Micromechanics and Inhomogeneity

The Toshio Mura 65th Anniversary Volume

Toshio Mura

G. J. Weng M. Taya H. Abé
Editors

Micromechanics and Inhomogeneity

The Toshio Mura 65th Anniversary Volume

With 211 Illustrations

Springer-Verlag
New York Berlin Heidelberg
London Paris Tokyo Hong Kong

G. J. Weng
Department of Mechanics and Materials Science
Rutgers University
Piscataway, NJ 08903
U.S.A.

M. Taya
Department of Mechanical Engineering
University of Washington
Seattle, WA 98195
U.S.A.

H. Abé
Department of Mechanical Engineering
Tohoku University
Sendai 980
Japan

Library of Congress Cataloging in Publication Data
Micromechanics and inhomogeneity, the Toshio Mura 65th anniversary volume / G.J. Weng, M. Taya, H. Abé, editors.
p. cm.
Selected papers from the Symposium on Micromechanics and Inhomogeneity held during the 1989 Winter Annual Meeting of the American Society of Mechanical Engineers in San Francisco. The symposium was co-sponsored by the Constitutive Equations Committee et al.
ISBN-13:978-1-4613-8921-7
1. Micromechanics—Congresses. 2. Dislocations in metals—Congresses. 3. Mura, Toshio, 1925– . I. Mura, Toshio, 1925– II. Weng, G.J. III. Taya, Minoru. IV. Abé, H. (Hiroyuki), 1936– . V. Symposium on Micromechanics and Inhomogeneity (1989: San Francisco, Calif.) VI. American Society of Mechanical Engineers. Winter Meeting (1989: San Francisco, Calif.) VII. American Society of Mechanical Engineers. Constitutive Equations Committee.
QC176.8.M5M53 1989 89-21898
620.1′63—dc20

Printed on acid-free paper.

Typeset by Asco Trade Typesetting Ltd., Hong Kong.

9 8 7 6 5 4 3 2 1

ISBN-13:978-1-4613-8921-7 e-ISBN-13:978-1-4613-8919-4
DOI: 10.1007/978-1-4613-8919-4

Editors' Preface

Toshio Mura has written extensively on micromechanics over the years, and in part due to his writings and many others in the field, micromechanics has gradually emerged as a recognized discipline in the study of mechanics of materials. The idea is to bring both the mechanics and physics on the microscopic level to the macroscopic scale, so that the deformation and fracture processes of materials can be better understood. While much apparently remains to be done, this approach has already shed new light on certain selected topics and has proved to be fruitful. It is indeed a happy occasion to celebrate both Toshio's upcoming 65th birthday and the emergence of this young science at the same time.

The volume contains thirty-seven original articles on the related topics of micromechanics and inhomogeneity; it is presented to Toshio by his friends, colleagues, and admirers as a wish for his good health and continuing productivity. The contributors belong to both the applied mechanics and the materials communities, all with a common belief that micromechanics is an indispensable area of research. It is hoped that this somewhat balanced structure will make the volume more useful to a wider range of readers, and that in the meantime it will still reflect more or less the spectrum of Toshio's lifelong works. As Editors we have at the outset set the highest possible standards for the book, with a keen anticipation that the volume will be widely circulated for many years to come.

We consider it an honor to have the opportunity to edit this volume. In the process of preparation, Toshio has been most helpful in providing all the needed information. We are grateful to the individual authors for their ready cooperation, and especially to Professor Mori for his efforts in writing the Toshio Mura Biography. The planning of this publication is also accompanied by a Symposium on Micromechanics and Inhomogeneity during the 1989 Winter Annual Meeting of the American Society of Mechanical Engineers in San Francisco. A total of sixty papers, including these thirty-seven, will be presented there. The other twenty-three papers, due to the constraints of time and other factors, regrettably could not be included here. A list of these twenty-three papers is appended at the end of the Preliminary matter; they are available directly from the authors to whom we also extend our appreciation. In this regard Professors M. Eisenberg, H. Kobayashi, T. Mori, and E. Tsuchida have also served on the Organizing Committee. The symposium was cosponsored by the Constitutive Equations Committee, the Applied Mechanics Division and Materials Division of ASME, and the Japanese Society of Mechanical Engineers.

Our special thanks are due to Springer-Verlag, New York, Inc., and its most capable and professional staff for their editing and production. The financial support of the Urakami Foundation for some overseas participants to the Mura Symposium is also gratefully acknowledged. Finally, we would like to express our appreciation to Mr. Shoji of Tohoku University for his help in the indexing of this book.

Piscataway, New Jersey
Seattle, Washington
Sendai, Japan

G. J. Weng
M. Taya
H. Abé

March 29, 1989

Contents

Biography of Toshio Mura

Toshio Mura, second son of Shinzo and Chie Fujii, was born in Ono, a small port village of Kanazawa, the capital of Ishikawa Prefecture, Japan, on December 7, 1925. Among the locals, the Fujiis are well known as brewers having a long history in the area. Kanazawa is an old city on the coast of the Sea of Japan, where traditional culture is proudly maintained and appreciated. The center of education on the coast, its fine arts and crafts, such as exquisite silk (Kaga Yuzen) and china (Kutani), are also famous. Ono adjoins Kanaiwa, the birthplace of Gohei Zeniya who was a very well-known entrepreneur, banker, and trader of the nineteenth century. It is said that Gohei quietly but bravely conducted foreign trade which was prohibited by the government at that time. It can be imagined that during his boyhood Mura often looked up at the huge bronze statue of Gohei, in a nearby park, which commands a distant view of the sea.

After graduation from the Kanazawa Second Middle School (Kanazawa Ni-chu) in 1941, Mura entered the Fourth Imperial High School (Shi-ko). Although militarism prevailed during this period, students in imperial high schools felt a traditionally observed liberal atmosphere there. Mura must have enjoyed this atmosphere, although at the same time feeling the pressure required to prepare for higher education. The imperial high schools, which were terminated after the conclusion of World War II, had been established to rear those chosen young who would be leaders in society.

In 1944, during the most difficult time of the war, Mura went to the Imperial University of Tokyo to read Aeronautical Engineering. After the war, his department was dissolved and changed to the Department of Applied Mathematics at the University of Tokyo. Fascinated by applied mathematics and with the encouragement of faculty members, he pursued his graduate study under the supervision of Tsuyoshi Hayashi in 1949. At that time the department had many outstanding scientists. Among those who had a particular influence upon Mura, either consciously or subconsciously, Kazuo Kondo and Sigeiti Moriguti must be mentioned. The title of his Ph.D. dissertation was "Study on Thermal Stresses." His work in the dissertation turned out to be one of the earliest papers on the dynamic wave of thermal stresses.

As a graduate student, Mura also began his teaching career as a mathematics professor at Meiji University, where he met and worked with his lifelong friend, Nobuo Kinoshita. Their joint paper, "On the boundary value problem of elasticity," which was published during his tenure at Meiji University (1956), agitated some Russian mathematicians in the field of integral equations. Had this work been extended, it would have led to the powerful computational technique now known as the boundary element method. Mura's subject of current research interest, the inverse problem, which he began with his student, Zhanjun Gao, has its origin in his work with Kinoshita.

At the graduate school, Mura was introduced to his future wife, Sawa, by her sister, Sumi, who had worked in the Department of Aeronautical Engineering. Sawa was the second daughter of Tetsuichi and Hanae Ozaki. Tetsuichi Ozaki was a career government official in the Ministry of Education at that time. During the courtship, Mura often visited the Ozaki's and Sumi fondly recalls that he praised Sawa's cooking. They married in 1953 and their first daughter, Miyako, was born in 1955.

The early 1950s were an exciting time when theoretical papers on dislocations, mostly by British scientists, were published. Because times were hard after the war, universities in Japan could not subscribe to a sufficient number of foreign journals. For example, the *Philosophical Magazine* was found in only one university library in Tokyo; Mura recalls going to this library to read the journal. Good papers were fascinating and inspiring, but his frustration accumulated. As a lion longs for a due hunting field, he wished to shorten the geographical and psychological distance to his unseen peers. Seizing an opportunity, Mura went to Northwestern University's Department of Materials Science, Evanston, Illinois, to work with John O. Brittain in 1958.

While at this department, Mura conceived the idea of the Periodic Distribution of Dislocations, which was documented in a paper and published later in the *Proceedings of the Royal Society of London* as a communication by A. H. Cottrell and R. E. Peierls (1964). In this paper, for the first time, the Fourier method was used to obtain the elastic field of dislocations. As seen in his later publications, the Fourier method became Mura's favorite tool to analyze elastic fields. The period of conception for this work coincided with the birth of his second daughter, Nanako, in early 1961. Mura claimed that too much excitement casued by these two factors hindered him from good sleep.

In 1961 Mura joined the Department of Civil Engineering at Northwestern University as an assistant professor. As a young lion, free in a savanna, roars and runs as he pleases, Mura actively and enthusiastically pursued the theory of dislocations. The pleasant but stimulating atmosphere, brewed by his colleagues, John Dundurs and Leon M. Keer, also encouraged him. Dundurs and Mura obtained the elastic fields of dislocations parallel to a cylindrical inhomogeneity (1964). Keer and Mura analyzed a penny-shaped crack with a plastic zone by solving an integral equation, Mura's first paper concerned with a crack (1963).

In 1963 Mura succeeded in expressing the elastic field of a curved dislocation in a line integral, now known as Mura's Formula (1963). The line integral is along the dislocation and contains only the state quantities that characterize the dislocation. This solution was later extended by John R. Willis, who gave the field of a dislocation segment in the form of algebraic equations, which required the solution of sextic equations (1970). Mura further explored the line integral expression to obtain an equation, readily applicable to any medium without using Green's function, requiring no differentiation and involving only computable integration. In his book, *Micromechanics of Defects in Solids* (first edition, 1982), Mura has shown that this expression can also reproduce the formulas developed by L. M. Brown, J. Lothe, R. J. Asaro, D. M. Barnett, and J. P. Hirth. The Brown *et al.* formulas are also seen to have beauty but contain second derivatives, which are not necessarily reliable when used in numerical computations. Mura and his student, D. R. J. Owen, applied the line integral formula to obtain stress fields of often observed but complexly shaped dislocations (1967). The paper in 1963 is also noteworthy for introducing the concept of a dislocation flux tensor, which is useful when the dynamic motion of dislocations is examined. The period, during which Mura's Formula was found, coincided with his promotion to Associate Professor of Civil Engineering.

Mura's Formula can be traced to the paper conceived earlier, "Periodic Distribution of Dislocations." In this paper Mura also introduced the concept of eigendistortion, although E. Kröner's earlier terminology of plastic distortion (1958) was used. Mura later coined the names, "eigendistortion and eigenstrain," to include all possible nonelastic distortions and strains.

The dislocation density and flux tensors were applied to continuum plasticity theory. Believing that a stress appearing within the framework of continuum plasticity was the sum of external and dislocation stresses, Mura published a series of papers, in the late 1960s, along these lines that emphasized the distribution and stress of dislocations (1967, 1968). He demonstrated his contention by solving classical problems of continuum plasticity through examination of the stress field, due to that distribution of dislocations that conformed to a deformation field of plasticity. Mura ingeniously discovered and used the concept of impotent dislocations, which are defined from the antisymmetric part of eigendistortions and yield neither displacement, strain, nor stress (1968). Prior to this work, a bold theory was proposed in which the von Mises yield criterion was interpreted within the context of the dislocation theory (1965).

In 1967 Mura became Professor of Civil Engineering. At that time Mura and J. G. Kuang, his student, obtained the solutions for a pile-up of edge dislocations against the interfacial boundary between different materials (1968). The Wiener–Hopf technique was applied to solve the Hilbert type integral equation for the pile-up in a closed form. Mura was very pleased with the use of modern mathematics for this solution.

The pioneering work of J. D. Eshelby, his beloved peer, appears to have inspired and stimulated Mura, as seen in his studies of static and dynamic fields of dislocations in anisotropic media and in dislocation pile-ups. As can be inferred from the preface to his book, *Micromechanics of Defects in Solids,* Mura regards Eshelby's work on inclusions and inhomogeneities (1957) as being the most important and fundamental. In 1970 Kinoshita joined Mura at Northwestern University. Using the Fourier integration method, they extended the celebrated ellipsoidal inclusion theory of Eshelby's to an anisotropic medium, and proved that the stress and strain fields inside an ellipsoidal inclusion are uniform (1971). The Eshelby tensors for an anisotropic medium were also given in the form of computable integrable formulas. R. J. Asaro and D. M. Barnett (1975) and T. Mura and N. Kinoshita (1978) further elaborated the mechanics of ellipsoidal inclusions with nonuniform eigenstrains. Mura and his student, P. C. Cheng, derived the stress and strain fields outside an ellipsoidal inclusion with general eigenstrains in an anisotropic medium (1977). In this work they recovered, in an elementary manner, the expressions for the jumps of distortion and stress at an interface with discontinuously changing eigenstrains, the expressions that were first obtained by R. Hill and later discussed by L. J. Walpole in conjunction with the jumps of polyharmonic potentials.

To Mura the evaluation of the disturbance in elastic fields due to elastic inhomogeneities is the most interesting application of the theory of inclusions. For example, Z. A. Moschovides and Mura solved the stress field caused by two inhomogeneities by applying the equivalent inclusion method with polynomial eigenstrains (1975). A computer, performing the numerical calculations, complained that the matrices involved for linear equations were singular. Moschovides looked for the bugs that might have caused this complaint, but no bugs were found. The linear equations were carefully examined analytically and the cause of the complaint was found. There existed certain distributions of eigenstrains that yielded no elastic field. Rozo Furuhashi, a visiting scholar, and Mura later generalized this finding and showed that impotent inclusions exist in a general sense (1979). The impotent inclusions have eigenstrains defined by derivatives of a continuous vector (displacement) that vanishes at the boundary of the inclusions. This anecdote illustrates Mura's teachings: "Study and examine a specific subject carefully. If there is anything strange and exciting, you can later generalize it in a broader sense."

Mura was trapped by the beauty of ellipsoidal inclusions, and extended the subject to see the elastic field when the boundary of an inclusion was allowed to slide until the tangential traction on the boundary vanished (sliding inclusion). Working with Furuhashi, Mura found that with only shear eigenstrains ellipsoidal inclusions, excluding the spheroidal case, yield no stress (1984). Although this finding caused controversy at the time, it is now believed to be true. The mechanics of sliding inclusions was further studied by his associate, Eiichiro Tsuchida and his students, Iwona M. Jasiuk and Dimitris A. Kouris (1985, 1988, 1989).

Mura also interacted with experimentalists, who eagerly sought his advice and aid on issues of mathematics and mechanics. In particular, Morris E. Fine, and his students in Northwestern's Department of Materials Science and Engineering, benefited from this interaction in their studies of the fatigue of alloys. Mura also gained insight into material properties and structures by the interactions with these materials scientists. His own student, Carl R. Vilman, developed a theory of crack growth in fatigue under Mura's supervision (1981). Plastic zones, which formed at a crack tip, were analyzed on the basis of "dislocations in plasticity." Mura is still pursuing the mechanics of fatigue. Very recently, he found negligibly potent dislocation walls that account for the dislocation structure of fatigued materials (1989). The negligibly potent dislocations are a discrete version of the impotent dislocations discovered earlier.

Because of his attachment to his native country, Mura has invited many visiting scholars from Japan. Some were taught, educated, and trained in mechanics; some continued in Japan the research begun at Northwestern. For all of them, Toshio and Sawa's house in Wilmette, Illinois, is their second home. To some, Toshio is a big brother and Sawa a big sister. Their daughters, Miyako and Nanako, and their daughters' husbands, Steve Izzo and Vince Kwasniewski, join lively gatherings at weekends, designed by Sawa for visiting scholars. Toshio is often the life and soul of the party at these gatherings. However, in recent years, Toshio seems most content when showing-off or playing with his two grandchildren, Courtney (1984) and Stephanie (1986).

Mura has been invited to several institutions as a visiting scientist: the National Bureau of Standards (1969–1970), the Atomic Energy Research Establishment at Harwell (1972), the Institute of Space and Astronautical Science (1983), the University of Tokyo (1984), and the Industrial Research Institute of Ishikawa (1987, 1988). Interaction with the scientists in these institutions also resulted in new developments for his studies. Among them is the concept of eigendistortions, introduced into lattice theory from a discussion with R. Bullough at Harwell (1978).

In 1986 Mura was elected to membership in the National Academy of Engineering, U.S.A., with the citation, "For Initiating and Promoting Micromechanics to Bridge the Gap Between Metal Physics and Engineering Mechanics." This was due valuation. His friends and disciples were delighted to know that Mura was now recognized, officially and publicly. During the same year, he was appointed Walter P. Murphy Professor in the Technological Institute at Northwestern University.

Besides numerous published research papers, Mura wrote three books, all of which were well received. In particular, his third book, *Micromechanics of Defects in Solids* (second revised edition, 1987), is highly regarded and often cited, and his original research and derivations of often-used formulas in mechanics are to be found in this treatise. The book is full of Mura's flavor. Mura writes his papers in commanding tones, using short and clear sentences. Although this may be partly due to English being his second language, this style is certainly due to the clarity and unambiguity that characterize his research. While reading his papers, what and how he thought can often be visualized. Mura encourages his readers to stretch to the advanced level of the subjects covered in his books. An example of this skill is shown in the preface to his first book, *Calculus of Variations* (1958). At the same time, he has shown his confidence and command of the subject in his second book, *Maikuro–Mekanikkusu* (1976).

In Japan, given names often reflect what parents wish their children to be or to become. "Toshio" is written in three characters. "To" stands for outward or abroad, "shi" determination or will, and "o" a man. Thus, "Toshio" means a man who is not confined and sets his eyes on the world. Surely, Mura has not disappointed his parents.

A man is as old as he feels. Mura is an outdoors man, playing tennis and swimming regularly. He is increasingly active in his research, full of curiosity and never tiring. He is also a good and kind teacher in class. I have mentioned that Mura was once inspired by the papers of his peers to which he had difficult access. Looking back, we can say that Mura has inspired young mechanicians and materials scientists. He will continue to play such a leading role, introducing exciting and novel ideas into Micromechanics.

This he can do.

Nagatsuta
January 1, 1989

T. Mori

List of Publications by Toshio Mura

A. Books

T. Hayashi and T. Mura, *Calculus of Variations*, Colona Col., Tokyo, 1958.

T. Mura, The continuum theory of dislocations, in *Advances in Materials Research*, Vol. 3, edited by H. Herman, 1968, pp. 1–108.

T. Mura (Editor), *Mathematical Theory of Dislocations*, Proceedings of the ASME Symposium, Northwestern University, June 1969, American Society of Mechanical Engineers, New York, 1969.

T. Mura and T. Mori, *Micromechanics*, Baifukan, Tokyo, 1976.

T. Mura (Editor), *Mechanics of Fatigue*, AMD, Vol. 47. Proceedings of the ASME Symposium, November 1981, American Society of Mechanical Engineers, New York, 1981.

T. Mura, *Micromechanics of Defects in Solids*, Martinus Nijhoff, The Hague, 1982.

T. Mura, *Micromechanics of Defects in Solids*, second revised edition, Martinus Nijhoff, The Hague, 1987.

B. Papers

1. T. Mura, Internal stresses in a solid steel cylinder due to quenching, *J. Appl. Mech. Japan*, **5** (1952), 16–19.
2. T. Mura, Interal stresses in a hollow cylinder due to cooling from the surfaces, *Trans. Japan Soc. Mech. Engr.*, **18** (1952), 16–22.
3. T. Mura, Thermal strains and stresses in transient state, *Proc. Japan Nat. Congr. Appl. Mech.*, **2** (1952), 9–14.
4. T. Mura, Unstable plastic yielding of a hollow cylinder by an internal pressure under a stational temperature distribution, *Trans. Japan Soc. Mech. Engr.*, **19** (1953), 19–87.
5. T. Mura, Buckling deformations of thin plates due to welding, *Proc. Japan Nat. Congr. Appl. Mech.*, **3** (1953), 103–106.
6. T. Mura and N. Kinoshita, Stefan-like problem of a cylinder, *Proc. Japan Nat. Congr. Appl. Mech.*, **4** (1955), 345–348.
7. T. Hattori and T. Mura, Residual stresses in bearing rings, *Trans. Japan Soc. Metal. Engr.*, **19** (1955), 282–286.

8. T. Mura, Extremum principles of the thermal elasto-plastic stresses, *J. Japan Soc. Aero.*, **3** (1955), 215–220.
9. T. Mura, Residual stress in quenched rings, *Proc. Japan Nat. Congr. Appl. Mech.*, **5** (1955), 49–52.
10. T. Mura, Extremum principles of thermal elasto-plastic problems, *Res. Rep. Faculty of Engng. Meiji Univ.*, **6** (1955–2), 1.
11. N. Kinoshita and T. Mura, On the process of ice formation, *Res. Rep. Faculty of Engng. Meiji Univ.*, **7** (1956–1), 1.
12. N. Kinoshita and T. Mura, Problem of the Stefan type, *Sugaku*, **8** (1957), 216–218.
13. T. Mura, Buckling type deformation of thin plates due to welding, *Res. Rep. Faculty of Engng. Meiji Univ.*, **7** (1956–1), 1.
14. T. Mura, Dynamical thermal stresses due to thermal shocks, *Res. Rep. Faculty of Engng. Meiji Univ.*, **8** (1956–2), 1.
15. N. Kinoshita and T. Mura, On a boundary value problem of elasticity, *Res. Rep. Faculty of Engng. Meiji Univ.*, **8** (1956–2), 56–82.
16. M. Ueno and T. Mura, Effects of tempering temperatures on initial stresses of ring-type test pieces, *Iron and Steel, Japan*, **42** (1956), 29–34.
17. T. Mura, Measurements of quenching stresses in rings, *Proc. Japan Nat. Congr. Appl. Mech.*, **6** (1956), 111–115.
18. T. Mura, Unstable plastic yield of a hollow cylinder under internal pressure and thermal stresses, *Res. Rep. Faculty of Engng. Meiji Univ.*, **9** (1957–1), 1.
19. T. Mura, Residual stresses due to thermal treatments, *Res. Rep. Faculty of Engng. Meiji Univ.*, **10** (1957–2), 1.
20. T. Mura and N. Kinoshita, Expression of initial stresses based on Green's functions, *J. Japan Soc. Aero.*, **5** (1957), 7–10.
21. T. Mura and H. Yoshimoto, Measurements of quenching stresses in bearing ring by interference fringes, *J. Appl. Phys.*, **29** (1958), 115–119.
22. T. Mura and J. O. Brittain, Contribution of dislocation line tension and the density of the solute atmosphere to the yield point in strain-aged ingot iron, *Acta Metallurgica*, **8** (1960), 767–772.
23. T. Mura, I. Tamura, and J. O. Brittain, On the internal friction of cold worked and quenched martensitic iron and steel, *J. Appl. Phys.*, **32** (1961), 92–98.
24. T. Mura, E. P. Lautenschlager, and J. O. Brittain, Segregation of solute atoms during strain aging, *Acta Metallurgica*, **9** (1961), 453–458.
25. T. Mura, Theory of continuous dislocations and its applications, AFOSR, No. 581 (1961).
26. I. Tamura, T. Mura, and J. O. Brittain, The influence of alloying elements on the internal friction of cold worked and quenched, martensitic iron and steel, *Trans. Metall. Soc. AIME*, **221** (1961), 1158–1162.
27. T. Mura and J. O. Brittain, Reply to R. Bullough's and R. C. Newman's comments on a paper by Mura, Lautenschlager, and Brittain entitled: Segregation of solute atoms during strain aging, *Acta Metallurgica*, **10** (1962), 973.
28. T. Mura, Continuous distribution of moving dislocations, *Phil. Mag.*, **8** (1963), 843–857.
29. T. Mura, On dynamic problems of continuous distribution of dislocations, *Int. J. Engng. Sci.*, **1** (1963), 371–381.
30. T. Mura and S. L. Lee, Application of variational principles to limit analysis, *Quart. Appl. Math.*, **21** (1963), 243–248.

31. J. S. Kao, T. Mura, and S. L. Lee, Limit analysis of orthotropic plates, *J. Mech. Phys. Solids*, **11** (1963), 429–436.
32. T. Mura, Periodic distributions of dislocations, *Proc. Roy. Soc. London*, **A280** (1964), 528–544.
33. J. Kiusalaas and T. Mura, On the elastic field around an edge dislocation with application to dislocation vibration, *Phil. Mag.*, **9** (1964), 1–7.
34. J. Dundurs and T. Mura, Interaction between an edge dislocation and a circular inclusion, *J. Mech. Phys. Solids*, **12** (1964), 177–189.
35. T. Mura, J. S. Kao, and S. L. Lee, Limit analysis of circular orthotropic plates, *Proc. ASCE, J. Engng. Mech. Div.*, **90** EM 5 (1964), 375–395.
36. J. Kiusalaas and T. Mura, On the motion of a screw dislocation, in *Recent Advances in Engineering Science*, Vol. 1, edited by A. C. Eringen, Gordon & Breach, New York, 1964, pp. 545–564.
37. T. Mura, Miscellaneous thoughts in America, *J. Japan Soc. Aero Space Sci.*, **13** (1965), 21–22.
38. T. Mura and A. Otsuka, Application of the slab analogy to the study of stress fields induced by imperfections in crystals, *Int. J. Solids Structures*, **1** (1965), 179–188.
39. T. Mura, Continuous distribution of dislocations and the mathematical theory of plasticity, *Phys. Stat. Sol.*, **10** (1965), 447–453.
40. T. Mura, Continuous distribution of dislocations and the mathematical theory of plasticity II, *Phys. Stat. Sol.*, **11** (1965), 683–688.
41. T. Mura, W. H. Rimawi, and S. L. Lee, Extended theorems of limit analysis, *Quart. Appl. Math.*, **23** (1965), 171–179.
42. Y. C. Hung and T. Mura, The elastic field of an ellipsoidal inclusion in an orthotropic medium, Proceedings of the 9th Midwestern Mechanics Conference, Wisconsin (1965). In *Developments in Mechanics*, Vol. 3, Part 1, *Solid Mechanical Materials*, edited by T. C. Huang and M. W. Johnson, Wiley, New York, 1967, pp. 81–90.
43. L. M. Keer and T. Mura, Stationary cracks and continuous distributions of dislocations, *Proceedings of the International Conference of Mechanics, Sendai, Japan, 1965*, **1** (1966), 99–115.
44. W. H. Rimawi, S. L. Lee, and T. Mura, Limit analysis of notched tension specimens, *Trans. ASCE, J. Engng. Mech. Div.*, **92** EM 1 (1966), 11–24.
45. T. Mura and W. C. Lyons, Continuous distribution of dislocations and energy dissipation in metals, *J. Acoust. Soc. Amer.*, **39** (1966), 527–531.
46. T. Mura and S. L. Lee, A variational method of limit analysis, *Proc. Fifth U.S. Nat. Congr. Appl. Mech.*, **577** (1966), 577–579.
47. T. Mura, Continuum theory of plasticity and dislocations, *Int. J. Engng. Sci.*, **5** (1967), 341–351.
48. S. L. Lee, T. Mura, and J. S. Kao, A variational method for the limit analysis of anisotropic plates, *Quart. Appl. Math.*, **24** (1967), 323–330.
49. D. R. J. Owen and T. Mura, Periodic dislocation distributions in a half-space, *J. Appl. Phys.*, **38** (1967), 1999–2009.
50. D. R. J. Owen and T. Mura, Dislocation configurations in cylindrical coordinates, *J. Appl. Phys.*, **38** (1967), 2818–2825.
51. S. L. Sass, T. Mura, and J. B. Cohen, Diffraction contrast from non-spherical distortions—in particular a cuboidal inclusion, *Phil. Mag.*, **16** (1967), 679–690.

52. T. Mura, S. L. Lee, R. H. Bryant, and W. H. Rimawi, Limit analysis by direct method of variation, *Proc. ASCE, J. Engng. Mech. Div.*, **93**, EM5 (1967), 67–78.
53. T. Mura, A. Otsuka, W. S. Fu, and J. James, Approach to inelasticity through dislocations and extended slab analogy, *Proceedings of the Third Southeastern Conference, Developments in Theoretical and Applied Mechanics*, **3** (1967), 35–55.
54. W. H. Rimawi, T. Mura, and S. L. Lee, Extended theorems of limit analysis of anisotropic solids, *Proceedings of the Third Southeastern Conference, Developments in Theoretical and Applied Mechanics*, **3** (1967), 57–71.
55. J. G. Kuang and T. Mura, Dislocation pile-up in two-phase materials, *J. Appl. Phys.*, **39** (1968), 109–120.
56. T. Mura, Continuum theory of dislocations and plasticity, in *IUTAM Symposium on Mechanics of Generalized Continua*, edited by E. Kroner, Springer-Verlag, Berlin, 1968, pp. 269–278.
57. T. Mura, The continuum theory of dislocations, in *Advances in Materials Research*, Vol. 3, edited by H. Herman, Interscience, New York, 1968, pp. 1–108.
58. T. Mura, Generalized stress–strain laws developed by dislocation theory, in *Work Hardening*, edited by J. P. Hirth and J. Weertman, *Metall. Soc. Conf.*, **46** (1968) 141–149.
59. T. Mura, Line integral expressions of interaction energy of dislocation loops in anisotropic materials, *J. Appl. Phys.*, **40** (1969), 73–75.
60. R. Hibbeler and T. Mura, Viscous creep ratchetting of nuclear reactor fuel elements, *Nuclear Engng. Design*, **9** (1969), 131–143.
61. R. H. Bryant, S. L. Lee, and T. Mura, Mises limited loads for simply supported conical sandwich shells under internal pressure, *Ingenieur-Archiv*, **37** (1969), 281–287.
62. T. Mura, Method of continuously distribution dislocations, in *Mathematical Theory of Dislocations*, edited by T. Mura, ASME, New York, 1969, pp. 25–45.
63. E. S. Pacheco and T. Mura, Interaction between a screw dislocation and a bimetallic interface, *J. Mech. Phys. Solids*, **17** (1969), 163–170.
64. E. S. Pacheco and T. Mura, Influence of shear moduli and lattice parameters on the equilibrium of a screw dislocation, *J. Composite Materials*, **3** (1969), 664–675.
65. J. G. Kuang and T. Mura, Dislocation pile-up in half-space, *J. Appl. Phys.*, **40** (1969), 5017–5021.
66. T. Mura, S. L. Lee, and W. H. Rimawi, A variational method for limit analysis of anisotropic and nonhomogeneous soilds, in *Developments in Theoretical and Applied Mechanics*, Vol. 4, edited by D. Frederick, Pergamon, Oxford, 1970, pp. 541–549.
67. T. Mura, Individual dislocations and continuum mechanics, in *Inelastic Behavior of Solids*, edited by M. F. Kanninen, W. F. Adler, A. R. Rosenfield, and R. I. Jaffee, McGraw-Hill, New York, 1970, pp. 211–229.
68. W. Huang and T. Mura, Elastic fields and energies of a circular edge disclination and a straight screw disclination, *J. Appl. Phys.*, **41** (1970), 5175–5179.
69. T. Mura, The elastic field of moving dislocations and disclinations, in *Fundamental Aspects of Dislocation Theory*, Vol. 2, edited by J. A. Simmons, R. deWit, and R. Bullough, NBS Special Publication 317, Conference Proceedings, April 21–25, 1969, Government Printing Office, Washington, DC, 1970, pp. 977–996.
70. T. Mura, Stress and velocity field produced by uniformly moving dislocations in anisotropic media, *Phil. Mag.*, **23** (1971), 235–237.

71. N. Kinoshita and T. Mura, Elastic fields of inclusions in anisotropic media, *Phys. Stat. Sol.* (a) **5** (1971), 759–768.
72. T. Mura and N. Kinoshita, Inclusions in anisotropic media, *Proceedings of the Fourth Symposium on Composite Materials*, International Composite Materiais Research Committee, JUSE, 1971, pp. 153–159.
73. T. Mura and N. Kinoshita, Green's functions for anisotropic elasticity, *Phys. Stat. Sol.* (b) **47** (1971), 607–618.
74. T. Mura, Displacement and plastic distortion fields produced by dislocations in anisotropic media, *J. Appl. Mech.*, **38** (1971), 865–868.
75. T. Mura and J. Dundurs, Deformational modes and microstructure, in *Structure, Solid Mechanics and Engineering Design*, Part 1, edited by M. Te'eni, Wiley-Interscience, New York, 1971, pp. 3–12.
76. K. Saito, R. O. Bozkurt, and T. Mura, Dislocation stresses in a thin film due to the periodic distributions of dislocations, *J. Appl. Phys.*, **43** (1972), 182–188.
77. W. Huang and T. Mura, Elastic energy of an elliptical-edge disclination, *J. Appl. Phys.*, **43** (1972), 239–241.
78. H. H. Kuo and T. Mura, Elastic field and strain energy of a circular wedge disclination, *J. Appl. Phys.*, **48** (1972), 1454–1457.
79. T. Mura, semi-microscopic plastic distortion and disclinations, *Archiwum Mechaniki Stosowanej*, **24** (1972), 449–456.
80. T. Mura, Dynamic response of dislocations in solids, in *Dynamic Response of Structures*, edited by G. Hermann and N. Perrone, Proc. Symp. Stanford Univ., Pergamon, Oxford, 1972, pp. 345–370.
81. H. H. Kuo and T. Mura, Circular disclinations and interface effects, *J. Appl. Phys.*, **43** (1972), 3936–3942.
82. T. Mura, A variational method for micromechanics of composite materials, in *Mechanical Behavior of Materials*, Proc. Int. Conf. Kyoto, Vol. V, The Society of Materials Science, Japan, Kyoto, Japan, 1972, pp. 12–18.
83. T. Mura, Why is dislocation theory necessary? *J. Materials Sci. Soc. Japan*, **9** (1972), 266–270.
84. H. H. Kuo, T. Mura, and J. Dundurs, Moving circular twist disclination loop in homogeneous and two-phase materials, *Int. J. Engng. Sci.*, **11** (1973), 193–201.
85. T. Mura and S. C. Lin, Elastic fields of inclusions in anisotropic media (II), *Phys. Stat. Sol.* (a) **15** (1973), 281–285.
86. S. C. Lin, T. Mura, M. Shibata, and T. Mori, The work-hardening behavior of anisotropic media by non-deforming particles of fibres, *Acta Metallurgica*, **21** (1973), 505–516.
87. S. C. Lin and T. Mura, Long-range elastic interaction between a dislocation and an ellipsoidal inclusion in cubic crystals, *J. Appl. Phys.*, **44** (1973), 1508–1514.
88. P. Wheeler and T. Mura, Dyname equivalence of composite material and eigenstrain problems, *J. Appl. Mech.*, **40** (1973), 498–502.
89. H. H. Kuo and T. Mura, Circular twist disclination in viscoelastic materials, *J. Appl. Phys.*, **44** (1973), 3586–3588.
90. M. Taya and T. Mura, Dynamic plastic behavior of structures under impact loading investigated by the extended Hamilton's principle, *Int. J. Solids Structures*, **10** (1974), 197–209.
91. T. Mura and S. C. Lin, Thin inclusions and cracks in anisotropic media, *J. Appl. Mech.*, **41** (1974), 209–214.

92. T. Mura, On mechanical engineering education in the U.S.A., *J. Japan Soc. Mech. Engng.*, **77** (1974), 384–387.
93. T. Mura, Foundation and application of continuous distribution of dislocations, *Trans. Japan Inst. Metals*, **13** (1974), 733–740.
94. T. Mura, Recent study on dislocation dynamics, *Proceedings of the 11th Annual Meeting of the Society of Engineering Science*, edited by G. J. Dvorak, Duke University, Durham, NC, 1974, pp. 72–73.
95. T. Mura and C. T. Lin, Theory of fatigue crack growth for work hardening materials, *Int. J. Fracture*, **10** (1974), 284–287.
96. T. Mura, A. Novakovic, and M. Meshii, A mathematical model of cyclic creep acceleration, *Materials Sci. Engng.*, **17** (1975), 221–225.
97. T. B. Edil, T. Mura, and R. J. Krizek, A micromechanistic formulation for stress–strain response of clay, *Int. J. Engng. Sci.*, **13** (1975), 831–840.
98. T. Mura, A note on the strain field of a dislocation line in anisotropic media, *Phys. Stat. Sol.* (b) **70** (1975), K1–K6.
99. D. K. Shetty, T. Mura, and M. Meshii, Analysis of creep deformation under cyclic loading conditions, *Materials Sci. Engng.*, **20** (1975), 261–266.
100. Z. A. Moschovidis and T. Mura, Two-ellipsoidal inhomogeneities by the equivalent inclusion method, *J. Appl. Mech.*, **42** (1975), 847–852.
101. T. Mura, L. M. Keer, and H. Abé, Analytical study of crack growth and shape by hydraulic fracturing of rocks, in *Geothermal Reservoir Engineering*, edited by P. Kruger and H. J. Ramey, Jr., Stanford Geothermal Program Workshop Report No. SGP-TR-12, Stanford University, Stanford, CA, 1975, pp. 180–184.
102. T. Mura and T. Mori, Elastic fields produced by dislocations in anisotropic media, *Phil. Mag.*, **33** (1976), 1021–1027.
103. T. Mori and T. Mura, Slip morphology in dispersion hardened materials, *Materials Sci. Engng.*, **26** (1976), 89–98.
104. H. Abé, T. Mura, and L. M. Keer, Growth rate of a penny-shaped crack in hydraulic fracturing of rocks, *J. Geophys. Res.*, **81** (1976), 5335–5340.
105. L. M. Keer, C. T. Lin, and T. Mura, Fracture analysis of adhesively bonded sheets, *J. Appl. Mech.*, **43** (1976), 652–656.
106. H. Abé, L. M. Keer, and T. Mura, Growth rate of a penny-shaped crack in hydraulic fracturing of rock, 2, *J. Geophys. Res.*, **81** (1976), 6292–6298.
107. T. Mori, P. C. Cheng, and T. Mura, Interaction among interstitial solute atoms in iron and tetragonal ordering, *Proceedings of the First JIM International Symposium on New Aspects of Martensitic Transformation*, Japan Institute of Metals, Tokyo, 1976, pp. 281–286.
108. T. Mura, T. Mori, and M. Kato, The elastic field caused by a general ellipsoidal inclusion and the application to martensite formation, *J. Mech. Phys. Solids*, **24** (1976), 305–313.
109. E. N. Mastrojannis, T. Mura, and L. M. Keer, Stress field of a planar elliptical dislocation loop, *Phil. Mag.*, **4** (1977), 1137–1139.
110. T. Mura and P. C. Cheng, An equivalent inclusion method for a three-dimensional lens-shaped crack in anisotropic media, *Fracture 1977*, ICF4, Waterloo, **3** (1977), 191–196.
111. T. Mura and P. C. Cheng, The elastic field outside an ellipsoidal inclusion, *J. Appl. Mech.*, **44** (1977), 591–594.
112. T. Mura, Eigenstrains in lattice theory, in *Continuum Models in Discrete Systems*, Proceedings of the 2nd International Conference, Mont Gabriel, Canada, edited

by J. W. Provan and H. H. E. Leipholz, Study No. 12, Solid Mechanics Division, University of Waterloo Press, 1978, pp. 503–519.

113. N. Funabashi, T. Mura, and L. M. Keer, Fatigue crack initiation and propagation in bonded sheets, *J. Adesion*, **9** (1978), 229–235.
114. T. Mori and T. Mura, Calculation of back stress decrease caused by climb of Orowan loops in a dispersion-hardened alloy, *Acta Metallurgica*, **26** (1978), 1199–1204.
115. T. Mori, P. C. Cheng, M. Kato, and T. Mura, Elastic strain energies of precipitates and periodically distributed inclusions in anisotropic media, *Acta Metallurgica*, **26** (1978), 1435–1441.
116. T. Mura and N. Kinoshita, The polynomial eigenstrain problem for an anisotropic ellipsoidal inclusion, *Phys. Stat. Sol.* (a) **48** (1978), 447–450.
117. Y. H. Kim, T. Mura, and M. E. Fine, Fatigue crack initiation and microcrack growth in 4140 steel, *Metall. Trans.*, **9A** (1978), 1679–1683.
118. T. Mura, Recent results in micromechanics, in *IUTUM Symposium on High Velocity Deformation of Solids*, edited by K. Kawata and J. Shioiri, Springer-Verlag, Berlin, 1978, pp. 295–304.
119. H. Abé, L. M. Keer, and T. Mura, Theoretical study of hydraulically fractured penny-shaped cracks in hot, dry rocks, *Int. J. Numer. Anal. Methods Geomech.*, **3** (1979), 79–96.
120. Y. H. Kim, M. E. Fine, and T. Mura, Plastic yielding at the tip of a blunt notch during static and fatigue loading, *Engng. Fracture Mech.*, **11** (1979), 653–660.
121. H. Sekine and T. Mura, The elastic field around an elliptical crack in an anisotropic medium under an applied stress of polynomial forms, *Int. J. Engng. Soc.*, **17** (1979), 641–649.
122. H. Sekine and T. Mura, Weakening of an elastic solid by a periodic array of penny-shaped cracks, *Int. J. Solids Structures*, **15** (1979), 493–502.
123. E. N. Mastrojannis, L. M. Keer, and T. Mura, Stress intensity factor for a plane crack under normal pressure, *Int. J. Fracture*, **15** (1979), 247–258.
124. T. Mura, On the extended J-integral, in *Recent Research on Mechanical Behavior of Solids*, University of Tokyo Press, Tokyo, 1979, pp. 237–243.
125. H. Sekine and T. Mura, A dislocation dipole in an anisotropic medium, *Phil. Mag.*, **40** (1979), 183–191.
126. K. Seo and T. Mura, The elastic field in a half-space due to ellipsoidal inclusions with uniform dilatational eigenstrains, *J. Appl. Mech.*, **46** (1979), 568–572.
127. R. Furuhashi and T. Mura, On the equivalent inclusion method and impotent eignstrains, *J. Elast.*, **9** (1979), 263–270.
128. T. Mura, Discussion for the paper, On the process of subsurface fatigue crack initiation in Ti-6A1-4V, by J. Ruppen, P Bhowal, E. Eylon and A. J. McEvily, in *Fatigue Mechanisms*, edited by K. T. Fong, ASTH, STP 675, 1979, pp. 65–67.
129. H. Sekine and T. Mura, Thermal stresses around an elastic ribbonlike inclusion with good thermal conductivity, *J. Thermal Stresses*, **2** (1979), 475–489.
130. C. Vilmann and T. Mura, Fracture related to a dislocation distribution, *J. Appl. Mech.*, **46** (1979), 817–820.
131. T. Mura, On the extended *J*-integral, in *Recent Research on Mechanical Behavior of Solids*, edited by The Committee on Recent Research on Mechanical Behavior of Solids, University of Tokyo Press, Tokyo, 1979, pp. 237–243.
132. E. N. Mastrojannis, L. M. Keer, and T. Mura, Growth of planar cracks induced by hydraulic fracturing, *Int. J. Number. Methods Engng.*, **15** (1980), 41–54.

133. T. Mori, M. Okabe, and T. Mura, Diffusional relaxation around a second phase particle, *Acta Metallurgica*, **28** (1980), 319–325.
134. H. Sekine and T. Mura, Characterization of a penny-shaped reservoir in a hot dry rock, *J. Geophys. Res.*, **85**, No. B7 (1980), 3811–3816.
135. R. Castles and T. Mura, Fracture criterion of flat ellipsoidal cracks in anisotropic bodies, *Phys. Stat. Sol.* (a) **60**, No. K1 (1980), 37–44.
136. M. Inokuti, H. Sekine, and T. Mura, General use of the Lagrange multiplier in nonlinear mathematical physics, in *Variational Methods in the Mechanics of Solids*, edited by S. Nemat-Nasser, Proceedings of the IUTAM Symposium, September 1978, Evanston, Illinois, Pergamon, New York, 1980, pp. 156–162.
137. R. Furuhashi, N. Kinoshita, and T. Mura, Periodic distributions of inclusions, *Int. J. Engng. Sci.*, **19** (1981), 231–236.
138. K. Tanaka and T. Mura, A dislocation model for fatigue crack initiation, *J. Appl. Mech.*, **48** (1981), 97–103.
139. Y. Izumi, M. E. Fine, and T. Mura, Energy considerations in fatigue crack propagation, *Int. J. Fracture*, **17** (1981), 15–25.
140. T. Mura, R. Furuhashi, and K. Tanaka, Equivalent inclusion method in composite materials, in *Composite Materials*, edited by K. Kawata and T. Akasaka, Proceedings of Japan–U.S. Conference, Tokyo, 1981, pp. 71–77.
141. T. Mura, Energy release rate and the J-integral, in *Three-Dimensional Constitutive Relations and Ductile Fracture*, edited by S. Nemat-Nasser, North-Holland, Amsterdam, 1981, pp. 147–153.
142. M. Taya and T. Mura, On stiffness and strength of an aligned short-fiber reinforced composite containing fiber-end cracks under uni-axial applied stress, *J. Appl. Mech.*, **48** (1981), 361–367.
143. M. Okabe, T. Mori, and T. Mura, Internal friction caused by diffusion around a second-phase particle A–Si alloy, *Phil. Mag.*, **44** (1981), 1–12.
144. T. Mura, Dislocation and inelasticity of solids, in *Dislocation Modelling of Physical Systems*, edited by M. F. Ashby, R. Bullough, C. S. Hartley, and J. P. Hirth, Pergamon, Oxford, 1981, pp. 357–367.
145. M. Wnuk and T. Mura, Comparative study of models for tensile quasi-static fracture, *Int. J. Engng. Sci.*, **19** (1981), 1517–1527.
146. M. Wnuk and T. Mura, Stability of a disc-shaped geothermal reservoir subjected to hydraulic and thermal ladings, *Int. J. Fracture*, **17** (1981), 493–517.
147. T. Mura and K. Tanaka, Dislocation dipole models for fatigue crack initiation, in *Mechanics of Fatigue*, edited by T. Mura, ASME, AMD, 47, 1981, pp. 111–131.
148. K. Tanaka and T. Mura, A theory of fatigue crack initiation at inclusions, *Metall. Trans.*, **13A** (1982), 117–123.
149. K. Tanaka and T. Mura, A micromechanical theory of fatigue crack initiation from notches, *Mech. Mater.*, **1** (1982), 63–73.
150. L. S. Fu and T. Mura, Volume integrals of ellipsoids associated with the inhomogeneous Helmholtz equation, *Wave Motion*, **4** (1982), 141–149.
151. H. Kobayshi and T. Mura, Fatigue crack closure and mechanical consideration, *Proceedings of the 27th JAPAN National Symposium on Strength, Fracture and Fatigue*, **27** (1982), 97–126.
152. T. Mura, Accumulation of elastic strain energy during cyclic loading, *Scripta Metall.*, **16** (1982), 811–814.
153. T. Mura and C. Vilmann, Fatigue crack propagation related to a dislocation

distribution, in *Defects and Fracture*, edited by G. C. Sih and H. Zorski, Martinus Nijhoff, The Hague, 1982, pp. 81–90.

154. H. Kobayashi and T. Mura, On the blunting line in the J_{IC} test—the comparison of theories and experiments, *Proceedings of the ICF International Symposium on Fracture Mechanics*, Science Press, Beijing, China, 1983, pp. 517–522.
155. T. Mori, M. Koda, R. Monzen, and T. Mura, Particle blocking in grain boundary sliding and associated internal friction, *Acta Metallurgica*, **31** (1983), 275–283.
156. E. N. Mastrojannis and T. Mura, On the problem of two coplanar cracks inside an infinite isotropic elastic solid, *Int. J. Numer. Methods Engng.*, **19** (1983), 27–35.
157. N. Anmadi, L. M. Keer, and T. Mura, Non-Hertzian contact stress analysis for an elastic half space—normal and sliding contact, *Int. J. Solids Structures*, **19** (1983), 357–373.
158. L. M. Keer, J. C. Lee, and T. Mura, Stress distributions for a quarter plane containing an arbitrarily oriented crack, *J. Appl. Mech.*, **50** (1983), 43–49.
159. L. M. Keer, J. C. Lee, and T. Mura, Hetenyi's elastic quarter space problem revisited, *Int. J. Solids Structures*, **19** (1983), 497–508.
160. M. P. Wnuk and T. Mura, Effect of microstructure on the upper and lower limits of material toughness in elastic–plastic fracture, *Mech. Mater.*, **2** (1983), 33–46.
161. E. Tsuchids and T. Mura, On the stress concentration around a spherical inclusion, *J. Reinforced Plastics and Composites*, **2** (1983), 29–33.
162. L. S. Fu and T. Mura, The determination of the elastodynamic fields of an ellipsoidal inhomogeneity, *J. Appl. Mech.*, **50** (1983), 390–396.
163. M. P. Wnuk and T. Mura, Extension of a stable crack at a variable growth step, in *Fracture Mechanics: Fourteenth Symposium–Theory and Analysis*, edited by J. C. Lewis and G. Sines, ASTM STP 791, 1983, pp. 96–127.
164. E. N. Mastrojannis, L. M. Keer, and T. Mura, Numerical solution of a three-part mixed boundary value problem of linear elastostatics, *Comput. Methods Appl. Mech. Engng.*, **39** (1983), 93–101.
165. T. Mura, H. Shirai, and J. R. Weertman, The elastic strain energy of dislocation structures in fatigued metals, in *Defects, Fracture and Fatigue*, edited by G. C. Sih and J. W. Provan, Martinus Nijhoff, The Hague, 1983, pp. 65–73.
166. N. Yamashita and T. Mura, Contact fatigue crack initiation under repeated oblique force, *Wear*, **91** (1983), 235–250.
167. E. Tsuchida and T. Mura, The stress field in an elastic half-space having a spheroidal inhomogeneity under all-around tension parallel to the plane boundary, *J. Appl. Mech.*, **50** (1983), 807–816.
168. N. Yamashita, T. Mura, and H. S. Cheng, Effect of stresses induced by a spherical asperity on surface pitting in elastohydrodynamic contacts, *ASLE Preprint*, 1983, Preprint No. 83-LC-2C-1.
169. E. N. Mastrojannis, T. Mura, and L. M. Keer, An axisymmetric Neumann potential problem for the circular annulus, *Computers and Structures*, **18** (1984), 365–368.
170. Y. Hirose and T. Mura, Nucleation mechanism of stress corrosion cracking from notches, *Engng. Fracture Mech.*, **19** (1984), 317–329.
171. T. Mura and R. Furuhashi, The elastic inclusion with a sliding interface, *J. Appl. Mech.*, **51** (1984), 308–310.
172. C. Gomez and T. Mura, Stresses caused by expansive cement in borehole, *J. Engng. Mech. ASCE*, **110** (1984), 1001–1005.

173. T. Mura and Y. Hirose, A dislocation model for crack initiation and propagation by stress corrosion in high-strength steel, in *Dislocations in Solids: Some Recent Advances*, edited by X. Markenscoff, ASME, AMD, 63, New York, 1984, pp. 59–68.
174. N. Kinoshita and T. Mura, Eigenstrain problems in a finite body, *SIAM J. Appl. Math.*, **44** (1984), 524–535.
175. Y. Hirose and T. Mura, Growth mechanism of stress corrosion cracking in high strength steel, *Engng. Fracture Mech.*, **19** (1984), 1057–1067.
176. L. M. Keer, J. C. Lee, and T. Mura, A contact problem for the elastic quarter space, *Int. J. Solids Structures*, **20** (1984), 513–524.
177. Y. Hirose and T. Mura, Effect of loading history on stress corrosion cracking in high strength steel, *Mech. Mater.*, **3** (1984), 95–110.
178. H. S. Cheng, L. M. Keer, and T. Mura, Analytical modelling of surface pitting in simulated gear-teeth contacts, *Society of Automotive Engineers. The Engineering Resource for Advancing Modility*, SP584—*Gear Design and Performance*. 1984, pp. 27–35.
179. T. Mura and J. Weertman, Dislocation models for threshold fatigue crack growth, in *Fatigue Crack Growth Threshold Concepts*, edited by D. Davidson and S. Suresh, Metall. Soc. AIME, New York, 1984, pp. 531–549.
180. L. M. Keer, N. Ahmadi, and T. Mura, Tangential loading of elastic bodies in contact, *Computer & Structures*, **19** (1984), 93–101.
181. M. Kato, S. Onaka, T. Mori, and T. Mura, Statistical consideration of plastic strain accumulation in cycle deformation and fatigue crack initiation, *Scripta Metallurgica*, **18** (1984), 1323–1326.
182. K. Tanaka and T. Mura, Fatigue crack growth along planer slip bands, *Acta Metallurgica*, **32** (1984), 1731–1740.
183. T. Mori and T. Mura, An inclusion model for crack arrest in fiber reinforced materials, *Mech. Mater.*, **3** (1984), 193–198.
184. N. Yamshita, T. Mura, and H. S. Cheng, Effect of stresses induced by a spherical asperity on surface pitting in elastohydrodynamic contacts, *ASLE Transactions*, **28** (1985), 11–20.
185. R. R. Castles and T. Mura, The analysis of eigenstrains outside an ellipsoidal inclusion, *J. Elasticity*, **15** (1985), 27–34.
186. T. Mura, General theory of inclusions, in *Fundamentals of Deformation and Fracture*, edited by B. A. Bilby K. J. Miller, and J. R. Willis, Eshelby Memorial Symposium, Cambridge University Press, Cambridge, 1985, pp. 75–89.
187. T. Mura, Boundary problems for dislocations, in *Dislocations in Solids*, edited by H. Suzuki, T. Ninomiya, K. Sumino, and Shin Takeuchi, Yamada Conference IX, University of Tokyo Press, Tokyo, 1985, pp. 17–23.
188. T. Mura, Sliding inclusions, in *The Mechanics of Dislocations*, edited by E. C. Aifantis and J. P. Hirth, International Symposium on Mechanics of Dislocations, American Society of Metals, 1985, pp. 77–79.
189. T. Mura and M. Taya, Residual stresses in and around a short fiber in metal matrix composites due to temperature change, in *Recent Advances in Composites in the United States and Japan*, edited by J. R. Vinson and M. Taya, ASTM STP 864, 1985, pp. 209–224.
190. Y. Hirose and T. Mura, Crack nucleation and propagation of corrosion fatigue in high-strength steel, *Engng. Fracture Mech.*, **22** (1985), 859–870.

191. T. Mura, A note on the crack opening displacement, *Mech. Mater.*, **4** (1985), 213–214.
192. T. Mura, I. Jasiuk, and B. Tsuchida, The stress field of sliding inclusion, *Int. J. Solids Structures*, **21** (1985), 1165–1179.
193. T. Mura, A new NDT: Evaluation of plastic strains in bulk from displacements on surfaces, *Mech. Res. Commun.*, **12** (1985), 243–248.
194. Y. Hirose and T. Mura, The effect of prior austinite grain size on the stress corrosion cracking susceptibility of AISI 4340 steel, in *Predictive Capabilities Environmentally Assisted Cracking*, edited by R. Rungta, ASME PVP 99, 1985, pp. 245–257.
195. M. R. Lin, M. E. Fine, and T. Mura, Fatigue crack initiation on slip bands: Theory and experiment, *Acta Metallurgica*, **34** (1986), 619–628.
196. E. Tsuchida, T. Mura, and J. Dundurs, The elastic field of an elliptic inclusion with a slipping interface, *J. Appl. Mech.*, **53** (1986), 103–107.
197. N. Kinoshita and T. Mura, An ellipsoidal inclusion with polynomial eigenstrains, *Quart. Appl. Math.*, **44** (1986), 195–199.
198. Y. Hirose, Z. Yajima, and T. Mura, X-ray fractography on fatigue fracture surfaces of AISI 4340 steel, in *Advances in X-Ray Analysis*, edited by C. S. Barrett, J. B. Cohen, J. Faber, R. Jenkins, D. E. Leyden, J. C. Russ, and P. K. Predecki, Plenum, New York, 29, 1986, pp. 63–70.
199. T. Mura, B. Cox, and Z. Gao, Computer-aided nondestructive measurements of plastic strains from surface displacements, in *Computational Mechanics '86. Theory and Applications*, edited by G. Yagawa and S. N. Atluri, Springer-Verlag, Tokyo, 1986, pp. II43–II48.
200. D. A. Kouris, E. Tsuchida, and T. Mura, An anomaly of sliding inclusions, *J. Appl. Mech.*, **53** (1986), 724–726.
201. T. Mura, I. M. Jasiuk, D. A. Kouris, and R. Furuhashi, Recent results of the equivalent inclusion method applied to composite materials, in *Composite '86 Recent Advances in Japan and the United States*, edited by K. Kawata, S. Umekawa, and A. Kobayashi, Proc. Japan–U.S., OCM-III, Tokyo, 1986, pp. 169–177.
202. E. N. Mastrojannis, L. M. Keer, and T. Mura, Thin circular plate under temperature loading in adhesive contact with an elastic half-space, *J. Thermal Stresses*, **10** (1987), 71–81.
203. D. A. Sotiropoulous and T. Mura, Torsion of an elasto-plastic bar via a dislocation method, *J. Appl. Mech.*, **54** (1987), 226–227.
204. S. J. Chang and T. Mura, Inclined pileup of screw dislocations at the crack tip with a dislocation-free zone, *Int. J. Engng. Sci.*, **25** (1987), 561–576.
205. R. R. Castles and T. Mura, Theory and application of harmonic eigenstrains, *Quart. J. Appl. Math.*, **40** (1987), 169–188.
206. T. Mura, The eigenstrains method applied to fracture and fatigue mechanics, in *Role of Fracture Mechanics in Modern Technology*, edited by G. C. Sih, H. Nishitani, and T. Ishihara, North-Holland Amsterdam, 1987, pp. 145–152.
207. T. Mura, R. Furuhashi, and T. Mori, Sliding ellipsoidal inhomogeneities under shear, in *Advanced Composite Materials and Structures*, edited by G. C. Sih and S. E. Hsu, VNU Science Press, The Netherlands 1987, pp. 113–122.
208. M. Morinaga, N. Yukawa, H. Adachi, and T. Mura, Electronic stability effect on local strain in martensite, *J. Phys. F: Met. Phys.*, **17** (1987), 2147–2162.

209. T. Mori and T. Mura, Blocking effect of inclusions on grain boundary sliding: spherical grain approximation, *J. Mech. Phys. Solids*, **35** (1987), 631–641.
210. N. Ahmadi, L. M. Keer, T. Mura, and V. Vithoontien, The interior stress field caused by tangential loading of a rectangular patch on an elastic half-space, *J. Tribology* (1987), 627–629.
211. I. Jasiuk, E. Tsuchida, and T. Mura, The sliding inclusion under shear, *Int. J. Solids Structures*, **23** (1987), 1373–1385.
212. T. Mura, Inclusion problems, *Appl. Mech. Rev.*, **41** (1988), 15–20.
213. I. Jasiuk, T. Mura, and E. Tsuchida, Thermal stresses and thermal expansion coefficients of short fiber composites with sliding interfaces, *J. Engng. Materials Technology (Trans. ASME)*, **110** (1988), 105–100.
214. B. N. Cox, D. B. Marshall, D. Kouris, and T. Mura, Surface displacement analysis of the transformed zone in magnesia partially stabilized zirconia, *J. Engng. Materials Technology (Trans. ASME)*, **110** (1988), 105–109.
215. M. Morinaga, N. Yukawa, H. Adachi, and T. Mura, Electronic state of interstitial atoms (C, N, O) in FCC Fe, *J. Phys. F: Met. Phys.*, **18** (1988), 923–934.
216. T. Mura, Advancement of micromechanics, *Japan Soc. Precision Engng.*, **54** (1988), 1040–1045.
217. A. Sato, Y. Watanabe, and T. Mura, Octahedral defects in a b.c.c. lattice examined by lattice theory, *J. Phys. Chem. Solids*, **49** (1988), 529–540.
218. E. N. Mastrojannis, L. M. Keer, and T. Mura, Axisymmetrically loaded thin circular plate in adhesive contact with an elastic half-space, *Comput. Mech.*, **3** (1988), 283–298.
219. I. B. Kwon, M. B. Fine, and T. Mure, Elastic strain energy analysis of the dislocation structures in fatigue, *Acta Metallurgica*, **36** (1988), 2605–2614.
220. S. Shibata, T. Mori, and T. Mura, Crack arrest by strong short fibers in a composite, in *Mechanical and Physical Behavior of Metallic and Ceramic Composites*, edited by S. I. Andersen, H. Lilholt, and O. B. Pedersen, 9th Riso International Symposium on Metallurgy and Materials Science, 1988, pp. 469–474.
221. T. Mori, K. Saito, and T. Mura, An inclusion model for crack arrest in a composite reinforced by sliding fibers, *Mech. Mater.*, **7** (1988), 49–58.
222. Y. Murakami, T. Mura, and M. Kobayashi, Change of dislocation structures and macroscopic conditions from initial state to fatigue crack nucleation, in *Basic Questions in Fatigue*, Vol. I, edited by J. T. Fong and R. J. Fields, American Society for Testing and Materials, 1988, pp. 39–63.
223. T. Mura, N. Yamashita, T. Mishima, and Y. Hirose, A dislocation model for hardness indentation problems—I, *Int. J. Engng. Sci.*, **27** (1989), 1–4.
224. K. Tanaka, H. Hoguchi, and T. Mura, A dislocation model for hardness indentation problems—II, *Int. J. Engng. Sci.*, **27** (1989), 11–28.
225. T. Mura, Impotent dislocation walls, *Mater. Sci. Engng. A.* **113** (1989), 149–152.
226. Z. Gao and T. Mura, Nondestructive evaluation of interfacial damages in composite materials, *Int. J. Solids Structures*, **7** (1989), 500–512.
227. Z. Gao and T. Mura, On the inversion of residual stresses from surface displacements, *J. Appl. Mech.*, **56** (1989), 530–537.
228. T. Mura and Z. Gao, Inverse problems in plasticity, '89, *2nd International Symposium on Plasticity and Its Current Applications*, edited by A. S. Khan and M. Tokuda, Pergamon Press, Oxford, 1989, pp. 573–576.
229. T. Mura and Y. Nakasone, A theory of fatigue crack initiation in solids, *J. Appl. Phys.*, **56** (1989), 300–306.

List of Contributors

HIROYUKI ABÉ, Department of Mechanical Engineering, Tohoku University, Sendai 980, Japan.

E. C. AIFANTIS, Department of Mechanical Engineering and Engineering Mechanics, Michigan Technological University, Houghton, MI 49931, U.S.A.

DAVID H. ALLEN, Aerospace Engineering Department, Texas A&M University, College Station, TX 77843, U.S.A.

MARC-HENRI AMBROISE, Laboratoire PMTM, CNRS, Université Paris-Nord, Villetaneuse, France.

P. F. BECHER, Metals and Ceramics Division, Oak Ridge National Laboratory, Oak Ridge, TN 37831-8051, U.S.A.

Y. BENVENISTE, Department of Civil Engineering, Rensselaer Polytechnic Institute, Troy, NY 12180-3590, U.S.A.

BRUNO A. BOLEY, Department of Civil Engineering and Engineering Mechanics, Columbia University, New York, NY 10027, U.S.A.

THIERRY BRETHEAU, Laboratoire PMTM, CNRS, Université Paris-Nord, Villetaneuse, France.

S.-J. CHANG, Engineering Technology Division, Oak Ridge National Laboratory, Oak Ridge, TN 37831-8051, U.S.A.

X. Y. CHEN, Department of Civil Engineering, University of California, Los Angeles, CA 90024-1593, U.S.A.

TSU-WEI CHOU, Center for Composite Materials, and Department of Mechanical Engineering, University of Delaware, Newark, DE 19716, U.S.A.

JOHN DUNDURS, Departments of Civil Engineering and Mechanical Engineering, Northwestern University, Evanston, IL 60208, U.S.A.

G. J. DVORAK, Department of Civil Engineering, Rensselaer Polytechnic Institute, Troy, NY 12180-3590, U.S.A.

MARTIN A. EISENBERG, Department of Aerospace Engineering, Mechanics and Engineering Science, University of Florida, Gainesville, FL 32611, U.S.A.

Alan D. Freed, NASA–Lewis Research Center, Cleveland, OH 44135, U.S.A.

G. Herrmann, Division of Applied Mechanics, Stanford University, Stanford, CA 94305-4040, U.S.A.

T. Honein, Division of Applied Mechanics, Stanford University, Stanford, CA 94305-4040, U.S.A.

M. Hori, Center of Excellence for Advanced Materials, Department of Applied Mechanics and Engineering Sciences, University of California, San Diego, La Jolla, CA 92093, U.S.A.

H. Horii, Department of Civil Engineering, University of Tokyo, Bunkyo-ku, Tokyo, Japan.

M. Iwamoto, Department of Mechanical and System Engineering, Kyoto Institute of Technology, Matugasaki, Sakyo-ku, Kyoto 606, Japan.

Iwona Jasiuk, Department of Metallurgy, Mechanics and Materials Science, Michigan State University, East Lansing, MI 48824-1226, U.S.A.

William C. Johnson, Carnegie Mellon University, Department of Metallurgical Engineering and Materials Science, Pittsburgh, PA 15213-3890, U.S.A.

Eric H. Jordan, University of Connecticut, Storrs, CT 06268, U.S.A.

Yutaka Kanoh, Department of Mechanical Engineering, Tohoku University, Sendai 980, Japan.

L. M. Keer, Department of Civil Engineering, Northwestern University, Evanston, IL 60208, U.S.A.

H. Koguchi, Department of Mechanical Engineering, Nagaoka University of Technology, Tomiokacho 1603-1, Nagaoka 940-21, Japan.

Demitris Kouris, Department of Mechanical and Aerospace Engineering, Arizona State University, Tempe, AZ 85287-6106, U.S.A.

E. Kröner, Institut für Theoretische und Angewandte Physik der Universität Stuttgart, Pfaffenwaldring 57, 7000 Stuttgart 80, Federal Republic of Germany.
Max-Planck-Institut für Metallforschung, Heisenbergstraße 1, 7000 Stuttgart 80, Federal Republic of Germany.

Jong-Won Lee, Aerospace Engineering Department, Texas A&M University, College Station, TX 77843, U.S.A.

J. K. Lee, Department of Metallurgical Engineering, Michigan Technological University, Houghton, MI 49931, U.S.A.

W. Lin, Department of Civil Engineering, Northwestern University, Evanston, IL 60208, U.S.A.

T. H. Lin, Department of Civil Engineering, University of California, Los Angeles, CA 90024-1593, U.S.A.

Shen-Yi Luo, Center for Composite Materials, and Department of Mechanical Engineering, University of Delaware, Newark, DE 19716, U.S.A.

H. Mei, Department of Mechanical Engineering and Applied Mechanics, University of Rhode Island, Kingston, RI 02881, U.S.A.

SITIRO MINAGAWA, The University of Electro Communications, Chofu, Tokyo 182, Japan.

T. MORI, Department of Materials Science and Engineering, Tokyo Institute of Technology, 4259 Nagatsuta, Midori-ku, Yokohama 227, Japan.

Y. MURAKAMI, Department of Mechanics and Strength of Solids, Faculty of Engineering, Kyushu University, Fukuoka 812, Japan.

T. NAKAMURA, Technical Research Laboratory, Toyo Umpanki, Co., Ltd., 3-Banchi, Ryugasaki, Ibaraki 301, Japan.

S. NEMAT-NASSER, Center of Excellence for Advanced Materials, Department of Applied Mechanics and Engineering Sciences, University of California, San Diego, La Jolla, CA 92093, U.S.A.

Y. NOMURA, Department of Mechanical and System Engineering, Kyoto Institute of Technology, Matugasaki, Sakyo-ku, Kyoto 606, Japan.

A. N. NORRIS, Department of Mechanics and Materials Science, College of Engineering, Rutgers University, Piscataway, NJ 08855-0909, U.S.A.

HIROSHI OGATA, The University of Electro Communications, Chofu, Tokyo 182, Japan.

O. B. PEDERSEN, Metallurgy Department, Risø National Laboratory, DK-4000 Roskilde, Denmark.

M. H. SADD, Department of Mechanical Engineering and Applied Mechanics, University of Rhode Island, Kingston, RI 02881, U.S.A.

K. SAHASAKMONTRI, Department of Civil Engineering, University of Tokyo, Bunkyo-ku, Tokyo, Japan.

K. SAITO, Department of Mechanical and System Engineering, Kyoto Institute of Technology, Matugasaki, Sakyo-ku, Kyoto 606, Japan.

H. SEKINE, Department of Engineering Science, Tohoku University, Sendai 980, Japan.

A. SHUKLA, Department of Mechanical Engineering and Applied Mechanics, University of Rhode Island, Kingston, RI 02881, U.S.A.

Y. SUMI, Department of Naval Architecture and Ocean Engineering, Yokohama National University, Tokiwadai Hodogaya-ku, Yokohama 240, Japan.

K. TANAKA, Department of Mechanical Engineering, Nagaoka University of Technology, Tomiokacho 1603-1, Nagaoka 940-21, Japan.

MINORU TAYA, Department of Mechanical Engineering, University of Washington, Seattle, WA 98195, U.S.A.

T. C. T. TING, Department of Civil Engineering, Mechanics and Metallurgy, University of Illinois at Chicago, Chicago, IL 60680, U.S.A.

THOMAS TSAKALAKOS, Department of Mechanics and Materials Science, College of Engineering, Rutgers University, Piscataway, NJ 08855-0909, U.S.A.

EIICHIRO TSUCHIDA, Department of Mechanical Engineering, Saitama University, 255 Shimo-Okubo, Urawa 338, Japan.

K. Wakashima, Research Laboratory of Precision Machinery and Electronics, Tokyo Institute of Technology, 4259 Nagatsuta, Midori-ku, Yokohama 227, Japan.

D. Walgraef, Department of Mechanical Engineering and Engineering Mechanics, Michigan Technological University, Houghton, MI 49931, U.S.A.
Permanent address: Service de Chimie–Physique, Université Libre de Bruxelles, CP 231, B-1050 Bruxelles.

Kevin P. Walker, Engineering Science Software, Inc., Smithfield, RI 02917, U.S.A.

L. J. Walpole, School of Mathematics, University of East Anglia, Norwich NR4 7TJ, U.K.

J. Weertman, Department of Materials Science and Engineering, Department of Geological Sciences, Northwestern University, Evanston, IL 60208, U.S.A.

G. J. Weng, Department of Mechanics and Materials Science, Rutgers University, New Brunswick, NJ 08903, U.S.A.

J. R. Willis, School of Mathematical Sciences, University of Bath, Bath BA2 7AY, U.K.

André Zaoui, Laboratoire PMTM, CNRS, Université Paris-Nord, Villetaneuse, France.

Y. H. Zhao, Department of Mechanics and Materials Science, Rutgers University, New Brunswick, NJ 08903, U.S.A.

C. Y. Zhu, Department of Mechanical Engineering and Applied Mechanics, University of Rhode Island, Kingston, RI 02881, U.S.A.

List of Additional Papers Presented

Statistical Continuum Theory for Inelastic Behavior of Heterogeneous Polycrystalline Media

B. L. Adams, Department of Mechanical Engineering, Yale University, New Haven, CT 06520, U.S.A.

On Elastic–Plastic Response of Metal–Matrix Composites

A. Agah-Tehrani, Department of Theoretical and Applied Mechanics, University of Illinois, Urbana, IL 68120, U.S.A.

Compaction of Micro Voids and Generation of Cracks in Compressive Field

T. Aizawa and J. Kihara, Department of Materials Science and Engineering, University of Tokyo, Japan

Factors Influencing Behavior of Composite Micromechanical Models Applied to Biological Soft Tissues

H. K. Ault and A. H. Hoffman, Worcester Polytechnic Institute, Worcester, MA 01609, U.S.A.

The Cut-Independence of the Energy of Volterra Dislocations

D. M. Barnett, Department of Materials Science and Engineering, Stanford University, Stanford, CA 94305, U.S.A.

Free Boundary Inclusion Problem. Application to Some Metal Plasticity Problems

M. Berveiller, Laboratoire de Physique et Mecanique des Materiaux, Universite de Metz, F-57045 Metz, France

A Simplified Stress Analysis in a Discontinuous Fiber Composite

C. R. Chiang, Department of Power Mechanical Engineering, National Tsing Hua University, Hsinchu 30043, Taiwan

A Computer Simulation Study of Anisotropic Flow of Polycrystalline Aluminum Undergone Combined Loadings at Elevated Temperature

P. H. Dluzewski, Institute of Fundamental Technological Research, Swietokrzyska 21, 00-049 Warszawa, Poland

Voids Evolution and Micro–Macro Transition

J. H. FAN and X. G. ZHENG, Department of Engineering Mechanics, Chongqing University, Chongqing 63004, China

The Cylindrical Dislocation

T. N. FARRIS and I. DEMIR, School of Aeronautics and Astronautics, Purdue University, West Lafayette, IN 47907, U.S.A.

Modeling the Crush Energy of Fiber-Reinforced Composites

S. F. HOYSAN and P. S. STEIF, Department of Mechanical Engineering, Carnegie Mellon University, Pittsburgh, PA 15213, U.S.A.

Cavitation Instability in Elastic–Plastic Solids

J. W. HUTCHINSON, Division of Applied Sciences, Harvard University, Cambridge, MA 02138, U.S.A.

Constrained Sintering of Thin Films

A. JAGOTA, R. K. BORDIA, and G. W. SCHERER, E356/347, Experimental Station, Central Research and Development Department, E. I. duPont de Nemours and Co., Wilmington, DE 19880, U.S.A.

A Model for the Evaluation of Effective Properties in Multi-Phase Composite Materials

G. C. JOHNSON and M. FERRARI, Department of Mechanical Engineering, University of California, Berkeley, CA 94720, U.S.A.

An Incremental Plastic Analysis of Multiphase Materials

Y. Y. LI and Y. CHEN, Department of Mechanics and Materials Science, Rutgers University, New Brunswick, NJ 08903, U.S.A.

The Surface Dislocation Method of Analyzing Incompatibility Stress

K. NISHIOKA, Department of Mechanical Engineering, University of Tokushima, Tokushima 770, Japan; and Y. ARIMITSU, Department of Mechanical Engineering, Kochi National College of Technology, Nankoku, Kochi 783, Japan

A Numerical Viewpoint on the Vectorial and Tensorial Damage Concepts

D. R. J. OWEN, Department of Civil Engineering, University College of Swansea, Swansea SA2 8PP, U.K.

Dislocation Nucleation From Crack Tips

J. R. RICE, Division of Applied Sciences, Harvard University, Cambridge, MA 02138, U.S.A.

Some Aspects of Polycrystalline Plasticity Solved by the Finite Element Method

M. TOKUDA, Department of Mechanical Engineering, Mie University, Tsu, Mie 514, Japan; and F. HAVLICEK, National Research Institute for Machine Design, Prague, Czechoslovakia

Improved Bounds on Transport and Mechanical Properties of Composite Media

S. TORQUATO, Department of Mechanical and Aerospace Engineering, North Carolina State University, Raleigh, NC 27695, U.S.A.

Some Aspects of Mechanical Properties of Composite Materials

S. S. WANG, Department of Theoretical and Applied Mechanics, University of Illinois, Urbana, IL 68120, U.S.A.

Constitutive Equations for Elastic–Viscoplastic Materials Based on Local Random Yielding of Microelements

P. W. WHALEY, Department of Engineering Mechanics, University of Nebraska–Lincoln, Lincoln, NE 68588, U.S.A.

Differential Schemes for Predicting the Effective Elastic Moduli of Heterogeneous Materials

R. W. ZIMMERMAN, Earth Sciences Division, Lawrence Berkeley Laboratory, University of California, Berkeley, CA 94720, U.S.A.

An Inverse Problem in Nondestructive Inspection of a Crack in a Plate with an Inhomogeneity by Means of the Electrical Potential Method

HIROYUKI ABÉ and YUTAKA KANOH
Department of Mechanical Engineering, Tohoku University,
Sendai 980, Japan

Abstract

A method was presented for the nondestructive evaluation of a crack on the back wall of an infinite strip having an inhomogeneity. The weldment is a typical inhomogeneous region as far as the electrical resistivity is concerned. The method was based on the distribution of d.c. electrical potential. The inhomogeneity problem was solved to determine the location, size, and inclination of the crack in the following way. First, a homogeneity problem was solved for a cracked strip, where the crack was modeled as a continuous distribution of singularities of the electrical potential. Next, the potential difference distribution was found for a plate without crack. By combining these two results, the crack geometry in the inhomogeneous strip was determined.

Application examples showed agreement between the evaluation and the actual crack observed in the weldment, and verified the applicability of the present method.

1. Introduction

The quantitative nondestructive evaluation of a crack is required for the integrity assessment of structural components based on fracture mechanics.

Generally, structural components are recognized as having inhomogeneity. The weldment is a typical inhomogeneity. Most of the defects in the components occur in the welds and some of these are regarded as cracks.

This paper describes a method of using d.c. electrical potential for nondestructive evaluation of a crack in a plate having inhomogeneity. This is a method utilizing direct current. The d.c. electrical potential method has the advantage of easy use for industry, compared with other techniques such as ultrasonic and radiography. So far many researchers have investigated the method (Futayama and Kamata, 1979; Prater *et al.*, 1985; Miyoshi and Nakano, 1986; Hayashi *et al.*, 1986; Kubo *et al.*, 1988; Abé *et al.*, 1988; Kanoh and Abé, 1989). Most of these have treated the components, including cracks, as homogeneous materials as far as electrical resistivity is concerned.

2. Evaluation of a Crack in Homogeneous Material

The d.c. electrical potential method consists of measuring the potential distribution on the surface of the cracked component, to which constant current is applied, and determining the crack geometry based on the potential difference due to the crack.

A potential distribution on the surface of a cracked component accords with a specific crack geometry. If the relation between the crack geometry and the potential distribution is known, the crack geometry can be determined from this relation by measuring potential differences.

In the laboratory this method has been used frequently for monitoring only the crack length in the fatigue and fracture specimens (Clark and Knott, 1975). However, a crack is generally identified with plural parameters, such as location, length, and inclination. In order to determine all of these parameters an appropriate approach should be developed. As a basis for treating the inhomogeneity problem, first let us consider the problem of a crack in a homogeneous medium. This problem will be solved in Section 2, where three parameters characterizing the crack will be evaluated.

2.1. Potential Distribution on Measuring Surface

Consider a two-dimensional problem as shown in Fig. 1. The constant d.c. is applied to an infinite strip having a surface crack. The electrical potential differences are measured on the lower surface. The crack lies on another surface, i.e., the back wall of the strip. Both upper and lower surfaces are electrically insulated except for the current input and output positions on the lower surface. The crack geometry is expressed in terms of the position α, the length l, and the inclination β.

In order to provide the theoretical relation between these parameters and the potential difference distribution on the measuring surface, the crack is modeled as a continuous distribution of singularities of the electrical potential. The singular integral equation is developed and solved to determine the distribution of electrical gaps.

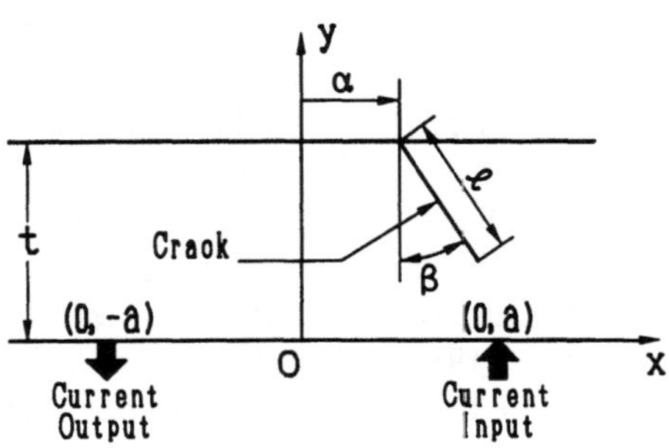

FIG. 1. A cracked infinite strip to which the constant current is applied.

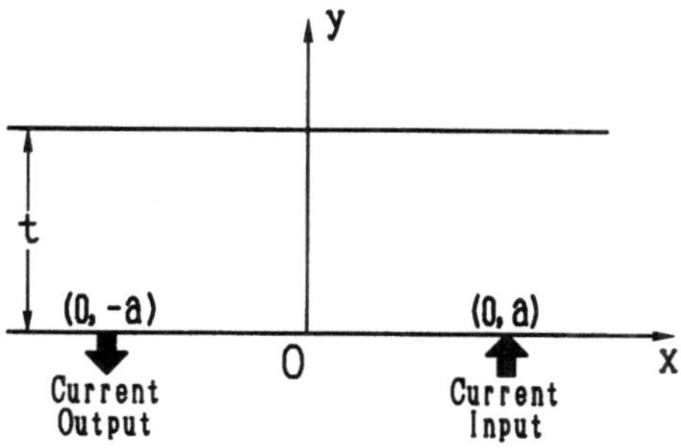

FIG. 2. An infinite strip with both upper and lower surfaces insulated except for current input and output positions, expressed by $\psi_1(z)$.

This analytical method does not need a long time for calculation compared with numerical methods such as FEM. Also it is easy to prepare the data for calculation.

The electrical potential satisfies the following Laplace equation:

$$\nabla^2\Phi(x, y) = 0, \qquad \left[\nabla^2 = \frac{\partial^2}{\partial x^2} + \frac{\partial^2}{\partial y^2}\right], \tag{2.1}$$

where x and y are the Cartesian coordinates taken as shown in Fig. 1. The solution of (2.1) is expressed by using the real part of an analytical function $\psi(z)$ of the complex variable z $(=x + iy)$ as

$$\Phi(x, y) = \mathrm{Re}[\![\psi(z)]\!]. \tag{2.2}$$

Let us divide the problem of Fig. 1 into two subproblems as shown in Figs. 2 and 3.

Let $\psi_1(z)$ be the solution for the problem of Fig. 2. Constant current is applied to the uniform strip with the same input and output positions as the problem of Fig. 1. Both the upper and lower surfaces are electrically insulated except for the current input and output positions on the lower surface. The function $\psi_1(z)$ is given by using the well-known formula in fluid mechanics

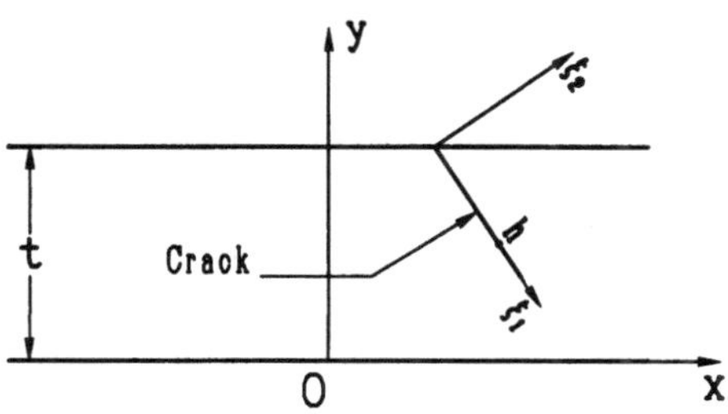

FIG. 3. A cracked strip with both upper and lower surfaces insulated, expressed by $\psi_2(z)$.

as follows:

$$\psi_1(z) = \frac{\rho m}{\pi} \ln\left\{\sinh \frac{\pi(z+a)}{2t}\right\} - \frac{\rho m}{\pi} \ln\left\{\sinh \frac{\pi(z-a)}{2t}\right\}, \tag{2.3}$$

where ρ is the electrical resistivity, m is the amount of applied current, $2a$ is the distance between the current input and output positions, and t is the thickness of the strip. The first and second terms on the right-hand side of (2.3) express the electrical flow obtained by the current input and output, respectively.

Next the problem of Fig. 3 will be treated. Both upper and lower surfaces of the strip are insulated electrically. The same amount of current as that obtained from the first problem (Fig. 2) is passed in the opposite direction from the crack surface. The boundary conditions are expressed as

$$-\frac{1}{\rho}\frac{\partial \Phi_2}{\partial y} = 0, \qquad \text{on } y = 0 \quad \text{and} \quad y = t, \tag{2.4}$$

$$-\frac{1}{\rho}\frac{\partial \Phi_2}{\partial n} = \frac{1}{\rho}\frac{\partial \Phi_1}{\partial n}, \qquad \text{on the crack surface,} \tag{2.5}$$

where Φ_1 and Φ_2 are the electrical potentials for the problems of Figs. 2 and 3, respectively, where $\Phi = \Phi_1 + \Phi_2$, and n is the normal to the crack surface.

In order to solve the problem of Fig. 3, first let us consider the problem that a singularity of electrical potential exists at position H of the semi-infinite domain characterized by the ξ, η-coordinate system (Fig. 4(a)). By setting an opposite singularity at the image position $\bar{H}$, the analytical function $\Psi_2(\zeta)$ of the complex variable ζ $(=\xi + i\eta)$ for this problem is given by

$$\Psi_2(\zeta) = \frac{ib}{2\pi}\ln(\zeta - H) - \frac{ib}{2\pi}\ln(\zeta - \bar{H}), \tag{2.6}$$

where b is the extent of the gap of the electrical potential. The function mapping the semi-infinite domain (Fig. 4(a)) into the infinite strip (Fig. 4(b)) conformally is given by (Batchelor, 1967)

$$z = \frac{t}{\pi}\ln \zeta. \tag{2.7}$$

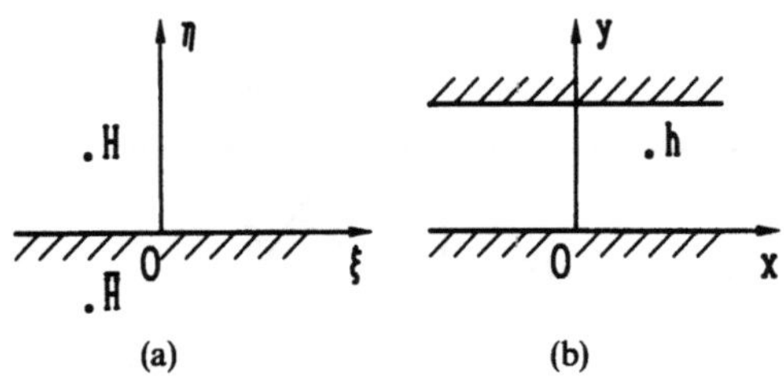

FIG. 4. Two domains with singularities of electrical potential: (a) semi-infinite domain; (b) infinite strip.

Therefore the analytical function $\psi_2(z)$ describing the solution for a problem, such that a singularity is put at the position $h\,(=(t/\pi \ln H)$ of the infinite strip with the surfaces insulated (Fig. 4(b)), is given by (2.6) and (2.7) as

$$\psi_2(z) = \frac{ib}{2\pi}\ln(e^{(\pi z/t)} - e^{(\pi h/t)}) - \frac{ib}{2\pi}\ln(e^{(\pi z/t)} - e^{(\pi \bar{h}/t)}), \tag{2.8}$$

where $\psi_2(z) = \Psi_2(e^{(\pi z/t)})$. By expressing the crack as the continuous distribution of the singularities of the electrical potential, the analytical function for the problem of Fig. 3 can be obtained.

By superposing the two problems of Figs. 2 and 3, we can find $\psi(z)$. Denoting the gap of the electrical potential on the line element $d\xi_1$ by $b(\xi_1)\,d\xi_1$, we have

$$\psi(z) = \psi_1(z) + \int_0^l \psi_2(z, \xi_1)\,d\xi_1, \tag{2.9}$$

where ξ_1 and ξ_2 are the Cartesian coordinates taken as shown in Fig. 3.

The problem is reduced to the determination of the unknown distribution density $b(\xi_1)$ from the condition for electrical insulation on the surface of the crack.

$$\frac{d}{dx}\psi(\chi) - \overline{\frac{d}{dx}\psi(\chi)} = 0, \tag{2.10}$$

where χ is a given point on the crack surface ($0 \leqq \chi \leqq l$). By substituting (2.9) into (2.10), the singular integral equation for the distribution density is given by

$$\frac{1}{\rho m\pi}\int_0^l b(\xi_1)\cdot M(\xi_1, \chi)\,d\xi_1 = P(\chi), \tag{2.11}$$

where

$$M(\xi_1, \chi) = \frac{e^{2\omega}\cos\beta - e^{\omega+\tau}\cos(\mu - \lambda + \beta)}{e^{2\omega} + e^{2\tau} - 2e^{\omega+\tau}\cos(\mu - \lambda)} - \frac{e^{2\omega}\cos\beta - e^{\omega+\tau}\cos(\mu + \lambda + \beta)}{e^{2\omega} + e^{2\tau} - 2e^{\omega+\tau}\cos(\mu + \lambda)}, \tag{2.12}$$

$$P(\chi) = \frac{1}{\pi}\left[\frac{(e^{\omega - a\pi/t} - e^{-\omega + a\pi/t})\sin\beta - 2\sin\mu\cos\beta}{e^{\omega - a\pi/t} + e^{-\omega + a\pi/t} - 2\cos\mu} - \frac{(e^{\omega + a\pi/t} - e^{-\omega - a\pi/t})\sin\beta - 2\sin\mu\cos\beta}{e^{\omega + a\pi/t} + e^{-\omega - a\pi/t} - 2\cos\mu}\right], \tag{2.13}$$

with

$$\omega = \frac{\chi\cos\beta + \alpha}{t}\pi, \qquad \mu = \frac{\chi\sin\beta + t}{t}\pi,$$

$$\tau = \frac{\xi_1\cos\beta + \alpha}{t}\pi, \qquad \lambda = \frac{\xi_1\sin\beta + t}{t}\pi. \tag{2.14}$$

The singular integral equation is solved by using the technique developed by Erdogan and Gupta (1972). The singular behavior of $b(\xi_1)$ at $\xi_1 = l$ can be described by $1/\sqrt{l-\xi_1}$ (Saka and Abé, 1983). This mathematical procedure is similar to that in two-dimensional elastic crack problems, see Fujino *et al.* (1984), for example.

The potential differences are measured on the lower surface ($y = 0$). Once the unknown distribution $b(\xi_1)$ is determined by solving (2.11), substitution of (2.9) into (2.2) gives the potential distribution on the measuring surface as

$$\Phi(x, 0) = \frac{\rho m}{2\pi} \ln \frac{\cosh \frac{x+a}{t}\pi - 1}{\cosh \frac{x-a}{t}\pi - 1} - \frac{1}{2\pi} \int_0^l b(\xi_1) \left[\tan^{-1} \left\{ \frac{-e^{\tau} \sin \lambda}{e^{(x/t)\pi} - e^{\tau} \cos \lambda} \right\} - \tan^{-1} \left\{ \frac{e^{\tau} \sin \lambda}{e^{(x/t)\pi} - e^{\tau} \cos \lambda} \right\} \right] d\xi_1. \tag{2.15}$$

2.2. *Evaluation Procedure*

The potential distribution on the measuring surface involves all information with respect to the unknown parameters characterizing the crack, as can be seen in (2.15). Since it is not easy to determine these parameters directly from the measured values of the distribution, an evaluation procedure, consisting of three steps, is presented as shown in Fig. 5.

Before determination of these three parameters, we consider estimating an approximate crack position. This estimation makes it possible to set the

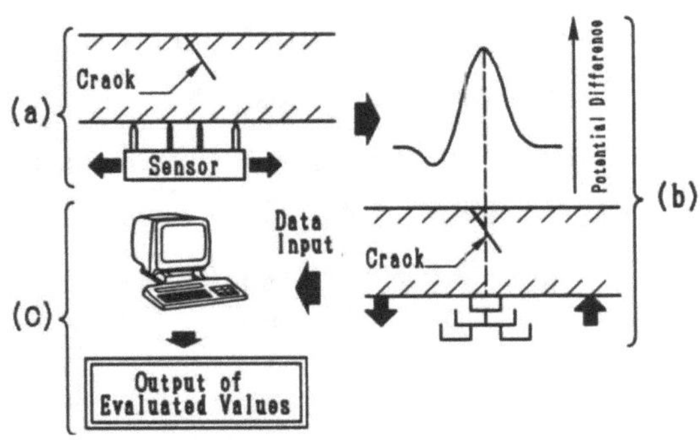

FIG. 5. Evaluation procedure for a surface crack in the homogeneous material: (a) estimation of approximate position; (b) detailed measurements of potential differences; (c) quantitative determination of crack geometry.

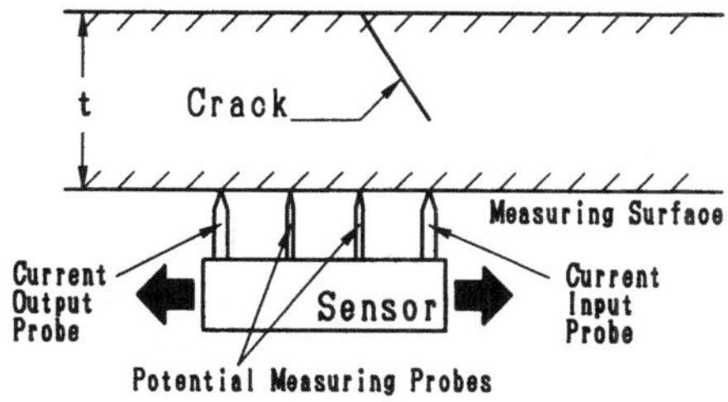

FIG. 6. A sensor for estimating approximate crack position.

current input and output probes on suitable positions, so as to measure the potential differences with adequate influence on the crack for the determination of its geometry.

Let the sensor, which has both measurement probes and the current input and output probes, be moved along the measuring surface, as shown in Fig. 6. Figure 7 shows the variation of the nondimensional potential difference $\Delta\phi$ with the position X_s of the sensor, where the potential difference is divided by that obtained for the uniform strip. It is easily shown that the position of the crack nearly corresponds to the location where the distribution attains its maximum. Similar results have been obtained in all kinds of examined cracks. Therefore, by detecting the location for the maximum value of the distribution, we can find the approximate position of the crack, α_m.

For the purpose of evaluating the crack position accurately, together with the other parameters, first set the current input and output probes on the measuring surface in such a way that the approximate position is found in the center between these two probes, as shown in Fig. 8. Let $\Delta\Phi_{mi}$ be the potential difference between the measuring probes, where i denotes the position of the left-hand side probe. Then, detailed measurements $\Delta\Phi_{mi}$ are

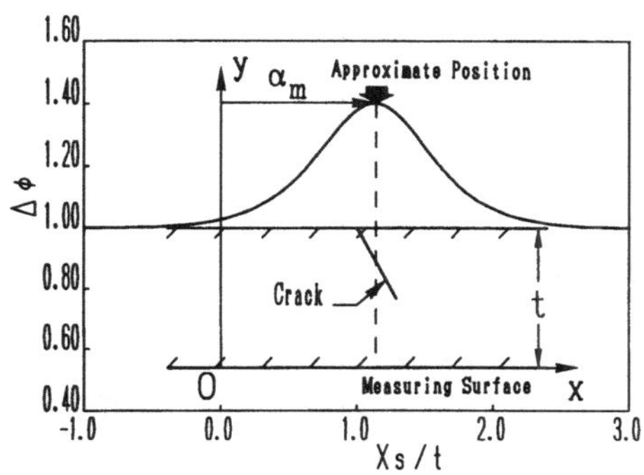

FIG. 7. The maximum location of the potential distribution and approximate crack position.

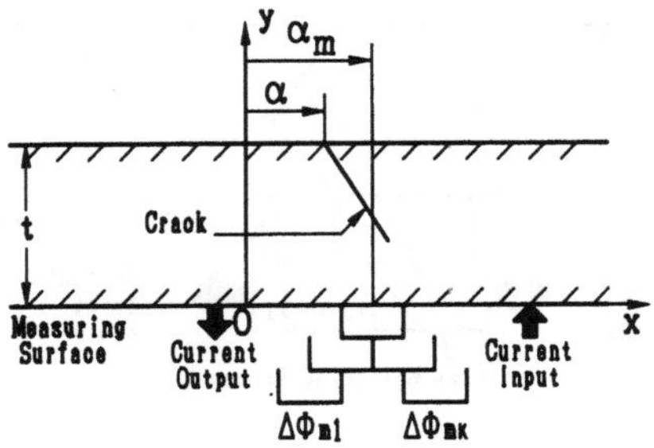

FIG. 8. The relation of positions between current probes and measurement probes in the detailed measurement.

carried out on a number of points on the surface near the approximate crack position.

Let F be defined as follows:

$$F = \sum_{i=1}^{K} (\Delta\Phi_{mi} - \Delta\Phi_i)^2, \tag{2.16}$$

where $\Delta\Phi_i$ are the potential differences obtained analytically using (2.15) at position i of the measuring probe, and K is the total number of measurement points.

An inverse problem with respect to $\Delta\Phi_i$ is solved by using the procedure for optimization. The optimization compares the measured potential differences $\Delta\Phi_{mi}$ with the analytical results $\Delta\Phi_i$. Both potential differences $\Delta\Phi_{mi}$ and $\Delta\Phi_i$ are divided by the corresponding values obtained at α_m, respectively. The unknown parameters α^* $(=\alpha - \alpha_m)$, l, and β are determined by means of the optimization procedure developed by Powell (1964), so as to minimize the function F.

2.3. *Numerical Simulation*

In order to investigate the accuracy of the present evaluation method, the potential difference distribution on the measuring surface obtained from the numerical simulation (FEM) is used, instead of the measured distribution by the experiment.

The result is shown in Fig. 9. The solid line indicates the objective geometry to be determined. The evaluation procedure was started by assuming an initial geometry indicated by a broken line and finally the evaluated geometry, indicated by a dot–dash line, was obtained through optimization. The number of total measurement points K was taken as 25. The evaluated crack was found almost near the objective crack, and it was shown that the present method is useful for the quantitative sizing of a crack.

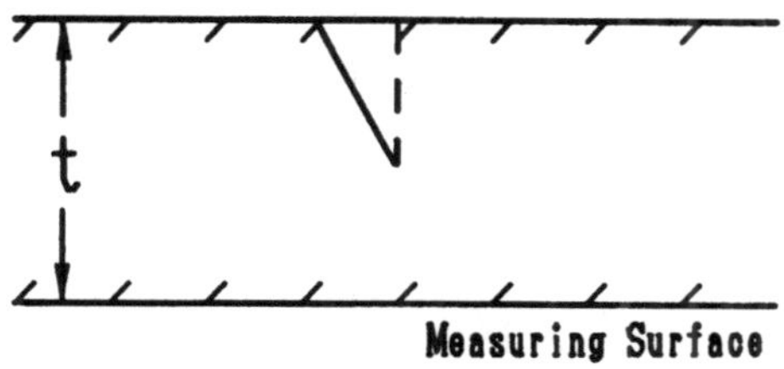

	α^*/t	l/t	β	Mark
Objective geometry	−0.300	0.577	30.0°	———
Initial geometry	0.000	0.500	0.000°	----
Evaluated geometry	−0.301	0.578	30.1°	———

$t = 20$ mm, $K = 25$.

FIG. 9. Evaluated result by numerical simulation.

3. Evaluation of a Crack in an Inhomogeneous Region

Let us consider the quantitative sizing of a crack in an inhomogeneous region with respect to electrical resistivity. The calculated potential difference distribution should be calibrated, taking account of the effect of electrical inhomogeneity. The crack geometry can be evaluated by applying the calibration to the homogeneous problem stated in Section 2. The weldment is a typical example of inhomogeneity. Some defects which are found in the weld can be regarded as cracks. These are divided into two types. One appears in welding processes as lack of fusion or incomplete penetration. The other is a fatique crack or a stress corrosion crack encountered in service. Of course, the former crack often spreads to the latter one.

3.1. Effect of Electrical Inhomogeneity on Potential Difference

Electrical resistivity of the weld differs from that of the base metal, and the component including the weld has an electrical inhomogeneity. Therefore, the quantitative sizing of a crack requires calibrating the effect of the inhomogeneity on the potential difference.

The test specimen having thickness 13 mm, width 9 mm, and length 110 mm is made from austenitic stainless steel. The specimen contains a metal inert gas arc weld in its center, as shown in Fig. 10. The crack is simulated by a machined slit (width 0.3 mm) which lies in the heat-affected zone. A current

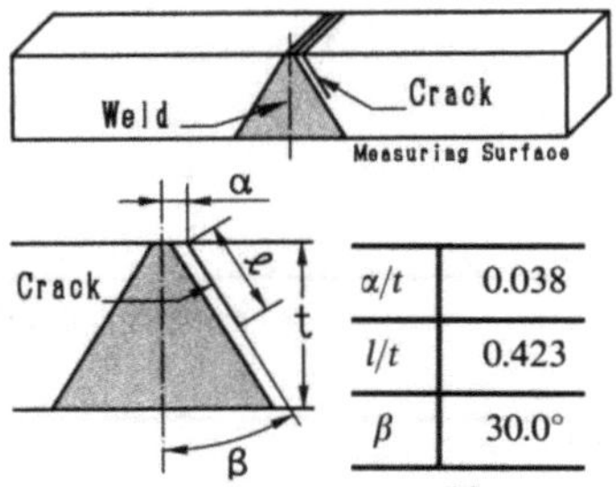

α/t	0.038
l/t	0.423
β	30.0°

$t = 13$ mm

FIG. 10. Test specimen.

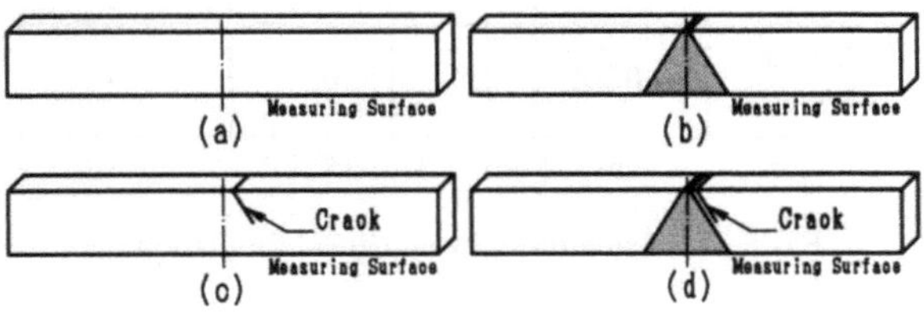

FIG. 11. Four problems to observe the effect of the inhomogeneity: (a) uniform problem; (b) weld problem; (c) crack problem; (d) crack and weld problem.

of 4 A is applied by a constant current source and the potential differences are measured by a microvoltmeter. The distance between the current input and output probes is 70 mm and that of the measurement probes is 10 mm.

Consider four problems, as shown in Fig. 11, in observing the change in the electrical potential distribution due to the weld as well as to the crack. Both problems (a) and (b) have no crack. On the other hand, both problems (c) and (d) have cracks with the same geometry. The weld shown in problem (b) is the same as that in (d). Figure 12 shows the electrical potential difference $\Delta\Phi$

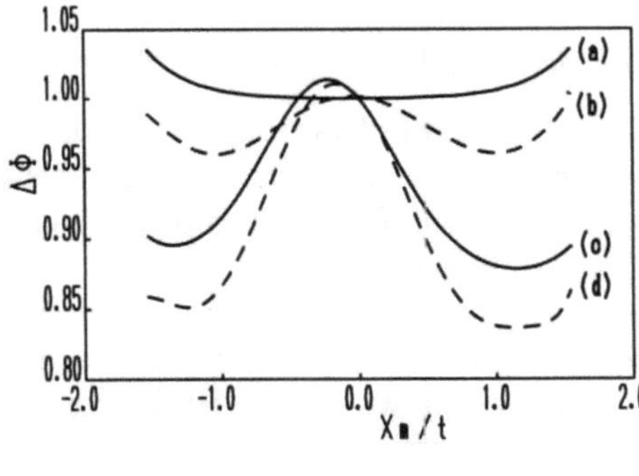

FIG. 12. Potential difference change caused by the weld.

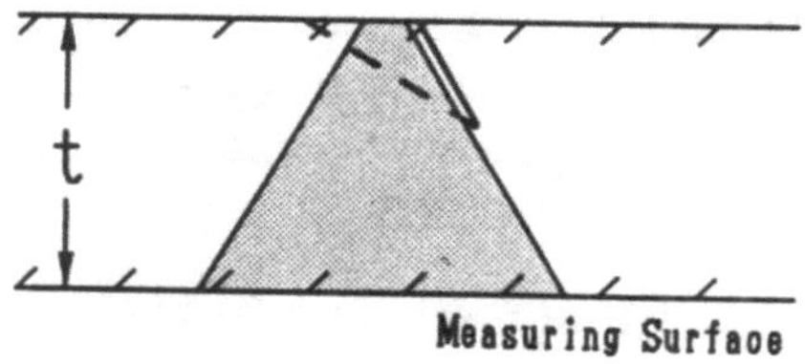

	α/t	l/t	β	Mark
Actual slit	0.038	0.423	30.0°	———
Evaluated geometry	−0.377	0.761	58.8°	----

$t = 13$ mm, $K = 25$.

FIG. 13. Evaluated result without calibration.

obtained from these problems, where the current input and output probes are set in such a way that the center of the specimen is found between these probes. Measurement is carried out near the center of the specimen. Let X_m be the distance from the center of the specimen to the center of the measurement probes. As for the nondimensionalization of the potential differences, position α_m has been replaced by X_m, being zero since the center of the specimen ($X_m = 0$) was almost the same as the approximate position of the crack (α_m). The distributions in (a) and (c) are calculated by using (2.15), and those in (b) and (d) are obtained from the measurement.

Compare (b) having the weld with the uniform specimen in (a). The distribution in (b) differs from that in (a), showing the effect of the weld on the potential difference distribution. The degree of the effect varies with X_m. In problem (b) the original dimensional values of the potential differences have been large in the weld compared with those in the base metal. The distribution in (d) differs from that in (c) in like manner. Now let us make an attempt to evaluate the crack geometry using the distribution in (d), without taking account of the effect of the weld on the potential differences. Figure 13 shows the result obtained in this way. The solid line indicates the geometry of the actual slit and the broken line represents the evaluated geometry. Even though the position of the evaluated crack tip almost corresponds to the actual one, the evaluation carried out for the length and inclination are not acceptable. Thus calibration is necessary where the effect of the weld is taken into account.

3.2. *Calibration Technique by Taking Account of the Effect of Inhomogeneity*

Let us consider the effect of the inhomogeneity on the potential differences by comparing the measured values in (d) with the distribution in (c) in Fig. 12.

Let $N(X_m)$ be $\Delta\Phi$ measured for (a) and let $\Delta C(X_m)$ be the change of the potential difference distribution caused by the crack. The distribution in (c), $C(X_m)$, obtained for the homogeneous material with a crack, is given in terms of $N(X_m)$ and $\Delta C(X_m)$ as follows:

$$C(X_m) = N(X_m) + \Delta C(X_m). \tag{3.1}$$

In the same way the potential differences in (b), denoted by $W(X_m)$, which describe the problem with the inhomogeneity having no crack, can be expressed as

$$W(X_m) = N(X_m) + \Delta W(X_m), \tag{3.2}$$

where $\Delta W(X_m)$ indicates the variation obtained by the inhomogeneity.

The change of distribution in (d) from that in (a) is considered to be caused by both the crack and the inhomogeneity. So the change may be expressed in terms of both the superposition and product of $\Delta C(X_m)$ and $\Delta W(X_m)$. The distribution in (d), denoted by $WC(X_m)$, may be put in the form

$$WC(X_m) = N(X_m) + \Delta C(X_m) + \Delta W(X_m) + \frac{\Delta C(X_m)\cdot\Delta W(X_m)}{A(X_m)}, \tag{3.3}$$

where $1/A(X_m)$ is the coefficient of the product term. It is assumed that $A(X_m)$ equals $N(X_m)$. Substitution of (3.1) and (3.2) into (3.3) gives

$$\frac{WC(X_m)}{C(X_m)} = \frac{W(X_m)}{N(X_m)}. \tag{3.4}$$

Equation (3.4) shows that, for any value of X_m, the potential change ratio for the case of having a weld to the case of having no weld is independent of the existence of the crack. It is noted that Futayama and Kamata (1979) have used the same expression as (3.4) for $X_m = 0$. The validity of (3.4) will be shown in Section 3.3.

Upon putting $WC(X_m)$, as the calibrated potential difference distribution,

$$\text{calibrated potential difference} = \frac{W(X_m)}{N(X_m)} C(X_m), \tag{3.5}$$

where $C(X_m)$ is the distribution calculated in Section 2. To use this equation both $N(X_m)$ and $W(X_m)$ must be prepared in advance.

The procedure for evaluating the crack in the inhomogeneous region is shown in Fig. 14, where the step which calibrates the calculated values by means of (3.5) is added.

3.3. Experimental Result

The potential distribution calculated for the homogeneous problem was indicated by a solid line in Fig. 15(a). The calculated result after the calibration was also shown. The calibrated potential distribution was denoted by a dot–dash line. The dot–dash line agrees well with the broken line. The

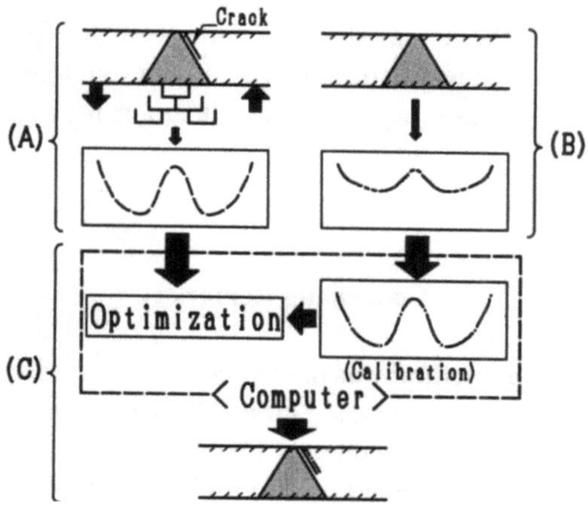

FIG. 14. Evaluation procedure for a surface crack in the inhomogeneous region: (A) detailed measurement near the weld; (B) evaluation of the effect of weld; (C) introduction of the effect of weld and quantitative determination of crack geometry.

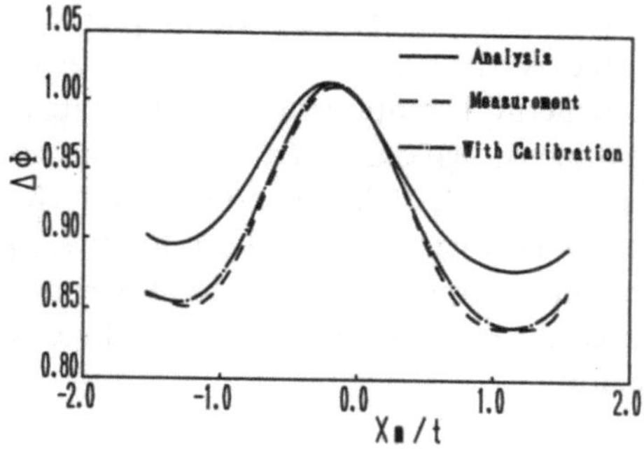

(a)

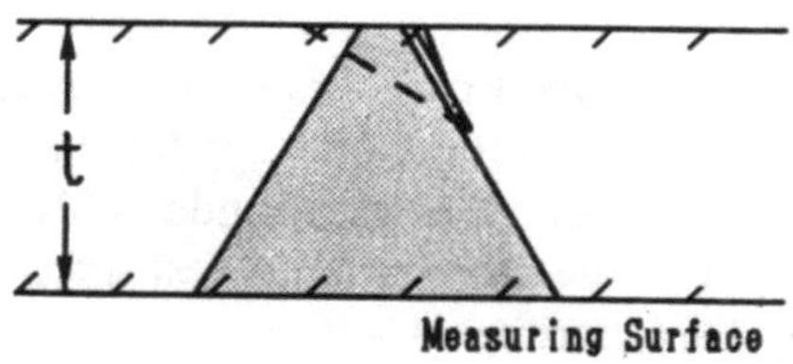

	α/t	l/t	β	Mark
Actual slit	0.038	0.423	30.0°	———
Evaluated geometry	−0.377	0.761	58.8°	- - - -
With calibration	0.077	0.408	21.6°	—·—

$t = 13$ mm, $K = 25$.

(b)

FIG. 15. Experimental result: (a) calibrated result; (b) evaluated result.

evaluated geometry was shown in Fig. 15(b). The evaluation geometry was found to approximate the actual slit with high accuracy for the length and inclination, as well as for the position. It was shown that (3.4) is valid, so that the present calibration technique is useful for the quantitative sizing of a crack in the weldment.

4. Concluding Remarks

A d.c. electrical potential method was presented for the nondestructive evaluation of a crack in an inhomogeneous region. The weld in an infinite strip has been taken as a typical inhomogeneous problem with respect to electrical resistivity. This problem was solved by combining the procedure for evaluating a crack in a homogeneous material and the calibration technique for electrical potential difference distribution obtained from a plate having inhomogeneity.

Two remarks, derived from the obtained result, are described. First, it can reasonably be expected that the present technique is applicable to other inhomogeneous problems with respect to electrical resistivity. Next, all of the actual cracks in the structural components are three-dimensional cracks. The geometry of a three-dimensional crack in a homogeneous body was evaluated in a previous paper (Abé *et al.*, 1988). Although the technique in this paper has been confined to two-dimensional cracks, it follows, by a combination of the method in the previous paper with the present technique, that the evaluation procedure can easily be extended to a three-dimensional crack in the imhomogeneous region.

Acknowledgments

The authors wish to acknowledge Professors H. Takahashi, T. Shoji and H. Niitsuma, and Dr. M. Saka of Tohoku University for their useful discussions and advice. The authors also express their gratitude to Messrs. K. Itagaki and M. Aihara for their help with the preparation of test specimens and to Mr. T. Suzuki for his help with the experiments. Appreciation is also extended to Mr. T. Shōji for his help in the preparation of the manuscript.

References

Abé, H., Saka, M., Wachi, T., and Kanoh, Y. (1988), An inverse problem in nondestructive inspection of a crack in a hollow cylinder by means of the electrical potential method, *Computational Mechanics '88*, **1**, 12.ii.2–12.ii.4.

Batchelor, G. K. (1967), *An Introduction to Fluid Dynamics*, 2nd edn., Cambridge University Press, Cambridge, p. 615.

Clark, G. and Knott, F. (1975), Measurement of fatigue cracks in notched specimen by means of theoretical electrical potential calibrations, *J. Mech. Phys. Solids*, **23**, 265–276.

Erdogan, F. and Gupta, G. D. (1972), On the numerical solution of singular integral equations, *Quart. Appl. Math.*, **29**, 525–534.

Fujino, K., Sekine, H., and Abé, H. (1984), Analysis of an edge crack in a semi-infinite composite with a long reinforced phase, *Int. J. Fract.*, **25**, 81–94.

Futayama, Y. and Kamata, H. (1979), Study on pipe inside-crack growth monitor by electric resistance method, *J. Japan Welding Soc.*, **48**, 820–824.

Hayashi, M., Ohtaka, M., Enomoto, K., Sasaki, T., and Kikuchi, T. (1986), Detection of configuration ot surface fatigue crack in inside of stainless steel pipe by DC potential drop method, *J. Soc. Mats. Sci., Japan*, **35**, 936–941.

Kanoh, Y. and Abé, H. (1989), A DC electrical potential method for discrimination and sizing of an embedded crack and a surface crack on the back wall, *Trans. JSME*, in press.

Kubo, S., Sakagami, T., Ohji, K., Hashimoto, T., and Matsumuro, Y. (1988), Quantitative measurement of three-dimensional surface crack by the electrical potential CT method, *Trans. JSME*, **54**, 218–225.

Miyshi, T. and Nakano, S. (1986), A study of determination of surface crack shape by the electric potential method, *Trans. JSME*, **52**, 1097–1104.

Powell, M. J. D. (1964), An efficient method for finding the minimum of a function of several variables without calculating derivatives, *Comput. J.*, **7**, 155–162.

Prater, T. A., Catlin, W. R., and Coffin, L. F. (1985), Application of the reversing DC electrical potential technique to monitoring crack growth in pipes, *Proceedings of the Second International Atomic Energy Agency Specialist's Meeting on Subcritical Crack Growth, Sendai, Japan.*

Saka, M. and Abé, H. (1983), A path-independent integral for 2-dimensional cracks in homogeneous isotropic conductive plate, *Int. J. Engng. Sci.*, **21**, 1451–1457.

The Effective Thermoelastic Properties of Whisker-Reinforced Composites as Functions of Material Forming Parameters

DAVID H. ALLEN and JONG-WON LEE
Aerospace Engineering Department, Texas A&M University,
College Station, TX 77843, U.S.A.

Abstract

A simple analytical method is proposed for predicting the effective thermoelastic properties of a composite material, reinforced by either continuous or discontinuous randomly oriented fibers subjected to different material forming processes. The elastic properties of an aligned fiber composite are predetermined from the equivalent inclusion method. The thermal expansion coefficients of an aligned fiber composite and the entire history of material forming processes of the same composite with initially random fiber orientation are assumed to be known *a priori*. Three combinations of hot pressing, extrusion, and rolling are investigated to quantitatively determine the effect of the forming sequence on the effective material properties. A number of special cases of the present model are compared with other theoretical models and experimental data appearing in the literature.

Nomenclature

a_{ij}	direction cosines
c^0_{ijkl}	stiffness tensor of the matrix material
c^1_{ijkl}	stiffness tensor of the inhomogeneity
C_{ijkl}	stiffness tensor of an aligned fiber composite
$(C_{ijkl})^{-1}$	compliance tensor of an aligned fiber composite
C'_{ijkl}	stiffness tensor in spherical coordinates
$\bar{C}_{ijkl}$	stiffness tensor after material forming
S_{ijkl}	Eshelby's tensor
D_a, D_b	diameters after and before extrusion
f	fiber volume fraction
t_a, t_b	thicknesses after and before hot pressing or rolling
w_a, w_b	widths after and before rolling which follows another forming
α	fiber aspect ratio

α_{ij}	thermal expansion coefficients of an aligned fiber composite
$\bar{\alpha}_{ij}$	thermal expansion coefficients after material forming
ε_{ij}	strain tensor
$\varepsilon_{ij}^0 = C_{ijkl}^0 \sigma_{kl}^0$	
ε_{ij}^1	strain disturbance due to a single inhomogeneity
$\tilde{\varepsilon}_{ij}$	average strain disturbance due to all inhomogeneities
ε_{ij}^*	eigenstrain within the inhomogeneity
σ_{ij}	stress tensor
σ_{ij}^0	applied stress
ν_0	matrix Poisson's ratio

1. Introduction

Considerable research efforts have been focused on the prediction of the effective thermoelastic properties of composite materials reinforced by either aligned or randomly oriented fibers. However, a systematic procedure has not been previously developed to model the properties of a randomly oriented fiber composite subjected to an arbitrary material forming history.

Mura (1982) discussed the usefulness of the equivalent inclusion method proposed by Eshelby (1957) for characterizing the redistribution of stress and strain due to inhomogenities in a matrix. This method has been generalized by Taya and Mura (1981) for predicting the effective stiffness and strength of a composite with aligned short fibers and fiber-end cracks, and by Taya and Chou (1981) for the effective Young's modulus of a hybrid composite containing two kinds of ellipsoidal inhomogeneities. However, the equivalent inclusion method does not include the effect of misaligned fibers which are common in fibrous composites. Although the variational method of Hashin and Rosen (1964) is capable of predicting the effective elastic constants of an aligned continuous fiber composite, it is very cumbersome to generalize this method for a short fiber composite because of the complex boundary conditions at the fiber–matrix interface. Papazian *et al.* (1987) compared a number of experimental data for SiC whiskers and particulate composites with finite element method solutions, but their finite element model is applicable only for certain specific fiber packing arrays.

A simple integration scheme was utilized by Christensen and Waals (1972) and Craft and Christensen (1984) for determining the effective elastic properties and thermal expansion coefficients, respectively, of two-dimensional or three-dimensional random continuous fiber composites. A similar integration scheme was utilized by Berthelot (1982) to determine the effect of fiber misalignment in a two-dimensional domain on the elastic properties of continuous/discontinuous fiber composites. Takao *et al.* (1982) predicted the effective longitudinal Young's modulus of misoriented short fiber composites utilizing the equivalent inclusion method together with an integration scheme.

However, the integration schemes for the possible fiber orientation mentioned above cannot be used as a general tool for a short fiber or whisker composite subjected to typical material forming processes, such as hot pressing, extrusion, and rolling.

Lee and Allen (1989) have developed an integration model representing fiber reorientation due to the three separate material forming processes mentioned above. In their study, the reoriented fiber distribution after each material forming was graphically described by a portion of a sphere. In general, a whisker composite is subjected to more than two consecutive material forming processes, for example, hot pressing followed by rolling. Therefore, the authors propose herein a systematic integration model which represents an extension of the previous model for determining the effect of combined material forming processes on the effective thermoelastic properties.

After a brief review of the equivalent inclusion method, from which the elastic constants of an aligned fiber composite are determined, a simple analytical model is presented for predicting the effective moduli and thermal expansion coefficients of a short fiber or whisker composite subjected to a number of different material forming processes.

2. Model Formulation

Consider a representative volume element of a composite reinforced by perfectly aligned fibers with finite length, as illustrated in Fig. 1(a). If the matrix thickness is assumed to be constant, the solution for the volume element will reduce to that for a continuous fiber when the fiber aspect ratio approaches

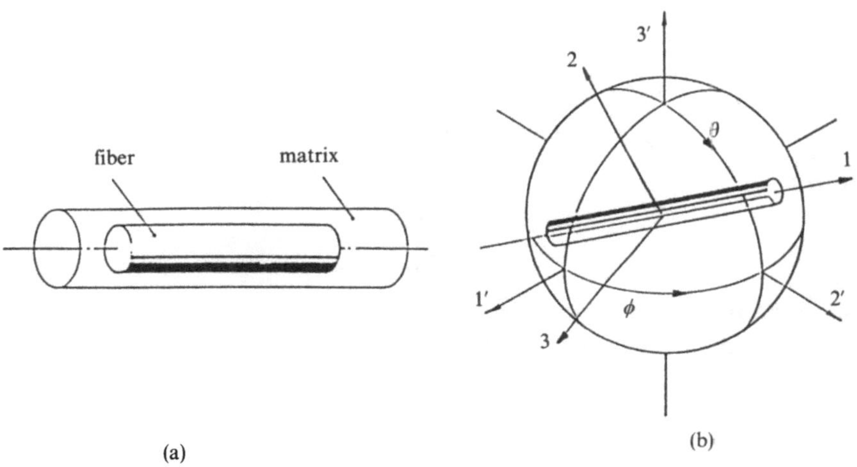

FIG. 1. Representative volume element: (a) aligned short fiber; (b) randomly oriented fiber.

infinity. Thus, due to the axial symmetry of the geometry, only five independent elastic constants and two thermal expansion coefficients of the representative composite element are required. To determine these seven constants which are equivalent to those of a perfectly aligned short fiber composite, the equivalent inclusion method discussed by Mura (1982) can be utilized as a first approximation even though the actual fiber is a cylinder of finite length rather than an ellipsoid. The present study utilizes the equivalent inclusion method developed by Taya and Mura (1981) and Taya and Chou (1981) to determine the five elastic properties of an aligned short fiber composite as input constants to the present model formulation, whereas the thermal expansion coefficients of the aligned fiber composite are assumed to be known, even though they can be determined from the equivalent inclusion method briefly discussed below.

2.1. Equivalent Inclusion Method

After Eshelby (1957) proposed the equivalent inclusion method for determining the elastic response of an ellipsoidal inclusion, as well as that of an ellipsoidal inhomogeneity, Mori and Tanaka (1973) extended the applicability of the equivalent inclusion method by including the back stress analysis which accounts for the mutual interaction among inclusions and external boundaries. The back stress analysis together with Eshelby's result was applied by Taya and Mura (1981) to aligned short fiber composites containing fiber-end cracks. Similar application of the back stress analysis to a hybrid composite was studied by Taya and Chou (1981). In the present section, this generalized equivalent inclusion method is briefly reviewed.

Consider a composite containing a single phase of ellipsoidal inhomogeneities. When all inhomogeneities are aligned in one direction, the following two equations (Taya and Mura, 1981) are obtained:

$$C^0_{ijkl}[\tilde{\varepsilon}_{kl} + f(\varepsilon^1_{kl} - \varepsilon^*_{kl})] = 0, \tag{2.1}$$

$$C^0_{ijkl}(\varepsilon^0_{kl} + \tilde{\varepsilon}_{kl} + \varepsilon^1_{kl} - \varepsilon^*_{kl}) = C^1_{ijkl}(\varepsilon^0_{kl} + \tilde{\varepsilon}_{kl} + \varepsilon^1_{kl}), \tag{2.2}$$

where

$$\varepsilon^1_{kl} = S_{klmn}\varepsilon^*_{mn}, \tag{2.3}$$

$$S_{ijkl} = S_{jikl} = S_{ijlk}. \tag{2.4}$$

Since the two different ways for expressing the strain energy stored in the composite should give identical results (Taya and Chou, 1981),

$$\tfrac{1}{2}(C_{ijkl})^{-1}\sigma^0_{ij}\sigma^0_{kl} = \tfrac{1}{2}(C^0_{ijkl})^{-1}\sigma^0_{ij}\sigma^0_{kl} + \frac{f}{2}\sigma^0_{ij}\varepsilon^*_{ij}. \tag{2.5}$$

Substituting (2.3) into (2.2) gives

$$-(C^1_{ijkl} - C^0_{ijkl})\varepsilon^0_{kl} = (C^1_{ijkl} - C^0_{ijkl})(S_{klmn}\varepsilon^*_{mn} + \tilde{\varepsilon}_{kl}) + C^0_{ijkl}\varepsilon^*_{kl}. \tag{2.6}$$

Substituting (2.1) into (2.6) gives

$$-(C^1_{ijkl} - C^0_{ijkl})\varepsilon^0_{kl} = \Xi_{ijkl}\varepsilon^*_{kl}, \tag{2.7}$$

where

$$\Xi_{ijkl} = (1-f)C^0_{ijkl} + (1-f)(C^1_{ijmn} - C^0_{ijmn})S^1_{mnkl} + fC^1_{ijkl}. \tag{2.8}$$

Inverting (2.7) gives

$$\begin{aligned}\varepsilon^*_{ij} &= -(\Xi_{ijkl})^{-1}(C^1_{klmn} - C^0_{klmn})\varepsilon^0_{mn}\\ &= -(\Xi_{ijkl})^{-1}(C^1_{klmn} - C^0_{klmn})(C^0_{mnpq})^{-1}\sigma^0_{pq}.\end{aligned} \tag{2.9}$$

From this point, the present authors propose a fully tensorial expression. Substituting (2.9) into (2.5) gives

$$\begin{aligned}\tfrac{1}{2}(C_{ijkl})^{-1}\sigma^0_{ij}\sigma^0_{kl} &= \tfrac{1}{2}(C^0_{ijkl})^{-1}\sigma^0_{ij}\sigma^0_{kl}\\ &\quad -\frac{f}{2}\sigma^0_{ij}(\Xi_{ijkl})^{-1}(C^1_{klmn} - C^0_{klmn})(C^0_{mnpq})^{-1}\sigma^0_{pq}.\end{aligned} \tag{2.10}$$

The effective compliance tensor of an aligned fiber composite, $(C_{ijkl})^{-1}$, is obtained by differentiating (2.10) twice with respect to the stress tensor.

$$(C_{ijkl})^{-1} = (C^0_{ijkl})^{-1} - \frac{f}{2}(C^1_{mnpq} - C^0_{mnpq})[(\Xi_{ijmn})^{-1}(C^0_{pqkl})^{-1} - (\Xi_{klmn})^{-1}(C^0_{pqij})^{-1}]. \tag{2.11}$$

When the shape of the inclusion is approximated by a prolate spheroid representing short fibers or whiskers, the nonvanishing terms of Eshelby's tensor (S_{ijkl}) are given by the following equations:

$$S_{1111} = \frac{1}{2(1-\nu_0)}\left[1 - 2\nu_0 + \frac{3\alpha^2 - 1}{\alpha^2 - 1} - \left\{1 - 2\nu_0 + \frac{3\alpha^2}{\alpha^2 - 1}\right\}g\right], \tag{2.12}$$

$$S_{2222} = S_{3333} = \frac{1}{4(1-\nu_0)}\left[\frac{3\alpha^2}{2(\alpha^2-1)} + \left\{1 - 2\nu_0 - \frac{9}{4(\alpha^2-1)}\right\}g\right], \tag{2.13}$$

$$S_{1122} = S_{1133} = \frac{1}{2(1-\nu_0)}\left[-(1-2\nu_0) - \frac{1}{\alpha^2-1} + \left\{1 - 2\nu_0 + \frac{3}{2(\alpha^2-1)}\right\}g\right], \tag{2.14}$$

$$S_{2211} = S_{3311} = \frac{1}{4(1-\nu_0)}\left[\frac{-2\alpha^2}{\alpha^2-1} - \left\{(1-2\nu_0) - \frac{3\alpha^2}{\alpha^2-1}\right\}g\right], \tag{2.15}$$

$$S_{2233} = S_{3322} = \frac{1}{4(1-\nu_0)}\left[\frac{\alpha^2}{2(\alpha^2-1)} - \left\{(1-2\nu_0) + \frac{3}{4(\alpha^2-1)}\right\}g\right], \tag{2.16}$$

$$S_{1212} = S_{1313} = \frac{1}{4(1-\nu_0)}\left[-2\nu_0 - \frac{2}{\alpha^2-1} + \left\{1 + \nu_0 + \frac{3}{\alpha^2-1}\right\}g\right], \tag{2.17}$$

$$S_{2323} = \frac{1}{4(1-\nu_0)}\left[\frac{\alpha^2}{2(\alpha^2-1)} + \left\{(1-2\nu_0) - \frac{3}{4(\alpha^2-1)}\right\}g\right], \tag{2.18}$$

where

$$g = \frac{\alpha}{(\alpha^2 - 1)^{3/2}} \{\alpha(\alpha^2 - 1)^{1/2} - \cosh^{-1} \alpha\}. \tag{2.19}$$

When the reinforcing fibers are continuous, the nonvanishing terms of Eshelby's tensor are given by

$$S_{2222} = S_{3333} = \frac{5 - 4\nu_0}{8(1 - \nu_0)}, \tag{2.20}$$

$$S_{2233} = S_{3322} = \frac{-1 + 4\nu_0}{8(1 - \nu_0)}, \tag{2.21}$$

$$S_{2211} = S_{3311} = \frac{1}{2(1 - \nu_0)}, \tag{2.22}$$

$$S_{2323} = \frac{3 - 4\nu_0}{8(1 - \nu_0)}, \tag{2.23}$$

$$S_{1212} = S_{1313} = \tfrac{1}{4}. \tag{2.24}$$

More detailed discussion of Eshelby's tensor for an ellipsoidal inclusion is given in Mura (1982).

Equation (2.11) gives the five independent elastic constants of a composite with aligned continuous/discontinuous reinforcing constituents. The five elastic constants are used as input data to a simple integration model of a composite subjected to a number of different material forming processes. In the next section, the reorientations of reinforcing fibers due to typical material forming processes are discussed with the aid of graphical representations which look like squashed dandelions.

2.2. *Dandelion Model*

If a piece of isotropic material is reinforced by randomly oriented short fibers, the resulting composite is also isotropic and the two independent elastic constants and one thermal expansion coefficient may be determined by integrating the stiffnesses over a spherical domain as shown in Fig. 1(b). Any material forming process, such as extrusion or rolling of the initially isotropic material, causes a permanent fiber reorientation which changes the material properties from isotropic to either transversely isotropic or orthotropic. The possible fiber orientation after a certain type of forming process can be graphically represented by a portion of a sphere if the fiber orientation density is assumed to be spatially homogeneous within the subdomain. For example, the reoriented fibers after isotropic extrusion fall within the spherical cone shown in Fig. 2(a). This fiber redistribution is radially isotropic and defined as "θ rotation." On the other hand, the dimension in the width direction is fixed during the rolling process resulting in "ϕ rotation" of the fibers as shown in Fig. 2(b). The present study includes a number of examples of combined

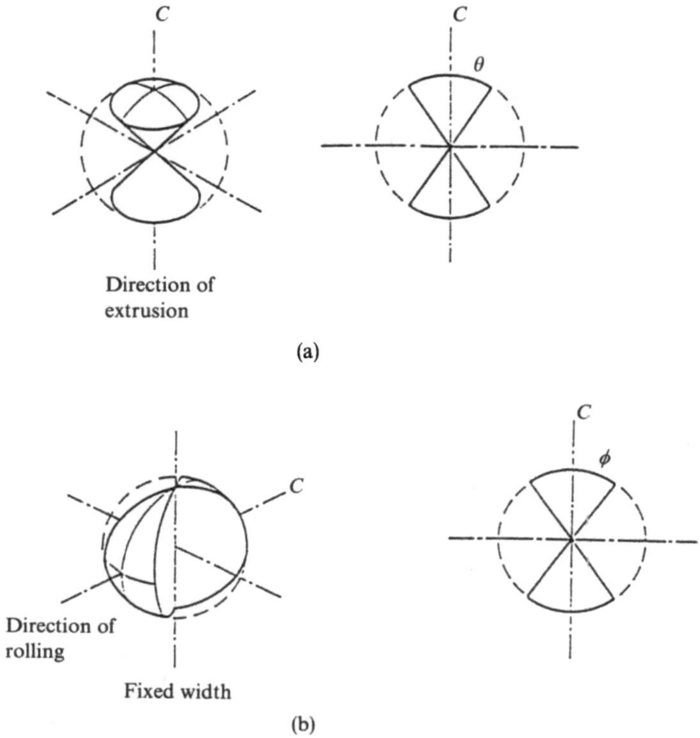

FIG. 2. Fiber reorientation due to extrusion and rolling: (a) θ rotation due to isotropic extrusion; (b) ϕ rotation due to rolling with fixed width.

fiber rotations and the change of effective thermoelastic properties of a short fiber reinforced composite due to combined material forming processes.

By combining hot pressing, extrusion, and rolling, which are the most commonly used forming processes for whisker reinforced composites, the same final shape can be obtained via a number of different forming histories. For example, a cube can be formed into a rectangular bar through the following forming processes resulting in the same dimensions.

(1) $\frac{1}{3}$ hot pressing followed by $\frac{3}{4}$ normal rolling, or $\frac{3}{4}$ hot pressing followed by $\frac{1}{3}$ normal rolling.
(2) $\frac{1}{2}$ by $\frac{1}{2}$ extrusion followed by $\frac{2}{3}$ transverse rolling.
(3) $\frac{1}{3}$ rolling with fixed width followed by $\frac{3}{4}$ rolling with fixed thickness, or $\frac{3}{4}$ rolling with fixed width followed by $\frac{1}{3}$ rolling with fixed thickness.

The graphical representations for these three cases are illustrated in Fig. 3(a), (b), and (c). Another possible forming process for the same final shape is an orthotropic extrusion where the fiber orientation is bounded within an elliptic cone. However, this last case is not discussed herein.

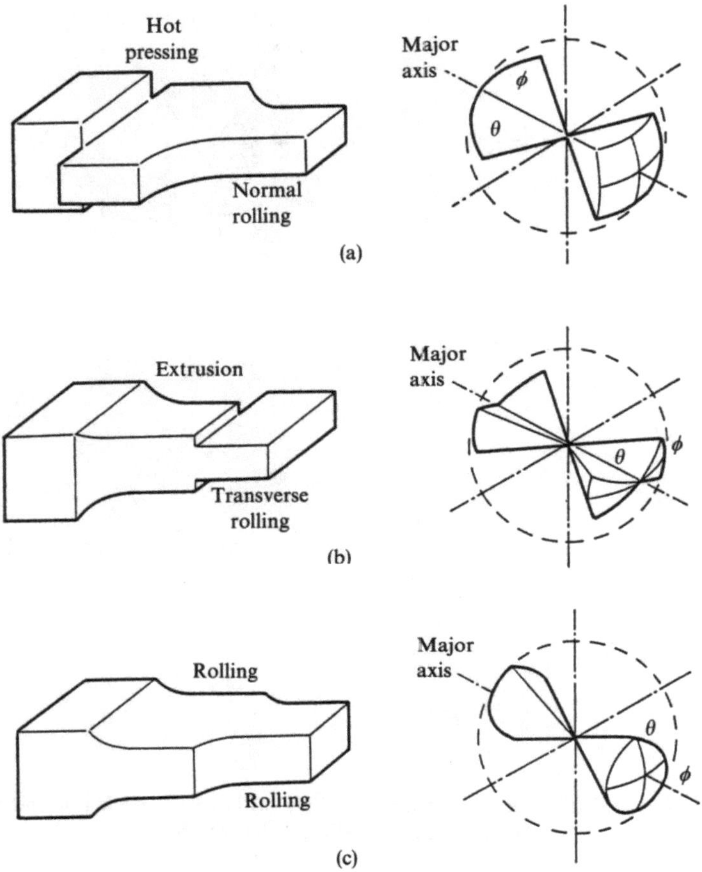

FIG. 3. Combined material forming processes: (a) hot pressing—normal rolling; (b) extrusion—transverse rolling; (c) rolling—rolling.

The general expression for the effective elastic moduli (C_{ijkl}) after any material forming process described above can be calculated by assuming statistical homogeneity and integrating the stiffnesses determined from the equivalent inclusion method in the spherical coordinate system over the relevant volume segment of a sphere. The stiffnesses in the spherical coordinate system are given by

$$C'_{ijkl} = a_{ip}a_{jq}a_{kr}a_{ls}C_{pqrs}, \tag{2.25}$$

where C_{ijkl} is the stiffness tensor determined from the equivalent inclusion method. And the direction cosines, a_{ij}, shown in Fig. 1(b), can be defined as

$$a_{ij} = \begin{bmatrix} \sin\theta\cos\phi & -\cos\theta\cos\phi & \sin\phi \\ \sin\theta\sin\phi & -\cos\theta\sin\phi & -\cos\phi \\ \cos\theta & \sin\theta & 0 \end{bmatrix}. \tag{2.26}$$

The general expression of the effective elastic moduli after an arbitrary material forming history becomes

$$\bar{C}_{ijkl} = \frac{1}{V} \int_{-\phi'}^{\phi'} \int_{-\theta'}^{\theta'} C'_{ijkl} f(\theta) \, d\theta \, d\phi, \tag{2.27}$$

where ϕ' and θ' are the ranges of integration in the ϕ and θ directions, respectively, and V is the volume of the relevant segment of a sphere with unit radius.

The stress tensor under uniform temperature change becomes

$$\begin{aligned} \sigma_{ij} &= \frac{1}{V} \int_{-\phi'}^{\phi'} \int_{-\theta'}^{\theta'} \sigma'_{ij} f(\theta) \, d\theta \, d\phi \\ &= \bar{C}_{ijkl}(\varepsilon_{kl} - \bar{\alpha}_{kl} \Delta T). \end{aligned} \tag{2.28}$$

Inverting (2.28) gives the effective thermal expansion coefficient tensor, $\bar{\alpha}_{ij}$.

For a combined material forming of longitudinal rolling and transverse rolling, it can be shown that

$$f(\theta) = \cos\left[\frac{(\theta - \pi/2)\pi}{2\eta}\right]. \tag{2.29}$$

For other combined material forming processes

$$f(\theta) = \sin \theta. \tag{2.30}$$

It should be noted that three main assumptions are required by the present integration model:

(1) The fiber distribution density is assumed to be spatially homogeneous within the subdomains illustrated in Fig. 3.
(2) The material forming processes mentioned herein are possible only through the plastic deformation of the matrix material. Plastic strain may not be spatially homogeneous in the matrix material due to stress concentrations near the reinforcing constituents. Also, it is well known that even an isotropic homogeneous material becomes transversely isotropic or orthotropic after the subject material forming processes. These plasticity effects are neglected in the present study.
(3) Although the material forming processes may cause defects, such as broken fibers, fiber–matrix debonding, etc., the present study assumes a defect-free and statistically homogeneous composite.

2.2.1. Combination of Hot Pressing and Rolling

From Fig. 3(a), the stiffness tensor after a combined forming of hot pressing and rolling becomes

$$\bar{C}_{ijkl} = \frac{1}{4\psi \sin \eta} \int_{-\psi}^{\psi} \int_{\pi/2-\eta}^{\pi/2+\eta} C'_{ijkl} \sin \theta \, d\theta \, d\phi. \tag{2.31}$$

Integrating gives

$$\begin{matrix}\bar{C}_{1111}\\ \bar{C}_{2222}\end{matrix} = C_{1111}\left[\frac{3}{4} \pm \frac{\sin 2\psi}{2\psi} + \frac{\sin 4\psi}{16\psi}\right]\left[\frac{1}{2} - \frac{\sin^2 \eta}{3} + \frac{\sin^4 \eta}{10}\right]$$

$$+ C_{2222}\left[\frac{3}{8} \mp \frac{\sin 2\psi}{4\psi} + \frac{\sin 4\psi}{32\psi} + \left\{1 - \frac{\sin 4\psi}{4\psi}\right\}\frac{\sin^2 \eta}{12}\right.$$

$$\left. + \left\{\frac{3}{4} \pm \frac{\sin 2\psi}{2\psi} + \frac{\sin 4\psi}{16\psi}\right\}\frac{\sin^4 \eta}{10}\right]$$

$$+ (2C_{1122} + 4C_{1212})\left[\left\{1 \pm \frac{\sin 2\psi}{2\psi}\right\}\left\{\frac{1}{2} - \frac{\sin^2 \eta}{6}\right\}\right.$$

$$- \left\{\frac{3}{4} \pm \frac{\sin 2\psi}{2\psi} + \frac{\sin 4\psi}{16\psi}\right\}$$

$$\left. \times \left\{\frac{1}{2} - \frac{\sin^2 \eta}{3} + \frac{\sin^4 \eta}{10}\right\}\right], \tag{2.32}$$

$$\bar{C}_{3333} = \frac{\sin^4 \eta}{5} C_{1111} + C_{2222}\left[1 - \frac{2\sin^2 \eta}{3} + \frac{\sin^4 \eta}{5}\right]$$

$$+ (2C_{1122} + 4C_{1212})\left[\frac{\sin^2 \eta}{3} - \frac{\sin^4 \eta}{5}\right], \tag{2.33}$$

$$\bar{C}_{1122} = (C_{1111} + C_{2222} - 4C_{1212})\left[\frac{1}{4} - \frac{\sin 4\psi}{16\psi}\right]\left[\frac{1}{2} - \frac{\sin^2 \eta}{3} + \frac{\sin^4 \eta}{10}\right]$$

$$+ C_{1122}\left[1 - \frac{\sin^2 \eta}{3} - \left\{\frac{1}{4} - \frac{\sin 4\psi}{16\psi}\right\}\left\{1 - \frac{2\sin^2 \eta}{3} + \frac{\sin^4 \eta}{5}\right\}\right]$$

$$+ \frac{\sin^2 \eta}{3} C_{2233}, \tag{2.34}$$

$$\begin{matrix}\bar{C}_{1133}\\ \bar{C}_{2233}\end{matrix} = (C_{1111} + C_{2222} - 4C_{1212})\left[1 \pm \frac{\sin 2\psi}{2\psi}\right]\left[\frac{\sin^2 \eta}{6} - \frac{\sin^4 \eta}{10}\right]$$

$$+ C_{1122}\left[\frac{1}{2} - \frac{\sin^2 \eta}{6} + \frac{\sin^4 \eta}{5} \pm \frac{\sin 2\psi}{2\psi}\left\{\frac{1}{2} - \frac{\sin^2 \eta}{2} + \frac{\sin^4 \eta}{5}\right\}\right]$$

$$+ C_{2233}\left\{1 \mp \frac{\sin 2\psi}{2\psi}\right\}\left\{\frac{1}{2} - \frac{\sin^2 \eta}{6}\right\}, \tag{2.35}$$

$$\begin{matrix}\bar{C}_{2323}\\ \bar{C}_{3131}\end{matrix} = (C_{1111} - 2C_{1122})\left\{1 \mp \frac{\sin 2\psi}{2\psi}\right\}\left\{\frac{\sin^2 \eta}{6} - \frac{\sin^4 \eta}{10}\right\}$$

$$+ C_{2222}\left[\frac{1}{4} + \frac{\sin^2 \eta}{12} - \frac{\sin^4 \eta}{10} \pm \frac{\sin 2\psi}{2\psi}\left\{\frac{1}{4} - \frac{\sin^2 \eta}{4} + \frac{\sin^4 \eta}{10}\right\}\right]$$

$$+ C_{1212}\left[\frac{1}{2} - \frac{\sin^2\eta}{2} + \frac{2\sin^4\eta}{5} \mp \frac{\sin 2\psi}{2\psi} \times \left\{\frac{1}{2} - \frac{5\sin^2\eta}{6} + \frac{2\sin^4\eta}{5}\right\}\right]$$

$$- C_{2233}\left\{1 \pm \frac{\sin 2\psi}{2\psi}\right\}\left\{\frac{1}{4} - \frac{\sin^2\eta}{12}\right\}, \tag{2.36}$$

$$\bar{C}_{1212} = (C_{1111} - 2C_{1122})\left\{1 - \frac{\sin 4\psi}{4\psi}\right\}\left\{\frac{1}{8} - \frac{\sin^2\eta}{12} + \frac{\sin^4\eta}{40}\right\}$$

$$+ C_{2222}\left[\frac{1}{8} + \frac{\sin^2\eta}{12} + \frac{\sin^4\eta}{40} - \frac{\sin 4\psi}{4\psi}\left\{\frac{1}{8} - \frac{\sin^2\eta}{12} + \frac{\sin^4\eta}{40}\right\}\right]$$

$$+ C_{1212}\left[\frac{1}{2} - \frac{\sin^4\eta}{10} + \frac{\sin 4\psi}{4\psi}\left\{\frac{1}{2} - \frac{\sin^2\eta}{3} + \frac{\sin^4\eta}{10}\right\}\right]$$

$$- \frac{\sin^2\eta}{6} C_{2233}. \tag{2.37}$$

where $\sin\eta = t_a/t_b$, and $\sin\psi = w_a/w_b$.

The effective thermal expansion coefficients become

$$\bar{\alpha}_{11} = (\bar{C}_{1111})^{-1}X + (\bar{C}_{1122})^{-1}Y + (\bar{C}_{1133})^{-1}Z, \tag{2.38}$$

$$\bar{\alpha}_{22} = (\bar{C}_{1122})^{-1}X + (\bar{C}_{2222})^{-1}Y + (\bar{C}_{2233})^{-1}Z, \tag{2.39}$$

$$\bar{\alpha}_{33} = (\bar{C}_{1133})^{-1}X + (\bar{C}_{2233})^{-1}Y + (\bar{C}_{3333})^{-1}Z, \tag{2.40}$$

$$\bar{\alpha}_{ij} = 0, \quad \text{otherwise}, \tag{2.41}$$

where

$$X = \frac{1}{2}\left[\left\{1 + \frac{\sin 2\psi}{2\psi}\right\}\left\{1 - \frac{\sin^2\eta}{3}\right\}C_{11kl} + \left\{1 + \frac{\sin 2\psi}{2\psi}\right\}C_{22kl} + \left\{1 - \frac{\sin 2\psi}{2\psi}\right\}C_{33kl}\right]\alpha_{kl}, \tag{2.42}$$

$$Y = \frac{1}{2}\left[\left\{1 - \frac{\sin 2\psi}{2\psi}\right\}\left\{1 - \frac{\sin^2\eta}{3}\right\}C_{11kl} + \left\{1 - \frac{\sin 2\psi}{2\psi}\right\}C_{22kl} + \left\{1 + \frac{\sin 2\psi}{2\psi}\right\}C_{33kl}\right]\alpha_{kl}, \tag{2.43}$$

$$Z = \left[\frac{\sin^2\eta}{3}C_{11kl} + \left\{1 - \frac{\sin^2\eta}{3}\right\}C_{22kl}\right]\alpha_{kl}. \tag{2.44}$$

2.2.2. Combination of Extrusion an Transverse Rolling

From Fig. 3(b), the stiffness tensor after a combined forming of extrusion and transverse rolling becomes

$$\bar{C}_{ijkl} = \frac{1}{2\psi(1-\cos\eta)}\int_{-\psi}^{\psi}\int_{0}^{\eta} C'_{ijkl}\sin\theta\, d\theta\, d\phi. \tag{2.45}$$

Integrating gives

$$\begin{aligned}
\begin{matrix}\bar{C}_{1111}\\ \bar{C}_{2222}\end{matrix} = C_{1111}&\left[\frac{3}{4} \pm \frac{\sin 2\psi}{2\psi} + \frac{\sin 4\psi}{16\psi}\right]\left[\frac{4}{15} - \frac{7\cos\eta}{30} - \frac{7\cos^2\eta}{30}\right.\\
&\left.\quad + \frac{\cos^3\eta}{10} + \frac{\cos^4\eta}{10}\right]\\
&+ C_{2222}\left[\frac{8}{15} + \frac{19\cos\eta}{120} + \frac{19\cos^2\eta}{120} + \frac{3\cos^3\eta}{40} + \frac{3\cos^4\eta}{40}\right.\\
&\quad \pm \frac{\sin 2\psi}{2\psi}\left\{-\frac{2}{5} + \frac{\cos\eta}{10} + \frac{\cos^2\eta}{10} + \frac{\cos^3\eta}{10} + \frac{\cos^4\eta}{10}\right\}\\
&\left.\quad + \frac{\sin 4\psi}{4\psi}\left\{\frac{1}{15} - \frac{7\cos\eta}{120} - \frac{7\cos^2\eta}{120} + \frac{\cos^3\eta}{40} + \frac{\cos^4\eta}{40}\right\}\right]\\
&+ (2C_{1111} + 4C_{1212})\left[\frac{2}{15} + \frac{\cos\eta}{120} + \frac{\cos^2\eta}{120} - \frac{3\cos^3\eta}{40} - \frac{3\cos^4\eta}{40}\right.\\
&\quad \pm \frac{\sin 2\psi}{2\psi}\left\{\frac{1}{15} + \frac{\cos\eta}{15} + \frac{\cos^2\eta}{15}\right.\\
&\quad \left. - \frac{\cos^3\eta}{10} - \frac{\cos^4\eta}{10}\right\}\\
&\quad - \frac{\sin 4\psi}{4\psi}\left\{\frac{1}{15} - \frac{7\cos\eta}{120} - \frac{7\cos^2\eta}{120}\right.\\
&\left.\left.\quad + \frac{\cos^3\eta}{40} + \frac{\cos^4\eta}{40}\right\}\right],
\end{aligned} \tag{2.46}$$

$$\begin{aligned}
\bar{C}_{3333} = &\frac{C_{1111}}{5}(1 + \cos\eta + \cos^2\eta + \cos^3\eta + \cos^4\eta)\\
&+ \frac{C_{2222}}{15}(1 - \cos\eta)^2(8 + 9\cos\eta + 3\cos^2\eta)\\
&+ \frac{(2C_{1122} + 4C_{1212})}{15}(1 - \cos\eta)(2 + 4\cos\eta + 6\cos^2\eta + 3\cos^3\eta),
\end{aligned} \tag{2.47}$$

$$\bar{C}_{1122} = (C_{1111} + C_{2222} - 4C_{1212})\left[1 - \frac{\sin 4\psi}{4\psi}\right]$$
$$\times \left[\frac{1}{15} - \frac{7\cos\eta}{120} - \frac{7\cos^2\eta}{120} + \frac{\cos^3\eta}{40} + \frac{\cos^4\eta}{40}\right]$$
$$+ C_{1122}\left[\frac{8}{15} - \frac{13\cos\eta}{60} - \frac{13\cos^2\eta}{60} - \frac{\cos^3\eta}{20} - \frac{\cos^4\eta}{20}\right.$$
$$\left. + \frac{\sin 4\psi}{4\psi}\left\{\frac{2}{15} - \frac{7\cos\eta}{60} - \frac{7\cos^2\eta}{60} + \frac{\cos^3\eta}{20} + \frac{\cos^4\eta}{20}\right\}\right]$$
$$+ \frac{C_{2233}}{3}(1 + \cos\eta + \cos^2\eta), \tag{2.48}$$

$$\begin{matrix}\bar{C}_{1133}\\ \bar{C}_{2233}\end{matrix} = (C_{1111} + C_{2222} - 4C_{1212})\left[1 \pm \frac{\sin 2\psi}{2\psi}\right]$$
$$\times \left[\frac{1}{15} + \frac{2\cos\eta}{15} + \frac{\cos^2\eta}{5} + \frac{\cos^3\eta}{10}\right](1 - \cos\eta)$$
$$+ C_{1122}\left[\frac{8}{15} + \frac{\cos\eta}{30} + \frac{\cos^2\eta}{30} + \frac{\cos^3\eta}{5} + \frac{\cos^4\eta}{5}\right.$$
$$\left. \pm \frac{\sin 2\psi}{2\psi}\left\{\frac{1}{5} - \frac{\cos\eta}{10} - \frac{2\cos^2\eta}{5} + \frac{\cos^3\eta}{5}\right\}(1 - \cos\eta)\right]$$
$$+ \frac{C_{2233}}{3}\left[1 \mp \frac{\sin 2\psi}{2\psi}\right]\left[1 + \frac{\cos\eta}{2}\right](1 - \cos\eta), \tag{2.49}$$

$$\begin{matrix}\bar{C}_{2323}\\ \bar{C}_{3131}\end{matrix} = (C_{1111} - 2C_{1122})\left\{1 \mp \frac{\sin 2\psi}{2\psi}\right\}$$
$$\times \left\{\frac{1}{15} + \frac{\cos\eta}{15} + \frac{\cos^2\eta}{15} - \frac{\cos^3\eta}{10} - \frac{\cos^4\eta}{10}\right\}$$
$$+ C_{2222}\left[\frac{7}{30} - \frac{\cos\eta}{60} - \frac{\cos^2\eta}{60} - \frac{\cos^3\eta}{10} - \frac{\cos^4\eta}{10}\right.$$
$$\left. \mp \frac{\sin 2\psi}{2\psi}\left\{-\frac{1}{10} + \frac{3\cos\eta}{20} + \frac{3\cos^2\eta}{20} - \frac{\cos^3\eta}{10} - \frac{\cos^4\eta}{10}\right\}\right]$$
$$+ C_{1212}\left[\frac{2}{5} - \frac{\cos\eta}{10} - \frac{\cos^2\eta}{10} + \frac{2\cos^3\eta}{5} + \frac{2\cos^4\eta}{5}\right.$$
$$\left. \mp \frac{\sin 2\psi}{2\psi}\left\{\frac{1}{15} - \frac{13\cos\eta}{30} - \frac{13\cos^2\eta}{30} + \frac{2\cos^3\eta}{5} + \frac{2\cos^4\eta}{5}\right\}\right]$$
$$- C_{2233}\left\{1 \pm \frac{\sin 2\psi}{2\psi}\right\}\left\{\frac{1}{6} - \frac{\cos\eta}{12} - \frac{\cos^2\eta}{12}\right\}, \tag{2.50}$$

$$\bar{C}_{1212} = (C_{1111} - 2C_{1122})\left\{1 - \frac{\sin 4\psi}{4\psi}\right\}$$
$$\times \left\{\frac{1}{15} - \frac{7\cos\eta}{120} - \frac{7\cos^2\eta}{120} + \frac{\cos^3\eta}{40} + \frac{\cos^4\eta}{40}\right\}$$
$$+ C_{2222}\left[\frac{7}{30} + \frac{13\cos\eta}{120} + \frac{13\cos^2\eta}{120} + \frac{\cos^3\eta}{40} + \frac{\cos^4\eta}{40}\right.$$
$$\left. - \frac{\sin 4\psi}{4\psi}\left\{\frac{1}{15} - \frac{7\cos\eta}{120} - \frac{7\cos^2\eta}{120} + \frac{\cos^3\eta}{40} + \frac{\cos^4\eta}{40}\right\}\right]$$
$$+ C_{1212}\left[\frac{2}{5} - \frac{\cos\eta}{10} - \frac{\cos^2\eta}{10} - \frac{\cos^3\eta}{10} - \frac{\cos^4\eta}{10}\right.$$
$$\left. + \frac{\sin 4\psi}{4\psi}\left\{\frac{4}{15} - \frac{7\cos\eta}{30} - \frac{7\cos^2\eta}{30} + \frac{\cos^3\eta}{10} + \frac{\cos^4\eta}{10}\right\}\right]$$
$$- \frac{C_{2233}}{6}(1 + \cos\eta + \cos^2\eta). \tag{2.51}$$

where $\sin\eta = D_a/D_b$ and $\sin\psi = w_a/w_b$.

The effective thermal expansion coefficients are given by (2.38), (2.39), and (2.40), where

$$X = \frac{1}{2}\left[\left\{1 + \frac{\sin 2\psi}{2\psi}\right\}\left\{\frac{2}{3} - \frac{\cos\eta}{3} - \frac{\cos^2\eta}{3}\right\}C_{11kl}\right.$$
$$+ \left\{1 + \frac{\sin 2\psi}{2\psi}\right\}\left\{\frac{1}{3} + \frac{\cos\eta}{3} + \frac{\cos^2\eta}{3}\right\}C_{22kl}$$
$$\left. + \left\{1 - \frac{\sin 2\psi}{2\psi}\right\}C_{33kl}\right]\alpha_{kl}, \tag{2.52}$$

$$Y = \frac{1}{2}\left[\left\{1 - \frac{\sin 2\psi}{2\psi}\right\}\left\{\frac{2}{3} - \frac{\cos\eta}{3} - \frac{\cos^2\eta}{3}\right\}C_{11kl}\right.$$
$$+ \left\{1 - \frac{\sin 2\psi}{2\psi}\right\}\left\{\frac{1}{3} + \frac{\cos\eta}{3} + \frac{\cos^2\eta}{3}\right\}C_{22kl}$$
$$\left. + \left\{1 + \frac{\sin 2\psi}{2\psi}\right\}C_{33kl}\right]\alpha_{kl}, \tag{2.53}$$

$$Z = \left[\left\{\frac{1}{3} + \frac{\cos\eta}{3} + \frac{\cos^2\eta}{3}\right\}C_{11kl} + \left\{\frac{2}{3} - \frac{\cos\eta}{3} - \frac{\cos^2\eta}{3}\right\}C_{22kl}\right]\alpha_{kl}. \tag{2.54}$$

2.2.2. Combination of Longitudinal Rolling and Transverse Rolling

From Fig. 3(c), the stiffness tensor after a combined forming of longitudinal rolling and transverse rolling becomes

$$\bar{C}_{ijkl} = \frac{\pi}{8\psi\eta} \int_{-\psi}^{\psi} \int_{\pi/2-\eta}^{\pi/2+\eta} C'_{ijkl} \cos\left[\frac{(\theta - \pi/2)}{2\eta}\right] d\theta \, d\phi. \tag{2.55}$$

Integrating and defining A and B as follows gives

$$A = 4\left(\frac{2\eta}{\pi}\right)^2 - 1, \tag{2.56}$$

$$B = 16\left(\frac{2\eta}{\pi}\right)^2 - 1, \tag{2.57}$$

$$\begin{aligned}
\begin{matrix}\bar{C}_{1111} \\ \bar{C}_{2222}\end{matrix} &= C_{1111}\left\{\frac{3}{4} \pm \frac{\sin 2\psi}{2\psi} + \frac{\sin 4\psi}{16\psi}\right\}\left\{\frac{3}{16} - \frac{\cos 2\eta}{4A} - \frac{\cos 4\eta}{16B}\right\} \\
&+ C_{2222}\left[\frac{41}{64} + \frac{5\cos 2\eta}{16A} - \frac{3\cos 4\eta}{64B}\right. \\
&\qquad \pm \frac{\sin 2\psi}{2\psi}\left\{-\frac{5}{16} + \frac{\cos 2\eta}{4A} - \frac{\cos 4\eta}{16B}\right\} \\
&\qquad \left. + \frac{\sin 4\psi}{16\psi}\left\{\frac{3}{16} - \frac{\cos 2\eta}{4A} - \frac{\cos 4\eta}{16B}\right\}\right] \\
&+ (2C_{1122} + 4C_{1212})\left[\frac{7}{64} - \frac{\cos 2\eta}{16A} + \frac{3\cos 4\eta}{64B}\right. \\
&\qquad \pm \frac{\sin 2\psi}{2\psi}\left\{\frac{1}{16} + \frac{\cos 4\eta}{16B}\right\} \\
&\qquad \left. - \frac{\sin 4\psi}{16\psi}\left\{\frac{3}{16} - \frac{\cos 2\eta}{4A} - \frac{\cos 4\eta}{16B}\right\}\right],
\end{aligned} \tag{2.58}$$

$$\begin{aligned}
\bar{C}_{3333} &= C_{1111}\left\{\frac{3}{8} + \frac{\cos 2\eta}{2A} - \frac{\cos 4\eta}{8B}\right\} + C_{2222}\left\{\frac{3}{8} - \frac{\cos 2\eta}{2A} - \frac{\cos 4\eta}{8B}\right\} \\
&+ (2C_{1122} + 4C_{1212})\left\{\frac{1}{8} + \frac{\cos 4\eta}{8B}\right\},
\end{aligned} \tag{2.59}$$

$$\begin{aligned}
\bar{C}_{1122} &= (C_{1111} + C_{2222} - 4C_{1212})\left[1 - \frac{\sin 4\psi}{4\psi}\right]\left[\frac{3}{64} - \frac{\cos 2\eta}{16A} - \frac{\cos 4\eta}{64B}\right] \\
&+ C_{1122}\left[\frac{13}{32} - \frac{3\cos 2\eta}{8A} + \frac{\cos 4\eta}{32B}\right. \\
&\qquad \left. + \frac{\sin 4\psi}{4\psi}\left\{\frac{3}{32} - \frac{\cos 2\eta}{8A} - \frac{\cos 4\eta}{32B}\right\}\right] \\
&+ C_{2233}\left[\frac{1}{2} + \frac{\cos 2\eta}{2A}\right],
\end{aligned} \tag{2.60}$$

$$\begin{aligned}\begin{matrix}\bar{C}_{1133}\\ \bar{C}_{2233}\end{matrix} &= (C_{1111} + C_{2222} - 4C_{1212})\left[1 \pm \frac{\sin 2\psi}{2\psi}\right]\left[\frac{1}{16} + \frac{\cos 4\eta}{16B}\right] \\ &\quad + C_{1122}\left[\frac{5}{8} + \frac{\cos 2\eta}{4A} - \frac{\cos 4\eta}{8B} \pm \frac{\sin 2\psi}{2\psi}\left\{\frac{1}{8} - \frac{\cos 2\eta}{4A} - \frac{\cos 4\eta}{8B}\right\}\right] \\ &\quad + C_{2233}\left[1 \mp \frac{\sin 2\psi}{2\psi}\right]\left[\frac{1}{4} - \frac{\cos 2\eta}{4A}\right],\end{aligned} \tag{2.61}$$

$$\begin{aligned}\begin{matrix}\bar{C}_{2323}\\ \bar{C}_{3131}\end{matrix} &= (C_{1111} - 2C_{1122})\left\{1 \mp \frac{\sin 2\psi}{2\psi}\right\}\left\{\frac{1}{16} + \frac{\cos 4\eta}{16B}\right\} \\ &\quad + C_{2222}\left[\frac{3}{16} - \frac{\cos 2\eta}{8A} + \frac{\cos 4\eta}{16B}\right. \\ &\qquad \left. \mp \frac{\sin 2\psi}{2\psi}\left\{-\frac{1}{16} + \frac{\cos 2\eta}{8A} + \frac{\cos 4\eta}{16B}\right\}\right] \\ &\quad + C_{1212}\left[\frac{1}{2} + \frac{\cos 2\eta}{4A} - \frac{\cos 4\eta}{4B} \pm \frac{\sin 2\psi}{2\psi}\left\{\frac{\cos 2\eta}{4A} + \frac{\cos 4\eta}{4B}\right\}\right] \\ &\quad - C_{2233}\left\{1 \pm \frac{\sin 2\psi}{2\psi}\right\}\left\{\frac{1}{8} - \frac{\cos 2\eta}{8A}\right\},\end{aligned} \tag{2.62}$$

$$\begin{aligned}\bar{C}_{1212} &= (C_{1111} - 2C_{1122})\left\{1 - \frac{\sin 4\psi}{4\psi}\right\}\left\{\frac{3}{64} - \frac{\cos 2\eta}{16A} - \frac{\cos 4\eta}{64B}\right\} \\ &\quad - \frac{C_{2233}}{4}\left\{1 + \frac{\cos 2\eta}{A}\right\} \\ &\quad + C_{2222}\left[\frac{19}{64} + \frac{3\cos 2\eta}{16A} - \frac{\cos 4\eta}{64B}\right. \\ &\qquad \left. + \frac{\sin 4\psi}{4\psi}\left\{-\frac{3}{64} + \frac{\cos 2\eta}{16A} + \frac{\cos 4\eta}{64B}\right\}\right] \\ &\quad + C_{1212}\left[\frac{5}{16} - \frac{\cos 2\eta}{4A} + \frac{\cos 4\eta}{16B} + \frac{\sin 4\psi}{4\psi}\left\{\frac{3}{16} - \frac{\cos 2\eta}{4A} - \frac{\cos 4\eta}{16B}\right\}\right],\end{aligned} \tag{2.63}$$

where $\sin \eta = t_a/t_b$ and $\sin \psi = w_a/w_b$.

The effective thermal expansion coefficients are given by (2.38), (2.39), and (2.40), where

$$\begin{aligned}X = \frac{1}{2}\Bigg[&\left\{1 + \frac{\sin 2\psi}{2\psi}\right\}\left\{\frac{1}{2} - \frac{\cos 2\eta}{2A}\right\}C_{11kl} \\ &+ \left\{1 + \frac{\sin 2\psi}{2\psi}\right\}\left\{\frac{1}{2} + \frac{\cos 2\eta}{2A}\right\}C_{22kl} \\ &+ \left\{1 - \frac{\sin 2\psi}{2\psi}\right\}C_{33kl}\Bigg]\alpha_{kl},\end{aligned} \tag{2.64}$$

$$Y = \frac{1}{2}\left[\left\{1 - \frac{\sin 2\psi}{2\psi}\right\}\left\{\frac{1}{2} - \frac{\cos 2\eta}{2A}\right\} C_{11kl} + \left\{1 - \frac{\sin 2\psi}{2\psi}\right\}\left\{\frac{1}{2} + \frac{\cos 2\eta}{2A}\right\} C_{22kl} + \left\{1 + \frac{\sin 2\psi}{2\psi}\right\} C_{33kl}\right]\alpha_{kl}, \tag{2.65}$$

$$Z = \left[\left\{\frac{1}{2} + \frac{\cos 2\eta}{2A}\right\} C_{11kl} + \left\{\frac{1}{2} - \frac{\cos 2\eta}{2A}\right\} C_{22kl}\right]\alpha_{kl}. \tag{2.66}$$

3. Results and Discussion

As mentioned earlier, the five elastic constants and two thermal expansion coefficients for a perfectly aligned fiber composite must be determined prior to the material forming processes discussed in the previous section. Among these seven independent constants, the five elastic constants are determined by the equivalent inclusion method. The remaining two thermal expansion coefficients are assumed to be known, even though they can be calculated from the equivalent inclusion method.

3.1. Aligned Fiber Orientation

The five elastic constants for an aligned short fiber composite determined by the equivalent inclusion method are illustrated in Fig. 4. Note that when the fiber volume fraction approaches one, the equivalent inclusion method recovers the fiber properties. In a real composite material which is generally treated as a statistically homogeneous material, the fiber volume fraction cannot exceed a certain value less than unity depending on the fiber packing array. Thus, the elastic constants determined by the equivalent inclusion method may not have any physical meaning for the fiber volume fraction close to one.

When the fiber aspect ratio approaches infinity, the equivalent inclusion method recovers the elastic constants of an aligned continuous fiber composite. The experimental data for the effective transverse Young's modulus in Tsai and Hahn (1980) and the effective shear modulus in Adams *et al.* (1967) are compared with the result from the equivalent inclusion method in Fig. 5(a) and (b), respectively. These results show that the equivalent inclusion method gives a reasonable prediction for the elastic constants of an aligned fiber composite.

3.2. Bounded Fiber Orientation

The five elastic constants for the aligned composite shown in Fig. 4 are used as input data for predicting the effective Young's modulus of the same compo-

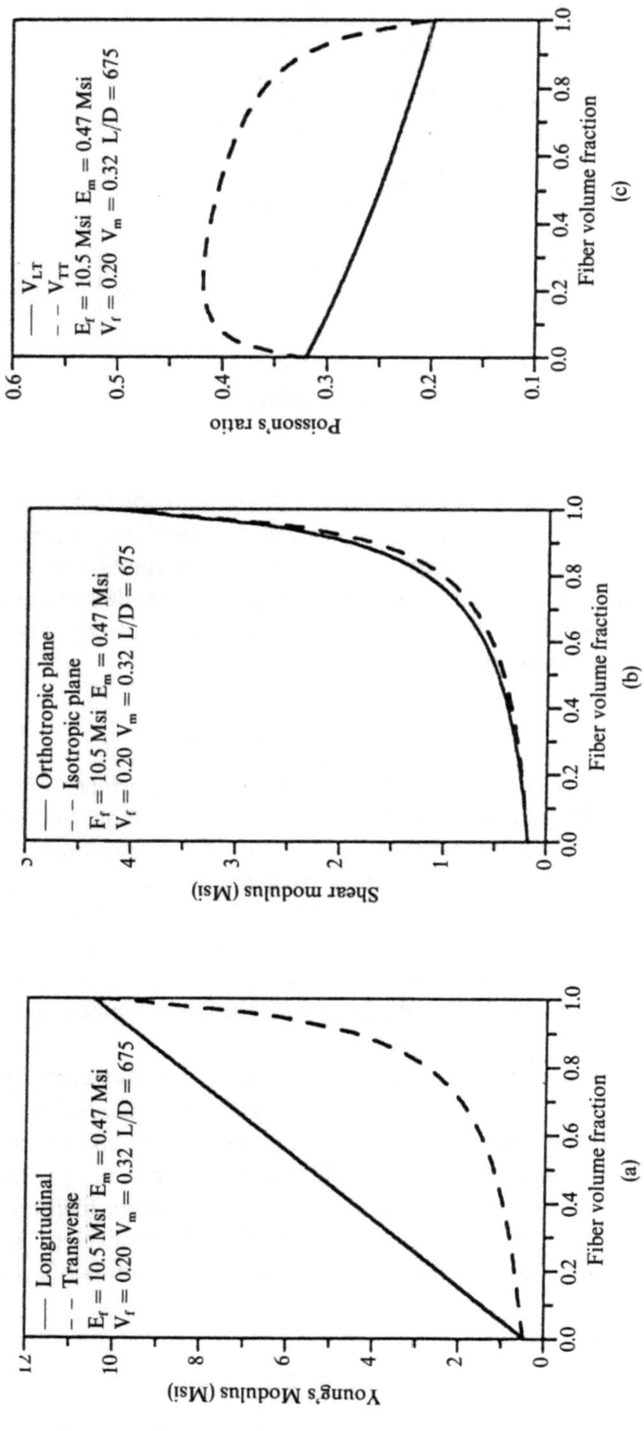

FIG. 4. Elastic constants of an aligned fiber composite: (a) Young's moduli; (b) shear moduli; (c) Poisson's ratios.

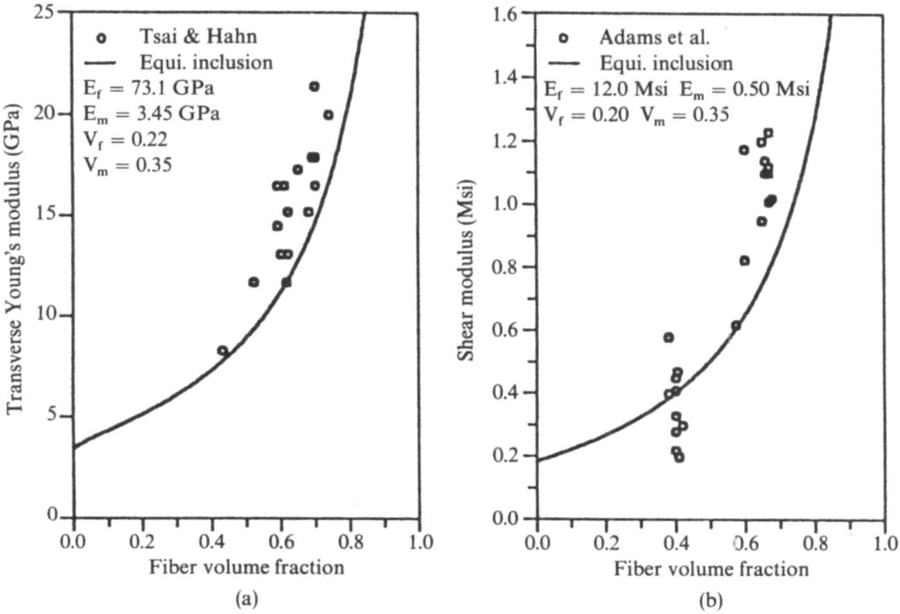

FIG. 5. Equivalent inclusion method versus experiental data: (a) transverse Young's modulus; (b) shear modulus.

site with randomly oriented fibers in a two-dimensional domain. The present analytic solutions for an extreme hot pressing with zero thickness ratio gives the effective elastic material properties of composite with two-dimensional random fibers. Figure 6 compares the present model prediction for the effective Young's modulus with the experimental result of the same composite in Lee (1969). When the thickness ratio of hot pressing is set to zero in the solutions given in Section 2.2.1, the result of Berthelot (1982) is recovered. Similarly, the present model simplifies to the solutions for the three-dimensional random fiber composite studied by Christensen and Waals (1972).

The effects of hot pressing, extrusion, and rolling on the axial Young's modulus are illustrated by Fig. 7. Consider an extreme case in which the thickness ratio is zero. In such a case the fiber orientation after hot pressing or rolling becomes planar. The material after hot pressing becomes transversely isotropic, whereas after rolling, it becomes orthotropic. When the material is subjected to an extreme extrusion in which the diameter ratio approaches zero, the material properties of a perfectly aligned fiber composite are retrieved. On the other hand, the material becomes three-dimensional isotropic when there is no dimension change due to material forming as shown in the same figure.

Figure 8 illustrates the effective axial Young's modulus and thermal expansion coefficient for the ideal composite in Fig. 7 after a number of different combined material forming processes resulting in the same final shape listed in Table 1. As shown in Fig. 8, the properties are significantly affected by the

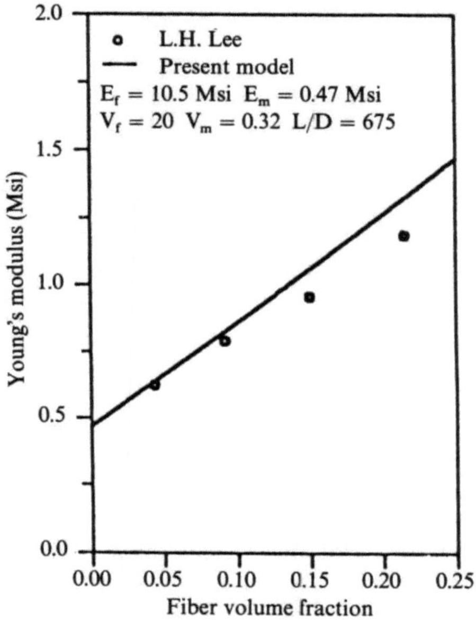

FIG. 6. Young's modulus of a two-dimensional random fiber composite.

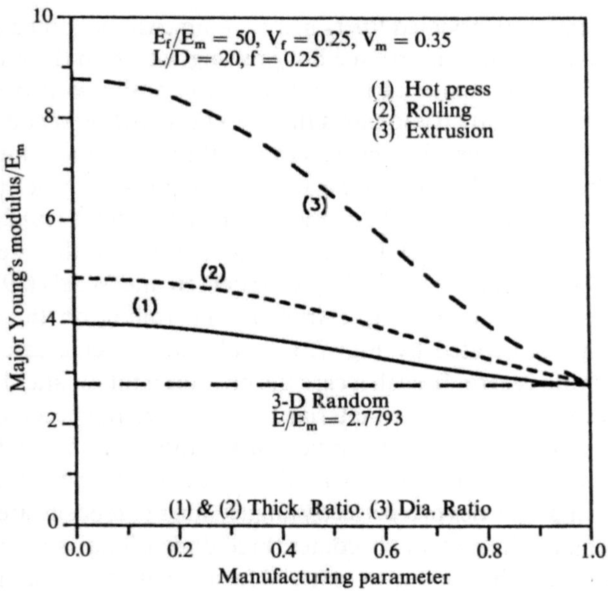

FIG. 7. Effective Young's modulus after each of hot pressing, extrusion, and rolling.

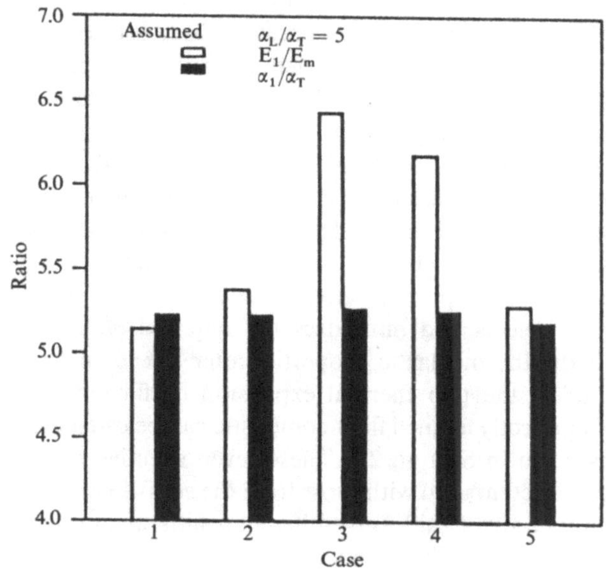

FIG. 8. Effective axial Young's modulus and thermal expansion coefficient after combined material forming.

TABLE 1. Combined material forming processes.

Case	Material forming sequence
1	$\frac{1}{3}$ hot pressing—$\frac{3}{4}$ normal rolling
2	$\frac{3}{4}$ hot pressing—$\frac{1}{3}$ normal rolling
3	$\frac{1}{2} \times \frac{1}{2}$ extrusion—$\frac{2}{3}$ transverse rolling
4	$\frac{1}{3}$ rolling with fixed thickness—$\frac{3}{4}$ rolling with fixed width
5	$\frac{3}{4}$ rolling with fixed thickness—$\frac{1}{3}$ rolling with fixed width
Initial dimension	$1 \times 1 \times 1$
Final dimension	$\frac{1}{3} \times \frac{3}{4} \times 4$

order and combination of material forming processes. Therefore, it is implied that the entire material forming history should be known *a priori* before conducting an experimental program to measure the thermoelastic properties of commercially available specimens.

4. Conclusions

Metal matrix composites reinforced by short fibers or whiskers are frequently subjected to a series of material forming processes discussed in Section 2.2. The present study utilizes a simple integration scheme for predicting the effect of material forming history on the thermoelastic properties of fiber reinforced composites. If the material forming processes discussed herein do not cause any defects, the utilization of the equivalent method or Halpin–Tsai equations (1969) may be reasonable for determining the thermoelastic constants of a perfectly aligned fiber composite as input data to the present integration model. However, broken fibers, local fiber clusters, and fiber–matrix debonding have been frequently observed (Riggs and Gillis, 1980; Allen *et al.*, 1987). It is very difficult to model these defects mathematically without painstaking experimental observation which is, in general, destructive.

Thus, the authors propose an alternative method which can include the effects of these defects and minimizes the required effort and number of specimens. If the thermoelastic properties after a known material forming history are given, the two thermal expansion coefficients and five elastic constants of a perfectly aligned fiber composite can be calculated by inverting the solutions given in Section 2.2. These seven effective thermoelastic constants then can be compared with those from the equivalent inclusion method. If the two methods give significantly different results, an internal state variable can be defined from the difference which represents the averaged effect of all possible defects before and during material forming. Then, the present integration model together with the internal state variable and the equivalent inclusion method can be utilized for predicting the thermoelastic response of the same composite after different material forming histories.

Acknowledgments

The present study has been supported by the Texas Advanced Research Program.

References

Adams, D. F., Doner, D. R., and Thomas, R. L. (1967), Mechanical behavior of fiber-reinforced composite materials, AFML-TR-67-96, May.

Allen, D. H., Harris, C. E., and Nottorf, E. W. (1987), A fractographic study of damage mechanisms in short-fiber metal matrix composites, in *Fractography of Modern Engineering Materials: Composites and Metals*, ASTM STP 948, edited by J. E. Masters and J. J. Au, American Society for Testing and Materials, Philadelphia, PA, pp. 189–213.

Berthelot, J. M. (1982), Effect of fibre misalignment on the elastic properties of oriented discontinuous fibre composites, *Fiber Sci. Technol.*, **17**, 25–39.

Christensen, R. M. and Waals, F. M. (1972), Effective stiffness of randomly oriented fibre composites, *J. Composite Materials*, **6**, 518–532.

Craft, W. J. and Christensen, R. M. (1984), Coefficient of thermal expansion for composites with randomly oriented fibers, in *Environmental Effects on Composite Materials*, Vol. 2, edited by G. S. Springer, Technomic, Lancaster, PA, pp. 331–347.

Eshelby, J. D. (1957), The determination of the elastic field of an ellipsoidal inclusion, and related problems, *Proc. Roy. Soc. London*, **A241**, 376–396.

Halpin, J. C. and Tsai, S. W. (1969), Effects of Environmental Factors on Composite Materials, AFML-TR-67-423, June.

Hashin, Z. and Rosen B. W. (1964), The elastic moduli of fiber-reinforced materials, *J. Appl. Mech.* **31**, 223–231.

Lee, J. W. and Allen, D. H. (1989), A model for predicting the effective elastic properties of randomly oriented fiber composites subjected to hot pressing, extrusion, and rolling, to appear in *Proc. AIAA/ASME*, 30th SDM Conference, Paper 89-1253.

Lee L. H. (1969), Strength-composition relationships of random short glass fiber–thermoplastics composites, *Polymer Engng. Sci.*, **9**, No. 3, 213–224.

Mori, T. and Tanaka, K. (1973), Average stress in matrix and average elastic energy of materials with misfitting inclusions, *Acta Metallurgica*, **21**, 571–574.

Mura, T. (1982), Micromechanics of defects in solids, in *Mechanics of Elastic and Inelastic Solids*, Vol. 3 edited by S. Nemat-Nassar, Martinus Nijhoff, The Hague.

Papazian, J. M., Levy, A., and Adler, P. N. (1987), Micro-mechanics of deformation in SiC/Al composites, AFOSR-TR-87-1658, August.

Riggs, D. M. and Gillis, P. (1980), The effect of mechanical working on SiC whisker-reinforced aluminum alloys, AMMRC TR 80-11.

Takao, Y., Chou, T. W., and Taya, M. (1982), Effective longitudinal Young's modulus of misoriented short fiber composites, *J. Appl. Mech.*, **49**, 537–540.

Taya, M. and Chou, T. W. (1981), On two kinds of ellipsoidal inhomogeneities in an infinite elastic body: An application to a hybrid composite, *Int. J. Solids Structures*, **17**, 553–563.

Taya, M. and Mura, T. (1981), On stiffness and strength of an aligned short-fiber reinforced composite containing fiber-end cracks under uniaxial applied stress, *J. Appl. Mech.*, **48**, 361–367.

Tsai, S. W. and Hahn, H. T. (1980), *Introduction to Composite Materials*, Technomic, Westport, CT 06880.

On the Specific Damaging Effects of Surface and Near-Surface Inclusions

MARC-HENRI AMBROISE, THIERRY BRETHEAU, and ANDRÉ ZAOUI
Laboratoire PMTM, CNRS, Université Paris-Nord, Villetaneuse, France

Abstract

The most significant results of an experimental study on the damaging effects of exogeneous ceramic inclusions in nickel-based superalloys are first reported. Attention is focused on some observations which conflict with a too-simple two-dimensional analysis of the problem. A more realistic theoretical treatment is then developed, which allows the calculation of configurational force and torque acting on a near-surface or surface inhomogeneity: the corresponding interfacial discontinuity of the elastostatic "energy–momentum" tensor is shown to be responsible for specific damaging effects of such inhomogeneities which could not be interpreted otherwise.

1. Introduction

For decades, inclusions have been associated with the general problem of fatigue failure of high strength alloys. Especially, oxide inclusions are known for their detrimental effect, mainly for specific fatigue processes such as surface and subsurface crack nucleation during uniaxial cyclic deformation of smooth specimens (Lankford, 1977a).

It is now generally accepted that the fatigue life of smooth specimens decreases with an increasing total inclusion content (Hauser and Wells, 1970; Lankford, 1977a): this observation is unambiguously related to the probability of finding a "critical" inclusion at sufficiently stressed places (Jablonski, 1981).

For hard inclusions with similar chemical composition, there exists a trend to an increasing propensity for nucleating cracks when the inclusion size increases (Hauser and Wells, 1970; Lankford, 1977a). In addition, the average minimum size of inclusions responsible for failure appears to increase with depth below the specimen surface (Hauser and Wells, 1970; Lankford, 1977b; Thompson *et al.*, 1979), so that surface or near-surface inclusions are generally more detrimental than deeper ones; at the same time, the size of surface inclusions plays a minor role on the number of cycles for crack initiation (Hauser and Wells, 1970; Law and Blackburn, 1980).

Thus, even a low concentration of small inclusions may induce a catastrophic response if ever *one* of them exists in a highly stressed surface or near-surface area. As long as a complete elimination of inclusions looks out of reach, efforts must be made to lower their intrinsic harmfulness due to a better knowledge of the inclusion/matrix mechanical interaction, particularly in the case of surface or near-surface inclusions. This paper reports both experimental and theoretical results in this field: due to original experimental techniques, allowing significant observations and measurements at the adequate scale and under realistic mechanical conditions, a careful micromechanical analysis of the inclusion/matrix interaction has been performed. While several observations were satisfactorily interpreted from a two-dimensional point of view directly associated with the surface investigations, some others clearly conflicted with such an approach. A theoretical analysis was then developed in order to derive the expression of configurational force and torque acting on a near-surface or surface elastic inhomogeneity in an elastic bounded medium: this allowed us to get a better understanding of the whole experimental investigation and to emphasize the specific damaging effects of such superficial inclusions.

2. Experimental

Most of the technical details concerning the experimental investigation are reported in Appendix A1. We report here the main information on the studied material and the experimental methodology.

2.1. Material

The material, a nickel-based superalloy, was processed by the powder metallurgy method. The Astroloy (NK 17 CDAT) powder was seeded with 300 ppm of 50–150 μm large alumina inclusions and then submitted to hot isostatic pressing and an isothermal forging at 1100°C. The microstructure induced by this thermomechanical treatment consisted of homogeneously distributed fine grains (<10 μm) without any structural or chemical inhomogeneity in the vicinity of the isolated ceramic inclusions (Fig. 1).

2.2. Mechanical Tests

Classical tensile tests were first performed at room temperature: observations of interesting areas around inclusions were possible only in a relaxed state, between step-by-step increments of the load. Such a procedure was found to be of limited efficiency and somewhat fallacious: the events occurring during the relaxation stages cannot be taken into account and wrong conclusions may be drawn on the actual sequence and location of the damaging processes leading to crack nucleations. That is why it was necessary to perform *in situ* tests inside a large-chamber scanning electron microscope equipped with a

FIG. 1. Microstructure of the matrix in the vicinity of an inclusion, as visualized by crystallographic contrast in the SEM. Neither a chemical reaction nor a grain size modification can be detected.

sufficiently high load capacity (more than 5 kN) microtensile machine, the characteristics of which are given in Appendix A1 (Bretheau *et al.*, 1988). Suffice it to say here that room temperature tensile and repeated tensile tests could be performed up to fracture on $15 \times 4 \times 0.8$ mm^3 specimens with simultaneous observation and recording at the scale of the inclusions, with a magnification ranging between 15 and 10,000. The superficial nature of such observations evidently fitted in quite well with that of the involved phenomena.

2.3. *Microextensometry*

One of the main parameters governing the damage processes in the studied material was found to be the interfacial strength at the matrix/inclusion boundary. Moreover, it was necessary to establish reliable correlations between the location of stress and strain concentrations and that of incipient microcracking, which was not possible from a qualitative visual estimate only. Consequently, microextensometric techniques should be used. This was performed by a microlithographic procedure (Attwood and Hazzledine, 1976) allowing the deposition, on about 1 mm^2 wide areas centered on emerging inclusions, of square fiducial microgrids made of 0.2–0.5 μm wide lines with a 2–5 μm pitch (Fig. 2(a)).

The very beginning and development of the decohesion at interfaces could then be analyzed easily and the flow pattern of the matrix around the inclusions determined. In-plane strain fields were derived either from a direct microcomputer-aided digitization of the grid nodes coordinates (Bretheau and Caldemaison, 1981, 1983), or from the generation of moiré patterns resulting from the interference between the deformed grids and the grating of the electron beam scanning, which is used for the observation of the inclusion and the surrounding matrix (Ambroise *et al.*, 1987) (Fig. 2(b), (c), (d)).

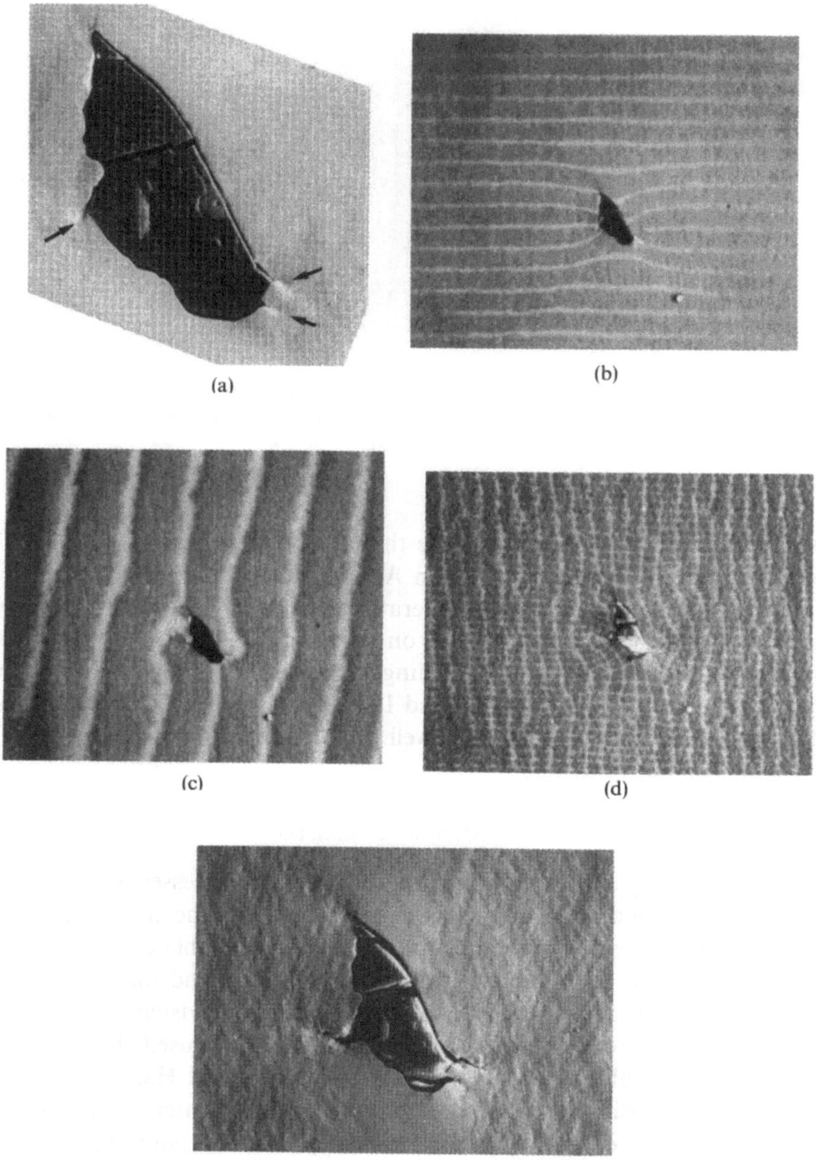

FIG. 2. Qualitative characterization of the strain field inhomogeneities (displacement field) around an inclusion. Initial grid pitch 3.5 μm; vertical tensile axis. (a) Interfacial decohesion, fracture of the inclusion, initiation of three microcracks (arrows). (Overall plastic strain $\varepsilon_p = 2\%$.) (b) Fringe moiré pattern showing the vertical displacement field. (c) Fringe moiré pattern showing the horizontal displacement field. (d) Cellular moiré pattern. (e) Evolution of the features observable on Fig. 2(a) ($\varepsilon_p = 8.7\%$).

3. Two-Dimensional Analysis

In order to identify and analyze the surface specific effects, a limited, purely two-dimensional, point of view is first adopted for the analysis of the matrix/inclusion interaction: this means that the observations are restricted to an in-plane view of the sample surface. The deficiencies of such an approach will then be appreciated and cured due to a more realistic three-dimensional analysis.

3.1. Governing Parameters

Thanks to the experimental techniques described above, the main parameters governing the nature and location of the damaging effects can be clearly brought out.

3.1.1. Cohesion at the Matrix/Inclusion Interface

The observation of any inclusion, as early as at the very beginning of the plastic flow, shows an interfacial decohesion starting from the "poles" (with respect to the tensile axis) of the inclusions (Fig. 2(a)). This means that the critical stress for decohesion is lower than the matrix yield stress. The corresponding strain field in the matrix is completely different from what it used to be in the case of a perfect fit at the interface: polar zones are almost undeformed even at quite large macroscopic strains, whereas "equatorial" zones are heavily strained in a compressive way (Fig. 2(b), (c), (d); Fig. 3(a), (b)). These strain concentrations can result in the nucleation of microcracks in the matrix, as well as in the fracture of the inclusions along planes roughly perpendicular to the tensile axis (Fig. 2(a), (e); Fig. 3(a)).

3.1.2. The Inclusion Morphology

Both the *overall* shape and the *local* morphological details play an important role in the detrimental character of the inclusion. Thus, when the global morphology does not respect the axial symmetry of the applied tensile load (Fig. 4(a)), the inclusion undergoes a rotation (Fig. 4(b)) that can induce local strain concentrations resulting in the nucleation of cracks (Fig. 4(c)). The local morphology effects are especially operative when salient angles are present in the equatorial area: the strain concentrations which already exist here, due to interfacial decohesion, may still be accentuated by cuspidal points, which results in crack nucleations (Fig. 2(a); Fig. 3(a); Fig. 7).

The damage initiation around an inclusion is now clearly seen to be ruled by the interface decohesion and the inclusion morphology. Nevertheless, the actual sequence of events resulting in a microcrack nucleation, as well as the effective detrimental effects of such microcracks, are still not elucidated. This needs under-stress observations and repeated tensile tests.

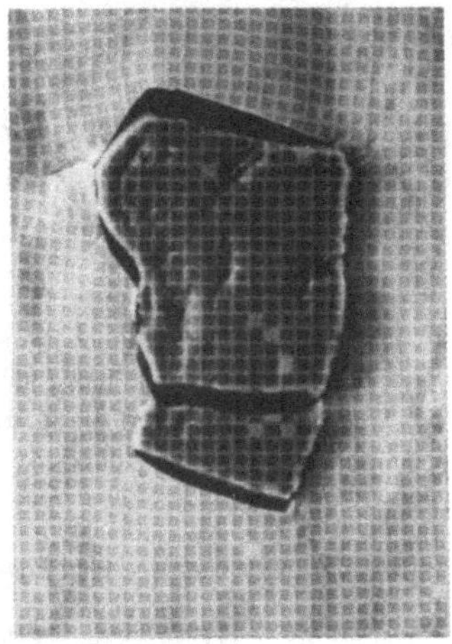

(a)

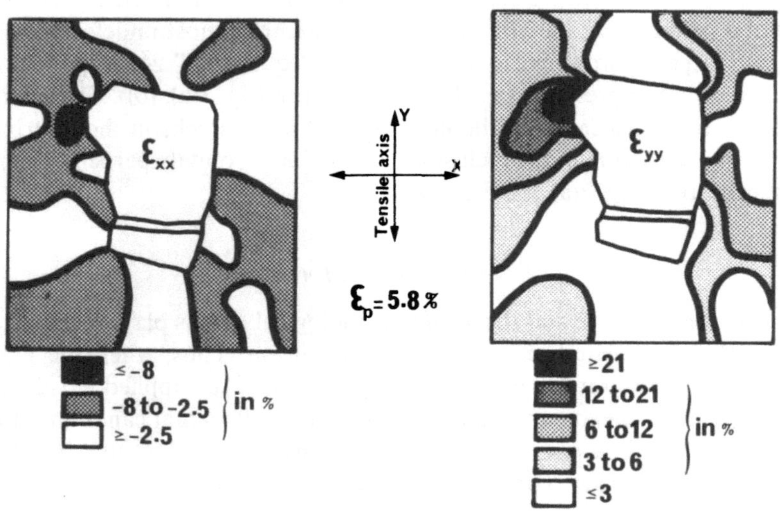

(b)

FIG. 3. Quantitative characterization of the strain field inhomogeneities around an inclusion. Initial grid pitch 3.5 μm; vertical tensile axis; $\varepsilon_p = 5.8\%$. (a) SEM micrograph. (b) Strain maps derived from direct measurements.

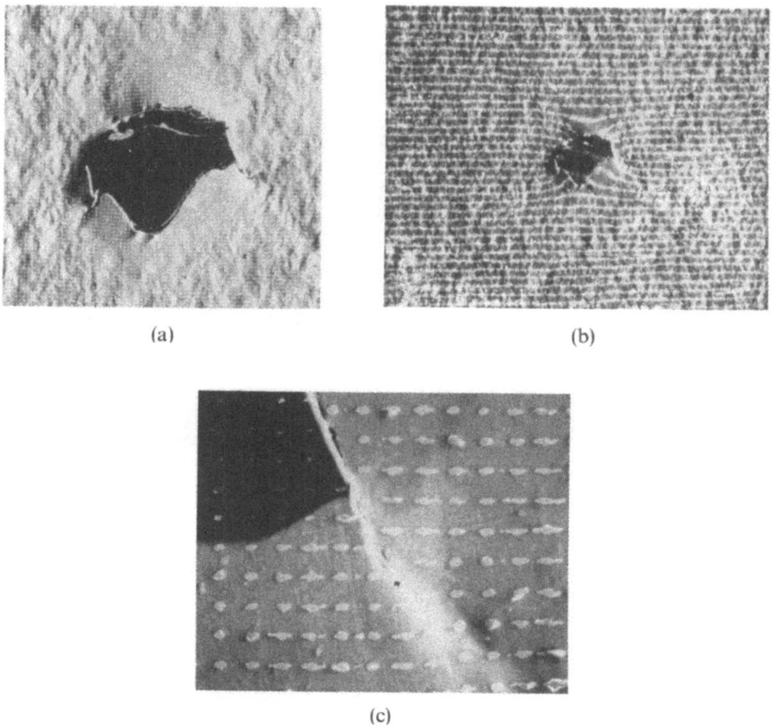

FIG. 4. Influence of the inclusion global morphology. (a) Direct SEM observation. (b) Fringe moiré pattern; the in-plane rotation of the inclusion is shown, thanks to the rotation of the fringes on the inclusion. (Vertical tensile axis; $\varepsilon_p = 8.7\%$.) (c) Initiation of a microcrack with a strong shear component in prolongation of a debonded interface. $\varepsilon_p = 2\%$.

3.2. Chronology of Damage Process

During the elastic stage of a monotonic tensile test (the linear part of the stress–strain curve), only very scarce events can be detected in the matrix areas which are far from any inclusion: a few slip lines appear in favorably oriented grains at stress levels close to the yield strength. On the contrary, the situation is much more animated in and around the ceramic inclusions (Fig. 5).

A decohesion nucleation at the interface can occur in the pole regions under a macroscopic tensile stress as low as 0.8 times the yield stress (Fig. 5(b)); if the load is then relaxed, the decohesion disappears completely (Fig. 5(c)): this reversibility indicates that even the localized matrix strain allowing the decohesion is mostly elastic at this stage.

Slip lines are statistically more frequent in the vicinity of the inclusion equator, which is in agreement with the location of stress concentrations.

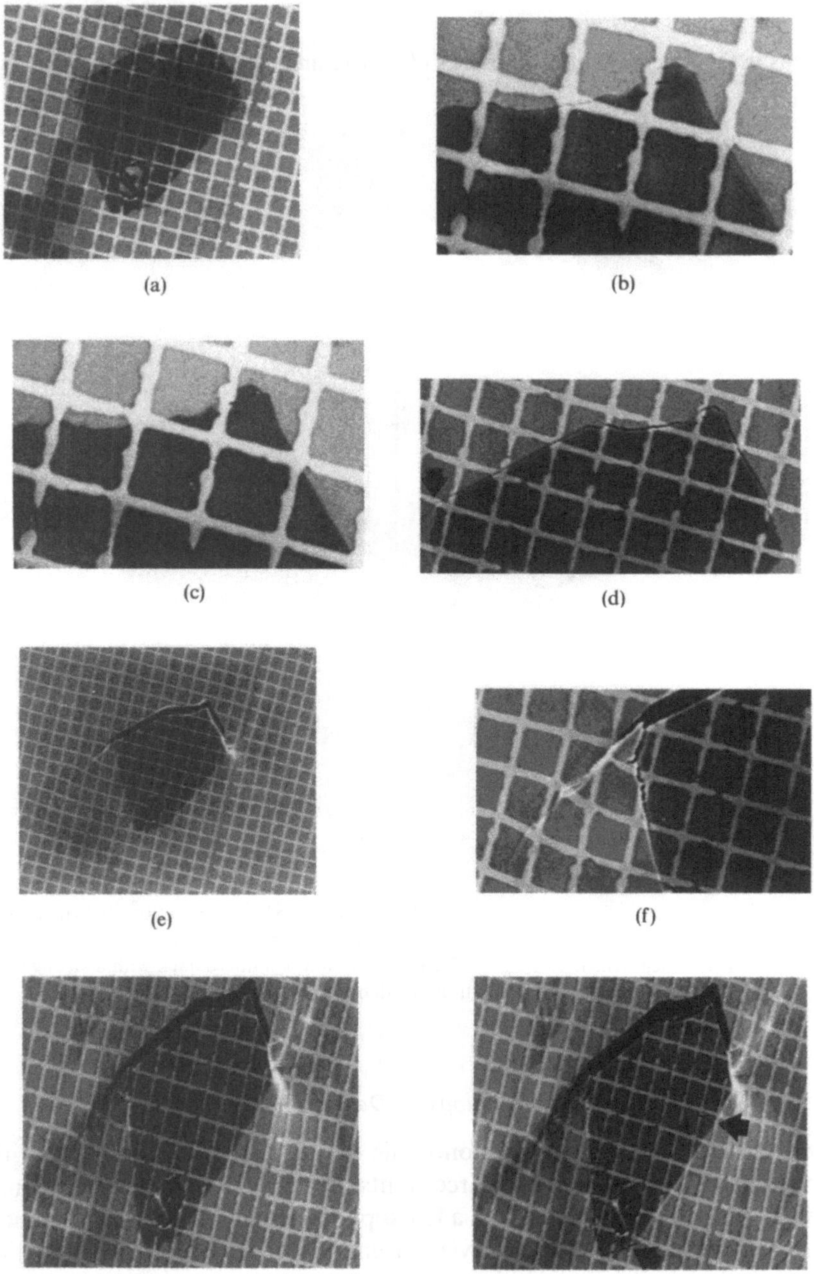

FIG. 5. Chronology of the damaging events around an inclusion. Vertical tensile axis; grid pitch 3.5 μm. (a) Initial state ($\sigma = 0$) (stage 1 on Fig. 5(i)). (b) First debonding ($\sigma = 900$ MPa) (2). (c) Relaxed state: the decohesion can no longer be observed (1). (d) The arrow points to the first damaging effect in the matrix ($\sigma = 1000$ MPa) (3). (e) Decohesion and damage at the macroscopic yield stress ($\sigma = 1200$ MPa) (4). (f) Beginning of crack propagation ($\sigma = 1300$ MPa; $\varepsilon_p \simeq 2\%$) (5). (g) The sample has been tilted by 50° around the tensile axis ($\sigma = 0$) (6). (h) Damaging effect due to a reloading ($\sigma = 1250$ MPa) (7). (i) Tensile stress versus strain curve showing the loading history of the sample.

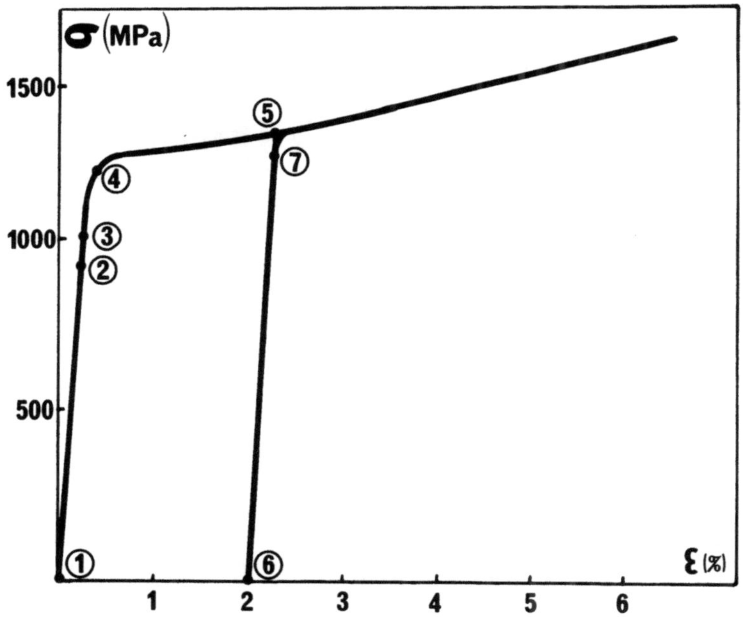

(i)

FIG. 5 *(continued)*

The first damaging events occur within the elastic stage (Fig. 5(d)). When the yield stress is reached (Fig. 5(e)), the prominent features associated with the inclusion/matrix interaction are already settled. Afterwards, when the monotonic tensile load is increasing, the existing defects enlarge or extend into the plastic regime, but very few new events occur.

Finally, it must be noted that interfacial cracks seem to be stable since they propagate only if the applied stress is increased (Fig. 6).

But, under monotonic tensile conditions, an inclusion could hardly be proved to be the original source of the subsequent breakdown of a sample. Since surface or near-surface inclusions are known to be especially operative under cyclic conditions, repeated *in situ* tensile tests were performed later on.

3.3. Repeated Tensile Tests

If a sample is unloaded after a small amount of plastic deformation (even smaller than that reported in Fig. 5) and then reloaded without any further plastic flow, several damaging events can occur, such as interfacial decohesion or inclusion fragmentation (Fig. 5(g), (h)). Thus, even only one elastic cycle performed after a slight plastic deformation can induce damage. When several cycles of repeated tension are imposed under quasi-static conditions, their effects are similar to those associated with a stress increase: the damage in and

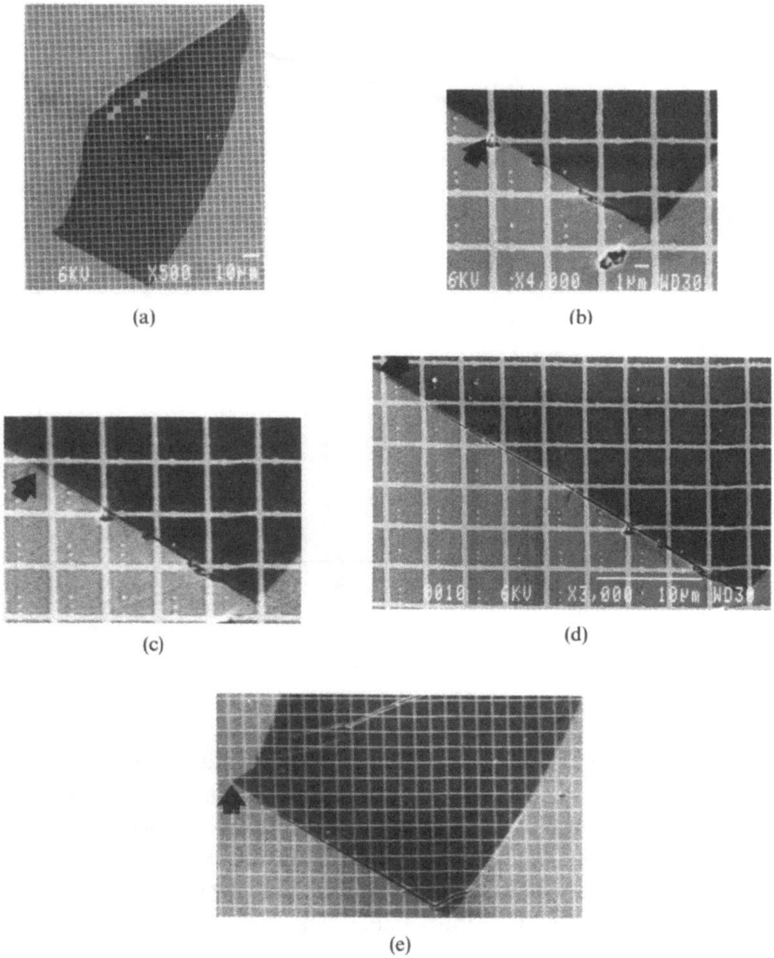

FIG. 6. Interfacial crack propagation in the macroscopically elastic state. Vertical tensile axis; grid pitch 4.5 μm. The arrows point to the crack front. (a) Initial state ($\sigma = 0$). (b) Decohesion nucleation ($\sigma = 915$ MPa). (c) Propagation ($\sigma = 1130$ MPa). (d) Propagation ($\sigma = 1225$ MPa). Macroscopic yield stress. (e) End of the propagation at the interface ($\sigma = 1260$ MPa).

around the inclusion slowly develops with more cracks in the inclusion, more interfacial decohesions, and a more pronounced plastification of the matrix in the equatorial area. At variance with the case of monotonic tests, fatigue cracks are now *nucleating* and *propagating easily* from the stress and strain concentration zones so that this can result in the ultimate breakdown of the sample (Fig. 7).

3.4. *Numerical Analysis*

In order to check the interpretation of the observations, concerning the influence of the inclusion shape and of the mechanical interface conditions on

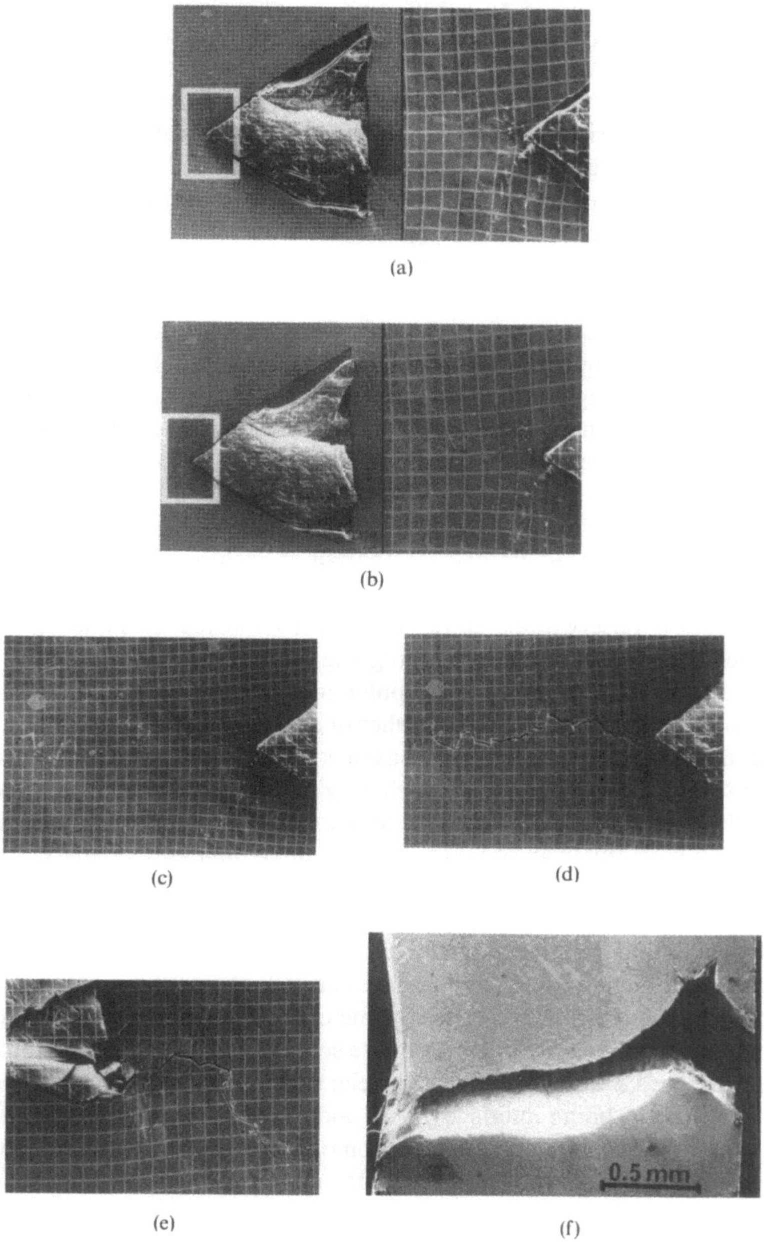

FIG. 7. Nucleation and propagation of a fatigue crack in repeated tensile tests. Vertical tensile axis; grid pitch 4.5 μm. (a) First cycle: $\varepsilon \simeq 2\%$; $\sigma = 1260$ MPa (observed in the relaxed state). (b) 2100 cycles: a microcrack has nucleated (relaxed state). (c) 7600 cycles: the crack propagates (relaxed state). (d) 7600 cycles: same as (c) under a macrostress of 1150 MPa. (e) 7600 cycles: same as (d). Crack initiated in the low right corner of the inclusion. (f) Sample broken under monotonic tensile conditions.

the stress level and space variation, several finite element calculations have been performed in the elastic case. Only one typical situation is reported here, concerning plane stress calculations for a polygonal inclusion with one salient angle in the equatorial region and two extreme interface conditions of maximal and minimal cohesion (with perfect sliding in the latter case).

The results of Fig. 8 clearly show that the most stressed areas transfer from a polar (Fig. 8(a): perfect fitting) to an equatorial (Fig. 8(b): no cohesion at all) location. Moreover, in the latter case, a stress overconcentration is induced by the sharp corner in the equatorial zone.

The experimental strain field of Fig. 3 and the calculated stress field of Fig. 8(b) are quite consistent with each other. Thus, as a first approximation, the two-dimensional approach can be considered as a right tool for understanding and even predicting the damaging effects of a quasi-rigid inclusion in a plasticized matrix in a number of cases. Nevertheless, this approach has been shown to fail in several other significant cases.

4. Three-Dimensional Analysis of Surface Effects

Several SEM observations show microcracks located at some quite unexpected places, according to the foregoing point of view. For example, Fig. 9(a) shows a crack nucleated in the polar zone instead of the expected equatorial zone, and Fig. 9(b) shows another one at too large a distance from the observable inclusion; Fig. 9(c) shows a crack which is far too long with respect to the size of the inclusion and which may be thought to be responsible for its nucleation, and Fig. 9(d) shows another crack which does not seem to be associated with any inclusion. Obviously, a three-dimensional analysis must be developed in order to take better account of the surface or near-surface position of such situations.

Any analytical treatment of the actual problem, including plastic flow, decohesion, inhomogeneity, and surface effects looks unobtainable, even for simplified geometries. Nevertheless, some of these aspects have received repeated attention separately. Let us quote some of the valuable contributions of T. Mura and his coworkers in this field; such as the problem of an elastic inclusion with a sliding interface (Mura and Furuhashi, 1984); the case of an elastic half-space with a spheroidal inhomogeneity under all-around tension parallel to the plane boundary (Tsuchida and Mura, 1983); or the case with ellipsoidal inclusions undergoing uniform dilatational eigenstrains (Seo and Mura, 1979). For the present purpose, a more global approach, which does not aim at deriving explicit stress or strain fields for given particular cases, but could draw some general conclusions concerning surface effects, looks more adequate. An example of what could be searched for is given, again, by Mura's derivation of the force acting on a subsurface inclusion, due to the free surface of a half-space, from the variation of the elastic strain energy with depth (Mura, 1982). The conclusion that, "the free surface attracts the inclu-

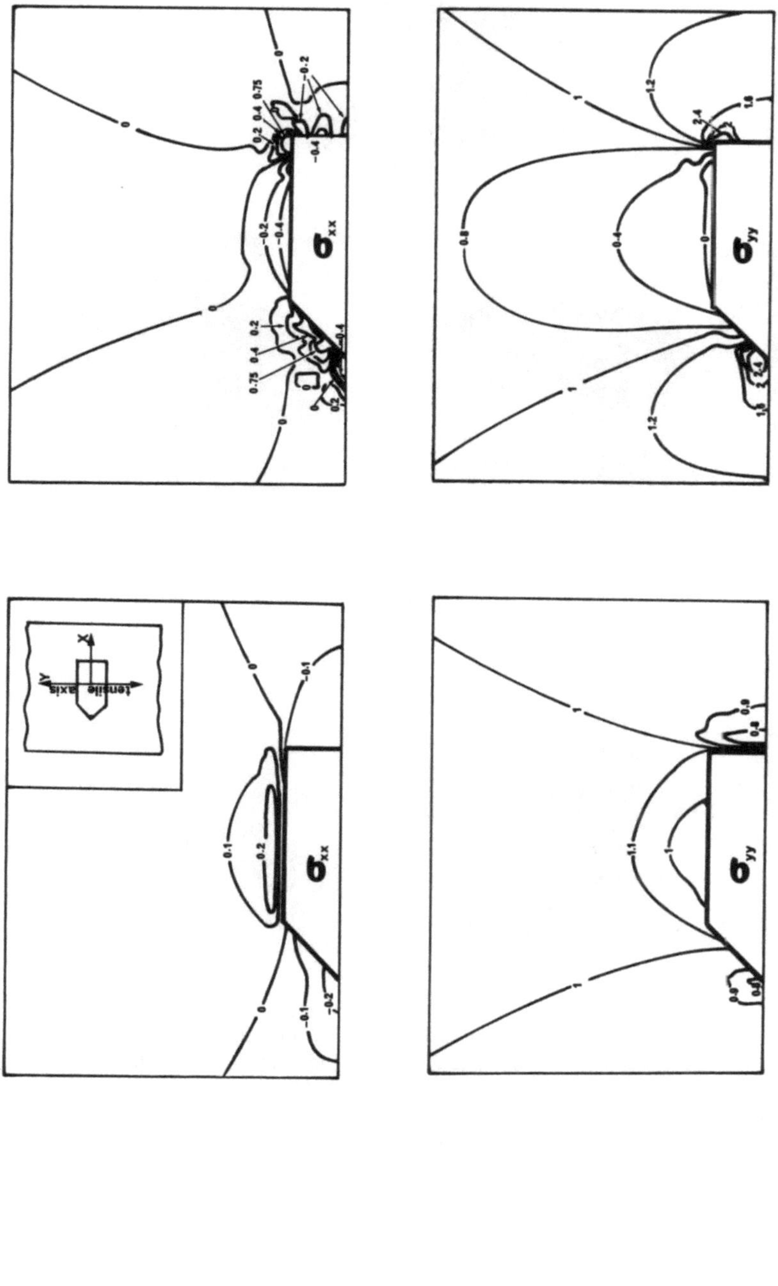

FIG. 8. Stress concentration maps (ratio of the local stress to the applied tensile stress) around a schematical penshape inclusion (compare with Fig. 3(b)). (a) Perfect interfacial cohesion. (b) No cohesion.

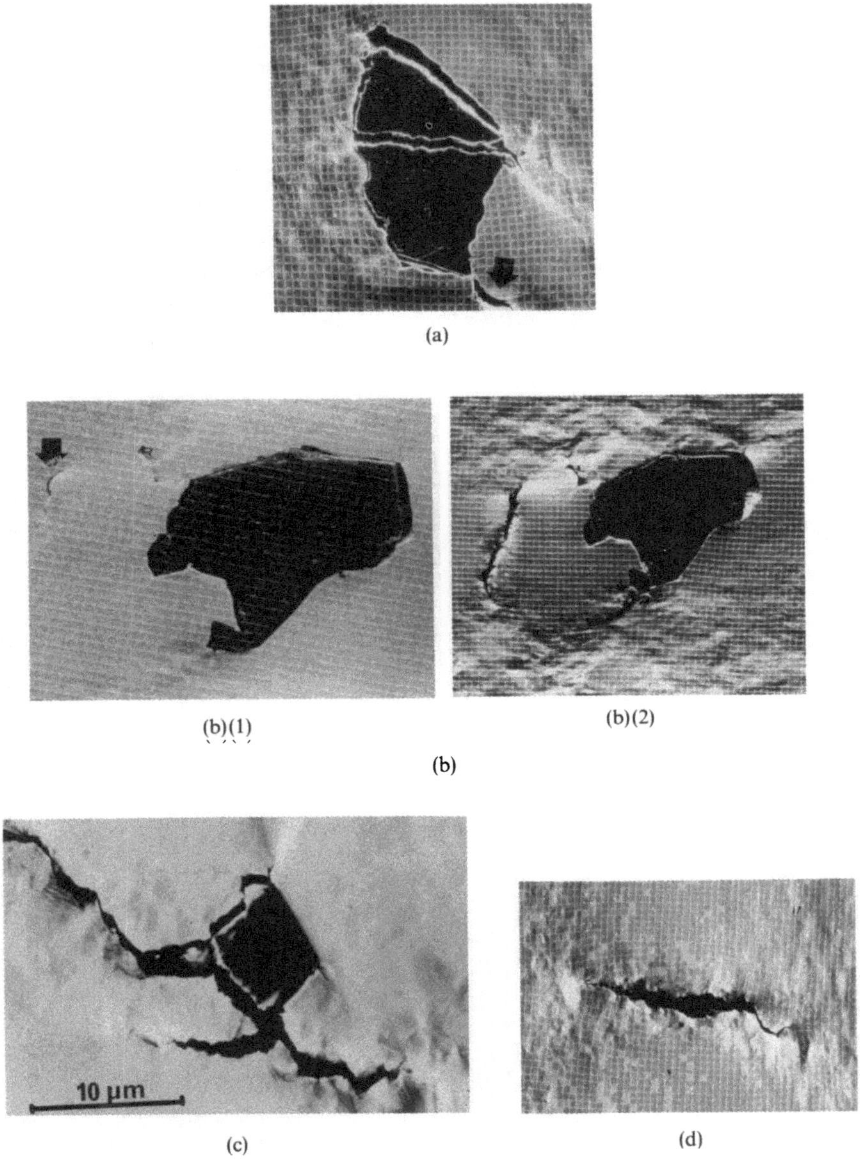

FIG. 9. Unexpected situations according to a two-dimensional analysis. (a) Crack nucleated in a polar zone. Vertical tensile axis; $\varepsilon_p = 8.7\%$; grid pitch 3.5 μm. (b) Crack nucleated far from an inclusion. Horizontal tensile axis; grid pitch 3.5 μm. bl: $\varepsilon_p = 2\%$; b2: $\varepsilon_p = 8.7\%$. (c) Long crack from an apparently small inclusion. Vertical tensile axis; $\varepsilon_p = 2\%$. (d) Crack nucleated without apparent inclusion. Vertical tensile axis; $\varepsilon_p = 2\%$.

sion," in this case, agrees with Asaro's analysis of the forces acting on internal stress sources (Asaro, 1972), and both of these studies can be related to Eshelby's celebrated treatment of the force acting on an elastic singularity (Eshelby, 1951). For such problems, which also include the case of elastic inhomogeneities, Eshelby has shown the advantage of dealing with the elastic analogue of the "energy–momentum" tensor of the classical theory of fields; this idea has been developed further by Eshelby (1975), Hill (1986), and others, and it will be shown here that it may also be used for a better understanding of some of the surprising observations reported above. First, the simpler case of a subsurface elastic perfectly bound inhomogeneity is addressed. Some extensions and conclusions are then suggested.

4.1. *Internal Inhomogeneity*

Let the body V, with the exterior surface S, consist of the elastic homogeneous matrix D and the perfectly bound inhomogeneity Ω with the surface Σ. V is not submitted to body forces but to given surface tractions $\mathbf{T}^g$ on S_T and given displacements $\mathbf{u}^g$ on S_u, with $S = S_T + S_u$ and $S_T \cap S_u = \phi$ (Fig. 10(a)). We can calculate the configurational force $\mathbf{F}$ and torque $\mathbf{C}$, associated with the matrix/inhomogeneity interaction, by expressing the virtual change $\delta\Phi$ of the total potential of the system, when the inhomogeneity undergoes a virtual infinitesimal rigid-body motion with the translation vector $\delta\boldsymbol{\xi}$ and the rotation vector $\delta\boldsymbol{\theta}$ around the origin, as

$$\delta\Phi = -(\mathbf{F}\cdot\delta\boldsymbol{\xi} + \mathbf{C}\cdot\delta\boldsymbol{\theta}). \tag{4.1}$$

The total potential Φ is defined by

$$\Phi = \int_V w\,dV - \int_{S_T} \mathbf{T}^g\cdot\mathbf{u}\,dS, \tag{4.2}$$

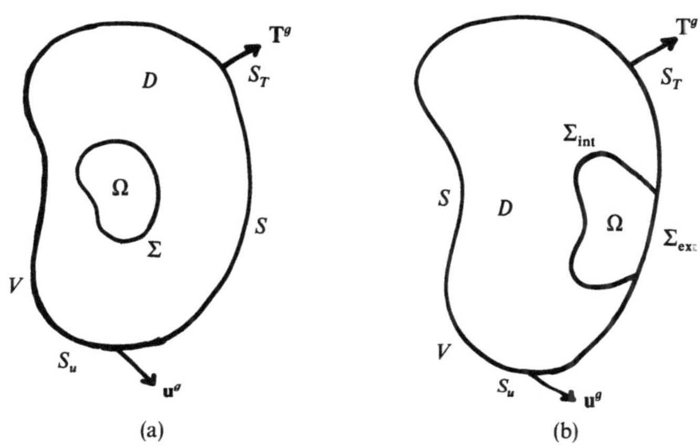

FIG. 10. Notations for (a) an internal and (b) a surface inhomogeneity.

where w is the elastic strain energy density such that the stress σ_{ij} is

$$\sigma_{ij} = \frac{\partial w}{\partial u_{i,j}}. \tag{4.3}$$

Using Bui's method (Bui, 1978) to derive the Rice J integral at a crack tip, we have to consider that Ω remains fixed and that D is moving according to a velocity field $\mathbf{v}$, obeying the following conditions:

$$\begin{cases} \mathbf{v} = 0, & \text{on } \Sigma, \\ \mathbf{v} = -(\dot{\boldsymbol{\xi}} + \dot{\boldsymbol{\theta}} \wedge \mathbf{r}) & \text{on } S \quad (\mathbf{r} \in S). \end{cases} \tag{4.4}$$

Except for (4.4), and the property of being continuous everywhere in V, $\mathbf{v}$ is arbitrary. Furthermore, we must state that the boundary conditions on S remain unchanged with respect to S. Consequently, we have

$$\begin{cases} \dfrac{\delta \mathbf{T}^g}{\delta t} = -\dot{\boldsymbol{\theta}} \wedge \mathbf{T}^g & \text{on } S_T, \\ \dfrac{\delta \mathbf{u}^g}{\delta t} = -\dot{\boldsymbol{\theta}} \wedge \mathbf{u}^g & \text{on } S_u, \end{cases} \tag{4.5}$$

where $\delta/\delta t$ denotes the total time derivative according to the velocity field $\mathbf{v}$. In this respect, we note that

$$\frac{\delta \mathbf{n}}{\delta t} = -\dot{\boldsymbol{\theta}} \wedge \mathbf{n}, \tag{4.6}$$

where $\mathbf{n}$ is the unit outwards normal vector of S. $\mathbf{F}$ and $\mathbf{C}$ will be derived from the equation

$$\frac{\delta \Phi}{\delta t} = \frac{\delta}{\delta t}\left[\int_V w\, dV - \int_{S_T} \mathbf{T}^g \cdot \mathbf{u}\, dS\right] = -\frac{\delta \boldsymbol{\xi}}{\delta t} \cdot \mathbf{F} - \frac{\delta \boldsymbol{\theta}}{\delta t} \cdot \mathbf{C}, \tag{4.7}$$

with $\dot{\boldsymbol{\xi}} = \delta \boldsymbol{\xi}/\delta t$ and $\dot{\boldsymbol{\theta}} = \delta \boldsymbol{\theta}/\delta t$. Referring to Appendix A2 for more details, we get, according to Germain (1986),

$$\frac{\delta \Phi}{\delta t} = \int_V \frac{\partial w}{\partial t}\, dV + \int_S w \mathbf{v} \cdot \mathbf{n}\, dS - \int_{S_T} \frac{\delta}{\delta t}(\mathbf{T}^g \cdot \mathbf{u})\, dS. \tag{4.8}$$

From (4.3) and the definition

$$\frac{\delta u_i}{\delta t} = \frac{\partial u_i}{\partial t} + u_{i,j} v_j, \tag{4.9}$$

the first term of (4.8) becomes

$$\int_V \frac{\partial w}{\delta t}\, dV = \int_V \sigma_{ij} \left(\frac{\partial u_i}{\partial t}\right)_{,j} dV = \int_V \sigma_{ij} \left(\frac{\delta u_i}{\delta t} - u_{i,k} v_k\right)_{,j} dV, \tag{4.10}$$

where it can be proved that the last integrand suffers no discontinuity on Σ. So, since σ_{ij} is divergence free,

$$\int_V \frac{\partial w}{\partial t}\, dV = \int_S \sigma_{ij} \left(\frac{\delta u_i}{\delta t} - u_{i,k} v_k\right) n_j\, dS. \tag{4.11}$$

Now, we get

$$\frac{\delta\Phi}{\delta t} = \int_S \sigma_{ij}\frac{\delta u_i}{\delta t} n_j \, dS + \int_S (wv_j - \sigma_{ij}u_{i,k}v_k)n_j \, dS$$
$$- \int_{S_T} \left(\frac{\delta \mathbf{T}^g}{\delta t}\cdot \mathbf{u} + \mathbf{T}^g \cdot \frac{\delta \mathbf{u}}{\delta t}\right) dS. \tag{4.12}$$

Since the first integral is

$$\int_S \sigma_{ij}\frac{\delta u_i}{\delta t} n_j \, dS = \int_{S_T} \mathbf{T}^g \cdot \frac{\delta \mathbf{u}}{\delta t} \, dS + \int_{S_u} \mathbf{T} \cdot \frac{\delta \mathbf{u}^g}{\delta t} \, dS, \tag{4.13}$$

we are left with

$$\frac{\delta\Phi}{\delta t} = \int_S (w\delta_{jk} - \sigma_{ij}u_{i,k})v_k n_j \, dS + \int_{S_u} \mathbf{T} \cdot \frac{\delta \mathbf{u}^g}{\delta t} \, dS - \int_{S_T} \frac{\delta \mathbf{T}^g}{\delta t} \cdot \mathbf{u} \, dS$$
$$= \int_S P_{jk} v_k n_j \, dS - \int_{S_u} \mathbf{T} \cdot (\dot{\boldsymbol{\theta}} \wedge \mathbf{u}^g) \, dS + \int_{S_T} \mathbf{u} \cdot (\dot{\boldsymbol{\theta}} \wedge \mathbf{T}^g) \, dS, \tag{4.14}$$

where the elastic energy–momentum tensor $P_{jk} = w\delta_{jk} - \sigma_{ij}u_{i,k}$ has been introduced and (4.5) used. With (4.4) and (4.7), we finally get

$$\begin{cases} F_i = \displaystyle\int_S P_{ji} n_j \, dS, \\ C_i = \epsilon_{ijk} \displaystyle\int_S (P_{mk}x_j + \sigma_{mk}u_j)n_m \, dS, \end{cases} \tag{4.15}$$

where ϵ_{ijk} is the completely antisymmetric permutation tensor. In (4.15) can be recognized the Knowles and Sternberg (1972) invariants, which allows us to replace S in both integrals, when D is homogeneous and isotropic, with any closed surface S' including Ω, such as Σ itself. Alternative expressions for $\mathbf{F}$ and $\mathbf{C}$ can be derived by using the discontinuity $[\mathbf{P}]$ of $\mathbf{P}$ across Σ, namely

$$F_i = \int_\Sigma [P_{ji}] n_j \, dS \qquad C_i = \epsilon_{ijk} \int_\Sigma [P_{mk}] x_j n_m \, dS, \tag{4.16}$$

which proves that the superficial discontinuity of the energy–momentum tensor across the interface is the very source of the configurational force and torque $\mathbf{F}$ and $\mathbf{C}$ (Ambroise *et al.*, 1987a).

4.2. Extensions and Discussion

The foregoing analysis may be applied without change when the inhomogeneity increasingly approaches the surface. Strictly speaking, the treatment is failing when it emerges on S with a finite part of Σ, Σ_{ext}, say (Fig. 10(b)). Nevertheless, since δt may be made arbitrarily small, and provided that the stress singularity at the line boundary between Σ_{ext} and S is weak enough (of the order of $r^{-\alpha}$ with $\alpha < 0.5$, according to a two-dimensional classical argument), formulae (4.15) and (4.16) are still valid, with Σ restricted to Σ_{int} and with the conditions $\delta\mathbf{T}^g/\delta t = 0$ on $\Sigma_{\text{ext}} \cap S_T$ and $\delta\mathbf{u}^g/\delta t = 0$ on $\Sigma_{\text{ext}} \cap S_u$.

At this stage, it must be admitted that, in the case of a well-bound inhomogeneity, **F** and **C** are quite fictitious vectors which have no more reality than the force at the tip of a fixed crack. But, since they express the way the total potential could change if the velocity **v** was an actual field, they indicate which force and torque would move the inhomogeneity if it was no longer bound up with the matrix. This is useful information for the case when the interfacial cohesion is weak enough to allow a complete debonding when the applied stresses are sufficiently high.

In order to state the fact that the configurational force would push the inhomogeneity towards the free surface, as it is known to do for an eigenstrained inclusion, we can stress two points:

First, the sign of **F** and **C** is unchanged when $\mathbf{T}^g$ and $\mathbf{u}^g$ are changed into their opposites: this is due to the expression of **P** which is the same if $\boldsymbol{\sigma}$ and **u** are changed into $-\boldsymbol{\sigma}$ and $-\mathbf{u}$. Note that the same was true in the case of Mura's eigenstrained inclusion where the attractive force, which decreases with depth, depends on ε^{*2}, if ε^* is the dilatational (or contractional) eigenstrain.

Second, according to Eshelby's (1951) analysis, the stress field $\boldsymbol{\sigma}$ may be written as

$$\boldsymbol{\sigma} = \boldsymbol{\sigma}^{\mathrm{A}} + \boldsymbol{\sigma}^{\mathrm{s}} \qquad \text{with} \quad \boldsymbol{\sigma}^{\mathrm{s}} = \boldsymbol{\sigma}^{\infty} + \boldsymbol{\sigma}^{\mathrm{I}}, \tag{4.17}$$

where $\boldsymbol{\sigma}^{\mathrm{A}}$ is the "applied" stress field, namely, the one the given surface tractions and displacements would produce in the body if it were homogeneous throughout, and $\boldsymbol{\sigma}^{\mathrm{s}}$ is the disturbance produced by the inhomogeneity. $\boldsymbol{\sigma}^{\mathrm{s}}$ may be considered as the self-stress associated with a singularity which is induced in the inhomogeneity by the applied forces. As for a real singularity, we can write $\boldsymbol{\sigma}^{\mathrm{s}} = \boldsymbol{\sigma}^{\infty} + \boldsymbol{\sigma}^{\mathrm{I}}$, where $\boldsymbol{\sigma}^{\infty}$ is the disturbance which would take place if the body was infinite, and $\boldsymbol{\sigma}^{\mathrm{I}}$ is the image stress field which allows the condition $(\boldsymbol{\sigma}^{\infty} + \boldsymbol{\sigma}^{\mathrm{I}})\cdot\mathbf{n} = 0$ on S to be fulfilled. Consequently, and just as for a real singularity, the resulting force exerted on the inhomogeneity is, with evident notations,

$$\mathbf{F} = \mathbf{F}(A, \infty) + \mathbf{F}(I, \infty) + \mathbf{F}(\infty, \infty). \tag{4.18}$$

Since $\mathbf{F}(\infty, \infty)$ vanishes, this means that the total force results from the interaction between the applied and image fields and the field which corresponds to an infinite body. If $\boldsymbol{\sigma}^{\mathrm{A}}$ is uniform, only the image field is operative: as for the inclusion, it originates from the free surface condition and results in the attraction of inhomogeneity by the free surface; note that the contribution $\mathbf{C}(A, \infty)$ to the total torque does not necessarily vanish in this case (Ambroise *et al.*, 1987b).

The reality of the effects of the configurational torque for adequate conditions was already proved by Fig. 4(b), where the moiré pattern clearly indicates

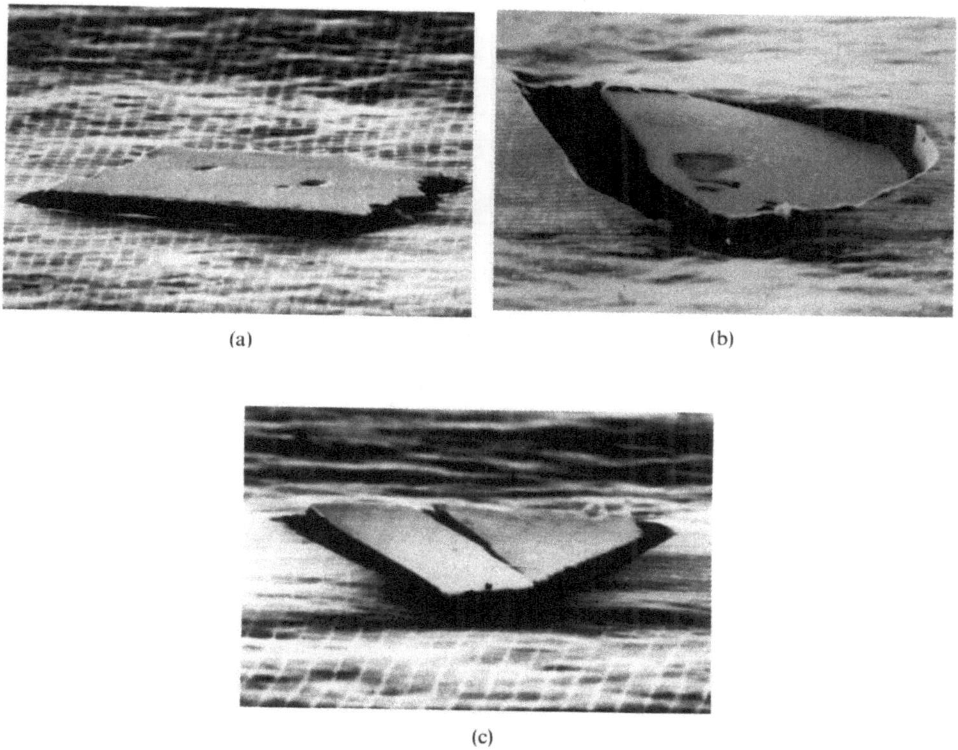

FIG. 11. Three-dimensional effects on surface inclusions. Horizontal tensile axis; $\varepsilon_p = 9\%$. (a) Exclusion. (b) Tilting. (c) Tilting of a broken inclusion.

an in-plane rotation of the inclusion with respect to the tensile axis. Additional illustrations of surface effects are given in Fig. 11, where exclusion and tilting of debonded surface inhomogeneities are clearly visible.

5. Conclusion

The foregoing comments may be checked in the experimental cases, which were reported at the beginning of Section 4, as conflicting with a two-dimensional interpretation. They strongly suggest that emerged or submerged inhomogeneities and image forces pushing them outwards could be responsible for the unexpected observations. Figure 12(a) is concerned with the situation of Fig. 9(a) where a crack is nucleated at an unexpected place: a chemical dissolution of the matrix exhibits a sharp edge just beneath this place, and which acts as a knife pushing on the thin matrix ligament. Figure 12(b) refers to the situation of Fig. 9(b): the SEM observation under a large tilt angle shows the out-of-plane movement of the inhomogeneity that pushes the

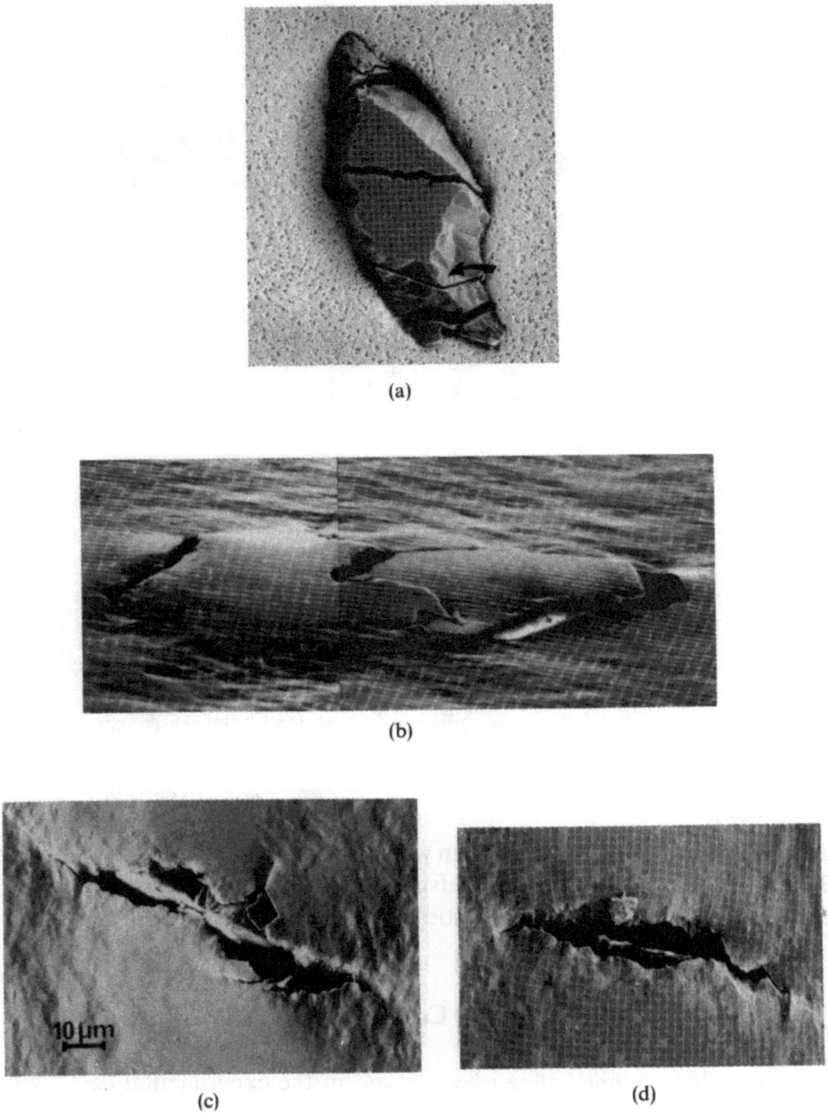

FIG. 12. The cracks of Fig. 9 revisited. (a) Fig. 9(a) after a chemical dissolution of the matrix: the arrow points to an internal indentor. (b) Fig. 9(b) observed under a tilt angle of 70°. (c) Fig. 9(c) after a strain increment. (d) Fig. 9(d) after a strain increment (the internal inhomogeneity is now visible).

ligament out. Similarly, Fig. 12(c) and (d) reveal, after a strain increment, the near-surface inclusions responsible for the crack nucleation of Fig. 9(c) and (d), respectively.

These cases clearly demonstrate the origin of the specific damaging effects of surface and near-surface inclusions, especially when the configurational force and torque can be converted into active ones due to the weak interfacial cohesion between the matrix and the inclusions. Several practical conclusions could be drawn from the foregoing on technological grounds: without going too far in this field, it can be suggested that the harmfulness of debonded ceramic inclusions could be significantly lowered, either by modifying the chemical composition of the matrix and the inclusions in order to increase the cohesion of their interfaces, or by performing adequate surface treatments (shot peening, etc.) in order to put the inclusions into a compressive residual stress state.

Acknowledgments

The authors wish to express grateful thanks to DRET for financial support and to SNECMA for supplying the material.

References

Ambroise, M. H., Bretheau, T., and Viaris de Lesegno, P. (1987), Surface microextensometry by means of moiré interferometry with a scanning electron microscope, *J. Appl. Mech.*, **54**, 237–239.

Ambroise, M. H., Bretheau, T., and Zaoui, A. (1987a), Sur une notion de "torseur de configuration" caractérisant les interactions élastiques entre inclusion et matrice, *C. R. Acad. Sci. Paris*, **304**, II, 245–250.

Ambroise, M. H., Bretheau, T., and Zaoui, A. (1987b), A note on configurational force and torque expressing the interaction between a superficial inhomogeneity and a matrix, *Phil. Mag. Lett.*, **56**, 7–11.

Asaro, R. J. (1972), Forces on internal stress sources in terms of elastic chemical potentials, *Script. Metall.*, **6**, 547–550.

Attwood, D. G. and Hazzledine, P. M. (1976), A fiducial grid for high-resolution metallography, *Metallography*, **9**, 483–501.

Bretheau, T. and Caldemaison, D. (1981), Test of mechanical interaction models between polycrystal grains by local strain measurements. *Proceedings of the 2nd Risø International Symposium on Metallurgy and Materials Science*, pp. 157–161.

Bretheau, T. and Caldemaison, D. (1983), Study of inclusion–matrix interaction by means of local strain measurements. *Proceedings 4th Risø International Symposium on Metallurgy and Materials Science*, pp. 173–178.

Bretheau, T., Caldemaison, D., and Ambroise, M. H. (1988), Inclusion/matrix mechanical interaction: an in situ study by tensile and fatigue tests in the S.E.M. *Proceedings ICSMA 8*, Tampere, Finland, pp. 1051–1056.

Bui, H. D. (1978), *Mécanique de la rupture fragile*, Masson, Paris.

Eshelby, J. D. (1951), The force on an elastic singularity, *Phil. Trans.*, **A244**, 87–112.

Eshelby, J. D. (1975), The elastic energy-momentum tensor, *J. Elasticity*, **5**, 321–335.

Germain, P. (1986), *Mécanique*, T1, Ellipses, Ecole Polytechnique, Paris.

Hauser, J. J. and Wells, M. G. H. (1970), Inclusions in high strength and bearing steels. Air Force Materials Laboratory Technical Report, 69-339.

Hill, R. (1986), Energy–momentum tensors in elastostatics: some reflexions on the general theory, *J. Mech. Phys. Solids*, **34**, 305–317.

Jablonski, D. A. (1981), The effect of ceramic inclusions on the low cycle fatigue life of low carbon Astroloy subjected to hot isostatic pressure, *J. Mat. Sci. Engng.*, **48**, 189–198.

Knowles, J. K. and Sternberg, E. (1972), On a class of conservation laws in linearized and finite elastostatics, *Arch. Rat. Mech. Anal.*, **44**, 187–211.

Lankford, J. (1977a), Effect of oxide inclusions on fatigue fracture, *Int. Met. Rev.*, Sept., 221–228.

Lankford, J. (1977b), Initiation and early growth of fatigue cracks in high strength steel, *Engng. Fracture Mech.*, **9**, 617–624.

Law, C. C. and Blackburn, M. J. (1980), Effects of ceramic inclusions on fatigue properties of a powder metallurgical nickel-base superalloy. *Proceedings of the International Powder Metallurgy Conference and Exhibition, Washington*, Vol. 12–14.

Mura, T. (1982), *Micromechanics of Defects in Solids*, Martinus Nihoff, the Hague.

Mura, T. and Furuhashi, R. (1984), The elastic inclusion with a sliding interface, *J. Appl. Mech.*, **51**, 308–310.

Seo, K. and Mura, T. (1979), The elastic field in a half-space due to ellipsoidal inclusions with uniform dilatational eigenstrains, *J. Appl. Mech.*, **46**, 568–572.

Thompson, F. A., Cutler, C. P., and Siddall, R. J. (1979), The influence of inclusions on the low cycle fatigue properties of the nickel-based superalloy APK1 produced via the powder-route. Henry Wiggin and Co. Ltd., Technical Report 3184.

Tsuchida, E. and Mura, T. (1983), The stress-field in an elastic half-space having a spheroidal inhomogeneity under all-around tension parallel to the plane boundary, *J. Appl. Mech.*, **50**, 807–816.

Appendix

A1. Experimental Techniques

A1.1. Tensile Tests in the SEM

A pertinent experimental study of the damaging effects of inhomogeneities demands surface observations of under-stress specimens. Two kinds of techniques can be used:

(i) The observation by SEM of plastic replicas of the surface for different stress or strain states (Lankford and Kusenberger, 1973; Lankford, 1976).

(ii) The under-stress direct observation of the real surface. Using a tensile stage mounted on the SEM.

The first technique gives no other observation than indirect and incremental. In order to benefit from a direct, real time, continuous observation, we used the second technique.

In situ tensile tests were performed, using a Raith tensile stage (Bretheau *et al.* 1988) designed specifically for a JEOL 845 SEM; the load capacity is higher than 5 kN, which allows us to test macroscopical specimens of high-strength materials; the cross-head velocity can be varied continuously between 0.1 and 60 μm/s. The load cell has exhibited very good linearity in calibration experiments, and a transducer measures the displacement of the moving grip: the load versus displacement curve is recorded continuously. The sample can be tilted around the tensile axis.

During the mechanical tests, the sample surface can be observed over an area of 20×20 mm^2 with a magnification ranging between 15 and 10,000. In order to improve the image quality, an on-line filtering and image summation are performed using a Crystal system; the treated images are recorded on a Sony video tape recorder.

In situ tensile test specimens with gauge dimensions $15 \times 4 \times 0.8$ mm^3 were spark cut from the hipped billets; one face was mechanically polished to a 1 μm diamond making the inclusions clearly visible. No damage due to surface preparation could be observed either in the matrix, the inclusion, or at the interface, except occasionally a very weak grooving on some parts of the interface; but it did not seem to affect the mechanical behavior.

A1.2. Microextensometry

Square fiducial microgrids made of 0.2–0.5 μm wide lines and with a 2–5 μm pitch have been deposited on emerging inclusions (Fig. 2).

The grids of evaporated metal were produced by a photo-resist technique using the electron beam of the scanning electron microscope for irradiation (Attwood and Hazzledine, 1976). In-plane strain fields have then been derived in the grid areas, due to two distinct measurement analyses:

The first (Bretheau and Caldemaison, 1981; Bretheau and Caldemaison, 1983) consists in direct digitization of the grid node coordinates using a microcomputer; it is a quantitative method that allows us to draw strain maps at the very local scale of the inhomogeneities; but it is also tedious and time consuming.

The second method (Ambroise *et al.*, 1987) directly uses iso-displacement lines of a moiré pattern, resulting from the interference between the deformed grids and the grating formed by the electron beam scanning which allows us to photograph the inclusion and the surrounding matrix; it is a much more qualitative and less local method, but it shows in real time the evolution of the strain inhomogeneities during a tensile test.

A2. Calculation of **F** *and* **C** *(Section* 4)

A2.1. Proof of (4.8)

$$\frac{\delta}{\delta t}\int_{S_T} \mathbf{T}^g\cdot\mathbf{u}\,dS = \frac{\delta}{\delta t}\int_{S_T} \sigma_{ij}u_i n_j\,dS$$

$$= \int_{S_T}\left(\frac{\delta}{\delta t}(\sigma_{ij}u_i)\cdot n_j + \sigma_{ij}u_i v_{k,k}n_j - \sigma_{ik}u_i v_{j,k}n_j\right)dS. \quad \text{(A2.1)}$$

But, according to (4.4), $v_{k,k} = 0$ on S. From (4.6), we get

$$\begin{cases} v_{j,k} = -v_{k,j} = \epsilon_{jkm}\dot{\theta}_m, \\ \dfrac{\delta n_i}{\delta t} = -\epsilon_{ijk}\dot{\theta}_j n_k = v_{i,k}n_k = -v_{k,i}n_k, \\ -\sigma_{ik}u_i v_{j,k}n_j = \sigma_{ik}u_i\dfrac{\delta n_k}{\delta t}. \end{cases} \quad \text{(A2.2)}$$

Consequently, we have

$$\frac{\delta}{\delta t}\int_{S_T} \mathbf{T}^g\cdot\mathbf{u}\,dS = \int_{S_T}\left(\frac{\delta}{\delta t}(\sigma_{ij}u_i)n_j + \sigma_{ik}u_i\frac{\delta n_k}{\delta t}\right)dS$$

$$= \int_{S_T}\frac{\delta}{\delta t}(\sigma_{ij}u_i n_j)\,dS = \int_{S_T}\frac{\delta}{\delta t}(\mathbf{T}^g\cdot\mathbf{u})\,dS. \quad \text{(A2.3)}$$

A2.2. Proof of (4.11)

$$\left[\sigma_{ij}\left(\frac{\delta u_i}{\delta t} - u_{i,k}v_k\right)n_j\right] = \left[T_i\frac{\delta u_i}{\delta t}\right] - [T_i u_{i,k}v_k]$$

$$= T_i\left[\frac{\delta u_i}{\delta t}\right] - T_i[u_{i,k}]v_k, \quad \text{(A2.4)}$$

since T_i is continuous across any surface lying between Σ and S, and the same is true for δu_i. Furthermore, $[u_{i,k}]$ is not zero only on Σ, but v_k vanishes here. So, the total studied discontinuity is zero.

Additional References

Lankford, J. (1976), Inclusion-matrix debonding and fatigue crack initiation in low alloy steel, *Int. J. Fracture*, **12**, 155–157.

Lankford, J. and Kusenberger, F. N. (1973), Initiation of fatigue cracks in 4340 steels, *Met. Trans.*, **4**, 553–559.

On a Correspondence Between Mechanical and Thermal Effects in Two-Phase Composites

Y. Benveniste* and G. J. Dvorak
Department of Civil Engineering, Rensselaer Polytechnic Institute, Troy, NY 12180-3590, U.S.A.

Abstract

This paper considers the thermomechanical loading problem of binary composites with any anisotropic elastic constituents and arbitrary phase geometry, subjected to homogeneous traction or displacement boundary conditions and uniform temperature change. It is shown that the solution of the thermomechanical problem is uniquely determined by the solution of the purely mechanical problem corresponding to zero temperature change. This result is used to obtain explicit relations between the effective thermal strain (or stress) coefficient tensor and the effective mechanical properties. The correspondence between thermomechanical and purely mechanical loads is also used to establish an important consistency property of the Mori–Tanaka model in the context of thermomechanical problems. Extensions of the results to composite systems with temperature-dependent properties is discussed.

1. Introduction

In recent papers, Dvorak (1983, 1986) has shown that for certain binary composites subjected to combined thermomechanical loading, the local thermal strain and stress concentration factors can be related in an exact way to the corresponding mechanical concentration factors. The considered systems were effectively isotropic binary composites with elastically isotropic phase but arbitrary phase geometry, and fibrous composites with elastically isotropic or transversely isotropic constituents of any cross section but of cylindrical geometry. The correspondence established in that paper between the concentration factors allows us to write expressions for the effective thermal expansion coefficients once the effective mechanical properties are known. Moreover, the derivation is made in a manner which makes the results applicable to inelastic systems.

The present paper generalizes the results obtained by Dvorak (1986) to

---

* On sabbatical leave from Tel-Aviv University.

binary composites with general anisotropic constituents and arbitrary phase geometry. The established results are then used to prove an important consistency property of the Mori–Tanaka micromechanics model in the context of thermomechanical problems.

In the second section of the paper the correspondence relations between thermomechanical problems and pure mechanical problems are obtained in a closed and simple form. Two types of loadings are considered:

(a) A combined thermomechanical loading with homogeneous traction boundary conditions and uniform temperature change (Equation (2.4)).
(b) A combined thermomechanical loading with homogeneous displacement boundary conditions and uniform temperature change (Equation (2.5)).

The purely mechanical problems are those corresponding to a zero temperature change.

The third section of the paper is concerned with evaluation of the tensor of effective thermal strain coefficients (thermal expansion) and the tensor of effective thermal stress coefficients, a subject which has drawn considerable interest in the literature in the last two decades. In a well-known paper, Levin (1967) has shown that in two-phase materials with arbitrary phase geometry the effective thermal expansion coefficients can be related to the effective elastic properties. This line of inquiry was extended by Schapery (1968), who derived bounds on thermal expansion coefficients of multiphase composites with isotropic phases, while Rosen and Hashin (1970) reviewed and extended Levin's result to general anisotropic phases. Laws (1973), on the other hand, has given a different treatment of the subject based on thermostatic considerations. We finally mention Craft and Christensen (1981) who considered the thermal expansion of composites with randomly oriented fibers. We show in the third section of the paper that the principle established in the second section allows a straightforward and elegant derivation of the results of Rosen and Hashin (1970) and Laws (1973). A dual formulation corresponding to zero traction or zero displacement boundary conditions is presented, resulting in expressions for the effective thermal strain and stress coefficient tensors.

The fourth section of the paper is concerned with the Mori–Tanaka (1973) model of composites in the context of thermomechanical problems. There exist several papers in the literature which predict the effective thermal coefficients of particulate composites by using the average matrix stress (or strain) concept of Mori and Tanaka (1973) (see Wakashima *et al.*, 1974; Takahashi *et al.*, 1985; Takao, 1985; Takao and Taya, 1985). These works base their derivation directly on eigenstrain concepts and the equivalent inclusion idea of Eshelby (1957), and do not make use of the results of Levin (1967) and Rosen and Hashin (1970). This section of the paper gives a concise derivation of the Mori–Tanaka micromechanics problem in the context of the thermomechanical properties, in the spirit of the exposition of this theory by Benveniste (1987) which dealt with the purely mechanical case. The correspondence relations established in the previous sections are then used to prove an important consistency property of this micromechanical mode. Specifically, we show that the direct application of the model to the prediction of the effective thermal coefficients produces results which are in agreement with those that would have been obtained if the relations of Levin (1967) and Rosen and Hashin (1970) were used.

The paper concludes with some comments on applications of the results to composite systems with temperature-dependent properties.

2. Correspondence of Overall Mechanical and Thermomechanical Loading

Consider a two-phase composite consisting of perfectly bonded anisotropic constituents of arbitrary phase geometry such that the orientation of the respective material axes in each of the phases is fixed throughout the aggregate. Let the thermoelastic constitutive relations of the homogeneous phases, $r = 1, 2$, be given by

$$\boldsymbol{\sigma}_r = \mathbf{L}_r \boldsymbol{\varepsilon}_r + \mathbf{l}_r \theta, \tag{2.1}$$

$$\boldsymbol{\varepsilon}_r = \mathbf{M}_r \boldsymbol{\sigma}_r + \mathbf{m}_r \theta, \tag{2.2}$$

where $\mathbf{L}_r$ and $\mathbf{M}_r = (\mathbf{L}_r)^{-1}$ are the phase stiffness and compliance tensors, $\mathbf{m}_r$ is the thermal strain vector (of expansion coefficients), and $\mathbf{l}_r$ the thermal stress vector, such that

$$\mathbf{l}_r = -\mathbf{L}_r \mathbf{m}_r. \tag{2.3}$$

Define the following thermomechanical loading problems for a representative volume V of the composite aggregate:

Problem I

$$\boldsymbol{\sigma}_n(S) = \boldsymbol{\sigma}_0 \mathbf{n}, \qquad \theta(S) = \theta_0. \tag{2.4}$$

Problem II

$$\mathbf{u}(S) = \boldsymbol{\varepsilon}_0 \mathbf{x}, \qquad \theta(S) = \theta_0, \tag{2.5}$$

where $\boldsymbol{\sigma}_n(S)$ and $\mathbf{u}(S)$ are the traction and displacement vectors at the external boundary S of V, with an outer unit normal $\mathbf{n}$, $\boldsymbol{\sigma}_0$ and $\boldsymbol{\varepsilon}_0$ are constant uniform overall stress and strain fields, and $\mathbf{x}$ denotes a Cartesian coordinate; $\theta(S)$ is the thermal change at S, and θ_0 is a constant quantity.

Note first that the uniform field $\theta(\mathbf{x}) = \theta_0$ in the volume V is the stationary temperature distribution that satisfies the boundary conditions $((2.4_2), (2.5_2))$. The local stresses and strains in the phase, which are caused, respectively, by the prescribed boundary conditions (2.4) and (2.5), can be written in the form

$$\boldsymbol{\sigma}_r(\mathbf{x}) = \mathbf{B}_r(\mathbf{x})\boldsymbol{\sigma}_0 + \mathbf{b}_r(\mathbf{x})\theta_0, \tag{2.6}$$

and

$$\boldsymbol{\varepsilon}_r(\mathbf{x}) = \mathbf{A}_r(\mathbf{x})\boldsymbol{\varepsilon}_0 + \mathbf{a}_r(\mathbf{x})\theta_0. \tag{2.7}$$

In the above equations, $\mathbf{A}_r(\mathbf{x})$ and $\mathbf{B}_r(\mathbf{x})$ are fourth-order tensors; their phase volume averages in a representative volume, $\mathbf{A}_r$ and $\mathbf{B}_r$ are usually called the mechanical strain and stress concentration factor tensors. The second-order tensors $\mathbf{a}_r(\mathbf{x})$ and $\mathbf{b}_r(\mathbf{x})$, have representative phase volume averages $\mathbf{a}_r$ and $\mathbf{b}_r$ which are called the thermal strain and stress concentration factors. These tensors without the argument $\mathbf{x}$ will denote average quantities in the sequel of the paper.

We will now show that for statistically homogeneous two-phase composites with distinct anisotropic constituents and completely arbitrary phase geometry, the tensor $\mathbf{b}_r(\mathbf{x})$ can be uniquely determined in terms of $\mathbf{B}_r(\mathbf{x})$, $\mathbf{M}_r$, and $\mathbf{m}_r$. A similar relation will be derived between $\mathbf{a}_r(\mathbf{x})$ and the tensors $\mathbf{A}_r(\mathbf{x})$, $\mathbf{L}_r$, and $\mathbf{l}_r$. These results will be obtained from the decomposition procedure proposed by Dvorak (1986, Sect. 3), which is extended here to systems with arbitrary phase anisotropy.

Consider first Problem I, Equation (2.4). By linearity, the effect of θ_0 can be determined separately, and is in fact represented by $\mathbf{b}_r(\mathbf{x})$ in (2.6). This tensor can be evaluated from the following decomposition scheme:

(a) The phases are separated from each other and subjected to a uniform temperature rise θ_0 which causes the uniform strains

$$\boldsymbol{\varepsilon}_1 = \mathbf{m}_1\theta_0, \qquad \boldsymbol{\varepsilon}_2 = \mathbf{m}_2\theta_0, \tag{2.8}$$

and zero stresses.

(b) Certain unknown tractions derived from an auxiliary uniform stress field

$$\hat{\boldsymbol{\sigma}}_1(\mathbf{x}) = \hat{\boldsymbol{\sigma}}_2(\mathbf{x}) = \hat{\boldsymbol{\sigma}}, \tag{2.9}$$

are applied to each phase such that the uniform strains caused by (2.8) and (2.9) make the phases compatible. This condition is met by demanding that

$$\mathbf{M}_1\hat{\boldsymbol{\sigma}} + \mathbf{m}_1\theta_0 = \mathbf{M}_2\hat{\boldsymbol{\sigma}} + \mathbf{m}_2\theta_0. \tag{2.10}$$

Therefore,

$$\hat{\boldsymbol{\sigma}} = (\mathbf{M}_1 - \mathbf{M}_2)^{-1}(\mathbf{m}_2 - \mathbf{m}_1)\theta_0, \tag{2.11}$$

providing that the inverse exists. A discussion of this proviso appears in Appendix A.

(c) In the final step of the decomposition procedure, overall tractions $-\hat{\boldsymbol{\sigma}}\mathbf{n}$ are applied at S to cancel the tractions introduced there in step (b). By superposition and with regard to the definition (2.6) of $\mathbf{B}_r(\mathbf{x})$, the tensor $\mathbf{b}_r(\mathbf{x})$ can now be written in the desired form

$$\mathbf{b}_r(\mathbf{x}) = [\mathbf{I} - \mathbf{B}_r(\mathbf{x})](\mathbf{M}_1 - \mathbf{M}_2)^{-1}(\mathbf{m}_2 - \mathbf{m}_1), \tag{2.12}$$

where $\mathbf{I}$ is the fourth-order unit tensor defined by

$$I_{ijkl} = \tfrac{1}{2}(\delta_{ik}\delta_{jl} + \delta_{il}\delta_{jk}), \tag{2.13}$$

with δ_{ik} being the Kronecker delta.

Consider next the thermal loading Problem II, Equation (2.5). The solution follows again from the above decomposition which remains unchanged except for the conditions (2.9) and (2.10) which are now replaced by the forms

$$\hat{\boldsymbol{\varepsilon}}_1(\mathbf{x}) = \hat{\boldsymbol{\varepsilon}}_2(\mathbf{x}) = \hat{\boldsymbol{\varepsilon}}, \tag{2.14}$$

$$\mathbf{L}_1\hat{\boldsymbol{\varepsilon}} + \mathbf{l}_1\theta_0 = \mathbf{L}_2\hat{\boldsymbol{\varepsilon}} + \mathbf{l}_2\theta_0. \tag{2.15}$$

The uniform auxiliary strain field then follows as

$$\hat{\boldsymbol{\varepsilon}} = (\mathbf{L}_1 - \mathbf{L}_2)^{-1}(\mathbf{l}_2 - \mathbf{l}_1)\theta_0. \tag{2.16}$$

The last step (c) of the procedure is now implemented by applying on S the displacement field

$$\mathbf{u}_r(S) = -\hat{\boldsymbol{\varepsilon}}\mathbf{x}. \tag{2.17}$$

That leads to the final expression for $\mathbf{a}_r(\mathbf{x})$

$$\mathbf{a}_r(\mathbf{x}) = [\mathbf{I} - \mathbf{A}_r(\mathbf{x})](\mathbf{L}_1 - \mathbf{L}_2)^{-1}(\mathbf{l}_2 - \mathbf{l}_1). \tag{2.18}$$

For the special case of a binary composite made of isotropic constituents, (2.12) and (2.18) can be reduced to the expressions for thermal stress and strain concentration factors given, respectively, by Dvorak (1986), equations (38) and (40). For fibrous composites with two transversely isotropic phases, the original decomposition procedure allows the imposition of an additional relation between phase stress or strain averages, such as equation (11) in the 1986 paper. Such additional relations cannot be prescribed in the present case of arbitrary phase geometry. We note, however, that the original and the present procedures coincide if we choose the additional constraint to be in agreement with (2.9) above, i.e., as $d\sigma_2^f = d\sigma_2^m$ (using the notation of the 1986 paper). We also note that (2.12) and (2.18) are analogous to the relations between mechanical and thermal microstress fields in fibrous composites consisting of three cylindrical transversely isotropic phases, which were recently derived by Dvorak and Chen (1988).

3. Effective Thermal Expansion Coefficients

We now utilize the above decomposition procedure in a derivation of the overall thermal expansion coefficients of binary composites with anisotropic phases.

The effective constitutive law of a heterogeneous thermoelastic medium can be written in the form

$$\bar{\boldsymbol{\varepsilon}} = \mathbf{M}\bar{\boldsymbol{\sigma}} + \mathbf{m}\bar{\theta}, \qquad \bar{\boldsymbol{\sigma}} = \mathbf{L}\bar{\boldsymbol{\varepsilon}} + \mathbf{l}\bar{\theta}, \tag{3.1}$$

with

$$\mathbf{l} = -\mathbf{L}\mathbf{m}, \qquad \mathbf{M} = \mathbf{L}^{-1}, \tag{3.2}$$

where the overbars denote representative volume averages of the stress or strain fields, $\mathbf{M}$, $\mathbf{L}$ are the effective overall compliance and stiffness tensors given by Hill (1963) as

$$\mathbf{M} = c_1\mathbf{M}_1\mathbf{B}_1 + c_2\mathbf{M}_2\mathbf{B}_2, \qquad \mathbf{L} = c_1\mathbf{L}_1\mathbf{A}_1 + c_2\mathbf{L}_2\mathbf{A}_2, \tag{3.3}$$

where c_r are the constituent volume fractions, and $\mathbf{m}$ and $\mathbf{l}$ in (3.1) and (3.2) are the overall thermal strain and thermal stress tensors. It is again recalled that $\mathbf{B}_r$ and $\mathbf{A}_r$ are representative phase volume averages of the fields $\mathbf{B}_r(\mathbf{x})$, $\mathbf{A}_r(\mathbf{x})$ defined in (2.6) and (2.7), respectively.

Consider first a special form of Problem I, Equation (2.4), in which $\boldsymbol{\sigma}_0 = \mathbf{0}$. Since in this case $\bar{\boldsymbol{\sigma}} = \mathbf{0}$ and $\bar{\theta} = \theta_0$, we find from (3.1_1) that

$$\mathbf{m}\theta_0 = \bar{\boldsymbol{\varepsilon}} = c_1\boldsymbol{\varepsilon}_1 + c_2\boldsymbol{\varepsilon}_2, \tag{3.4}$$

where $\boldsymbol{\varepsilon}_1$, $\boldsymbol{\varepsilon}_2$ are phase volume averages of local strains $\boldsymbol{\varepsilon}_1(\mathbf{x})$ and $\boldsymbol{\varepsilon}_2(\mathbf{x})$. Substitution of (2.12) into (2.6) with $\boldsymbol{\sigma}_0 = \mathbf{0}$, and the result in (2.2) provides

$$\boldsymbol{\varepsilon}_r = \mathbf{M}_r[\mathbf{I} - \mathbf{B}_r](\mathbf{M}_1 - \mathbf{M}_2)^{-1}(\mathbf{m}_2 - \mathbf{m}_1)\theta_0. \tag{3.5}$$

Equations (3.5), (3.4), and (3.3), finally give the overall thermal strain tensor $\mathbf{m}$

$$\mathbf{m} = c_1\mathbf{m}_1 + c_2\mathbf{m}_2 + (\mathbf{M} - c_1\mathbf{M}_1 - c_2\mathbf{M}_2)(\mathbf{M}_1 - \mathbf{M}_2)^{-1}(\mathbf{m}_1 - \mathbf{m}_2), \tag{3.6}$$

which is precisely equation (2.2) in Rosen and Hashin (1970). It is interesting to observe that in this process of substitution, $\mathbf{B}_1$ and $\mathbf{B}_2$ combine in the manner of (3.3_1) into the overall property $\mathbf{M}$. Equivalent forms of (3.6) which can be arrived at after some manipulation are

$$\mathbf{m} = \mathbf{m}_1 + (\mathbf{M}_1 - \mathbf{M})(\mathbf{M}_2 - \mathbf{M}_1)^{-1}(\mathbf{m}_1 - \mathbf{m}_2), \tag{3.7}$$

and the symmetric form

$$\mathbf{m} = (\mathbf{M} - \mathbf{M}_2)(\mathbf{M}_1 - \mathbf{M}_2)^{-1}\mathbf{m}_1 + (\mathbf{M} - \mathbf{M}_1)(\mathbf{M}_2 - \mathbf{M}_1)^{-1}\mathbf{m}_2. \tag{3.8}$$

We now turn to Problem II, Equation (2.5), and seek the solution for $\boldsymbol{\varepsilon}_0 = \mathbf{0}$. Equation (3.1_2) provides

$$\mathbf{l}\theta_0 = \boldsymbol{\sigma} = c_1\boldsymbol{\sigma}_1 + c_2\boldsymbol{\sigma}_2, \tag{3.9}$$

where $\boldsymbol{\sigma}_1$ and $\boldsymbol{\sigma}_2$ are phase volume averages of $\boldsymbol{\sigma}_1(\mathbf{x})$ and $\boldsymbol{\sigma}_2(\mathbf{x})$, respectively. This, together with (2.18) and (2.7) at $\boldsymbol{\varepsilon}_0 = \mathbf{0}$, and (2.1) with (3.3_1) yields

$$\mathbf{l} = c_1\mathbf{l}_1 + c_2\mathbf{l}_2 + (\mathbf{L} - c_1\mathbf{L}_1 - c_2\mathbf{L}_2)(\mathbf{L}_1 - \mathbf{L}_2)^{-1}(\mathbf{l}_1 - \mathbf{l}_2). \tag{3.10}$$

The equivalent forms are

$$\mathbf{l} = \mathbf{l}_1 + (\mathbf{L}_1 - \mathbf{L})(\mathbf{L}_2 - \mathbf{L}_1)^{-1}(\mathbf{l}_1 - \mathbf{l}_2), \tag{3.11}$$

$$\mathbf{l} = (\mathbf{L} - \mathbf{L}_2)(\mathbf{L}_1 - \mathbf{L}_2)^{-1}\mathbf{l}_1 + (\mathbf{L} - \mathbf{L}_1)(\mathbf{L}_2 - \mathbf{L}_1)^{-1}\mathbf{l}_2. \tag{3.12}$$

Laws (1973) derived this last formula in a different way.

Using (2.3) and (3.2_2), we can verify that $\mathbf{m}$ and $\mathbf{l}$ as given by (3.6) and (3.10), or by their equivalent forms, satisfy (3.2_1).

Special forms of (3.6) and (3.10) for fibrous and particulate composites were given by equations (26), (29), (37), and (39) in Dvorak's (1986) paper.

The direct derivation of the overall thermal strain and stress tensors **m** and **l**, from the above relations between thermal and mechanical concentration factors, implies that these tensors can be found once the effective mechanical properties have been predicted by a certain micromechanical model. However, a valid question to ask is whether the model, if used to estimate the effective thermal strain and stress tensors directly, would give the predictions that coincide with those found from (3.6) and (3.10). The following section establishes such consistency for the Mori–Tanaka model.

4. Application of the Mori–Tanaka Method to Thermoelastic Problems

4.1. Review of the Method and Principal Results

We start by giving a concise summary of this model, in the framework of its application to purely mechanical problems, as presented by Benveniste (1987).

Consider the loading configuration of Problem I, Equation (2.4), with $\theta_0 = 0$. In this case, $\bar{\boldsymbol{\sigma}} = \boldsymbol{\sigma}_0$, hence we have

$$c_1\mathbf{B}_1 + c_2\mathbf{B}_2 = \mathbf{I}, \tag{4.1}$$

which, when combined with (3.3_1), results in

$$\mathbf{M} = \mathbf{M}_1 + c_2(\mathbf{M}_2 - \mathbf{M}_1)\mathbf{B}_2, \tag{4.2}$$

so that we need to know $\mathbf{B}_2$ (or $\mathbf{B}_1$) to determine $\mathbf{M}$.

Similarly, in Problem II, Equation (2.5), with $\theta_0 = 0$ there is $\bar{\boldsymbol{\varepsilon}} = \boldsymbol{\varepsilon}_0$, hence

$$c_1\mathbf{A}_1 + c_2\mathbf{A}_2 = \mathbf{I}, \tag{4.3}$$

with

$$\mathbf{L} = \mathbf{L}_1 + c_2(\mathbf{L}_2 - \mathbf{L}_1)\mathbf{A}_2. \tag{4.4}$$

In all results obtained so far it was possible to regard both phases on equal footing. In contrast, the Mori–Tanaka method makes a clear distinction between the continuous matrix and the discrete fibrous or particulate reinforcement. Therefore, it is necessary to designate the phases by numbers, which we select as 1 for the matrix and 2 for the reinforcement. We furthermore assume here that the particulate phase is represented by ellipsoidal inclusions of similar shape but which can be, however, of different size. The model can be applied to inclusions which are aligned, may have a certain orientation distribution, or are of random orientation. For simplicity of the exposition we chose to deal here with the case of aligned inclusions.

The approximate evaluation of the strain concentration factor $\mathbf{A}_2$ will illustrate the method. Under dilute conditions, $\mathbf{A}_2$ would be found from strains in a single inclusion embedded in an infinite matrix subjected to the uniform boundary strains (2.5) with $\theta = 0$. The solution of this problem is

$$\mathbf{A}_2 = \mathbf{T}, \tag{4.5}$$

where

$$\mathbf{T} = [\mathbf{I} + \mathbf{S}\mathbf{L}_1^{-1}(\mathbf{L}_2 - \mathbf{L}_1)]^{-1}, \tag{4.6}$$

and **S** is the Eshelby (1957) tensor.

Of course, in the case of a single inclusion the average strain in the infinite matrix is not affected by the presence of the inclusion and is thus equal to $\boldsymbol{\varepsilon}_0$. In contrast, when many inclusions are present, the magnitude of the average strain in the matrix, and in the inclusions, is influenced by their interaction. The Mori–Tanaka method assumes that the average strain $\boldsymbol{\varepsilon}_r$ in the interacting inclusions can be approximated by that of a single inclusion embedded in an infinite matrix subjected to the uniform average matrix strain $\boldsymbol{\varepsilon}_1$. This is illustrated in Fig. 1(a) which shows Problem I with $\theta_0 = 0$, which must now be solved for a single inclusion in a certain large volume V' which is enclosed

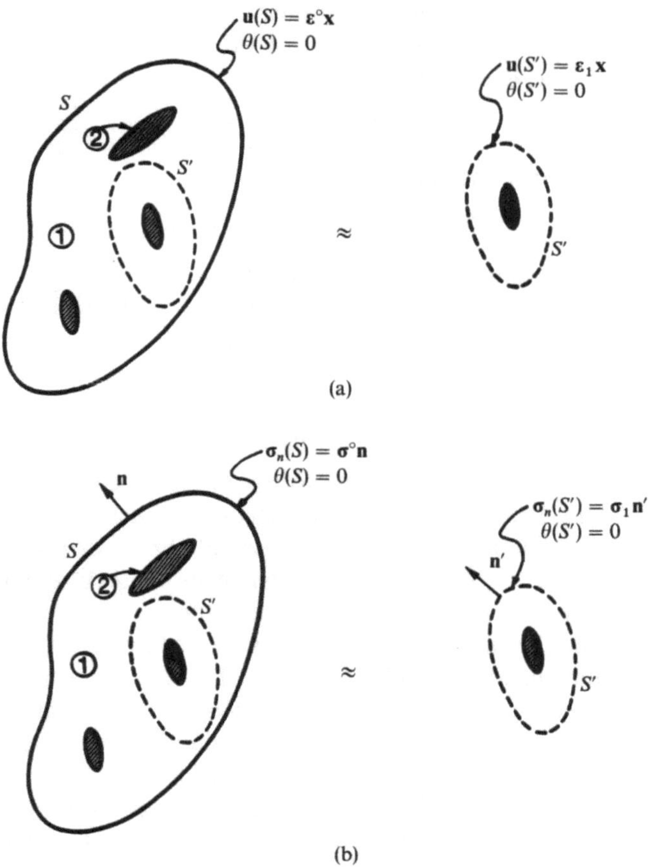

FIG. 1. A schematic representation of the Mori–Tanaka model for the case of mechanical loading.

by a surface S', and subjected to the boundary condition

$$\mathbf{u}(S') = \boldsymbol{\varepsilon}_1 \mathbf{x}. \tag{4.7}$$

The solution is

$$\boldsymbol{\varepsilon}_2 = \mathbf{T}\boldsymbol{\varepsilon}_1, \tag{4.8}$$

and as clarified by Benveniste (1987), it represents the essential assumption of the Mori–Tanaka method. It also implies that

$$\mathbf{A}_2 = \mathbf{T}\mathbf{A}_1. \tag{4.9}$$

Using (4.3), we obtain the estimates of the matrix (1) and reinforcement (2) strain concentration factors

$$\mathbf{A}_1 = (c_1\mathbf{I} + c_2\mathbf{T})^{-1}, \tag{4.10}$$

$$\mathbf{A}_2 = \mathbf{T}(c_1\mathbf{I} + c_2\mathbf{T})^{-1}, \tag{4.11}$$

which can be substituted into (4.4) to provide an estimate of the overall stiffness

$$\mathbf{L} = \mathbf{L}_1 + c_2(\mathbf{L}_2 - \mathbf{L}_1)\mathbf{T}(c_1\mathbf{I} + c_2\mathbf{T})^{-1}. \tag{4.12}$$

A similar procedure with the boundary conditions (2.4) and $\theta_0 = 0$ and (4.1) and (4.2) yields an estimate of the overall compliance

$$\mathbf{M} = \mathbf{M}_1 + c_2(\mathbf{M}_2 - \mathbf{M}_1)\mathbf{W}(c_1\mathbf{I} + c_2\mathbf{W})^{-1}, \tag{4.13}$$

where $\mathbf{W}$ denotes the stress concentration factor tensor of an isolated inclusion and is given by

$$\mathbf{W} = \mathbf{L}_2\mathbf{T}\mathbf{M}_1. \tag{4.14}$$

Benveniste (1987) had shown that the results (4.12) and (4.13) are consistent in the sense that $\mathbf{M} = \mathbf{L}^{-1}$. We note that these equations cannot be reduced to a symmetric form because the two phases do not enter on equal footing.

Turning next to the effective thermal strain and stress tensors, we substitute (4.12) and (4.13) into (3.6) and (3.10), respectively, to obtain

$$\mathbf{m} = \mathbf{m}_1 + c_2(\mathbf{M}_2 - \mathbf{M}_1)\mathbf{W}(c_1\mathbf{I} + c_2\mathbf{W})^{-1}(\mathbf{M}_2 - \mathbf{M}_1)^{-1}(\mathbf{m}_2 - \mathbf{m}_1), \tag{4.15}$$

$$\mathbf{l} = \mathbf{l}_1 + c_2(\mathbf{L}_2 - \mathbf{L}_1)\mathbf{T}(c_1\mathbf{I} + c_2\mathbf{T})^{-1}(\mathbf{L}_2 - \mathbf{L}_1)^{-1}(\mathbf{l}_2 - \mathbf{l}_1). \tag{4.16}$$

Since (3.6) and (3.10) satisfy (3.2_1) and, as we just concluded, $\mathbf{M} = \mathbf{L}^{-1}$, it follows that the $\mathbf{m}$ and $\mathbf{l}$ obey the relation $\mathbf{l} = -\mathbf{L}\mathbf{m}$.

4.2. *Proof of Consistency*

The Mori–Tanaka method has been used previously to predict the effective thermal expansion coefficients of particulate composites (Wakashima *et al.*, 1974; Takahashi *et al.*, 1980; Takao, 1985; Takao and Taya, 1985). In these works, the method was not implemented through (3.6) and (3.10). Instead, it was applied directly to the thermal case by using the equivalent inclusion idea of Eshelby (1957) and the eigenstrain concept. In what follows we rederive the Mori–Tanaka results in a different way, and then prove that the results are consistent with (4.15) and (4.16).

Consider first the thermal loading problem (2.4) with $\boldsymbol{\sigma}_0 = \mathbf{0}$. As in Section 4.1, we write the solution of this problem for the single inclusion case in the form

$$\boldsymbol{\sigma}_1^s = \mathbf{0}, \qquad \boldsymbol{\sigma}_2^s = \mathbf{w}\theta_0, \tag{4.17}$$

where $\mathbf{w}$ is the average stress in a single inclusion embedded in the matrix and subjected to unit temperature and zero traction at the remote boundary S'. From the volume average of (2.12) we find the average stress in the single inclusion as

$$\mathbf{w} = (\mathbf{I} - \mathbf{W})(\mathbf{M}_1 - \mathbf{M}_2)^{-1}(\mathbf{m}_2 - \mathbf{m}_1). \tag{4.18}$$

For a finite concentration of the reinforcement, the solution is obtained again from the Mori–Tanaka assumption that the average stress in each inclusion is equal to that found for the dilute case with the boundary conditions shown in Fig. 2(a)

$$\theta(S') = \theta_0, \qquad \boldsymbol{\sigma}_{n'}(S') = \boldsymbol{\sigma}_1 \mathbf{n}', \tag{4.19}$$

where $\boldsymbol{\sigma}_1$ is the unknown average stress in the matrix at finite concentration and $\mathbf{n}'$ is the outside normal to S'.

To find $\boldsymbol{\sigma}_1$, we note that the boundary conditions (4.19) cause the average stress in each inclusion

$$\boldsymbol{\sigma}_2 = \mathbf{w}\theta_0 + \mathbf{W}\boldsymbol{\sigma}_1, \tag{4.20}$$

where the second term accounts for particle interaction and $\mathbf{W}$ is given by (4.14). Recall now (3.4) and write the strains as in (2.2)

$$\mathbf{m}\theta_0 = c_1(\mathbf{M}_1\boldsymbol{\sigma}_1 + \mathbf{m}_1\theta_0) + c_2(\mathbf{M}_2\boldsymbol{\sigma}_2 + \mathbf{m}_2\theta_0). \tag{4.21}$$

Also recall that for $\boldsymbol{\sigma}_0 = \mathbf{0}$

$$\bar{\boldsymbol{\sigma}} = c_1\boldsymbol{\sigma}_1 + c_2\boldsymbol{\sigma}_2 = \mathbf{0}. \tag{4.22}$$

The last three equations lead to the expressions

$$\mathbf{m}\theta_0 = [\mathbf{m}_1 + c_2(\mathbf{m}_2 - \mathbf{m}_1)]\theta_0 + (\mathbf{M}_1 - \mathbf{M}_2)c_1\boldsymbol{\sigma}_1, \tag{4.23}$$

$$\boldsymbol{\sigma}_1 = -c_2(c_1\mathbf{I} + c_2\mathbf{W})^{-1}\mathbf{w}\theta_0. \tag{4.24}$$

Substitution of (4.18) into (4.24) and then into (4.23) yields

$$\begin{aligned}\mathbf{m} = \mathbf{m}_1 + c_2[\mathbf{I} - (\mathbf{M}_1 - \mathbf{M}_2)c_1(c_1\mathbf{I} + c_2\mathbf{W})^{-1}\\ \cdot(\mathbf{I} - \mathbf{W})(\mathbf{M}_1 - \mathbf{M}_2)^{-1}](\mathbf{m}_2 - \mathbf{m}_1),\end{aligned} \tag{4.25}$$

which can be written as

$$\begin{aligned}\mathbf{m} = \mathbf{m}_1 + c_2(\mathbf{M}_1 - \mathbf{M}_2)(c_1\mathbf{I} + c_2\mathbf{W})^{-1}\\ \cdot[(c_1\mathbf{I} + c_2\mathbf{W})(\mathbf{M}_1 - \mathbf{M}_2)^{-1} - c_1(\mathbf{I} - \mathbf{W})(\mathbf{M}_1 - \mathbf{M}_2)^{-1}](\mathbf{m}_2 - \mathbf{m}_1).\end{aligned} \tag{4.26}$$

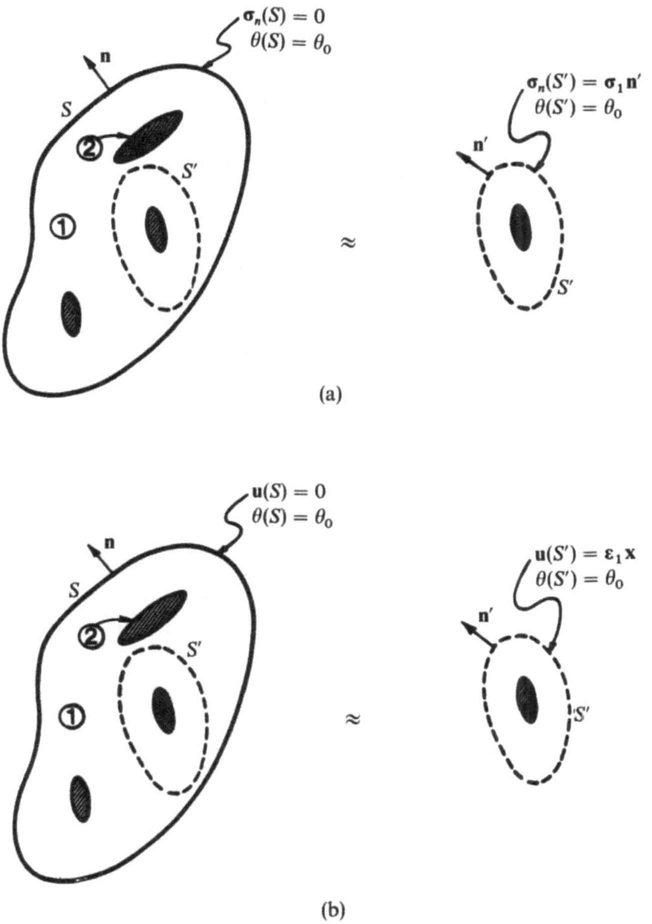

FIG. 2. A schematic representation of the Mori–Tanaka model for the case of thermal loadings.

Factoring out $(\mathbf{M}_1 - \mathbf{M}_2)^{-1}$ in the middle expression provides

$$\mathbf{m} = \mathbf{m}_1 + c_2(\mathbf{M}_2 - \mathbf{M}_1)(c_1\mathbf{I} + c_2\mathbf{W})^{-1}\mathbf{W}(\mathbf{M}_2 - \mathbf{M}_1)^{-1}(\mathbf{m}_2 - \mathbf{m}_1), \quad (4.27)$$

which, together with the identity

$$(c_1\mathbf{I} + c_2\mathbf{W})^{-1}\mathbf{W} = \mathbf{W}(c_1\mathbf{I} + c_2\mathbf{W})^{-1}, \quad (4.28)$$

shows that (4.27) is identical to (4.15). In addition, we demonstrate in Appendix B that the overall $\mathbf{m}$ given by (4.27) is actually identical to that derived in a different way by Takao and Taya (1985).

A similar proof of consistency can be given for the overall thermal stress vector $\mathbf{l}$, (4.16). Here we consider Problem II, with the boundary conditions (2.5) and $\boldsymbol{\varepsilon}_0 = \mathbf{0}$. In the spirit of the Mori–Tanaka method, we first take a single inclusion in a large matrix volume V' and specify that on S', $\mathbf{u}(S') = \boldsymbol{\varepsilon}_1 \mathbf{x}$, the average matrix strain at finite concentration, Fig. 2(b). The inclusion strain is given by the counterpart of (4.20)

$$\boldsymbol{\varepsilon}_2 = \mathbf{t}\theta_0 + \mathbf{T}\boldsymbol{\varepsilon}_1, \tag{4.29}$$

where $\mathbf{t}$ follows from the volume average of (2.18) as

$$\mathbf{t} = (\mathbf{I} - \mathbf{T})(\mathbf{L}_1 - \mathbf{L}_2)^{-1}(\mathbf{l}_2 - \mathbf{l}_1). \tag{4.30}$$

Here, the first term denotes the average strain in an isolated particle embedded in an infinite, stress-free matrix medium and subjected to a uniform thermal change θ_0, whereas the second term, with $\mathbf{Y}$ given by (4.6), accounts for particle interaction. We also write

$$\bar{\boldsymbol{\varepsilon}} = c_1\boldsymbol{\varepsilon}_1 + c_2\boldsymbol{\varepsilon}_2 = \mathbf{0}, \tag{4.31}$$

and substitute from (2.1) and (3.1_1), with $\bar{\boldsymbol{\varepsilon}} = \mathbf{0}$, to find

$$\mathbf{l}\theta_0 = c_1(\mathbf{L}_1\boldsymbol{\varepsilon}_1 + \mathbf{l}_1\theta_0) + c_2(\mathbf{L}_2\boldsymbol{\varepsilon}_2 + \mathbf{l}_2\theta_0). \tag{4.32}$$

Due to the similar structure of ((4.29), (4.30), (4.31), (4.32)) ((4.20), (4.18), (4.22), (4.21)), respectively, it is clear that the former set will result in $\mathbf{l}$ given by (4.16), in the same manner that the latter set resulted in (4.15).

5. Temperature-Dependent Material Properties

In many cases of practical interest, and particularly in high-temperature applications of composite materials, the magnitudes of certain material properties such as stiffness and thermal expansion coefficients depend on temperature. Typically, elastic moduli or compliances are experimentally measured at specific temperatures, and the coefficients of thermal expansion are obtained as derivatives of strain–temperature records taken in a certain temperature interval. In any case, the temperature dependence of thermoelastic coefficients can be represented by suitable functions which approximate the experimental data with sufficient accuracy.

We recall that all the results obtained in the preceding sections were derived from solutions of either Problem I or II, Equations (2.4) and (2.5), which were formulated for linear thermoelastic materials with temperature-independent properties. These results certainly remain valid for infinitesimal thermal changes. For example, consider Problem I, and assume that at a given $\boldsymbol{\sigma}_0$ and θ_0 the tensors $\mathbf{B}_r$, $\mathbf{b}_r$, $\mathbf{M}_r$, and $\mathbf{m}_r$ are known, and are now functions of θ_0. For an increase in temperature and stress denoted by $d\theta_0$ and $\boldsymbol{\sigma}_0$, we can now write

$$d\boldsymbol{\sigma}_r = \mathbf{B}_r(\mathbf{x}; \theta_0)\, d\boldsymbol{\sigma}_0 + \mathbf{b}_r(\mathbf{x}; \theta_0)\, d\theta_0, \tag{5.1}$$

with

$$\mathbf{b}_r(\mathbf{x}; \theta_0) = [\mathbf{I} - \mathbf{B}_r(\mathbf{x}; \theta_0)][\mathbf{M}_1(\theta_0) - \mathbf{M}_2(\theta_0)]^{-1}[\mathbf{m}_2(\theta_0) - \mathbf{m}_1(\theta_0)]. \tag{5.2}$$

It is therefore clear that with the temperature-dependent properties represented in a step-wise constant manner, an incremental implementation of the results derived in the present paper becomes possible.

Acknowledgments

The first author is indebted to Professor G. J. Dvorak for the visiting appointment at RPI during 1987/1988. Support from the Office of Naval Research and the DARPA-HiTASC program at RPI is gratefully acknowledged.

References

Benveniste, Y. (1987), A new approach to the application of Mori–Tanaka's theory in composite materials, *Mech. Materials*, **6**, 147–157.

Dvorak, G. J. (1983), Metal matrix composites: Plasticity and fatigue, in *Mechanics of Composite Materials—Recent Advances*, edited by Z. Hashin and C. T. Herakovich, Pergamon Press, New York, pp. 73–92.

Dvorak, G. J. (1986), Thermal expansion of elastic–plastic composite materials, *J. Appl. Mech.*, **53**, 737–743.

Dvorak, G. J. and Chen, T., (1988), Thermal expansion of three-phase composite materials, to be published.

Eshelby, J. D. (1957), The determination of the elastic field of an ellipsoidal inclusion and related problems, *Proc. Roy. Soc. London*, **A241**, 376–396.

Hill, R. (1963), Elastic properties of reinforced solids: Some theoretical principles, *J. Mech. Phys. Solids*, **11**, 357–372.

Laws, N. (1973), On the thermostatics of composite materials, *J. Mech. Phys. Solids*, **21**, 9–17.

Laws, N. (1974), The overall thermoelastic moduli of transversely isotropic composites according to the self-consistent method, *Int. J. Engng. Sci.*, **12**, 79–87.

Levin, V. M. (1967), Thermal expansion coefficients of heterogeneous materials, *Mekhanika Tverdogo Tela*, **2**, 88–94.

Mori, T. and Tanaka, K. (1973), Average stress in matrix and average elastic energy of materials with misfitting inclusions, *Acta Metallurgica*, **21**, 571–574.

Rosen, B. W. and Hashin, Z. (1970), Effective thermal expansion coefficients and specific heats of composite materials, *Int. J. Engng. Sci.* **8**, 157–173.

Schapery, R. A. (1968), Thermal expansion coefficients of composite materials based on energy principles, *J. Composite Materials*, **2**, 380–404.

Takahashi, K., Harakawa, K., and Sakai, T. (1980), Analysis of the thermal expansion coefficients of particle filled polymers, *J. Composite Materials, Suppl.*, **14**, 144–159.

Takao, Y. (1985), Thermal expansion coefficients in misoriented short-fiber composites, in *Recent Advances in Composites in the United States and Japan*, ASTM STP 864, 685–689.

Takao, Y. and Taya, M. (1985), Thermal expansion coefficients and thermal stresses in an aligned short fiber composite with application to a short carbon fiber/aluminum, *J. Appl. Mech.*, **52**, 806–810.

Wakashima, K., Otsuka, M., and Umekawa, S. (1974), Thermal expansion of heterogeneous solids containing aligned ellipsoidal inclusions, *J. Composite Materials*, **8**, 391–404.

Appendix A

This appendix presents a discussion of the decomposition scheme for the degenerate cases in which $(\mathbf{M}_1 - \mathbf{M}_2)^{-1}$ or $(\mathbf{L}_1 - \mathbf{L}_2)^{-1}$ fail to exist. The two important cases of isotropic and transversely isotropic constituents will be analyzed.

Consider first the case of isotropic constituents for which the tensors $\mathbf{M}_i$ and $\mathbf{m}_i$ can be written in component form as follows:

$$(M_{pqrs})_i = \frac{1}{9\kappa_i}\delta_{pq}\delta_{rs} + \frac{1}{4\mu_i}(\delta_{pr}\delta_{qs} + \delta_{ps}\delta_{qr} - \tfrac{2}{3}\delta_{pq}\delta_{rs}), \tag{A1.1}$$

$$(m_{rs})_i = \alpha_i\delta_{rs}, \tag{A1.2}$$

where κ_i, μ_i are the bulk and shear moduli of the phases and α_i are the thermal expansion coefficients. For the sake of brevity we limit ourselves to (2.11), a similar discussion applies for (2.16).

The difference $(\mathbf{M}_1 - \mathbf{M}_2)^{-1}$ can now be written in component form

$$[(\mathbf{M}_1 - \mathbf{M}_2)^{-1}]_{pqrs} = \left(\frac{1}{\kappa_1} - \frac{1}{\kappa_2}\right)^{-1}\delta_{pq}\delta_{rs} + \left(\frac{1}{\mu_1} - \frac{1}{\mu_2}\right)^{-1}(\delta_{pr}\delta_{qs} + \delta_{ps}\delta_{qr} - \tfrac{2}{3}\delta_{pq}\delta_{rs}). \tag{A1.3}$$

Use of (A1.2) and (A1.3) in (2.11) yields

$$\hat{\sigma}_{pq} = 3\left(\frac{1}{\kappa_1} - \frac{1}{\kappa_2}\right)^{-1}(\alpha_1 - \alpha_2)\delta_{pq}, \tag{A1.4}$$

which means that even though $(\mathbf{M}_1 - \mathbf{M}_2)^{-1}$ becomes singular for $\mu_1 = \mu_2$, the decomposition scheme continues to be well defined. The decomposition fails, however, if $\kappa_1 = \kappa_2$.

Equation (2.11), and the specific result of (A1.4) in (24), give the well-known Levin's formula

$$\alpha = c_1\alpha_1 + c_2\alpha_2 + (\alpha_1 - \alpha_2)\left(\frac{1}{\kappa_1} - \frac{1}{\kappa_2}\right)^{-1}\left[\frac{1}{\kappa} - \frac{c}{\kappa_1} - \frac{c_2}{\kappa_2}\right], \tag{A1.5}$$

which also becomes singular when $\kappa_1 = \kappa_2$. It should be noted, however, that this difficulty can be circumvented if the more general Levin's result, based on the concentration factors $\mathbf{B}_r$ (see equation (2.17) in Rosen and Hashin (1970)), is used instead of (A1.5). It is of interest to note that these degenerate cases of Levin's result have not been dealt before in the literature.

We consider next the case of transversely isotropic constituents and choose this time to illustrate the analysis with (2.16). The tensors **L**, and **l** can now be denoted using the scheme used by Hill (1963), Walpole (1969), and Laws (1974) (see the last reference for a comprehensive exposition of this notation)

$$\mathbf{L}_i = (2k_i, l_i, l_i, n_i, 2m_i, 2p_i), \tag{A1.6}$$

$$\mathbf{l}_i = (\beta_T^{(i)}, \beta_L^{(i)}), \tag{A1.7}$$

where k_i is the plane strain bulk modulus for lateral dilatation without longitudinal extension, n_i is the modulus for longitudinal uniaxial straining, l_i is the associated cross modulus, m_i is the shear modulus for shearing in any transverse direction, and n_i is the modulus for longitudinal shearing; finally $\beta_T^{(i)}$ and $\beta_L^{(i)}$ denote, respectively, the thermal stress coefficients in the transverse and longitudinal direction.

In the notation of (A1.6), $(\mathbf{L}_1 - \mathbf{L}_2)$ becomes

$$(\mathbf{L}_1 - \mathbf{L}_2) = [2(k_1 - k_2), (l_1 - l_2), (l_1 - l_2), (n_1 - n_2), (2m_1 - 2m_2), (2p_1 - 2p_2)], \tag{A1.8}$$

so that $(\mathbf{L}_1 - \mathbf{L}_2)^{-1}$ is given by

$$(\mathbf{L}_1 - \mathbf{L}_2) = \left(\frac{n_1 - n_2}{2\lambda}, -\frac{l_1 - l_2}{2\lambda}, -\frac{l_1 - l_2}{2\lambda}, \frac{k_1 - k_2}{\lambda}, \frac{1}{2m_1 - 2m_2}, \frac{1}{2p_1 - 2p_2}\right), \tag{A1.9}$$

where λ is defined as

$$\lambda = (k_1 - k_2)(n_1 - n_2) - (l_1 - l_2)^2. \tag{A1.10}$$

The product $(\mathbf{L}_1 - \mathbf{L}_2)^{-1}(\mathbf{l}_1 - \mathbf{l}_2)$ therefore becomes

$$(\mathbf{L}_1 - \mathbf{L}_2)^{-1}(\mathbf{l}_1 - \mathbf{l}_2) = (p, q), \tag{A1.11}$$

with p and q given by

$$p = \frac{n_1 - n_2}{2\lambda}(\beta_T^{(1)} - \beta_T^{(2)}) - \frac{l_1 - l_2}{2\lambda}(\beta_L^{(1)} - \beta_L^{(2)}), \tag{A1.12}$$

$$q = -2\frac{l_1 - l_2}{2\lambda}(\beta_T^{(1)} - \beta_T^{(2)}) + \frac{k_1 - k_2}{2\lambda}(\beta_L^{(1)} - \beta_L^{(2)}), \tag{A1.13}$$

so that the product in (A1.11), and thus the decomposition scheme fail to exist when λ, as given by (A1.10), vanishes.

The reduction of the present results to the case of isotropic constituents can be readily carried out by noting that for isotropic phases

$$\mathbf{L}_i = [2(\kappa_i + \tfrac{1}{3}\mu_i), (\kappa_i - \tfrac{2}{3}\mu_i), (\kappa_i - \tfrac{2}{3}\mu_i), \kappa_i + \tfrac{4}{3}\mu_i, 2\mu_i, 2\mu_i], \tag{A1.14}$$

$$\mathbf{l} = [\beta, \beta], \tag{A1.15}$$

thus reducing λ to

$$\lambda = 3(\mu_1 - \mu_2)(\kappa_1 - \kappa_2), \tag{A1.16}$$

and p and q to

$$p = q = \frac{1}{3(\kappa_1 - \kappa_2)}(\beta_1 - \beta_2). \tag{A1.17}$$

Therefore, in the case of isotropic constituents, the decomposition scheme fails only when $\kappa_1 = \kappa_2$.

Appendix B

In this Appendix we will prove the equivalence between our result (4.25) and that obtained by Takao and Taya (1985). The approach used by these authors is based on the eigenstrain concepts and equivalent inclusion formalism. Four equations in Takao and Taya (1985) determine the effective thermal expansion tensor. We will reproduce them here and show that they lead, in fact, to the relatively compact closed form of (4.27).

We first note that the effective thermal expansion tensor is denoted in the Takao and Taya paper by $\boldsymbol{\alpha}_c$, whereas the symbol $\boldsymbol{\alpha}^*$ is used there to denote the thermal strain due to the difference between $\mathbf{m}_1$ and $\mathbf{m}_2$ under temperature change θ.

Equations (3), (6), (7), and (13) of that work are,

$$\boldsymbol{\alpha}^* = (\mathbf{m}_2 - \mathbf{m}_1)\theta_0, \tag{B1.1}$$

$$\mathbf{L}_1[\bar{\mathbf{e}} + (\mathbf{S} - \mathbf{I})\boldsymbol{\alpha}^* + (\mathbf{S} - \mathbf{I})\mathbf{e}^*] = \mathbf{L}_2[\bar{\mathbf{e}} + (\mathbf{S} - \mathbf{I})\boldsymbol{\alpha}^* + \mathbf{S}\mathbf{e}^*], \tag{B1.2}$$

$$\bar{\mathbf{e}} + c_2(\mathbf{S} - \mathbf{I})(\boldsymbol{\alpha}^* + \mathbf{e}^*) = 0, \tag{B1.3}$$

$$\mathbf{m} = \mathbf{m}_1 + \frac{c_2(\mathbf{e}^* + \boldsymbol{\alpha}^*)}{\theta_0}. \tag{B1.4}$$

We have also used $\mathbf{m}_1, \mathbf{m}_2$ for $\boldsymbol{\alpha}_1, \boldsymbol{\alpha}_2$; $\mathbf{L}_1$ and $\mathbf{L}_2$ for $\mathbf{C}_m$ and $\mathbf{C}_f$; θ_0 for Δt; and c_2 has been used instead of f in Takao and Taya. In that work $\mathbf{e}^*$ denotes the fictitious strain called "eigenstrain" or "transformation strain" and $\bar{\mathbf{e}}$ is the volume-averaged disturbance of strain in the matrix. The tensor $\mathbf{S}$ stands again for the Eshelby tensor appearing in our equation (4.6).

We start by substituting (B1.1) into (B1.4) .

$$\mathbf{m} = c_1\mathbf{m}_1 + c_2\mathbf{m}_2 + \frac{c_2\mathbf{e}^*}{\theta_0}. \tag{B1.5}$$

Use of (B1.3) and (B1.1) in (B1.2) gives for $\mathbf{e}^*$

$$\frac{1}{c_1}[c_1(\mathbf{L}_2 - \mathbf{L}_1)\mathbf{S} + c_1\mathbf{L}_1 + c_2\mathbf{L}_2]\mathbf{e}^* = (\mathbf{L}_1 - \mathbf{L}_2)(\mathbf{S} - \mathbf{I})(\mathbf{m}_2 - \mathbf{m}_1)\theta_0, \tag{B1.6}$$

which, when substituted in (B1.5) provide Takao and Taya's principal result

$$\mathbf{m} = c_1\mathbf{m}_1 + c_2\mathbf{m}_2 + (c_2c_1)[c_1(\mathbf{L}_2 - \mathbf{L}_1)\mathbf{S} + c_1\mathbf{L}_1 + c_2\mathbf{L}_2]^{-1} \cdot(\mathbf{L}_1 - \mathbf{L}_2)(\mathbf{S} - \mathbf{I})(\mathbf{m}_2 - \mathbf{m}_1). \tag{B1.7}$$

We will now show that this last expression is identical to (4.27) or to an equivalent form which is obtained by substituting (4.13) into (3.6).

After some manipulation we obtain

$$\mathbf{m} = c_1\mathbf{m}_1 + c_2\mathbf{m}_2 + c_1c_2(\mathbf{M}_1 - \mathbf{M}_2)(\mathbf{I} - \mathbf{W}) \cdot(\mathbf{c}_1\mathbf{I} + c_2\mathbf{W})^{-1}(\mathbf{M}_2 - \mathbf{M}_1)^{-1}(\mathbf{m}_2 - \mathbf{m}_1) \tag{B1.8}$$

this implies that (B1.7) is identical to (B1.8) if the following equality holds:

$$[c_1(\mathbf{L}_2 - \mathbf{L}_1)\mathbf{S} + c_1\mathbf{L}_1 + c_2\mathbf{L}_2]^{-1}(\mathbf{L}_1 - \mathbf{L}_2)(\mathbf{S} - \mathbf{I}) = (\mathbf{M}_1 - \mathbf{M}_2)(\mathbf{I} - \mathbf{W})(c_1\mathbf{I} + c_2\mathbf{W})^{-1}(\mathbf{M}_2 - \mathbf{M}_1)^{-1}. \tag{B1.9}$$

Using (4.6) and (4.14) we note first that

$$\mathbf{I} - \mathbf{W} = [\mathbf{M}_2 + \mathbf{S}(\mathbf{M}_1 - \mathbf{M}_2)]^{-1}(\mathbf{S} - \mathbf{I})(\mathbf{M}_1 - \mathbf{M}_2), \tag{B1.10}$$

$$(c_1\mathbf{I} + c_2\mathbf{W})^{-1} = [c_2\mathbf{M}_1 + c_1\mathbf{M}_2 + c_1\mathbf{S}(\mathbf{M}_1 - \mathbf{M}_2)]^{-1}\cdot[\mathbf{M}_2 + \mathbf{S}(\mathbf{M}_1 - \mathbf{M}_2)]. \tag{B1.11}$$

Next, taking into account the equality

$$(\mathbf{I} - \mathbf{W})(c_1\mathbf{I} + c_2\mathbf{W})^{-1} = (c_1\mathbf{I} + c_2\mathbf{W})^{-1}(\mathbf{I} - \mathbf{W}), \tag{B1.12}$$

it is easy to show that (B1.9) is in fact valid. This proves the equivalence between our relation (B1.8) and that resulting from Takao and Taya's work.

On Atomic Spacing in Large Regular Cubic Lattices

BRUNO A. BOLEY
Department of Civil Engineering and Engineering Mechanics,
Columbia University, New York, NY 10027, U.S.A.

Abstract

Atomic spacing in an infinite regular lattice, which is *a priori* indeterminate because of symmetry, is determined here by considering the infinite array as the limiting case of a sequence of finite ones. A simple approximate formula for the spacing is derived.

1. Introduction

The force between atomic particles is usually expressed as the difference between two forces, one attractive and one repulsive, each commonly described by an inverse-power law. Thus, for a separation distance r,

$$F = \frac{A}{r^m} - \frac{B}{r^n} \tag{1.1}$$

corresponding to a potential

$$V = \frac{Br^{1-n}}{n-1} - \frac{Ar^{1-m}}{m-1}, \tag{1.2}$$

where m and n are integers ($n > m > 1$), and A and B are constants. In any given lattice the position of each particle corresponding to equilibrium can then be calculated, and subsequently, if desired, the internal force distribution as well. This is often done by starting with an undisturbed array, and then determining the aberrations caused in it by the specific characteristics of the body being analyzed, be they surfaces, cracks, or internal defects (e.g., Kanzaki, 1957). In such a procedure, the original spacing in the undisturbed lattice may be of interest, but it so happens that in certain cases it may not immediately be known. For example, in an infinite array possessing sufficient symmetry (e.g., a face-centered cubic lattice), the spacing in question is undetermined, because symmetry insures that equilibrium holds with *any* constant spacing. It is the purpose of the present work to determine the spacing in such an infinite lattice, by considering it as the limiting case of a sequence of finite ones.

Before proceeding to that task, it should be mentioned that if one is interested only in the gross properties of the body, such as the material constants of the corresponding continuum, and the relation between them and the quantities in (1.1), then the actual spacing is irrelevant. For example, Boley (1978) showed that Poisson's ratio can be written (after correcting a typographical error, and changing to the present notation) as

$$\nu = \begin{cases} \dfrac{n+1}{3n-1} & \text{for} \quad n > 3, \text{ and any } m < n, \\[2ex] \dfrac{m-1}{m+1} & \text{for} \quad m < n < 3, \end{cases} \tag{1.3}$$

independently of the value of the constant spacing.

Note also that the constants m and n (as here defined) differ from the more conventional ones; the latter are obtained by replacing these by $(m + 1)$ and $(n + 1)$, respectively. The current notation is a little more convenient for the analysis which follows.

2. Finite Arrays

If only *two* particles are involved, equilibrium occurs at the distance

$$r_0 = \left(\frac{B}{A}\right)^{1/(m-n)}, \tag{2.1a}$$

while the maximum force F_{max} would occur at the distance

$$r_{max} = \left(\frac{n}{m}\right)^{1/(n-m)} r_0. \tag{2.1b}$$

In dimensionless terms, with

$$\rho = \frac{r}{r_0}; \qquad \rho_{max} = \frac{r_{max}}{r_0} = \left(\frac{n}{m}\right)^{1/(n-m)}, \tag{2.2a}$$

we then have

$$\frac{F}{F_{max}} = \frac{\rho^{-m} - \rho^{-n}}{\rho_{max}^{-m} - \rho_{max}^{-n}}. \tag{2.2b}$$

We may proceed to determine the equilibrium spacing for any number N of particles, simply by setting to zero the sum of the forces exerted on each particle by all others, or

$$\sum_{j=1}^{N} F_{ij} = 0, \qquad i = 1, 2, \ldots, N, \tag{2.3}$$

where $F_{ij} = -F_{ji}$ is the force exerted on particle i by particle j, the distance between them being $r_{ij} = r_{ji}$.

Let the particles be numbered consecutively from the top (see Table 1);

TABLE 1

	$N = 2$	3	4	5	6	7	8	∞
ρ_{12}	1	.983	.979	.978	.977	.977	.977	.977
ρ_{23}		.983	.963	.959	.957	.957	.956	.956
ρ_{34}			.979	.959	.954	.952	.952	.951
ρ_{45}				.978	.957	.952	.951	.949
ρ_{56}					.977	.957	.952	.948
ρ_{67}						.977	.956	.947
ρ_{78}							.977	.947
ρ_{ctr}								.946

$m = 4, \quad n = 7$

$\rho = \dfrac{r}{r_0}$

then symmetry prevails about particle $(N + 1)/2$ if N is odd, and midway between particles $(N/2)$ and $(N + 2)/2$ if N is even. Symmetry also gives $r_{ij} = r_{N-i+1,N-j+1}$.

For $N = 3$ we obtain

$$\rho_{12} = \rho_{23} = \left(\frac{1 + 2^{-n}}{1 + 2^{-m}}\right)^{1/(n-m)}, \tag{2.4}$$

but for any larger value of N no similarly simple closed-form result can be found, and numerical calculation must be resorted to. Clearly, this can only be done for specific values of m and n. The results collected in Table 1 for the special case of $m = 4$, $n = 7$, are illustrative of the outcome of such calculations.

3. Infinite Array

The question we have posed is, What are the limiting values of spacings such as those listed in Table 1, as N is allowed to become infinite? As we examine Table 1, we see that the numbers begin to approach the expected distribution, namely a large constant central area unaffected by the receding surface, and a thin surface-tension layer near the boundary itself. We shall seek limits for both regions; in particular, however, we are interested in the value approached by the central spacing as $N \to \infty$, since, as we have indicated, we shall take this limit to represent the spacing in an infinite array.

To calculate the limits, we develop an approximate method of calculation, which exploits the obvious fact that intensity of interaction decreases rapidly with distance, and that therefore errors made in estimating separations between distant particles are relatively unimportant. We then proceed by isolating in (2.3) the terms corresponding to near-neighbor interaction; then

$$\sum_{j=1}^{i-2} F_{ij} + F_{i,i-1} + F_{i,i+1} + \sum_{j=i+2}^{N} F_{ij} = 0, \qquad i = 1, 2, \ldots, N. \tag{3.1}$$

Summing now the first i of these equations, we obtain

$$F_{i,i+1} = -\sum_{k=1}^{i} \left(\sum_{j=1}^{k-2} F_{kj} + \sum_{j=k+2}^{N} F_{kj} \right), \qquad i = 1, 2, \ldots, N. \tag{3.2}$$

The quantity $F_{i,i+1}$ appears on the left-hand side only, so that it can be calculated directly; if the terms on the right-hand side are known or (as in what follows) approximated. These terms, however, depend exclusively on further-than-near neighbor interactions, and therefore errors introduced in them will have a relatively small effect on the calculated value of $F_{i,i+1}$. Should more accurate approximations be desired, they could be obtained by a straightforward extension of the present analysis. This would involve isolating, in (2.3), terms involving the effect of second-nearest neighbors rather than nearest neighbors only; the equations analogous to (3.2) would then require simultaneous solution for two unknowns rather than for a single one as here. This will, however, not be done here for simplicity of presentation, and because the approximation presented appears adequate to yield three-figure accuracy for most values of m and n, provided that, as in the present example, they are large enough.

The approximation we use for these particular calculations consists in assuming that the right-hand side of (3.2) can be calculated by assuming equal values for the distances between more-than-successive particles. It can then be shown that

$$\rho_{12}^{n-m} = \frac{\Sigma_n}{\Sigma_m}, \tag{3.3a}$$

$$\rho_{23}^{n-m} = \frac{2\Sigma_n - 1 - (N-1)^{-n}}{2\Sigma_m - 1 - (N-1)^{-m}}, \tag{3.3b}$$

$$\rho_{34}^{n-m} = \frac{3\Sigma_n - 2 - 2^{-n} - (N-2)^{-n} - 2(N-1)^{-n}}{3\Sigma_m - 2 - 2^{-m} - (N-2)^{-m} - 2(N-1)^{-m}}, \tag{3.3c}$$

$$\rho_{45}^{n-m} = \frac{4\Sigma_n - 3 - 2^{1-n} - 3^{-n} - (N-3)^{-n} - 2(N-2)^{-n} - 3(N-1)^{-n}}{4\Sigma_m - 3 - 2^{1-m} - 3^{-m} - (N-1)^{-m} - 2(N-2)^{-m} - 3(N-1)^{-m}}, \tag{3.3d}$$

and so forth. For the center of the array, we have (considering, for simplicity, only the case of N even)

$$\rho_{N/2,(N+1)/2}^{n-m} = \frac{\Sigma_{n-1} + \sum_{i=(N/2)+1}^{N-1} (N-2i)/i^n}{\Sigma_{m-1} + \sum_{i=(N/2)+1}^{N-1} (N-2i)/i^m}. \tag{3.4}$$

In all of the above, the notation

$$\Sigma_p = \sum_{i=1}^{N-1} \left(\frac{1}{i}\right)^p, \tag{3.5}$$

has been used.

4. Limiting Case

We now, finally, proceed to evaluate the limits of the above expressions as $N \to \infty$. We note first that

$$\lim_{N\to\infty} \Sigma_p = \sum_{i=1}^{\infty} \left(\frac{1}{i}\right)^p = \zeta(p), \tag{4.1}$$

where $\zeta(p)$ stands for the Riemann zeta function. Then, *in the limit*, we have

$$\rho_{12}^{n-m} = \frac{\zeta(n)}{\zeta(m)}, \tag{4.2a}$$

$$\rho_{23}^{n-m} = \frac{2\zeta(n) - 1}{2\zeta(m) - 1}, \tag{4.2b}$$

$$\rho_{34}^{n-m} = \frac{3\zeta(n) - 2 - (2)^{-n}}{3\zeta(m) - 2 - 2^{-m}}, \tag{4.2c}$$

$$\rho_{45}^{n-m} = \frac{4\zeta(n) - 3 - 2(2)^{-n} - (3)^{-n}}{4\zeta(m) - 3 - 2(2)^{-m} - 3^{-m}}, \tag{4.2d}$$

$$\rho_{56}^{n-m} = \frac{5\zeta(n) - 4 - 3(2)^{-n} - 2(3)^{-n} - (4)^{-n}}{5\zeta(m) - 4 - 3(2)^{-m} - 2(3)^{-m} - (4)^{-m}}, \tag{4.2e}$$

and so forth. Finally, from (3.4), the limit value of the center spacing is simply

$$\rho_{\text{ctr}}^{n-m} = \frac{\zeta(n-1)}{\zeta(m-1)}. \tag{4.3}$$

Since $\zeta(1) = \infty$, it is clear that the above developments are valid only for $m, n > 2$.

Numerical values illustrating (4.2) and (4.3) are included in Table 1. They clearly show that surface tension is restricted to a very thin layer, and that, for the major portion of the body, the spacing is essentially that prevailing at the center and given by (4.3). This is then the result we were looking for, namely, the spacing pertaining to an infinite body, which we had agreed to consider as the limiting case of a sequence of finite ones.

References

Boley, B. A. (1978), Some consideration of elastic analyses of discrete models of solids, *Computers & Structures*, **8**, 345–347.

Kanzaki, H. (1957), Point defects in face-centred cubic lattice—I. Distorsions around defects, *J. Phys. Chem. Solids*, **2**, 24–36.

Crack Tip Toughening by Inclusions with Pairs of Shear Transformations

S.-J. Chang* and P. F. Becher†
*Engineering Technology Division,
†Metals and Ceramics Division,
Oak Ridge National Laboratory, Oak Ridge, TN 37831-8051, U.S.A.

Abstract

The deviatoric transformation strain of an inclusion is modeled by applying an equivalent distribution of dislocations along a surface which exhibits a discontinuous change in the transformation strains. This method is applied to qualitatively model the twin structures generated in transformation toughened ceramics. For this case, the transformation shear strain of the inclusion is assumed to consist of a number of symmetrical pairs of (twinning) shears in a rectangular grain. The elastic energy is derived and expressed in terms of elementary functions. With one pair of shears, the inclusion induced toughening effect in the presence of a crack is calculated by applying a recent solution of the crack–dislocation interaction problem. Numerical results show that the toughening due to the inclusion (as compared to that due to dilatation) is not negligible if the inclusion is located within a distance equal to several grain sizes from the crack tip. Moreover, the toughening depends strongly on the orientation of the inclusion relative to the crack.

1. Introduction

The martensitic transformation of tetragonal zirconia particles or grains has been shown to increase the toughness of both monolithic zirconia and ceramics containing zirconia inclusions (McMeeking and Evans, 1982; Evans and Cannon, 1986; Becher *et al.*, 1987; Rose, 1987). Analysis of the toughening behavior indicated that the dilatational or volumetric expansion of the transformed particles is a significant contributor to the toughening. Experimental observations of the martensitic transformation of ZrO_2 show that twin structures with different variants are generated (Muddle and Hannink, 1986). The condition required to trigger the transformation seems to remain a subject of investigation, although the transformation mechanism has been shown to be closely related to the local shear (Chen and Morel, 1986). Questions have been raised (McMeeking and Evans, 1982), as to how the orientation of the twins may interact with the crack tip to provide further reduction in the stress intensity.

In the present study, we shall not discuss the criteria for the nucleation of the transformation, but only suggest a method for calculating the toughening contribution as a result of the inclusions which have been subjected to pairs of shear transformations. The method is based on the distribution of dislocations introduced to match the misfit or incompatible deformation along the interfaces between an inclusion or twin and the matrix (Bilby *et al.*, 1955; Mura, 1987). The twin structure is assumed to be subjected to pairs of shear transformations in a rectangular grain. The energy of the deformation twinning is then calculated and the results show that the elastic twin energy per unit volume decreases as the number of pairs of twins increases. The transformation, therefore, leads to a reduction of the total elastic energy of the system. The stress is found to increase logarithmically as the tip is approached from the matrix side. Moreover, the toughening contribution of the twin structures is analyzed by applying a recent solution of the dislocation–crack interaction problem. The numerical results show that if the twins are located near the crack tip, their contribution to the toughening depends strongly on their orientation, and is not negligible compared to that due to the dilatation toughening.

2. Dislocation Density Description for Inelastic Inclusion

A dislocation density tensor is used to represent the incompatible or misfit deformation across the boundary between an inclusion and the surrounding matrix. It is also used to model the nonuniform inelastic transformation strain within the inclusion. This formulation enables us to solve the inclusion problem with nonuniform transformation strain in terms of the solution of an equivalent dislocation problem.

Let u_k^{e} and u_k^{p} denote Cartesian components of the displacements due to elastic and plastic deformations, respectively. The elastic distortion β_{ik}^{e} and plastic distortion β_{ik}^{p} are defined, respectively, as the gradients of the displacements

$$\beta_{ik}^{\mathrm{e}} = \frac{\partial u_k^{\mathrm{e}}}{\partial x_i}, \tag{2.1}$$

$$\beta_{ik}^{\mathrm{p}} = \frac{\partial u_k^{\mathrm{p}}}{\partial x_i}, \tag{2.2}$$

where x_i $(i = 1, 2, 3)$ are the Cartesian coordinates. Since the plastic distortion appears repeatedly in the following text, we shall use the notation β_{ik} without superscript for brevity. The Burgers vector b is defined as a contour integral along a Burgers circuit L with respect to a line segment of a dislocation loop

$$b_k = \oint_L \beta_{ik}\, dx_i. \tag{2.3}$$

The above line integral can be transformed to a surface integral according to

$$\oint_L \beta_{ik}\, dx_i = \int_S e_{ilm} \frac{\partial \beta_{mk}}{\partial x_l}\, dS_i = \int_S e_{ilm} \frac{\partial^2 u_k^p}{\partial x_m\, \partial x_l}\, dS_i, \tag{2.4}$$

where e_{ilm} is the permutation tensor. If u_k^p were continuous and differentiable, then we should have

$$\frac{\partial^2 u_k^p}{\partial x_m\, \partial x_l} = \frac{\partial^2 u_k^p}{\partial x_l\, \partial x_m}, \tag{2.5}$$

and (2.4) would be zero. However, (2.5) is not satisfied everywhere within S due to the disturbance of the u_i^p field induced by the dislocation loop which passes through S. In order to satisfy (2.3) for a single dislocation line, we must have

$$e_{ilm} \frac{\partial \beta_{mk}}{\partial x_l} = -b_k v_i \delta(\xi). \tag{2.6}$$

In the above equation, $\delta(\xi)$ is the two-dimensional δ function and v_i is the normal direction of the surface S such that

$$\int_S v_i \delta(\xi)\, dS_i = 1. \tag{2.7}$$

In (2.6), the discrete Burgers vector b_k may be viewed as a distribution of Burgers vectors with the distribution function $b_k\delta(\xi)$. Therefore, given more dislocation lines passing through S, we may use (2.6) to define a distribution of dislocations. This yields

$$\alpha_{ik} = v_i b_k = -e_{ilm} \frac{\partial \beta_{mk}}{\partial x_l}, \tag{2.8}$$

where α_{ik} represents the kth component b_k of the total Burgers vector per unit area that has a unit normal v_i along the ith direction. α_{ik} is a tensor quantity because it is composed of two vectors b and v. A physical view of the distribution of dislocations is to smear out the discrete dislocation lines and to replace them by the equivalent continuous distribution.

It is seen from the above equation, (2.8), that α_{ik} vanishes if every derivative of the plastic distortion β_{mk} does. Suppose that the derivatives of β_{mk} are nonzero only along certain surfaces in three-dimensional space, or lines in two-dimensional space. This means that the inelastic strain is uniform everywhere except along these surfaces. Then it is more convenient to represent the inelastic strain field by the discontinuities of β's across these surfaces. This is shown in the following.

In (2.8), α_{ik} may be interpreted as the divergence of a flux vector $e_{ilm}\beta_{mk}$ with $l = 1, 2, 3$ as its three components. Suppose that the flux vector is discontinuous across a surface where n_l is the direction cosine and the quantity $[\beta_{mk}]$ is the discontinuity across the surface. We may prescribe a thin layer Δt across the surface and apply the divergence theorem to obtain

$$\alpha'_{ik} = \lim_{\Delta t \to 0} \alpha_{ik} \Delta t = -e_{ilm}[\beta_{mk}] n_l. \tag{2.9}$$

From the above equation, α'_{ik} is defined as the surface dislocation density as a result of the discontinuity $[\beta_{mk}]$. Therefore, an equivalent distribution of surface dislocations can be used to represent the discontinuous change of β_{mk} across the surfaces, and the solution of the inclusion problem with nonuniform transformation strain can be reduced to that of an equivalent dislocation problem. We shall formulate the problem more specifically in the next section.

3. Two-Dimensional Inclusion

For a two-dimensional inclusion, a direct application of (2.9) leads to

$$\begin{bmatrix} \alpha'_1 \\ \alpha'_2 \end{bmatrix} = -\begin{bmatrix} [\beta_{11}] & [\beta_{21}] \\ [\beta_{12}] & [\beta_{22}] \end{bmatrix} \begin{bmatrix} s_1 \\ s_2 \end{bmatrix}, \tag{3.1}$$

where s is the tangent vector along the interface of discontinuity $[\beta]$. For example, the change in the β_{11} component becomes

$$[\beta_{11}] = \beta_{11}^{+} - \beta_{11}^{-}, \tag{3.2}$$

if β_{11}^{+} and β_{11}^{-} are the plastic distortions of the matrix and the inclusion, respectively.

The above representation suggests that we may define a complex surface dislocation density α' by

$$\alpha' = \alpha'_1 + i\alpha'_2, \tag{3.3}$$

and rewrite (3.1) as

$$2\alpha' = -([\beta_{11}] + [\beta_{22}])\frac{dz}{ds} - ([\beta_{11}] - [\beta_{22}])\frac{d\bar{z}}{ds} - i([\beta_{12}] - [\beta_{21}])\frac{dz}{ds} - i([\beta_{12}] + [\beta_{21}])\frac{d\bar{z}}{ds}, \tag{3.4}$$

where $z = x_1 + ix_2$. The first two terms in (3.4) are the effect due to uniform stretching, and the last two terms simple shear. By using the complex dislocation density, (3.4), we have an advantage in calculating some of the contour integrals.

The above description of surface dislocations can be used to calculate the stress distribution as a result of the misfit deformation along a certain surface provided that we know the stress functions due to one single edge dislocation. For an edge dislocation at z_1 in an isotropic and elastic medium with shear modulus μ and Poisson's ratio v, the stress potentials are

$$\phi(z) = B\alpha\frac{1}{z - z_1}, \tag{3.5}$$

$$\psi(z) = -B\bar{\alpha}\frac{1}{z - z_1} + B\alpha\frac{\bar{z}_1}{(z - z_1)^2}, \tag{3.6}$$

where B is a material constant

$$B = \frac{\mu}{4\pi i(1 - \nu)}. \tag{3.7}$$

The solutions of two-dimensional inclusion problems can be expressed in terms of the line integrals of (3.5) and (3.6) along the inclusion surface with appropriate boundary conditions. Functions $\phi(z)$ and $\psi(z)$ for an inclusion of a general shape in an infinite medium are shown in Appendix A.

These expressions for ϕ and ψ have been applied to circular and elliptic inclusions and the well-known solutions of the inclusion problems (Mura, 1987) were confirmed.

4. Twin Structure Idealized as Pairs of Shear Transformations

It is assumed that the twin structure generated by the phase transformation can be represented qualitatively by pairs of inelastic shear deformations in a rectangular grain, as shown schematically in Fig. 1.

Each pair consists of a positive shear S at the right-half and a negative shear

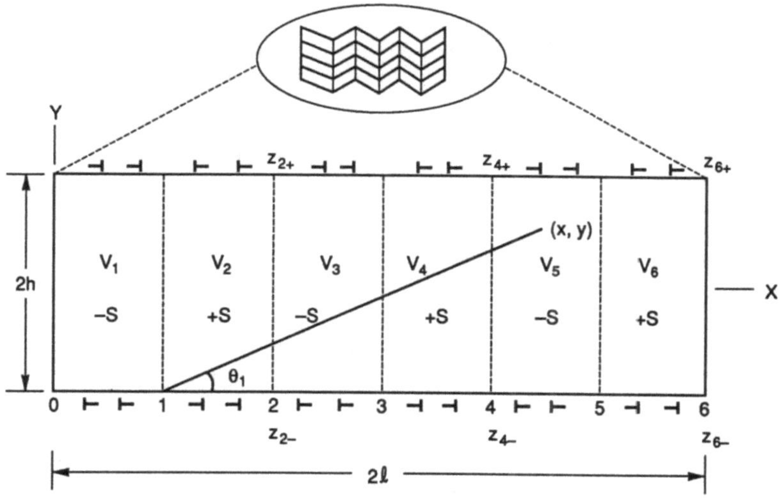

FIG. 1. Schematic representation of the deformation twinning for an inclusion consisting of $N/2$ pairs of shear deformations. Each pair has negative shear $-S$ at left and positive shear $+S$ at right. An equivalent description of the inelastic deformation is represented by a distribution of edge dislocations located along the top and bottom boundaries between the inclusion and the surrounding matrix. Points at the ends of the twin boundaries are labeled by $1+, 1-, \ldots, N+, N-$, with the complex coordinates $Z_{1+}, Z_{1-}, \ldots, Z_{N+}, Z_{N-}$, respectively.

$-S$ at the left

$$\beta_{12} = \frac{\partial u_2}{\partial x_1} = S \qquad \text{right-half}$$

$$= -S \quad \text{left-half.} \tag{4.1}$$

The other components of β are zero, including β_{21}. The matrix is assumed to deform elastically. By applying (3.4), the misfit strain between the matrix and the inclusion can be used to derive the equivalent distribution of edge dislocations on the top and bottom part of the twin. It is noted that, although there exist changes in plastic strain along the twin boundaries, there is no distribution of dislocations along these boundaries.

Elementary calculation shows that the solution in terms of the complex stress functions for the twinning deformation is

$$\frac{i}{B\beta_{12}}\phi(z) = 2\sum_{i=0}^{N}\left[(-1)^i \log\frac{z_{i-} - z}{z_{i+} - z}\right] - \log\frac{z_{0-} - z}{z_{0+} - z} - \log\frac{z_{N-} - z}{z_{N+} - z}, \tag{4.2}$$

$$\frac{i}{B\beta_{12}}\psi(z) = 2\sum_{i=0}^{N}\left[(-1)^i\left(\frac{\bar{z}_{i-}}{z_{i-} - z} - \frac{\bar{z}_{i+}}{z_{i+} - z}\right)\right] - \left(\frac{\bar{z}_{0-}}{z_{0-} - z} - \frac{\bar{z}_{0+}}{z_{0+} - z}\right)$$
$$- \left(\frac{\bar{z}_{N-}}{z_{N-} - z} - \frac{\bar{z}_{N+}}{z_{N+} - z}\right). \tag{4.3}$$

From the potential functions $\phi(z)$ and $\psi(z)$, shown in the above equations, the stress components are

$$\sigma_1 + \sigma_2 = \frac{2B\beta_{12}}{i} 2 \operatorname{Re}\left[2\sum_{i=0}^{N}(-1)^i \log\frac{z_{i-} - z}{z_{i+} - z}\right.$$
$$\left. - \log\frac{(z_{0-} - z)(z_{N-} - z)}{(z_{0+} - z)(z_{N+} - z)}\right], \tag{4.4}$$

$$\sigma_2 - \sigma_1 + 2i\sigma_{12}$$
$$= \frac{2B\beta_{12}}{i}\left[2\sum_{i=0}^{N}(-1)^i\left(\frac{\bar{z}_{i-} - \bar{z}}{z_{i-} - z} - \frac{\bar{z}_{i+} - \bar{z}}{z_{i+} - z}\right)\right.$$
$$\left. - \left(\frac{\bar{z}_{0-} - \bar{z}}{z_{0-} - z} - \frac{\bar{z}_{0+} - \bar{z}}{z_{0+} - z}\right) - \left(\frac{\bar{z}_{N-} - \bar{z}}{z_{N-} - z} - \frac{\bar{z}_{N+} - \bar{z}}{z_{N+} - z}\right)\right]. \tag{4.5}$$

5. Stress Intensity

From the stress distribution, (4.4) and (4.5), we observe that at the ends of each twinning plane the stress tends to infinity according to

$$\sigma_1 = \sigma_2 \to 4Bi\beta_{12} \log|z - z_{i+}| \to -\infty \qquad \text{for } z \to z_{i+}, \qquad i = \text{odd},$$
$$\to +\infty \qquad \text{for } z \to z_{i+}, \qquad i = \text{even}. \tag{5.1}$$

Therefore the stress intensity factor is

$$K = 4Bi\beta_{12} = \frac{\mu\beta_{12}}{\pi(1-\nu)}. \tag{5.2}$$

It is observed that K is linearly proportional to the shear β_{12} deformed within the twin and is independent of the dimension of the transformed inclusion.

6. Total Energy

The total elastic energy due to the transformation can be calculated by applying the relation

$$W = -\frac{1}{2}\int_v \sigma_{ij}\varepsilon_{ij}^* \, dv, \tag{6.1}$$

where ε_{ij}^* is the inelastic transformation strain and v is the volume of the inclusion prior to the transformation. In the present case it has a rectangular shape, and the transformation consists of several pairs of shears. It has been shown recently (Mura *et al.*, 1985) that the above equation is a valid expression even if the interface between the inclusion and the matrix has a sliding contact boundary condition. The derivation is shown in Appendix B.

After some elementary calculations, we obtain the following expression of energy for the present problem

$$W = \frac{\mu\beta_{12}^2}{2\pi(1-\nu)}\sum_{j=1}^{N}\sum_{i=1}^{N}(-1)^{i+j+1}\int_{v_j}\sin 2\theta_i \, dx\, dy, \tag{6.2}$$

where θ_i is defined as the angle between x axis and the ray from z_{i-} to the point (x, y). As shown in Appendix B, (6.2) can be further simplified to logarithmic functions.

The numerical result shows that the energy of the two-dimensional case is less than that of the three-dimensional counterpart (Mura *et al.*, 1976). However, the energy of the two-dimensional case qualitatively described the physical nature exhibited by its three-dimensional counterpart. In addition, the present transformation strain for the case of one pair of shears is obtained via an approximate polynomial transformation (Asaro and Barnett, 1975). Finally, a numerical calculation for the total energy was then carried out and the result is plotted as the bottom curve in Fig. 2. This curve shows a continuous decrease in energy as the number of pairs increases, provided that both the dimension of the inclusion and the magnitude of the shear are fixed.

As discussed earlier (Evans and Cannon, 1986), the criterion for the nucleation of the twin structure may be related to the minimum total energy of the system. It is possible that an additional energy contribution may come from other sources, in which the surface energy along the interface of the discontinuous shear deformation may be a dominant quantity. It is not known how large the surface energy should be. We calculated a series of curves by

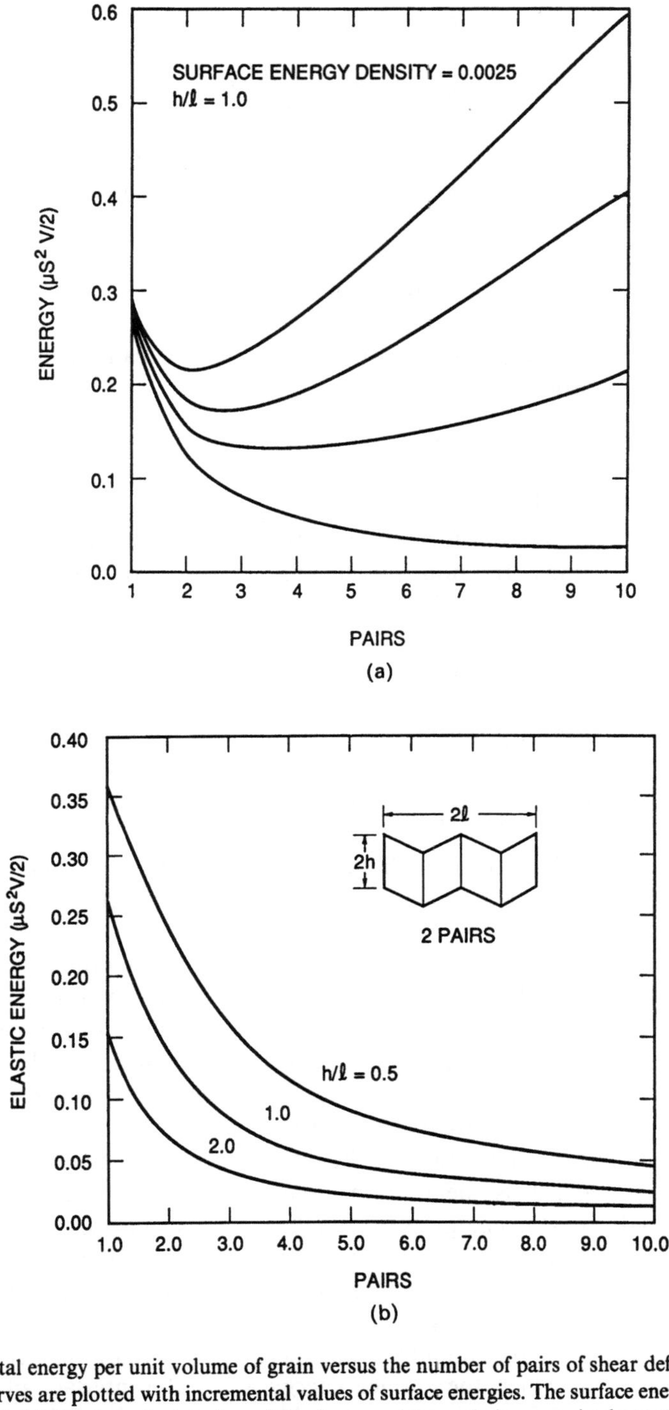

FIG. 2. Total energy per unit volume of grain versus the number of pairs of shear deformations. (a) The curves are plotted with incremental values of surface energies. The surface energy density increment is 0.0025 along the interface of the shears and the grain is square in shape with volume V. (b) The grain is rectangular in shape.

assigning a range of surface energies. The results are plotted as a series of curves in Fig. 2. Each of these curves has a minimum value. The number of pairs which are required to reach the minimum decreases as the magnitude of the assigned surface energy increases.

7. Dislocation and Crack Interaction

In the preceding sections, we have discussed the use of dislocations to model the misfit created by the inelastic transformation of the inclusion against the matrix. The stress field generated as a result of the misfit, however, depends on the constitutive properties of the inclusion as well as the matrix. At present, we shall assume that both the inclusion and the matrix have the same isotropic elastic constants.

In order to calculate the toughening effect, we need an explicit solution of the dislocation–crack interaction problem. There are several ways to derive the solution (Thomson, 1986). At present, a two-dimensional solution is derived by the author in Appendix C. The problem is formulated by invoking a distribution of (virtual) dislocations to model the displacement of the crack surface. The edge dislocation which induces the solution is located near the crack with the complex coordinate β. The problem is schematically shown in Fig. 3, in which a complex coordinate is chosen so that the negative x axis coincides with the semi-infinite crack surface and the crack tip is located at the origin. Any point in the two-dimensional body is represented by the coordinate $z = x + iy$.

As shown in Appendix C, the complex stress intensity K ($K = K_1 + iK_2$), induced by an edge dislocation which is located at the complex coordinate β with a complex Burgers vector b, has the expression

$$K = -\frac{\sqrt{2\pi}}{\sqrt{\beta}}\left[\bar{a}\left(1 + \frac{\sqrt{\beta}}{\sqrt{\bar{\beta}}}\right) + \frac{a}{2}\left(1 - \frac{\bar{\beta}}{\beta}\right)\right], \tag{7.1}$$

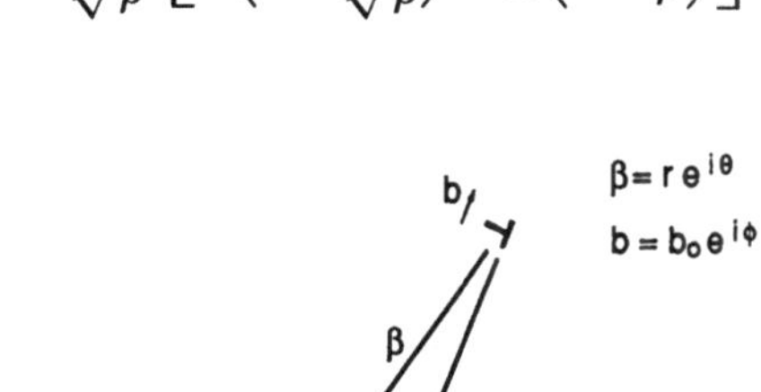

$$K = -\frac{\mu i}{(1-\nu)\sqrt{2\pi r}}\left[e^{-i\phi} - \frac{1}{2}e^{i(\phi-\theta)}\left(1 - e^{-i\theta}\right)\right]\cos\frac{\theta}{2}$$

FIG. 3. An edge dislocation located at β and the Burgers vector b relative to a semi-infinite crack which coincides with the negative real axis.

where K_1 and K_2 are the mode I and mode II stress intensities, respectively, and the complex constant a is defined as $B\alpha$.

The distribution function, also complex-valued, is

$$f(x) = -\frac{1}{A\pi\sqrt{-x}}\left[-\bar{a}\left(\frac{\sqrt{\beta}}{x-\beta} - \frac{\sqrt{\bar{\beta}}}{x-\bar{\beta}}\right) + \frac{a}{2}\left(1 - \frac{\bar{\beta}}{\beta}\right)\frac{(x+\beta)\sqrt{\beta}}{(x-\beta)^2}\right], \quad (7.2)$$

from which we obtain K by taking the limit

$$K = \lim_{x\to 0} \sqrt{2\pi}\, A\pi\sqrt{-x}\, f(x). \quad (7.3)$$

The real part of the function $f(x)$ denotes the mode I component of the distribution function, and the imaginary part of $f(x)$, the mode II component. The complex displacement for the crack surface is

$$u = b_0 \int_x^0 f(x)\, dx, \quad (7.4)$$

where b_0 is the magnitude of the Burgers vector for the (virtual) dislocation. The above expressions are valid for b not necessarily parallel to β. For $b \| \beta$, K reduces to

$$K = \frac{bu}{2\sqrt{2\pi r}(1-\nu)} \cos\frac{\theta}{2}[3 \sin\theta + i(3\cos\theta - 1)]. \quad (7.5)$$

8. Inclusion Induced Toughness

It has been illustrated earlier that the transformation strain for an inclusion consisting of pairs of shears can be represented by a distribution of dislocations along the inclusion boundary. In the following calculations, the deformation misfit is modeled approximately by a discrete number of edge dislocations.

At present, we only use four dislocations to represent the transformation of one pair of shears. The arrangement of the dislocations is schematically shown in Fig. 4. It represents an inclusion which has a rectangular shape of $2\Delta x$ by Δy in dimension and has been subjected to a pair of shears. For the purpose of comparison, a dilatational transformation is approximately represented by a circular arrangement of four dislocations.

The stress intensity K_1 induced by a symmetric pair of inclusions, each consisting of a single pair of shear deformations, is calculated and the result is shown in Fig. 5, where the directions of the shears to the horizontal axis are $\pi/2$ and π. The induced K_1 which has a negative value indicates a toughening effect for the crack tip region.

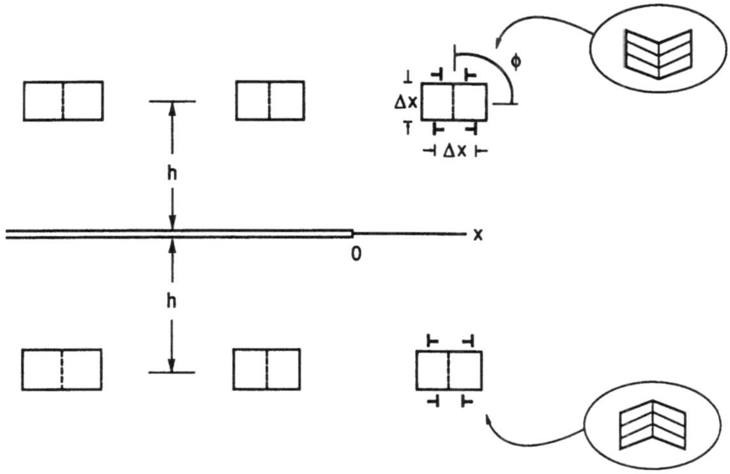

FIG. 4. Two lines of inclusions located parallel to the crack surface.

To estimate the overall K effect on the crack growth, two lines of inclusions are assumed as shown in Fig. 4. The arrangement is similar to that of some known models (Weertman *et al.*, 1983; Rose, 1987). The double line model shows a change in toughness as the crack grows. It approximately represents the R curve, or the crack growth resistance curve. The numerical values of these R curves are plotted in Fig. 6. It is shown that each of the curves varies significantly only near the crack tip region and that the effect is short-ranged. However, since the transformations are induced by the crack tip stress and the density is relatively large near the crack tip region, the local influence on the crack propagation may not be totally discounted. Moreover, results in Fig. 5 also indicate that the toughening effect due to the inclusions consisting of single pair of shears depends strongly on their orientations.

For the purpose of comparison, the numerical results corresponding to dilatational inclusions are plotted in Fig. 7. The dilatational inclusions in most of the locations only result in a toughening effect, except in regions in front of the crack tip where they may cause an increase in crack tip stress intensity and enhance the crack propagation. The R curve always shows a negative limiting value which contributes to the resistance to the crack growth.

The numerical value of K for an inclusion consisting of single pair of shears has been expressed in terms of a nondimensional constant

$$C_1 = \frac{\mu b}{(1-\nu)\sqrt{h}}\left(\frac{\Delta x}{h}\right)^2. \tag{8.1}$$

In the above equation, μ is the shear modulus, ν is Poisson's ratio, b is the Burgers vector, h is a distance, and Δx is the distance between the

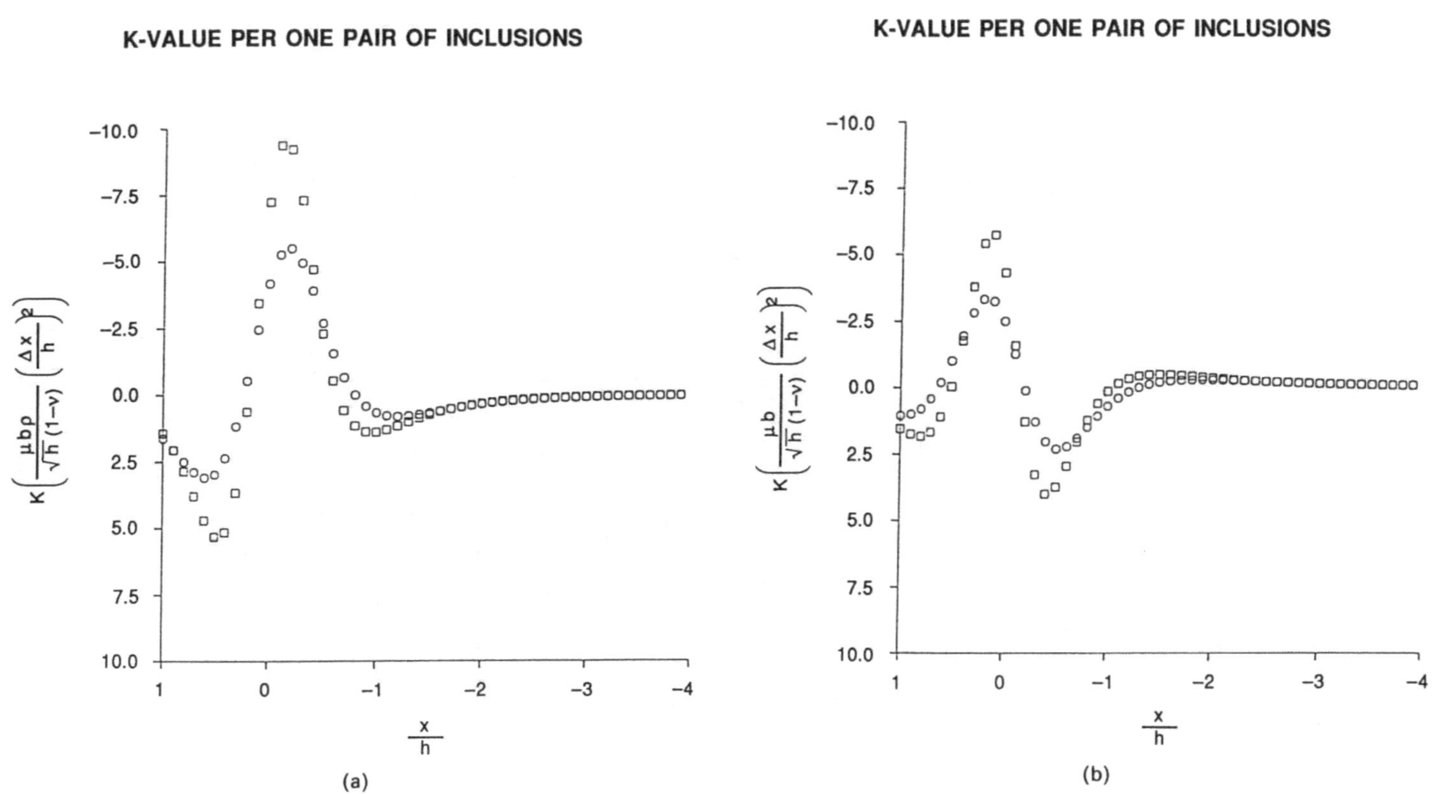

FIG. 5. Toughening by two symmetrically located inclusions each consisting of one pair of shear transformations located at h and $0.8h$ from the crack: (a) direction of shear $\phi = \pi/2$; (b) direction of shear $\phi = \pi$.

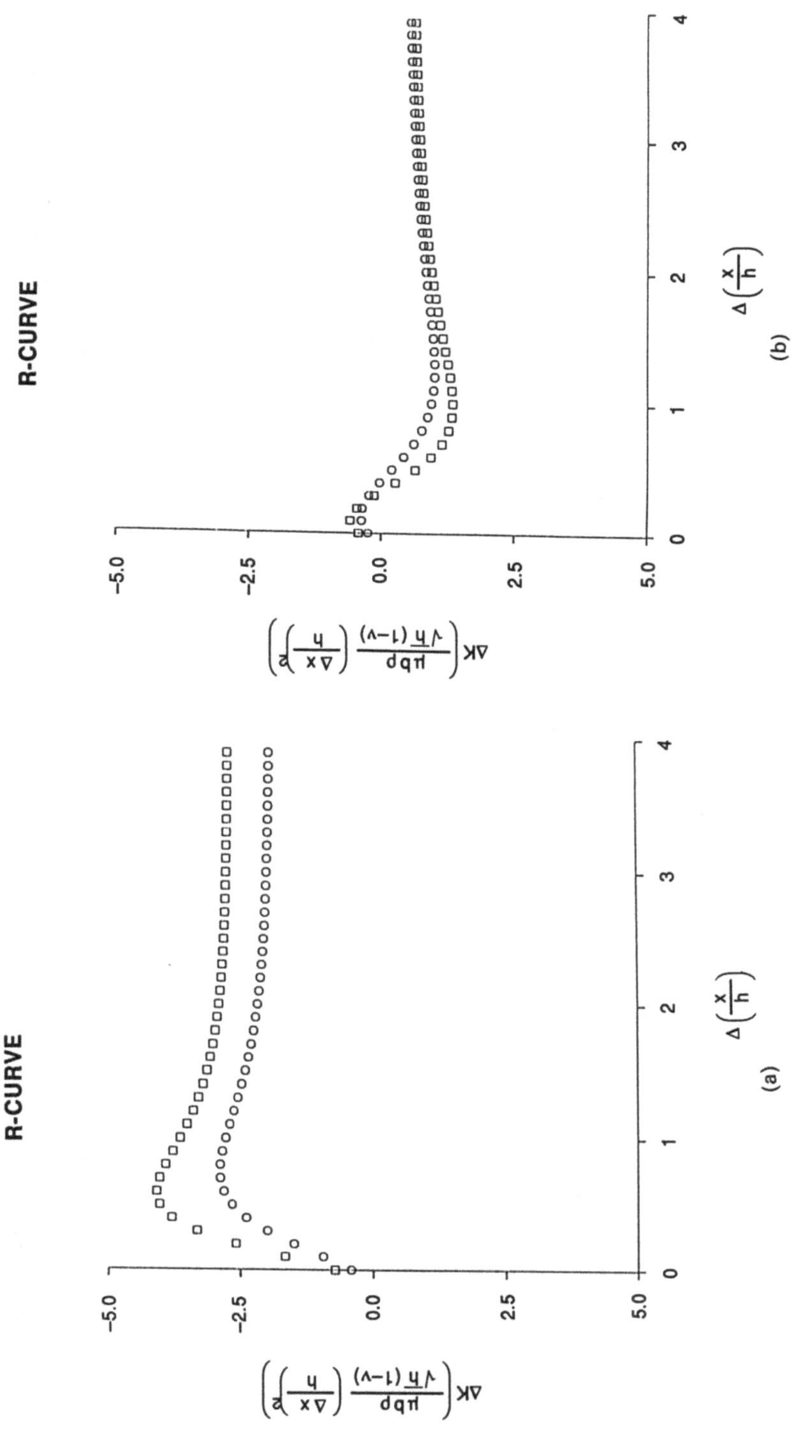

FIG. 6. *R* curves generated by two parallel lines of inclusions. Each inclusion has one pair of shear transformations: (a) direction of shear $\phi = \pi/2$; (b) direction of shear $\phi = \pi$.

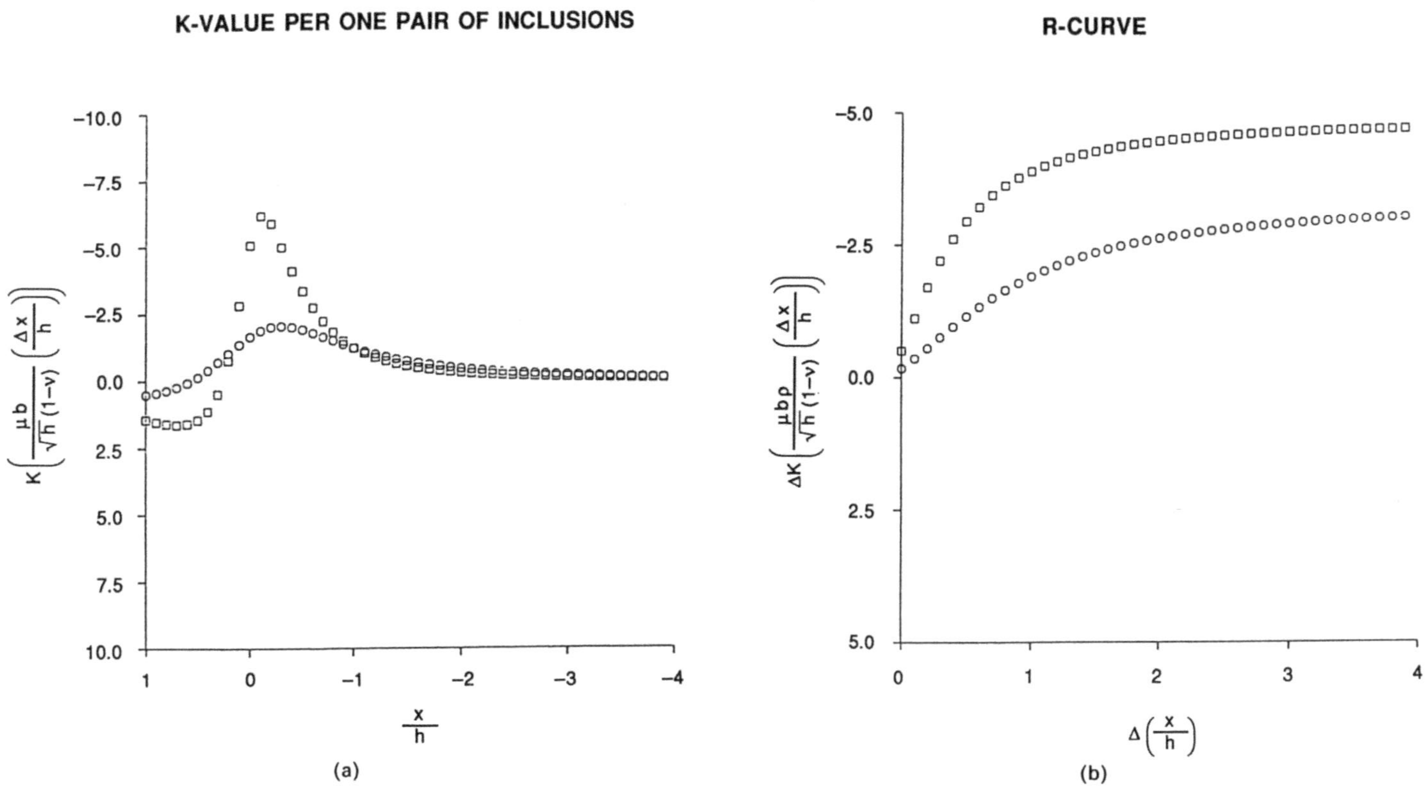

FIG. 7. (a) Toughening by two symmetrically located inclusions of dilatation, located at h and $0.5h$ from the crack. (b) The corresponding R curves.

two dislocations (or the grain size). For the R curve calculation, the nondimensional constant is $C_1 \rho$ where ρ is the number of inclusions per h along the longitudinal direction of the line of inclusions. The K value for the case of pairs of dilatations is plotted with the nondimensional constant

$$C_2 = \frac{\mu b}{(1 - \nu)\sqrt{h}} \left(\frac{\Delta x}{h} \right), \tag{8.2}$$

which depends only on the first power of $\Delta x/h$. The numerical value of K, due to either a pair of shears or dilatation, has the same order of magnitude, but the nondimensional constants are different by a factor of $\Delta x/h$. It implies that the influence of a pair of shears is less than the influence of dilatation by the ratio $\Delta x/h$. In other words, the ratio of the stress intensities has the same order of magnitude as that of the grain size versus the distance from the crack tip. We conclude, therefore, that if a grain with a pair of shear transformations is located near the crack tip within several grain sizes, its influence on the K value should not be neglected compared to that due to a grain with a dilatation transformation.

9. Conclusion and Discussion

The present investigation shows that it is convenient to use the method of continuous distribution of dislocations to calculate the general inclusion–crack interaction problem. Since the method does not restrict the shape of the inclusion, it can be programmed into an efficient numerical routine such that the more general microstructure–crack interaction problems can be handled.

This paper only suggests a method of calculation. The more difficult and fundamental problem concerning the nucleation of transformation twins remains intact. Important problems such as the determination of the transformation boundary and the density of the transformed grains remain unsolved. Recently, a more-than-additive increase in toughness was observed for some composites if the transformation-toughened ceramic was supplemented with whisker reinforcement (Becher and Tiegs, 1987). This problem also seems related to the criteria of transformation by which the transformation boundary changes substantially as a result of the second toughening mechanism.

Acknowledgments

Research sponsored by the Division of Materials Sciences. U.S. Department of Energy, under contract DE-AC05-840R21400 with Martin Marietta Energy Systems, Inc. The authors thank T. Mura, T. Mori, and C. H. Hsueh for helpful discussions.

References

Asaro, R. J. and Barnett, D. M. (1975), The non-uniform transformation strain problem for an anisotropic ellipsoidal inclusion, *J. Mech. Phys. Solids*, **23**, 77–83.

Becher, P. F., Swain, M. V., and Sōmiya, S. (1987), *Advanced Structural Ceramics*, MRS, Vol. 78, Materials Research Society, Pittsburgh.

Becher, P. F. and Tiegs, T. N. (1987), Toughening behavior involving multiple mechanisms: Whisker reinforcement and zirconia toughening, *J. Amer. Ceram. Soc.*, **70**, 651–54.

Bilby, B. A., Bullough, R., and Smith, E. (1955), Continuous distribution of dislocations, *Proc. Roy. Soc. London*, **A231**, 263–273.

Chen, I.-W. and Morel, P. E. (1986), Implications of transformation plasticity in ZrO_2-containing ceramics, *J. Amer. Ceram. Soc.*, **69**, 181–89.

Evans, A. G. and Cannon, R. M. (1986), Toughening of brittle solids by martensitic transformations, *Acta Metallurgica*, **34**, 761–800.

McMeeking, R. and Evans, A. G. (1982), Mechanics of transformation toughening in brittle materials, *J. Amer. Ceram. Soc.*, **65**, 242–45.

Muddle, B. C. and Hannink, R. H. J. (1986), Crystallography of the tetragonal to monoclinic transformation in MgO partially stabilized zirconia, *J. Amer. Ceram. Soc.*, **69**, 547–55.

Mura, T. (1987), *Micromechanics of Defects in Solids*, 2nd ed., Martinus Nijhoff, Dordrecht.

Mura, T., Jasiuk, I., and Tsuchida, B. (1985), The stress field of a sliding inclusion, *Int. J. Solids Structures*, **21**, 1165–1179.

Mura, T., Mori, T., and Kato, M. (1976), The elastic field caused by a general ellipsoidal inclusion and the application to martensite formulation, *J. Mech. Phys. Solids*, **24**, 305–18.

Rose, L. R. F. (1987), The mechanics of transformation toughening, *Proc. Roy. Soc. London*, **A412**, 169–97.

Thomson, R. (1986), Physics of fracture, in *Solid State Physics*, Vol. 39, edited by H. Ehrenreich and D. Turnbull, Academic Press, New York, pp. 1–129.

Weertman, J., Lin I.-H., and Thomson, R. (1983), Double ship plane crack model, *Acta Metallurgica*, **31**, 473–486.

Appendix A

The potentials for the inclusion problem are

$$\frac{2}{B}\phi(z) = -([\beta_{11}] + [\beta_{22}])\int \frac{dz_1}{z - z_1} - ([\beta_{11}] - [\beta_{22}])\int \frac{d\bar{z}_1}{z - z_1} - i([\beta_{12}] - [\beta_{21}])\int \frac{dz_1}{z - z_1} - i([\beta_{12}] + [\beta_{21}])\int \frac{d\bar{z}_1}{z - z_1}, \tag{A1.1}$$

$$\frac{2}{B}\psi(z) = ([\beta_{11}] + [\beta_{22}])\int \frac{d\bar{z}_1}{z - z_1} + ([\beta_{11}] - [\beta_{22}])\int \frac{dz_1}{z - z_1}$$

$$- i([\beta_{12}] - [\beta_{21}]) \int \frac{d\bar{z}_1}{z - z_1} - i([\beta_{12}] + [\beta_{21}]) \int \frac{dz_1}{z - z_1}$$

$$- ([\beta_{11}] + [\beta_{22}]) \int \frac{\bar{z}_1 \, dz_1}{(z - z_1)^2} - ([\beta_{11}] - [\beta_{22}]) \int \frac{\bar{z}_1 \, d\bar{z}_1}{(z - z_1)^2}$$

$$- i([\beta_{12}] - [\beta_{21}]) \int \frac{\bar{z}_1 \, dz_1}{(z - z_1)^2} - i([\beta_{12}] + [\beta_{21}]) \int \frac{\bar{z}_1 \, d\bar{z}_1}{(z - z_1)^2}. \tag{A1.2}$$

By straightforward application of (A1.1) and (A1.2), the results for inclusion problems of circular and elliptic inclusions subjected to uniform eigenstrains were confirmed.

Appendix B

The elastic strain energy w of the inclusion and the matrix due to plastic deformation of the inclusion with possible debonding is

$$W = \frac{1}{2} \int_D \sigma_{ij}(u_{i,j} - \varepsilon_{ij}^*) \, dv. \tag{B1.1}$$

The inclusion domain D is the sum of that for the matrix $D - \Sigma$ and inclusion Σ. The potential energy for the matrix is the equation when expressed in terms of boundary traction

$$\int_{D-\Sigma} \sigma_{ij} u_{i,j} \, dv = \int_{\partial D} \sigma_{ij} n_j u_i \, ds - \int_{\partial \Sigma} \sigma_{ij} n_j u_i \, (\text{out}) \, ds. \tag{B1.2}$$

Also, we have the elastic energy for the inclusion

$$\int_\Sigma \sigma_{ij} u_{i,j} \, dv = \int_{\partial \Sigma} \sigma_{ij} n_j u_i \, (\text{in}) \, ds. \tag{B1.3}$$

The sum of the above two equations gives

$$\int_D \sigma_{ij} u_{i,j} \, dv = - \int_{\partial \Sigma} \sigma_{ij} n_j [u_i] \, ds, \tag{B1.4}$$

where $\sigma_{ij} n_j = 0$ on ∂D and $[u_i] = u_i\,(\text{out}) - u_i\,(\text{in})$.

This discontinuous $[u_i]$ is zero for a bounded interface. For the case of a sliding inclusion, the traction force is continuous and has no shear along $\partial \Sigma$. For this case, $\sigma_{ij} n_j$ has no tangential shear component. The term $\sigma_{ij} n_j [u_i]$ vanishes along the interface. Therefore, for both cases, (B1.1) becomes

$$W = -\frac{1}{2} \int_\Sigma \sigma_{ij} \varepsilon_{ij}^* \, dv. \tag{B1.5}$$

Appendix C

From (4.5), we can prove that

$$\sigma_{12} = Bi\beta_{12}\left[2\sum_{i=0}^{N}(-1)^i(\sin 2\theta_{i-} - \sin 2\theta_{i+}) - (\sin 2\theta_{0-} - \sin 2\theta_{0+}) - (\sin 2\theta_{N-} - \sin 2\theta_{N+})\right]. \quad \text{(C1.1)}$$

Substituting the above equation into (6.1), the elastic energy becomes

$$\begin{aligned} W &= -\frac{1}{2}\sum_{j=1}^{N}(-1)^j\int_{v_j}\sigma_{12}\beta_{12}\,dv \\ &= -\tfrac{1}{2}\beta_{12}\sum_{j=1}^{N}(-1)^j\left[\int_{v_j}4Bi\beta_{12}\sum_{i=0}^{N}(-1)^i\sin 2\theta_{i-}\,dx\,dy - \int_{v_j}2Bi\beta_{12}(\sin 2\theta_{0-} + \sin 2\theta_{N-})\,dx\,dy\right] \\ &= 2Bi\beta_{12}^2\sum_{j=1}^{N}\sum_{i=1}^{N}(-1)^{i+j+1}\int_{v_j}\sin 2\theta_{i-}\,dx\,dy, \end{aligned} \quad \text{(C1.2)}$$

where we have made use of the following relation:

$$\sum_{j=1}^{N}(-1)^j\int_{v_j}\sin 2\theta_{0-}\,dx\,dy = \sum_{j=1}^{N}(-1)^j\int_{v_j}\sin 2\theta_{N-}\,dx\,dy. \quad \text{(C1.3)}$$

The integral in (C1.2) can be further reduced to

$$\begin{aligned} \int_{v_j}\sin 2\theta_{i-}\,dx\,dy &= \tfrac{1}{2}(x_j - x_i)^2\log\left[1 + \frac{(2h)^2}{(x_j - x_i)^2}\right] \\ &\quad - \tfrac{1}{2}(x_{j-1} - x_i)^2\log\left[1 + \frac{(2h)^2}{(x_{j-1} - x_i)^2}\right] \\ &\quad + \tfrac{1}{2}(2h)^2\log\left[\frac{(x_j - x_i)^2 + (2h)^2}{(x_{j-1} - x_i)^2 + (2h)^2}\right]. \end{aligned} \quad \text{(C1.4)}$$

Appendix D

The dislocation–crack interaction problem is solved in this appendix. For one edge dislocation located at β, as shown in Fig. 3, we formulate the pileup integral equation as

$$A\int_{-\infty}^{0}\frac{f(x')\,dx'}{x - x'} + \sigma_y + i\sigma_{xy} = 0, \quad \text{(D1.1)}$$

where $f(x)$ is the (virtual) distribution function, A is a constant, and

$$\sigma_y + i\sigma_{xy} = \frac{\bar{a}}{x - \beta} + \frac{\bar{a}}{x - \bar{\beta}} + \frac{a(-\beta + \bar{\beta})}{(x - \beta)^2}. \quad \text{(D1.2)}$$

The solution to the integral equation is

$$f(x) = -\frac{1}{A\pi^2\sqrt{-x}} \int_{-\infty}^{0} \frac{\sqrt{-x'}[\sigma_y(x') + i\sigma_{xy}(x')]\,dx'}{x' - x}$$

$$= -\frac{1}{A\pi\sqrt{-x}}\left[-\bar{a}\left(\frac{\sqrt{\beta}}{x-\beta} + \frac{\sqrt{\bar{\beta}}}{x-\bar{\beta}}\right) + \frac{a}{2}\left(1 - \frac{\bar{\beta}}{\beta}\right)\frac{(x+\beta)\sqrt{\beta}}{(x-\beta)^2}\right]. \tag{D1.3}$$

The complex stress intensity K $[K \underset{x \to 0}{=} \sqrt{2\pi}\, A\pi\sqrt{-x}\, f(x)]$ is

$$K = -\frac{\sqrt{2\pi}}{\sqrt{\beta}}\left[\bar{a}\left(1 + \frac{\sqrt{\beta}}{\sqrt{\bar{\beta}}}\right) + \frac{a}{2}\left(1 - \frac{\bar{\beta}}{\beta}\right)\right]. \tag{D1.4}$$

Boundary Conditions at Interfaces

JOHN DUNDURS
Departments of Civil Engineering and Mechanical Engineering,
Northwestern University, Evanston, IL 60208, U.S.A.

Abstract

It is suggested that new insight can be gained in the mechanical conditions at an interface by thinking in terms of stretch strains and curvature changes. For simplicity, the discussion is restricted to elasticity and two dimensions, but particular attention is paid to eigenstrains.

1. Introduction

The conventional way to state the boundary conditions at a bonded interface between two phases is to say that tractions and displacements are to be continuous. Continuity of tractions is nothing else but Newton's third law and should not be touched. Continuity of displacements is another matter, as it can be replaced with conditions on strains and curvature change. This idea has many facets and is best introduced by a simple example. I have chosen for this purpose the interface between elastic and isotropic phases in two dimensions.

2. A Curvature Relation

Take a flat sheet and draw on it a smooth curve C with the curvature κ. Suppose now that the sheet is strained, with e_{ij} $(i, j = 1, 2)$ denoting the small strains in the sheet. The curvature of C then changes to $\kappa + \Delta\kappa$. The curvature change in terms of the strains is (Dundurs, 1989)

$$\Delta\kappa = 2\frac{\partial e_{ns}}{\partial s} - \frac{\partial e_{ss}}{\partial n} - \kappa e_{nn}, \tag{2.1}$$

where the strain components are referred to the tangential and normal directions. The sign convention is the following: Assign the curve C a direction. The arc coordinate s points in the direction of the curve, and the normal n points to the left when marching along C in the assigned direction. The curvature κ is reckoned positive when the center of curvature is on the side in which n points. The sign convention for the strain components e_{ss}, e_{ns}, and e_{nn} is the customary one applied to the (s, n) system. The derivation of (2.1) is tedious when approached formally (Dundurs, 1989), but can be greatly simplified when one has insight (Shield, 1988). It is also possible to visualize the individual terms in (2.1) by drawing some appropriate sketches.

Hooke's law, in two dimensions, is

$$e_{nn} = \frac{1}{2\mu}\{\sigma_{nn} - \tfrac{1}{4}(3-k)(\sigma_{nn}+\sigma_{ss})\} + \varepsilon_{nn} + \eta\varepsilon_{zz}, \tag{2.2}$$

$$e_{ns} = \frac{1}{2\mu}\sigma_{ns} + \varepsilon_{ns}, \tag{2.3}$$

$$e_{ss} = \frac{1}{2\mu}\{\sigma_{ss} - \tfrac{1}{4}(3-k)(\sigma_{nn}+\sigma_{ss})\} + \varepsilon_{ss} + \eta\varepsilon_{zz}, \tag{2.4}$$

where e is used to denote the total strain and ε the eigenstrain (Mura, 1982). The eigenstrain ε_{zz} is the normal strain perpendicular to the plane. Moreover, with ν denoting Poisson's ratio

$$k = 3 - 4\nu, \qquad \eta = \nu, \tag{2.5}$$

for plane strain, and

$$k = \frac{3-\nu}{1+\nu}, \qquad \eta = 0, \tag{2.6}$$

for plane stress. Substituting the strains into formula (2.1) for the curvature change

$$\Delta\kappa = \frac{1}{8\mu}\left\{8\frac{\partial\sigma_{ns}}{\partial s} - (k+1)\frac{\partial\sigma_{ss}}{\partial n} + (3-k)\frac{\partial\sigma_{nn}}{\partial n} + \kappa[(3-k)\sigma_{ss} - (k+1)\sigma_{nn}]\right\}$$
$$+ 2\frac{\partial\varepsilon_{ns}}{\partial s} - \frac{\partial\varepsilon_{ss}}{\partial n} - \kappa\varepsilon_{nn} - \eta\left(\frac{\partial\varepsilon_{zz}}{\partial n} + \kappa\varepsilon_{zz}\right). \tag{2.7}$$

This expression can be simplified by using the equilibrium condition

$$\frac{\partial\sigma_{ns}}{\partial s} + \frac{\partial\sigma_{nn}}{\partial n} + \kappa(\sigma_{ss} - \sigma_{nn}) = 0. \tag{2.8}$$

The result is

$$\Delta\kappa = \frac{1}{8\mu}\left\{(5+k)\frac{\partial\sigma_{ns}}{\partial s} - (k+1)\frac{\partial\sigma_{ss}}{\partial n} - 2\kappa(k-1)\sigma_{nn}\right\}$$
$$+ 2\frac{\partial\varepsilon_{ns}}{\partial s} - \frac{\partial\varepsilon_{ss}}{\partial s} - \kappa\varepsilon_{nn} - \eta\left(\frac{\partial\varepsilon_{zz}}{\partial n} + \kappa\varepsilon_{zz}\right), \tag{2.9}$$

3. Bonded Interface

The boundary conditions at a geometrically smooth interface between two bonded phases are

$$\sigma_{nn}^{(1)} = \sigma_{nn}^{(2)}, \tag{3.1}$$

$$\sigma_{ns}^{(1)} = \sigma_{ns}^{(2)}, \tag{3.2}$$

$$e_{ss}^{(1)} = e_{ss}^{(2)}, \tag{3.3}$$

$$\Delta\kappa_1 = \Delta\kappa_2, \tag{3.4}$$

where σ_{nn} and σ_{ns} are the stress components in the (s, n) system mentioned before. Clearly (3.1) and (3.2) imply continuity of tractions. Conditions (3.3) and (3.4) state that line elements at the interface must have equal stretch strains and curvature changes.

The next step is to substitute (2.4) and (2.9) into (3.3) and (3.4). The resulting expressions contain the three elastic constants μ_2/μ_1, k_1, and k_2. A reduction in the dependence on these constants can be achieved by dividing through with $(\mu_2/\mu_1)(k_1 + 1) + k_2 + 1$, and introducing the parameters

$$\alpha = \frac{\mu_2(k_1 + 1) - \mu_1(k_2 + 1)}{\mu_2(k_1 + 1) + \mu_1(k_2 + 1)}, \qquad \beta = \frac{\mu_2(k_1 - 1) - \mu_1(k_2 - 1)}{\mu_2(k_1 + 1) + \mu_1(k_2 + 1)}, \tag{3.5}$$

which are measures for the mismatches in uniaxial and bulk compliances in two dimensions (Dundurs, 1969). Thus (3.3) and (3.4) yield

$$\begin{aligned}(1 - \alpha)\sigma_{ss}^{(2)} - (1 + \alpha)\sigma_{ss}^{(1)} + 2(\alpha - 2\beta)\sigma_{nn} \\ = 16m\{\varepsilon_{ss}^{(2)} - \varepsilon_{ss}^{(1)} + \eta_2\varepsilon_{zz}^{(2)} - \eta_1\varepsilon_{zz}^{(1)}\},\end{aligned} \tag{3.6}$$

$$\begin{aligned}(1 - \alpha)\frac{\partial\sigma_{ss}^{(2)}}{\partial n} - (1 + \alpha)\frac{\partial\sigma_{ss}^{(1)}}{\partial n} + 2(3\alpha - 2\beta)\frac{\partial\sigma_{ns}}{\partial s} - 4\kappa\beta\sigma_{nn} \\ = 16m\left\{2\left(\frac{\partial\varepsilon_{ns}^{(2)}}{\partial s} - \frac{\partial\varepsilon_{ns}^{(1)}}{\partial s}\right) - \left(\frac{\partial\varepsilon_{ss}^{(2)}}{\partial n} - \frac{\partial\varepsilon_{ss}^{(1)}}{\partial n}\right) - \kappa(\varepsilon_{nn}^{(2)} - \varepsilon_{nn}^{(1)})\right. \\ \left. - \left(\eta_2\frac{\partial\varepsilon_{zz}^{(2)}}{\partial n} - \eta_1\frac{\partial\varepsilon_{zz}^{(1)}}{\partial n}\right) - \kappa(\eta_2\varepsilon_{zz}^{(2)} - \eta_1\varepsilon_{zz}^{(1)})\right\},\end{aligned} \tag{3.7}$$

where

$$m = \frac{\mu_1\mu_2}{\mu_2(k_1 + 1) + \mu_1(k_2 + 1)}. \tag{3.8}$$

Equations (3.6) and (3.7) are boundary conditions that enforce continuity of deformations. They can also be viewed as jump conditions: If the stress fields are known in phase 1, (3.6) yields the hoop stress $\sigma_{ss}^{(2)}$ and (3.7) its normal derivative $\partial\sigma_{ss}^{(2)}/\partial n$ in phase 2 on the other side of the interface. Jump conditions on stresses have been discussed before (Hill, 1961; Dundurs, 1969; Hill, 1972; Mura, 1982), but (3.6) is more explicit for the case discussed, and (3.7) appears to be new. Moreover, the jump conditions are seen to evolve from the simple idea about conformity of deformations expressed by (3.3) and (3.4), and it takes only a few steps in the derivations to get them.

4. Slipping Interface

The boundary conditions for a frictionless slipping interface are

$$\sigma_{nn}^{(1)} = \sigma_{nn}^{(2)}, \tag{4.1}$$

$$\sigma_{ns}^{(1)} = \sigma_{ns}^{(2)} = 0, \tag{4.2}$$

$$\Delta\kappa_1 = \Delta\kappa_2. \tag{4.3}$$

The last of these conditions (4.3) now yields

$$(1-\alpha)\frac{\partial\sigma_{ss}^{(2)}}{\partial n} - (1+\alpha)\frac{\partial\sigma_{ss}^{(1)}}{\partial n} - 4\kappa\beta\sigma_{nn} = 16m\left\{\quad\right\}, \tag{4.4}$$

where the brace on the right-hand side is the same as in (3.7). Since (3.3) was discarded, there is no condition on the jump in stresses, and only a condition on the normal derivatives of the hoop stresses survives.

5. Interface with Friction

The boundary conditions of a frictionless slipping interface (4.1)–(4.4) may be all right on a very small scale, when the interface can transmit some tension and the shear tractions disappear by some kind of relaxation mechanism. On larger scales, when there is no bonding between the phases, we would expect a much more complicated behavior influenced by friction.

A critical feature of a problem with friction is that the deformation state reached depends on the load path. This requires in essence that the loading process be followed in time.

Adopting Coulomb's law as the simplest model for friction, the boundary conditions in a stick zone are

$$g = 0, \tag{5.1}$$

$$V = 0, \tag{5.2}$$

$$\sigma_{nn} < 0, \tag{5.3}$$

$$|\sigma_{ns}| < -\rho\sigma_{nn}, \tag{5.4}$$

where g denotes the gap between the two phases and V is the slip velocity (counted positive if the phase on the positive side of n slips in the positive direction of s with respect to the other phase). Moreover, ρ denotes the coefficient of friction (in static elasticity, it is not possible to distinguish between static and kinetic friction). Condition (3.3) is lost, because the present stick zone may have been a slip zone at some previous time. However, (5.2) can be replaced with the condition

$$\dot{e}_{ss}^{(1)} = \dot{e}_{ss}^{(2)} \tag{5.5}$$

if the slip velocity V vanishes at one point. Hence (3.6) holds in a stick zone for the time derivatives of the quantities involved. If the gap g vanishes and the slopes match at one point, (3.4) implies (5.1), and consequently, (3.7) remains valid in a stick zone.

In an active slip zone, the boundary conditions are

$$g = 0, \tag{5.6}$$

$$\sigma_{ns} = -\rho\sigma_{nn}\,\mathrm{sgn}\,V, \tag{5.7}$$

$$\sigma_{nn} < 0. \tag{5.8}$$

The sign of V in (5.7) ensures that friction dissipates mechanical energy. Now there is no condition on the stretch strains, and (3.6) is lost. However, (5.6) again implies (3.7). Using (5.7), we can also replace the tangential derivative of the shear stress in (3.7) with that of the normal stress.

The boundary conditions in a separation zone are

$$\sigma_{nn} = 0, \tag{5.9}$$

$$\sigma_{ns} = 0, \tag{5.10}$$

$$g > 0. \tag{5.11}$$

Neither (3.6) nor (3.7) hold now.

It is known (Saeedvafa and Dundurs, 1988) that there are two fundamentally different kinds of dependence on the load path, which may be called loose and strict. For loading from a given initial state, loose dependence implies that the same deformation state is reached, regardless of what the actual path is, provided that the direction of the forward tangent at every point on the load path falls within certain bounds. Under the conditions of strict dependence, the deformation state reached depends on all details of the load path, and the direction of the forward tangent must be fully specified in the analysis. Loose dependence is associated with all slip zones expanding into stick zones. Strict dependence follows when a stick zone grows at the expense of either a slip or a separation zone. The importance of recognizing loose versus strict dependence ahead of time becomes particularly clear if we think about the numerical solution of problems: Loose dependence requires the solution of only one boundary value problem for a specified initial state and final loading conditions. In contrast, strict dependence requires the solution of a sequence of boundary value problems in an incremental formulation.

References

Dundurs, J. (1969), Discussion of a paper by D. B. Bogy, *J. Appl. Mech.*, **36**, 650–652.
Dundurs, J. (1989), Cavities vis-à-vis rigid inclusions and some general results in plane elasticity, *J. Appl. Mech.*, in press.
Hill, R. (1961), Discontinuity relations in mechanics of solids, in *Progress in Solid Mechanics*, Vol. 2, edited by I. N. Sneddon and R. Hill, North-Holland, Amsterdam.
Hill, R. (1972), An invariant treatment of interfacial discontinuities in elastic composites, in *Continuum Mechanics and Related Problems of Analysis*, edited by L. I. Sedov, Nauka, Moscow.
Mura, T. (1982), *Micromechanics of Defects in Solids*, Martinus Nijhoff, The Hague.
Saeedvafa, M. and Dundurs, J. (1988), An example for load-path dependence in elasticity with friction, *J. Mècan. thèor. appl.*, **7**, 211–226.
Shield, R. T. (1988), Private communication.

On Viscoplasticity and Continuum Dislocation Theory

MARTIN A. EISENBERG
Department of Aerospace Engineering,
Mechanics and Engineering Science, University of Florida,
Gainesville, FL 32611, U.S.A.

Abstract

Concepts of dislocation force and flux or velocity tensors introduced by Mura are related to external and self-equilibrating eigenstress fields and to lattice and dislocation-related deformations. Constitutive assumptions for a dislocation-based viscoplasticity theory are introduced and the results compared with macroscopic continuum formulations. Kinematic and kinetic decompositions of the deformation gradients are carefully discussed in the context of thermodynamical restrictions on the constitutive theory. The insights gained from the dislocation-theoretic approach are used to provide both a physical interpretation and means to evaluate properties derivable from the macroscopic theory but not easily understood without recourse to dislocation concepts.

1. Introduction

Continuum macroscopic constitutive theories and theories of microscale dislocation behavior have tended to be developed without the benefit of valuable insights derivable from the closely related disciplines. To be sure, there are notable exceptions to this trend, but instances of cross-fertilization are indeed exceptional. The continuum theory of dislocation behavior provides a natural bridge between the two disciplines.

Herein, concepts of dislocation force and flux or velocity tensors introduced by Mura (1963) are related to external and self-equilibrating eigenstress fields and to lattice and dislocation-related deformations. Constitutive assumptions for a dislocation-based viscoplasticity theory are introduced and the results compared with macroscopic continuum formulations. Kinematic and kinetic decompositions of the deformation gradients are carefully discussed in the context of thermodynamical restrictions on the constitutive theory. The insights gained from the dislocation-theoretic approach are used to provide both a physical interpretation and means to evaluate properties derivable from the macroscopic theory, but not easily understood without recourse to dislocation concepts.

2. Description of the Dislocated Continuum

Let the body under consideration undergo a motion which is described by the continuous displacement field, $u_i(x_j)$. For simplicity, we assume that the *displacement gradients* or *distortions* measured with respect to some convenient reference configuration

$$\beta_{ij} = \frac{\partial u_j}{\partial x_i}, \tag{2.1}$$

are small. Consider a small region of the body and let l_{ij} represent the outer product of the jth component of the length of an element of dislocation line which has crossed an area with unit normal in the i direction. The dislocation line is assumed to have a Burgers vector b_k associated with it. We define the *dislocation movement tensor*, N_{ijk}, by

$$N_{ijk} = \sum l_{ij} b_k, \tag{2.2}$$

where the summation is to be taken over all dislocations which have crossed the elemental surface area. This definition of N_{ijk} (Kröner and Rieder, 1956; Kröner, 1958; Nabarro, 1967) is seen to be an integral over time of the *dislocation velocity tensor*, V_{ijk} (Kröner, 1958; Mura, 1963, 1965a, b, 1968). In Mura's notation

$$V_{ijk} = \sum \mu V_i v_j b_k, \tag{2.3}$$

where μ is the density of a family of dislocations with a line direction v_j, a Burgers vector b_k, and a velocity V_i, and the summation is taken over all families of dislocations at the point under consideration. Since N_{ijk} may be obtained from V_{ijk} only by a path-dependent integration of

$$\dot{N}_{ijk} = V_{ijk}, \tag{2.4}$$

and since we shall subsequently write constitutive relations for V_{ijk}, it will become apparent that N_{ijk} depends upon the history of deformation in the region under consideration.

We now introduce the *state of dislocation tensor* (Nye, 1953) or the *dislocation density* α_{ij} as the sum of all b_j components of dislocations passing through the unit area normal to x_i. Mura (1968) has shown that

$$\begin{aligned} \dot{\alpha}_{ij} &= \varepsilon_{imn}\varepsilon_{pqn} V_{pqj,m} \\ &= V_{imj,m} - V_{mij,m}, \end{aligned} \tag{2.5}$$

where ε_{imn} is the permutation tensor. The second term in (2.5) is the negative of the divergence of V_{mij} with respect to x_m and represents an accumulation of dislocations due to transport into and out of the region under consideration, and the first term represents the creation of dislocations due to rotational motion of the dislocation lines. We may now define a tensor quantity

$$\dot{\beta}_{ij}^{\mathrm{D}} = -\varepsilon_{imn} V_{mnj}, \tag{2.6}$$

so that

$$\dot{\alpha}_{ij} = -\varepsilon_{imn}\dot{\beta}^{\mathrm{D}}_{nj,m}. \tag{2.7}$$

From (2.3) and (2.6) it follows that $\dot{\beta}^{\mathrm{D}}_{ij}$ represents the ith component of the vector product of the dislocation velocity with the dislocation line direction times the jth component of the Burgers vector. That is, $\dot{\beta}^{\mathrm{D}}_{ij}$ is the rate of plastic gliding of the x_i plane in the x_j direction or the *rate of change of dislocation distortion*. Consequently, we may interpret β_{ij} as

$$\beta_{ij} = \beta^{\mathrm{L}}_{ij} + \beta^{\mathrm{D}}_{ij}, \tag{2.8}$$

where β^{D}_{ij} is the dislocation distortion due to N_{ijk} and β^{L}_{ij} is the *lattice distortion*. The total displacement gradient may then be split into symmetric and antisymmetric parts

$$\beta_{ij} = \beta_{(ij)} + \beta_{[ij]}, \tag{2.9}$$

where $\beta_{(ij)}$ is the total strain ε_{ij} and $\beta_{[ij]}$ is the total rotation ω_{ij} of a material point.

Some simplification in the kinematic relations is possible if we consider certain special cases of interest. The dislocation velocity tensor may be written as

$$V_{ijk} = V_{(ij)k} + V_{[ij]k}, \tag{2.10}$$

when referred to the coordinate system x_i and, under coordinate rotation to x'_1, $V_{(ij)k} \to V'_{(ij)k}$ and $V_{[ij]k} \to V'_{[ij]k}$, since symmetry properties of tensors are preserved under coordinate transformation. Nabarro (1967) has observed that no physical significance may be attached to the components of V_{ijk} for which $i = j$, since they represent translation of a dislocation along its own line. If a coordinate system x'_i can be found such that $V_{(ij)k}$ vanished for $i \neq j$, then we may assert, without loss of generality, that V_{ijk} must be antisymmetric in i and j. In general, no such transformation exists; however, if V_{ijk} can be represented as

$$V_{ijk} = A_{ij}B_k = \sum_{\mu} (C^{(\mu)}_{ij})B_k \quad \text{or} \quad A_{ij}\sum_{\mu} D^{(\mu)}_k, \tag{2.11}$$

the desired coordinate system is the principal coordinate system associated with $A_{(ij)}$. It follows that if either the directions of the Burgers vectors of all families of dislocations are the same, or if the velocity and line directions of all dislocation families are the same but the Burgers vectors vary, then V_{ijk} can be assumed to be antisymmetric. In any event, the dislocation distortion rate depends only upon $V_{[ij]k}$, but the relation between α_{ij} and V_{ijk} (2.5) assumes the simple form

$$\dot{\alpha}_{ij} = 2V_{imj,m},$$

when the above conditions are satisfied.

The continuously dislocated continuum undergoes the described deformation under the influence of a symmetric stress field σ_{ij}. Thus the rate of plastic work is given by

$$\sigma_{ij}\dot{\varepsilon}_{ij}^{\mathrm{D}} = \sigma_{ij}\dot{\beta}_{ij}^{\mathrm{D}} = -\sigma_{ij}\varepsilon_{imn}V_{mnj}$$
$$= p_{mnj}V_{mnj},$$

where

$$p_{mnj} = -\sigma_{ij}\varepsilon_{imn}. \tag{2.12}$$

The tensor p_{mnj} represents the component of force per unit volume in the x_m direction, acting on dislocations whose lines are in the x_n direction and whose Burgers' vectors are in the x_j direction.

3. A Dislocation-Based Viscoplasticity Theory

We assume that there exists a range of values of *forces acting on each material element* of the body for which the deformations will be accomplished solely by distortion of the lattice structure. That is, $V_{ijk} = 0$ and hence $\dot{\beta}_{ij}^{\mathrm{D}} = 0$. *For such ranges of force*, changes in these lattice distortions are assumed to be related uniquely to changes in the force and temperature field. Beyond this range of *forces acting on each material element* the configuration of dislocations changes in response to the externally applied loads. Note that the *forces and associated stresses*[1] experienced locally at each material element consist of two components:

(1) An *externally imposed stress field*, σ_{ij}^{E}, that satisfies the nontrivial force boundary conditions for the body; and
(2) An *internal stress field*, σ_{ij}^{P}, that satisfies traction-free boundary conditions on the surface of the body and represents self-equilibrating stresses that are associated with internal eigenstrain (Mura, 1987) sources, such as discrete and/or continuously distributed dislocation fields associated with plastic flow mechanisms.

That is,

$$\sigma = \sigma^{\mathrm{E}} + \sigma^{\mathrm{P}}. \tag{3.1}$$

During plastic flow events we may envisage the externally applied stresses, σ^{E}, to generate changes in dislocation structure, β^{D}, and their corresponding internal stress fields, σ^{P}, which, in turn, create additional lattice distortions, β^{I}. Thus, during plastic flow events

$$\dot{\beta} = \dot{\beta}^{\mathrm{E}} + \dot{\beta}^{\mathrm{P}}, \tag{3.2}$$

where, in turn,

$$\dot{\beta}^{\mathrm{P}} = \dot{\beta}^{\mathrm{I}} + \dot{\beta}^{\mathrm{D}}. \tag{3.3}$$

[1] Where the contextual meaning is apparent we shall use the term "stresses" to refer to both forces and stresses for brevity in the ensuing discussion. Also, subscripts will *not* be explicitly written except where they contribute to clarity of the expression.

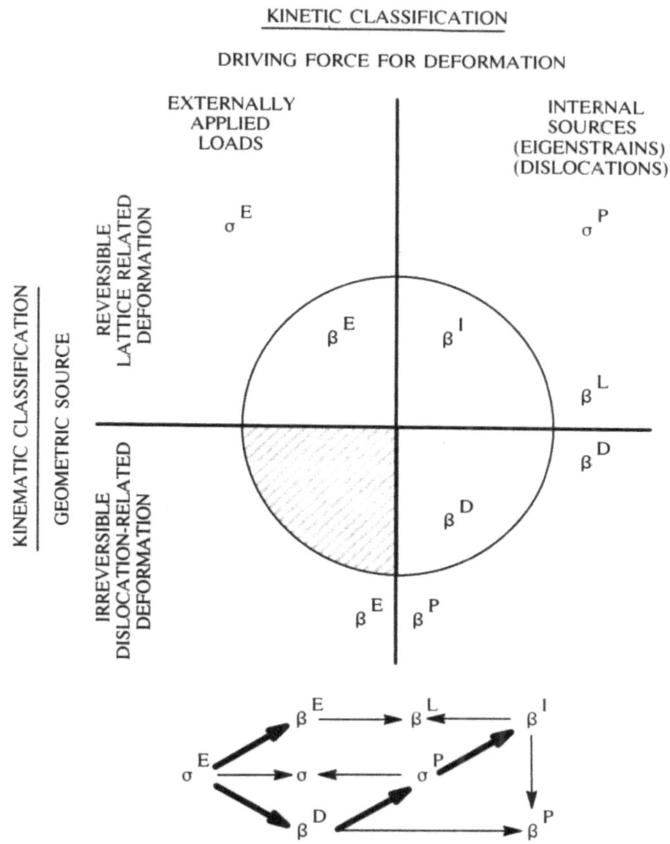

FIG. 1

The above concepts are summarized graphically in Fig. 1, and the *kinematic relations* in (2.8) and (3.3), and the *kinetic relations* in (3.2) above and (3.4) below

$$\dot{\beta}^{L} = \dot{\beta}^{E} + \dot{\beta}^{I}. \tag{3.4}$$

Note that any three of the relations (2.8), (3.2)–(3.4) may be considered fundamental, the fourth, a derivable corollary.[2]

To describe the range of forces for which the material behaves elastically, we introduce a yield function

$$f = f[p - p_0(N, \alpha, k^{(i)}, T)] \tag{3.5}$$

[2] It is essential that the fundamental but subtle differences among these deformation decompositions be clearly understood. Unfortunately, many if not most authors (including the present author) have not consistently distinguished these differences and myriad nomenclatures have contributed to confusion. We shall elaborate on these comments in the next section.

and corresponding critical value $\kappa(N, \alpha, k^{(i)}, T)$, which are assumed to be continuously differentiable functions of their variables. Here, T is the absolute temperature and p, N, and α are the dislocation force, dislocation movement, and dislocation density tensors, respectively, and $k^{(i)}$ are a finite number of such other parameters as may be necessary to characterize the effects of prior deformation history on the material response. The parameters, $k^{(i)}$, are assumed to be governed by evolution-type equations of the form

$$\dot{k}^{(i)} = g^{(i)}(p, \varepsilon^{\mathrm{P}}, k^{(i)}, T, \dot{\varepsilon}^{\mathrm{P}}), \tag{3.6}$$

where $g^{(i)}$ are assumed to be homogeneous of degree one in $\dot{\varepsilon}^{\mathrm{P}}$. Note that N and α can be subsumed in the list $k^{(i)}$. They are mentioned explicitly to highlight the role of the dislocation motion tensor and of the possible nonlocal[3] nature of the theory, since α is given by the gradient of N (see (2.5) and (2.7)).

For all values of f less than κ the deformation is assumed to be completely elastic. We now consider a 28-dimensional vector space whose positions are described by an orthogonal Cartesian coordinate system with axes corresponding to the 27 components of the dislocation force tensor f and the temperature T. The equation

$$f = f[p - p_0(N, \alpha, k^{(i)}, T)] = \kappa, \tag{3.7}$$

then represents a hypersurface in this force–temperature space, the interior of which describes the totality of conditions for which the material responds elastically. The location, shape, and size of this hypersurface or *yield surface* depends in general on the current values of N, α, $k^{(i)}$, and κ. Note that κ defines the size of the elastic domain while p_0 defines the location of the center of the domain. That is, in the limit as $\kappa \to 0$, $p \to p_0$.

Let there exist a point on the yield surface, p^* defined by the intersection of (3.7) and the radial line

$$p^* - p = c^2(p - p_0), \tag{3.8}$$

which connects p_0 and the current loading point p. The yield surface is assumed to be convex and thus *a fortiori* star-shaped with respect to p_0. Thus c and p^* are determined uniquely by (3.7) and (3.8). That is, to every state of dislocation force and temperature (p, T) there exists a corresponding force, temperature pair (p^*, T) which, for reasons that will become obvious, we denote the *quasi-static loading state.*

Let the distance between p and p^* in the hyperspace be denoted by the Euclidean norm

$$H = |p - p^*|. \tag{3.9}$$

From the definition of the yield surface, for all p within the surface ($f < \kappa$),

---

[3] In Section 4 additional sources of nonlocality will become evident.

there is no dislocation motion. We assume that for all p outside the yield surface ($f > \kappa$) the distance in dislocation force space, H, governs the magnitude of the rate of change of dislocation state, and that the yield function f serves as a potential function to define the direction of dislocation motion. That is, there exists an associated flow rule

$$\dot{N} = V = \gamma\langle\Phi(H)\rangle n(p^*), \tag{3.10}$$

where $n(p^*)$ is the unit external normal to the yield surface in the hyperspace defined by the current temperature, T, and γ and Φ are material functions to be determined, and $\langle\ \rangle$ represents Macaulay brackets. The quantity H represents a dislocation driving force over and above a critical threshold value. As $H \to 0$, $p \to p^*$, and $V \to 0$, hence the designation of (p^*, T) as the *quasi-static loading state*.

Equation (3.10) is the analog of equation (3) of Eisenberg and Yen's (1981) macroscopic theory of anisotropic viscoplasticity. Both relations represent generalizations of Malvern's (1951) uniaxial overstress theory and of the isotropic viscoplasticity theories based on the works of Bingham (1922), Hohenemser and Prager (1932), and Perzyna (1963). The direction of the dislocation velocity tensor, V, as determined by the normal to (3.7) at a point defined by (3.8), is identical to that predicted by an anisotropic generalization of Perzyna's theory. Chaboche and Rousselier's (1983) theory would predict V to be in the direction defined directly by line (3.8). By an alternative but similar constitutive assumption to (3.8), we may predict dislocation velocity directions in agreement with Phillips and Wu (1973). All the dislocation velocity tensor directions coincide for many loading paths and detailed constitutive hypotheses.

The behavior described by (3.10) in terms of dislocation velocity and force tensors, V and p, also represents a generalization of the uniaxial concept of relaxation boundary. Eisenberg *et al.* (1977) demonstrate that phenomena of relaxation to a deformation history-defined natural state, primary and secondary creep, and strain rate-dependent constitutive response are all subsumed in (3.10) and its analog (Eisenberg and Yen, 1981, Eqn. (3)).

4. Thermodynamics, Deformation Decomposition, and Dislocations

Eisenberg (1970) discussed the relation between continuum macroscopic rate-independent plasticity theories and dislocation theory. In Section 3, the rudiments of a viscoplasticity theory based on dislocation-theoretic concepts have been introduced and related to continuum theories of rate-dependent plastic flow. Further elaboration of the micro/macromechanical analogies for rate-dependent flow follow by arguments similar to those for the rate-independent case. Among the concepts discussed by Eisenberg (1970) were thermodynamic restrictions that led to the definition of a thermodynamic reference stress, σ^0, or thermodynamic reference dislocation force, p^0.[4]

Eisenberg *et al.* (1977) discussed the thermodynamic reference state and thermodynamic restrictions in much greater detail in the context of macroscopic thermoplasticity theories of both rate-dependent and rate-independent type. They demonstrated that there are two alternative strain decompositions that are suggested by the theory, but without explicit recourse to dislocation-theoretic concepts made an infelicitous choice of nomenclature that failed to reveal the physical significance of the alternative concepts.[5] Later, Eisenberg and Hartley (1982), making explicit use of the dislocation-theoretic formulation, succeeded in providing a more satisfying interpretation of the results derived in 1977. By summarizing these results below, in the context of more consistent nomenclature graphically represented in Fig. 1 and in the context of the constitutive hypotheses of the previous section, we hope to clarify further these subtle but important theoretical concepts.

Let there exist functions

$$\psi = \psi(\varepsilon^{\mathrm{E}}, \varepsilon^{\mathrm{P}}, k^{(i)}, T), \tag{4.1}$$

$$s = s(\varepsilon^{\mathrm{E}}, \varepsilon^{\mathrm{P}}, k^{(i)}, T), \tag{4.2}$$

$$q_k = q_k(\varepsilon^{\mathrm{E}}, \varepsilon^{\mathrm{P}}, k^{(i)}, T, T_{,k}), \tag{4.3}$$

where ψ is the specific Helmholtz free energy, s is the specific entropy, q_k is the outward heat flux vector, and $T_{,k}$, is the spatial temperature gradient. From the Clausius–Duhem inequality we conclude that

$$s = -\frac{\partial\psi}{\partial T}, \tag{4.4}$$

$$\sigma = \rho\frac{\partial\psi}{\partial\varepsilon^{\mathrm{E}}}. \tag{4.5}$$

Since incremental elastic response is governed by the fundamental lattice structure, barring massive disruption of this structure by dislocation motion, the incremental stress, elastic strain, and temperature response should be insensitive to plastic flow or dislocation motion. It can be shown (Eisenberg *et al.*, 1977) that under these conditions

$$\psi = \psi^{\mathrm{E}}(\varepsilon^{\mathrm{E}}, T) + \psi^{\mathrm{P}}(\varepsilon^{\mathrm{P}}, k^{(i)}) + \varepsilon^{\mathrm{E}}\phi(\varepsilon^{\mathrm{P}}, k^{(i)}) - [T - T^{a}(\varepsilon^{\mathrm{P}}, k^{(i)})]s^{\mathrm{P}}(\varepsilon^{\mathrm{P}}, k^{(i)}). \tag{4.6}$$[6]

---

[4] The dislocation force state, p_0, the quasi-static loading state, p^*, the thermodynamic reference state, p^0, and their corresponding stresses defined by (2.12) represent forces and stresses that are related to forces and stresses that have been variously described by the adjectives, "equilibrium," "back," "drag," ... and discussed at some length by Krempl (1987).

[5] The quantities identified as ε^{E} and ε^{P} in 1977 should more appropriately have been denoted ε^{L} and ε^{D} in our present notation.

[6] Hartley and Eisenberg define ϕ as the negative of the ϕ defined herein.

Note that there are four fundamental components to equation (4.6):

(1) an elastic contribution;
(2) a plastic contribution;
(3) an elastic–plastic coupling term, ϕ; and
(4) a temperature–plastic coupling term, s^{P}.

Only the first two terms were considered by Eisenberg (1970). Lubliner (1974) and Naghdi and Trapp (1974) derived relations neglecting the third and fourth terms, respectively. Lubliner attributed to s^{P} the physical interpretation of Cottrell's (1953) "configurational entropy," but no interpretation for ϕ was offered by Naghdi and Trapp.

If we further assume that the incremental thermoelastic response is linear, then

$$\dot{\varepsilon}^{\mathrm{E}} = S\dot{\sigma} - S\eta\dot{T} + \rho S\left(\frac{\delta\phi}{\delta\varepsilon^{\mathrm{P}}}\right)\dot{\varepsilon}^{\mathrm{P}}, \tag{4.7}$$

where

$$\frac{\delta}{\delta\varepsilon^{\mathrm{P}}} = \frac{\partial}{\partial\varepsilon^{\mathrm{P}}} + \frac{\partial}{\partial k^{(i)}}\frac{\partial k^{(i)}}{\partial\varepsilon^{\mathrm{P}}}. \tag{4.8}$$

From Fig. 1

$$\dot{\varepsilon}^{\mathrm{E}} = S\dot{\sigma}^{\mathrm{E}} - S\eta\dot{T}, \tag{4.9}$$

and

$$\dot{\varepsilon}^{\mathrm{I}} = S\dot{\sigma}^{\mathrm{P}}, \tag{4.10}$$

thus

$$\dot{\varepsilon}^{\mathrm{L}} = S\dot{\sigma} - S\eta\dot{T}. \tag{4.11}$$

Here S and η represent elastic compliance and thermoelastic coefficients, respectively. Note that it is somewhat arbitrary as to whether to assign the $S\eta\dot{T}$ term to $\dot{\varepsilon}^{\mathrm{E}}$ or to $\dot{\varepsilon}^{\mathrm{I}}$, for we can think of the temperature as external driving force or as an eigenstrain source. Similarly, it should be noted that in deriving (2.7) we omitted consideration of contributions to α by the plastic rotations associated with disclination fields (Mura, 1987). Restriction of the eigenstress field to dislocations better suits our current purposes.

From (4.7) and (4.11)

$$\dot{\varepsilon}^{\mathrm{E}} = \dot{\varepsilon}^{\mathrm{L}} + \rho S\left(\frac{\delta\phi}{\delta\varepsilon^{\mathrm{P}}}\right)\dot{\varepsilon}^{\mathrm{P}}. \tag{4.12}$$

From (3.4), (4.10), and (4.12)

$$\dot{\varepsilon}^{\mathrm{I}} = -\rho S\left(\frac{\delta\phi}{\delta\varepsilon^{\mathrm{P}}}\right)\dot{\varepsilon}^{\mathrm{P}}, \tag{4.13}$$

and

$$\sigma^{\mathrm{P}} = -\rho\left(\frac{\delta\phi}{\delta\varepsilon^{\mathrm{P}}}\right)\dot{\varepsilon}^{\mathrm{P}}. \tag{4.14}$$

Thus, both the "lattice strain," ε^{L}, and the "elastic strain," ε^{E}, are entitled to the sobriquet, "elastic"—the former in the hypoelastic, the latter in the hyperelastic sense. When the dislocation velocity tensor, V, vanishes, both $\dot{\beta}^{\mathrm{D}}$ and $\dot{\beta}^{\mathrm{I}}$ vanish, so there is no distribution between the ε^{L} and ε^{E} in that important case.

In principle, evaluation of the elastic–plastic coupling follows from the identification of a Green's function consistent with the dislocation velocity distribution and the computation from it of a projection operator (Simmons and Bullough, 1970) from which we may, in turn, evaluate the lattice strain associated with the internal eigenstress field. By quadratures we may compute the contribution to the Helmholtz potential which assumes the form

$$\varepsilon^{\mathrm{E}}\phi = \frac{1}{\rho}\bar{K}\varepsilon^{\mathrm{E}}\varepsilon^{\mathrm{P}}, \tag{4.15}$$

where

$$\bar{K}_{rspq}\varepsilon^{\mathrm{P}}_{pq} = \int K_{rspq}\delta(\varepsilon^{\mathrm{P}}_{pq}), \tag{4.16}$$

and

$$K_{rspq} = -\rho\left[\frac{\delta\phi_{rs}}{\delta\varepsilon^{\mathrm{P}}_{pq}}\right]. \tag{4.17}$$

Details of the derivation are presented in Hartley and Eisenberg (1982).

Finally, it should be noted that the Clausius–Duhem inequality imposes certain inequality constraints on the constitutive equations. Specifically,

$$(\sigma - \sigma^0)\dot{\varepsilon}^{\mathrm{P}} \geqq 0, \tag{4.18}$$

where

$$\sigma^0 = \sigma^a - (T - T^a)C + K(\varepsilon^{\mathrm{E}} - \varepsilon^{\mathrm{E}a}), \tag{4.19}$$

$$C = \rho\cdot\frac{\delta s^{\mathrm{P}}}{\delta\varepsilon^{\mathrm{P}}}, \tag{4.20}$$

$$\sigma^{\mathrm{a}} = \rho\left[\frac{\delta\psi^{\mathrm{P}}}{\delta\varepsilon^{\mathrm{P}}} + \varepsilon^{\mathrm{Ea}}\frac{K}{\rho} + \frac{\delta T^a}{\delta\varepsilon^{\mathrm{P}}}s^{\mathrm{P}}\right], \tag{4.21}$$

where σ^a and $\varepsilon^{\mathrm{E}a}$ denote the stresses and elastic strains at the high-temperature limit of the elastic domain in stress–temperature space. Eisenberg (1970), based on a simplified theory in which both s^{P} and ϕ vanished, argued for the plausible identification of σ^0 with the internal stress, σ^{P}, and with the back stress. The situation is less clear in the more general theory. Still, with the aid of dislocation-theoretic concepts we can gain some valuable insights. Neglecting, for the time being, the configurational entropy, note that the last term in (4.19) is the product of K and the difference in elastic strain from the high-temperature limit. The quantity K represents the rate of change of σ^{P} with plastic strain and is thus of the order of a plastic tangent modulus, E_{t}. The strain difference is bounded by a quantity of the order of the size of the elastic domain, D, divided by a quantity of the order of the elastic modulus, E. Thus, the correction term to σ^0 is of the order of $DE_t/E \ll D$. Thus, there is reason to believe that the earlier arguments in favor of identifying the thermodynamic reference stress with the yield surface center, the back stress, and the internal eigenstress appear still to be qualitatively reasonable.

5. Closure

The continuum dislocation-theoretic formalism has been coupled with the theory of continuum thermoviscoplasticity and shown to provide useful insights and heuristic guidance in the formulation of the macroscopic theory. Two closely allied yet distinct definitions of "elastic" strain have been examined, and their proper respective roles relative to the constitutive formulation have been critically assessed. The implications of an elastic/plastic coupling term in the Helmholtz potential have been interpreted. Although the relation between this coupling term and the internal eigenstress has been established, it remains to carry out quantitative evaluations for particular dislocation configurations in order to realize the full synergistic potential of the merging of the two mainstreams of macroscopic and microscopic analysis of plastic flow.

References

Bingham, E. C. (1922), *Fluidity and Plasticity*, McGraw-Hill, New York.

Chaboche, J. L. and Rousselier, G. (1983), On the plastic and viscoplastic constitutive equations, Parts I and II, *Trans. ASME J. Pressure Vessel Technol.*, **105**, 153–164.

Cottrell, A. (1953), *Dislocations and Plastic Flow in Crystals*, Oxford University Press, London.

Eisenberg, M. A. (1970), On the relation between continuum plasticity and dislocation theories, *Int. J. Engng. Sci.*, **8**, No. 3, 261–371.

Eisenberg, M. A., Lee, C. W., and Phillips, A. (1977), Observations on the theoretical and experimental foundations of thermo-plasticity, *Int. J. Solids Structures*, **13**, 1239–1255.

Eisenberg, M. A. and Yen, C. F. (1981), A theory of multiaxial anisotropic viscoplasticity, *J. Appl. Mech.*, **48**, 276–284.

Eisenberg, M. A. and Hartley, C. S. (1982), Phenomenological plasticity and dislocation models of deformation, in *Proceedings, Acta/Scripta Metallurgica International Conference on Dislocation Modelling of Physical Systems*, edited by M. F. Ashby *et al.*, Pergamon Press, Gainesville, FL., April 1980.

Hohenemser, K. and Prager, W. (1932), Über die Ansätze der Mechanik isotroper Kontinua, *Z. Angew. Math. Mech.*, **12**, 216–226.

Krempl, E. (1987), Models of viscoplasticity, some comments on equilibrium (back) stress and drag stress, *Acta Mechanica*, **69**, No. 5. 1–4, 25–42.

Kröner, E. and Rieder, G. (1956), Kontinuumsteorie der Verstezungen, *Z. Physik*, **145**, 424–429.

Kröner, E. (1958), *Kontinuumsteorie der Versetzungen und Eigenspannungen*, Springer-Verlag, Berlin.

Lubliner, J. (1974), A simple theory of plasticity, *Int. J. Solids Structures*, **10**, 313–319.

Malvern L. E. (1951), The propagation of longitudinal waves of plastic deformation in a bar of material exhibiting a strain-rate effect, *ASME J. Appl. Mech.*, **18**, 203–208.

Mura, T. (1963), On dynamic problems of continuous distribution of dislocations, *Int. J. Engng. Sci.*, **1**, 371–381.

Mura, T., (1965a), Continuous distribution of dislocations and the mathematical theory of plasticity, *Physica Status Solidi*, **10**, 447–453.

Mura, T. (1965b), Continuous distribution of dislocations and the mathematical theory of plasticity (II), *Physica Status Solidi*, **11**, 683–688.

Mura, T. (1968), Continuum theory of dislocations and plasticity, in *Proceedings IUTAM Symposium on Mechanics of Generalized Continua*, edited by E. Kröner, Springer-Verlag, Berlin, pp. 269–278.

Mura, T. (1987), *Micromechanics of Defects in Solids*, 2nd ed., Martinus Nijhoff, Dordrecht.

Nabarro, F. R. N. (1967), *Theory of Crystal Dislocations*, Chap. 1, Oxford University Press, London.

Naghdi, P. M. and Trapp, J. A. (1974), On finite elastic–plastic deformation of metals, *J. Appl. Mech.*, **41**, 254–260.

Nye, J. F. (1953), Some geometrical relations in dislocated crystals, *Acta Metallurgica*, **1**, 153–162.

Perzyna, P. (1963), The constitutive equations for rate sensitive plastic materials, *Quart. Appl. Math.*, **20**, 321–332.

Phillips, A. and Wu, H. C. (1973), A theory of viscoplasticity, *Int. J. Solids Structures*, **9**, 15–30.

Simmons, J. A. and Bullough, R. (1970), *Fundamental Aspects of Dislocation Theory, NBS Special Publication* 317, *Volume I*, National Bureau of Standards, Washington, DC, p. 89.

A Circular Inclusion with Slipping Interface in Plane Elastostatics

T. HONEIN and G. HERRMANN
Division of Applied Mechanics, Stanford University, Stanford, CA 94305-4040, U.S.A.

Abstract

It is shown that the solution, in plane elastostatics, for an infinite domain subjected to arbitrary loading and into which a circular inclusion of a different elastic material has been inserted (heterogeneous problem) may be obtained, in the case of a slipping interface, from the solution of the corresponding homogeneous problem (i.e., when there is no inclusion) merely by a single quadrature and simple algebraic manipulations. This novel procedure of heterogenization is illustrated by several specific examples.

1. Introduction

The purpose of present paper is to apply the procedure of heterogenization for the solution of an infinite domain (matrix) subjected to arbitrary loading and into which a circular inclusion of a different elastic material has been inserted, assuming a slipping interface between the inclusion and the matrix.

This novel procedure consists in expressing the solution of the heterogeneous problem in terms of that of the homogeneous one, i.e., the solution for the complete infinite plane occupied by the matrix material and subjected to the same loading. This heterogenization procedure is based, for problems with circular or straight boundaries, on a recent paper by Honein and Herrman (1988a). There it was shown that the problem of plane elastostatics for a circular disk and that of an infinite region with a circular hole, subjected either to the same traction or same displacement along the common boundary, are related by an involution. Also, it was shown in the same paper that the solution of either problem, with singularities within the region, is related to the corresponding problem for the complete infinite region (subjected to the same singularities). Thus, knowing the solution for an infinite region, the solution for an infinite region with a circular hole can be obtained by mere substitution into a simple algebraic expression, assuming, of course, that all singularities lie outside the prospective hole. If all singularities, assumed now self-equilibrating, reside within a circular region then the solution for the disk is obtained by substitution into the same expression.

In a subsequent paper (Honein and Herrmann, 1988b), the authors have extended the above-mentioned results to the case when, instead of a hole, a circular elastic inclusion perfectly bonded to the matrix is present. It was shown there, that the transformation for the hole remains essentially valid provided the different terms involved are weighted by some coefficients which depend only on the material moduli. Thus the procedure of heterogenization consists, in this instance, in performing essentially some algebraic manipulations.

In contrast with the previous cases, the more complicated heterogenization procedure for a circular inclusion with a slipping interface will involve, in addition to some essentially algebraic manipulations, a single quadrature.

It is found, in formulating the aforementioned problems, most convenient to employ the Kolosov–Muskhelishvili representation, allowing thus for lucidity and compactness. This formalism is reviewed very briefly in Section 2. Then in Section 3, after a brief review of some notions introduced in Honein and Herrmann (1988a), we proceed to derive the transformation relating the solution, for an infinite domain subjected to arbitrary loading and into which a circular elastic inclusion with a slipping interface has been inserted, to that of the homogeneous problem. Several specific examples to illustrate this new procedure are presented in Section 4 and finally some conclusions are drawn in Section 5.

2. Review of Fundamental Relations of Plane Elastostatics

In this section we review very briefly the basic relations of plane elastostatics, using the Kolosov–Muskhelishvili formalism. For details the reader is referred to Muskhelishvili (1953) or Sokolnikoff (1983).

We consider a linear elastic body (homogeneous and isotropic) occupying an open region Ω of $\mathbf{R}^2$ with piecewise smooth boundary $\partial\Omega$ consisting of a finite number of contours Γ_j, $j = 1, 2, \ldots, n$. A point on one of these contours will be specified by prescribing the arc length s measured from a reference point on that contour.

Let $t_\alpha(s)$ $(\alpha = 1, 2)$ denote the components of traction at a point of the boundary with respect to a Cartesian system of coordinates Ox_1x_2. If we define (see Sokolnikoff, 1983)

$$f_1 + if_2 = i\int_0^s [t_1(\lambda) + it_2(\lambda)]\, d\lambda,$$

where the sense of integration is always such that the material is on the left as λ increases, then the first boundary value problem of plane elastostatics is reduced to the determination of the complex Kolosov–Muskhelishvili potentials $\phi(z)$, $\psi(z)$. These are subjected to the boundary condition

$$\phi(z) + z\overline{\phi'(z)} + \overline{\psi(z)} = f_1 + if_2 + \text{constant} \quad \text{on } \partial\Omega,$$

where the constants have different values on each of the contours forming the boundary $\partial\Omega$. The components of stress in polar coordinates are related to ϕ and ψ by

$$\sigma_{rr} + \sigma_{\theta\theta} = 2[\phi'(z) + \overline{\phi'(z)}] = 4\,\mathrm{Re}[\phi'(z)], \tag{2.1a}$$

$$\sigma_{\theta\theta} - \sigma_{rr} + 2i\sigma_{r\theta} = 2[\bar{z}\phi''(z) + \psi'(z)]e^{2i\theta}, \tag{2.1b}$$

where Re designates the real part of the argument.

The components of displacement u_α in Cartesian coordinates are given by

$$2\mu(u_1 + iu_2) = \kappa\phi(z) - z\overline{\phi'(z)} - \overline{\psi(z)},$$

and in polar coordinates by

$$2\mu(u_r + iu_\theta) = e^{-i\theta}(\kappa\phi(z) - z\overline{\phi'(z)} - \overline{\psi(z)}), \tag{2.2}$$

where μ is the shear modulus and

$$\kappa = 3 - 4\nu, \qquad \text{for plane strain, and}$$

$$\kappa = \frac{3 - \nu}{1 + \nu}, \qquad \text{for plane stress,}$$

ν being Poisson's ratio.

3. Derivation of the Transformation

We consider now a circular elastic inclusion, with a slipping interface placed into an infinite matrix, subjected to an elastic field whose sources are in the matrix (including infinity), so that the elastic field is free of singularities inside or on the boundary of the circular inclusion.

We propose to express the solution to this problem in terms of the corresponding much simpler problem of an infinite homogeneous body subjected to the same loading. It will be shown that the transformation leading from one solution to the other is obtained by performing a single quadrature and some essentially algebraic operations. This procedure is quite systematic and can be completely automated.

The elastic moduli of the inclusion are denoted by μ_{I}, κ_{I} while those of the matrix are designated by μ_{M}, κ_{M}, μ being the shear modulus and κ as defined in the previous section.

We recall here briefly some notations which were introduced in the paper by Honein and Herrmann (1988a). Let u be a function of polar coordinates r, θ. The Kelvin transformation in the plane is defined by

$$\tilde{u}(r, \theta) = u\left(\frac{a^2}{r}, \theta\right).$$

The importance of this transformation lies in the fact that if u is a harmonic function on the disk $r < a$, then $\tilde{u}$ is a harmonic function on the region $r > a$, including the point at infinity and vice versa. If u is regarded as the real part of a holomorphic function

$$f(z) = u(r, \theta) + iv(r, \theta),$$

then the "hat" transformation is defined by

$$\hat{f}(z) = \overline{f\left(\frac{a^2}{\bar{z}}\right)} = u\left(\frac{a^2}{r}, \theta\right) - iv\left(\frac{a^2}{r}, \theta\right) = \tilde{u}(r, \theta) - i\tilde{v}(r, \theta).$$

The hat transformation maps holomorphic functions defined on $r < a$ into holomorphic functions defined on $r > a$, including the point at infinity and vice versa.

Let the Kolosov–Muskhelishvili complex potentials for the homogeneous problem be denoted by ϕ, ψ, then the most general solution which does not alter the singularities, after the inclusion has been introduced, can be written as

$$\phi_{\mathrm{I}} = \phi + g,$$
$$\psi_{\mathrm{I}} = \psi + k,$$
$$\phi_{\mathrm{M}} = \phi + \hat{f},$$
$$\psi_{\mathrm{M}} = \psi + \hat{h},$$

where g, k, f, and h are holomorphic inside the disk $r < a$.

The additional terms g, k, f, and h must introduce the same tractions along the common boundary and thus must be related by the involution correspondence (Honein and Herrmann, 1988a) i.e., we must have

$$f = k + \frac{a^2}{z}(g'(z) - g'(0)),$$

$$h = g + z\overline{g'(0)} + \frac{z^3}{a^2} f'.$$

The remaining conditions which determine g and k are:

(1) The shear stress is zero along the interface; and
(2) The normal displacement u_r is continuous across the interface.

The net result is recorded in the following theorem whose detailed proof in a more general setting will be presented in a forthcoming paper.

THEOREM. *The solution of the inclusion problem with a slipping interface, when all singularities are in the matrix, is related to that of the corresponding homo-*

geneous problem (ϕ, ψ) *(i.e., when the complete region is occupied by the matrix material) by*

$$\phi_I = \phi + g,$$
$$\psi_I = \psi + k,$$
$$\phi_M = \phi + \hat{f},$$
$$\psi_M = \psi + \hat{h},$$

where g is determined by

$$(1+\eta)zg' - 2g = 2\alpha\phi - (\alpha + \beta\eta)z\phi' - (1-\alpha)\left[z^2\phi'' + \frac{z^3}{a^2}\psi' + \gamma\phi'(0)z\right], \tag{3.1}$$

and k, f, and h are given in terms of g by

$$k = -\beta\psi + \frac{1}{1+\eta}\frac{a^2}{z^2}\left\{\frac{1-\eta}{\eta}[g + \alpha\phi + (1-\alpha)z\phi'] + (1-\alpha)\left(z^2\phi'' + \frac{z^3}{a^2}\psi'\right) - \gamma(1-\beta)\phi'(0)z\right\},$$

$$f = k + \frac{a^2}{z}(g' - g'(0)),$$

$$h = g + zg'(0) + \frac{z^3}{a^2}f',$$

where ϕ *is chosen such that* $\phi'(0)$ *is real.*

Here α, β, η, *and* γ *are the following combinations of the material moduli:*

$$\alpha = \frac{\mu_M\kappa_I - \mu_I\kappa_M}{\mu_I + \mu_M\kappa_I},$$
$$\beta = \frac{\mu_M - \mu_I}{\mu_M + \mu_I\kappa_M},$$
$$\eta = \frac{\mu_M + \mu_I\kappa_M}{\mu_I + \mu_M\kappa_I} = \frac{1-\alpha}{1-\beta},$$
$$\gamma = \frac{\mu_M(\kappa_I - 1) - \mu_I(\kappa_M - 1)}{2\mu_I + \mu_M(\kappa_I - 1)} = \frac{\alpha - \beta}{(1-\beta) - \beta(1-\alpha)},$$

a being the radius of the inclusion, and the origin is taken to be at the center of the inclusion.

The problem reduces to the determination of g which is governed by a first-order equidimensional (Euler) inhomogeneous ordinary differential equation whose solution can be written down explicitly and involves a single quadrature, namely,

$$g = \frac{\xi}{2} z^{\xi} \int^{z} t^{-\xi-1} R(t)\, dt,$$

where the constant of integration must be chosen such as to make g analytic in the disk $r < a$. Here ξ, is defined by

$$\xi = \frac{2}{1+\eta},$$

and R designates the right-hand side of (3.1).

It must be noted that, for certain types of loading, the normal stress may be positive along some parts of the circular interface which is not admissible in the present formulation. This situation may be avoided by superimposing on the previous solution the one of an insertion of a sufficiently large radius. For the case of identical materials, the foregoing equations simplify considerably and we obtain

$$\begin{aligned}
g &= -\frac{z}{2}\left(\phi' + \frac{z}{a^2}\psi - \frac{1}{a^2}\chi\right), \\
k &= \tfrac{1}{2}(a^2\phi'' + z\psi'), \\
f &= -\frac{a^2}{2z}\left(\phi' + \frac{z}{a^2}\psi - \frac{1}{a^2}\chi\right), \\
h &= -\frac{z^2}{2}\left(\phi'' + \frac{z}{a^2}\psi'\right),
\end{aligned}$$

where χ is the antiderivative of ψ (i.e. $\chi' = \psi$) satisfying $\chi(0) = a^2\phi'(0)$.

4. Examples

4.1. Elastic Inclusion Under Remote Uniform Traction

As our first example we consider a circular elastic inclusion with a slipping interface placed into a matrix which is subjected to uniform traction of intensity T and cT at infinity along the x and y axes, respectively.

The solution of the corresponding homogeneous problem is trivially given as

$$\phi(z) = \frac{1+c}{4} Tz, \qquad \psi(z) = \frac{-1+c}{2} Tz.$$

An immediate application of the formulas given in the previous section yields the solution as

$$\phi_I(z) = (1-\gamma)\frac{1+c}{4}Tz + \frac{1-\alpha}{1+3\eta}\frac{1-c}{2}T\frac{z^3}{a^2},$$

$$\psi_I(z) = \frac{3(1-\alpha)}{1+3\eta}(-1+c)Tz,$$

$$\phi_M(z) = \frac{1+c}{4}Tz + \frac{[3(\eta+\alpha)-2](1-c)}{2}T\frac{a^2}{z},$$

$$\psi_M(z) = \frac{-1+c}{2}Tz - \gamma\frac{1+c}{2}T\frac{a^2}{z}.$$

For $c = 0$, this corresponds to the same solution obtained by Muskhelishvili (1953, pp. 215–217), essentially by guessing.

4.2. *Center of Dilatation or Concentrated Moment*

We consider the complex potentials given by

$$\phi(z) = 0, \qquad \psi(z) = \frac{\Gamma}{z - x_0},$$

where $x_0 > a$.

This corresponds to a center of dilatation or a concentrated moment acting at $(x_0, 0)$ if Γ is real or purely imaginary, respectively (Fig. 1).

In this case, we have

$$R(t) = \frac{(1-\alpha)\Gamma}{a^2}\frac{t^3}{(t-x_0)^2}.$$

In order to carry out the integration explicitly, we expand $R(t)$ in power series yielding

$$R(t) = \frac{(1-\alpha)\Gamma}{a^2}\sum_{n=1}^{+\infty}\frac{nt^{n+2}}{x_0^{n+1}}.$$

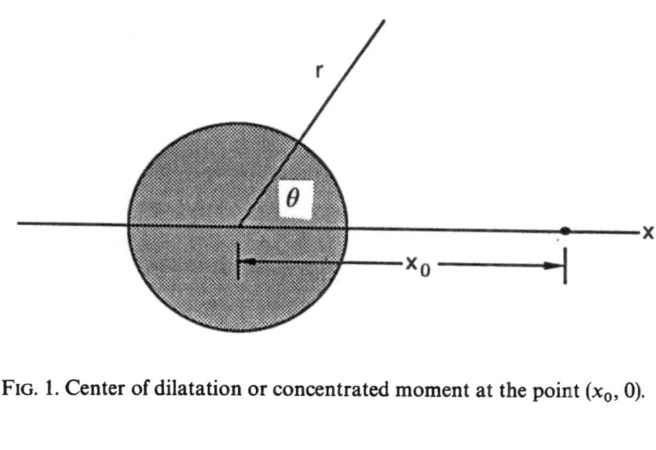

FIG. 1. Center of dilatation or concentrated moment at the point $(x_0, 0)$.

Then the solution is easily obtained as

$$g = \frac{\xi(1-\alpha)\Gamma}{2a^2} \sum_{n=1}^{+\infty} \frac{nz^{n+2}}{(n-\xi+2)x_0^{n+1}},$$

$$k = \frac{-\beta\Gamma}{z-x_0} - \frac{(1-\alpha)\xi\Gamma z}{2(z-x_0)^2} + (\alpha-\beta)\frac{\xi^2}{4}\Gamma \sum_{n=1}^{+\infty} \frac{nz^n}{(n-\xi+2)x_0^{n+1}},$$

$$f = \frac{-\beta\Gamma}{z-x_0} - \frac{(1-\alpha)\xi\Gamma z}{2(z-x_0)^2} + \frac{\xi^2}{4}(\alpha-\beta)\Gamma \sum_{n=1}^{+\infty} \frac{nz^n}{(n-\xi+2)x_0^{n+1}}$$

$$+ \frac{\xi}{2}(1-\alpha)\Gamma \sum_{n=1}^{+\infty} \frac{n(n+2)z^n}{(n-\xi+2)x_0^{n+1}},$$

$$h = \frac{\beta\Gamma z^3}{a^2(z-x_0)^2} + \frac{(1-\alpha)\xi\Gamma(z+x_0)z^3}{2a^2(z-x_0)^3} + \frac{\xi}{2}\frac{(1-\alpha)}{a^2}\Gamma \sum_{n=1}^{+\infty} \frac{n(n+1)^2 z^{n+2}}{(n-\xi+2)x_0^{n+1}}$$

$$+ \frac{\xi^2}{4a^2}(\alpha-\beta)\Gamma \sum_{n=1}^{+\infty} \frac{n^2 z^{n+2}}{(n-\xi+2)x_0^{n+1}},$$

from which the stresses can be obtained as Fourier series without great difficulty.

4.3. Point Force

As our final example, we consider a point force of intensity F parallel to the x axis and acting at the point $(-x_0, 0)$ (Fig. 2). In order to simplify the results, we treat the case of two identical materials.

The solution of the homogeneous problem is given by

$$\phi(z) = -\frac{F}{2\pi(1+\kappa)}\log(z+x_0),$$

$$\psi(z) = \frac{\kappa F}{2\pi(1+\kappa)}\log(z+x_0) - \frac{x_0 F}{2\pi(1+\kappa)(z+x_0)}.$$

We choose a branch cut of the logarithm that does not intersect the

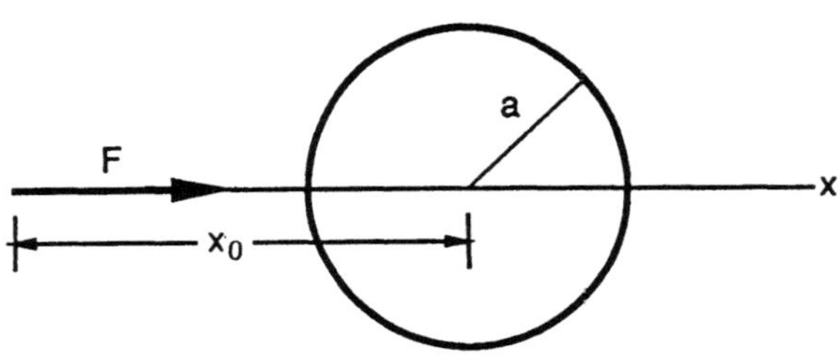

FIG. 2. Point force acting at $(-x_0, 0)$.

inclusion (e.g., the principal branch), and we obtain the solution as

$$g = -\frac{zF}{4\pi(1+\kappa)}\left((1-\kappa)\frac{x_0}{a^2}\ln\left(1+\frac{z}{x_0}\right)-\frac{1}{z+x_0}\left(1+\frac{x_0}{a^2}z\right)+\frac{\kappa z}{a^2}+\frac{1}{x_0}\right),$$

$$k=\frac{F}{4\pi(1+\kappa)}\left(\frac{x_0 z+a^2}{(z+x_0)^2}+\frac{\kappa z}{z+x_0}\right),$$

$$f=\frac{a^2}{z^2}g,$$

$$h=-\frac{z^2}{a^2}k.$$

The jump in the tangential displacement along the interface is given by

$$[u_\theta] = \frac{F}{4\pi\mu}\left\{(1-\kappa)\varepsilon\,\mathrm{Arctan}\left(\frac{\sin\theta}{\varepsilon+\cos\theta}\right)+\left(\kappa+\frac{1-\varepsilon^2}{1+\varepsilon^2+2\varepsilon\cos\theta}\right)\sin\theta\right\},$$

where $[u_\theta]$ is defined by

$$[u_\theta] = u_\theta(a^+,\theta) - u_\theta(a^-,\theta),$$

and ε is the ratio x_0/a.

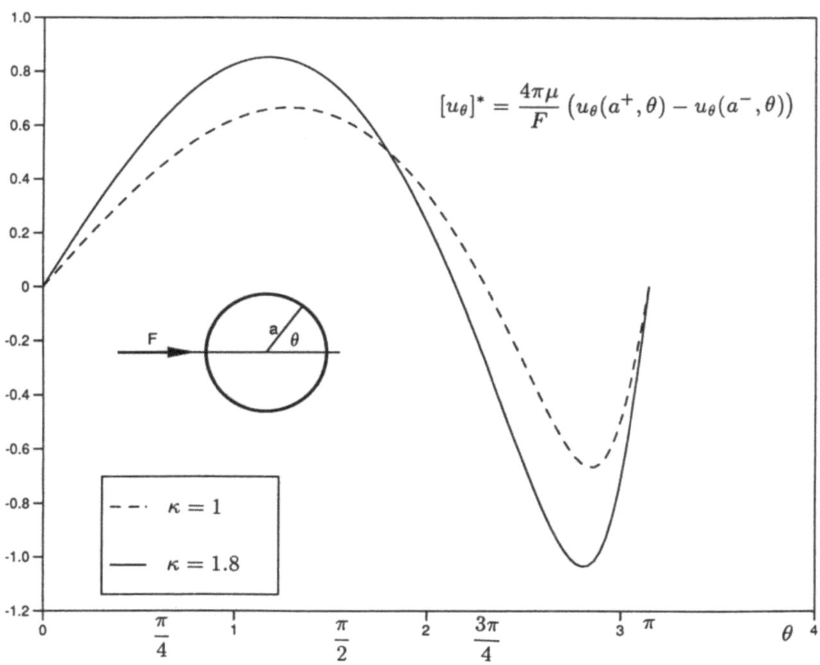

FIG. 3. General behavior of the circumferential displacement discontinuity along the circular interface for a point force.

The dimensionless quantity $[u_\theta]^* = (4\pi\mu/F)[u_\theta]$ is sketched in Fig. 3 for $\varepsilon = 1.5$ and $\kappa = 1, 1.8$. Note that $\kappa = 1$ corresponds to incompressible materials in plane strain.

5. Conclusions

In this paper, a systematic procedure has been derived allowing the expression, of the solution of a circular elastic inclusion with a slipping interface inserted into an infinite matrix and subjected to arbitrary loading, in terms of the solution of the corresponding homogeneous problem (i.e., when there is no inclusion).

This heterogenization procedure involves a single quadrature and simple algebraic manipulations. Several illustrative examples have been given.

The solution for a single inclusion advanced here can be used as a building block for the problem of several inclusions using, for example, the Schwarz alternating method, see Sokolnikoff (1983).

The relevance of the present analysis and the results obtained may be found in phenomena of grain boundary sliding, relaxation, and related processes, see Ghahremani (1980).

Acknowledgments

This work was performed with the support of the U.S. Department of Energy and the National Science Foundation. This support is gratefully acknowledged.

References

Ghahremani, F. (1980), Effect of grain boundary sliding on anelasticity of polycrystals, *Int. J. Solids Structures*, **16**, 825–845.

Honein, T. and Herrmann, G. (1988a), The involution correspondence in plane elastostatics for regions bounded by a circle, *J. Appl. Mech.*, **55**, 566–573.

Honein, T. and Herrmann, G. (1988b), On bonded inclusions with circular or straight boundaries in plane elastostatics, to be published in *J. Appl. Mech.*

Muskhelishvili, N. I. (1953), *Some Basic Problems of the Mathematical Theory of Elasticity*, Noordhoff, Groningen, Holland.

Sokolnikoff, I. S. (1983), *Mathematical Theory of Elasticity*, Krieger, Florida.

Mechanical Properties of Cracked Solids: Validity of the Self-Consistent Method

H. Horii and K. Sahasakmontri
Department of Civil Engineering, University of Tokyo, Bunkyo-ku, Tokyo, Japan

Abstract

The evaluation of overall elastic moduli of materials containing inhomogeneities is one of the classical problems in micromechanics. Crack is one of the typical examples of inhomogeneities. The overall properties of solids containing cracks depend on the density, orientation, and distribution of cracks. To describe the overall response of such solids under applied stresses, some quantities are necessary that characterize the arrangement of cracks, and also the relationship between the quantities and the overall properties needs to be found.

Instead of dealing with a general problem, the problem of solids containing a doubly periodic rectangular array of cracks is considered as a special case. The analysis is made based on the method of pseudotractions, and a first-order approximate but explicit solution is derived. This study solves the puzzle in the problem of doubly periodic cracks. In terms of the obtained solution, discussion is made on the essential characteristic of the required quantities, and also on the sufficiency of the fabric tensor (or the crack tensor), which is proposed to describe the effect of crack arrays on the overall properties of a cracked solid.

Then, the problem of solids containing randomly distributed cracks is investigated in an attempt to examine the validity of the self-consistent method, which incorporates indirectly the interaction effects into the estimation of the overall moduli of cracked solids. The method of pseudotractions is again employed in the analysis, and results are compared with those estimated by the self-consistent method and the solutions of the problem of doubly periodic cracks.

1. Introduction

The behavior of solids containing cracks has for decades been a subject of interest of numerous researches, especially those in the field of geomechanics. The mechanical response of the cracked solids depends not only on the properties of the matrix surrounding the cracks, but also on the size, shape, orientation, and distribution of the embedded cracks. There have been considerable efforts directed at estimating the overall response of solids containing cracks of various shapes and arrangements; for example, Walsh (1965a, b),

Vakulenko and Kachanov (1971), Garbin and Knopoff (1973), Salganik (1973), Vavakin and Salganik (1975), Hudson (1980), Budiansky and O'Connell (1976), Eimer (1978, 1979), Hoenig (1979), Kachanov (1980), Leguillon and Sanchez-Palencia (1982), Horii and Nemat-Nasser (1983), Oda (1982, 1983), Oda *et al.* (1984).

In general, it is difficult to establish a boundary value problem if the orientation and distribution of cracks are allowed to be arbitrary. Therefore, some restrictions have to be made regarding the arrangement and scale of the cracks to render the problem mathematically tractable. The investigation of the behavior of solids containing cracks is then directed to those special problems concerning cracks arranged in regular patterns, among which the simplest form is the doubly periodic rectangular array. Although the problem may lose generalities, such simplified models can disclose, at least, some general trends, and in fact they may be of some interest on their own account.

In connection with the problem of a doubly periodic array of cracks, Delameter *et al.* (1975) investigated the stress intensity factors for a doubly periodic rectangular array of cracks under mode I and mode II deformation. Representing each crack by distributed dislocations, they obtained singular integral equations for the dislocation density from the boundary conditions at the traction-free crack faces. Extending the work of Delameter *et al.* (1975), Karihaloo (1978, 1979) examined the stress relaxation process from the tips of the cracks in doubly periodic (rectangular and diamond-shaped) arrays in all modes of deformation. Isida (1972) and Isida *et al.* (1981) made extensive analyses on a single crack in a rectangular finite plate by employing the boundary collocation method. By choosing appropriate boundary conditions, they obtained numerical solutions to the problem of doubly periodic rectangular array of cracks under mode I. The numerical results reported by Isida (1972) and Isida *et al.* (1981) are seen to be different from those given for mode I by Delameter *et al.* (1975). Noting the difference, Delameter *et al.* (1977) published the corrected results and reasoned that the errors were caused by an incorrect evaluation of an infinite series. It will later be shown that a divergent, doubly infinite series—of which the limit depends on the way of taking the summation—always arises if the superposition principle is used to account for the interaction of the doubly periodic cracks. In this connection it is of interest to note that the divergence of the summation which appears in the formulation for periodic distribution of inclusions is discussed by Furuhashi *et al.* (1981); see also Mura (1986).

For the geomaterials such as rocks, the overall behavior always exhibits anisotropy, partly due to the complicated configuration and the preferred orientation of the cracks, and partly due to the load-induced anisotropy caused by crack-closure. To describe the overall response of such solids under applied stresses, some quantities are necessary that characterize the arrangement of cracks, and also the relationship between the quantities and the overall properties needs to be found. For this purpose, Oda (1982) has proposed the so-called fabric tensor, taking into account:

(1) volume density of cracks;
(2) dimension of cracks; and
(3) orientation of cracks.

The usefulness of the fabric tensor in characterizing the anisotropy of crack patterns has been recognized in many experiments. As an example, Oda *et al.* (1984) have shown that the supersonic wave velocity propagating through cracked bodies is closely related to the fabric tensor. This seems to suggest the possibility of estimating the fabric tensor of *in situ* rock masses by field measurement of the wave velocity. However, since the distribution of cracks in space is not included in the definition of the fabric tensor, it is possible that cracked bodies, having the crack geometries being described by the same fabric tensor, may have different overall moduli. We may therefore wonder whether or not the fabric tensor is sufficient in describing the effect of cracks on the mechanical properties of cracked solids. In this study, a discussion is made on the sufficiency of the fabric tensor, using the closed-form solution for the doubly periodic cracks.

Provided that the statistical distribution of cracks is random, use can be made of the so-called "self-consistent method" to include the interaction effects in estimating the overall properties of the body. This method was originally proposed for aggregates of crystals and was later applied to solids with randomly distributed inhomogeneities, such as solid inclusions or voids; see Willis (1981) and Walpole (1981), who give extensive lists of the literature. Application of the method to crack problems was pioneered by Budiansky and O'Connel (1976). They considered a body containing randomly distributed penny-shaped cracks, and estimated analytically its effective elastic moduli. Since then, this method has been extensively exploited in the analyses of the macroscopic properties of the body with cracks of various sizes and configurations; see, for example, Hoenig (1979), Kachanov (1980), Horii and Nemat-Nasser (1983). Although the method has found its application in a wide variety of problems, it evaluates the effect of interaction between neighboring inhomogeneities in an indirect manner. In the present work, the overall moduli of the solid containing randomly distributed, undirectionally oriented cracks are evaluated using the self-consistent method. The results are compared, in an attempt to examine the validity of the self-consistent method, to those computed using the method of pseudotractions as well as those obtained for the doubly periodic, rectangular array of cracks.

2. A Doubly Periodic Rectangular Array of Cracks

2.1. Mathematical Formulation

An infinite elastic plane containing a doubly periodic rectangular array of cracks is shown in Fig. 1. Each crack in the rectangular array is of length $2c$, and is separated from other cracks by a distance H vertically and a distance

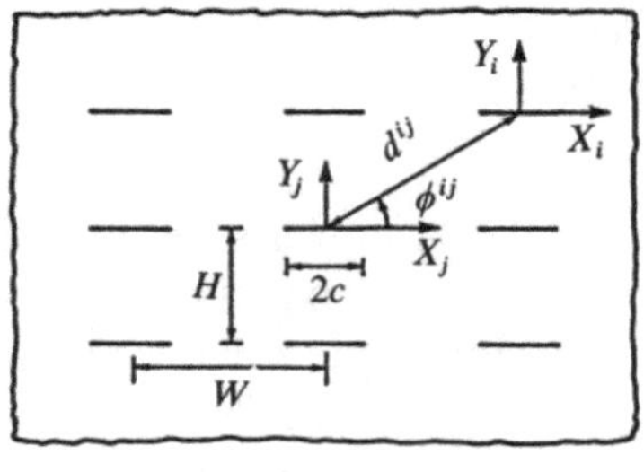

FIG. 1. A doubly periodic rectangular array of cracks.

W horizontally. The quantities which are used hereafter are also shown in Fig. 1.

The problem of an infinitely extended elastic solid containing a doubly periodic rectangular array of cracks under far-field stresses is referred to as the "original problem." (In the doubly periodic problems the far-field stresses are not well defined. In this study, they are introduced through the following superposition.) The solution to the problem is given by superposition of the solutions of a homogeneous problem and a subsidiary problem, as shown in Fig. 2. The homogeneous problem contains no cracks and is subjected to uniform stresses at infinity, σ_y^∞ and τ_{xy}^∞. In the subsidiary problem, the same constant stresses with opposite sign are prescribed on the surfaces of all the cracks with no stresses at infinity. The boundary conditions along any crack k are given by

$$\sigma_y^k = -\sigma_y^\infty, \qquad \tau_{xy}^k = -\tau_{xy}^\infty, \qquad -c \leqq x^k \leqq c, \qquad y^k = 0. \tag{2.1}$$

Since the solution to the homogeneous problem is known, only the subsidiary problem is to be solved. To apply the method of pseudotractions, proposed by Horii and Nemat-Nasser (1985), all the cracks are numbered and the subsidiary problem is decomposed into infinite number of subproblems, as shown in Fig. 3. In the subproblem j, the boundary conditions along the crack surfaces of the crack j are given by

$$\sigma_y^j = -(\sigma_y^\infty + \sigma_y^{\text{P}}), \qquad \tau_{xy}^j = -(\tau_{xy}^\infty + \tau_{xy}^{\text{P}}), \qquad -c \leqq x^j \leqq c, \qquad y^j = 0. \tag{2.2}$$

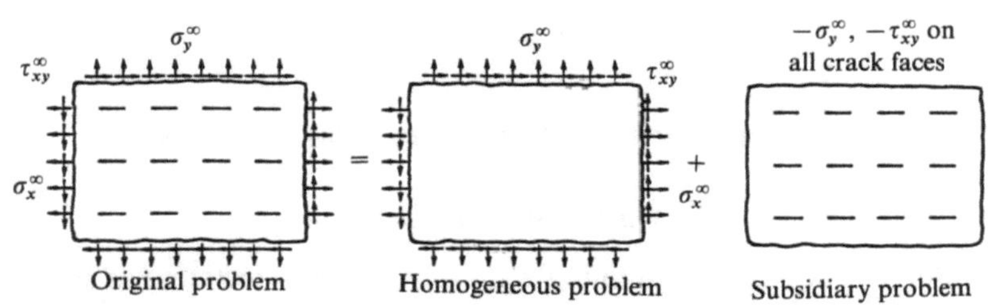

FIG. 2. Decomposition of an original problem into a homogeneous problem and a subsidiary problem.

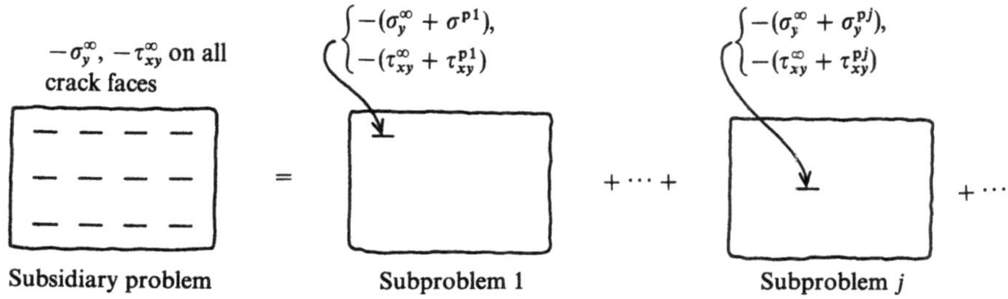

FIG. 3. Decomposition of the subsidiary problem into subproblems.

The quantities σ_y^{p}, τ_{xy}^{p}, being called the "pseudotractions," are unknown tractions which must be determined such that the boundary conditions (2.1) are satisfied when all the subproblems are superimposed. Note that the pseudotractions are the same for all cracks because of symmetry of the problem.

In the present analysis, we employ Muskhelishvili's complex stress functions $\phi(z)$ and $\psi(z)$ (see Muskhelishvili, 1953). Stress and displacement components are given in terms of these potentials by

$$\begin{aligned} \sigma_x(z) + \sigma_y(z) &= 2[\Phi(z) + \overline{\Phi(z)}], \\ \sigma_y(z) - \sigma_x(z) + 2i\tau_{xy}(z) &= 2[\bar{z}\Phi'(z) + \Psi(z)], \\ 2G[u(z) + iv(z)] &= \kappa\phi(z) - z\overline{\phi'(z)} - \overline{\psi(z)}, \end{aligned} \tag{2.3}$$

where $\Phi(z) = \phi'(z)$, $\Psi(z) = \psi'(z)$, $z = x + iy$, $i = \sqrt{-1}$, G is the shear modulus, $\kappa = 3 - 4\nu$ for plane strain, $\kappa = (3 - \nu)/(1 + \nu)$ for plane stress, ν is the Poisson ratio, the overbar denotes a complex conjugate, and the prime stands for differentiation with respect to the argument.

For the subproblem j, the stress functions are given by (see Muskhelishvili, 1953)

$$\begin{aligned} \Phi^j(z^j) &= -\frac{1}{2\pi i(z^{j^2} - c^2)^{1/2}} \int_{-c}^{c} \frac{(x^2 - c^2)^{1/2}}{x - z^j} [(\sigma_y^\infty + \sigma_y^{\mathrm{p}}) - i(\tau_{xy}^\infty + \tau_{xy}^{\mathrm{p}})]\, dx, \\ \Psi^j(z^j) &= \overline{\Phi^j(\overline{z^j})} - \Phi^j(z^j) - z^j\Phi^{j\prime}(z^j), \quad z^j = x^j + iy^j, \quad j = 1, 2, \ldots. \end{aligned} \tag{2.4}$$

The requirement that the sum of the subproblems must be equivalent to the subsidiary problem leads to

$$\begin{aligned} \sigma_y^{\mathrm{p}} - i\tau_{xy}^{\mathrm{p}} &= \sum_k^{\mathrm{All}} [\Phi^k(z^k) + \overline{\Phi^k(z^k)} + z^k\overline{\Phi^{k\prime}(z^k)} + \overline{\Psi^k(z^k)}], \\ z^k &= d^{jk}e^{i\phi^{jk}} + x^j, \qquad -c \leqq x^j \leqq c, \end{aligned} \tag{2.5}$$

where d^{jk}, ϕ^{jk}, x^j are defined in Fig. 1, and $\sum_k^{\mathrm{All}}$ denotes the summation for all the cracks except for the crack j under consideration. Note that $\sum_k^{\mathrm{All}}$ represents a doubly infinite summation in the present problem. The right-hand side of (2.5) represents the sum of tractions on the crack j caused by all other cracks.

Equations (2.4) and (2.5) form an integral equation for the pseudotractions, which are functions of x^j. Although it is not possible to solve this integral equation explicitly, this integral equation can be reduced to a system of algebraic equations in the following manner.

We expand the pseudotractions into a Taylor series as

$$\sigma_y^{\mathrm{p}}(x^j) - i\tau_{xy}^{\mathrm{p}}(x^j) = \sum_{n=0}^{\infty} (P_n - iQ_n)\left(\frac{x^j}{c}\right)^n. \tag{2.6}$$

Substituting (2.6) into (2.4) and (2.5), the integral equation (2.5) is reduced to a system of algebraic equations for P_n and Q_n. Horii and Nemat-Nasser (1985) showed that P_n and Q_n are of the order of $(c/d)^{n+2}$, where d denotes distance between cracks. Neglecting terms of order higher than $(c/d)^{N+1}$, we have $2N$ linear algebraic equations for $P_0, \ldots, P_{N-1}, Q_0, \ldots, Q_{N-1}$. The first-order approximate but explicit solutions are obtained by taking $N = 1$. Equation (2.6) then becomes

$$\sigma_y^{\mathrm{p}}(x^j) - i\tau_{xy}^{\mathrm{p}}(x^j) = P_0 - iQ_0, \tag{2.7}$$

implying that the pseudotractions are constant. Substituting (2.7) into (2.4), we obtain the stress function $\Phi^j(z^j)$ for the subproblem j as

$$\Phi^j(z^j) = \tfrac{1}{2}[(\sigma_y^{\infty} + P_0) - i(\tau_{xy}^{\infty} + Q_0)]\left\{\frac{1}{(z^{j2} - c^2)^{1/2}} - 1\right\}. \tag{2.8}$$

It follows from Horii and Nemat-Nasser (1985) that

$$P_0 = \left[\sum_k^{\mathrm{ALL}} A_{00}^{jk}\right]\sigma_y^{\infty} + \left[\sum_k^{\mathrm{ALL}} B_{00}^{jk}\right]\tau_{xy}^{\infty}, \quad Q_0 = \left[\sum_k^{\mathrm{ALL}} C_{00}^{jk}\right]\sigma_y^{\infty} + \left[\sum_k^{\mathrm{ALL}} D_{00}^{jk}\right]\tau_{xy}^{\infty}, \tag{2.9}$$

where

$$A_{00}^{jk} = [\cos(2\phi^{jk}) - \tfrac{1}{2}\cos(4\phi^{jk})](c/d^{jk})^2,$$
$$B_{00}^{jk} = \tfrac{1}{2}[-\sin(2\phi^{jk}) + \sin(4\phi^{jk})](c/d^{jk})^2,$$
$$C_{00}^{jk} = B_{00}^{jk}, \qquad D_{00}^{jk} = \tfrac{1}{2}\cos(4\phi^{jk})(c/d^{jk})^2.$$

The stress functions of the subproblem j are given by (2.8) and (2.9), and the stress and displacement fields are known through (2.3). The stress and displacement fields of the original problem are then obtained from superposition. Furthermore, following Horii and Nemat-Nasser (1985), the stress intensity factors for mode I and II are given by

$$\frac{K_{\mathrm{I}}}{\sigma_y^{\infty}\sqrt{\pi c}} = 1 + \frac{P_0}{\sigma_y^{\infty}}, \qquad \frac{K_{\mathrm{II}}}{\tau_{xy}^{\infty}\sqrt{\pi c}} = 1 + \frac{Q_0}{\tau_{xy}^{\infty}}. \tag{2.10}$$

The doubly infinite series in (2.9) represent the interaction effects on the crack j from all other cracks. These summations are numerically investigated by considering the limits of the partial sums for a finite number of cracks. If the doubly infinite series are convergent, the way to take the partial sums does not affect the values of the limits. Here, we use a rectangular array of R rows by C columns of cracks for the partial sums. Keeping the ratio C/R constant, we increase C and R. Then the doubly infinite series in (2.9) are expressed as

$$\sum_k^{\text{ALL}} A_{00}^{jk} = \lim_{R,C\to\infty}\left(\frac{c}{W}\right)^2\left[-\frac{3}{\left(\frac{H}{W}\right)^2}\sum_{m=1}^{R/2}\left(\frac{1}{m^2}\right)+\sum_{n=1}^{C/2}\left(\frac{1}{n^2}\right)\right.$$

$$\left.+2\sum_{m=1}^{R/2}\sum_{n=1}^{C/2}\left\{\frac{-3m^4\left(\frac{H}{W}\right)^4+6m^2n^2\left(\frac{H}{W}\right)^2+n^4}{\left(m^2\left(\frac{H}{W}\right)^2+n^2\right)^3}\right\}\right],$$

$$\sum_k^{\text{ALL}} B_{00}^{jk} = \sum_k^{\text{ALL}} C_{00}^{jk} = 0,$$

$$\sum_k^{\text{ALL}} D_{00}^{jk} = \lim_{R,C\to\infty}\left(\frac{c}{W}\right)^2\left[\frac{1}{\left(\frac{H}{W}\right)^2}\sum_{m=1}^{R/2}\left(\frac{1}{m^2}\right)+\sum_{n=1}^{C/2}\left(\frac{1}{n^2}\right)\right.$$

$$\left.+2\sum_{m=1}^{R/2}\sum_{n=1}^{C/2}\left\{\frac{m^4\left(\frac{H}{W}\right)^4-6m^2n^2\left(\frac{H}{W}\right)^2+n^4}{\left(m^2\left(\frac{H}{W}\right)^2+n^2\right)^3}\right\}\right]. \qquad (2.11)$$

The convergence of the infinite series in (2.11) is numerically examined and found to be fast. As an example, for $H/W = 1$, $c/W = 0.1$ with $C/R = 1$, an array of 10×10 cracks gives $\sum_k^{\text{ALL}} A_{00}^{jk} = -0.01036$ and $\sum_k^{\text{ALL}} D_{00}^{jk} = +0.01036$, while the final values are -0.01039 and $+0.01039$, respectively. Varying the ratio C/R, the limits of the series are investigated and shown in Figs. 4 and 5 for $H/W = 1$, $c/W = 0.1$. The ratio C/R represents the shape of the array that is used to approach the limits of the sums. The results reveal that as we increase the number of cracks under summation, the value of each sum approaches a different limit for different values of C/R. Therefore, it is concluded that each of the doubly infinite series in (2.11) diverges because the limiting value depends on the way of taking the summation. This is the essential difficulty of the problem of doubly periodic cracks, and always arises when we use the superposition principle to account for the influence from all the cracks. The superposition always leads to the doubly infinite series of which limits depend on the way of taking the summation.

2.2. Average Stresses and Double Periodicity

The reason for the difficulty shown in the preceding subsection is given as follows. We evaluate the infinite series as the limits of finite sums. This

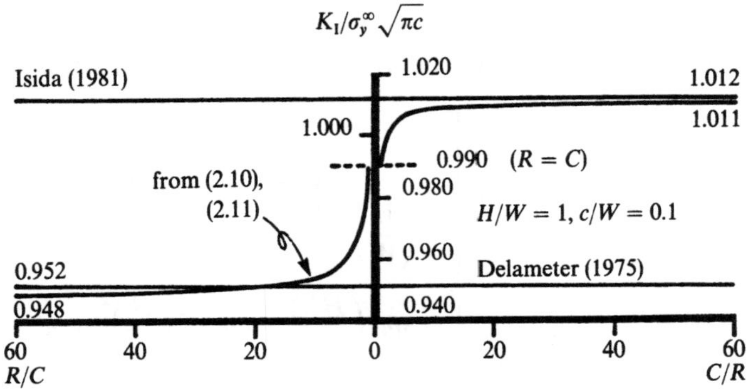

FIG. 4. Limits of $K_I/\sigma_y^\infty\sqrt{\pi c}$.

corresponds to solving a problem of an infinite plane with cracks of finite number. Since the solid is infinitely extended, no matter how we increase the number of cracks, we always have an uncracked area surrounding the cracked region. For each value of C/R the crack array has a particular shape and the stress and strain fields have different patterns. The average stresses within the crack array are then, as will be shown in this section, different from the applied remote stresses.

Now we evaluate the average stresses along the middle part of the cracked portion of the plane, i.e., along line AA' in Fig. 6. By superposition, the average stresses $\hat{\sigma}_y$, $\hat{\tau}_{xy}$ are found to be

$$\hat{\sigma}_y - i\hat{\tau}_{xy} = \sigma_y^\infty - i\tau_{xy}^\infty + \frac{4}{W}\sum_{m=1}^{R/2}\sum_{n=1}^{C/2}\left[\int_{(n-1)W}^{nW}\{\sigma_y(x, y_m) - i\tau_{xy}(x, y_m)\}\,dx\right],$$

$$y_m = H(m - \tfrac{1}{2}), \tag{2.12}$$

where $\sigma_y(x, y_m)$ and $\tau_{xy}(x, y_m)$ are the stresses along the segment AA' in the subproblems.

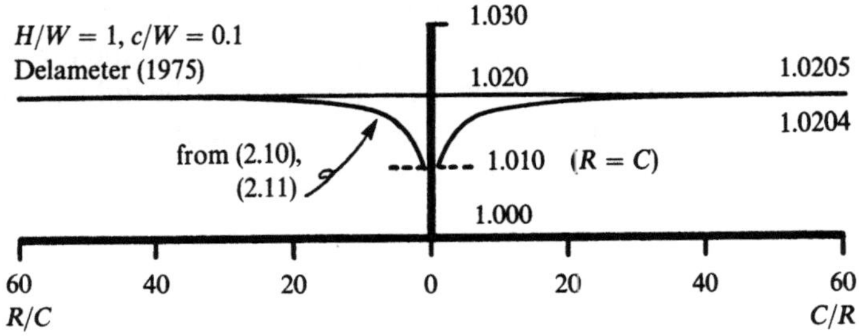

FIG. 5. Limits of $K_{II}/\tau_{xy}^\infty\sqrt{\pi c}$.

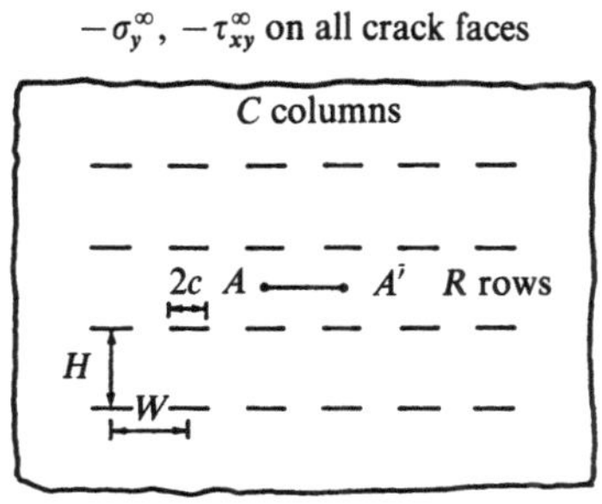

FIG. 6. A solid with an array of R rows by C columns of cracks.

With the help of (2.3), (2.8), and (2.9), we get

$$\hat{\sigma}_y - i\hat{\tau}_{xy} = \lim_{R,C\to\infty}\left[\sigma_y^\infty\left\{1 + 2\left(\frac{c}{W}\right)^2 \sum_{m=1}^{R/2}\sum_{n=1}^{C/2} C_{mn}^{\mathrm{I}}\right\} - i\tau_{xy}^\infty\left\{1 + 2\left(\frac{c}{W}\right)^2 \sum_{m=1}^{R/2}\sum_{n=1}^{C/2} C_{mn}^{\mathrm{II}}\right\}\right], \tag{2.13}$$

where

$$C_{mn}^{\mathrm{I}} = \frac{n^3(n-1)^3 + \left(\frac{y_m}{W}\right)^2 n(n-1)(7n^2-7n-3) + \left(\frac{y_m}{W}\right)^4 (3n^2-3n-1) - 3\left(\frac{y_m}{W}\right)^6}{\left[(n-1)^2 + \left(\frac{y_m}{W}\right)^2\right]^2\left[n^2 + \left(\frac{y_m}{W}\right)^2\right]^2},$$

$$C_{mn}^{\mathrm{II}} = \frac{n^3(n-1)^3 - \left(\frac{y_m}{W}\right)^2\left[\left(\frac{y_m}{W}\right)^2 + n(n-1)\right](5n^2-5n+1) + \left(\frac{y_m}{W}\right)^6}{\left[(n-1)^2 + \left(\frac{y_m}{W}\right)^2\right]^2\left[n^2 + \left(\frac{y_m}{W}\right)^2\right]^2}.$$

As example, the results for $H/W = 1$, $c/W = 0.1$ are shown in Figs. 7 and 8. It is seen that for different C/R, both summations tend to different limits as the number of cracks in the solid is increased.

To have the solution for the doubly periodic cracks, all the field variables must be expressed in terms of the stresses averaged within the cracked region instead of the applied remote stresses. In the sequel, we show that being expressed in terms of the average stresses, the stress intensity factors are uniquely determined.

From (2.10), (2.11), and (2.13), the stress intensity factors, normalized by the average stresses instead of the stresses at infinity, are given by

$$\begin{aligned} \frac{K_{\mathrm{I}}}{\hat{\sigma}_y\sqrt{\pi c}} &= 1 + \left(\frac{c}{W}\right)^2 F_{\mathrm{I}}(\phi) = 1 + \left(\frac{c}{H}\right)^2 G_{\mathrm{I}}(\phi) = 1 + \frac{c^2}{HW} H_{\mathrm{I}}(\phi), \\ \frac{K_{\mathrm{II}}}{\hat{\tau}_{xy}\sqrt{\pi c}} &= 1 + \left(\frac{c}{W}\right)^2 F_{\mathrm{II}}(\phi) = 1 + \left(\frac{c}{H}\right)^2 G_{\mathrm{II}}(\phi) = 1 + \frac{c^2}{HW} H_{\mathrm{II}}(\phi), \end{aligned} \tag{2.14}$$

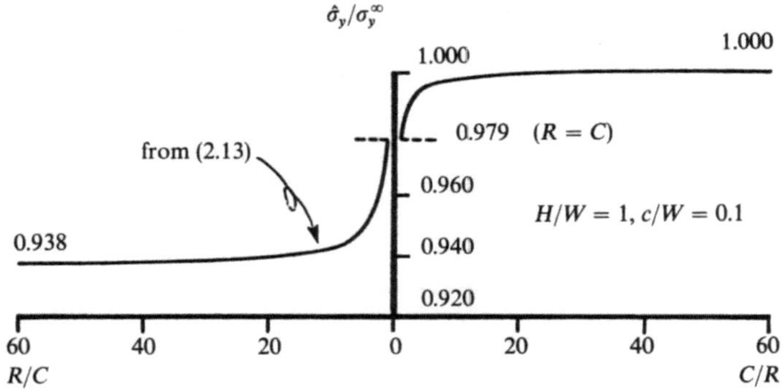

FIG. 7. Limits of $\hat{\sigma}_y/\sigma_y^{\infty}$.

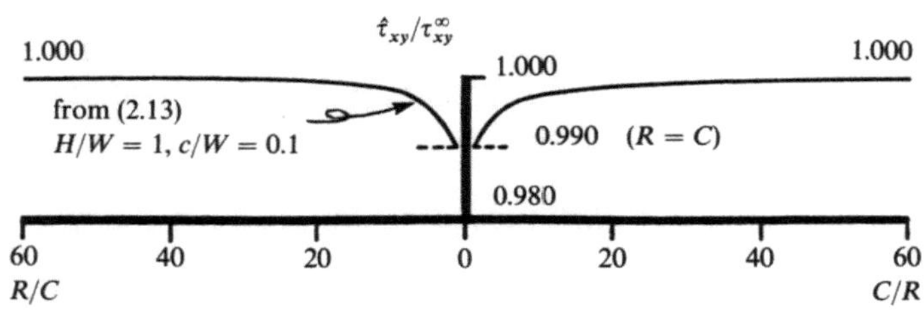

FIG. 8. Limits of $\hat{\tau}_{xy}/\tau_{xy}^{\infty}$.

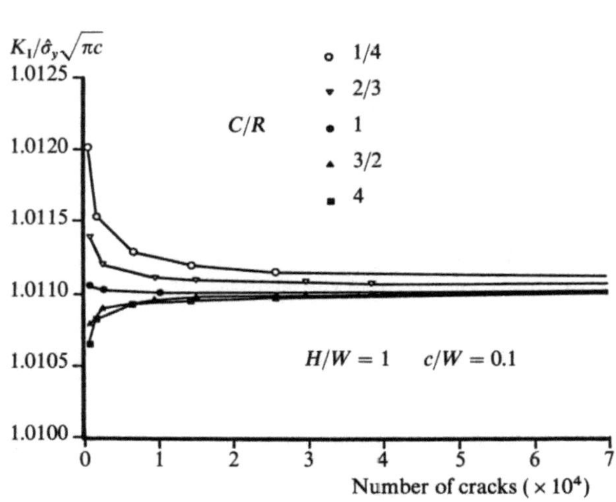

FIG. 9. Convergence of $K_I/\hat{\sigma}_y\sqrt{\pi c}$.

where

$$F_{\mathrm{I}}(\phi) = \lim_{R,C\to\infty}\left[-\frac{3}{\left(\frac{H}{W}\right)^2}\sum_{m=1}^{R/2}\left(\frac{1}{m^2}\right)+\sum_{n=1}^{C/2}\left(\frac{1}{n^2}\right)\right.$$

$$+2\sum_{m=1}^{R/2}\sum_{n=1}^{C/2}\left\{\frac{-3m^4\left(\frac{H}{W}\right)^4+6m^2n^2\left(\frac{H}{W}\right)^2+n^4}{\left(m^2\left(\frac{H}{W}\right)^2+n^2\right)^3}\right.$$

$$\left.\left.-\frac{n^3(n-1)^3+\left(\frac{y_m}{W}\right)^2 n(n-1)(7n^2-7n-3)+\left(\frac{y_m}{W}\right)^4(3n^2-3n-1)-3\left(\frac{y_m}{W}\right)^6}{\left[(n-1)^2+\left(\frac{y_m}{W}\right)^2\right]^2\left[n^2+\left(\frac{y_m}{W}\right)^2\right]^2}\right\}\right],$$

$$F_{\mathrm{II}}(\phi) = \lim_{R,C\to\infty}\left[\frac{1}{\left(\frac{H}{W}\right)^2}\sum_{m=1}^{R/2}\left(\frac{1}{m^2}\right)+\sum_{n=1}^{C/2}\left(\frac{1}{n^2}\right)\right.$$

$$+2\sum_{m=1}^{R/2}\sum_{n=1}^{C/2}\left\{\frac{m^4\left(\frac{H}{W}\right)^4-6m^2n^2\left(\frac{H}{W}\right)^2+n^4}{\left(m^2\left(\frac{H}{W}\right)^2+n^2\right)^3}\right.$$

$$\left.\left.-\frac{n^3(n-1)^3-\left(\frac{y_m}{W}\right)^2\left[\left(\frac{y_m}{W}\right)^2+n(n-1)\right](5n^2-5n+1)+\left(\frac{y_m}{W}\right)^6}{\left[(n-1)^2+\left(\frac{y_m}{W}\right)^2\right]^2\left[n^2+\left(\frac{y_m}{W}\right)^2\right]^2}\right\}\right],$$

$$G_{\mathrm{I}}(\phi) = \left(\frac{H}{W}\right)^2 F_{\mathrm{I}}(\phi), \qquad H_{\mathrm{I}}(\phi) = \frac{H}{W}F_{\mathrm{I}}(\phi),$$

$$G_{\mathrm{II}}(\phi) = \left(\frac{H}{W}\right)^2 F_{\mathrm{II}}(\phi), \qquad H_{\mathrm{II}}(\phi) = \frac{H}{W}F_{\mathrm{II}}(\phi),$$

$$\tan\phi = \frac{H}{W}.$$

The numerical results obtained from (2.14) for the case of $H/W = 1$, $c/W = 0.1$ are shown in Figs. 9 and 10. It is seen that a single limit exists for each mode, which means that the doubly infinite series in (2.14) are convergent series and the stress intensity factors are uniquely determined.

For $H/W = 1$, $c/W = 0.1$, $K_{\mathrm{I}}/\hat{\sigma}\sqrt{\pi c}$ is given by (2.14) to be 1.0110, which is very close to the value of $K_{\mathrm{I}}/\sigma_0\sqrt{\pi c}$ reported by Isida (1972) and Isida *et*

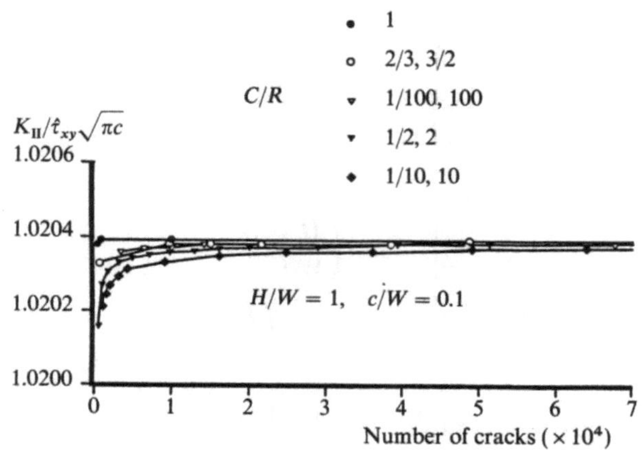

FIG. 10. Convergence of $K_{II}/\hat{t}_{xy}\sqrt{\pi c}$.

al. (1981), being 1.012. (Note that the present solution is an approximate solution.) However, the numerical value of $K_I/\sigma\sqrt{\pi c}$ for the present case is shown by Delameter *et al.* (1975) to be 0.9517. The numerical solutions given by Isida (1972) and Isida *et al.* (1981) are for a rectangular plate with a single crack, with the appropriate boundary conditions of the doubly periodic array of cracks. On the other hand, the method used by Delameter *et al.* (1975) has its basis on the superposition. In their formulation, Delameter *et al.* (1975) represented each crack by a certain distribution of dislocations, and employed the superposition principle to obtain the stress field due to the applied stresses and the dislocations. This leads to a term with doubly infinite summation of the contribution of all other cracks. He evaluated the doubly infinite summations in the vertical direction first and set up a numerical scheme to solve the resulting integral equations. If the ratio of C/R is set to be very small, which is equivalent to taking the summation in the vertical direction first, (2.10) gives $K_I/\sigma_y^\infty\sqrt{\pi c}$ the value of 0.9482, see also Fig. 4. Hence, it seems that the stress intensity factors obtained by Delameter *et al.* (1975) correspond to (2.10) with a very small value of C/R, which differs from Isida's and our solution (2.14). Noticing the discrepancies of the results, Delameter *et al.* (1977) published the corrected results and explained that the errors were due to an incorrect evaluation of an infinite series. We would like to point out that without introducing the average stresses, the doubly infinite series resulting from superposition is divergent, like those in (2.11). Although, we may be able to get the correct numerical results from these divergent series by changing the way of taking the summation (see Fig. 4), this actually requires *a priori* knowledge to choose the way of taking the summation such that $\hat{\sigma}_y/\sigma_y^\infty$ is equal to one (C/R being very large, see Fig. 7).

The accuracy of our first-order approximate solution is examined by comparing the values of K_I given by (2.14) to those reported by Isida *et al.* (1981).

TABLE 1. $K_I/\sigma_y\sqrt{\pi c}$ for several values of H/W and c/W.

H/W \ c/W	0	0.05	0.10	0.15	0.20	Remark
0.5	1.0	0.982	0.928	0.838		
1.0	1.0	1.003	1.011	1.025		(2.14)
1.5	1.0	1.004	1.016	1.036	1.064	
0.5	1.0	0.983	0.948	0.922		from
1.0	1.0	1.003	1.012	1.031		Isida *et al.*
1.5	1.0	1.004	1.017	1.039	1.074	(1981)

Table 1 shows the values of $K_I/\hat{\sigma}_y\sqrt{\pi c}$ computed from (2.14), and those obtained by Isida *et al.* (1981) for various values of H/W and c/W. Good agreement between both results are observed for small values of c/W.

For mode II, it can be shown from (2.13) that, as C/R decreases, the value of $\hat{\tau}_{xy}/\tau_{xy}^{\infty}$ approaches 1.0, making the values of $K_{II}/\tau_{xy}^{\infty}\sqrt{\pi c}$ of (2.10) approach those of $K_{II}/\hat{\tau}_{xy}\sqrt{\pi c}$ of (2.14). (See also Figs. 5 and 8.) Accordingly, the numerical results for mode II, reported by Delameter *et al.* (1975), are expected to agree with our first-order approximate solution. Table 2 shows the comparison of the values of $K_{II}/\hat{\tau}_{xy}\sqrt{\pi c}$ obtained through (2.14) with those in parentheses computed by Delameter *et al.* (1975) for several values of H/W

TABLE 2. $K_{II}/\tau_{xy}\sqrt{\pi c}$ for several values of H/W and $c/\min(H/W)$.

H/W \ $c/\min(H, W)$	0.100	0.125	0.167	0.200	0.250	Remark
0.0	1.0164	1.0257	1.0457		1.1028	
0.4					1.1028	
0.6			1.0460			(2.14)
0.8		1.0274			1.1094	
1.0	1.0204					
1.2				1.0713		
1.6				1.0664		
2.0				1.0659		
∞	1.0164			1.0658		
0.0	1.0160	1.0246	1.0424		1.0881	
0.4					1.0881	from
0.6			1.0429			Delameter
0.8		1.0267			1.1022	*et al.*
1.0	1.0205					(1975)
1.2				1.0787		
1.6				1.0757		
2.0				1.0753		
∞	1.0160			1.0753		

and $c/\min(H, W)$. The agreement is observed to be satisfactory for small values of $c/\min(H, W)$.

When H/W goes to infinity, (2.14) yields the solution for an infinite row of collinear cracks which agrees with the expansion of the exact solution by Westergaard (1939), neglecting higher-order terms, i.e.,

$$\lim_{H/W\to\infty} \frac{K_{\rm I}}{\hat{\sigma}_y\sqrt{\pi c}} = 1 + \frac{\pi^2}{6}\left(\frac{c}{W}\right)^2, \quad \lim_{H/W\to\infty} \frac{K_{\rm II}}{\hat{\tau}_{xy}\sqrt{\pi c}} = 1 + \frac{\pi^2}{6}\left(\frac{c}{W}\right)^2. \qquad (2.15)$$

The solution for a single stack of cracks is obtained from (2.14) if H/W goes to zero. Γhe solution (2.16), shown below, compares well with Fig. 2.29 (p. 123) of Isida (1973) and corresponds to the first-order approximate solutions of Horii and Nemat-Nasser (1985).

$$\lim_{H/W\to 0} \frac{K_{\rm I}}{\hat{\sigma}_y\sqrt{\pi c}} = 1 - \frac{\pi^2}{2}\left(\frac{c}{H}\right)^2, \quad \lim_{H/W\to 0} \frac{K_{\rm II}}{\hat{\tau}_{xy}\sqrt{\pi c}} = 1 + \frac{\pi^2}{6}\left(\frac{c}{H}\right)^2. \qquad (2.16)$$

In Figs. 11 and 12, the values of the stress intensity factors computed using (2.14) are shown for various values of $c/\min(H, W)$ and ϕ. (Note that the angle

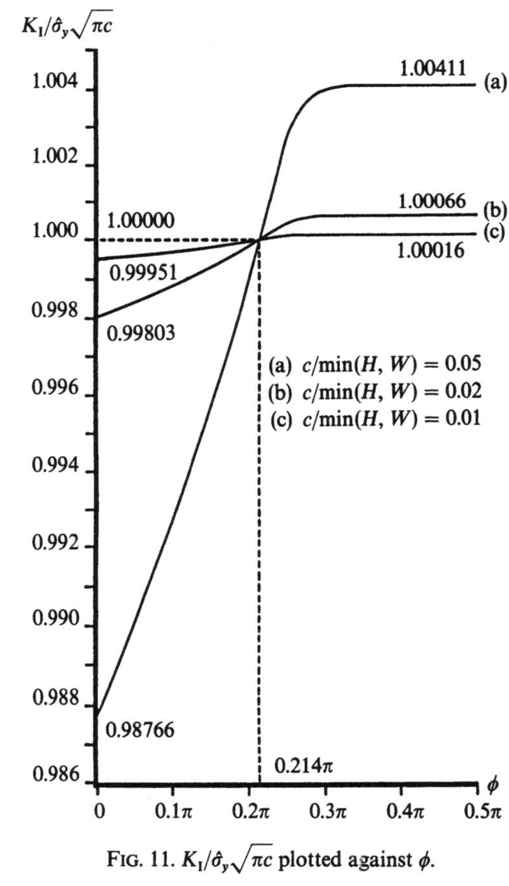

FIG. 11. $K_{\rm I}/\hat{\sigma}_y\sqrt{\pi c}$ plotted against ϕ.

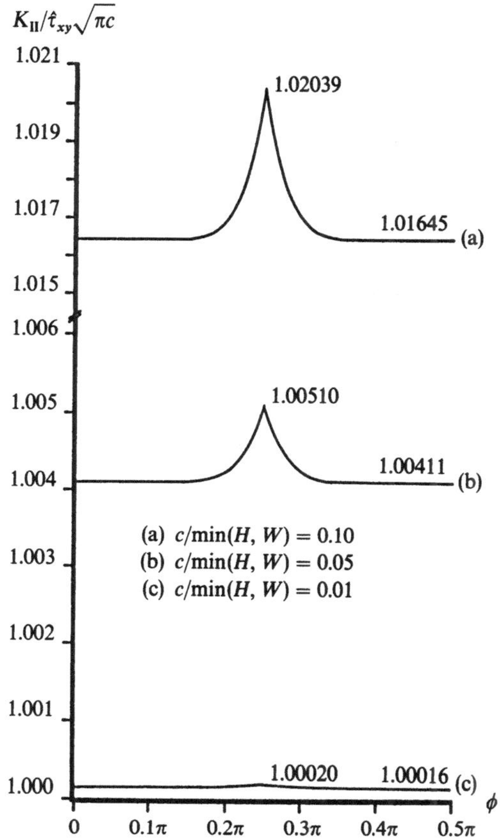

FIG. 12. $K_{II}/\hat{t}_{xy}\sqrt{\pi c}$ plotted against ϕ.

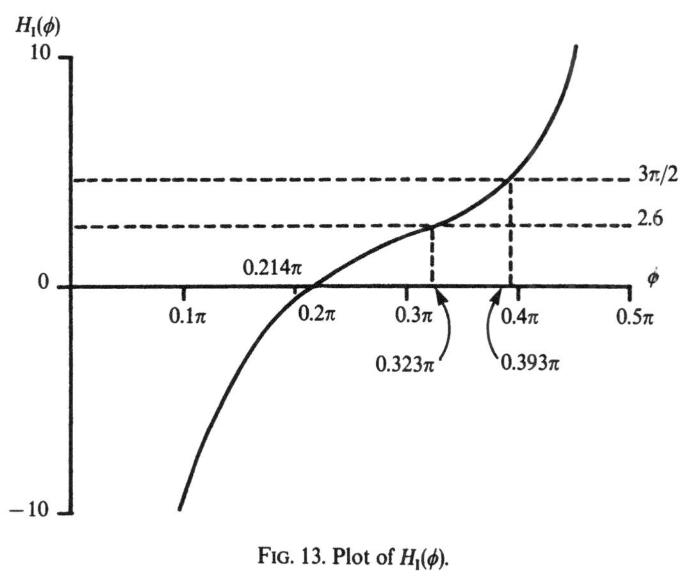

FIG. 13. Plot of $H_I(\phi)$.

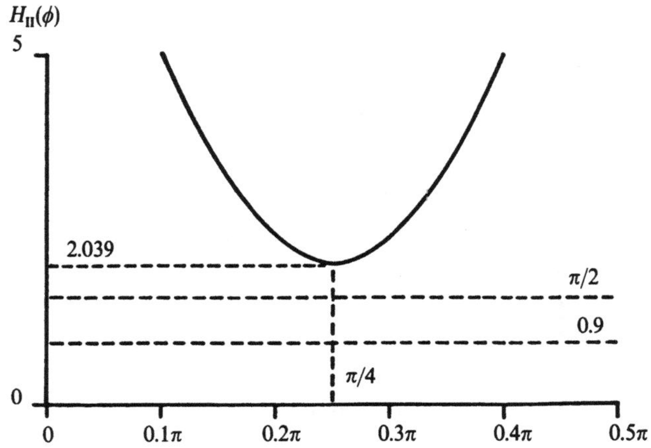

FIG. 14. Plot of $H_{II}(\phi)$.

ϕ defines the crack arrangement; $\tan \phi = H/W$.) It is seen that for ϕ less than 0.214π, the values of $K_I/\hat{\sigma}_y\sqrt{\pi c}$ are less than one. For mode II, the values of $K_{II}/\hat{\tau}_{xy}\sqrt{\pi c}$ are seen to be always larger than one.

From (2.14) it is seen that the variation of the stress intensity factors with the crack arrangement at a constant crack density, c^2/HW, is represented by functions $H_I(\phi)$ and $H_{II}(\phi)$. The functions $H_I(\phi)$ and $H_{II}(\phi)$ are plotted against ϕ in Figs. 13 and 14, respectively.

3. The Overall Compliance

The relationship between the average strain $\hat{\varepsilon}_{ij}$ and the average stress $\hat{\sigma}_{ij}$ in a cracked solid has been shown by Horii and Nemat-Nasser (1983) to be

$$\begin{aligned}\hat{\varepsilon}_{ij} &= D_{ijkl}\hat{\sigma}_{kl} + \frac{1}{V}\int_S \tfrac{1}{2}([u_i]n_j + [u_j]n_i)\, dS = (D_{ijkl} + H_{ijkl})\hat{\sigma}_{kl} \\ &= \bar{D}_{ijkl}\hat{\sigma}_{kl}, \qquad i, j, k, l = 1, 2, 3,\end{aligned} \tag{3.1}$$

where $[u_i]$ denotes the displacement gap along the crack surface with unit normal vector n_i, D_{ijkl} is the elastic compliance of the matrix, $\bar{D}_{ijkl}$ is the overall compliance of the cracked solid, and the integration is carried over the crack surface S contained in the solid of volume V. Once the displacement gap is known, the tensor H_{ijkl} and then $\bar{D}_{ijkl}$ are readily obtained.

For a two-dimensional problem, the stress–strain relation reduces to

$$\begin{Bmatrix} \hat{\varepsilon}_x \\ \hat{\varepsilon}_y \\ 2\hat{\varepsilon}_{xy} \end{Bmatrix} = \begin{bmatrix} \bar{D}_{11} & \bar{D}_{12} & \bar{D}_{16} \\ \bar{D}_{21} & \bar{D}_{22} & \bar{D}_{26} \\ \bar{D}_{61} & \bar{D}_{62} & \bar{D}_{66} \end{bmatrix} \begin{Bmatrix} \hat{\sigma}_x \\ \hat{\sigma}_y \\ \hat{\tau}_{xy} \end{Bmatrix}, \tag{3.2}$$

with $\bar{D}_{ij} = D_{ij} + H_{ij}$ and $i, j = 1, 2, 6$.

3.1. Without Interaction

If the interaction between neighboring cracks is neglected, we can utilize the solution of a single crack in an infinite plane for the evaluation of the tensor H_{ij}. It can be easily shown that in this case the overall compliance $\bar{D}_{ij}$ can be expressed as $\bar{D}_{ij}(\mathbf{F})$, $\mathbf{F}$ being the second-rank fabric tensor (see Oda, 1982) whose components are given by

$$F_{ij} = \rho \langle 4c^2 \rangle \langle n_i n_j \rangle, \tag{3.3}$$

in which

$$\langle 4c^2 \rangle = 4 \int_0^\infty c^2 f(c)\, dc,$$

$$\langle n_i n_j \rangle = \int_0^{2\pi} n_i n_j e(\hat{n})\, d\beta, \qquad i, j = 1, 2,$$

where $f(c)$ is a probability density function of c, representing the distribution of the crack of length $2c$; $e(\hat{n})$ is a probability density function of $\hat{n}$, representing the distribution of the crack having the unit normal vector $\hat{n}$; $n_1 = \sin\beta$, $n_2 = \cos\beta$ are the components of $\hat{n}$; and ρ is the number of cracks per unit volume. Here, we assume that $\hat{n}$ and c are mutually independent and the solid is of unit thickness. For plane stress, the overall compliance $[\bar{D}_{ij}(\mathbf{F})]$ is given by

$$[\bar{D}_{ij}(\mathbf{F})] = \frac{1}{E}\begin{bmatrix} 1 + \frac{\pi}{2}F_{11} & -\nu & \frac{\pi}{2}F_{12} \\ -\nu & 1 + \frac{\pi}{2}F_{22} & \frac{\pi}{2}F_{12} \\ \frac{\pi}{2}F_{12} & \frac{\pi}{2}F_{12} & 2(1+\nu) + \frac{\pi}{2}(F_{11} + F_{22}) \end{bmatrix}. \tag{3.4}$$

As an example, we consider the solid containing a periodic array of cracks shown in Fig. 1, it follows from (3.3) that the fabric tensor is

$$[F_{ij}] = \begin{bmatrix} 0 & 0 \\ 0 & 4\rho^* \end{bmatrix}, \qquad i, j = 1, 2, \tag{3.5}$$

where ρ^* is the normalized crack–density parameter defined by $\rho^* = c^2\rho$, and for the present case $\rho = 1/HW$. From (3.4) and (3.5) the overall moduli of the solid are found to be

$$E_y = \frac{E}{1 + 2\pi\rho^*}, \qquad G_{xy} = \frac{G}{1 + 2\frac{G}{E}\pi\rho^*}, \tag{3.6}$$

where G is the shear modulus of the matrix surrounding the cracks; E_y and G_{xy} are, respectively, the overall tensile modulus in the direction normal to the crack faces, and the overall shear modulus of the cracked solid.

3.2. With Interaction

It is seen in the previous subsection that the fabric tensor is the only parameter needed to characterize the influence of crack pattern on the overall compliance of the solid provided that the interaction effect is ignored. However, if the interaction effect is taken into account, the higher-order terms in the crack-density parameter will enter the compliance matrix. This can be exemplified as follows by considering the solid containing a doubly periodic array of cracks.

Using the solution obtained in the previous section, the components of the tensor H_{ij}, for the solid containing a doubly periodic set of cracks, are given by

$$H_{22} = \frac{(\kappa + 1)}{4G}\pi\rho^*[1 + \rho^* H_{\text{I}}(\phi)], \quad H_{66} = \frac{(\kappa + 1)}{4G}\pi\rho^*[1 + \rho^* H_{\text{II}}(\phi)], \tag{3.7}$$

otherwise $H_{ij} = 0$, in which $H_{\text{I}}(\phi)$ and $H_{\text{II}}(\phi)$ are given by (2.14). From (3.2) and (3.7) the overall compliance of the solid for the case of plane stress is then given by

$$[\bar{D}_{ij}] = \frac{1}{E}\begin{bmatrix} 1 & -\nu & 0 \\ -\nu & 1+2\pi\rho^*[1+\rho^* H_{\text{I}}(\phi)] & 0 \\ 0 & 0 & 2(1+\nu)+2\pi\rho^*[1+\rho^* H_{\text{II}}(\phi)] \end{bmatrix}. \tag{3.8}$$

from which the overall moduli are found to be

$$E_y = \frac{E}{1 + 2\pi\rho^*[1 + \rho^* H_{\text{I}}(\phi)]}, \qquad G_{xy} = \frac{G}{1 + 2\dfrac{G}{E}\pi\rho^*[1 + \rho^* H_{\text{II}}(\phi)]}. \tag{3.9}$$

The same results are obtained by evaluating the difference between the strain energy of the body with and without cracks subjected to the same applied stresses. To evaluate the change in strain energy, the amount of energy required to close all the cracks in a unit volume is calculated, either by integrating the energy release rate or by integrating the crack opening displacements.

It is seen from (3.6) and (3.9) that including the interaction, the overall moduli of the cracked body become dependent on the shape of the crack array or, in other words, the distribution of cracks in the solid. The fabric tensor alone cannot, in this case, completely describe the effect of the crack array geometry on the overall moduli of the solid (even though a higher rank fabric tensor is employed), and additional parameters $[H_{\text{I}}(\phi), H_{\text{II}}(\phi)]$ are needed that take into account the spacial distribution of the cracks. It is important to notice that this new parameter depends on the loading condition and is obtained by considering the interaction effect based on micromechanics.

3.3. The Self-Consistent Method

The overall elastic moduli of solids containing randomly distributed, unidirectional cracks (see Fig. 15) are evaluated by the self-consistent method in the

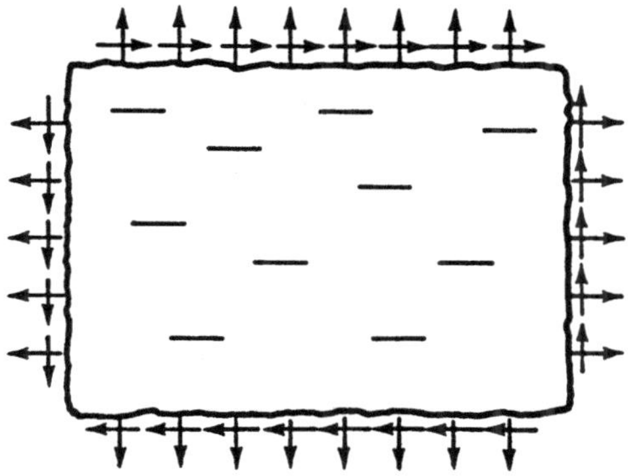

FIG. 15. Randomly distributed unidirectional cracks.

following manner. Consider a single crack embedded in the infinite solid whose compliance is set to be the overall compliance $\bar{D}_{ij}$ of the cracked body, which is the unknown to be determined. (Note that for a solid containing randomly distributed, unidirectional cracks, the overall response of the solid is anisotropic.) The displacement gap along the crack surface is expressed in terms of the unknown overall compliance by using the solution of a single crack in an anisotropic material given by Sih *et al.* (1965). According to (2.17) and (2.18), the displacement gap is integrated and then the overall compliance is obtained as

$$\begin{aligned}\bar{D}_{22} &= D_{22} + \pi\rho^*\sqrt{\bar{D}_{22}(2D_{12} + \bar{D}_{66} + 2\sqrt{D_{11}\bar{D}_{22}})},\\ \bar{D}_{66} &= D_{66} + \pi\rho^*\sqrt{\bar{D}_{11}(2D_{12} + \bar{D}_{66} + 2\sqrt{D_{11}\bar{D}_{22}})}.\end{aligned} \tag{3.10}$$

This equation is called the consistency condition for the unknown overall compliance $\bar{D}_{ij}$. An approximate solution is obtained through asymptotic expansions of the unknown variables with respect to a small parameter ρ^* as

$$\begin{aligned}\bar{D}_{22} &= \frac{1}{E} + \frac{2}{E}\pi\rho^*[1 + \rho^* H_{\mathrm{I}}].\\ \bar{D}_{66} &= 2\frac{(1+\nu)}{E} + \frac{2}{E}\pi\rho^*[1 + \rho^* H_{\mathrm{II}}].\end{aligned} \tag{3.11}$$

for plane stress, in which $H_{\mathrm{I}} = 3\pi/2$ and $H_{\mathrm{II}} = \pi/2$.

3.4. Random Distribution and the Method of Pseudotractions

In this subsection, the overall moduli of a solid containing randomly distributed, unidirectional cracks are evaluated using the method of pseudotractions. The position of the center of each crack is randomly generated using the random number generator. To simplify the analysis, the infinite plane is divided into doubly periodic rectangular cells of width W and height H, and the cracks are randomly placed inside the cell. The distribution of the cracks in the whole plane is consequently not random in the rigorous sense but has a large periodicity.

The analysis for the present case is the same as the one previously given for the doubly periodic array of cracks, except that the values of P_0 and Q_0 are different for each crack. Since the crack distribution is doubly periodic on a large scale, we calculate the average stresses inside the cell and express all the field variables in terms of the average stresses, in order to avoid the difficulty due to the divergent series. The overall compliance of the cracked body is then computed using the method shown at the beginning of this section, and the overall moduli are given by (3.9) with $H_I(\phi)$ and $H_{II}(\phi)$ being replaced by H_I and H_{II}, respectively.

The calculation for the same number of cracks in a cell is repeated and the mean values of H_I and H_{II} are computed. It is found that the value of H_{II} is less scattered than that of H_I, and as the number of the cracks in a cell is increased the scatter of the values of H_I and H_{II} decreases. With $H/W = 1.0$ and $N = 500$, H_I and H_{II} are found to be 2.6 and 0.9, respectively.

Comparing the above results with (3.11) it is seen that values of the interaction terms H_I and H_{II} estimated by the self-consistent method are larger, 2.6 and 0.9 compared with $3\pi/2$ and $\pi/2$. This seems to agree with the argument pointed out by Henyey and Pomphrey (1982), that the self-consistent method overestimates the interaction effect. Figures 13 and 14 compare the interaction terms of randomly distributed cracks with those of doubly periodic cracks. We can see that for mode II the results of randomly distributed cracks are the lower bound of those of doubly period cracks, but for mode I no such bound exists. From Figs. 13 and 14, we can get an array shape that produces the same interaction as the random cracks. The square-shape array gives the nearest value for mode II, but for mode I the array is found to be the one with ϕ of about 58°. From the same figures, the same interaction as the self-consistent method is obtained with the crack array with ϕ around 71° for mode I. For mode II, $\phi = 45°$ gives the nearest interaction. It is seen that the interaction behavior under mode I and II is completely different. This phenomenon stems from the fact that the disturbance in the stress field due to the presence of another crack, i.e., the interaction effect, depends on the relative position (as specified by ϕ^{ij} in Fig. 1), and the dependency differs for mode I and II. This can be seen by examining the terms A_{00}^{jk} and D_{00}^{jk} in (2.9) which, respectively, represent the interaction of the kth crack on the jth crack. From Figs. 16 and 17 which, respectively, show the variations of A_{00}^{ij} and D_{00}^{ij} with ϕ^{ij}, we can see that the interaction behavior under mode I and II is different, that of mode I having a stronger direction-dependence. Regarding the argument saying that the self-consistent method may not provide a good estimate when the interaction effect has a strong direction-dependence, further study is necessary to clarify this point.

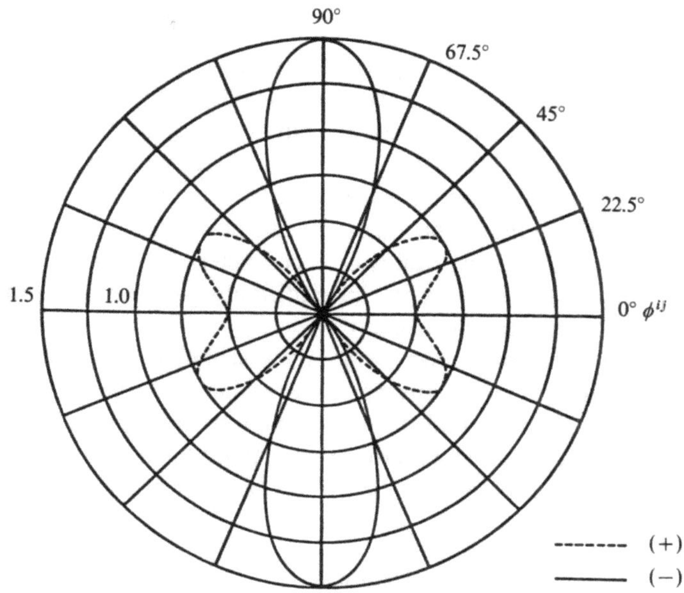

FIG. 16. Variations of $A_{00}^{ij}/(c/d^{ij})^2$ with ϕ^{ij}.

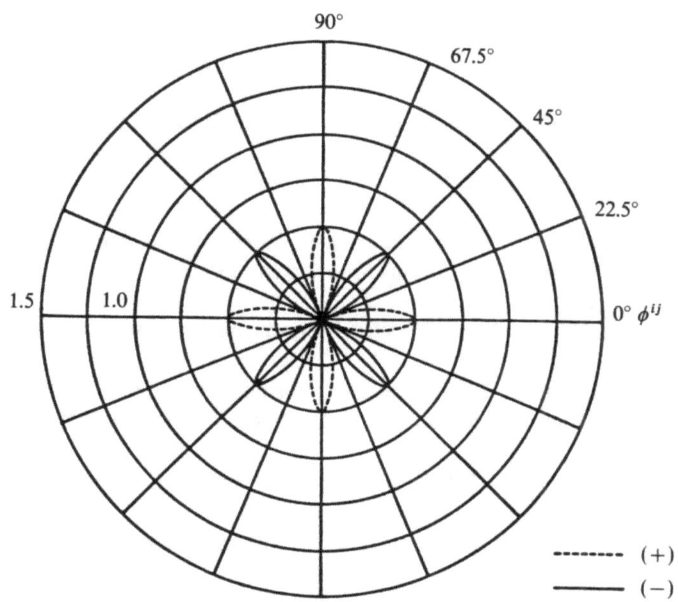

FIG. 17. Variations of $D_{00}^{ij}/(c/d^{ij})^2$ with ϕ^{ij}.

4. Conclusions

The present study reveals that the difficulty in solving the problem of doubly periodic cracks arises from the superposition principle, and is overcome by considering the average stress. An approximate but explicit solution is derived. The stress intensity factors and the overall compliance are explicitly obtained as functions of the crack density and a parameter describing the geometry of the crack array. It is shown that if the interaction effect is neglected, the effect of a crack array on the overall moduli can be characterized with the fabric tensor alone. Including the interaction, an additional parameter is found necessary and its characteristic is presented.

The problem of solids containing randomly distributed unidirectional cracks is also investigated in an attempt to examine the validity of the self-consistent method. The method of pseudotractions is employed in the analysis, and results are compared with those estimated by the self-consistent method and the solution of the problem of doubly periodic array of cracks.

Acknowledgment

This study was supported in part by the Grant-in-Aid for Scientific Research, from the Japanese Ministry of Education, Science, and Culture.

References

Budiansky, B. and O'Connel, R. J. (1976), Elastic moduli of a cracked solid, *Int. J. Solids Structures*, **12**, 81–97.

Delameter, W. R., Herrmann, G., and Barnett, D. M. (1975), Weakening of an elastic solid by a rectangular array of cracks, *J. Appl. Mech.*, **42**, 74–80.

Delameter, W. R., Herrmann, G., and Barnett, D. M. (1977), *Erratum, J. Appl. Mech., March*, **44**, 190.

Eimer, Cz. (1978), Elasticity of cracked medium, *Arch. Mech.*, **30**, No. 6, 827–836.

Eimer, Cz. (1979), Bulk constitutive relations for cracked materials, *Arch. Mech.*, **31**, No. 4, 519–532.

Furuhashi, R., Kinoshita, N., and Mura, T. (1981), Periodic distributions of inclusions, *Int. J. Engng. Sci.*, **19**, 231–236.

Garbin, H. D. and Knopoff, L. (1973), The compressional modulus of a material permeated by a random distribution of circular cracks, *Quart. Appl. Math.*, **30**, 453–464.

Henyey, F. S. and Pomphrey, N. (1982), Self-consistent moduli of a cracked solid, *Geophys. Res. Lett.*, **9**, 903–906.

Hoenig, A. (1979), Elastic moduli of a non-randomly cracked body, *Int. J. Solids Structures*, **15**, 137–154.

Horii, H. and Nemat-Nasser, S. (1983), Overall moduli of solids with microcracks: Load-induced anisotropy, *J. Mech. Phys. Solids*, **31**, No. 2, 155–177.

Horii, H. and Nemat-Nasser, S. (1985), Elastic fields of interacting inhomogeneities, *Int. J. Solids Structures*, **21**, No. 7, 731–745.

Hudson, J. A. (1980), Overall properties of a cracked solid, *Math. Proc. Camb. Philos. Soc.* (1975), **88**, 371–384.

Isida, M. (1972), Effects of specimen geometry and loading conditions on the crack tip plastic zone, in *Mechanical Behaviour of Materials*, Vol. 1, Society of Materials and Science, Japan, pp. 394–407.

Isida, M. (1973), Laurent series expansion for internal crack problems, in *Method of Analysis and Solutions of Crack Problems*, Chap. 2, edited by G. C. Sih, Noordhoff, Leyden.

Isida, M., Ushijima, N., and Kishine, N. (1981), Rectangular plates, strips and wide plates containing internal cracks under various boundary conditions, *Trans. Japan Soc. Mech. Engrs., A*, **47**, No. 413, 27–35. (In Japanese.)

Kachanov, M. (1980), Continuum model of medium with cracks, *J. Engng. Mech. Division, ASCE*, **106**, EM5, 1039–1051.

Karihaloo, B. L. (1978), Fracture characteristics of solids containing doubly-periodic arrays of cracks, *Proc. Roy. Soc. London*, **A360**, 373–387.

Karihaloo, B. L. (1979), Fracture of solids containing arrays of cracks, *Engng. Fracture Mech.*, **12**, 49–77.

Leguillon, D. and Sanchez-Palencia, E. (1982), On the behaviour of a cracked elastic body with (or without) friction, *J. Mécan. Appl.*, **1**, No. 2, 195–209.

Mura, T. (1986), *Micromechanics of Defects*, 2nd rev. ed., Martinus Nijhoff, The Netherlands.

Muskhelishvili, N. I. (1953), *Some Basic Problems in the Mathematical Theory of Elasticity*, (translated from the Russian, edited by J. R. M. Radok, Noordhoff, Groningen.

Oda, M. (1982), Fabric tensor for discontinuous geological materials, *Soils and Foundations*, **22**, No. 4, 96–108.

Oda, M. (1983), A method for evaluating the effect of crack geometry on the mechanical behavior of cracked rock mass, *Mech. Mater.*, **2**, No. 2, 163–171.

Oda, M., Suzuki, K., and Maeshibu, T. (1984), Elastic compliance for rock like materials with random cracks, *Soils and Foundations*, **24**, No. 3, 27–40.

Salganik, R. L. (1973), Mechanics of bodies with many cracks, *Izv. AN SSSR. MTT*, **8**, No. 4, 149–158. (English translation, *Mechanics of Solids*, **8**, No. 4, 135–143.)

Sih, G. C., Paris, P. C., and Irwin, G. R. (1965), On cracks in rectilinearly anisotropic bodies, *Int. J. Fracture Mech.*, **1**, 189–203.

Vakulenko, A. A. and Kachanov, M. L. (1971), Continual theory of a medium with cracks, *Izv. AN SSSR. MTT*, **6**, No. 4, 159–166. (English translation, *Mechanics of Solids*, **6**, No. 4, 145–151.)

Vavakin, A. S. and Salganik, R. L. (1975), Effective characteristics of nonhomogeneous media with isolated nonhomogeneities", *Izv. AN SSSR. MTT*, **10**, No. 3, 65–75. (English translation, *Mechanics of Solids*, **10**, No. 3, 58–66.)

Walpole, L. J. (1981), Elastic behavior of composite materials: Theoretical foundations, *Adv. Appl. Mech.*, **21**, 169–242.

Walsh, J. B. (1965a), The effect of cracks on the compressibility of rock. *J. Geophys. Res.*, **70**, No. 2, 381–389.

Walsh, J. B. (1965b), The effect of cracks on the uniaxial elastic compression of rocks, *J. Geophys. Res.*, **70**, No. 2, 399–411.

Westergaard, H. M. (1939), Bearing pressures and cracks, *J. Appl. Mech.*, **6**, 49–53.

Willis, J. R. (1981), Variational and related methods for the overall properties of composites, *Adv. Appl. Mech.*, **21**, 1–78.

The Elastic and Diffusional Interaction of Spherical Inhomogeneities in a Uniaxial Stress Field

William C. Johnson
Carnegie Mellon University, Department of Metallurgical Engineering and Materials Science, Pittsburgh, PA 15213-3890, U.S.A.

Abstract

The influence of an applied uniaxial stress field on the coarsening kinetics of two elastically and diffusionally interacting coherent spherical particles (inhomogeneities) is examined. The particles do not possess an eigenstrain but have elastic constants different from those of the matrix phase, and it is assumed that compositional inhomogeneity in the matrix does not engender stress. Under these assumptions, the elastic fields are obtained to first order in the difference between the elastic constants of the particle and matrix phases. The coupled equations of diffusion and elasticity are then solved and expressions obtained for the matrix composition field, the local normal velocities of the particles, the isotropic particle growth rates, and the velocity of the particles' centers of mass. The coarsening kinetics are independent of the sign of the elastic inhomogeneity to first order in the difference in elastic constants, when the far-field composition is chosen such that mass is only exchanged between particles. There are large ranges of the thermophysical parameters that allow for inverse coarsening, i.e., the smaller particle grows at the expense of the larger particle. The solutions obtained herein provide a basis for examining the influence of an external stress on microstructural evolution in the late stages of the precipitation process (coarsening).

1. Introduction

Professor Mura has made numerous contributions to the fields of mechanics and materials science. One of his important contributions to the study of diffusional phase transformations has been the development of techniques to calculate elastic fields associated with misfitting inclusions (second phase domains that possess an eigenstrain or stress-free transformation strain), and elastic inhomogeneities in an external stress field (domains that do not possess an eigenstrain but which have elastic constants different from those of the matrix phase). In this work, we show how these techniques can be used to understand the influence of an applied stress field on the kinetics of the late stages of a first-order diffusional phase transformation. In particular, we examine how a uniaxial stress field changes the coarsening kinetics of two elastically and diffusionally interacting spherical inhomogeneities.

A first-order diffusional phase transformation occurs when an initially, compositionally homogeneous binary alloy is cooled to a temperature for which the equilibrium thermodynamic state consists of two phases of different composition. In the early stages of the phase separation or precipitation process, composition fluctuations give rise to discrete domains or particles of one phase (the precipitate or minority phase) embedded in a connected majority phase (the matrix). These particles grow via mass diffusion, completing with one another for the diminishing solute in the matrix phase, and resulting in a dispersion of second-phase particles. Significant internal stresses may result during the precipitation process if the precipitate and matrix possess different lattice parameters or crystal structures (eigenstrains).

During the late stages of the precipitation process, the supersaturation of the matrix phase is small and much of the driving force for precipitation has been dissipated. The compositions and volume fractions of the phases are approaching their equilibrium values. In the absence of elastic stress, the two-phase system is still not in equilibrium owing to the interfacial energy associated with the large particle–matrix interfacial area. Increasing the size scale of the particles, at fixed volume fraction, decreases the total interfacial area and, hence, lowers the interfacial and system energies. This stage of the precipitation process is termed coarsening or Ostwald ripening (Ostwald, 1901; Lifshitz and Slyozov, 1961; Wagner, 1961). When elastic stresses are present as a result of the eigenstrains or an externally applied stress field, the energy of the system depends upon the absolute size and spatial arrangement of the particles as well as the interfacial energy. This is because of the elastic interaction between particles. As the elastic interaction energy of two equal-sized particles increases as the cube of the particle radius, while the interfacial energy increases as the square of the particle radius, an increase in the size scale of the system need not decrease the system energy (Ardell and Nicholson, 1966; Johnson, 1984). Rather, as recent kinetic analyses of elastically and diffusionally interacting particles have shown, the system energy may be decreased via particle migrations (Geguzin and Krivoglaz, 1973; Voorhees and Johnson, 1988) or inverse coarsening (Johnson, 1984; Schmeizer and Gutzow, 1986; Kawasaki and Enomoto, 1988; Johnson *et al.*, 1988).

The application of an external stress during the precipitation process may significantly alter the microstructure of two-phase coherent solids. Well-documented examples of the influence of an applied stress include changes in the morphology of isolated particles (Tien and Copley, 1971) and the development of rafted structures in which interacting particles align perpendicular to the direction of the applied elastic field (Mackay and Ebert, 1983; Nathal and Ebert, 1983). Furthermore, like an external pressure, an applied uniaxial stress may change the equilibrium compositions and volume fractions of the coherent phases from that expected in the absence of an external stress. If the external stress is applied during the precipitation process, solute fluxes and changes in the particle volume fraction result with significant effects on the particle coarsening kinetics (Tien and Copley, 1971).

Here, the influence of an applied uniaxial stress field on the coarsening kinetics of two elastically and diffusionally interacting coherent particles (inhomogeneities) is examined. By limiting this initial analysis to two particles, it is hoped that some of the relevant physics associated with the elastic stress can be identified. The field equations and boundary conditions necessary to analyze the coupled elasticity and diffusion problem are presented in Section 2 along with the assumptions employed in the model. In Section 3, a solution to the elastic field associated with two coherent spherical inhomogeneities in an uniaxial stress field is presented. These elastic fields are then used in the solution of the diffusion equation to obtain expressions for the matrix composition field, isotropic particle growth rates, and particle translation rates assuming that the matrix supersaturation is small. Results of the calculations are presented in Section 4 and the kinetic analysis is compared to elastic energy calculations in Section 5.

2. Theoretical Formulation

2.1. Physical Assumptions

We consider a two-phase substitutional binary alloy system of components A and B. In the absence of an external stress, the system is one of coherent second-phase spherical particles (the β phase) in an infinite (α) matrix. The two phases possess identical crystal lattices and lattice parameters so that no stresses are engendered by transformation strains or eigenstrains. The partial molar volumes of the solute and solvent are assumed to be equal so that no stresses are generated as a result of concentration gradients within each phase. Interfacial stresses, arising from the excess forces associated with a coherent interface, are assumed to be negligible. The particle elastic constants, c^{β}_{ijkl}, may be different from those of the matrix, c^{α}_{ijkl}, so that nonuniform stresses result when an applied field is present.

The two particles are of radii r_- and r_+, respectively, as depicted in Fig. 1. The relative position of the particles, as measured with respect to the crystallographic axes, is given by the vector $\mathbf{d}$ directed from the particle of radius r_- to that of radius r_+. The applied uniaxial field is directed along the x_3 axis and the two particles are assumed to lie in the x_1–x_3 plane as the present analysis is limited to elastically isotropic systems. The stress field approaches that of the applied field far from the particles while the concentration field approaches a constant, c_∞. As shown in Fig. 1, there are two Cartesian coordinate systems that can be associated with this system. One Cartesian system is based on the crystallographic axes with the origin located at the center of the particle of radius r_- while the other is based on an orthogonal, bispherical coordinate system. In the bispherical coordinate system, the origin lies along the vector connecting the two particle centers with the x_3 axis lying parallel to $\mathbf{d}$.

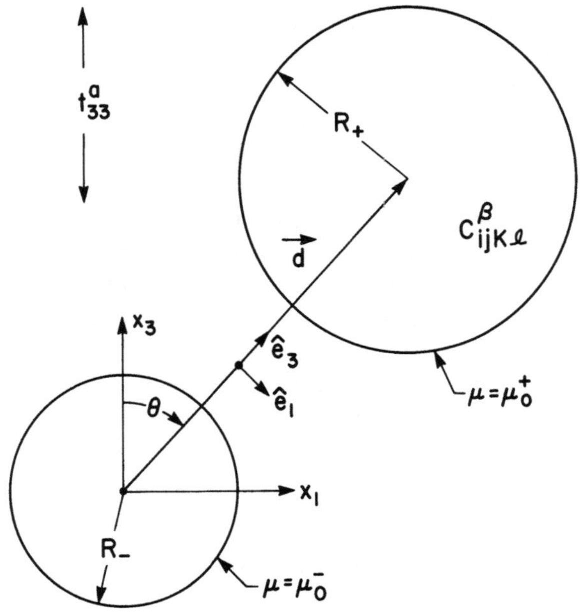

FIG. 1. Schematic diagram illustrating the two-particle system. The only source of stress is the uniaxial stress field directed along the x_3 axis, which is perturbed by the two spherical inhomogeneities.

In the presence of the applied stress field, the morphology of the particles departs from sphericity owing to the nonspherically symmetric elastic and composition fields. Studying the temporal evolution of these systems thus entails solving a two-body, free-boundary value problem wherein both the diffusion and elastic fields surrounding arbitrarily shaped particles must be solved self-consistently. As these problems would involve intensive numerical calculations, the problem is posed here in terms of an initial value problem; the particle growth and migration rates are determined at time $t = 0$ when the particles are spherical.

2.2. Field Equations and Boundary Conditions

In this section the governing field equations and boundary conditions necessary to determine the composition and elastic fields, the local normal interfacial velocities, the isotropic particle growth rates, and the instantaneous velocity of the particles' centers of mass (migration rate) are presented. Lower- and uppercase letters are used to denote dimensional and nondimensional quantities, respectively. A superscript is used to denote the α or β phase. This development parallels that employed for misfitting particles in the absence of an external stress field (Johnson *et al.*, 1988). These equations are applicable to one, two, or many interacting inhomogeneities.

2.2.1. *Elasticity*

In what follows, the reference state for the measurement of deformation is the unstressed matrix phase. Elastic equilibrium requires that the stress, t_{ij}, satisfies Cauchy's relation in each phase

$$t_{ij,j} = 0, \tag{2.1}$$

where the comma denotes differentiation with respect to x_j and repeated indices are summed. The strain, e_{ij}, and stress are assumed to be linearly related by

$$t^{\alpha}_{ij} = c^{\alpha}_{ijkl} e^{\alpha}_{kl}, \tag{2.2}$$

and

$$t^{\beta}_{ij} = c^{\beta}_{ijkl} e^{\beta}_{kl}. \tag{2.3}$$

The coherency constraint requires both the continuity of traction and the continuity of displacement across the particle–matrix interface

$$t^{\alpha}_{ij} n^{\alpha}_j + t^{\beta}_{ij} n^{\beta}_j = 0, \tag{2.4}$$

$$u^{\alpha}_i = u^{\beta}_i. \tag{2.5}$$

$\hat{n}$ is the outward pointing unit normal of the designated phase and $\mathbf{u}$ is the displacement. The final boundary condition requires that the stress field approaches that of the external field far from the particles.

2.2.2. *Diffusion*

The local flux of solute, $\mathbf{j}$, is also expressed with respect to the unstressed matrix (reference state). Since the solute and solvent are assumed to possess identical partial molar volumes, the diffusion field for the isotropic system must obey (Larche and Cahn, 1985)

$$\frac{\partial c}{\partial t} = D_c \nabla^2 c, \tag{2.6}$$

where the diffusion coefficient, D_c, is assumed independent of composition and c is the solute (B) composition measured in mole fraction.

The boundary conditions for diffusion are obtained by assuming local thermodynamic equilibrium at the particle–matrix interface. The local interfacial equilibrium concentrations, c^{α} and c^{β}, are position-dependent as the elastic field is position-dependent. They can be expressed in terms of the unstressed equilibrium compositions, c^{α}_i and c^{β}_i, i.e., those compositions that would be obtained from the (unstressed) phase diagram, as (Voorhees and Johnson, 1986)

$$c^{\alpha}(\mathbf{x}) = c^{\alpha}_i + \frac{[g(\mathbf{x}) + 2\sigma/r]}{\chi^{\alpha}(c^{\beta}_i - c^{\alpha}_i)}, \tag{2.7}$$

and

$$c^{\beta}(\mathbf{x}) = c^{\beta}_i + \frac{[g(\mathbf{x}) + 2\sigma/r]}{\chi^{\beta}(c^{\beta}_i - c^{\beta}_i)}, \tag{2.8}$$

where σ is the interfacial energy and r is the radius of the relevant particle. $g(\mathbf{x})$ is a function of the elastic field at a point $\mathbf{x}$ on the particle–matrix interface

$$g(\mathbf{x}) = (e_{ij}^{\alpha} - e_{ij}^{\beta})t_{ij}^{\alpha} + \tfrac{1}{2}(e_{ij}^{\beta} - e_{ij}^{\mathrm{T}})t_{ij}^{\beta} - \tfrac{1}{2}t_{ij}^{\alpha}e_{ij}^{\alpha}, \tag{2.9}$$

and

$$\chi = \frac{\rho_0 k\theta}{c_i(1 - c_i)}\left[1 + \frac{\partial \ln \gamma}{\partial \ln c}\right], \tag{2.10}$$

where ρ_0 is the density of lattice sites in the reference state, θ is the abolute temperature, k is Boltzmann's constant, and γ is the activity coefficient of the solute in solution. Equations (2.7) and (2.8) result from assuming that the change in equilibrium composition upon application of the external stress is small.

The final diffusion boundary condition assumes that the composition field approaches a constant far from the two particles. This far-field composition, c_∞, can be arbitrarily specified or chosen such that mass is exchanged only between the particles and not with the surroundings.

2.2.3. Particle Growth Rates

The local normal interfacial velocity, v_{n}, is obtained from a flux conservation condition at the particle–matrix interface. When mass flow occurs only in the matrix phase, this flux condition is

$$(c^{\beta} - c^{\alpha})v_{\mathrm{n}} = -D_{\mathrm{c}}\nabla c^{\alpha}\cdot \hat{n}^{\alpha}, \tag{2.11}$$

where the normal interfacial velocity is measured with respect to the reference state.

The rate of change of the particle radius with time, or the isotropic particle growth rate, dr/dt, is the surface average of v_{n}. When $\chi^{\alpha} = \chi^{\beta}$, the jump in composition across the interface is independent of position and $c^{\beta} - c^{\alpha} = c_i^{\beta} - c_i^{\alpha}$. The isotropic growth rate then becomes

$$\frac{dr}{dt} = \frac{-D_{\mathrm{c}}}{4\pi r^2(c_i^{\beta} - c_i^{\alpha})}\iint_s \nabla c\cdot \hat{n}^{\alpha}\, ds, \tag{2.12}$$

where s is the particle–matrix interface.

2.2.4. Particle Migration Rates

The instantaneous velocity of the center of mass of a spherical particle, $\mathbf{v}_{\mathrm{cm}}$, can be expressed formally as

$$\mathbf{v}_{\mathrm{cm}} = \frac{3}{4\pi r^2}\iint_s v_n\hat{n}\, ds, \tag{2.13}$$

where the local normal growth rate is given by (2.11).

2.3. Nondimensionalization

The number of independent thermophysical variables can be significantly reduced through nondimensionalization. In what follows, the nondimensional quantities are denoted using uppercase letters. A scaled concentration, C, is defined as

$$C = \frac{c - c_i^\alpha}{c_i^\beta - c_i^\alpha}, \tag{2.14}$$

where c_i^α is the equilibrium matrix composition of an unstressed system consisting of an isolated spherical particle of β in an infinite matrix. c_i^α is the unstressed solvus.

A nondimensional stress, T_{ij}, is defined in terms of the applied stress. If in the absence of the spherical inhomogeneities the only nonzero component of the stress tensor is $t_{33} = \tau^a$, a convenient nondimensional stress is

$$T_{ij} = \frac{\tau^a t_{ij}}{\chi c_{44}^\alpha (c_i^\beta - c_i^\alpha)^2}. \tag{2.15}$$

A normalized strain, E_{ij}, is likewise defined as

$$E_{ij} = \frac{e_{ij} c_{44}^\alpha}{\tau^a}, \tag{2.16}$$

where c_{44}^α is the shear modulus of the matrix phase. The applied elastic field, in the absence of the perturbation induced by the elastic inhomogeneity, is denoted by a superscript "a."

For the two-particle problem, there are two possible characteristic length scales; the capillary length, l_c, and a length that depends upon the size of the particles, q. The capillary length is defined in the usual way as

$$l_c = \frac{2\sigma}{\chi (c_i^\beta - c_i^\alpha)^2}. \tag{2.17}$$

The other characteristic length, q, is here defined as

$$q = \left[\frac{3(v_- + v_+)}{8\pi}\right]^{1/3}, \tag{2.18}$$

where v_- and v_+ are the volumes of the respective particles. Either characteristic length scale can be employed in the nondimensionalization. For ease in computing the elastic fields, q is chosen as the characteristic length for the two-particle problem. The nondimensional lengths are then of the form $R_+ = r_+/q$. As it is the ratio of these characteristic lengths that determines the coarsening behavior of the system, a relative capillary length, L_c, is also defined (Johnson *et al.*, 1988)

$$L_c = \frac{l_c}{q} = \frac{2\sigma}{\chi (c_i^\beta - c_i^\alpha)^2 q}. \tag{2.19}$$

Finally, a dimensionless time, τ, is defined as

$$\tau = \frac{tD_c}{q^2}. \tag{2.20}$$

It is convenient to express the relative size of the particles in terms of a parameter R as

$$R = \frac{R_+ - R_-}{R_+ + R_-}. \tag{2.21}$$

When both particles are of the same size, $R_- = R_+ = 1$ and $R = 0$.

When expressed in terms of the nondimensional variables, the governing equations of elasticity, (2.1)–(2.5), become

$$T_{ij,j} = 0, \tag{2.22}$$

$$T_{ij}^{\alpha} = C_{ijkl}^{\alpha} E_{kl}^{\alpha}, \tag{2.23}$$

$$T_{ij}^{\beta} = C_{ijkl}^{\beta} E_{kl}^{\beta}, \tag{2.24}$$

$$T_{ij}^{\alpha} + T_{ij}^{\beta} n_j^{\beta} = 0, \tag{2.25}$$

$$U_i^{\alpha} = U_i^{\beta}. \tag{2.26}$$

C_{ijkl} are dimensionless elastic constants defined by

$$C_{ijkl} = \frac{(\tau^{a})^2 c_{ijkl}^{\alpha}}{\chi (c_{44}^{\alpha})^2 (c_i^{\beta} - c_i^{\alpha})^2}, \tag{2.27}$$

and

$$C_{ijkl}^{\beta} = \frac{(\tau^{a})^2 c_{ijkl}^{\beta}}{\chi (c_{44}^{\alpha})^2 (c_i^{\beta} - c_i^{\alpha})^2}. \tag{2.28}$$

The diffusion equation becomes

$$\frac{\partial C}{\partial \tau} = \nabla^2 C. \tag{2.29}$$

The position-dependent diffusional boundary conditions at the particle–matrix interface for each of the phases are

$$C^{\alpha}(\mathbf{X}) = G(\mathbf{X}) + \frac{L_c}{R}, \tag{2.30}$$

$$C^{\beta}(\mathbf{X}) = 1 + G(\mathbf{X}) + \frac{L_c}{R}, \tag{2.31}$$

where

$$G(\mathbf{X}) = (E_{ij}^{\alpha} - E_{ij}^{\beta}) T_{ij}^{\alpha} + \tfrac{1}{2} T_{ij}^{\beta} E_{ij}^{\beta} - \tfrac{1}{2} T_{ij}^{\alpha} E_{ij}^{\alpha}. \tag{2.32}$$

In the absence of an applied stress, the stress field vanishes and $G(\mathbf{X}) = 0$. In this case, the diffusional boundary conditons are determined only by the capillary contribution.

The nondimensional local normal velocity is

$$V_{\mathrm{n}} = -\nabla C \cdot \hat{n}^{\alpha}. \tag{2.33}$$

The nondimensional isotropic particle growth rates are

$$\frac{dR_+}{d\tau} = \dot{R}_+ = \frac{-1}{4\pi R_+^2} \iint_{S_+} \nabla C \cdot \hat{n}^{\alpha}\, dS, \tag{2.34}$$

and

$$\frac{dR_-}{d\tau} = \dot{R}_- = \frac{-1}{4\pi R_-^2} \iint_{S_-} \nabla C \cdot \hat{n}^{\alpha}\, dS. \tag{2.35}$$

The dimensionless velocity of each particle's center of mass is

$$\mathbf{V}_{\mathrm{cm}}^{+} = \frac{-3}{4\pi R_+^2} \iint_{S_+} V_{\mathrm{n}} \hat{n}^{\alpha}\, dS, \tag{2.36}$$

and

$$\mathbf{V}_{\mathrm{cm}}^{-} = \frac{-3}{4\pi R_-^2} \iint_{S_-} V_{\mathrm{n}} \hat{n}^{\alpha}\, dS. \tag{2.37}$$

3. Solutions

When solute and solvent possess identical partial molar volumes, compositional inhomogeneity does not engender stress. The elastic field depends only upon the sign and magnitude of the applied stress field, the difference in elastic constants between particle and matrix, and the relative position and size of the particles. This allows the elastic field to be calculated independently of the concentration field. Once the elastic fields have been determined, they can be used to calculate the local equilibrium compositions at the particle–matrix interface. This provides the boundary conditions necessary to solve the diffusion problem. In Section 3.1, an approximate solution to the elastic field of two spherical inhomogeneities perturbing an uniaxial stress is presented. In Section 3.2, the elastic fields are used to calculate the interfacial boundary conditions from which the matrix composition field, isotropic particle growth rates, and particle migration rates are then determined.

3.1. Elastic Fields

When the system is elastically inhomogeneous ($C_{ijkl}^{\alpha} \neq C_{ijkl}^{\beta}$), and no misfit strains are present, the strain field is known to satisfy the following integral equation (Chen and Young, 1977; Johnson *et al.*, 1980):

$$u_{m,n} = u_{m,n}^{\mathrm{a}} + \Delta c_{ijkl} \iiint_{V_+} G_{ml,kn}(\mathbf{x} - \mathbf{x}') u_{i,j}\, d^3x'$$
$$+ \Delta c_{ijkl} \iiint_{V_-} G_{ml,kn}(\mathbf{x} - \mathbf{x}') u_{i,j}\, d^3x', \tag{3.1}$$

where u_m^a would be the displacement field resulting from the applied stress in the absence of the elastic inhomogeneities, the integration is performed over the volume of each precipitate, G_{ml} is the infinite space elastic Green's function for an elastically *homogeneous* system, and

$$\Delta c_{ijkl} = c_{ijkl}^{\beta} - c_{ijkl}^{\alpha}. \tag{3.2}$$

If the difference between the elastic constants of the particle and those of the matrix is small, i.e.,

$$\left|\frac{\Delta c_{ijkl}}{c_{ijkl}^{\alpha}}\right| \ll 1, \tag{3.3}$$

the elastic field in the presence of the inhomogeneities can be approximated to first order in the difference in the elastic constants using the Born approximation (Johnson, 1983). If $u_{m,n}^a$ is spatially uniform, as would result from a uniform applied stress field, the displacement gradients can be approximated as

$$u_{m,n} \approx u_{m,n}^a + u_{m,n}^+ + u_{m,n}^-, \tag{3.4}$$

where

$$u_{m,n}^+ = \Delta c_{ijkl} u_{i,j}^a \iiint_{V_+} G_{ml,kn}(\mathbf{x} - \mathbf{x}')\, d^3x', \tag{3.5}$$

and

$$u_{m,n}^- = \Delta c_{ijkl} u_{i,j}^a \iiint_{V_-} G_{ml,kn}(\mathbf{x} - \mathbf{x}')\, d^3x'. \tag{3.6}$$

We now limit the analysis to isotropic systems in which the Poisson ratio, ν, of each phase is equal, i.e.,

$$\nu^{\alpha} = \nu^{\beta} = \nu. \tag{3.7}$$

We further define

$$\frac{c_{12}^{\beta}}{c_{12}^{\alpha}} = \frac{c_{44}^{\beta}}{c_{44}^{\alpha}} = \delta. \tag{3.8}$$

When $\delta = 1$, the system is elastically homogeneous. The difference between the elastic constants of the particles and matrix can be expressed in terms of δ as

$$\Delta c_{ijkl} = (\delta - 1) c_{ijkl}^{\alpha}. \tag{3.9}$$

It is the term $(\delta - 1)$ that is assumed to be small in this analysis, $|\delta - 1| \ll 1$.

For an isotropic system, the infinite-space elastic Green's function is

$$G_{ml}(\mathbf{x} - \mathbf{x}') = \frac{\delta_{ml}}{4\pi c_{44}^{\alpha}|\mathbf{x} - \mathbf{x}'|} - \frac{1}{16\pi c_{44}^{\alpha}(1 - \nu)} \frac{\partial^2}{\partial x_m\, \partial x_l}|\mathbf{x} - \mathbf{x}'|. \tag{3.10}$$

When the applied stress is a uniaxial stress field directed along the x_3 axis,

$$t_{ij}^a = \tau^a \delta_{i3} \delta_{j3}, \tag{3.11}$$

and

$$u_{ij}^{\mathrm{a}} = \frac{\tau^{\mathrm{a}}}{2c_{44}^{\alpha}}\left\{\frac{-\nu}{(1+\nu)}\delta_{ij} + \delta_{i3}\delta_{j3}\right\}. \tag{3.12}$$

Substituting (3.12) into (3.5) and (3.6) and simplifying yields the following dimensional expressions:

$$u_{m,n}^{-}(\mathbf{x}) = \frac{\tau^{\mathrm{a}}(\delta-1)}{4\pi c_{44}^{\alpha}}\left\{\delta_{m3}\phi_{,n3}(\mathbf{x}) - \frac{1}{4(1-\nu)}\psi_{,mn33}(\mathbf{x})\right\}, \tag{3.13}$$

$$u_{m,n}^{+}(\mathbf{x}) = \frac{\tau^{\mathrm{a}}(\delta-1)}{4\pi c_{44}^{\alpha}}\left\{\delta_{m3}\phi_{,n3}(\mathbf{x}-\mathbf{d}) - \frac{1}{4(1-\nu)}\psi_{,mn33}(\mathbf{x}-\mathbf{d})\right\}. \tag{3.14}$$

d is vector locating the center of the second inhomogeneity with respect to the first, and ϕ and ψ are the harmonic and biharmonic potential functions, respectively (Eshelby, 1957). Equations (3.13) and (3.14) provide an analytic approximation of the elastic strain field to first order in the difference between the shear moduli of the particle and matrix phases. In this limit, the strain fields superimpose and the method can be extended to many particles. When the difference between the particle and matrix shear moduli is large, techniques such as those developed by Moschovidis and Mura (1975) must be employed.

3.2. Diffusion Boundary Conditions

The strain fields determined above are now used to calculate the equilibrium compositions at the particle–matrix interface. The matrix equilibrium composition is used as the boundary condition in the solution of the diffusion problem. Substituting (2.32), (3.4), and (3.12)–(3.14) into (2.30) and then simplifying yields the following expression for the local equilibrium interfacial composition in nondimensional form for the particle of radius R_-:

$$G(\mathbf{X}) = G^s(\mathbf{X}) + G^i(\mathbf{X}) + \frac{L_c}{R_-}, \tag{3.15}$$

where

$$G^s(\mathbf{X}) = \frac{(\delta-1)C_{44}^{\alpha}}{4(1+\nu)} + \frac{(\delta-1)^2(10\nu-7)C_{44}^{\alpha}}{30(1-\nu)} + \frac{(\delta-1)^2 c_{44}^{\alpha}}{4(1-\nu)}\{(1-\nu)X_3^2 - X_3^4\}, \tag{3.16}$$

and

$$G^i(\mathbf{X}) = \frac{(\delta-1)^2 C_{44}^{\alpha}}{4\pi}\left\{\phi_{,33}(\mathbf{X}-\mathbf{D}) - \frac{1}{4(1-\nu)}\psi_{,3333}(\mathbf{X}-\mathbf{D})\right\}. \tag{3.17}$$

A similar expression exists for the particle of radius R_+. $G^s(\mathbf{X})$ is the contribution to the interfacial composition that is independent of the presence of the other precipitate. This term is position-dependent and therefore results in changes in precipitate morphology (Johnson, 1987). $G^i(\mathbf{X})$ is the contribution to the interfacial composition that results from the presence of the second inhomogeneity. It is this term that induces particle migration. The interfacial compositions can be determined analytically as the harmonic and biharmonic potential functions are known analytically for spherical particles. These interfacial compositions are now used as boundary conditions to compute the matrix composition fields.

3.3. Composition Fields

As the concern here is with the late stages of the precipitation process, both the supersaturation of and the concentration gradients in the matrix phase are small. This permits use of the quasi-stationary solution to the diffusion equation to approximate the composition field in the matrix phase; the time derivative of the concentration is set equal to zero in (2.29). Diffusion within the particles is assumed to be negligibly slow. These approximations allow analytic series solutions to be derived for the isotropic particle growth rates and the particle migration rates in terms of the formal expressions for the boundary conditions of diffusion, (2.30) and (2.31).

Laplace's equation for the concentration field can be solved analytically in the orthogonal bispherical coordinate system. This system consists of the three coordinates, μ, η, and ϕ. When the particles lie along the x_3 axis, the coordinate μ locates spheres of radius A csch $|\mu|$ with centers at the points (0, 0, A coth μ). Thus the surface of the particles of radii R_+ and R_- can be defined by μ_0^+ and μ_0^-, respectively. A is the dimensionless interfocal distance that is obtained by solving

$$D = [R_-^2 + A^2]^{1/2} + [R_+^2 + A^2]^{1/2}. \tag{3.18}$$

The coordinate η designates an azimuthal position on the surface of the sphere defined by μ while ϕ provides an axial position. If the system possesses axial symmetry, as would be the case when the particles are aligned along the x_3 axis, the composition field displays no ϕ dependence. The variables assume the values, $-\infty < \mu < \infty$, $0 \le \eta \le \pi$, and $0 \le \phi \le 2\pi$. The Cartesian coordinate system associated with the bispherical coordinate system need not coincide with that of the crystallographic axes and proper accounting of the translation and rotation between the two coordinate systems is essential.

The composition field of the matrix is

$$\begin{aligned} C(\mu, \eta, \phi) = C_\infty &+ (\cosh\mu - \cos\eta)^{1/2} \sum_{l=0}^{\infty} \{A_{l0} \cosh[(l + \tfrac{1}{2}\mu] \\ &+ B_{l0} \sinh[(l + \tfrac{1}{2})\mu]\} Y_{l0} \\ &+ (\cosh\mu - \cos\eta)^{1/2} \sum_{l=0}^{\infty} \sum_{m=1}^{l} \{\{C_{lm}^{+} \cosh[(l + \tfrac{1}{2})\mu] \\ &+ D_{lm}^{+} \sinh[(l + \tfrac{1}{2})\mu]\} Y_{lm}^{+} \\ &+ \{C_{lm}^{-} \cosh[(l + \tfrac{1}{2})\mu] + D_{lm}^{-} \sinh[(l + \tfrac{1}{2})\mu]\} Y_{lm}^{-}\}, \end{aligned} \tag{3.19}$$

where the Y_{lm} are the spherical harmonics as defined in the Appendix.

The coefficients of the eigenfunction expansion for the composition field, A_{l0}, B_{l0}, C_{lm}^{+}, C_{lm}^{-}, D_{lm}^{+}, and D_{lm}^{-}, are determined from the local equilibrium interfacial concentrations, given by (3.15), and the orthogonality conditions for the spherical harmonics. This is accomplished by setting $\mu = \mu_0^+$ in (3.19), equating the resultant expression to the equilibrium interfacial composition, and then using the orthogonality condition for spherical harmonics to obtain one equation for two of the unknown coefficients. The second equation is obtained by setting $\mu = \mu_0^-$ in (3.19), equating the resultant expression to the equilibrium interfacial composition of the particle with radius R_-, and again invoking orthogonality. Solving the two linear simultaneous equations for each set of eigenfunction coefficients gives, after some algebra:

$$A_{l0} = \frac{\{Q_{l0}^{+} \sinh[(l+\frac{1}{2})\mu_0^-] - Q_{l0}^{-} \sinh[(l+\frac{1}{2})\mu_0^+]\}}{\sinh[(l+\frac{1}{2})(\mu_0^- - \mu_0^+)]}, \tag{3.20}$$

$$B_{l0} = \frac{\{Q_{l0}^{-} \cosh[(l+\frac{1}{2})\mu_0^+] - Q_{l0}^{+} \cosh[(l+\frac{1}{2})\mu_0^-]\}}{\sinh[(l+\frac{1}{2})(\mu_0^- - \mu_0^+)]}, \tag{3.21}$$

$$C_{lm}^{+} = \frac{\{R_{lm}^{+} \sinh[(l+\frac{1}{2})\mu_0^-] - R_{lm}^{-} \sinh[(l+\frac{1}{2})\mu_0^+]\}}{\sinh[(l+\frac{1}{2})(\mu_0^- - \mu_0^+)]}, \tag{3.22}$$

$$C_{lm}^{-} = \frac{\{S_{lm}^{+} \sinh[(l+\frac{1}{2})\mu_0^-] - S_{lm}^{-} \sinh[(l+\frac{1}{2})\mu_0^+]\}}{\sinh[(l+\frac{1}{2})(\mu_0^- - \mu_0^+)]}, \tag{3.23}$$

$$D_{lm}^{+} = \frac{\{R_{lm}^{-} \cosh[(l+\frac{1}{2})\mu_0^+] - R_{lm}^{+} \cosh[(l+\frac{1}{2})\mu_0^-]\}}{\sinh[(l+\frac{1}{2})(\mu_0^- - \mu_0^+)]}, \tag{3.24}$$

$$D_{lm}^{-} = \frac{\{S_{lm}^{-} \cosh[(l+\frac{1}{2})\mu_0^+] - S_{lm}^{+} \cosh[(l+\frac{1}{2})\mu_0^-]\}}{\sinh[(l+\frac{1}{2})(\mu_0^- - \mu_0^+)]}, \tag{3.25}$$

where

$$Q_{l0}^{+} = \int_0^{2\pi}\int_0^{\pi} \frac{[C(\mu_0^+, \eta, \phi) - C_\infty] Y_{l0} \sin\eta \, d\eta \, d\phi}{[\cosh\mu_0^+ - \cos\eta]^{1/2}}, \tag{3.26}$$

$$Q_{l0}^{-} = \int_0^{2\pi}\int_0^{\pi} \frac{[C(\mu_0^-, \eta, \phi) - C_\infty] Y_{l0} \sin\eta \, d\eta \, d\phi}{[\cosh\mu_0^- - \cos\eta]^{1/2}}, \tag{3.27}$$

$$R_{lm}^{+} = \int_0^{2\pi}\int_0^{\pi} \frac{[C(\mu_0^+, \eta, \phi) - C_\infty] Y_{lm}^{+} \sin\eta \, d\eta \, d\phi}{[\cosh\mu_0^+ - \cos\eta]^{1/2}}, \tag{3.28}$$

$$R_{lm}^{-} = \int_0^{2\pi}\int_0^{\pi} \frac{[C(\mu_0^-, \eta, \phi) - C_\infty] Y_{lm}^{+} \sin\eta \, d\eta \, d\phi}{[\cosh\mu_0^- - \cos\eta]^{1/2}}, \tag{3.29}$$

$$S_{lm}^{+} = \int_0^{2\pi}\int_0^{\pi} \frac{[C(\mu_0^+, \eta, \phi) - C_\infty] Y_{lm}^{-} \sin\eta \, d\eta \, d\phi}{[\cosh\mu_0^+ - \cos\eta]^{1/2}}, \tag{3.30}$$

$$S_{lm}^{-} = \int_0^{2\pi}\int_0^{\pi} \frac{[C(\mu_0^-, \eta, \phi) - C_\infty] Y_{lm}^{-} \sin\eta \, d\eta \, d\phi}{[\cosh\mu_0^- - \cos\eta]^{1/2}}. \tag{3.31}$$

The local normal interfacial velocities can be determined directly from (2.33) and (3.19) as

$$V_n^+ = \nabla C \cdot \hat{n}^\alpha = \nabla C \cdot \hat{e}_\mu = \frac{\partial C}{\partial \mu} \qquad (\mu = \mu_0^+), \tag{3.32}$$

and

$$V_n^- = \nabla C \cdot \hat{n}^\alpha = -\nabla C \cdot \hat{e}_\mu = -\frac{\partial C}{\partial \mu} \qquad (\mu = \mu_0^-), \tag{3.33}$$

where $\hat{e}_\mu$ is the unit μ basis vector in bispherical coordinates.

Integrating the total normal growth rate over the surface of each sphere gives for the nondimensional isotropic particle growth rates:

$$\frac{dR_+}{d\tau} = \dot{R}_+ = \frac{-A}{\sqrt{8\pi}\, R_+^2} \sum_{l=0}^{\infty} (2l+1)^{1/2}(A_{l0} + B_{l0}), \tag{3.34}$$

and

$$\frac{dR_-}{d\tau} = \dot{R}_- = \frac{A}{\sqrt{8\pi}\, R_-^2} \sum_{l=0}^{\infty} (2l+1)^{1/2}(B_{l0} - A_{l0}). \tag{3.35}$$

The isotropic growth rates depend only on the eigenfunction coefficients A_{l0} and B_{l0} and assume the same functional form as for the case of misfit strains in the absence of an applied stress (Johnson *et al.*, 1988).

The instantaneous velocity of the center of mass of the particles can be expressed, after significant algebraic manipulation, as

$$\mathbf{V}_{\mathrm{cm}}^- = \frac{-\sinh^2 \mu_0^-}{2\sqrt{\pi}\, R_-} \sum_{l=0}^{\infty} [l(l+1)(2l+1)]^{1/2}\{T_l^1 \hat{e}_1 + T_l^2 \hat{e}_2 + T_l^3 \hat{e}_3\} \tag{3.36}$$

and

$$\mathbf{V}_{\mathrm{cm}}^+ = \frac{\sinh^2 \mu_0^+}{2\sqrt{\pi}\, R_+} \sum_{l=0}^{\infty} [l(l+1)(2l+1)]^{1/2}\{T_l^1 \hat{e}_1 + T_l^2 \hat{e}_2 + T_l^3 \hat{e}_3\}. \tag{3.37}$$

$\hat{e}_1$, $\hat{e}_2$, and $\hat{e}_3$ are the Cartesian coordinate basis vectors associated with the bispherical coordinate system and

$$T_l^1 = C_{l1}^+ \operatorname{sgn}(\mu_0)[e^{-(2l+1)|\mu_0|} - 2] - D_{l1}^+[e^{-(2l+1)|\mu_0|} + 2], \tag{3.38}$$

$$T_l^2 = C_{l1}^- \operatorname{sgn}(\mu_0)[e^{-(2l+1)|\mu_0|} - 2] - D_{l1}^-[e^{-(2l+1)|\mu_0|} + 2], \tag{3.39}$$

$$T_l^3 = \frac{[2l+1-\operatorname{sgn}(\mu_0)\coth \mu_0]}{[2l(l+1)]^{1/2}}\{A_{l0}[e^{-(2l+1)|\mu_0|} - 2] - \operatorname{sgn}(\mu_0) B_{l0}[e^{-(2l+1)|\mu_0|} + 2]\}. \tag{3.40}$$

Equations (3.38)–(3.40) apply to each particle with $\mu_0 = \mu_0^+$ or $\mu_0 = \mu_0^-$ as appropriate. If the system is axisymmetric, $C_{lm}^+ = C_{lm}^- = D_{lm}^+ = D_{lm}^- = 0$, and particle migrations are along the axis of symmetry. The translations can be expressed in terms of the crystal Cartesian coordinate system using vector rotation.

4. Results

In this section, the influence of an applied uniaxial stress field directed along the x_3 axis on particle growth rates and velocities of the particles' centers of mass is examined. Owing to the complex spatial dependence of the interfacial concentrations, the coefficients for the eigenfunction expansion for the composition field, (3.20)–(3.25), are determined numerically. In doing so, the nondimensional elastic constants, $\nu = \frac{1}{3}$ and $C_{44}^{\alpha} = 10^{-3}$, are used. The far-field composition is chosen such that net mass is exchanged only between the particles and not with the surroundings. The value of the nondimensional far-field potential, C_{∞}^{0}, that conserves the mass of the two particles is given by (Johnson *et al.*, 1988)

$$C_{\infty}^{0} = \frac{S_1}{\sqrt{8\pi}\, S_2}, \tag{4.1}$$

where

$$S_1 = \sum_{n=0}^{\infty} \frac{(2n+1)^{1/2}\{I_n^+ \sinh[(n+\frac{1}{2}\mu_0^-] - I_n^- \sinh[(n+\frac{1}{2}\mu_0^+]\}}{\sinh[(n+\frac{1}{2})(\mu_0^- - \mu_0^+)]}, \tag{4.2}$$

$$S_2 = \sum_{n=0}^{\infty} \left\{1 + \frac{\cosh[(n+\frac{1}{2})(\mu_0^- - \mu_0^+)] - \cosh[(n+\frac{1}{2})(\mu_0^- + \mu_0^+)]}{\sinh[(n+\frac{1}{2})(\mu_0^- - \mu_0^+)]}\right\}, \tag{4.3}$$

$$I_n^+ = \sqrt{\pi} \int_0^{\pi} \frac{C^{\alpha}(\mu_0^+, \eta) P_n(\cos\eta)(2n+1)^{1/2} \sin\eta \, d\eta}{[\cosh\mu_0^+ - \cos\eta]^{1/2}}, \tag{4.4}$$

$$I_n^- = \sqrt{\pi} \int_0^{\pi} \frac{C^{\alpha}(\mu_0^-, \eta) P_n(\cos\eta)(2n+1)^{1/2} \sin\eta \, d\eta}{[\cosh\mu_0^+ - \cos\eta]^{1/2}}. \tag{4.5}$$

To first order in the difference in elastic constants between particle and matrix, the equilibrium interfacial composition of (3.15) is independent of the sign of the applied stress. This is because the nondimensional elastic constants are independent of the sign of the applied stress (2.27). The contribution of the stress field of one particle to the equilibrium interfacial composition around the other particle is also independent of the sign of the elastic inhomogeneity. Since particle translation rates arise only from particle–particle interactions, the magnitude and sign of the translation rates are independent of the sign of the elastic inhomogeneity, i.e., the translation rates are proportional to $(\delta - 1)^2$. This is not true for the particle growth rates which depend upon the sign of the elastic inhomogeneity, as can be seen from (3.16). However, when the far-field composition is chosen such that mass is exchanged only between the two particles, the growth rates also become independent of the sign of $(\delta - 1)$. As the results presented here are determined for such a far-field composition, they are independent of the sign of both the applied stress field and elastic inhomogeneity $(\delta - 1)$. The value $|\delta - 1| = 0.1$ is employed throughout.

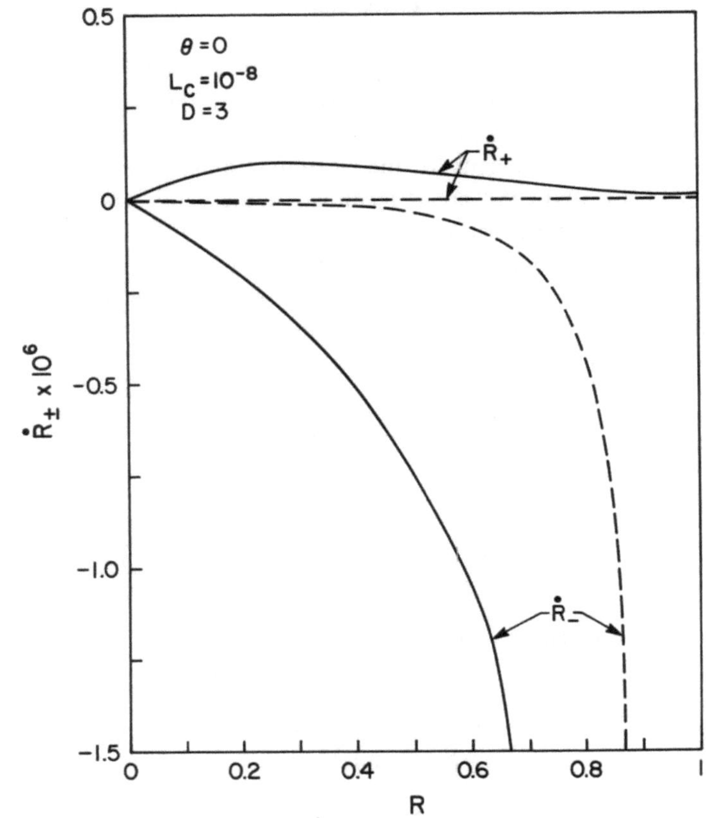

FIG. 2. Isotropic particle growth rates are plotted as a function of the relative particle size, R, when the particles are aligned along the direction of the applied stress. The growth rates are enhanced by over an order of magnitude with respect to the stress-free case (dashed lines).

4.1. Particle Growth Rates

Figures 2–4 depict the isotropic particle growth rates as a function of the relative size of the particles, R. The relative capillary length and intercenter distance are $L_c = 10^{-8}$ and $D = 3$. The broken lines give the particle growth rates in the absence of the applied stress field. In this case, all diffusion is capillarity induced. The solid lines give the isotropic growth rates upon application of the applied stress. In the absence of elastic stress, the larger particle always grows at the expense of the smaller particle and the magnitude of the growth rates scale with L_c.

Figure 2 corresponds to the case when the particles are aligned parallel to the direction of the uniaxial stress field ($\theta = 0$). The applied stress field does not change the qualitative behavior of the isotropic particle growth rates as

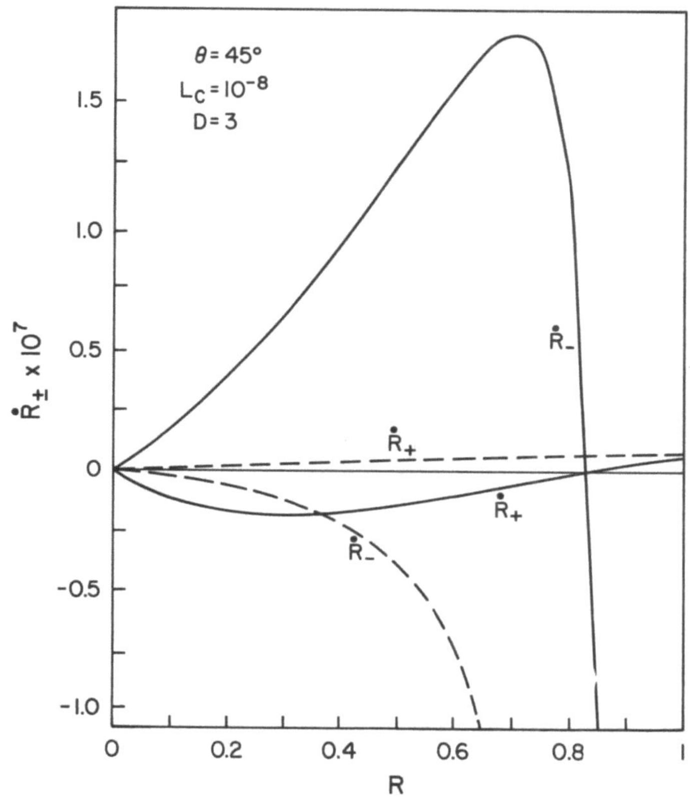

FIG. 3. Isotropic particle growth rates are plotted as a function of the relative particle size, R, when the particles are aligned at 45° to the direction of the applied stress. Regions of inverse coarsening exist for $0 < R < 0.8$.

a function of the relative particle size with respect to that predicted for the stress-free case; the larger particle always grows at the expense of the smaller particle. The magnitude of the isotropic growth rates, however, is increased by well over an order of magnitude.

As the orientation of the particles with respect to the applied stress is changed, the coarsening kinetics may change radically. Figure 3 shows the isotropic particle growth rates when the particles are oriented at $\theta = 45°$ to the applied stress field, all other parameters are the same as those of Fig. 2. The smaller particle now grows at the expense of the larger particle for $0 < R < 0.83$. This phenomenon is known as inverse coarsening (Johnson, 1984). For larger values of R, the small particle dissolves and the larger particle grows at its expense. The applied stress field alters both the sign and magnitude of the particle growth rates with respect to the classical case of capillarity-driven coarsening.

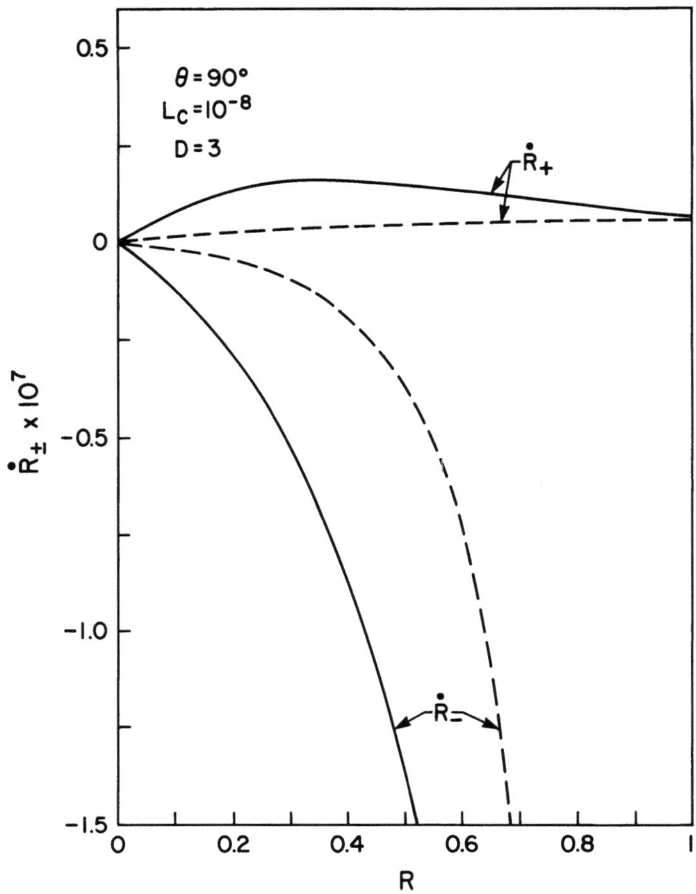

FIG. 4. Isotropic particle growth rates are plotted as a function of the relative particle size, R, when the particles are aligned in the direction perpendicular to the applied stress.

Figure 4 depicts the isotropic particle growth rates when the particles are aligned perpendicular to the applied stress field. The coarsening kinetics are qualitatively similar to those illustrated in Fig. 2 for $\theta = 0$. However, the magnitude of the isotropic growth rates is about a factor of 10 less than those depicted in Fig. 2. Figure 5 illustrates how the isotropic particle growth rates change as a function of the orientation of the particles with respect to the applied stress field for the relative particle size $R = 0.25$. Inverse coarsening is observed for $0.3 < \cos\theta < 0.8$ or $36° < \theta < 72°$.

Changing either the intercenter distance, D, or the relative capilarity length, L_c, changes both the qualitative and quantitative coarsening behavior of the system. As the intercenter distance is increased, the particle interactions decrease as D^{-3}. The range of inverse coarsening becomes smaller and eventually disappears while the magnitudes of the isotropic growth rates approach

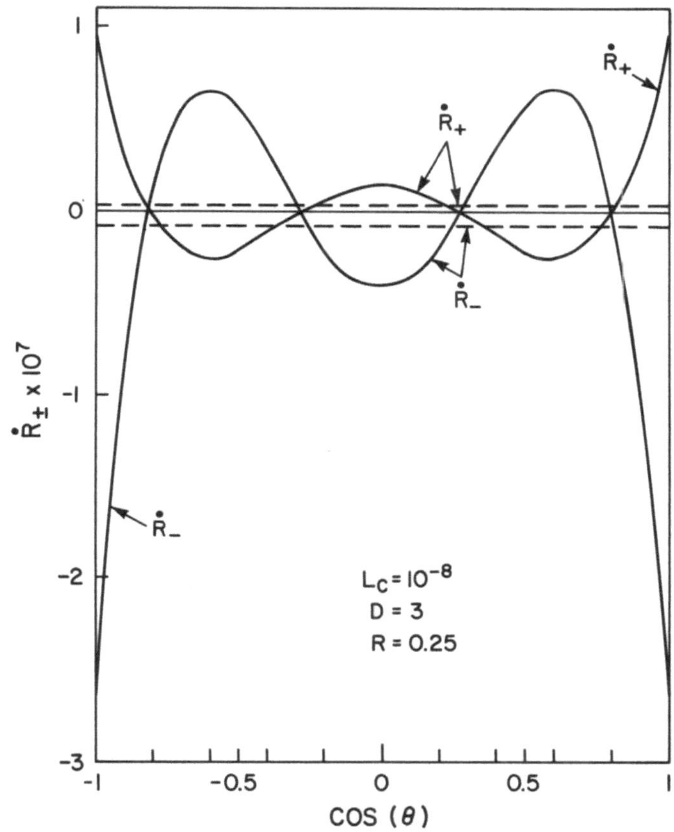

FIG. 5. Isotropic particle growth rates are plotted as a function of $\cos\theta$ and θ denotes the orientation of the particles with respect to the applied stress field. The relative particle size is $R = 0.25$. The dashed lines give the growth rates for the stress-free case.

those of the stress-free situation. Similar trends in the coarsening behavior are observed as the relative capillary length is increased. In this case, the contribution of elasticity to the thermodynamics, with respect to the capillarity term, becomes progressively smaller as L_c decreases.

4.2. Particle Translation Rates

The nonsymmetric elastic and diffusion fields associated with the interacting particles results in a translation of the particles' centers of mass. An illustration of the influence of the applied stress on the nondimensional particle translation rate is given in Fig. 6 as a function of the relative particle size, R. The solid and broken lines correspond to the translation rates of the large and small particles, respectively. The subscript refers to the component of the

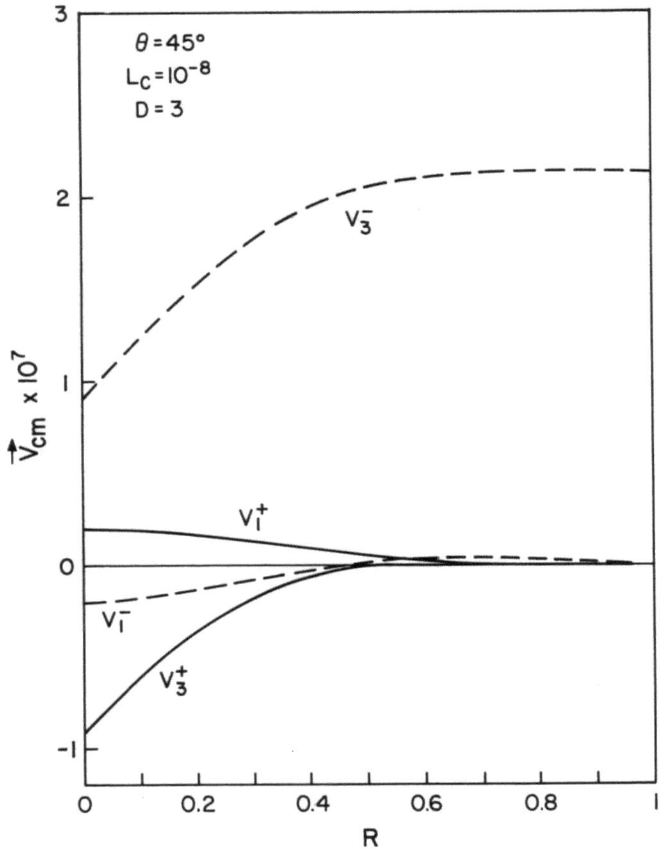

FIG. 6. The components of the velocity of the particles' centers of mass are plotted as a function of the relative particle size. Solid and dashed lines represent the large and small particles, respectively.

translation vector as measured with respect to the Cartesian basis associated with the crystal. Thus V_3^- is the component of the translation vector of the small particle parallel to the direction of the applied stress, or along the x_3 axis. Nondimensional translation rates of the particles in the absence of the applied stress field are on the order of 1×10^{-9} for identical conditions to those employed in Fig. 6 and are not shown. The nondimensional translation rate of the small particle in the applied stress field is approximately two orders of magnitude larger than the translation rate resulting from capilarity-induced diffusion. For this orientation, the centers of mass of the particles are approaching one another along the x_3 axis and moving apart more slowly along the x_1 direction. The distance between the centers of mass of the particles is decreasing while it is increasing in the absence of the applied stress. The nondimensional translation rates also depend strongly on the particle orientation with respect to the applied stress field, the relative capillary length, and the intercenter distance.

5. Discussion

A qualitative understanding of the influence of an applied stress field on the coarsening kinetics of a two-particle system can be obtained by considering the elastic interaction between two spherical inhomogeneities. For a uniaxial stress field aligned along the x_3 axis, the interaction energy to first order in the difference in elastic constants between particle and matrix is (Johnson, 1983)

$$E_{\text{int}} = \frac{\pi(\tau^\alpha)^2(\delta - 1)^2 r_+^3 r_-^3}{3c_{44}^\alpha(1 - \nu)d^3}\left\{\frac{4\nu - 1}{3} - \frac{3(r_+^2 + r_-^2)}{5d^2} - \left[2 + 4\nu - \frac{6(r_+^2 + r_-^2)}{d^2}\right]X^2 + \left[5 - \frac{7(r_+^2 + r_-^2)}{d^2}\right]X^4\right\}, \quad (5.1)$$

where $X = \cos\theta$ and it has been assumed that $\nu^\alpha = \nu^\beta$.

Figure 7 shows the functional dependence of the interaction energy on X for three different values of the relative particle size. The intercenter distance is $D = 3$ and the interaction energy has been normalized by

$$\frac{(\tau^a)^2(\delta - 1)^2(v_+ + v_-)}{4c_{44}^\alpha(1 + \nu)}.$$

The curve for $R = 0.25$ is the elastic interaction energy of the system for which the isotropic particle growth rates are calculated in Fig. 5.

The interaction energy is positive when the particles are aligned in the direction of, and perpendicular to, the applied stress field ($X = 1$ and $X = 0$, respectively). The interaction energy is negative for $0.05 < X < 0.75$. The magnitude of the interaction energy is greatest for equal-sized particles aligned parallel to the applied stress. When the interaction energy is positive, the two interacting spheres possess a higher elastic energy than when the two spheres are isolated or when there is just one large sphere of equivalent mass.

When the elastic energy is added to the interfacial energy, the lowest energy state for a system in which the mass of the particles is held fixed is one large sphere if the interaction energy is positive. Two spheres of equal size maximize both the elastic and interfacial energies. Consequently, when two spheres of unequal size interact both elastically and diffusionally and possess a positive interaction energy, the system tends to evolve towards a one-particle system (Johnson, 1984). When the interaction energy is negative, the one-particle system is the lowest energy state when the particles are small (the relative capillary length is large). At sufficiently large particle sizes, the elastic interaction energy begins to dominate over the interfacial energy and two particles

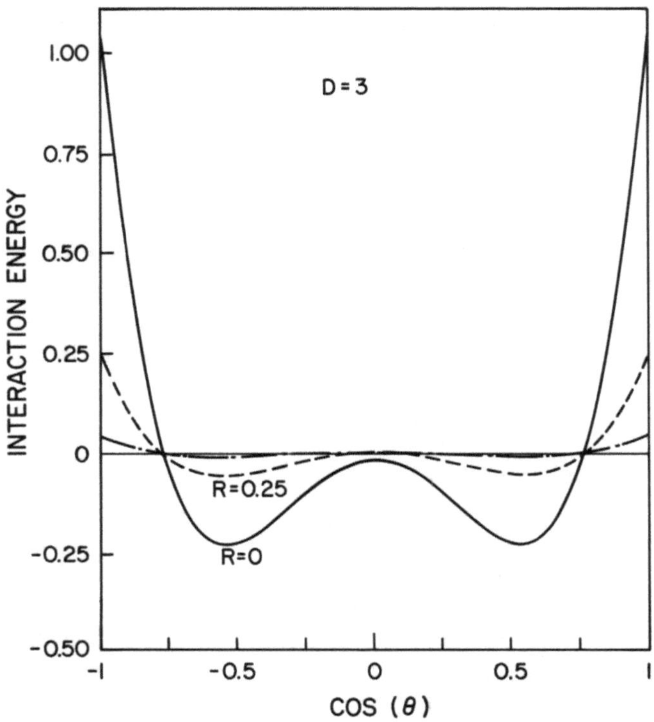

FIG. 7. The nondimensional elastic interaction energy of two spherical inhomogeneities as a function of cos θ where θ denotes the orientation of the particles with respect to the applied stress field. The energy curve for $R = 0.25$ depicts the situation displayed in Fig. 5.

of equal size give a lower system energy than one large particle. The decrease in the system energy owing to the elastic interaction energy more than offsets the increase in interfacial energy from the two particles. An energetic analysis including both the elastic and interfacial energies is given elsewhere (Johnson, 1984).

The *qualitative* relationship between the energy and kinetic analyses can be seen by comparing Figs. 5 and 7 for the relative particle size, $R = 0.25$. For $L_c = 10^{-8}$ (the value of the relative capillary length employed in Fig. 5), the larger particle grows at the expense of the smaller particle over roughly the same range of θ as the interaction energy is positive. Likewise, inverse coarsening is often observed where the interaction energy is negative. Decreasing the relative capillary length increases the range of inverse coarsening only slightly for this system, whereas increasing the relative capillary length will eliminate the region of inverse coarsening entirely. When $L_c = 10^{-8}$, the elasticity is controlling the kinetics; at larger values of L_c the capillarity begins to control the kinetics.

Although there is qualitative agreement between the energy and kinetic analyses, there is often significant quantitative disagreement. This is particularly true in determining the onset of inverse coarsening. The kinetic analysis usually predicts inverse coarsening at particle sizes much smaller than the energy analysis (Johnson, 1984). This is partially a result of the kinetic analysis allowing, in principle, for both particle shape changes and particle migrations.

The particle migrations induced by the applied elastic field are often one or two orders of magnitude larger (and of different sign) than those arising from capillarity induced diffusion. The ratio of the particle migration rate to the isotropic growth rate, the quantity that determines if the particle translation rates might be experimentally observed, is often greater than one in the presence of the applied stress as compared to less than one in the absence of the applied stress. Consequently, particle migrations induced by particle interactions in a uniform applied stress field may be an important mechanism in the development of two-phase microstructures.

The influence of elastic stress on the kinetics of a diffusional phase transformation requires the ability to calculate elastic fields associated with misfitting inclusions and elastic inhomogeneities. Professor Mura's contributions to this field have been tremendous and will grow in importance as our understanding of solid–solid phase transformations increases.

6. Conclusions

The effect of an applied uniaxial stress field on the kinetics of particle growth and particle translation was examined for two elastically and diffusionally interacting spherical inhomogeneities. The boundary conditions for the diffusion problem were obtained from the assumption of local thermodynamic equilibrium at the particle–matrix interface. The elastic field was determined to first order in the difference in elastic constants between particle and matrix phases using an integral equation method. The matrix composition field was approximated by Laplace's equation which was solved in the orthogonal bispherical coordinate system. The applied stress field, like misfit strains, can significantly after the kinetics of two-particle coarsening. In particular:

(1) The applied stress field may alter both the sign and magnitude of the coarsening kinetics with respect to that of capillarity-induced coarsening.
(2) For larger values of the relative capillary length, the system behavior is dominated by the capillary terms while for smaller values of L_c, elasticity dominates. Thus the qualitative coarsening behavior of the system under uniaxial stress should be expected to change with time as the particles grow.
(3) Inverse coarsening, or the growth of small particles at the expense of large particles, may occur over large ranges of the thermophysical parameters. This includes systems for which the particles are elastically harder than the matrix phase.

(4) The translation rate of a particle's center of mass is sufficiently large with respect to its isotropic growth rate, that in some cases it should be important in the development of microstructure in two-phase binary alloys under stress.

Acknowledgments

I wish to acknowledge a number of helpful discussions with T. A. Abinandanan and P. W. Voorhees. I also wish to thank the Division of Materials Research of the Department of Energy under grant DE-FG02-84ER45166 for support of this work.

References

Ardell, A. J. and Nicholson, R. B. (1966), On the modulated structure of aged Ni–Al alloys, *Acta Metallurgica*, **14**, 1295–1309.

Chen, F. C. and Young, K. (1977), Inclusions of arbitrary shape in an elastic medium, *J. Math. Phys.*, **12**, 1412–1416.

Doi, M. and Miyazaki, T. (1986), γ' percipitate morphology formed under the influence of elastic interaction energies in nickel-based alloys, *Mater. Sci. Engng.*, **78**, 87–95.

Eshelby, J. D. (1957), Determination of the elastic field of an ellipsoidal inclusion and related problems, *Proc. Roy. Soc., London*, **A252**, 376–396.

Geguzin, Y. E. and Krivoglaz, M. A. (1973), *Migration of Macroscopic Inclusions in Solids*, Plenum, New York, p. 133.

Johnson, W. C., Earmme, Y. Y., and Lee, J. K. (1980), Approximation of the strain field associated with an inhomogeneous precipitate, *J. Appl. Mech.*, **47**, 775–780.

Johnson, W. C. (1983), Elastic interaction of two precipitates subjected to an applied stress field, *Metall. Trans.*, **14A**, 2219–2227.

Johnson, W. C. (1984), On the elastic stabilization of precipitates against coarsening under applied load, *Acta Metallurgica*, **32**, 465–475.

Johnson, W. C. (1987), Precipitate shape evolution under applied stress—Thermodynamics and kinetics", *Metall. Trans.*, **18A**, 233–247.

Johnson, W. C., Voorhees, P. W., and Zupon, D. E. (1988), The effects of elastic stress on the kinetics of Ostwald ripening: The two-particle problem, *Metall. Trans.*, in press.

Kawasaki, K. and Enomoto, Y. (1988), Statistical theory of Ostwald ripening with elastic field interaction, *Physica A*, in press.

Larche, F. C. and Cahn, J. W. (1985), The interaction of composition and stress in crystalline solids, *Acta Metallurgica*, **33**, 331–358.

Lifshitz, I. M. and Slyozov, V. V. (1961), The kinetics of precipitation from supersaturated solid solutions, *J. Phys. Chem. Solids*, **19**, 35–50.

Mackay, R. A. and Ebert, L. J. (1983), The development of directional coarsening of the γ' precipitate in superalloy single crystals, *Scripta Metall.*, **17**, 1217–1222.

Moschovidis, Z. A. and Mura, T. (1975), Two ellipsoidal inhomogeneities by the equivalent inclusion method, *J. Appl. Mech.*, **42**, 847–851.

Nathal, M. V. and Ebert, L. J. (1983), Gamma prime shape changes during creep of a nickel-based superalloy, *Scripta Metall.*, **17**, 1151–1156.

Ostwald, W. (1901), *Z. Phys. Chem.*, **37**, 385.

Schmeizer, J. and Gutzow, I. (1986), On the kinetic description of Ostwald ripening in elastic media, *Z. Wihelm-Pieck-Uni.*, **35**, 5–12.

Tien, J. K. and Copley, S. M. (1971), The effect of uniaxial stress on the periodic morphology of coherent gamma prime precipitates in nickel-based superalloy crystals, *Metall. Trans.*, **2**, 215–219.

Voorhees, P. W. and Johnson, W. C. (1986), Interfacial equilibrium during a first-order phase transformation in solids, *J. Chem. Phys.*, **84**, 5108–5121.

Voorhees, P. W. and Johnson, W. C. (1988), On the development of spatial coorelations during diffusional late-stage phase transformations in stressed solids, *Phys. Rev. Lett.*, **61**, 2225–2228.

Wagner, C. (1961), Theorie der Laterung von Niederschlägen durch Umlösen, *Z. Elektrochemie*, **65**, 581–595.

Analysis of Cracks in Transversely Isotropic Media

L. M. KEER and W. LIN
Department of Civil Engineering, Northwestern University,
Evanston, IL 60208, U.S.A.

Abstract

The boundary integral method is applied to analyze cracks in transversely isotropic media. Since the Green's functions are written in tensor form, the integral equations can easily be constructed to provide the relationship between the crack opening displacements (COD) and the applied tractions. The kernels of the integral equations are expressed concisely in a closed form. The numerical scheme to solve the singular integral equations is presented. The method can be applied not only to nonelliptically shaped cracks and but also to nonplanar cracks. The final results contain the crack opening displacement and energy consumed by cracks.

1. Introduction

The analysis of flaws or cracks in anisotropic materials is of recent concern because of the wide use of fiber-reinforced composites and considerations of microcracking. The relevant literature for the analysis of inclusions or cracks in anisotropic materials over the past few decades can be found in Mura (1982). To analyze cracks in anisotropic materials, three related articles are given by Eshelby (1957), Willis (1968), and Hoenig (1978), and concern elliptically shaped cracks subjected to remote forces. The main conclusion of Eshelby and Willis is that the crack opening displacement for an elliptically shaped crack is also of elliptical shape. Hoenig applied the M and J integrals to evaluate the magnitude of the elliptically shaped crack opening displacement. In this paper, the boundary integral method is applied to evaluate the response of transversely isotropic media containing internal cracks. With this approach the crack need not be planar and elliptical shaped, but can possess a general shape.

To analyze cracks in transversely isotropic materials by the boundary element technique, the Green's functions derived by Pan and Chou (1976) are essential. In this paper, the Green's functions are taken in tensor forms from a recent paper by Lin and Keer (1989). Since the Green's functions are in tensor form, the kernels of the integral equation can immediately be obtained in a tensor form and the kernels will contain only two types of functions. In Section

2, the mathematical formulation is described and the kernels are constructed. In Section 3, a simple numerical scheme is proposed to solve the integral equation. The essence of the numerical scheme is the evaluation of the singular finite part integrals, by performing line integrals where no singular points are involved. The final section presents numerical results to show that the proposed numerical scheme works.

2. Mathematical Formulation

The main step in the boundary integral formulation is to construct the kernel of the integral equation in a manner that is convenient for numerical evaluation. The construction of the integral equation, to relate the crack opening displacement and the applied remote force, depends on the Green's function. For a transversely isotropic material, the Green's function has been obtained by Pan and Chou (1976). After a slight rearrangement, the Green's function is written as (see Lin and Keer, 1989):

$$G_{ij}(\mathbf{X}, \xi) = \sum_{\alpha=1}^{2} \left[\frac{\delta_{i3}\delta_{j3}P_\alpha}{R_\alpha} + \delta_{ia}\delta_{jb}Q_\alpha \left(\frac{x_a}{R_\alpha^*} \right)_{,b} - (\delta_{ia}\delta_{j3} + \delta_{ja}\delta_{i3})A_\alpha \left(\frac{x_a}{R_\alpha^*} \right)_{,3} \right] + De_{ia3}e_{jb3}\left(\frac{x_a}{R_3^*} \right)_{,b}, \tag{2.1}$$

where $G_{ij}(\mathbf{X}, \xi)$ is the Green's function and is interpreted as the displacement at $\mathbf{X}$ in the ith direction due to a unit point force applied at ξ in the jth direction, δ_{ia} is the Kronecker delta, and e_{ia3} is the permutation tensor. The constants P_α, Q_α, A_α, D, and ν_α depend on the transversely isotropic elastic constants (see Appendix), $x_a = X_a - \xi_a$, $R_\alpha = [x_1^2 + x_2^2 + (\nu_\alpha x_3)^2]^{1/2}$, $R_\alpha^* = R_\alpha + \nu_\alpha |x_3|$, the elastic properties are isotropic in the plane $X_3 = \text{const.}$, and the symbol $(\cdot)_{,b}$ represents $\partial(\cdot)/\partial X_b$. Repeated indices a and b imply summation over the range 1, 2, while other repeated indices imply summation over the range 1, 2, 3.

Since the Green's functions are known, the well-known representation integral is readily established, i.e.,

$$u_s(\xi) = C_{mnpq} \iint G_{ps,q}(\mathbf{X}, \xi) N_n(\mathbf{X}) \Delta u_m(\mathbf{X})\, dS(\mathbf{X}), \tag{2.2}$$

where C_{mnpq} are elastic constants, $\Delta u_m(\mathbf{X})$ is the jump of u_m at $\mathbf{X}$ and is equal to the displacement of u_m at S^+ minus the displacement at S^-, and N_n is the nth component of the normal vector from surface S^- to surface S^+. For a planar crack, N_n is independent of $\mathbf{X}$. After further application of the boundary conditions on the crack surfaces and the stress–strain relationship, the integral equation is obtained as follows:

$$-T_i(\xi) = \sigma_{ij}(\xi) N_j(\xi) = \text{f.p.} \iint K_{im}(\mathbf{X}, \xi) \Delta u_m(X)\, dS(\mathbf{X}), \tag{2.3}$$

where the kernel is

$$K_{im} = -C_{ijst}N_j(\xi)C_{mnpq}N_n(\mathbf{X})G_{ps,qt}(\mathbf{X}, \xi). \tag{2.4}$$

Here, $T_i(\xi)$ is the corresponding traction on the crack surface due to remote force, and "f.p." denotes that the area integral is a finite part integral. Finally, by substituting (2.1) into (2.4), the explicit form of the kernel is

$$K_{im} = \sum_{\alpha=1}^{2}\left[\left(\frac{1}{R_\alpha}\right)_{,qt} P_{im\alpha qt} + \left(\frac{x_a}{R_\alpha^*}\right)_{,bqt} Q_{im\alpha abqt} - \left(\frac{x_a}{R_\alpha^*}\right)_{,3qt} A_{im\alpha a3qt}\right] + \left(\frac{x_a}{R_3^*}\right)_{,bqt} D_{im3abqt}, \tag{2.5}$$

and

$$P_{im\alpha qt} = L_{istmpq}(\xi, \mathbf{X})\delta_{p3}\delta_{s3}P_\alpha, \tag{2.6a}$$

$$Q_{im\alpha abqt} = L_{istmpq}(\xi, \mathbf{X})\delta_{pa}\delta_{sb}Q_\alpha, \tag{2.6b}$$

$$A_{im\alpha a3qt} = L_{istmpq}(\xi, \mathbf{X})(\delta_{pa}\delta_{s3} + \delta_{sa}\delta_{p3})A_\alpha, \tag{2.6c}$$

$$D_{im3abqt} = L_{istmpq}(\xi, \mathbf{X})e_{pa3}e_{sb3}D, \tag{2.6d}$$

$$L_{istmpq}(\xi, \mathbf{X}) = -C_{ijst}N_j(\xi)C_{mnpq}N_n(\mathbf{X}). \tag{2.6e}$$

For a planar crack, functions $P_{im\alpha qt}$, $Q_{im\alpha abqt}$, $A_{im\alpha a3qt}$, and $D_{im3abqt}$ are independent of $\mathbf{X}$ and ξ.

3. Numerical Scheme

The first step in the numerical scheme is to write (2.3) in discrete form. Thus, the crack domain is divided into many triangular elements and a number j, is given for each element. Within the jth element a collocation point $\mathbf{X}_j$ is chosen, ξ_j is the same point as $\mathbf{X}_j$, and $\Delta u_m(\mathbf{X})$ is approximated by $w_j(\mathbf{X})\Delta U_m(\mathbf{X}_j)$. The shape function $w_j(\mathbf{X})$ equals $[2ae(\mathbf{X}) - e^2(\mathbf{X})]^{1/2}$, where $2a$ is the maximum distance between any two points within the entire crack domain, and $e(\mathbf{X})$ is the minimum distance from the point $\mathbf{X}$ to the crack tip. Even if the crack is nonplanar, $C_{mnpq}N_n(\mathbf{X})$ in (2.6e) is a constant for each triangular element because the normal vector

$$N_n(\mathbf{X}) = \frac{(\mathbf{p}_2 - \mathbf{p}_1) \times (\mathbf{p}_3 - \mathbf{p}_1)}{|(\mathbf{p}_2 - \mathbf{p}_1) \times (\mathbf{p}_3 - \mathbf{p}_1)|},$$

where $\mathbf{p}_1$, $\mathbf{p}_2$, and $\mathbf{p}_3$ are the three nodal points of the jth element. Thus, the integral equation has the following discrete form:

$$-T_i(\xi_n) = \sum_{j=1}^{N} \text{f.p.} \iint K_{im}(\mathbf{X}, \xi_n)w_j(\mathbf{X})\, dS_j(\mathbf{X})\Delta U_m(\mathbf{X}_j). \tag{3.1}$$

There are only two types of functions involved, i.e., $w_j(\mathbf{X})(1/R_\alpha)_{,qt}$ and $w_j(\mathbf{X})(x_a/R_\alpha^*)_{,bqt}$, since functions in 2.6(a)–(e) are constants for each discrete element.

For the case $n \neq j$, the kernel $K_{im}(\mathbf{X}, \xi_n)$ within the j'th element is finite and the integration can be performed by applying a low-order quadrature formula provided the area of j'th element, S_j, is small enough. If S_j is not small enough, the j'th element is further subdivided until a satisfactory convergent value is obtained for the integral over S_j.

When $n = j$, the kernel $K_{im}(\mathbf{X}, \xi_n)$ within the j'th element contains a singular point, and special treatment is required. The basic goal is to convert the area integrals in (2.3) into line integrals to avoid the singular integration points. To achieve this aim, a local coordinate system is built on the nodal points $\mathbf{p}_1$, $\mathbf{p}_2$, and $\mathbf{p}_3$ of the element where the three axes of this system are $\mathbf{e}_{1'} = N_n$, $\mathbf{e}_{2'} = (\mathbf{p}_2 - \mathbf{p}_1)/[(\mathbf{p}_2 - \mathbf{p}_1)]$, and $\mathbf{e}_{3'} = \mathbf{e}_{1'} \times \mathbf{e}_{2'}$. Physically, $\mathbf{e}_{1'}$ is the normal vector of the plane where the element is located, and $\mathbf{e}_{2'}$ is the vector from the first nodal point to the second nodal point. Thus

$$\left(\frac{1}{R_\alpha}\right)_{,qt} = C_{qq'}(\mathbf{X}_j)C_{tt'}(\mathbf{X}_j)\left(\frac{1}{R_\alpha}\right)_{,q't'},$$

$$\left(\frac{x_a}{R_\alpha^*}\right)_{,bqt} = C_{qq'}(\mathbf{X}_j)C_{tt'}(\mathbf{X}_j)\left(\frac{x_a}{R_\alpha^*}\right)_{,bq't'},$$

where $C_{qq'}$ is a rotation matrix. For the case that S_j is not around the crack tip, the shape function $w_j(\mathbf{X})$ may be set equal to $w_j(\mathbf{X}_j)$ and the two types of integrals to be performed are

$$\text{f.p.} \iint \left(\frac{x_a}{R_\alpha^*}\right)_{,bq't'} dX_{2'}\, dX_{3'} \qquad \text{and} \qquad \text{f.p.} \iint \left(\frac{1}{R_\alpha}\right)_{,q't'} dX_{2'}\, dX_{3'}.$$

The Gauss divergence theorem cannot be directly applied to the area integrals in (2.3) because q' and t' may be equal to $1'$. However, the area integrals can be converted into line integrals, easily, unless both q' and t' are equal to $1'$. The conversions are as follows:

$$\text{f.p.} \iint (\text{Function})_{,2'}\, dX_{2'}\, dX_{3'} = \int_{P2}^{P3} (\text{Function})\, dX_{3'} - \int_{P1}^{P3} (\text{Function})\, dX_{3'}, \tag{3.2}$$

and

$$\text{f.p.} \iint (\text{Function})_{,3'}\, dX_{3'}\, dX_{2'} = -\int_{P1}^{P2} (\text{Function})\, dX_{2'} - \int_{P2}^{P3} (\text{Function})\, dX_{2'} - \int_{P3}^{P1} (\text{Function})\, dX_{2'}. \tag{3.3}$$

Since the collocation point is chosen within the element, there is no singular point along the line integrals in (3.2) and (3.3). Thus, a computational problem arises only in the case that both q' and t' equals $1'$. To evaluate

$$\text{f.p.} \iint \left(\frac{1}{R_\alpha}\right)_{,1'1'} dX_{2'}\, dX_{3'},$$

the relations

$$\left(\frac{1}{R_\alpha}\right)_{,1'1'} + \left(\frac{1}{R_\alpha}\right)_{,2'2'} + \left(\frac{1}{R_\alpha}\right)_{,3'3'} = \left(1 - \frac{1}{v_\alpha^2}\right)\left(\frac{1}{R_\alpha}\right)_{,33}$$

$$= C_{3q'}(\mathbf{X}_j)C_{3t'}(\mathbf{X}_j)\left(1 - \frac{1}{v_\alpha^2}\right)\left(\frac{1}{R_\alpha}\right)_{,q't'} \tag{3.4}$$

are applied. By rearranging (3.4) so that the $(1/R_\alpha)_{,1'1'}$ term is moved to the left-hand side and the remaining terms to the right-hand side of the equation, the area integral of $(1/R_\alpha)_{,1'1'}$ can be evaluated by performing the line integrals in (3.2) and (3.3). To evaluate

$$\text{f.p.} \iint \left(\frac{x_a}{R_\alpha^*}\right)_{,b1'1'} dX_{2'}\, dX_{3'},$$

apply

$$\left(\frac{x_a}{R_\alpha^*}\right)_{,b1'1'} = -\left(\frac{x_a}{R_\alpha^*}\right)_{,b2'2'} - \left(\frac{x_a}{R_\alpha^*}\right)_{,b3'3'} + (1 - v_\alpha^2)\left(\frac{1}{R_\alpha}\right)_{,ab}$$

$$- 2v_\alpha \delta(x_3)(\log R_\alpha^*)_{,ab}. \tag{3.5}$$

Then, the integration of $(x_a/R_\alpha^*)_{,b2'2'}$ and $(x_a/R_\alpha^*)_{,b3'3'}$ over an area can also be converted into line integrals. The integration of $(1/R_\alpha)_{,ab}$ has been evaluated by (3.2)–(3.4). The integration of the last term in (3.5) is

$$\text{f.p.} \iint \delta(x_3)(\log R_\alpha^*)_{,ab}\, dX_{2'}\, dX_{3'} = -(\delta_{ab} - 2C_{a2''}C_{b2''})(1/r_1'' + 1/r_2''),$$

For $C_{a2''}$ and $C_{b2''}$, the $X_{2''}$ axis is normal to both the X_3 axis and the $X_{1'}$ axis. Also, r_1'' and r_2'' are the two distances from the point $(x_1, x_2, x_3) = (0, 0, 0)$ to the two points of interception on the plane $x_3 = 0$ and the contour line of the element.

For the case that the shape function is not a constant, such as $w_j(\mathbf{X}) = [2ae(\mathbf{X}) - e^2(\mathbf{X})]^{1/2}$, then numerical integration is performed by the following manner:

$$\text{f.p.} \iint K_{im}(\mathbf{X}, \xi_n) w_j(\mathbf{X})\, dS_j(\mathbf{X}) = w_j(\mathbf{X}_j)\, \text{f.p.} \iint K_{im}(\mathbf{X}, \xi_n)\, dS_j(\mathbf{X})$$

$$+ \text{f.p.} \iint K_{im}(\mathbf{X}, \xi_n)[w_j(\mathbf{X}) - w_j(\mathbf{X}_j)]\, dS_j(\mathbf{X}). \tag{3.6}$$

The first integral on the right-hand side of (3.6) has already been discussed. The second integral is a Cauchy-type integral in the polar coordinate system embedded on the element with the origin point at the collocation point, i.e.

$$\text{f.p.} \iint K_{im}(\mathbf{X}, \xi_n)[w_j(\mathbf{X}) - w_j(\mathbf{X}_j)]\, dS_j(\mathbf{X})$$
$$= \int [r''^3 K_{im}(\mathbf{X}, \xi_n)]\, d\theta'' \;\text{f.p.} \int \frac{w_j(\mathbf{X}) - w_j(\mathbf{X}_j)}{r''^2}\, dr''. \quad (3.7)$$

The function $r''^3 K_{im}(\mathbf{X}, \xi_n)$ is independent of r'' and there is no singularity along θ''. The finite-part integral along r'' has been discussed by Kutt (1975).

4. Numerical Results

As a numerical example, a penny-shaped crack in an infinite medium is considered. The geometry of the mesh for the crack is given in Fig. 1. The normal direction of the crack surface is along the $(-\sin\theta, 1, \cos\theta)$ direction. The traction on the crack surfaces are $-T_1$, $-T_2$, and $-T_3$. Four materials are considered:

(1) a nearly isotropic material with a Poisson ratio of 0.2;
(2) magnesium;
(3) zinc; and
(4) a fibre-reinforced composite material (Datta *et al.*, 1988).

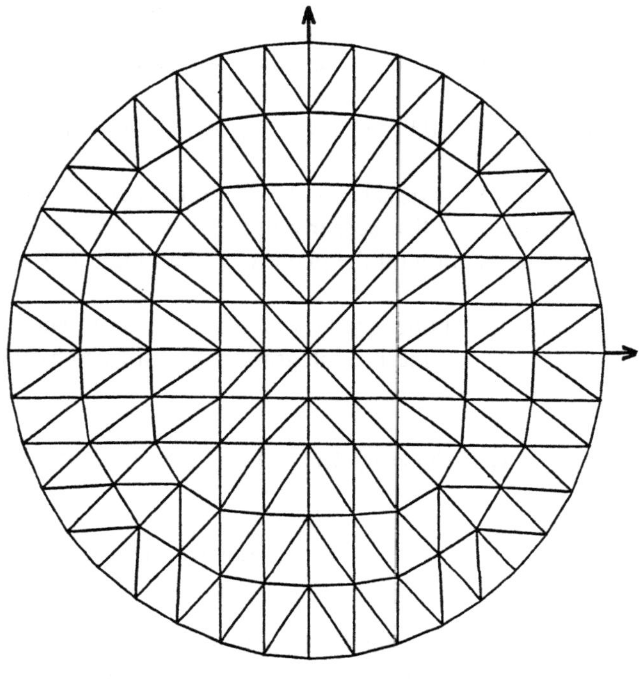

FIG. 1. After Lin and Keer (1989).

TABLE 1. Isotropic material.

θ	Δu_1	$\Delta u_2/f_2$	Δu_3
0	$1.03f_1$	1.15	$1.15f_3$
30	$1.06f_1 + 0.05f_3$	1.14	$1.12f_3 + 0.05f_1$

TABLE 2. Magnesium.

θ	$\Delta u_1/1.64$	$\Delta u_2/1.64f_2$	$\Delta u_3/1.64$
0	$0.532f_1$	0.630	$0.618f_3$
15	$0.537f_1 + 0.019f_3$	0.630	$0.609f_3 + 0.019f_1$
30	$0.552f_1 + 0.032f_3$	0.629	$0.588f_3 + 0.032f_1$
45	$0.574f_1 + 0.037f_3$	0.626	$0.562f_3 + 0.037f_1$
60	$0.597f_1 + 0.032f_3$	0.622	$0.538f_3 + 0.032f_1$
75	$0.617f_1 + 0.019f_3$	0.621	$0.521f_3 + 0.019f_1$
89	$0.625f_1$	0.625	$0.514f_3$
90	$0.618f_1$	0.618	$0.508f_3$ (exact)

TABLE 3. Zinc.

θ	$\Delta u_1/3.96$	$\Delta u_2/3.96f_2$	$\Delta u_3/3.96$
0	$0.189f_1$	0.199	$0.350f_3$
15	$0.193f_1 + 0.015f_3$	0.202	$0.353f_3 + 0.015f_1$
30	$0.205f_1 + 0.028f_3$	0.209	$0.360f_3 + 0.028f_1$
45	$0.221f_1 + 0.036f_3$	0.218	$0.367f_3 + 0.036f_1$
60	$0.234f_1 + 0.033f_3$	0.228	$0.369f_3 + 0.033f_1$
75	$0.238f_1 + 0.018f_3$	0.234	$0.369f_3 + 0.018f_1$
89	$0.238f_1$	0.238	$0.369f_3$
90	$0.235f_1$	0.235	$0.365f_3$ (exact)

The solved crack opening displacements are given in Tables 1–4, where $f_i = T_i(a^2 - R^2)^{1/2}/C_{44}$, a is the radius of the crack, and $R = (X_1^2 + X_2^2 + X_3^2)^{1/2}$.

For the nearly isotropic material, the relative stiffnesses are $C_{11} = 2.666$, $C_{13} = 0.667$, $C_{33} = 2.666$, $C_{44} = 1$, $C_{66} = 1$. The exact solutions are $\Delta u_1 = f_1(t \cos^2\theta + s \sin^2\theta) + f_3(t - s)\cos\theta \sin\theta$, $\Delta u_2 = sf_2$, and $\Delta u_3 = f_3(s\cos^2\theta + t\sin^2\theta) + f_1(t - s)\cos\theta\sin\theta$, where $t = 4(1 - \nu)/\pi = 1.02$ and $s = 8(1 - \nu)/(2 - \nu)\pi = 1.13$. For $\theta = 0°$, the exact solutions are $\Delta u = 1.02f_1$, $\Delta u_2 = 1.13f_2$, and $\Delta u_3 = 1.13f_3$. For $\theta = 30°$, the exact solutions are $\Delta u_1 = 1.05f_1 + 0.05f_3$, $\Delta u_2 = 1.13f_2$, and $\Delta u_3 = 0.05f_1 + 1.10f_3$. Thus, the numerical error in Table 1 is about 2%.

The elastic properties of magnesium and zinc are available from Payton (1983, p. 3) or Landolt-Bornstein (1969). The relative stiffnesses for magnesium are $C_{11} = 5.92$, $C_{13} = 2.14$, $C_{33} = 6.14$, $C_{44} = 1.64$, and $C_{66} = 1.675$. The relative stiffnesses for zinc are $C_{11} = 16.5$, $C_{13} = 5.0$, $C_{33} = 6.2$, $C_{44} = 3.96$, and $C_{66} = 6.7$. For fiber-reinforced composite material (Datta *et al.*, 1988),

TABLE 4. Fibre-reinforced composite.

θ	$\Delta u_1/0.0707$	$\Delta u_2/0.0707f_2$	$\Delta u_3/0.0707$
0	$18.5f_1$	24.7	$6.39f_3$
15	$18.5f_1 + 0.10f_3$	24.2	$6.04f_3 + 0.10f_1$
30	$18.4f_1 + 0.14f_3$	23.0	$5.48f_3 + 0.14f_1$
45	$18.4f_1 + 0.15f_3$	21.3	$4.98f_3 + 0.15f_1$
60	$18.5f_1 + 0.14f_3$	19.9	$4.62f_3 + 0.14f_1$
75	$18.7f_1 + 0.08f_3$	18.9	$4.41f_3 + 0.08f_1$
89	$18.7f_1$	18.7	$4.33f_3$
90	$18.6f_1$	18.6	$4.28f_3$ (exact)

the relative stiffnesses are $C_{11} = 0.1392$, $C_{13} = 0.0644$, $C_{33} = 1.6073$, $C_{44} = 0.0707$, and $C_{66} = 0.035$. For $\theta = 90°$, closed form solutions for the crack opening displacement are

$$\frac{\Delta u_1}{f_1} = \frac{\Delta u_2}{f_2} = \frac{(2/\pi^2 C_{44})}{Dv_3^2 - \sum_{\alpha=1}^{2} [P_\alpha + v_\alpha^2(2A_\alpha - Q_\alpha)]},$$

and

$$\frac{\Delta u_3}{f_3} = \frac{C_{44}}{\left[\pi^2 \sum_{\alpha=1}^{2} Q_\alpha(C_{33}h_\alpha - C_{13})^2\right]},$$

where $h_\alpha = -A_\alpha v_\alpha^2/Q_\alpha$.

From Tables 1–4, the energy dissipated (Mura, 1982, pp. 174–182, eqs. (25.17) and (25.46)) by the cracks can easily be obtained by integrating f_i over the crack area, i.e.,

$$g_i = 2\pi \int f_i r\, dr = \frac{2\pi a^3 T_i}{3C_{44}}.$$

Then, by replacing f_i by g_i in Tables 1–4 and multiplying the values with their corresponding tractions, the energy dissipation caused by the loadings and the crack is thereby obtained.

Acknowledgments

The authors are grateful to support from Amoco Production Company in this research. Particular thanks is given to Drs. R. W. Veatch and Z. Moschovidis.

References

Datta, S. K., Shah, A. H., Bratton, R. L., and Chakrabarty, T. (1988), Wave propagation in laminated composite plates, *J. Acoustical Soc. Amer.*, **83**, (6), 2020–2026.

Eshelby, J. D. (1957), The determination of the elastic field of an ellipsoidal inclusion, and related problems, *Proc. Roy. Soc.*, **A241**, 376–396.

Hoenig, A. (1978). The behavior of a flat elliptical crack in an anisotropic elastic body, *Int. J. Solids Structures*, **14**, 925–934.

Hosten B., Deschamps M., and Tittmann B. R. (1987), Inhomogeneous wave generation and propagation in lossy anisotropic solids. Application to the characterization of viscoelastic composite material, *J. Acoustical Soc. Amer.*, **82**, (5), 1763–1770.

Kutt, H. R. (1975), The numerical evaluation of principle value integrals by finite part integration, *Numer. Math.*, **24**, 205–210.

Landolt-Bornstein, (1969), Numerical data and functional relationships in science and technology, New Series, Group 3, Vol. I, article by R. F. S. Hearmon, Springer-Verlag, Berlin-Heidelberg-New York.

Lin, W. and Keer, L. M. (1989), Three-dimensional analysis of cracks in layered transversely isotropic media, *Proc. Roy. Soc. London* (in press).

Mura, T. (1982), *Micromechanics of Defects in Solids*, Martinus Nijhoff, The Hague.

Murakami, Y. and Nemat-Nasser, S. (1983), Growth and stability of interacting surface flaws of arbitrary shape, *Enging. Fracture Mech.*, **17**, 193–210.

Pan, Y. C. and Chou, T. W. (1976), Point force solution for an infinite transversely isotropic solid, *ASME J. Appl. Mech.*, **43**, 608–612.

Payton, R. G. (1983), *Elastic Wave Propagation in Transversely Isotropic Media*, Martinus Nijhoff, The Hauge.

Willis, J. R. (1968), The stress field around an elliptical crack in an anisotropic elastic medium, *Int. J. Engng. Sci.*, **6**, 253–263.

Appendix

Symbols

$$A_1 v_1 = -A_2 v_2 = -\frac{C_{13} + C_{44}}{4\pi C_{33} C_{44}(v_1^2 - v_2^2)},$$

$$D = 1/(4\pi C_{44} v_3),$$

$$P_\alpha = \frac{A_\alpha(C_{11} - C_{44} v_\alpha^2)}{C_{13} + C_{44}},$$

$$Q_\alpha = \frac{(-1)^\alpha (C_{44} - C_{33} v_\alpha^2)}{4\pi C_{33} C_{44}(v_1^2 - v_2^2) v_\alpha},$$

$$v_1 = \left\{\frac{[(C_{11}C_{33})^{1/2} - C_{13}][(C_{11}C_{33})^{1/2} + C_{13} + 2C_{44}]}{4C_{33}C_{44}}\right\}^{1/2} + \left\{\frac{[(C_{11}C_{33})^{1/2} + C_{13}][(C_{11}C_{33})^{1/2} - C_{13} - 2C_{44}]}{4C_{33}C_{44}}\right\}^{1/2},$$

$$v_2 = \left\{\frac{[(C_{11}C_{33})^{1/2} - C_{13}][(C_{11}C_{33})^{1/2} + C_{13} + 2C_{44}]}{4C_{33}C_{44}}\right\}^{1/2} - \left\{\frac{[(C_{11}C_{33})^{1/2} + C_{13}][(C_{11}C_{33})^{1/2} - C_{13} - 2C_{44}]}{4C_{33}C_{44}}\right\}^{1/2},$$

$$v_3 = \left(\frac{C_{66}}{C_{44}}\right)^{1/2}.$$

Modified Green Functions in the Theory of Heterogeneous and/or Anisotropic Linearly Elastic Media

E. KRÖNER
Institut für Theoretische und Angewandte Physik der Universität Stuttgart, Pfaffenwaldring 57, 7000 Stuttgart 80, Federal Republic of Germany
Max-Planck-Institut für Metallforschung, Heisenbergstraße 1, 7000 Stuttgart 80, Federal Republic of Germany

Abstract

The concept of the modified Green function, defined as certain second derivatives of the conventional tensor Green function, has been proved useful in solving elastic (and other) problems in the media of complicated constitution, such as heterogeneous or anisotropic continua. Emphasis is laid on infinite media. Contrary to the conventional approach, situations with finite stress and strain in the infinite regions are also investigated. Depending on the conditions four modified Green functions appear. These are limits of the four (modified) Green functions of finite bodies that grow beyond bounds. Certain properties of these functions are illustrated of the two basic boundary value problems of potential theory, assuming that the growing bodies are spheres. It is shown that the concept of modified Green functions is particularly useful in the elasticity theory of random media.

1. Introduction

One of the most powerful tools in physical field theories is the Green function, which is particularly useful in linear or linearized theories. Important contributions to the field of the Green functions are due to Mura (1963, 1982). It therefore seems appropriate to dedicate to Professor Mura this investigation, which among other things is concerned with certain peculiarities which the Green functions show in the case of so-called infinite medium. No real medium is infinite. Therefore it might appear sufficient to treat just finite media. However, the mathematical problem of solving the field-theoretical equations for infinite bodies is so much simpler than that for finite ones, that the wish to utilize this simplicity is only natural. For this reason the idealization of the infinite medium and, along with it, the use of the Green function for infinite media has been popular for more than 100 years.

Of course, the infinite medium is not a good concept in those problems, where the boundary of the medium is essential. Strictly speaking, the boundary is always essential, because there is no material medium without a boundary. In fact, the correct view is to model an infinite medium as a large, but finite medium and then solve the boundary value problem (BVP). In general, the state of the body will depend on its linear dimensions in all directions. There are certain situations where the state of the body does not change much, if the body is increasingly "blown up," so that the limit values for the infinite body have a meaning. This is investigated in the present work.

In Section 2 we introduce the modified Green functions starting from the ordinary (second rank) tensor Green functions of the two standard boundary value problems. There are four modified functions which we denote by Γ^{I}, Γ^{II}, Δ^{I}, and Δ^{II}. Two relate to the first boundary value problem (surface displacements given, superscript I), and two to the second boundary value problem (surface fractions given, superscript II). All these functions are fourth rank self-adjoint tensor functions of two points $\mathbf{r}$ and $\mathbf{r}'$. Following Gubernatis and Krumhansl (1975) we call the Γ's *strain Green functions* and the Δ's *stress Green functions*; for the reason, see below.

An important property of the modified Green functions for homogeneous bodies is that their volume integral equals the tensor of elastic moduli C (Δ^{I}), or that of elastic compliances $S = C^{-1}$ (Γ^{II}), or zero (Δ^{II}, Γ^{I}). This result does not apply to the infinite medium if there we replace the Γ's and Δ's by Γ^{∞} and Δ^{∞}, as derived from the ordinary (or displacement) Green function G^{∞} of the infinite medium (the Kelvin–Somigliana function in the case of the homogeneous, isotropic medium). To obtain the correct integrals, Γ^{∞} and Δ^{∞} have to be completed by the addition of an extra term for each. This term is different for Γ and Δ, and different for the two boundary value problems. How all this looks is illustrated, for the example of the potential theory, in Section 3.

Some important formulas for elasticity theory are given in Section 4. Section 5 contains four subsections with remarks concerning the applicability of the concept. In particular, Subsection 5.3 shows, how the modified Green functions of heterogeneous and anisotropic media can be obtained. In Subsection 5.4 we discuss application to the theory of random elastic media.

2. The Concept of the Modified Green Function

The important result of the theory of Green functions in elastic media is that, with their help, the displacement field $u = (u_i)$ can be represented as a sum of a volume (V) and a surface (S) integral, which contain beside the elastic moduli only the Green function, the volume density $f = (f_j)$ of external forces, and either the displacements $g = (g_i)$ on the surface (boundary value problem I) or the surface tractions $h = (h_i)$ (boundary value problem II). The Green function, also called the *displacement Green function*, of elasticity theory is a two-point second rank tensor function $G^{\mathrm{I}} = (G^{\mathrm{I}}_{ik})$ or $G^{\mathrm{II}} = (G^{\mathrm{II}}_{ik})$ for the respective boundary value problem.

The formulas in question are then

$$u_i(\mathbf{r}) = \int_V G^{\mathrm{I}}_{ik}(\mathbf{r}, \mathbf{r}')f_k(\mathbf{r}')\,dV' - \int_S g_k(\mathbf{r}')c_{klmn}(\mathbf{r}')\partial'_n G^{\mathrm{I}}_{mi}(\mathbf{r}', \mathbf{r})\,dS'_l, \tag{2.1}$$

where c_{klmn} are the (perhaps anisotropic and heterogeneous) elastic moduli, and

$$u_i(\mathbf{r}) = \int_V G^{\mathrm{II}}_{ik}(\mathbf{r}, \mathbf{r}')f_k(\mathbf{r}')\,dV' + \int_S G^{\mathrm{II}}_{ik}(\mathbf{r}, \mathbf{r}')h_k(\mathbf{r}')\,dS'. \tag{2.2}$$

Since in these formulas the u's are the actually prevailing displacements for given f and g or h, we can predict that the equilibrium conditions

$$\partial_i\sigma_{ij} = -f_j, \tag{2.3}$$

are satisfied, where σ_{ij} are the components of the stress tensor σ. Furthermore, the boundary conditions are satisfied, namely,

$$u_i(S) = g_i$$

for boundary value problem I (2.4)

$$n_i\sigma_{ij} = h_j$$

for boundary value problem II. (2.5)

If (2.3) is inserted into the volume integral of (2.2), and if then the partial integration is performed, we find that with (2.5) the surface integrals in (2.2) cancel. For the elastic strain tensor

$$\varepsilon_{ij} = \tfrac{1}{2}(\partial_i u_j + \partial_j u_i), \tag{2.6}$$

we find

$$\varepsilon_{ij}(\mathbf{r}) = \int_V \Gamma^{\mathrm{II}}_{ijkl}(\mathbf{r}, \mathbf{r}')\sigma_{kl}(\mathbf{r}')\,dV', \tag{2.7}$$

with

$$\Gamma^{\mathrm{II}}_{ijkl}(\mathbf{r}, \mathbf{r}') \equiv \partial_j\partial'_l G^{\mathrm{II}}_{ik}(\mathbf{r}, \mathbf{r}')|_{(ij)(kl)} = \Gamma^{\mathrm{II}}_{klij}(\mathbf{r}', \mathbf{r}). \tag{2.8}$$

Here the parentheses (ij) and (kl) indicate symmetrization in the concerned indices. For later use we rewrite (2.7) in the abridged notation

$$\varepsilon = \Gamma^{\mathrm{II}} * \sigma, \tag{2.9}$$

where, obviously, the star represents a kind of convolution. Since Γ^{II}, like Γ^{I} (see below) leads us to the strain, the Γ's are called the *strain* Green functions. Strictly speaking, they are not Green functions, because they have no inverse (see below). For this reason we also speak of the *modified* Green function.

In (2.7) any stress tensor satisfying the equilibrium and traction conditions gives us the correct strains. It is often quite simple to find stresses satisfying (2.3) and (2.5), if we do not demand that the strains related (by Hooke's law) to these stresses

$$\sigma_{ij} = c_{ijkl}\varepsilon_{kl}, \qquad \varepsilon_{ij} = s_{ijkl}\sigma_{kl}, \tag{2.10}$$

are compatible. For convenience we write (2.10), too, in abridged notation

$$\sigma = c * \varepsilon, \qquad \varepsilon = s * \sigma. \tag{2.11}$$

Here c and s are also understood as operators, namely, with kernels $c_{ijkl}\delta(\mathbf{r} - \mathbf{r}')$ and $s_{ijkl}\delta(\mathbf{r} - \mathbf{r}')$ for elastically local media. In the case of nonlocal elasticity, equations (2.11) keep their appearance, but c and s have more complex kernels. We shall not pursue the nonlocal case in this work.

As desired, ε on the left-hand side of (2.9) is compatible, ε on the right-hand side of (2.11) may not be. If we insert (2.11) in (2.9) to obtain

$$\varepsilon = \Gamma^{\mathrm{II}} * c * \varepsilon, \tag{2.12}$$

then the two ε's are the same if, and only if, ε on the right-hand side is also compatible. *Proof*: Imagine some compatible ε and the corresponding $\sigma = c * \varepsilon$ as given. By (2.3) and (2.5) we can derive the volume and surface forces which produce the state (σ, ε). But the correct strain for this state is also given by the left-hand side of (2.12). Thus, if a compatible ε is inserted on the right-hand side of (2.12), then the same compatible ε results on the left-hand side.

Since the decomposition of an arbitrary strain field into a compatible and incompatible part is unique, we see that the quantity $\Gamma * c$ acts as a projection operator, which projects out the compatible part from the general strain field. This property of $\Gamma * c$ has been discussed for the infinite medium by Kunin (1964) and by Simmons and Bullough (1970). Incidentally, the projection property of $\Gamma * c$ is the reason for the nonexistence of an inverse, also of Γ.

Because (2.7) is valid for all stresses satisfying (2.3) and (2.5) it is also true for constant stress, in which case

$$\varepsilon_{ij}(\mathbf{r}) = \int_V \Gamma^{\mathrm{II}}_{ijkl}(\mathbf{r}, \mathbf{r}')\, dV' \sigma_{kl}. \tag{2.13}$$

The (compatible) strain on the left-hand side is constant provided we consider a homogeneous (not necessarily isotropic) medium. Comparing (2.12) with (2.10_2) yields

$$\int_V \Gamma^{\mathrm{II}}_{ijkl}(\mathbf{r}, \mathbf{r}')\, dV' = S_{ijkl}, \tag{2.14}$$

as anticipated in Section 1. Here, and later, we use a capital letter for the elastic constants when these relate to a homogeneous medium.

Now consider the first boundary value problem for the special case of zero boundary displacements, $g = 0$. Integrating by parts the volume integral in (2.1), we see that due to the boundary condition for the Green function, $G^{\mathrm{I}} = 0$

on S, the newly arising surface integral also vanishes and we obtain

$$\varepsilon_{ij}(\mathbf{r}) = \int_V \Gamma^{\mathrm{I}}_{ijkl}(\mathbf{r}, \mathbf{r}')\sigma_{kl}(\mathbf{r}')\, dV', \tag{2.15}$$

with compatible strain on the left-hand side. Equation (2.15) is valid under the condition of no surface displacement. Another interesting case is that of vanishing volume forces ($f = 0$). Of course, (2.10_2) is always true. To exclude volume forces, we subtract from it (2.15) and obtain

$$\varepsilon = s * \sigma - \Gamma^{\mathrm{I}} * \sigma, \tag{2.16}$$

which is equivalent to

$$\sigma = \Delta^{\mathrm{I}} * \varepsilon, \tag{2.17}$$

where Δ^{I}, defined by

$$\Delta^{\mathrm{I}} = c - c * \Gamma^{\mathrm{I}} * c \qquad (\Gamma^{\mathrm{I}} = s - s * \Delta^{\mathrm{I}} * s) \tag{2.18}$$

is called the *stress* Green function of the first boundary value problem. Equations (2.16)–(2.18) are valid for all strains for which $f = 0$, therefore for constant strain also. In a similar manner as before we conclude that

$$\int_V \Delta^{\mathrm{I}}_{ijkl}(\mathbf{r}, \mathbf{r}')\, dV' = C_{ijkl} \tag{2.19}$$

is valid for homogeneous media. It also follows from (2.18_2) that for such media

$$\int_V \Gamma^{\mathrm{I}}_{ijkl}(\mathbf{r}, \mathbf{r}')\, dV' = 0. \tag{2.20}$$

Similar considerations lead us to

$$\int_V \Delta^{\mathrm{II}}_{ijkl}(\mathbf{r}, \mathbf{r}')\, dV' = 0, \tag{2.21}$$

valid again for homogeneous media. Here

$$\Delta^{\mathrm{II}} = c - c * \Gamma^{\mathrm{II}} * c \qquad (\Gamma^{\mathrm{II}} = s - s * \Delta^{\mathrm{II}} * s), \tag{2.22}$$

where, as in (2.18), s and c are the operators explained earlier.

Summary of results: We have obtained the following four equations with modified Green functions, which are valid under the indicated conditions:

$$\begin{aligned} \varepsilon &= \Gamma^{\mathrm{I}} * \sigma, && (\eta = 0, g = 0) \\ \varepsilon &= \Gamma^{\mathrm{II}} * \sigma, && (\eta = 0), \\ \sigma &= \Delta^{\mathrm{I}} * \varepsilon, && (f = 0), \\ \sigma &= \Delta^{\mathrm{II}} * \varepsilon, && (f = 0, h = 0), \end{aligned} \tag{2.23}$$

(see the systematics). Here η is the incompatibility tensor ($\eta = \nabla \times \varepsilon \times \nabla$), whose vanishing implies compatible strain. Obviously, the last two equations can be applied to incompatible situations as well. The pertaining stresses are known as internal, or eigenstresses, or residual stresses. They remain in the body when the external loads are removed.

3. Illustration: The Modified Green Function for the Potential Problem of a Large Sphere

The considerations of preceding section were rather abstract. Therefore, before coming to some application, let us illustrate the obtained results by an example. To make it simple, we shall verify some equations of the last section for the case of isotropic dielectric materials. In this case we have second rank instead of fourth rank material tensors, and vector fields instead of tensor fields. The whole formalism remains unchanged. If c_{ij} is the dielectric material tensor then, due to the isotropy, we have $c_{ij} = c\delta_{ij}$ with c as the dielectric constant. The relevant differential equation is Poisson's equation

$$c\nabla^2 U = -\rho, \tag{3.1}$$

where U is the electric potential and ρ is the charge density. The Green function for the infinite domain is a scalar, namely,

$$G^{\infty}(\mathbf{r}, \mathbf{r}') = \frac{1}{4\pi c|\mathbf{r} - \mathbf{r}'|}, \tag{3.2}$$

(Coulomb potential). For the modified Green function we easily find

$$\Gamma_{ij}^{\infty}(\mathbf{r}, \mathbf{r}') \equiv \partial_i \partial_j' G^{\infty}(\mathbf{r}, \mathbf{r}') = E_{ij}\delta(\mathbf{r} - \mathbf{r}') + \frac{F_{ij}(\theta, \varphi)}{|\mathbf{r} - \mathbf{r}'|^3}, \tag{3.3}$$

where

$$E_{ij} = \frac{1}{3c}\delta_{ij}, \qquad F_{ij} = \frac{1}{4\pi c}(\delta_{ij} - 3e_i e_j). \tag{3.4}$$

The e_i are the components of the unit vector $(\mathbf{r} - \mathbf{r}')/|\mathbf{r} - \mathbf{r}'|$ and θ, φ are the pertaining polar angles.

The integral of F_{ij}, taken over the volume of the body, vanishes for reasons of symmetry. Thus

$$\int_V \Gamma_{ij}^{\infty}(\mathbf{r}, \mathbf{r}')\, dV' = \int_V E_{ij}\delta(\mathbf{r} - \mathbf{r}')\, dV' = \frac{1}{3c}\delta_{ij}. \tag{3.5}$$

The equations which should be verified (see (2.14) and (2.20)) are

$$\int_V \Gamma_{ij}^{\mathrm{I}}(\mathbf{r}, \mathbf{r}')\, dV' = 0 \qquad \int_V \Gamma_{ij}^{\mathrm{II}}(\mathbf{r}, \mathbf{r}')\, dV' = \frac{1}{c}\delta_{ij}, \tag{3.6}$$

where the Γ's are the modified Green functions of the respective boundary value problems.

The disagreement between (3.5) and (3.6) shows us that Γ^{∞}, thus also G^{∞}, cannot be the complete Green function of the infinite medium. The question then arises as to whether all the infinite medium is a legitimate medium in the sense of continuum mechanics. We shall investigate this question by modeling the infinite medium as a large sphere whose radius R we let go to infinity at the appropriate moment. More precisely, we shall now calculate Γ^{I} and Γ^{II} for this sphere and see what remains when R goes to infinity.

The Green functions G^{I} and G^{II} for the sphere have been calculated, e.g., by the method of reciprocal radii (Frank and von Mises, 1930, pp. 748 and 759). Introduce the angle γ by

$$|\mathbf{r} - \mathbf{r}'|^2 = r^2 + r'^2 - 2rr' \cos \gamma, \tag{3.7}$$

the quantity A by

$$A \equiv R^4 + r^2r'^2 - 2R^2rr' \cos \gamma = R^4 + r^2r'^2 - R^2r^2 - R^2r'^2 + (\mathbf{r} - \mathbf{r}')^2, \tag{3.8}$$

and the quantity B by

$$B = R^2 - rr' \cos \gamma = R^2 - \tfrac{1}{2}(r^2 + r'^2) + |\mathbf{r} - \mathbf{r}'|^2. \tag{3.9}$$

r and r' are the distances of the field point and the source point from the center of the sphere.

Herewith the two Green functions can be written as

$$G^{\mathrm{I}}(\mathbf{r}, \mathbf{r}') = \frac{1}{4\pi c}\left[\frac{1}{|\mathbf{r} - \mathbf{r}'|} - RA^{-1/2}\right], \tag{3.10}$$

$$G^{\mathrm{II}}(\mathbf{r}, \mathbf{r}') = \frac{1}{4\pi c}\left[\frac{1}{|\mathbf{r} - \mathbf{r}'|} + RA^{-1/2} - \frac{1}{R}\ln(B + A^{1/2}) - \frac{1}{R} + \frac{1}{R}\ln(2R^2)\right]. \tag{3.11}$$

To calculate

$$\Gamma^{\mathrm{I}}_{ij}(\mathbf{r}, \mathbf{r}') = \partial_i\partial'_j G^{\mathrm{I}}(\mathbf{r}, \mathbf{r}'), \qquad \Gamma^{\mathrm{II}}_{ij}(\mathbf{r}, \mathbf{r}') = \partial_i\partial'_j G^{\mathrm{II}}(\mathbf{r}, \mathbf{r}'), \tag{3.12}$$

is just an exercise in differentiating. We do not need the full expressions, but only what remains in the limit $R \to \infty$. We find

$$\lim_{R\to\infty} \partial_i\partial_{j'}\left[\frac{1}{R}\ln(B + A^{1/2})\right] = R^{-3}\left[1 + O\left(\frac{r^2}{R^2}, \frac{r'^2}{R^2}\right)\right]\delta_{ij}, \tag{3.13}$$

$$\lim_{R\to\infty} \partial_i\partial'_j(RA^{-1/2}) = R^{-3}\left[1 + O\left(\frac{r^2}{R^2}, \frac{r'^2}{R^2}\right)\right]\delta_{ij}. \tag{3.14}$$

In all physically interesting situations it is allowed to consider $r/R \ll 1$ and $r'/R \ll 1$, and we have

$$\lim_{R\to\infty} \Gamma^{\mathrm{I}}_{ij} = \frac{1}{3c}\delta_{ij}\delta(\mathbf{r} - \mathbf{r}') + \frac{1}{4\pi c}(\delta_{ij} - 3e_ie_j)\frac{1}{|\mathbf{r} - \mathbf{r}'|^3} - \frac{1}{4\pi c}\delta_{ij}\frac{1}{R}. \tag{3.15}$$

Noting that the second term on the right-hand side of (3.15) does not contribute to the volume integral of Γ^{I}, and that $4\pi R^3/3$ is the volume V of the sphere we obtain

$$\lim_{R\to\infty}\int_V \Gamma^{\mathrm{I}}_{ij}(\mathbf{r},\mathbf{r}')\,dV' = \frac{1}{3c}\delta_{ij} - \frac{1}{3c}\delta_{ij} = 0. \tag{3.16}$$

The corresponding procedure applied to Γ^{II} yields

$$\lim_{R\to\infty}\Gamma^{\mathrm{II}}_{ij} = \frac{1}{3c}\delta_{ij}\delta(\mathbf{r}-\mathbf{r}') + \frac{1}{4\pi c}(\delta_{ij} - 3e_ie_j)\cdot\frac{1}{|\mathbf{r}-\mathbf{r}'|^3}$$
$$+ \frac{1}{4\pi c}\delta_{ij}\frac{1}{R^3} + \frac{1}{4\pi c}\delta_{ij}\frac{1}{R^3}, \tag{3.17}$$

from where

$$\lim_{R\to\infty}\int_V \Gamma^{\mathrm{II}}_{ij}(\mathbf{r},\mathbf{r}')\,dV' = \frac{1}{3c}\delta_{ij} + \frac{1}{3c}\delta_{ij} + \frac{1}{3c}\delta_{ij} = \frac{1}{c}\delta_{ij}. \tag{3.18}$$

With the results (3.15) and (3.17), the equations (3.6) are indeed verified.

It is now easy to find the displacement Green functions G^{I} and G^{II} of the infinite sphere. We only have to add to G^{∞} a term G^{01} (or G^{02}) such that the correct Γ^{I} (or Γ^{II}) results. Of course, G^{01} and G^{02} must satisfy Laplace's equation. Consider the harmonic functions

$$G^{01} = -\frac{1}{3cV}(xx' + yy' + zz'), \qquad G^{02} = \frac{2}{3cV}(xx' + yy' + zz'). \tag{3.19}$$

Then

$$\partial_i\partial'_j G^{01} = \Gamma^{01}_{ij} = -\frac{1}{3cV}\delta_{ij}, \tag{3.20}$$

and

$$\partial_i\partial'_j G^{02} = \Gamma^{02}_{ij} = \frac{2}{3cV}\delta_{ij}, \tag{3.21}$$

as it should be. It is possible to add expressions linear in x, y, z or x', y', z' to G^{01} and G^{02} without changing Γ^{01} and Γ^{02}. It follows from our derivation that parts of G which do not contribute to Γ have no physical meaning and, therefore, can be omitted.

The above results are true under the conditions $r/R \ll 1$ and $r'/R \ll 1$. The modified Green functions for finite spheres follow by adding additional terms on the right-hand sides of (3.15) and (3.17). Such terms, however, do not contribute to the integrals of the type occurring in (3.16) and (3.18) because these integrals must be independent of the size of the sphere.

Above we have introduced an infinite medium as a sphere with infinite radius. Such a medium is not different from the one obtained if any other convex form is expanded to infinity in all directions. In particular, (3.19)–(3.21) remain valid, so that at the end there is just one infinite medium.

4. Explicit Form of the Green Functions G^∞ and Γ^∞ for Homogeneous, Isotropic Elastic Media

In Section 3 we gave the explicit form of the Green functions G^∞ and Γ^∞ for an infinite homogeneous and isotropic medium in the case of potential theory. The analogous results for elasticity theory are, with κ as compression modulus and μ as shear modulus,

$$G_{ij}^{\infty}(\mathbf{r}, \mathbf{r}') = \frac{1}{8\pi\mu}\left(\frac{3\kappa + 7\mu}{3\kappa + 4\mu}\delta_{ij} + \frac{3\kappa + \mu}{3\kappa + 4\mu} e_i e_j\right)\frac{1}{|\mathbf{r} - \mathbf{r}'|}, \tag{4.1}$$

whereas $\Gamma_{ijkl}^{\infty}(\mathbf{r}, \mathbf{r}')$ has the form

$$\Gamma_{ijkl}^{\infty} = E_{ijkl}\delta(\mathbf{r} - \mathbf{r}') + F_{ijkl}(\theta, \varphi)\frac{1}{|\mathbf{r} - \mathbf{r}'|^3}. \tag{4.2}$$

As before, θ, φ are the directional angles of the vector $e = (\mathbf{r} - \mathbf{r}')/|\mathbf{r} - \mathbf{r}'|$. Equation (4.2) is also valid for anisotropic (homogeneous) media. In the isotropic case (e.g., Kröner, 1986)

$$\begin{aligned} E_{ijkl} &= \frac{1}{15\mu}\left(-\frac{3\kappa + \mu}{3\kappa + 4\mu} + 9\frac{\kappa + 2\mu}{3\kappa + 4\mu} I_{ijkl}\right), \\ I_{ijkl} &= \frac{\delta_{ik}\delta_{jl} + \delta_{il}\delta_{jk}}{2}, \end{aligned} \tag{4.3}$$

and

$$\begin{aligned} F_{ijkl} = {} & \frac{1}{8\pi\mu(3\kappa + 4\mu)}[6\mu I_{ijkl} - (3\kappa + \mu) \\ & \times (\delta_{ij}\delta_{kl} - 3e_i e_j\delta_{kl} - 3e_k e_l\delta_{ij} + 15e_i e_j e_k e_l) + 6(3\kappa - 2\mu)(e_i e_k \delta_{jl})_{(ij)(kl)}]. \end{aligned} \tag{4.4}$$

(Equation (5.37) of Kröner (1986) contains a sign mistake: replace "−" by "+" in front of 15.) Because of symmetry, the volume integral of the last term in (4.2) vanishes so that

$$\int_V \Gamma_{ijkl}^{\infty}(\mathbf{r}, \mathbf{r}')\, dV' = E_{ijkl}. \tag{4.5}$$

This result is also valid for anisotropic media, if the anisotropic E is used.

5. Applications

5.1. General Remarks

It is known that the Green function G^∞ of the infinite, homogeneous and isotropic elastic medium (the Kelvin–Somigliana function) solves problems of infinite media, provided stress and strain go to zero in the infinity. This is often, but not always, the case. For instance, we may be interested to know how the stress state in a sphere of radius R changes with R, given the surface tractions as functions of R and of the polar angles θ, φ. Of course, this problem is physically meaningful only when the stresses remain finite everywhere. In special cases, we might find that the stress state is independent of R, e.g., in the case of simple constant stress, where stress and strain are finite in the infinity.

Consider an infinite continuum in a state of, everywhere, finite stress. In a imaginery experiment cut out a large sphere, or a volume of any other form and measure the volume and surface forces necessary to keep the stress state in this volume unchanged. We can imagine cutting out larger and larger spheres and see that the concept of surface tractions still has a meaning. Therefore, the derivation of equations like (2.7) is still valid and the strain, given by

$$\varepsilon_{ij}(\mathbf{r}) = \int_V \Gamma^{\mathrm{II}}_{ijkl}(\mathbf{r}, \mathbf{r}')\sigma_{kl}(\mathbf{r}')\, dV', \tag{5.1}$$

is compatible, thus it is the correct strain prevailing in the (infinite or finite) medium.

Now consider a constant stress σ. In an elastically homogeneous medium the corresponding strain is indeed compatible and the whole problem is trivial. We may ask: What is the (compatible) strain state in a heterogeneous medium with the same stress sources that produce constant σ in a homogéneous medium? It follows from the heterogeneous (inverse) Hooke's law that the strain

$$s_{ijkl}(\mathbf{r})\sigma_{kl} \tag{5.2}$$

is, in general, incompatible, due to the position dependence of the compliances. In fact, in a medium with such a strain each volume element, when cut out, will relax differently, and the relaxed body will be a nonfitting assembly of volume elements. As explained in Section 2, (5.1) nevertheless gives us the correct compatible strain in the heterogeneous medium and therefore solves our problem not only in the case of finite media, but also of infinite media. From the strain, the stress in the heterogeneous medium follows immediately by application of the (heterogeneous) Hooke's law.

Since the analogus arguments apply to more general stress distributions, we can state that (5.1) solves the following problem:

Given a finite or infinite elastically homogeneous medium in a compatible strain state.
Wanted: The (compatible) strain state produced by the same stress sources in a heterogeneous elastic medium of the same shape.

The homogeneous medium is often called the comparison medium—to be compared with the real medium, which is of interest. The above statement also applies to media that are anisotropic instead of heterogeneous, or heterogeneous *and* anisotropic.

5.2. The Elastic Inclusion Problem

The method of the modified Green functions gives us quick access to the elastic inclusion problem, to which Mura also gave important contributions; see, e.g., Mura (1982) and Kinoshita and Mura (1971). Cut out a simply connected, not necessarily ellipsoidal, region (called an *inclusion*) from a homogeneous, in general, anisotropic infinite stress-free elastic medium. Give the inclusion a constant stress-free strain, say ε^*, e.g., by plastic deformation. Deform the inclusion back to its original shape by an elastic strain, insert it and glue it along the interface. The total strain is zero everywhere at this moment. Now take away the forces applied before to the surface of the inclusion to produce the elastic back-strain. This is equivalent to applying the opposite forces to the interface. Under these forces both matrix and inclusion will deform into the final state, thereby suffering an elastic strain ε^{T}, say. We are interested to know ε^{T}, given ε^* and the shape of the inclusion.

To apply the formalism of the strain Green function we need to know a stress field which is in equilibrium with the applied forces. These forces are those applied to the interface, opposite to the forces which before had produced the homogeneous stress state $-c_{ijkl}\varepsilon^*_{kl}$ in the inclusion. Thus

$$\varepsilon^{\mathrm{T}}_{ij}(\mathbf{r}) = \int_{\mathrm{incl}} \Gamma^{\infty}_{ijkl}(\mathbf{r} - \mathbf{r}')\, dV' c_{klmn}\varepsilon^*_{mn}, \tag{5.3}$$

and is valid for the interior and exterior of the inclusion. Instead of Γ^{∞} we could also use the Γ^{II} of the infinite medium, because the additional term would be of order V_{incl}/V $(V \to \infty)$ and therefore negligible. The integration extends over the inclusion volume, because only there ε^*_{mn} does not vanish. For this reason the integral in (5.3) is not in general equal to that in (5.1) with constant stress. It is equal to it for spherical inclusions, in which case

$$\varepsilon^{\mathrm{T}}_{ij} = E_{ijkl}c_{klmn}\varepsilon^*_{mn}. \tag{5.4}$$

Inside the inclusion ε^{T} depends on position except for the spherical and ellipsoidal form. Averaging (overbar) (5.3) over the inclusion volume gives us

$$\overline{\varepsilon^{\mathrm{T}}_{ij}} = \frac{1}{V_{\mathrm{incl}}} \int\!\!\int_{\mathrm{incl}} \Gamma^{\infty}_{ijkl}(\mathbf{r} - \mathbf{r}')\, dV\, dV' c_{klmn}\varepsilon^*_{mn} \equiv S^{\mathrm{Esh}}_{ijmn}\varepsilon^*_{mn}, \tag{5.5}$$

where S^{Esh}_{ijmn} is Eshelby's inclusion tensor, derived by him in a different way. For further results on the inclusion problem, see Mura (1982) and Kunin and Sosnina (1971).

5.3. Green Functions of Heterogeneous and Anisotropic Media

The discussed Green functions are by no means simple functions in heterogeneous or anisotropic media, and not even in homogeneous media with nonsimple shape. There exists, however, a systematic method of calculating anisotropic or heterogeneous Green functions when that of the simple comparison media, say $\overset{\circ}{\Gamma}$, is known. The relevant formula is an integral equation, originally developed in the quantum mechanical scattering theory by Lippmann, Schwinger, and Dyson (LSD) and introduced into elasticity theory by Dederichs and Zeller (1972). Similar developments are due to Rieder (1962). In our abridged notation the LSD equation has the form

$$(I + \overset{\circ}{\Gamma} * \delta \overset{\circ}{c}) * \Gamma = \overset{\circ}{\Gamma}, \qquad \delta \overset{\circ}{c} \equiv c - \overset{\circ}{C}, \tag{5.6}$$

where c and $\overset{\circ}{C}$ are the operators of the elastic moduli of the real medium and of the comparison medium. I is the unity operator. Equation (5.6) is valid for the Green functions Γ^{I} and Γ^{II} of the two boundary value problems, and also for Γ^{∞}.

From our former result the LSD equation is obtained as follows. The equation (2.12)

$$\varepsilon = \Gamma^{\mathrm{II}} * c * \varepsilon, \tag{5.7}$$

is true for arbitrary ε on the right-hand side, whereas the ε on the left-hand side is always the compatible part of the right-hand side ε. If for the latter, a compatible tensor field is inserted, then this is equal to the left-hand side ε. See the discussion after (2.12). Responsible for all this is the projection operator $\Gamma^{\mathrm{II}} * c$, which, as a genuine projection operator, satisfies

$$\Gamma^{\mathrm{II}} * c * \Gamma^{\mathrm{II}} * c = \Gamma^{\mathrm{II}} * c, \tag{5.8}$$

or

$$\Gamma^{\mathrm{II}} * c * \Gamma^{\mathrm{II}} = \Gamma^{\mathrm{II}}. \tag{5.9}$$

This is an integral equation for Γ^{II}, given c. The other modified Green functions satisfy corresponding equations. Note that we cannot shorten by Γ^{II} in (5.9), because Γ^{II}, like $\Gamma^{\mathrm{II}} * c$, has no inverse.

All these results can be proved directly by introducing the decomposition of ε,

$$\varepsilon = (\nabla u)^{\mathrm{sym}} + \nabla \times \iota \times \nabla, \tag{5.10}$$

into compatible and incompatible parts in (5.7). In (5.10) ι is an arbitrary symmetric tensor. In particular, (5.7) is formed to be true for any elastic medium, so that besides (5.9),

$$\Gamma = \overset{\circ}{\Gamma} * \overset{\circ}{C} * \Gamma, \qquad \overset{\circ}{\Gamma} = \Gamma * c * \overset{\circ}{\Gamma} = \overset{\circ}{\Gamma} * c * \Gamma, \tag{5.11}$$

also follows, where $\overset{\circ}{\Gamma}$ and $\overset{\circ}{C}$ may represent any elastic medium. We have chosen the capital C, because we shall be interested in a homogeneous medium to be taken as the comparison medium (see above). The last equation in (5.11) is valid because the Γ's and c are self-adjoint.

A combination of (5.11) immediately gives the LSD equation (5.6). This equation is usually solved in the form of a Neumann series, formally a geometric series in terms of multiple integrals of increasing order

$$\Gamma = \overset{\circ}{\Gamma} - \overset{\circ}{\Gamma} * \delta \overset{\circ}{c} * \overset{\circ}{\Gamma} + \overset{\circ}{\Gamma} * \delta \overset{\circ}{c} * \overset{\circ}{\Gamma} * \delta \overset{\circ}{c} * \overset{\circ}{\Gamma} \cdots. \tag{5.12}$$

5.4. Random Media

The LSD equation has important applications in the theory of random elastic media, whose distribution of elastic moduli is microscopically heterogeneous but macroscopically homogeneous (not necessarily isotropic). The interest is in average values of stress, strain, and other quantities. The theory utilizes ensemble averaging (symbol $\langle \cdot \rangle$), which commutes with differentiation and integration.

Let us return to (2.7), or (2.9), $\varepsilon = \Gamma^{\mathrm{II}} * \sigma$. There Γ^{II} is a purely internal quantity representing the material properties, whereas σ is determined by the external loads. Hence, as long as these are not correlated to the distribution of elastic moduli we can use the theorem, that (ensemble) averages of products equal the product of averages of the factors. Thus averaging of (2.7) yields

$$\langle \varepsilon_{ij}(\mathbf{r}) \rangle = \int_V \langle \Gamma^{\mathrm{II}}_{ijkl}(\mathbf{r}, \mathbf{r}' \rangle \langle \sigma_{kl}(\mathbf{r}') \rangle \, dV'. \tag{5.13}$$

Introducing an effective compliance tensor by

$$\langle \varepsilon_{ij}(\mathbf{r}) \rangle = S^{\mathrm{eff}}_{ijkl} \langle \sigma_{kl}(\mathbf{r}) \rangle, \tag{5.14}$$

and noting that (5.13) is also true for constant $\langle \sigma_{kl} \rangle$ we conclude that

$$S^{\mathrm{eff}}_{ijkl} = \int_V \langle \Gamma^{\mathrm{II}}_{ijkl}(\mathbf{r}, \mathbf{r}') \rangle \, dV'. \tag{5.15}$$

This means that although the effective elastic moduli and compliances are not the ensemble averages of the local moduli and compliances, the ensemble-averaged strain Green function is the correct Green function of the effective medium. Equation (5.15) gives us a possibility to calculate S^{eff}.

To obtain $\langle \Gamma \rangle$, which we need in (5.15), we average (5.12)

$$\langle \Gamma \rangle = \mathring{\Gamma} - \mathring{\Gamma} * \langle \delta \mathring{c} \rangle * \mathring{\Gamma} + \mathring{\Gamma} * \langle \delta \mathring{c} * \mathring{\Gamma} * \delta \mathring{c} \rangle * \mathring{\Gamma} \cdots, \tag{5.16}$$

where we have omitted the superscripts II for simplicity. Since the $\mathring{\Gamma}$'s are nonrandom, they have been put largely outside the brackets, so that the averaging creates so-called correlation functions $\langle \delta \mathring{c}(\mathbf{r}_1) \rangle$ (one-point), $\langle \delta \mathring{c}(\mathbf{r}_1) \delta \mathring{c}(\mathbf{r}_2) \rangle$ (two-point), $\langle \delta \mathring{c}(\mathbf{r}_1) \delta \mathring{c}(\mathbf{r}_2) \delta \mathring{c}(\mathbf{r}_3) \rangle$ (three-point), etc. The whole set of these functions contains the complete statistical information about the random elastic medium. Every quantity of interest can be calculated, at least in principle, if all these correlation functions are known.

By volume integration over (5.16) we obtain the effective compliances. In this case, the $\mathring{\Gamma}$'s are the completed Green functions (including the surface part). However, it suffices to calculate E^{eff}, because we then find the components of C^{eff} from (4.3) (assuming macroisotropy for the moment) when there we replace κ and μ by κ^{eff} and μ^{eff}. For this calculation we replace all the Green functions in (5.16) by their Γ^{∞} part and form $\int \Gamma^{\infty} \, dV'$.

Because the choice of $\mathring{\Gamma}$ in the LSD equation is arbitrary, it is often convenient to choose $\mathring{\Gamma} = \langle \Gamma \rangle \equiv \Gamma^{\text{eff}}$. Then (5.16) can be written

$$\langle \delta c^{\text{eff}}(I + \Gamma^{\text{eff}} * \delta c^{\text{eff}})^{-1} \rangle = 0. \tag{5.17}$$

It is shown, e.g., in Kröner (1986), that with the just made choice of $\mathring{\Gamma}$, there are possible situations in which F does not contribute. We may then replace Γ^{eff} by E^{eff} in (5.17), thus

$$\langle \delta c^{\text{eff}}(I + E^{\text{eff}} * \delta c^{\text{eff}})^{-1} \rangle = 0, \tag{5.18}$$

which effectively is an algebraic equation, because the operators E^{eff} and δc^{eff} have a delta kernel.

The C^{eff} calculated from this equation is called self-consistent, see the corresponding equations of Walpole (1966) or of Kröner (1958). It is valid under the mentioned condition for F (no contribution!) which means complete lack of order in the distribution of elastic moduli. This property is reflected in the correlation functions.

6. Conclusion

The introduction of the modified Green functions allows us simple formulations of the boundary value problems of elasticity theory and similar linearized field theories. Simple formulation does not mean that the solution also becomes practicable. In fact, it lies in the very nature of these problems that they are often extremely involved, and this is so because the Green functions are complex, except in certain simpler problems, e.g., those of the infinite medium. The simple formulations discussed in this work might nevertheless be useful because they show, in a systematic way, how to solve these problems. The important tool for this will often be the computer—perhaps in connection with finite element methods.

References

Dederichs, P. and Zeller, R. (1973), Variational treatment of the elastic constants of disordered materials, *Z. Physik*, **259**, 103–116.

Frank, P. and von Mises, R. (1930), *Die Differential- und Integralgleichungen der Mechanik und Physik*, 1. Band, 2. Vermehrte Auflage, Vieweg und Sohn, Braunschweig.

Gubernatis, J. E. and Krumhansl, J. A. (1975), Macroscopic engineering properties of polycrystalline materials: Elastic properties, *J. Appl. Phys.*, **46**, 1875–1883.

Kinoshita, N. and Mura, T. (1971), Elastic fields of inclusions in anisotropic media, *Phys. Stat. Sol.* (a), **5**, 759–768.

Kröner, E. (1958), Berechnung der elastischen Konstanten des Vielkristalls aus den Konstanten des Einkristalls, *Z. Physik*, **151**, 504–518.

Kröner, E. (1986), Statistical modelling, in *Modelling Small Deformation of Polycrystals*, edited by J. Gittus and J. Zarka, Elsevier Applied Science, London, New York, pp. 229–291.

Kunin, I. A. (1964), Green's tensor for anisotropic elastic medium with internal stress sources, *Dokl. Akad. Nauk SSSR*, **157**, 1319–1320.

Kunin, I. A. and Sosnina, E. G. (1971), An ellipsoidal inhomogeneity in an elastic medium, *Dokl. Akad. Nauk SSSR*, **199**, 571–574.

Mura, T. (1963), Continuous distribution of moving dislocations, *Phil. Mag.*, **8**, 843–857

Mura, T. (1982), *Micromechanics of Defects in Solids*, Nijhoff, The Hague.

Rieder, G. (1962), Iterationsverfahren und Operatorgleichungen in der Elastizitätstheorie, *Abh. Braunschweig. Wiss. Ges.*, **14** (2), 109–343.

Simmons, J. A. and Bullough, R. (1970), Internal stress and the incompatibility problem in anisotropic elasticity, in *Proceedings of the Conference on Fundamental Aspects of Dislocation Theory* (Gaithersburg), edited by J. A. Simmons, R. de Wit, and R. Bullough, U.S. Special Publication 317, Vol. 1, National Bureau of Standards, Washington, pp. 89–124.

Walpole, L.J. (1966), On bounds of the overall elastic moduli of inhomogeneous systems, I and II, *J. Mech. Phys. Solids*, **14**, 151–162, 289–301.

Inclusions With and Without Free Surfaces in a Plane Strain

J. K. LEE
Department of Metallurgical Engineering,
Michigan Technological University,
Houghton, MI 49931, U.S.A.

Abstract

The stress field and strain energy of inclusions with and without free surfaces in a plane strain condition are studied via an atomistic approach, in which a triangular lattice is considered with linear atomic interactions. The result of a polygonal inclusion in an infinite matrix shows that the stress concentrations at the polygonal vertices increase as the ratio of inclusion shear modulus to matrix shear modulus, μ^*/μ, increases. The strain energy is nearly proportional to $\mu^*/(2\mu^* + \mu)$, which is the proportionality constant for an elliptic inclusion with a dilatational eigenstrain. For an undergrowth whose one side is free of traction, a maximum principal stress analysis reveals stress concentrations in a tensile mode at its traction-free side, indicating a possibility of crack initiation for a brittle undergrowth. For an overgrowth whose three sides are free of traction, a number of interesting results are found. Unlike an undergrowth or inclusion, the stress is mostly confined in the neighborhood of the overgrowth–substrate interface. For a positive eigenstrain, the stress field starts with a strong compressive state at the interface area and diminishes as the distance from the interface increases. However, nearby the substrate, the traction-free perimter of an overgrowth is also found to be in a tensile stress mode. Quite different from the case of inclusions and undergrowths is that the total strain energy associated with an overgrowth approaches an asymptotic value as its thickness increases. That is, the strain energy density of an overgrowth is found to depend on its shape and also on its size.

1. Introduction

In joining ceramic to metal matrixes, there arise residual thermoelastic stresses due to differences in the thermal expansion coefficient between the ceramic and metal matrixes, and they often render a difficult engineering problem because cracks can be initiated at the stress concentrations. In growing a semiconductor thin film on a substrate, the coherency strain due to the difference in lattice parameter between thin film overgrowth and substrate often introduces misfit dislocations into the system, thus weakening its electronic transport quality. The ceramic part in the ceramic–metal composite and the overgrowth on a substrate can be broadly defined as inclusions having partially free surfaces. Following Eshelby's classic work on elastic inclusions (Eshelby, 1957), several investigations (Chiu, 1980; Tsuchida and Mura, 1983; Lee and Hsu, 1985; Mura, 1987) have been made on the stress field and strain energy of inclusions having such free surfaces. Because of complex boundary conditions associated with the problems, however, only a few cases have been solved and many remain to be solved. In this investigation, we study, by means of the atomistic approach, the stress field and strain energy of the following inclusions:

(A) Nonelliptical inclusions embedded in an infinite matrix. Rectanglar and hexagonal *inclusions* are examined.
(B) Rectangular inclusion whose three sides are surrounded by a half-space matrix and the remaining side is free of traction. This inclusion is termed an *undergrowth.*
(C) Rectangular inclusion whose one side is welded to a half-space matrix (substrate) and the other three edges are free of traction. This is inclusion is termed an *overgrowth.*

Particular attention is paid to a rectangular inclusion since continuum elasticity solutions are available for both cases (A) and (B), provided that elastic constants are assumed to be homogeneous and isotropic. Similarly, for an elliptic inclusion, solutions for homogeneous as well as inhomogeneous cases are available. However, the elliptic shapes are not amenable to the condition of traction-free boundaries, which are relevant to overgrowth thin films or ceramic–metal joint problems for which both cases (B) and (C) have direct implications. In this study, we will compare the results of the atomistic approach with those of continuum elasticity whenever they are possible. In section 2, the atomistic approach is explained, followed by the solutions of continuum elasticity in Section 3. Section 4 presents the results combined with discussions. Conclusions are given in Section 5.

2. Atomistic Approach with a Triangular Lattice

Hoover *et al.* (1974) have demonstrated that a two-dimensional triangular lattice can simulate an elasticity problem in a plane strain condition. Figure 1(a) depicts a rectangular inclusion with its side lengths equal to $2H$ and $2L$, embedded in an infinite triangular lattice matrix. Because of symmetry, only the first quarter is shown in the picture. If a traction-free condition is imposed along the x axis, the same figure also represents an undergrowth with side lengths equal to $2L$ and H, surrounded by a half-space triangular lattice matrix. Figure 1(b) represents an overgrowth with side lengths equal to $2L$ and H. For this case, the region of $y \geqq 0$ represents a substrate or half-space matrix. The potential energy of a triangular lattice having a uniform spring

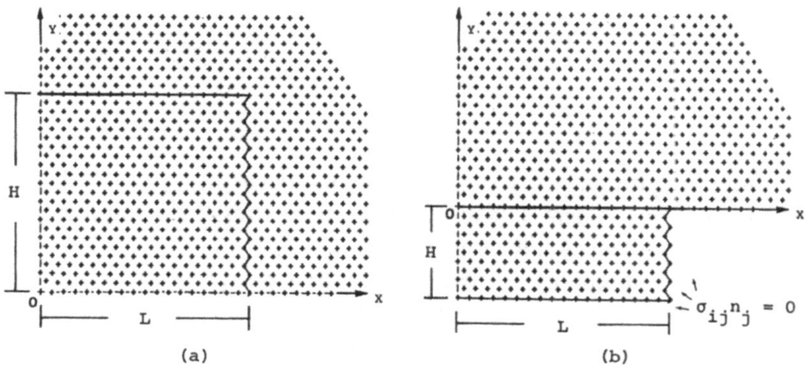

FIG. 1. Coordinates of various inclusions. (a) Depicts a rectangular *inclusion* with its side lengths equal to $2H$ and $2L$, embedded in an infinite triangular lattice matrix. Because of symmetry, only the first quarter is shown. If a traction-free condition is imposed along the x axis, (a) also represents an *undergrowth* with side lengths equal to $2L$ and H, surrounded by a half-space triangular lattice matrix. (b) Represents an *overgrowth* with lengths equal to $2L$ and H. The region of $y \geq 0$ represents a substrate or half-space matrix. All the coordinates are given in units of the matrix lattice parameter, d.

constant, κ, is given by†

$$\Phi = \frac{\kappa}{2} \sum_{i>j} (r_{ij} - d)^2, \tag{2.1}$$

where

$$r_{ij}^2 = x_{ij}^2 + y_{ij}^2, \qquad x_{ij} = x_i - x_j, \qquad y_{ij} = y_i - y_j, \tag{2.2}$$

and d is the triangular lattice parameter. For an infinitesimal displacement case, which is the basis of linear elasticity, the calculation of (2.1) is limited to the *first nearest neighbors for a given atom.* For this lattice, the Lamé constants, λ and μ are, under a plane strain condition, given by

$$\lambda = \mu = \frac{3^{1/2}\kappa}{4}. \tag{2.3}$$

The Poisson ratio, $\nu = \lambda/(2\lambda + 2\mu)$, becomes $\frac{1}{4}$. By calculating the force on an atom, we find the stress components of the ith atom to be

$$\sigma_{11} = \frac{\kappa}{3^{1/2}d^2} \sum_j \left(1 - \frac{d}{r_{ij}}\right) x_{ij}^2, \tag{2.4}$$

$$\sigma_{12} = \frac{\kappa}{3^{1/2}d^2} \sum_j \left(1 - \frac{d}{r_{ij}}\right) x_{ij} y_{ij}, \tag{2.5}$$

$$\sigma_{22} = \frac{\kappa}{3^{1/2}d^2} \sum_j \left(1 - \frac{d}{r_{ij}}\right) y_{ij}^2. \tag{2.6}$$

† The subindices i and j in x, y, z, and r indicate the coordinates of the ith and jth atoms, not those used for a tensor notation.

Within the region of an inclusion, use of the different spring constant, κ^*, and the lattice parameter, d^*, suffices the inclusion problem of present interest;

$$\kappa^* = \gamma\kappa, \tag{2.7}$$

$$d^* = (1 + \varepsilon)d. \tag{2.8}$$

If γ is equal to 1, it represents an elastically homogeneous system, otherwise an inhomogeneous system. ε is an eigenstrain or stress-free transformation strain, which represents a lattice mismatch.

The elastic strain energy is then the potential energy of an equilibrated system for a given ε, since the potential energy is zero when ε becomes zero. The equilibration of a system is accomplished via Powell's conjugate gradient method (1977). To minimize nonlinear interactions among the atoms, the eigenstrain, ε, is assigned with a small value, 0.5%, and the atoms are allowed to move until the average force per atom reaches 10^{-5} κ or less. Both free boundary and elastic boundary conditions for the atoms in the peripheral region are examined for Case (A), and it is found that if the system is sufficiently large, about 12,000 atoms for the present study, the difference in the numerical solutions of stress and strain energy is found to be within a few percent.

3. Continuum Linear Elasticity

3.1. *Inclusion in an Infinite Matrix; Case* (A)

For this case, both two- and three-dimensional analytical solutions have been obtained in the literature (Faivre, 1964; Lee and Johnson, 1977; Khachaturyan, 1983). For the purpose of comparison with the atomistic approach, the analytical solutions are reformulated here in the manner of Faivre (1964). Under a purely dilatational eigenstrain case, the elastic stress field of a rectangular parallelepiped with its edge lengths equal to $2L$, $2H$, and $2W$ are given by

$$\sigma_{11}(\vec{r}) = -2\mu\xi\varepsilon S(\vec{r}) + \frac{\mu\xi\varepsilon}{2\pi}\left[\tan^{-1}\frac{u_2u_3}{u_1U} + \frac{\nu}{(1-\nu)} \cdot\left(\tan^{-1}\frac{u_3u_1}{u_2U} + \tan^{-1}\frac{u_1u_2}{u_3U}\right)\right], \tag{3.1}$$

$$\sigma_{12}(\vec{r}) = -\frac{\mu\S\varepsilon}{2\pi}\left[\tanh^{-1}\frac{u_3}{U}\right], \tag{3.2}$$

where

$$[f(u_1, u_2, u_3)] = f(u_1, u_2, u_3)\Big|_{x-L,y-H,z-W}^{x+L,y+H,z+W},$$

$$U^2 = u_1^2 + u_2^2 + u_3^2, \qquad \vec{r} = (x, y, z),$$

$$\xi = \frac{1+\nu}{1-2\nu}, \qquad \S = \frac{1+\nu}{1-\nu},$$

and $S(\vec{r})$ is equal to 1 if $\vec{r}$ is inside the inclusion, otherwise it is zero. Other terms in $\sigma_{ij}(\vec{r})$ can be obtained through appropriate cyclic permutations among the subscripts, 1, 2, and 3.

3.2. *Undergrowth*: *Case* (B)

Several studies have been made for this case (Chiu, 1980; Lee and Hsu, 1985). The solutions of Lee and Hsu are reproduced here for the purpose of comparison. Using the coordinates shown in Fig. 1(a), the stress field of a rectangular parallelepiped with its lengths equal to $2L$, H, and $2W$ becomes

$$\sigma_{11}(\vec{r}) = -2\mu\S\varepsilon S(\vec{r}) + \frac{\mu\S\varepsilon}{2\pi}\left[\tan^{-1}\frac{u_2u_3}{u_1U} + 3\tan^{-1}\frac{v_2v_3}{v_1V} + 4\nu\tan^{-1}\frac{v_1v_2}{v_3V} - \frac{2yv_1}{V(V+v_3)}\right], \tag{3.3}$$

$$\sigma_{22}(\vec{r}) = -2\mu\S\varepsilon S(\vec{r}) + \frac{\mu\S\varepsilon}{2\pi}\left[\tan^{-1}\frac{u_3u_1}{u_2U} - \tan^{-1}\frac{v_3v_1}{v_2V} - \frac{2yv_3v_1(V^2+v_2^2)}{V(V^2-v_3^2)(V^2-v_1^2)}\right], \tag{3.4}$$

$$\sigma_{12}(\vec{r}) = -\frac{\mu\S\varepsilon}{2\pi}\left[\ln(U+u_3) + \ln(V+v_3) + \frac{2yv_2}{V(V+v_3)}\right], \tag{3.5}$$

$$\sigma_{31}(\vec{r}\,) = -\frac{\mu\S\varepsilon}{2\pi}\left[\ln(U+u_2) + (3-4\nu)\ln(V+v_2) + \frac{2y}{V}\right], \tag{3.6}$$

where

$$[f(u_1, u_2, u_3)] = f(u_1, u_2, u_3)|_{x-L,y-H,z-W}^{x+L,y,z+W},$$
$$[f(v_1, v_2, v_3)] = f(v_1, v_2, v_3)|_{x-L,y,z-W}^{x+L,y+H,z+W},$$
$$V^2 = v_1^2 + v_2^2 + v_3^2.$$

$\sigma_{33}(\vec{r})$ and $\sigma_{23}(\vec{r})$ can be obtained through appropriate cyclic permutations from (3.3) and (3.5), respectively.

By letting W approach infinity and dividing the results by a factor, $(1 + \nu)$ (due to the dimensionality difference), we can obtain the continuum solutions for the stress field of both Case (A) and Case (B), i.e., in a plane strain condition.

4. Numerical Results and Discussion

4.1. *Case* (A): *Inclusion*

To demonstrate the validity of the atomistic approach, a comparison is made in Fig. 2, in which the stress field along the y axis from the inclusion center is

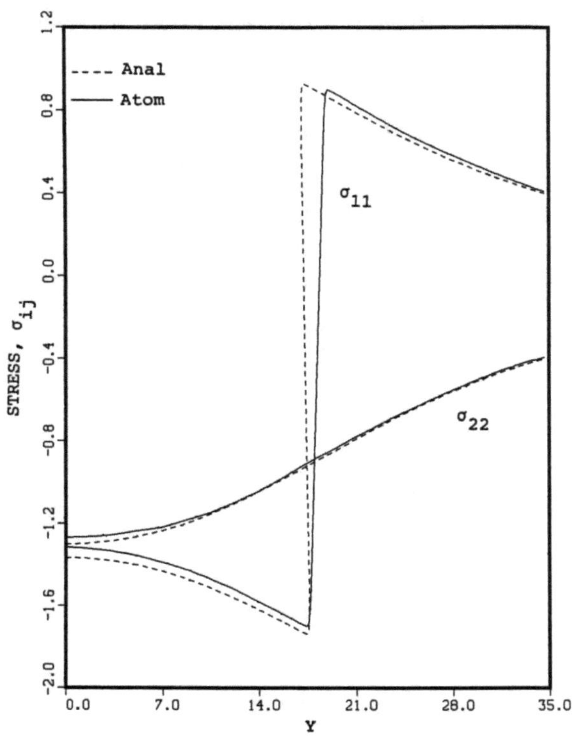

FIG. 2. σ_{11} and σ_{22} along the y axis from the center for an inclusion with $L = 18d$ and $H = 17.32d$. The stresses are given in units of $\mu\varepsilon$. The solid curves show the atomistic approach, while the dashed curves show the continuum solution.

displayed for an inclusion with $L = 18d$ and $H = 17.32d$. The solid curves are for the atomistic approach, while the dashed curves are for the continuum solution of (3.1). The stresses are given in units of $\mu\varepsilon$, and the coordinates are expressed in units of lattice parameter, d, throughout this study. The σ_{22} component is continuous across the inclusion–matrix interface at $y = 17.32$, and both methods are shown to be in excellent agreement. The σ_{11} component is discontinuous at the interface, and the continuum solution shows a usual stress concentration at the interface position (Faivre, 1964). We note, however, that the solution of the atomistic approach shows a stress relaxation at this position. The relaxation can be credited to a discrete structural effect (Ashurst and Hoover, 1976), termed "lattice trapping."

Figure 3 shows a maximum shear stress, τ_{12}^{m}, for two different shapes of a hard inclusion with $\mu^*/\mu = \gamma = 3$; a regular hexagon with $L = 18d$ in (a) and a rectangle with $L = 18d$ and $H = 17.32d$ in (b). The maximum shear stress is defined as

$$\tau_{12}^{m} = ((\sigma_{11} - \sigma_{22})^2 + 4\sigma_{12}^2)^{1/2}. \tag{4.1}$$

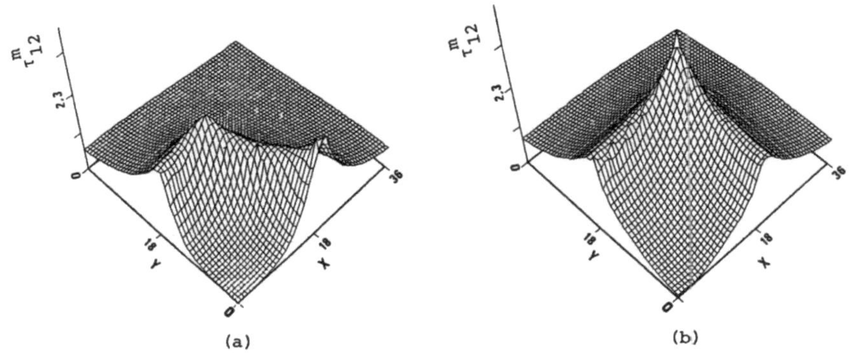

FIG 3. Maximum shear stress, τ_{12}^m, for two different shapes of a hard inclusion with $\mu^*/\mu = \gamma = 3$: (a) a regular hexagon with $L = 18d$, and (b) a rectangle with $L = 18d$ and $H = 17.32d$.

Note that the shear stress, τ_{12}^m, reveals remarkably well stress concentrations located at the vertices of the polygonal inclusions. As the shape becomes more rounded, the magnitude of such stress concentrations diminishes, and eventually, for a circular or an elliptic shape, no sharp stress concentrations exist. For a circular inclusion (Bhargava, 1961), the elastic strain energy is given by

$$\Phi = \frac{4 \text{ Area} \cdot \gamma\mu\varepsilon^2}{2\gamma + 1}. \tag{4.2}$$

For a regular hexagon, the atomistic approach yields

$$\Phi = \frac{3.96 \text{ Area} \cdot \gamma\mu\varepsilon^2}{2\gamma + 1}. \tag{4.3}$$

The small difference may manifest that because of stress relaxation at the vertices, the strain energy is slightly reduced for a polygonal inclusion in the atomistic approach.

4.2. *Case* (B): *Undergrowth*

We again start with comparisons between the atomistic approach and the Lee–Hsu analytical solution for an elastically homogeneous case, i.e., $\gamma = 1$. In Fig. 4, the σ_{11} component of an undergrowth with $L = 18d$ and $H = 17.32d$ is plotted along two different directions; 0° and 60° from the x axis. Along the x axis, i.e., the 0° direction, the atomistic result shows a significant stress relaxation across the undergrowth–matrix interface (at $x = 18$). Because this position is a corner, a strong structural effect, much higher than at a position in the middle of an interface, is clearly demonstrated. Contrary to the 0° case, the 60° case shows a very small difference between the atomistic and continuum approaches. In Fig. 5, a normalized strain energy, Φ/κ, is plotted as a

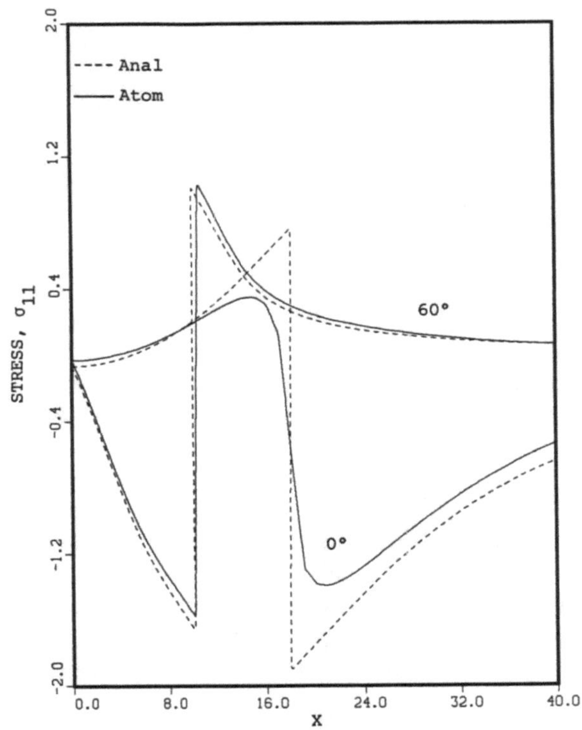

FIG. 4. σ_{11} along two different directions, 0° and 60° from the x axis, for an undergrowth with $L = 18d$ and $H = 17.32d$. Solid curves show the atomistic approach while dashed curves show the continuum elasticity theory. Note a significant stress relaxation across the undergrowth–matrix interface (at $x = 18$) for the atomistic case.

function of H for a rectangular undergrowth with $L = 18d$. Again, excellent agreement is shown between the two approaches. The difference becomes larger as the height, H, decreases. It can be attributed to the stress relaxation in the atomistic approach, though it is surprising to find the stress relaxation increasing the strain energy slightly rather than decreasing it.

In Fig. 6, a maximum principal stress, $\sigma(\mathrm{mp})$, is shown for an undergrowth with $L = 18d$, $H = 17.32d$, and $\mu^* = 2\mu$. The maximum principal stress is defined as

$$\begin{aligned} \sigma(\mathrm{mp}) &= 0.5(\sigma_{11} + \sigma_{22} + \tau_{12}^{\mathrm{m}}), \qquad \text{if} \quad \sigma_{11} + \sigma_{22} \geqq 0, \\ &= 0.5(\sigma_{11} + \sigma_{22} - \tau_{12}^{\mathrm{m}}), \qquad \text{if} \quad \sigma_{11} + \sigma_{22} < 0. \end{aligned} \tag{4.4}$$

The three-dimensional perspective reveals that due to the positive eigenstrain employed in this calculation, most of the undergrowth is under a compressive stress mode. However, in the proximity of the traction-free edge, i.e., along the x axis, we observe a tensile stress mode, the magnitude of which becomes

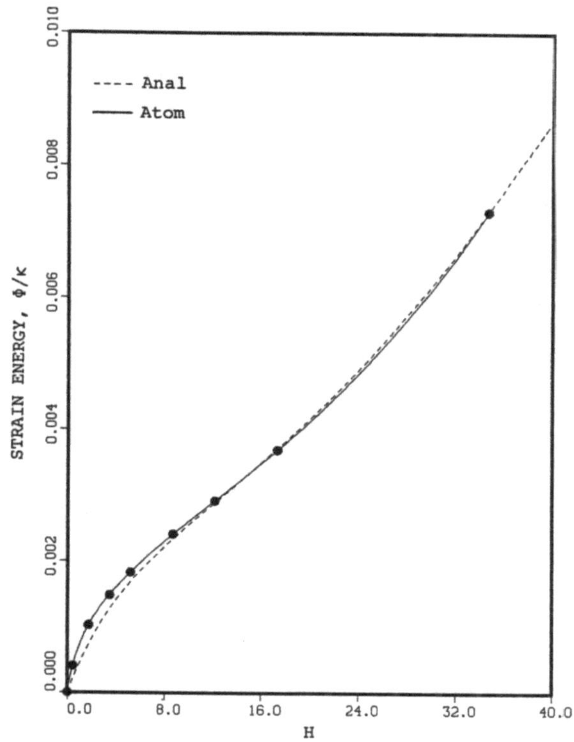

FIG. 5. Normalized strain energy, Φ/κ, versus height, H, for a rectangular undergrowth with $L = 18d$. The solid curve shows the atomistic approach while the dashed curve shows the continuum elasticity theory.

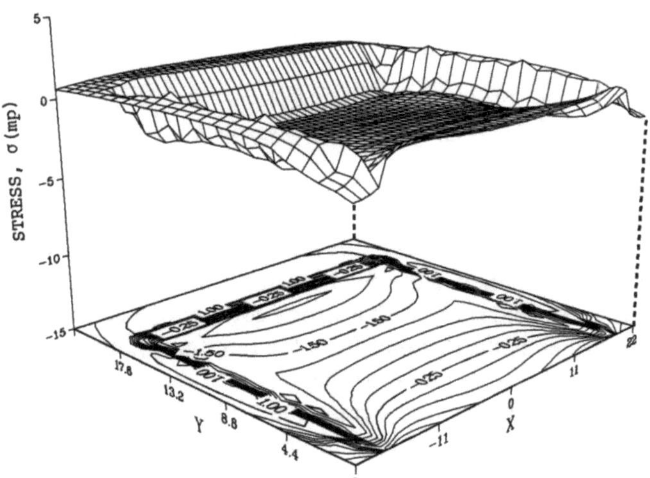

FIG. 6. Maximum principal stress, σ(mp), for an undergrowth with $L = 18d$, $H = 17.32d$, and $\mu^* = 2\mu$. Note a tensile stress mode in the proximity of the traction-free edge.

highest at the undergrowth–matrix interface. This finding may render a direct implication on the role of thermal residual stress in a ceramic : metal composite. Consider a ceramic part surrounded by a half-space metal matrix, as in the present case. Because of the difference in the thermal expansion coefficient between the ceramic part and the metal matrix, there arises a thermoelastic residual stress. Ceramic is usually harder than the metal matrix (hence a higher shear modulus) but prone to a tensile stress for possible cracks. The present stress analysis reinforces the usual observation that cracks originate at a corner located along a ceramic–metal joint.

The stress field of an undergrowth strongly depends on the H/L ratio. In Fig. 7(a), the σ_{11} component is plotted along the x axis, i.e., along the free edge from the center of an undergrowth with $L = 18d$ and $H = 1.732d$, while Fig. 7(b) shows the results of an undergrowth with $L = 18d$ and $H = 17.32d$. Both cases are examined for three different values of μ^*, 0.5μ, 1μ, and 2μ. As the H/L ratio approaches zero, the stress becomes of a one-dimensional nature, entirely accommodated with the undergrowth; Fig. 7(a) shows this trend. Note that in Fig. 7(a), as μ increases, σ_{11} becomes increasingly negative within the undergrowth ($x \leqq 18$), but its value rapidly approaches zero in the matrix ($x > 18$). As the H/L ratio increases, the stress field spreads over into the matrix side, as shown in Fig. 7(b), and the σ_{11} value becomes of a positive nature in some cases.

Strain energy, Φ/κ, is compared as a function of both the height, H, and the shear modulus, μ^*, in Fig. 8. As expected, the strain energy increases as both H and μ^* increase. According to (4.3) for a circular inclusion, a strain/energy ratio should yield that $\Phi(\mu^* = 2\mu)/\Phi(\mu^* = \mu) = 1.2$ and $\Phi(\mu^* = \mu/2)/\Phi(\mu^* = \mu) = 0.75$. For a rectangular undergrowth, the energy ratio depends on the height, H, and comes close to those of a circular inclusion as H increases; for example, when $L = 18d$ and $H = 17.32d$, $\Phi(\mu^* = 2\mu)/\Phi(\mu^* = \mu) = 1.26$ and $\Phi(\mu^* = \mu/2)/\Phi(\mu^* = \mu) = 0.74$.

4.3. *Case* (C) *Overgrowth*

In Fig. 9, a three-dimensional perspective is displayed for the maximum principal stress, $\sigma(\text{mp})$, of an overgrowth having $L = 18d$, $H = 17.32d$, and $\mu^* = 2\mu$. Unlike an undergrowth of the same size shown in Fig. 6, the stress is found to be mostly confined in the neighborhood of the overgrowth–substrate interface. The stress field starts with a strong compressive state at the interface area and diminishes as the distance from the interface, y, increases. However, careful examination reveals that near the substrate, the traction-free perimeter of the overgrowth is in a tensile stress mode. This result suggests that for a ceramic–metal joint, cracks can be initiated at joint corners unless some measure for stress relief is arranged. A detailed contour picture for σ_{11}, σ_{12}, and σ_{22} is shown in Fig. 10 for an overgrowth with $L = 18d$, $H = 8.66d$, and $\mu^* = 2\mu$. Because of the dilatational nature of the eigenstrain employed in this study, the trend of the two normal components, σ_{11} and σ_{22}, is found to be

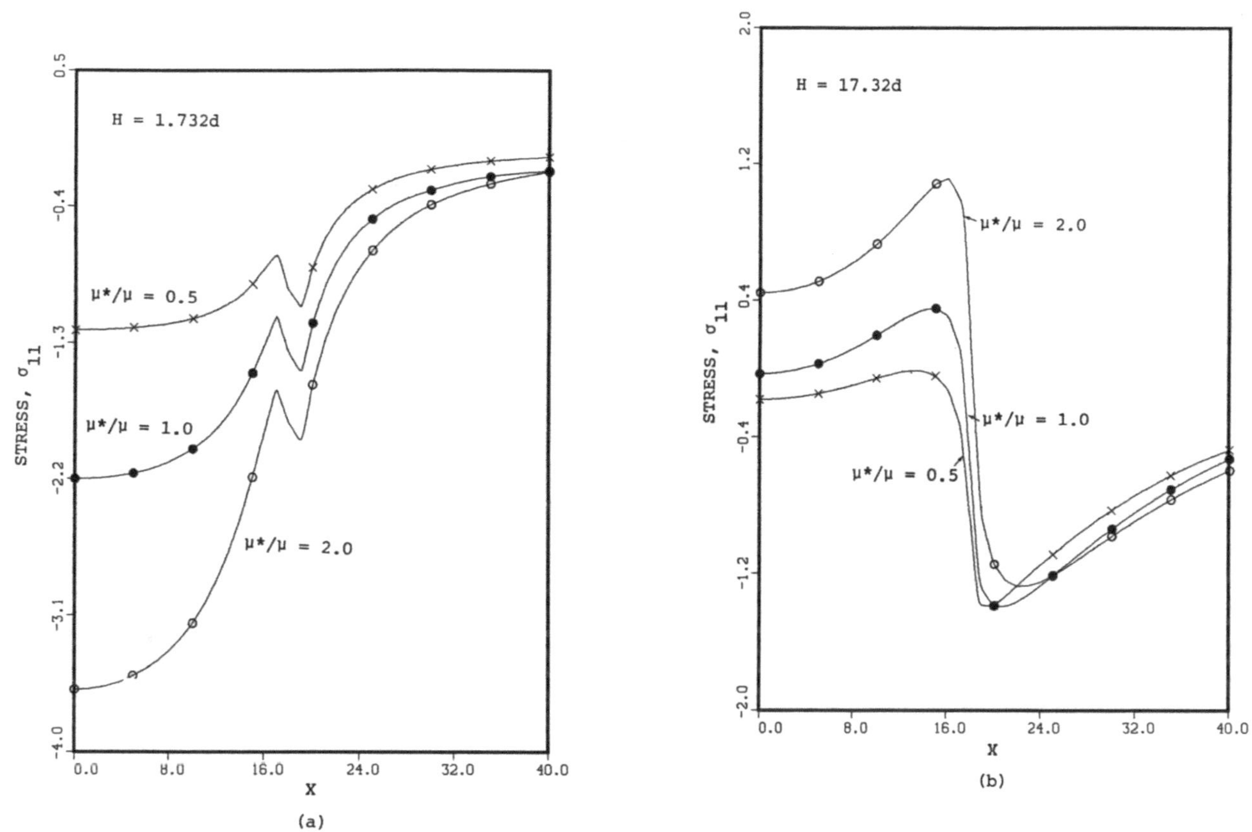

FIG. 7. σ_{11} along the x axis from the center of an undergrowth: (a) undergrowths with $L = 18d$ and $H = 1.732d$, and (b) undergrowths with $L = 18d$ and $H = 17.32d$. Both cases are examined for three different values of μ^*; 0.5μ, 1μ, and 2μ.

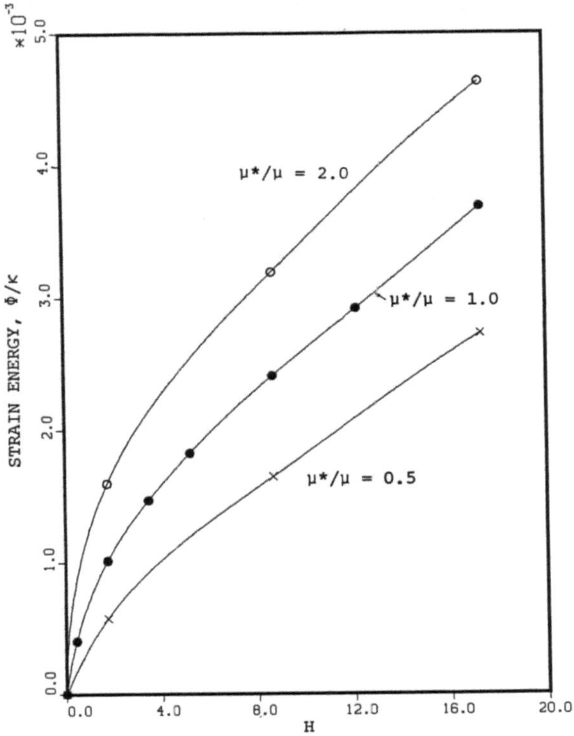

FIG. 8. Normalized strain energy, Φ/κ, versus height, H, for undergrowths with three different values of μ^*; 0.5μ, 1μ, and 2μ.

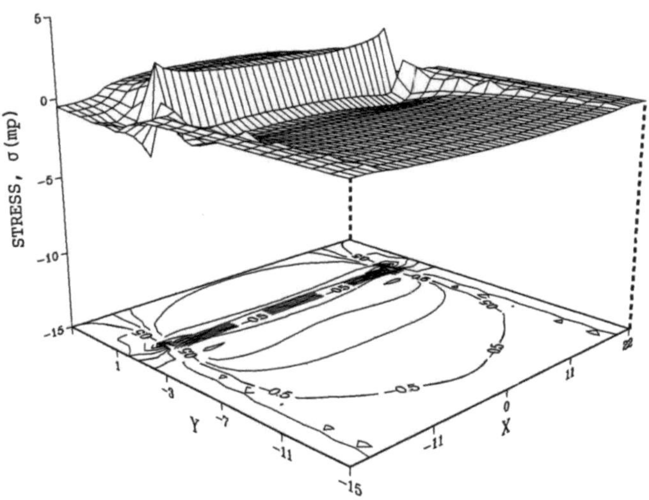

FIG. 9. Maximum principal stress, σ(mp), of an overgrowth having $L = 18d$, $H = 17.32d$, and $\mu^* = 2\mu$.

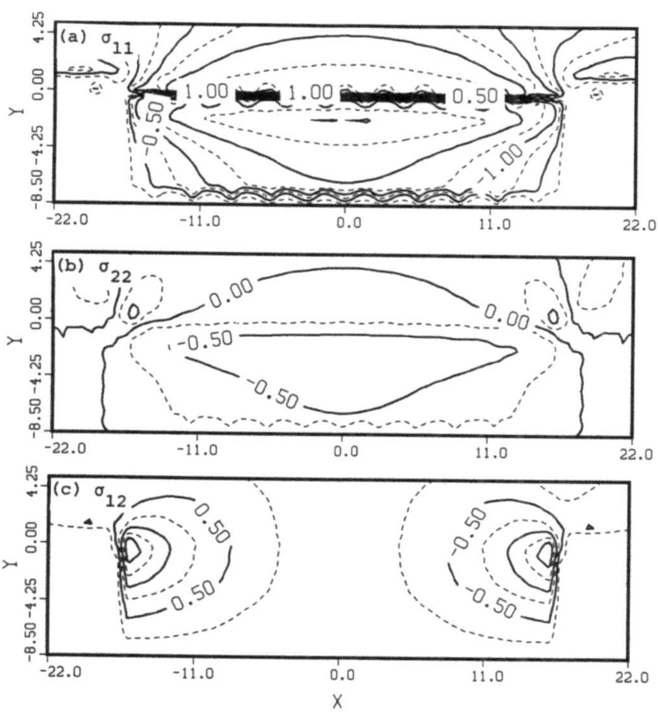

FIG. 10. (a) σ_{11}, (b) σ_{22}, and (c) σ_{12} for an overgrowth with $L = 18d$, $H = 8.66d$, and $\mu^* = 2\mu$.

similar, although σ_{11} is much more intense than σ_{22}. The shear component, σ_{12}, is localized in the two joint corners where severe distortion is undergone. The oscillatory behavior of the stress field near the free edge, $y = 8.5$, is due to the triangular–lattice structural effect.

Figure 11 examines σ_{11} as a function of height, H, for an overgrowth having $L = 18d$ and $\mu^* = 2\mu$. In Fig. 11(a), the σ_{11} component is depicted along the overgrowth–substrate interface, and in Fig. 11(b) the same component is shown along the y axis. We note that the magnitude of the stress on the substrate side (the region of $y \geqq 0$ including the x axis) is shown to increase initially with an increase in the overgrowth thickness, H, and eventually approach an asymptotic state at a larger thickness. On the other hand, the stress magnitude on the overgrowth side ($y < 0$), especially near the interface, Fig. 11(b), initially decreases as H increases. This behavior suggests a tendency that the stress becomes localized within an overgrowth (for a given length L) as the thickness, H, decreases.

The asymptotic nature of the stress field is well demonstrated in the plot of strain energy versus height in Fig. 12. The strain energy, Φ/κ, is given for three different cases of μ^*: 0.5μ, 1μ, and 2μ. In all, the strain energy is found to reach

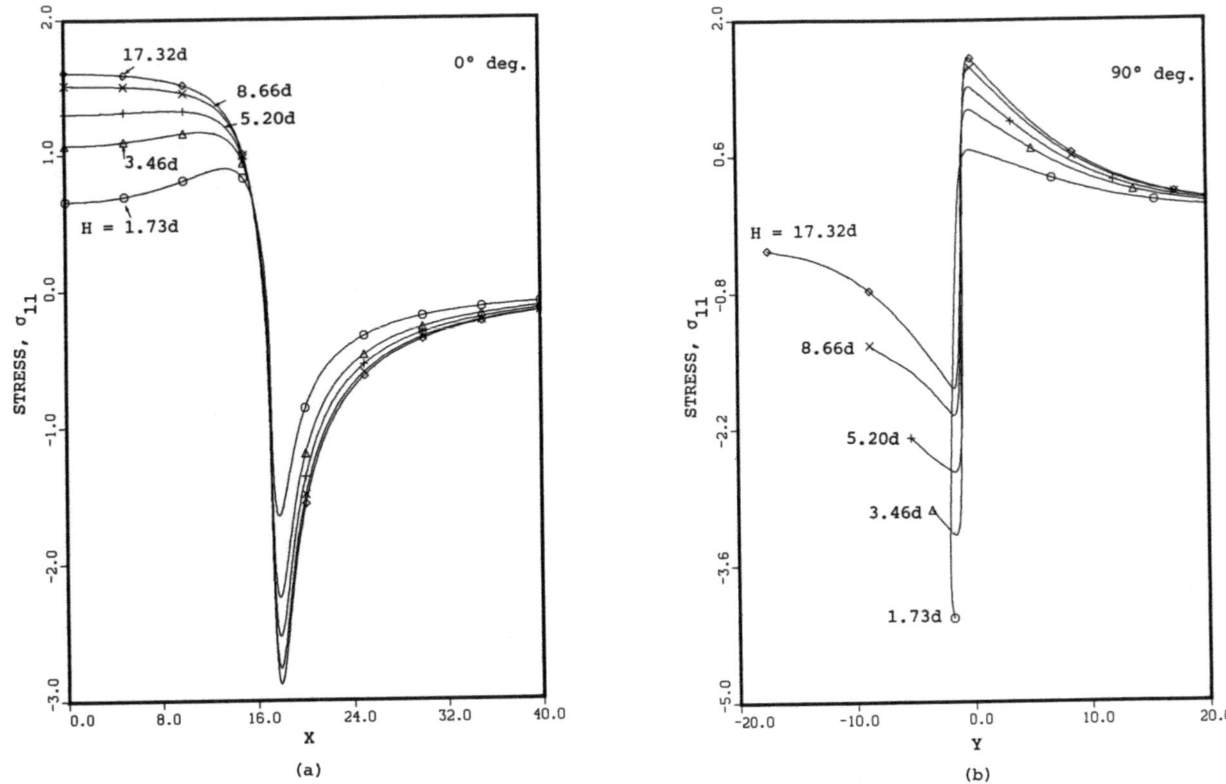

FIG. 11. σ_{11} as a function of thickness, H, for an overgrowth with $L = 18d$ and $\mu^* = 2\mu$: (a) along the x axis, i.e., along the overgrowth–substrate interface but on the substrate side, and (b) along the y axis.

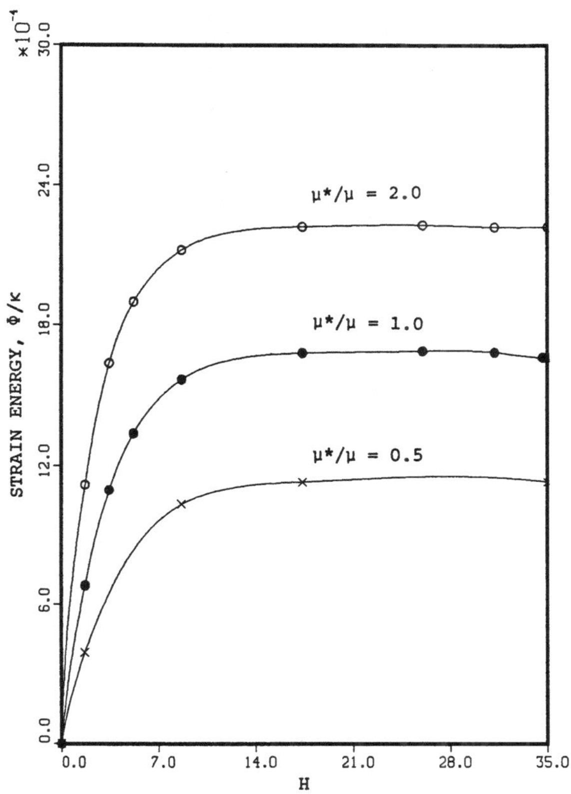

FIG. 12. Normalized strain energy, Φ/κ, as a function of overgrowth thickness, H, for three different cases of μ^*; 0.5μ, 1μ, and 2μ. Note that the strain energy reaches its plateau at the height of about $H = 18d$.

its plateau at a height of about $H = 18d$. Because of the extensive free surface associated with an overgrowth, the strain energy ratio, $\Phi(\mu_1^*)/\Phi(\mu_2^*)$, for overgrowths with different shear moduli, finds quite different values from those of inclusions. For example, when $L = 18d$ and $H = 17.32d$, $\Phi(\mu^* = 2\mu)/\Phi(\mu^* = \mu) = 1.32$ and $\Phi(\mu^* = \mu/2)/\Phi(\mu^* = \mu) = 0.67$. For *inclusions* and *undergrowths*, strain energy density depends only on the inclusion shape. In other words, the strain energy per unit area of inclusion is a constant for a fixed ratio of H/L. For overgrowths, however, this is not true as shown in Fig. 13. Here, the strain energy density, $\Phi/\kappa d^2$, is plotted as a function of H for overgrowths with $H/L = 0.866$ and $\mu^* = \mu$. The strain energy density initially increases as the size, H, increases, and approaches a maximum, followed by a decrease. The decrease in the strain energy density can be attributed to the fact that the stress of a large overgrowth is localized in the proximity of the overgrowth–substrate interface, and the areal fraction of the stressed zone decreases as its size increases.

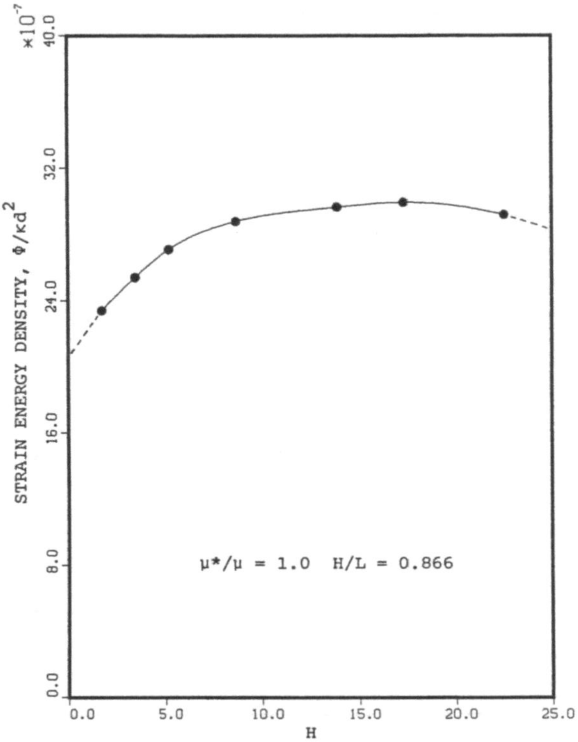

FIG. 13. Normalized strain energy density, $\Phi/\kappa d^2$, of an overgrowth versus thickness, H. All the overgrowths have $H/L = 0.866$ and $\mu^* = \mu$. Note that the strain energy density decreases as the H value becomes large.

5. Conclusion

For elastically homogeneous inclusions and undergrowths, the atomistic approach and linear elasticity theory are shown to be in excellent agreement. In fact, Ashurst and Hoover (1976) proved, in their fracture studies, an equivalence between the triangular–lattice atomistic approach and the finite element method. This is the basis for the success of the triangular–lattice atomistic approach in analyzing an inclusion problem.

The analysis of polygonal inclusions shows that the stress concentrations at the polygonal vertices increase as the ratio of the inclusion shear modulus to the matrix shear modulus, μ^*/μ, increases. The strain energy is nearly proportional to $\mu^*/(2\mu^* + \mu)$, which is proportionality constant for an elliptic inclusion with a dilatational eigenstrain. For an undergrowth, a maximum principal stress analysis reveals stress concentrations in a tensile mode at its traction-free side, indicating a possibility of crack initiation for a brittle undergrowth.

For an overgrowth inclusion, several interesting results are found. Unlike an undergrowth or inclusion, the stress is mostly confined in the neighborhood of the overgrowth–substrate interface. For a positive eigenstrain, the stress field starts with a strong compressive state at the interface area and diminishes as the distance from the interface increases. However, near the substrate, the traction-free perimeter of an overgrowth is found to be in a tensile stress mode, again suggesting that for a ceramic–metal joint, cracks can be initiated at joint corners unless some measure for stress relief is arranged. Quite different from the case of inclusions and undergrowths is that the total strain energy associated with an overgrowth approach an asymptotic value as its thickness increases. That is, the strain energy density of an overgrowth is found to depend on its shape and also on its size. This finding should shed some light on an interesting coherency strain problem; i.e., the use of the strain energy associated with an *inclusion* can be a gross approximation for an *overgrowth thin film*.

Acknowledgments

This paper is dedicated to Professor Tosio Mura in honor of his 65th birthday, whose teaching has greatly influenced the author's career. The work was supported by the U.S. Department of Energy (DE-FG02-87ER45315) (Dr. Jules Routbort, contract monitor), for which the author expresses his deep appreciation.

References

Ashurst, W. T. and Hoover, W. G. (1976), Microscopic fracture studies in the two-dimensional triangular lattice, *Phys. Rev.*, **B14**, 1465.

Bhargava, R. D. (1961), The inclusion problem, *Appl. Sci. Res.*, **A10**, 80.

Chiu, Y. P. (1980), On the stress field and surface domains in a half-space with a cuboidal zone in which initial strains are uniform, *J. Appl. Mech.*, **45**, 302.

Eshelby, J. D. (1957), The determination of the elastic field of an ellipsoidal inclusion, and related problems, *Proc. Roy. Soc. London*, **A241**, 376.

Faivre, G. (1964), Déformations de cohérence d'un précipite quardratique, *Phys. Stat. Sol.*, **35**, 249.

Hoover, W. G., Ashurst, W. T., and Olness, R. J. (1974), Two-dimensional computer studies of crystal stability and fluid viscosity, *J. Chem. Phys.*, **60**, 4043.

Khachaturyan, A. G. (1983), *Theory of Structural Transformation in Solids*, Wiley, New York p. 213.

Lee, J. K. and Johnson, W. C. (1977), Elastic strain energy and interactions of thin square plates which have undergone a simple shear, *Scripta Metallurgica*, **11**, 477.

Lee, S. and Hsu, C. C. (1985), Thermo-elastic stress due to surface parallelepiped inclusions, *J. Appl. Mech.*, **52**, 225.

Mura, T. (1987), *Micromechanics of Defects in Solids*, 2nd edition, Matinus Nijhoff, Dordrecht, p. 110.

Powell, M. J. D. (1977), Restart procedures for the conjugate gradient method, *Math. Programming*, **12**, 241.

Tsuchida, E. and Mura, T. (1983), The stress field in an elastic half-space having a spheroidal inhomogeneity under all-around tension parallel to the plane boundary, *J. Appl. Mech.*, **50**, 807.

Interaction of Slip Bands in High-Cycle Fatigue Crack Initiation

T. H. LIN and Q. Y. CHEN
Department of Civil Engineering, University of California, Los Angeles, CA 90024-1593, U.S.A.

Abstract

A number of fatigue bands spaced 3 microns apart are assumed to exist in a most favorably oriented crystal at a free surface of a polycrystal. Each fatigue band is taken to consist of three thin slices P, Q, and R with R sandwiched in P and Q. P and Q in each band are assumed to have equal and opposite initial resolved shear stresses. Plastic strain distributions in the fatigue bands under cyclic loading were calculated.

The maximum shear strain at this free surface is a measure of the height of extrusion or depth of intrusion, which in turn is taken as a measure of crack initiation and hence the level of fatigue damage. The applied cyclic shear stress versus the number of cycles to yield different levels of fatigue damage were calculated. The shape of these curves seems to be similar to that of Wohler's diagram and that of the Coffin–Mansan equations. It is hoped that this analysis gives some physical explanations to the Wohler and the Coffin–Mason relations.

1. Introduction

Fatigue failure is developed in two stages: fatigue crack initiation and fatigue crack propagation. Wohler's famous fatigue diagram, Fig. 1 (Kocanda, 1978), relating the alternate cyclic stress to the cycles of failure covers both initiation and propagation. Crack propagation has been much more extensively studied than crack initiation. In low-cycle fatigue, crack propagation predominates fatigue life. However, in high-cycle fatigue, initiation predominates. Both crack initiation and propagation are important in the reduction of fatigue failures.

Metals are subject to fatigue failure at temperatures as low as 1.7 K (McCammon and Rosenberg, 1957; MacCone *et al.*, 1959). This indicates that surface corrosion, gas adsorption, gas diffusion (into the metal or vacancy diffusion to form voids) are not necessary to fatigue crack initiation. This leaves localized plastic deformation to play a basic role in fatigue crack initiation. Following the clue—provided by the observation of extrusions and intrusions in fatigue specimens by Forsyth and Stubbington (1955), Hull

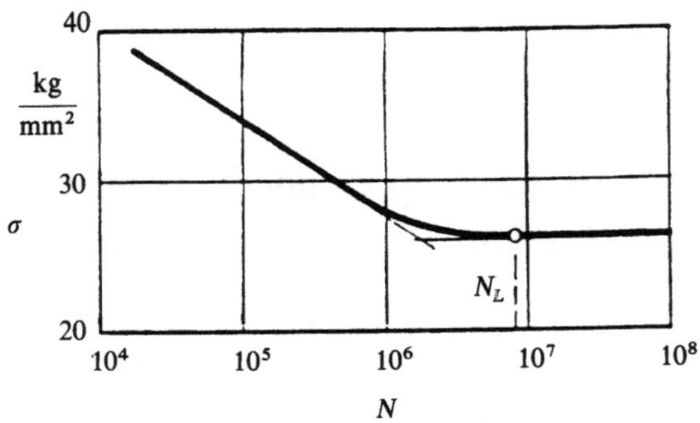

FIG. 1. The Wohler fatigue diagram in σ–log N coordinates.

(1958), Thompson (1959), Meke and Blockwitz (1980), and Mughrabi (1980)—a number of theories of fatigue crack initiation have been proposed by different distinguished investigators: Wood (1956), Cottrell and Hull (1957), Mott (1958), McEviley and Machlin (1959), Thompson (1959), and others. These theories consider mainly the paths of dislocation movement to produce extrusion or intrusion without analytically considering the resolved shear stress required to cause this movement. Lin and Ito (1969) calculated the resolved shear stress fields due to slip or the movement of dislocations, and developed a micromechanic theory of crack initiation. This seems to be the first quantitative theory of fatigue crack initiation. This theory will be used here to analyze the interaction effect of slip bands.

2. Ratchet Mechanism Provided by Microstress

All metals have defects. This initial resolved shear stress field caused by initial defects is denoted by τ^{I}. During loading, when the critical shear stress is exceeded in some region, slip occurs. After unloading this slip remains and causes a residual resolved shear stress field τ^{R}. The resolved shear stress caused by loading is denoted by τ^{A}. After reloading, the resolved shear stress is then

$$\tau = \tau^{\mathrm{I}} + \tau^{\mathrm{R}} + \tau^{\mathrm{A}}. \tag{2.1}$$

Lin and Ito (1969) considered a single slip band in a most favorably oriented crystal at a free surface of a polycrystal. For an extrusion to initiate, positive shear has to occur in a thin slice P and negative shear in a closely located slice Q, as shown in Fig. 2. The initial stress field favorable for this slip sequence is one having positive shear stress in P and negative shear stress in Q. Such an initial stress field, shown by Lin and Lin (1983), can be provided by an initial strain $e^{\mathrm{I}}_{\alpha\alpha}$ in the thin slice R. Lin and Ito (1969) suggested that this positive $e^{\mathrm{I}}_{\alpha\alpha}$ can be caused by an array of interstitial dislocation dipoles and negative $e^{\mathrm{I}}_{\alpha\alpha}$

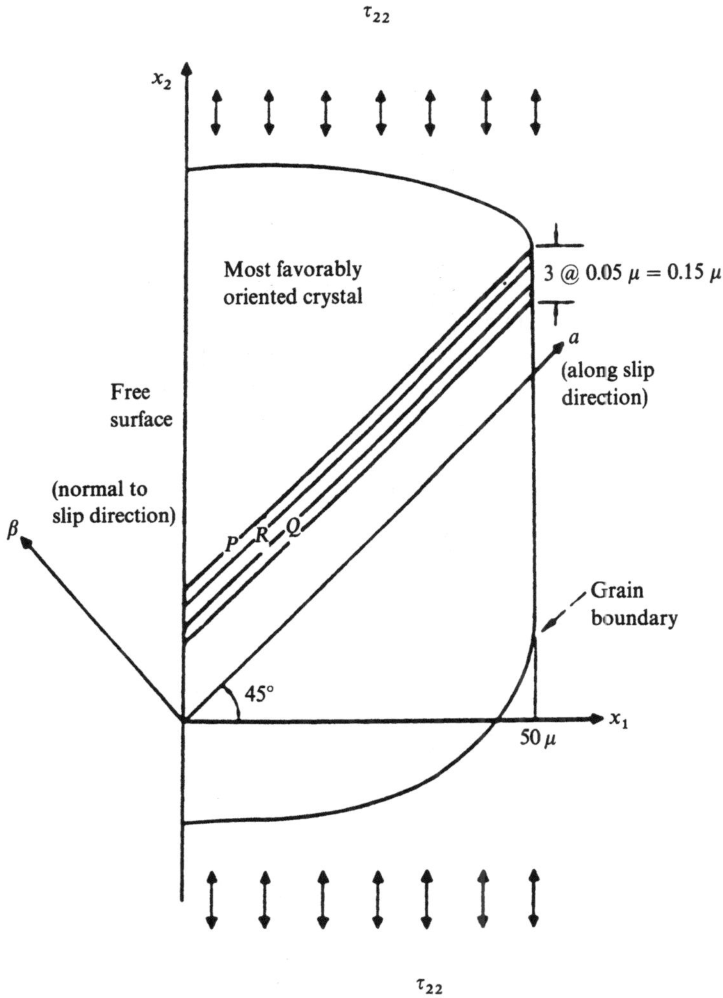

FIG. 2. Most favorably oriented crystal at free surface.

by vacancy dipoles. It has been recently pointed out by Antonopoulos *et al.* (1967) and Mughrabi *et al.* (1983) that the observed ladder structure in a persistent slip band (PSB) can be represented by an array of dislocation dipoles causing initial resolved shear stresses at the interfaces between the PSB and the matrix.

Assuming τ^{I} to be positive in *P* and negative in *Q*, as that caused by an array of interstitial dislocation dipoles in *R*, a tensile loading causing a positive resolved shear stress τ^{A} gives a shear stress $(\tau^{\mathrm{I}} + \tau^{\mathrm{A}})$ in *P* to reach first the critical shear stress τ^{C} and hence *P* slides first. After slip occurs, it produces a residual stress field τ^R. When the resolved shear stress given by (2.1) equals τ^C in some region, slip may initiate or continue in this region. Due to the continuity of the resolved shear stress fields and the closeness of Q to P, slip in P relieves not only the positive shear stress in P but also in the neighboring region including Q (Lin and Ito, 1969; Lin, 1977). Hence only P slides during forward loading. This slip increases the negative shear stress in Q to cause Q to slide more readily in the reverse loading. When the negative shear stress in Q reaches τ^C in the reversed loading, slip occurs in Q and a new residual stress field is produced. Similarly, the relief of negative resolved shear stress, not only confined in Q but also in P, thus makes P more ready to slide in the next forward loading. This process is repeated for every cycle, providing a natural ratchet mechanism to give monotonic build-up of local plastic strain in P and Q, pushing the slice R out and starting an extrusion. Interchanging the signs of the initial stresses in P and Q will initiate an intrusion instead of an extrusion. This theory of Lin and Ito (1969) has been verified by many metallurgical observations (Lin, 1977).

3. Decreasing Rate of Extrusion Growth

An initial tensile strain $e^I_{\alpha\alpha}$ in R causes an initial compressive stress $\tau^I_{\alpha\alpha}$ in R, which in turn induces an initial positive shear stress $\tau^I_{\alpha\beta}$ and P and a negative one in Q. This initial stress field causes P and Q to slide alternatively under cyclic loading. The build-up of slip strain $e^P_{\alpha\beta}$ in P and Q may be considered to be caused by $e^I_{\alpha\alpha}$ in R. If R were cut out, the free length of R would be longer than the slot, by an amount referred to as the "static extrusion" by Mughrabi *et al.* (1983). Under cyclic loading the extrusion grows and the thin slice R increases in length causing this initial compression to decrease. This decrease of compressive stress in R causes a reduction of $(\tau^I_{\alpha\beta} + \tau^R_{\alpha\beta})$ in P, hence Q induces a decreasing rate of extrusion growth as observed in experiments (Mughrabi *et al.*, 1983).

4. Effect of Secondary Slip on the Extent of Extrusion

There are twelve slip systems in an f.c.c. crystal. This change of $\tau_{\alpha\alpha}$ in R causes changes of resolved shear stresses in all slip systems. When the decrease of compression in R becomes large, its residual stress combined with the applied stress can cause a second slip system to have shear stress reaching the critical and slide. The plastic strain due to slip in this second slip system has a tensor component $e^P_{\alpha\alpha}$ just like $e^I_{\alpha\alpha}$ in causing the positive and negative shear stress $\tau^I_{\alpha\beta}$ in P and Q, respectively. Hence this secondary slip $e^P_{\xi\eta}$ can increase greatly the extent of extrusion and intrusion (Lin *et al.*, 1989).

5. Polycrystal Model

The previous studies have been confined to a single slip band in the most favorably oriented crystal at the free surface (Fig. 2). The present paper considers a number of fatigue bands in this crystal. Fatigue data of single copper crystals by Mughrabi (1978) show the variation of the resolved shear stress τ with the cumulative plastic shear strain at different plastic shear strain amplitudes. The applied resolved shear stress at a given plastic strain amplitude increases rapidly at the beginning, approaches a maximum, then decreases, i.e., softening and then approaching a saturation stress. Persistent slip bands were found at the beginning of saturation. The volume of PSB's increases with the plastic strain amplitude γ_{pl}. A number of PSB's can be present in the most favorably oriented crystal of a polycrystal. The crystals surrounding this most favorably oriented crystal are less favorably oriented and hence have lower resolved shears stress due to the applied load. In the early stages of a high-cycle fatigue, the residual resolved shear stresses in the neighboring crystals can be small compared to the decrease of the applied resolved shear stresses due to the change of orientation. Hence, in the early stages of high-cycle fatigue, slip in the most favorable oriented crystal can develop to some extent without significant slip in the neighboring crystals. For numerical study, we consider this most favorably oriented crystal to have eleven slip bands spaced at 3 microns apart as shown in Fig. 3. Each band is assumed to consist of three thin slices P, Q, and R as shown in Fig. 2. P and Q in each band are assumed to have equal and opposite initial resolved shear stresses of the amounts as indicated in Fig. 3. The polycrystal is taken to be of pure aluminum with a critical shear stress of 2.94 MPa (430 p.s.i.). The polycrystal is subject to a cyclic tension and compression in the x_2 direction. Let the excessive shear stress be defined as the initial shear stress τ^{I} plus the residual shear stress τ^{R}, and the maximum applied shear stress τ^{A} minus the critical shear stress τ^{C},

$$\tau^{\mathrm{E}} = (\tau^{\mathrm{I}} + \tau^{\mathrm{R}} + \tau^{\mathrm{A}} - \tau^{\mathrm{C}}). \tag{5.1}$$

The initial excessive shear stress $\tau^{\mathrm{E_I}}$ is the $\tau^{\mathrm{I}} + \tau^{\mathrm{A}} - \tau^{\mathrm{C}}$. During loading this excessive shear stress τ^{E} is to be relieved with slip in the different slip bands. Slip distributions in these bands are to be calculated.

6. Method of Calculating Slip Strain Distributions in Slip Bands

Consider the strain e_{ij} in an elastic–plastic body to be the sum of the elastic part e_{ij}^{E} and the plastic part e_{ij}^{P}. The elastic strain is the difference between the total strain and the plastic strain. For an elastically isotropic body, the stress–strain relation is written as

$$\tau_{ij} = \lambda\delta_{ij}(\theta - \theta^{\mathrm{P}}) + 2\mu(e_{ij} - e_{ij}^{\mathrm{P}}), \tag{6.1}$$

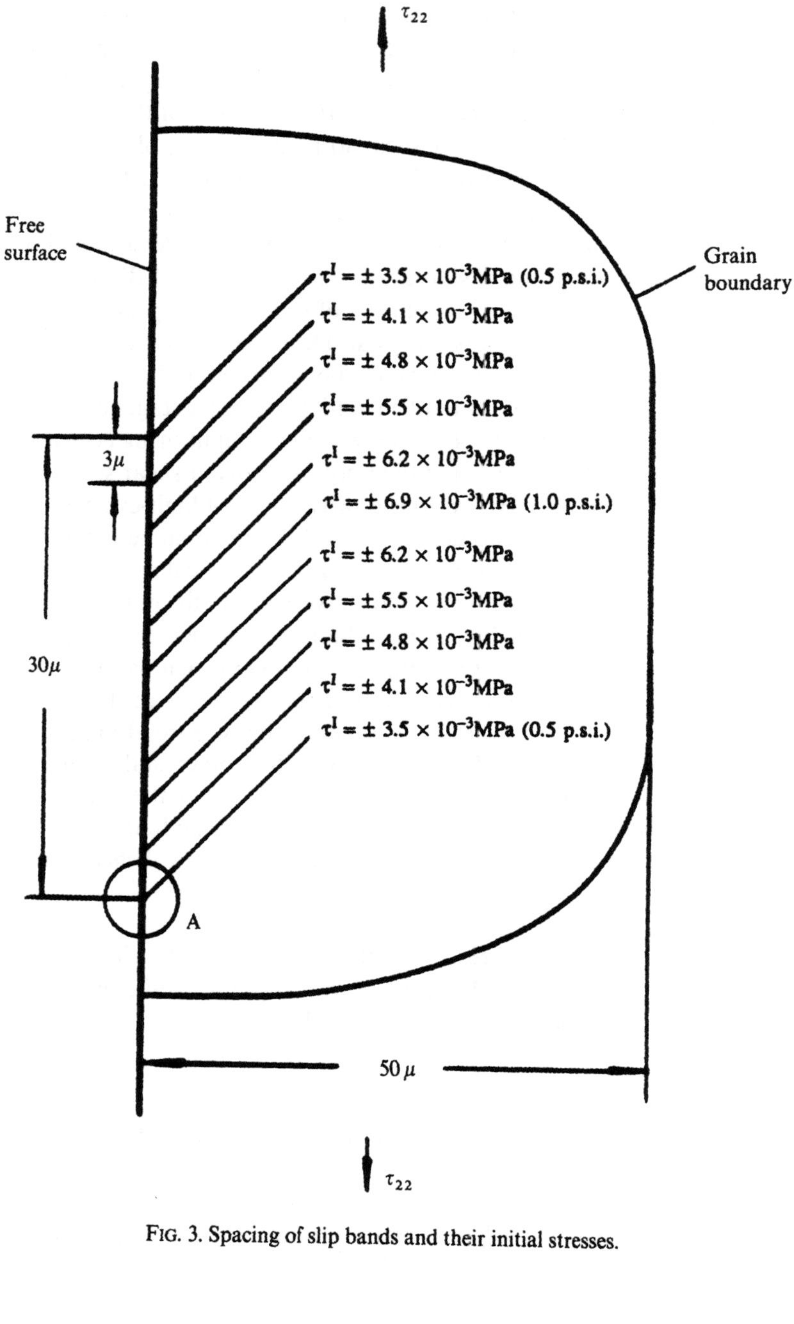

FIG. 3. Spacing of slip bands and their initial stresses.

where μ is the shear modulus, $\lambda = 2\mu v/(1-2v)$, θ is the dilatation, and v is Poisson's ratio. Substituting the above relation into the equilibrium conditions (Lin, 1968), we obtain

$$\lambda\delta_{ij}\theta_{,j} + 2\mu e_{ij,j} + (-\lambda\delta_{ij}\theta^{P}_{,j} - 2\mu e^{P}_{ij,j}) + F_i = 0 \tag{6.2}$$

in the interior of the body, and

$$S_i + (\lambda\delta_{ij}n_j\theta^{P} + 2\mu n_j e^{P}_{ij}) = \lambda\delta_{ij}n_j\theta + 2\mu n_j e_{ij} \tag{6.3}$$

on the boundary surface, where F_i denotes body force per unit volume along the x_i direction, S_i denotes traction along the x_i direction per unit area of the boundary, n_j denotes the cosine of the angle between the exterior normal n with the x_j axis, the subscript j after the comman denotes differentiation with respect to the j axis, and the repetition of the subscripts denotes summation from one to three. From (6.2) and (6.3) it is seen that the terms in parentheses are equivalent to F_i and S_i, respectively, in causing e_{ij}. These parenthetical terms are called, by Lin the equivalent body and surfaces forces and denoted by $\underline{F_i}$ and $\underline{S_i}$, respectively. This concept of equivalent forces is here used for slip analysis. The polycrystal considered is a fine-grain aggregate. The equivalent force produced by slip, causing plastic strain in the crystal at the surface, may be treated as an external force applied to a semi-infinite elastic medium. The thickness of the slices is much less than the length of the traces of slip bands on the surface. Consequently, plastic strain and the equivalent forces are taken to be constant along this slip band direction. This semi-infinite solid may hence be considered to be under plane deformation.

For numerical calculations, the slices in the slip bands are divided into a number of parallelogram grids along the length in the α direction (Fig. 2). Each grid is taken to have constant plastic strain. The corresponding stress is the average stress in the grid (Lin *et al.*, 1988). From the plane strain solution of a semi-infinite medium, the resolved shear stress field $\tau^{R}_{\alpha\beta}$ caused by a constant plastic strain $e^{P}_{\alpha\beta_n}$ in the nth grid was calculated. Since we refer to the same slip system, we drop the subscript $\alpha\beta$. The average residual stress τ^{R}_{m} over this mth grid is expressed as

$$\tau^{R}_{m} = -C_{mn}e^{P}_{n}, \tag{6.4}$$

where C_{mn} is the average resolved shear stress in the $\alpha\beta$ slip system in the mth grid, due to a unit uniform plastic strain e^{P}_{n} in grid n. As the cyclic loading proceeds, slip occurs in more grids in the slices. This total resolved shear stress in the mth grid is the sum of the initial, residual, and applied stresses

$$\tau_m = \tau^{I}_{m} + \tau^{A}_{m} - \sum C_{mn}e^{P}_{n}, \tag{6.5}$$

where "n" is summed over all grids with plastic strain.

The microscopic plastic strain in the thin slices is much larger than the macroscopic plastic strain of a crystal, so the strain hardening of the thin slices is much less than that of the metal and is neglected. This yields a constant critical shear stress τ^{C}. Sliding occurs in the grids with $\tau = \pm\tau^{C}$. For an incremental loading $\Delta\tau_m^A$, $\Delta\tau_m = 0$,

$$\Delta\tau_m^A = \sum C_{mn} \Delta e_n^P. \tag{6.6}$$

Intrusion at the free surface is considered to be a crack initiation. This depth of intrusion or the height of extrusion is taken as a measure of crack initiation and hence as a measure of fatigue damage.

7. Numerical Results

Using the method described above, each thin slice *P*, *Q*, and *R* of each band is divided along the α direction into three parallelogram grids, and the slip distributions in the different slices of the slip bands were calculated. The center fatigue band has the maximum initial shear stress and has been found to give the maximum plastic shear strain at the free surface. This shear strain is considered to be a measure of the fatigue damage. The variations of this surface strain with the number of cycles of loading are plotted at different values of initial excessive resolved shear stresses as shown in Fig. 4. For each value of this initial excessive shear stress, the numbers of active slip bands at different cycles of loading are shown in Fig. 5. From the curves in Fig. 4 we pick the number of cycles to yield the given value of the plastic resolved shear strain, $e_{\alpha\beta}^P$ at the free surface, thus we obtain the initial excessive shear stress τ^{E_i}

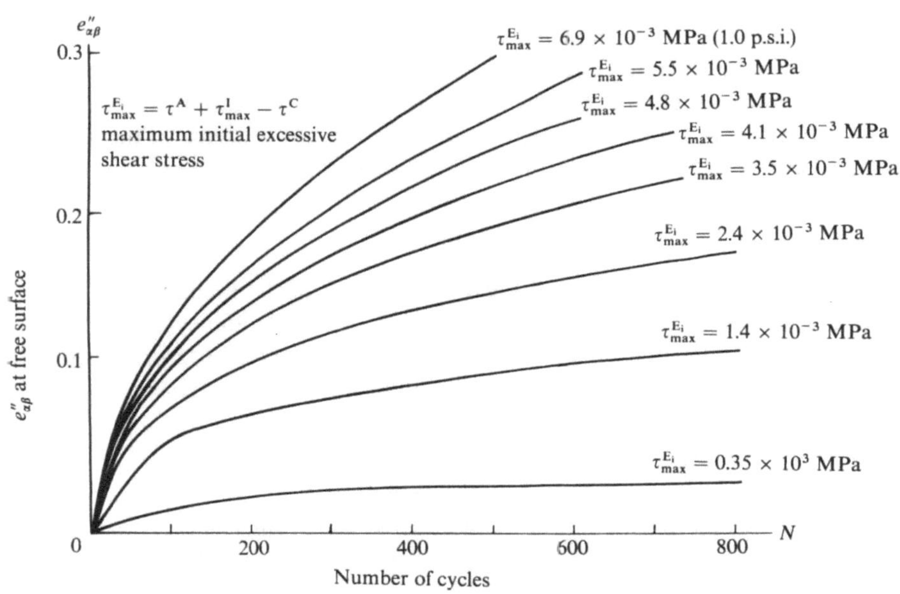

FIG. 4. Plastic strain $e''_{\alpha\beta}$ at free surface versus number of cycles at different applied shear stresses.

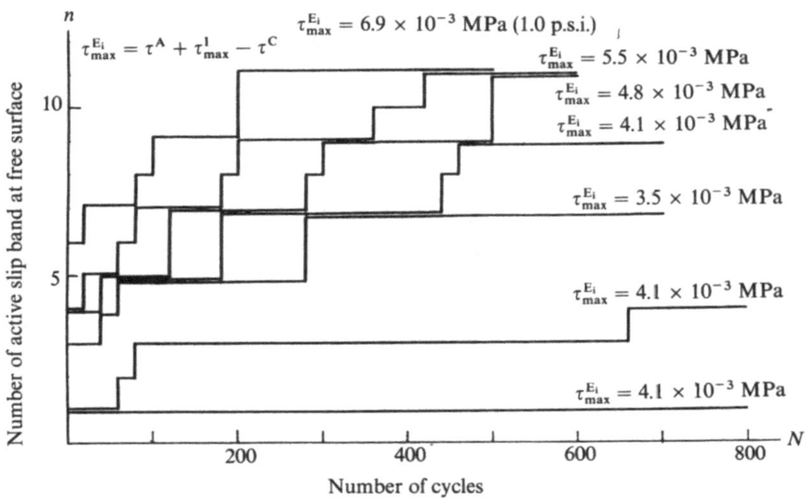

FIG. 5. Number of active slip bands at free surface versus number of cycles.

versus cycles of loading to yield the given surface plastic strain, which is a measure of the extent of extrusion or intrusion and hence the extent of crack initiation. This is shown in Fig. 6.

8. Conclusions

It is seen from Fig. 6 that the shape of the curves of τ^{E_I} versus N for different levels of fatigue damage is somewhat similar to Wohler's diagram. This shows the initial progressive stages of fatigue failure.

At the initial stage of fatigue loading, the slip in a single band is directly proportional to the excessive shear stress. The rate of incremental slip in P and Q per cycle depends on the positive and negative resolved shear stresses $(\tau^I + \tau^R)$. As loading proceeds, extrusion grows and this shear stress $(\tau^I + \tau^R)$ decreases and reduces the growth rate of extrusion. This has the effect of decreasing the slope of the $e^P_{\alpha\beta}$ at the free surface with N. It has been shown that this slope approaches zero when the extrusion approaches the "static extrusion" (Lin *et al.*, 1989). When more than one slip band becomes active, slip in one band relieves the shear stress not only in this band but also in other bands. This decreases the amount of slip in each band. This causes the curves $e^P_{\alpha\beta}$ versus N to bend further down. This has the effect of causing the τ^A versus N curve to give a specified value of surface strain to become more flat.

The above considers the stress-controlled fatigue. The coffin–Manson equation of the S–N curve is mainly for strain-controlled fatigue. The plastic strain amplitude per cycle is kept constant. The average incremental plastic

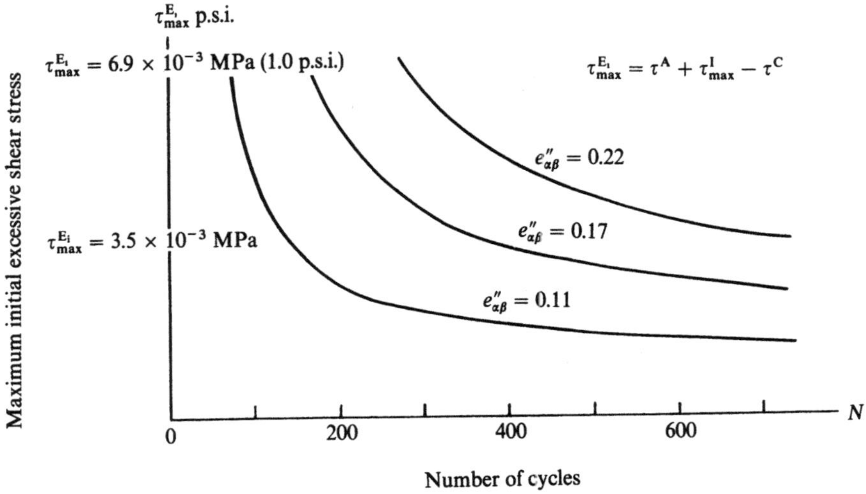

FIG. 6. Maximum initial excessive shear stress versus number of cycles. τ–N curve at different levels of fatigue damage.

strain $e^{P}_{\alpha\beta}$, in the most favorably oriented crystal per cycle, is taken to represent the plastic strain amplitude γ_{pl} of the polycrystal. The maximum surface slip strain $e^{P}_{\alpha\beta}$, to give a given γ_{pl} in the polycrystal, is less for the case of more active slip bands than that of the single active band. The decrease of the shear stress $(\tau^{I} + \tau^{R})$ in a single slip band with cycles of loading further decreases the number of cycles to yield a given $e^{P}_{\alpha\beta}$ at the free surface. Hence the interaction effect of slip bands causes the stress cycle, i.e., $\tau^{E_i} - N$, curve on reaching different levels of fatigue damage to become more flat. This present analysis seems to give some explanation to the shapes of Wohler's diagram and the form of the Coffin–Manson equations.

Acknowledgments

The support of the U.S. Office of Naval research Through Contract N00014-86-K-153 and the interest of the scientific officer, Dr. Yapa Rajapakse are greatfully acknowledged.

This research was sponsored by Office of Naval Research through Contract N00014-86-K0153.

References

Antonopoulos, J. G., Brown, L. M., and Winter, A. T. (1967), Vacancy dipoles in fatigued copper, *Phil Mag.*, **34**, 549.

Cottrell, A. H. and Hull, D. (1957), Extrusions and intrusions by cyclic slip in copper, *Proc. Roy. Soc. London*, **242A**, 211.

Forsyth, P. J. E. and Stubbington, C. A. (1955), The slip band extrusion effect observed in some aluminum alloys subjected to cyclic stress, *J. Inst. Metals*, **83**, 395.

Hull, D. (1958), Surface structure of slip bands on copper fatigued at 293, 90, 20, and 4.2 K, *J. Inst. Metals*, **86**, 425.

Kocanda, S. (1978), *Fatigue Failures of Metals*, Sijithoff & Noordhoff, p. 5.

Lin, T. H. (1968), *Theory of Inelastic Structures*, Wiley, New York, pp. 44–48.

Lin, T. H. (1977), Micromechanics of deformation of slip bans under monotonic and cyclic loadings, in *Reviews of the Deformation Behavior of Materials*, edited by P. Felham, Freund, Tel-Aviv.

Lin, T. H. and Ito, Y. M. (1969), Mechanics of a fatigue crack nucleation mechanism, *J. Mech. Phys. Solids*, **17**, 511–523.

Lin, T. H. and Ito, Y. M. (1969), Fatigue crack nucleation of metals, *Proc. U.S. Acad. Sci.*, **62**, No. 3, 631–635.

Lin, S. R. and Lin, T. H. (1983), Initial strain field and crack initiation mechanics, *J. Appl. Mech.*, **50**, 367–372.

Lin, T. H., Lin, S. R., Wu, X. Q., and Chen, Q. Y. (1988), Reciprocal theorem of residual stress and inelastic strain, UCLA School of Engineering and Applied Science, Civil Engineering Department, Technical Report UCLA-ENG-88-09.

Lin, T. H., Lin, S. R., and Wu, X. Q. (1989), Micromechanics of an extrusion in high cycle fatigue, UCLA School of Engineering and Applied Science, Civil Engineering Department, Technical Report UCLA-ENG-88-07, and to be published in *Philosophical Magazine*.

MacCone, R. K., McCammon, R. D., and Rosenberg, H. M. (1959), The fatigue of metals at 1.7 K, *Phil. Mag.*, **4**, 267.

McCammon, R. D. and Rosenberg, H. M. (1957), The fatigue and ultimate tensile strengths of metals between 4.2 and 293 K, *Proc. Roy. Soc. London*, **242A**, 203.

McEviley, J., Jr. and Machlin, E. S. (1959), Critical experiments on the nature of fatigue in crystalline materials, *Proceedings of the International Conference on the Atomic Mechanisms of Fracture*, Technology Press, MIT Cambridge, MA, and Wiley, New York.

Meke, K. and Blockwitz, C. (1980), Internal displacement of persistent slip bands in cyclically deformed nickel single crystals, *Phys. Stat. Sol.*, (a) **61**, 5.

Mott, N. F. (1958), Origin and fatigue cracks, *Acta Metallurgica*, **6**, 195.

Mughrabi, H. (1978), The cyclic hardening and saturation behavior of copper single crystals, *Mater. Sci. Engng.*, **33**, 207–223.

Mughrabi, H. (1980), *Microscopic Mechanisms of Metal Fatigue Strength of Metals and Alloys*, Vol. 3, Pergamon, Oxford and New York.

Mughrabi, H., Wang, R., Differet, K., and Essmann, U. (1983). Fatigue crack initiation by cyclic slip irreversibilities in high-cycle fatigue, Fatigue Mechanism, ASTM, STP 811, pp. 5–45.

Thompson, N. (1959), *Proceedings of the International Conference on the Atomic Mechanisms of Fracture*, Technology Press, MIT Cambridge, MA, and Wiley, New York.

Wood, W. A. (1956), Mechanisms of fatigue, in *Fatigue in Aircraft Structures*, edited by A. M. Freudental, Academic, New York, pp. 1–19.

Elastic Behavior of Laminated Flexible Composites Under Finite Deformation

Shen-Yi Luo and Tsu-Wei Chou
Center for Composite Materials, and Department of Mechanical Engineering, University of Delaware, Newark, DE 19716, U.S.A.

Abstract

This paper examines the nonlinear elastic behavior of laminated flexible composites under finite deformation. The constitutive relations have been derived by applying a model for unidirectional flexible composite lamina, which is based on the Lagrangian description and a strain–energy density referring to the initial principal material coordinates. The fabrication technique for flexible composite laminates is discussed. Experiments have been conducted to verify the constitutive relations in the finite deformation range. Good agreements have been found between the theory and experiments.

Nomenclature

a_{ij}	components of the orthogonal transformation matrix
C_{ij}, C_{ijk}, C_{ijkl}	elastic stiffness constants related to linear, bi-modulus, and nonlinear properties, respectively
c	$= \cos \theta_0$
E_{ij}	Lagrangian strain
g_{ij}	deformation gradient
N_{ij}	stress resultant
s	$= \sin \theta_0$
W	strain–energy density referring to the initial principal material coordinates
W_{ij}	$= \partial W/\partial E_{ij}$
X	Lagrangian coordinate
$\bar{X}$	initial principal material coordinate with $\bar{X}_1$ parallel to the fiber direction
θ_0	initial fiber orientation
Π_{ij}	Lagrangian stress (Piola–Kirchoff stress)
σ_{ij}	Cauchy stress (Eulerian stress)

– (overbar) for a quantity referring to the initial principal material coordinates $\bar{X}_1$–$\bar{X}_2$

1. Introduction

The classical lamination theory of composites (for instance, Jones, 1975; Vinson and Chou, 1975) is a relatively mature subject, and it is adequate in describing the elastic properties of composite laminates under the assumptions of linear stress–strain relations and infinitesimal deformation. Under finite deformation, the composite configuration change could be large, the fiber reorientation usually cannot be neglected, and the geometric and material nonlinearities may be quite significant. Thus the conventional lamination theory, developed for rigid composites, is no longer applicable for studying the elastic behavior of laminated flexible composites, and better understanding needs to be developed.

The term "flexible composites" is used here to identify composites based upon elastomeric polymers, of which the usable range of deformation is much larger than those of the conventional thermosetting or thermoplastic polymer-based composites (Chou and Takahashi, 1987; Luo and Chou, 1988a). The ability of flexible composites to sustain large deformation and fatigue loading, and still provide high load-carrying capacity has been mainly analyzed in cord/rubber composites. However, most of the existing analyses on the mechanics of cord/rubber composites (for instance, Akasaka, 1959–1964; Clark, 1980; Walter and Patel, 1978) are primarily based on the classical lamination theory for infinitesimal deformation. Chou (1988) has recently provided a review of the mechanics of flexible composites.

Adkins and Rivlin (1955) treated the nonlinear, anisotropic, and finite deformation problem of fiber reinforced elastomeric laminates by using the "ideal fiber reinforced material theory"; the assumptions of volume incompressibility and fiber inextensibility are basic to the analysis. Further developments of this theory can be found in the work of Rivlin (1964), Pipkin and Rogers (1971), and Spencer (1972). Difficulties often rise in applying this theory to composites with complicated fiber geometries and in cases where the extension of the fibers cannot be neglected.

In an effort to provide a rigorous treatment of the finite deformation problem of flexible composites, Luo and Chou (1988b, c) and Rivlin (1987) recently developed a nonlinear constitutive model for flexible composite lamina, which is based upon the Lagrangian description and a strain–energy density expressed in the principal material coordinates. In this paper, the model has been applied to study the finite nonlinear elastic behavior of laminated flexible composites. The fabrication technique for flexible composite laminates has been examined. Experiments have been conducted to verify the constitutive relations in the finite deformation range. Good agreements have been found between the theory and experiments.

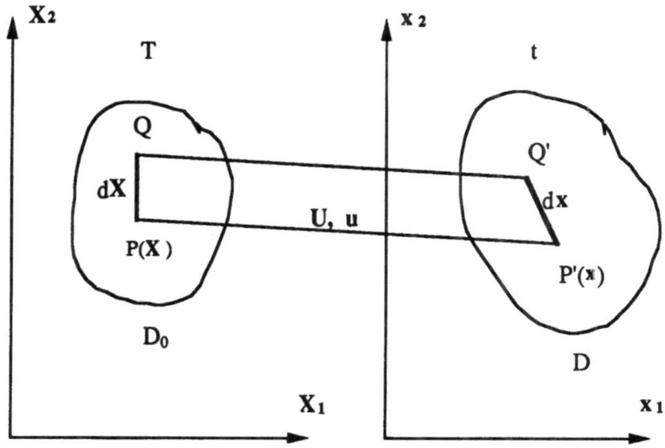

FIG. 1. The undeformed configuration D_0 and the deformed configuration D of an elastic medium.

2. Analysis

2.1. Background

The description of the relation between the undeformed and deformed configurations of a continuum can be considered as a "mapping" between domains D_0 and D (Fig. 1). To find the transformation relation, let X and x be two fixed rectangular Cartesian coordinates associated with the original and deformed configurations, respectively. The position of a generic particle P inside the domain D_0 is defined by the position vector $\mathbf{X}$ and coordinates X_j $(j = 1, 2, 3)$. After deformation, this particle assumes the location P' with the new position vector $\mathbf{x}$ and coordinates x_i $(i = 1, 2, 3)$. Then, the "mapping" or the deformation of the configuration can be described mathematically by the coordinate transformation between X_j and x_i (Fung, 1969; Rivlin, 1970). The definition of deformation, in which the independent variable is the particle position vector $\mathbf{X}$ in the original state, is known as the Lagrangian description. The reference system X is known as the Lagrangian coordinate.

The Lagrangian strain (E_{ij}, also known as the Green or St. Venant strain) tensor is defined as

$$E_{ij} = \tfrac{1}{2}(g_{ki}g_{kj} - \delta_{ij}), \tag{2.1}$$

where the dummy, or repeating, index denotes a summation with respect to that index over its range, δ_{ij} is the Kronecker delta, and

$$g_{ij} = \frac{\partial x_i}{\partial X_j}. \tag{2.2}$$

Equation (2.1) can be rewritten in terms of the displacement vector, $\mathbf{U} = \mathbf{x} - \mathbf{X}$ (Fig. 1), as

$$E_{ij} = \frac{1}{2}\left(\frac{\partial U_i}{\partial X_j} + \frac{\partial U_j}{\partial X_i} + \frac{\partial U_K}{\partial X_i}\frac{\partial U_K}{\partial X_j}\right). \tag{2.3}$$

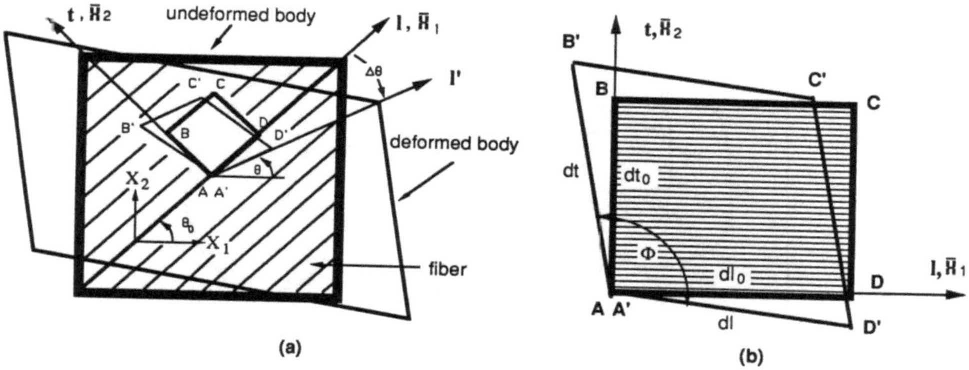

FIG. 2. A rectangular element of composite lamina before and after loading in the Lagrangian system.

If the displacement gradients are sufficiently small, the quadratic terms in (2.3) can be neglected, and the Lagrangian strain tensors are reduced to the infinitesimal strain tensors as expected.

Following Rivlin (1986–1988) the strain–energy density is assumed to be a function of the Lagrangian strain components referring to the initial principal material coordinate $\bar{X}$ with the axis $\bar{X}_1$ being parallel to the fiber direction (Fig. 2). The strain–energy per unit volume of the undeformed lamina is further assumed in the following fourth-order polynomial form (Luo and Chou, 1988c)

$$W = \tfrac{1}{2}C_{11}\bar{\mathbf{E}}_1^2 + \tfrac{1}{3}C_{111}\bar{\mathbf{E}}_1^3 + \tfrac{1}{4}C_{1111}\bar{\mathbf{E}}_1^4 + C_{12}\bar{\mathbf{E}}_1\bar{\mathbf{E}}_2 + \tfrac{1}{2}C_{22}\bar{\mathbf{E}}_2^2 + \tfrac{1}{3}C_{222}\bar{\mathbf{E}}_2^3 + \tfrac{1}{4}C_{2222}\bar{\mathbf{E}}_2^4 + \tfrac{1}{2}C_{66}\bar{\mathbf{E}}_6^2 + \tfrac{1}{4}C_{6666}\bar{\mathbf{E}}_6^4, \tag{2.4}$$

where C_{ij}, C_{ijk}, and C_{ijkl} are elastic constants. In the above equation, the short-handed notations are used, namely, $\bar{\mathbf{E}}_1 = \bar{\mathbf{E}}_{11}$, $\bar{\mathbf{E}}_2 = \bar{\mathbf{E}}_{22}$, and $\bar{\mathbf{E}}_6 = 2\bar{\mathbf{E}}_{12}$. A quantity with an overbar refers to the initial principal material coordinate $\bar{X}$.

In finite elasticity, the stress can be defined as the force per either the undeformed or deformed area. The former is known as the Piola–Kirchoff stress or the Lagrangian stress (Π_{ij}), where the latter is known as the Cauchy stress or the Eulerian stress (σ_{ij}). The general relations between the two stress descriptions are

$$\sigma_{ji} = (\det \mathbf{g})^{-1} g_{jk}\Pi_{ki}, \tag{2.5}$$

where (det $\mathbf{g}$) is the determinant of $[\mathbf{g}]$.

The two-dimensional Eulerian stress referring to the principal material coordinate system $\bar{X}_1$–$\bar{X}_2$, is given in terms of W (Rivlin, 1970),

$$\bar{\boldsymbol{\sigma}}_{ij} = \frac{1}{\det(\bar{\mathbf{g}})}\bar{\mathbf{g}}_{ip}\frac{\partial W}{\partial \bar{\mathbf{g}}_{jp}}. \tag{2.6}$$

To derive the general constitutive equations with reference axes other than the material principal directions, a two-dimensional rectangular Cartesian coordinate system X_1–X_2 is chosen in the plane of the lamina. The angle between X_i and $\bar{X}_i$ is θ_0 (Fig. 2). Let $[\mathbf{a}]$ be an orthogonal transformation matrix, $[\mathbf{X}] = [\mathbf{a}][\bar{\mathbf{X}}]$, and

$$[\mathbf{a}] = \begin{bmatrix} \cos\theta_0 & -\sin\theta_0 \\ \sin\theta_0 & \cos\theta_0 \end{bmatrix}. \tag{2.7}$$

Then, between coordinate systems X and $\bar{X}$, the transformation relations of deformation gradient, Lagrangian strain, and stress are given in matrix form, respectively, as

$$\begin{aligned} [\bar{\mathbf{g}}] &= [\mathbf{a}]^{\mathrm{T}}[\mathbf{g}][\mathbf{a}], \\ [\bar{\mathbf{E}}] &= [\mathbf{a}]^{\mathrm{T}}[\mathbf{E}][\mathbf{a}], \\ [\boldsymbol{\sigma}] &= [\mathbf{a}][\bar{\boldsymbol{\sigma}}][\mathbf{a}]^{\mathrm{T}}. \end{aligned} \tag{2.8}$$

Using (2.5)–(2.8), the following can be derived (Luo and Chou, 1988c)

$$\Pi_{ji} = g_{ip}\{a_{p1}a_{j1}W_{11} + a_{p2}a_{j2}W_{22} + \tfrac{1}{2}(a_{p1}a_{j2} + a_{p2}a_{j1})W_{12}\}, \tag{2.9}$$

where

$$\begin{aligned} W_{11} &= \frac{\partial W}{\partial \bar{\mathbf{E}}_{11}} = C_{11}\bar{\mathbf{E}}_1 + C_{111}\bar{\mathbf{E}}_1^2 + C_{1111}\bar{\mathbf{E}}_1^3 + C_{12}\bar{\mathbf{E}}_2, \\ W_{22} &= \frac{\partial W}{\partial \bar{\mathbf{E}}_{22}} = C_{22}\bar{\mathbf{E}}_2 + C_{222}\bar{\mathbf{E}}_2^2 + C_{2222}\bar{\mathbf{E}}_2^3 + C_{12}\bar{\mathbf{E}}_1, \\ W_{12} &= \frac{\partial W}{\partial \bar{\mathbf{E}}_{12}} = 2(C_{66}\bar{\mathbf{E}}_6 + C_{6666}\bar{\mathbf{E}}_6^3). \end{aligned} \tag{2.10}$$

With (2.7), (2.9) is expressed in the following explicit form:

$$\begin{aligned} \Pi_{11} &= [g_{11}c^2 + g_{12}cs]W_{11} + [g_{11}s^2 - g_{12}cs]W_{22} \\ &\quad + [-g_{11}cs + \tfrac{1}{2}g_{12}(c^2 - s^2)]W_{12}, \\ \Pi_{22} &= [g_{22}s^2 + g_{21}cs]W_{11} + [g_{22}c^2 - g_{21}cs]W_{22} \\ &\quad + [g_{22}cs + \tfrac{1}{2}g_{21}(c^2 - s^2)]W_{12}, \\ \Pi_{12} &= [g_{22}cs + g_{21}c^2]W_{11} + [g_{21}s^2 - g_{22}cs]W_{22} \\ &\quad + [-g_{21}cs + \tfrac{1}{2}g_{22}(c^2 - s^2)]W_{12}, \\ \Pi_{21} &= [g_{11}cs + g_{12}s^2]W_{11} + [g_{12}c^2 - g_{11}cs]W_{22} \\ &\quad + [g_{12}cs + \tfrac{1}{2}g_{11}(c^2 - s^2)]W_{12}, \end{aligned} \tag{2.11}$$

where $c = \cos\theta_0$ and $s = \sin\theta_0$.

Equation (2.11) gives the general constitutive equations for a composite lamina under finite deformation, where the deformation gradients, g_{ij}, represent the geometric nonlinearity induced by the configuration changes of the lamina. The nonlinear expressions of W_{ij} (referring to (2.10)) represent the material nonlinearity of the composites. If the deformation of the composite lamina is infinitesimal (i.e., $g_{ij} = \delta_{ij}$) and only the linear terms (i.e., C_{ij}) remain in the expression of W_{ij}, (2.11) can be easily reduced to the familiar linear stress–strain equation used for rigid composites.

2.2. *Constitutive Equation of Flexible Composite Laminate*

Now the constitutive model for composite laminae discussed above are applied to study the constitutive relations of laminated flexible composites (Fig. 3) under finite plane deformation. The stress resultant in Lagrangian description (N_{ij}) is defined as

$$N_{ij} = \int_{-h/2}^{h/2} \Pi_{ij}\, dz, \tag{2.12}$$

where h is the initial thickness of the laminate. The N_{ij} so defined gives the total force in the i direction per unit length of the undeformed laminate.

Assume that the laminate is composed of n layers of laminae. By neglecting the interlaminar shear deformation, the deformation gradient, g_{ij} (2.2), has the same value for all the layers, so does E_{ij} (2.1). For an arbitrary kth lamina within the laminate, let $\theta_0^{(k)}$ be the fiber orientation angle with respect to the coordinate X_1, and let $a_{ij}^{(k)}$ be the values given by (2.7) for $\theta_0 = \theta_0^{(k)}$. Also, for the kth lamina, let $\bar{\mathbf{E}}_{ij}^{(k)}$ be the values of $\bar{\mathbf{E}}_{ij}$ given by (2.8), and let $W_{ij}^{(k)}$ be the values of W_{ij} given by (2.10). Then, from (2.11) and (2.12) the following can be derived

$$N_{ji} = g_{ip} \sum_{k=1}^{n} (h_k - h_{k-1}) \times \{a_{p1}^{(k)} a_{j1}^{(k)} W_{11}^{(k)} + a_{p2}^{(k)} a_{j2}^{(k)} W_{22}^{(k)} + \tfrac{1}{2}(a_{p1}^{(k)} a_{j2}^{(k)} + a_{p2}^{(k)} a_{j1}^{(k)}) W_{12}^{(k)}\}. \tag{2.13}$$

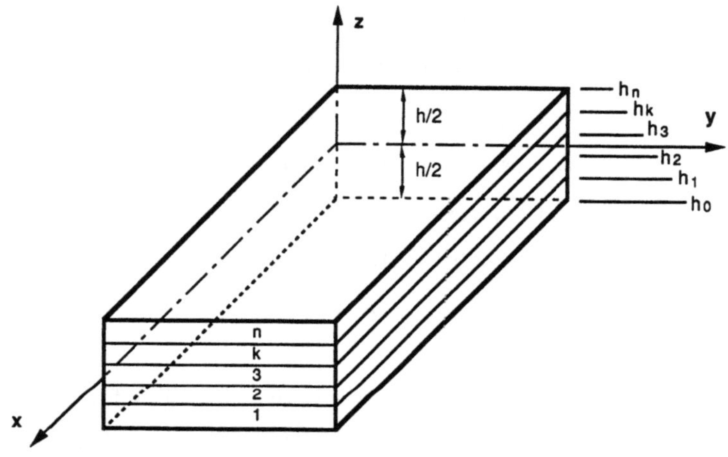

FIG. 3. A composite laminate.

If the laminae are identical (except the fiber orientation) with thickness t, then (2.13) can be rewritten as

$$N_{ji} = tg_{ip} \sum_{k=1}^{n} \{a_{p1}^{(k)} a_{j1}^{(k)} W_{11}^{(k)} + a_{p2}^{(k)} a_{j2}^{(k)} W_{22}^{(k)} + \tfrac{1}{2}(a_{p1}^{(k)} a_{j2}^{(k)} + a_{p2}^{(k)} a_{j1}^{(k)}) W_{12}^{(k)}\} \quad (2.14)$$

Equations (2.13) and (2.14) are the general constitutive equations for flexible composite laminates under finite deformations. The calculation procedure is further exemplified in the following.

2.3. Homogeneous Deformation

The homogeneous deformation of Fig. 4 can be defined as

$$\begin{aligned} x_1 &= \lambda_1 X_1, \\ x_2 &= \lambda_2 X_2, \end{aligned} \quad (2.15)$$

where λ_1 and λ_2 are the extension ratios in the X_1 and X_2 directions, respectively. Consequently, referring to (2.1) and (2.2), for all laminae

$$[\mathbf{g}] = \begin{bmatrix} \lambda_1 & 0 \\ 0 & \lambda_2 \end{bmatrix}, \qquad 2[\mathbf{E}] = \begin{bmatrix} \lambda_1^2 - 1 & 0 \\ 0 & \lambda_2^2 - 1 \end{bmatrix}. \quad (2.16)$$

The Lagrangian strains referring to the principal material coordinate, for the kth lamina, are obtained from (2.8)

$$\begin{aligned} \bar{\mathbf{E}}_{11}^{(k)} &= \tfrac{1}{2}(\lambda_1^2 \cos^2 \theta_0^{(k)} + \lambda_2^2 \sin^2 \theta_0^{(k)} - 1), \\ \bar{\mathbf{E}}_{22}^{(k)} &= \tfrac{1}{2}[(\lambda_1^2 \sin^2 \theta_0^{(k)} + \lambda_2^2 \cos^2 \theta_0^{(k)} - 1), \\ \bar{\mathbf{E}}_{12}^{(k)} &= \tfrac{1}{2}(\lambda_2^2 - \lambda_1^2) \sin \theta_0^{(k)} \cos \theta_0^{(k)}. \end{aligned} \quad (2.17)$$

Then the components of the Lagrangian stress resultant are obtained

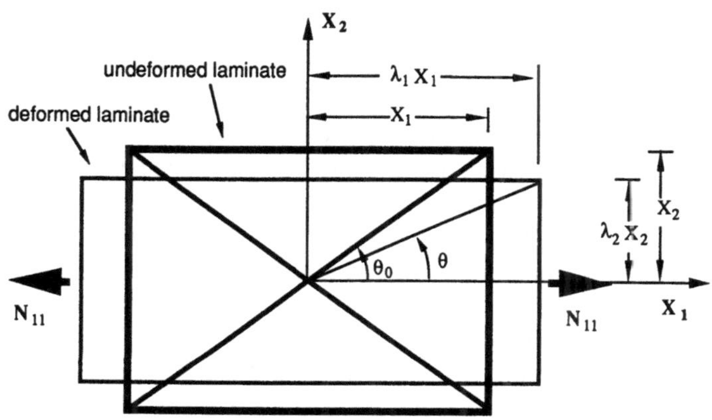

FIG. 4. A symmetric flexible composite laminate under a uniaxial load.

from (2.14)

$$
\begin{aligned}
N_{11} &= \frac{t}{2}\lambda_1 \sum_{k=1}^{n} \{W_{11}^{(k)}(1 + \cos 2\theta_0^{(k)}) + W_{22}^{(k)}(1 - \cos 2\theta_0^{(k)}) - W_{12}^{(k)} \sin 2\theta_0^{(k)}\},\\
N_{22} &= \frac{t}{2}\lambda_2 \sum_{k=1}^{n} \{W_{11}^{(k)}(1 - \cos 2\theta_0^{(k)}) + W_{22}^{(k)}(1 + \cos 2\theta_0^{(k)}) + W_{12}^{(k)} \sin 2\theta_0^{(k)}\},\\
N_{12} &= \frac{t}{2}\lambda_2 \sum_{k=1}^{n} \{(W_{11}^{(k)} - W_{22}^{(k)}) \sin 2\theta_0^{(k)} + W_{12}^{(k)} \cos 2\theta_0^{(k)}\},\\
N_{21} &= \frac{t}{2}\lambda_1 \sum_{k=1}^{n} \{(W_{11}^{(k)} - W_{22}^{(k)}) \sin 2\theta_0^{(k)} + W_{12}^{(k)} \cos 2\theta_0^{(k)}\},
\end{aligned}
\tag{2.18}
$$

where $W_{11}^{(k)}$, $W_{22}^{(k)}$, and $W_{12}^{(k)}$ are obtained from (2.10) and (2.17) with two variates λ_1 and λ_2.

3. Experiments

3.1. Materials

The specimens of Kevlar-49/silicone elastomer laminate, with a fiber orientation sequence of $+\theta_0/-\theta_0/-\theta_0/+\theta_0$, were fabricated by the authors. The fiber material is Kevlar-49, 1420 denier (1 denier = 1 gram/9,000 meters) with 1,000 filaments in a yarn, manufactured by the Du Pont Company. The SYLGARD 184 silicone elastomer matrix, supplied by the Dow Corning Company, is transparent and has an elongation up to 100%. This resin is supplied in two parts of liquid components: base and curing agent. It can be cured at room temperature.

3.2. Fabrication

A metal frame with a saw-shaped edge is machined first as shown in Fig. 5(a). The spacing between two adjacent teeth of the sawed edge is determined by the volume fraction of the specimens (or numbers of yarns per unit length perpendicular to the fiber). The yarn is then wound onto the frame according to the required fiber orientations (Figs. 5(b) and 5(c)). The frame with would fibers is placed in a vacuum container. When the inner pressure is down to 600–700 mm of mercury less than one atmospheric pressure, the premixed resin is injected through a pipe on the top of the container. During resin injection, vacuum is maintained to eliminate air bubbles in the composites. The curing process could take place in an autoclave according to the curing cycles. However, our experience has shown that for better fiber/matrix bonding and minimizing porosity, specimens should first be cured at room temperature for 24 hours, and then placed in an oven at a temperature of 65 °C for another 24 hours. The cured lamina can be cut by a sharp knife from the frame into specimens with the desired dimensions.

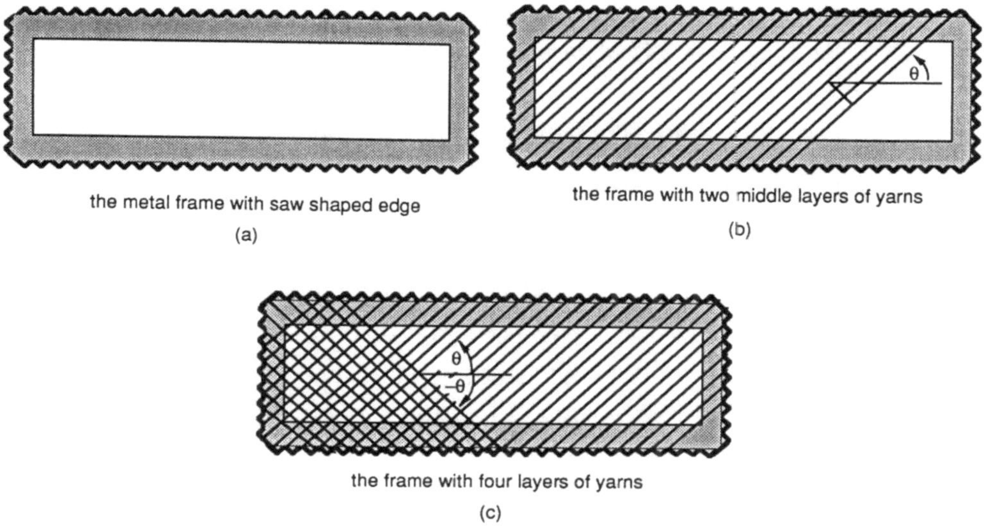

FIG. 5. The fiber winding sequence for making a symmetric laminate.

3.3. Tests

The experiments of flexible composite laminates under longitudinal tension have been conducted to verify the constitutive relations obtained in Section 2. The tensile tests are performed on an Instron tester at the cross-head speed of 5 mm/min. The fiber volume fraction of the specimen is 9%; and the specimen dimensions of the test section are 4 mm × 26 mm × 180 mm.

The conventional strain guaging technique is not suitable for measuring the large strains of flexible composites under finite deformation. Consequently, a photographic method is used (Luo and Chou, 1988a). At each predetermined strain level, a photograph is taken to record the deformed configuration of the grid premarked on the specimen. Measurements of the locations of the markers on the specimen determine the displacements. In order to eliminate the influence of stress relaxation of the material, the specimen is kept at a constant strain for 5 seconds before a photographic record is made. The measurement error of this method is believed to be small compared to the scale of deformation.

4. Numerical Examples

To verify the constitutive equations obtained in Section 2, the theoretical predictions are compared with the experimental results of $+\theta_0/-\theta_0/-\theta_0/+\theta_0$ laminates under unidirectional tension

$$N_{11} \neq 0, \qquad \text{and} \qquad N_{22} = N_{12} = N_{21} = 0. \tag{4.1}$$

The elastic constants of Kevlar-49/silicone elastomer are (Luo, 1988): $C_{11} =$

8,600 MPa, $C_{12} = -1.3$ MPa, $C_{22} = 2.77$ MPa, $C_{2222} = -12.5$ MPa, $C_{66} =$ 2.57 MPa, $C_{6666} = -2.45$ MPa, and $C_{111} = C_{1111} = C_{222} = 0$.

4.1. Tensile Stress–Strain Relation

The state of homogeneous deformation is assumed for a symmetric composite laminate with fiber orientation sequences of $+\theta_0/-\theta_0/-\theta_0/+\theta_0$ under unidirectional tension (Fig. 4). Because $\bar{\mathbf{E}}_{11}$ and $\bar{\mathbf{E}}_{22}$ are even functions of θ_0, and $\bar{\mathbf{E}}_{12}$ is an odd function of θ_0 (2.17), we have from (2.10)

$$\begin{aligned} W_{11}^{(\theta)} &= W_{11}^{(-\theta)}, \\ W_{22}^{(\theta)} &= W_{22}^{(-\theta)}, \\ W_{12}^{(\theta)} &= -W_{12}^{(-\theta)}. \end{aligned} \tag{4.2}$$

Equation (2.18) is then reduced to

$$\begin{aligned} N_{11} &= \frac{h}{2}\lambda_1\{W_{11}^{(\theta)}(1+\cos 2\theta_0) + W_{22}^{(\theta)}(1-\cos 2\theta_0) - W_{12}^{(\theta)}\sin 2\theta_0\}, \\ N_{22} &= \frac{h}{2}\lambda_2\{W_{11}^{(\theta)}(1-\cos 2\theta_0) + W_{22}^{(\theta)}(1+\cos 2\theta_0) + W_{12}^{(\theta)}\sin 2\theta_0\}, \\ N_{12} &= N_{21} = 0, \end{aligned} \tag{4.3}$$

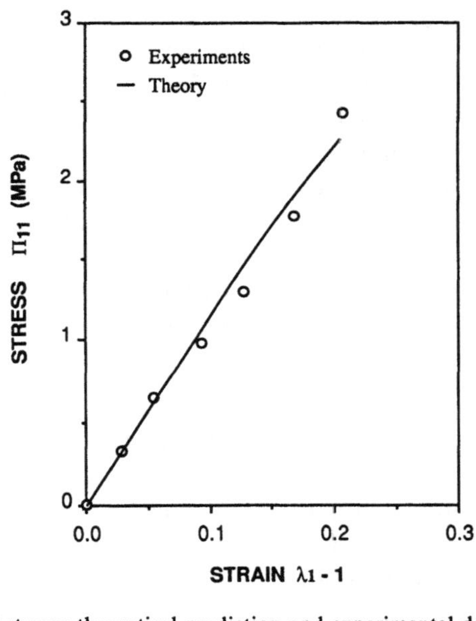

FIG. 6. Comparisons between theoretical prediction and experimental data of the stress–strain response of a $(\pm 45°)_S$ Kevlar/silicone elastomer composite laminate under uniaxial load.

where h is the thickness of the laminate; $W_{ij}^{(\theta)}$, obtained from (2.10) and (2.17), is a function of λ_1 and λ_2. With the stress condition of (4.1) and the values of elastic constants, the two unknowns, λ_1 and λ_2, in (4.3) can be solved.

For example, let $\theta_0 = 45°$, from (2.10), (2.17), and (4.3),

$$N_{11}/h = \lambda_1\{C_{66}(\lambda_1^2 - \lambda_2^2) + \tfrac{1}{4}C_{6666}(\lambda_1^2 - \lambda_2^2)^3\},$$

$$\begin{aligned} N_{22}/h &= 0 \\ &= (D - 4C_{66})\lambda_1^2 - (D + 4C_{66})\lambda_2^2 + \tfrac{1}{4}(C_{111} + C_{222})(\lambda_1^2 + \lambda_2^2 - 2)^2 \\ &\quad + \tfrac{1}{16}(C_{1111} + C_{2222})(\lambda_1^2 + \lambda_2^2 - 2)^3 + C_{6666}(\lambda_2^2 - \lambda_1^2)^3, \end{aligned} \tag{4.4}$$

where $D = C_{11} + 2C_{12} + C_{22}$. On the other hand, the relation of $(4.4)_1$ may be used to determine experimentally the shear constants C_{66} and C_{6666} by a curve fitting method, if the experimental data of N_{11}, λ_1, and λ_2 are given.

Figure 6 shows the comparison between the theoretical predictions and experimental results of the stress–strain relation of $[\pm 45°]_S$ Kevlar/silicone elastomer composite laminates under uniaxial load. Reasonable agreement has been found.

4.2. *Effective Poisson Ratio*

The Poisson ratio is defined as the negative ratio of the strain in the X_j direction to the strain in the X_i direction due to an applied stress in the X_i direction. The Poisson ratio for a symmetric composite laminate was derived by Posfalvi (1977) based upon the finite deformation consideration. However, the comparison of theory with experiments in this work is still limited to the small deformation range although the experimental results of large deformation were presented; the details of the experiments (specimen preparation, fiber volume fraction, test method, etc.) were not described.

From the analysis of Section 4.1, the effective Poisson ratio in the finite deformation range can be readily predicted. For example, $a + \theta_0/-\theta_0/-\theta_0/+\theta_0$ laminate under unidirectional load and the effective Poisson ratio at a given strain level can be determined from (2.16) as

$$\frac{E_{22}}{E_{11}} = \frac{(\lambda_2^2 - 1)}{(\lambda_1^2 - 1)}, \tag{4.5}$$

where the relation of λ_1 and λ_2 can be obtained from $(4.3)_2$ and (2.10) with $N_{22} = 0$.

The approximate order of the ratio E_{22}/E_{11} can be obtained by neglecting the nonlinear terms (i.e., C_{111}, C_{6666}, etc.) in the expressions of (2.10) for W_{ij}, then

$$-\frac{(\lambda_2^2 - 1)}{(\lambda_1^2 - 1)} = \frac{A}{B}, \tag{4.6}$$

where

$$A = C_{11}\cos^2\theta_0\sin^2\theta_0 + C_{12}(\sin^4\theta_0 + \cos^4\theta_0) + C_{22}\cos^2\theta_0\sin^2\theta_0 - 4C_{66}\cos^2\theta_0\sin^2\theta_0,$$

$$B = C_{11}\sin^4\theta_0 + 2C_{12}\cos^2\theta_0\sin^2\theta_0 + C_{22}\cos^4\theta_0 + 4C_{66}\cos^2\theta_0\sin^2\theta_0.$$

For example, let $\theta_0 = 45°$, (4.6) yields

$$\frac{A}{B} = \frac{(C_{11} + 2C_{12} + C_{22}) - 4C_{66}}{(C_{11} + 2C_{12} + C_{22}) + 4C_{66}}. \tag{4.7}$$

Since the shear modulus C_{66} for flexible composites is relatively small, it can be assumed $A/B \approx 1$. Then, (4.6) becomes

$$(\lambda_2^2 - 1) = -(\lambda_1^2 - 1). \tag{4.8}$$

Furthermore, if the flexible composite is very stiff in the fiber direction (i.e., $C_{11} \gg C_{ij}$, $ij \neq 11$) and $\theta_0 \neq 0$, (4.6) becomes

$$\frac{E_{22}}{E_{11}} = \frac{(\lambda_2^2 - 1)}{(\lambda_1^2 - 1)} = -\frac{\cos^2\theta_0}{\sin^2\theta_0}. \tag{4.9}$$

The results of (4.9) can also be derived by using the "ideal fiber reinforced material theory" (Adkins and Rivlin, 1955).

Figure 7 gives the comparison between the theoretical predictions and experimental results of the ratio λ_2/λ_1 for $(\pm\theta_0)_S$ Kevlar/silicone elastomer composite laminates under uniaxial load. The initial fiber orientations are 15°, 30°, and 45°. Very good agreement has been found.

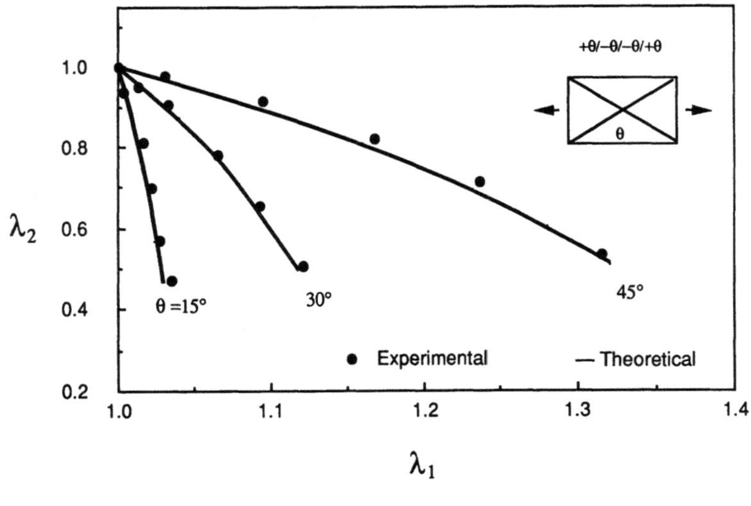

FIG. 7. Comparisons between theoretical predictions and experimental results of the ratio λ_2/λ_1 of $(\pm\theta)_S$ Kevlar/silicone elastomer composite laminates under uniaxial load.

Also, using the same definition of Posfalvi (1977), the current Poisson ratio at a given strain level can be derived from (4.6) as

$$\frac{d\lambda_2}{d\lambda_1} = \frac{\lambda_1}{\lambda_2}\frac{A}{B}. \tag{4.10}$$

The current fiber orientation, $\theta^{(k)}$ of the kth lamina, can be expressed in terms of λ_1, λ_2 and the initial fiber orientation $\theta_0^{(k)}$, referring to Fig. 4, as

$$\tan\theta^{(k)} = \frac{\lambda_2 \sin\theta_0^{(k)}}{\lambda_1 \cos\theta_0^{(k)}} = \frac{\lambda_2}{\lambda_1}\tan\theta_0^{(k)}, \tag{4.11}$$

where λ_1 and λ_2 are obtained by solving (4.3).

5. Conclusion

The nonlinear constitutive equations of laminated composites under finite deformation have been derived by applying a model for unidirectional flexible composite lamina, which is based on the Lagrangian description and a strain–energy density referring to the initial principal material coordinates. The assumption of negligible small interlaminar shear deformation is made in this analysis.

As an example, the constitutive relations are derived explicitly for the case of homogeneous deformation of $a + \theta_0/-\theta_0/-\theta_0/+\theta_0$ laminate under unidirectional load. Both the stress–strain relation and the effective Poisson ratio are discussed. Also, experiments have been conducted to verify the constitutive relations in the finite deformation range. Good agreements have been found between the theory and experiments.

References

Adkins, J. E. and Rivlin, R. S. (1955), Large elastic deformation of isotropic materials *X*. Reinforcement by inextensible cords, *Phil. Trans. Roy. Soc. London*, **248**, A, 944.

Akasaka, T. (1959–1964), *Various Reports/Bulletins, Faculty of Science and Engineering*, Chuo University, Tokyo.

Chou, T. W. (1989), Flexible composites, *J. Mater. Sci.*, **24**, 761–783.

Chou, T. W. and Takahashi, K. (1987), Nonlinear elastic behavior of flexible composites, *Composites*, **18**, No. 1, 25–34.

Clark, S. K. (1980), The role of textiles in pneumatic tires, in *Mechanics of Flexible Fiber Assemblies*, edited by J. W. S. Hearle, J. J. Thwaites and J. Amirbayat, Sijthoff and Noordhoff, The Netherlands, pp. 471.

Fung, Y. C. (1969), *A First Course in Continuum Mechanics*, Prentice-Hall, Englewood Cliffs, N.J.

Jones, R. M. (1975), *Mechanics of Composite Materials*, McGraw-Hill, New York.

Luo, S. Y., and Chou T. W. (1988a), Finite deformation and nonlinear elastic behavior of flexible composites, *J. Appl. Mech.*, **55**, 149–155.

Luo, S. Y., and Chou, T. W. (1988b), Constitutive relations of flexible composites under finite elastic deformation, in *Mechanics of Composite Materials—1988*, edited by G. J. Dvorak, and N. Laws, ASME, AMD, Vol. **92**, pp. 209–216.

Luo, S. Y., and Chou, T. W. (1988c), Finite deformation of flexible composites, submitted for publication.

Luo, Shen-Yi (1988), Theoretical modelling and experimental characterization of flexible composites under finite deformation, Ph.D. Dissertation, University of Delaware.

Pipkin, A. C., and Rogers, T. G. (1971), Plane deformation of incompressible fiber-reinforced materials, *J. Appl. Mech.*, 634–640.

Posfalvi, O. (1977), The Poisson ratio for rubber-cord composites, *Rubber Chem. Tech.*, **50**, 224–232.

Rivlin, R. S. (1964), Networks of inextensible cords, in *Nonlinear Problems of Engineering*, Academic Press, New York.

Rivlin, R. S. (1970), in *Centro Internazionale Matematico Estivo*, edited by II. Ciclo and R. S. Rivlin, Edizioni Cremonese, Roma.

Rivlin, R. S. (1986–1988), private communication.

Spencer, A. J. M. (1972), *Deformation of Fiber Reinforced Materials*, Oxford, Clarendon Press.

Vinson, J. R. and Chou, T. W. (1975), *Composite Materials and Their Use in Structures*, Elsevier–Applied Science, London.

Walter, J. D., and Patel, H. P. (1978), Approximate expressing for the elastic composites of cord–rubber laminates, *ASME Winter Annual Meeting*, San Francisco.

On the Basic Components of the Interaction Energy Between Two Infinitesimal Circular Defects in an Isotropic Elastic Body

SITIRO MINAGAWA and HIROSHI OGATA†
The University of Electro-Communications, Chofu, Tokyo 182, Japan

Abstract

The energy of interaction between two infinitesimal circular defects—dislocations and/or Frank disclinations—are given by the method of tensor analysis. The stress, incompatibility, and stress functions are expressed in terms of the dislocation and disclination density tensors. The interaction energy is given in terms of those tensors by means of the double-volume integrals with respect to the regions where there exist continuous distributions of defects. For discrete defects, the double-volume integrals are converted into the double-line integrals. The integrations are carried out for infinitesimal circular defects and the interaction energy between infinitesimal circular defects is given by a linear combination of certain basic components. Finally, the physical meaning of those components is given.

1. Introduction

There are two main currents in the investigations of the mechanics of dislocations and disclinations; i.e., the investigations based on the atomic (or discrete) models, and those by means of the continuum (or field) theories. For example, Blin (1955) obtained the formulas for the interaction energy between two closed dislocation loops, and Hirth and Lothe (1968) applied these formulas to estimate the energy of interaction between dislocation segments and the interaction forces acting between them. On the other hand, Kröner (1958) calculated the interaction energy between dislocations by means of the theory of continuous distributions of dislocations. DeWit (1971) and Kossecka and deWit (1977a, 1977b) advanced their work into the problem of disclinations, and Kondo (1955, 1958, 1962, 1968) extensively made his pioneering work on the field of dislocations and disclinations by the application of non-Riemannian geometry.

---

† Present address: The Tokyo Metropolitan Government, 1-2-28 Kounan, Minato-ku, Tokyo.

Mura (1987) advanced his comprehensive study on dislocations and disclinations by means of both discrete and continuum models; his special attention is focused at bridging the gap between the two currents of investigations. Readers who are interested in his investigations may consult his book, as well as his original publications in the List of Publications by T. Mura (p. xvii).

As is well known (e.g., p. 45 of Mura (1987)), the dislocation (line) is defined as a part of the boundary of a plane which is embedded in a material, and along which a translational slip is assumed. If a rotational slip is assumed in place of the translational one, a (Frank) disclination is given along the boundary of the same plane. In this paper, *defect* is a general term for dislocations and disclinations, as well as for their mixtures. Therefore, the *defect-line* is characterized by the Burgers and Frank vectors, and the axis of rotation, to make the Frank disclination. When we examine the problem from the standpoint of the continuum theory, *a continuous distribution of defects* should be taken into consideration; it is characterized by both the dislocation and disclination density tensors.

Section 2 will be devoted to a review of some basic equations for the field of stress and incompatibility, produced by a distribution of defects in terms of the dislocation and disclination density tensors. In Section 3 we shall give the tensorial expressions for the interaction energy between two aggregates of continuous distributions of defects. They will be converted into those for the interaction energy between two discrete defects. In Section 4 computations will be made for infinitesimal circular defects, which are placed in an arbitrary sense; it will be stated by a linear combination of certain basic components. The physical meaning of those components will be investigated. Throughout this paper, the formulas given by Kröner (1958), DeWit (1973a), Mura (1987), etc., will be used.

2. Mathematical Preliminaries

2.1. Displacement and Curvature

We assume an orthogonal Cartesian coordinate system, with respect to which the position of a material point is stated by x_i. Throughout this paper, lowercase indices i, j, k, ..., take the values of 1, 2, or 3, and Einstein's summation convention is used for indices appearing twice in one expression. Let $\mathbf{x}$ (or $\mathbf{x}'$) be the vector having x_i (or x_i') as its components, and a comma followed by an index (or indices) designates the derivative(s) with respect to the corresponding spatial coordinate(s).

Let $u_i(\mathbf{x})$ be the field of displacement. The fields of strain and rotation are defined by

$$e_{ij} = \tfrac{1}{2}(u_{i,j} + u_{j,i}), \qquad \phi_i = \tfrac{1}{2}\varepsilon_{ijk}u_{k,j}, \tag{2.1}$$

where ε_{ijk} is Eddington's permutation symbol, which is 1 or -1 according to whether ijk is an even or odd permutation of 123, or otherwise zero. The curvature tensor is given by the gradient of the field of rotation so that

$$\kappa_{ij} = \phi_{j,i}. \tag{2.2}$$

The strain and curvature are divided into the elastic and plastic parts such that

$$e_{ij} = e_{ij}^{\mathrm{E}} + e_{ij}^{\mathrm{P}}, \qquad \kappa_{ij} = \kappa_{ij}^{\mathrm{E}} + \kappa_{ij}^{\mathrm{P}}, \tag{2.3}$$

where e_{ij}^{E} and κ_{ij}^{E} are the elastic strain and curvature, while e_{ij}^{P} and κ_{ij}^{P} the plastic strain and curvature

2.2. *Continuous Distributions of Defects*

The dislocation and disclination density tensors are given by the plastic strain and curvature tensors through

$$\alpha_{ij} = -\varepsilon_{ipq}(e_{qj,p}^{\mathrm{P}} - \varepsilon_{qsj}\kappa_{ps}^{\mathrm{P}}), \qquad \theta_{ij} = -\varepsilon_{ipq}\kappa_{qj,p}^{\mathrm{P}}. \tag{2.4}$$

The incompatibility tensor is defined by

$$\eta_{ij} = \varepsilon_{ipq}\varepsilon_{jmn}e_{nq,pm}^{\mathrm{P}}. \tag{2.5}$$

The substitution of (2.1), (2.3), and (2.4) into (2.5) leads

$$\eta_{ij} = -\tfrac{1}{2}(\varepsilon_{iqs}\alpha_{js,q} + \varepsilon_{jqs}\alpha_{is,q}) - \tfrac{1}{2}(\theta_{ij} - \theta_{ji}). \tag{2.6}$$

On the other hand, the stress field is stated as follows:

$$\sigma_{ij} = -\varepsilon_{ipq}\varepsilon_{jmn}\chi_{nq,pm}, \tag{2.7}$$

where χ_{ij} is the stress function tensor. For the stress field produced by a continuous distribution of defects, we have

$$\begin{aligned}\chi_{ij}(\mathbf{x}) = \frac{G}{8\pi}\int\Bigg[&\left(\varepsilon_{ipq}\delta_{rj} + \varepsilon_{jpq}\delta_{ri} + \frac{2\nu}{1-\nu}\delta_{ij}\varepsilon_{rpq}\right)R_{,p}\alpha_{rq}(\mathbf{x}') \\ &+ \left(\delta_{ip}\delta_{jq} + \delta_{jp}\delta_{iq} + \frac{2\nu}{1-\nu}\delta_{ij}\delta_{pq}\right)R\theta_{pq}(\mathbf{x}')\Bigg]\,d\mathbf{x}',\end{aligned} \tag{2.8}$$

where R means $|\mathbf{x} - \mathbf{x}'|$, G is the shear modulus, ν is Poisson's ratio, and δ_{ij} is Kronecker's delta. The substitution of (2.8) into (2.7) leads to

$$\begin{aligned}\sigma_{ij}(\mathbf{x}) = \frac{G}{4\pi}\int\Bigg[&\left(\varepsilon_{iks}\delta_{jp} + \varepsilon_{jks}\delta_{ip} - \frac{2}{1-\nu}\delta_{ij}\varepsilon_{pks}\right)\left(\frac{1}{R}\right)_{,k} \\ &+ \frac{1}{1-\nu}\varepsilon_{pks}R_{,kij}\Bigg]\alpha_{ps}(\mathbf{x}')\,d\mathbf{x}' \\ + \frac{G}{4\pi}\int\Bigg[&\left(\delta_{ip}\delta_{jq} + \delta_{jp}\delta_{iq} - \frac{2}{1-\nu}\delta_{ij}\delta_{pq}\right)\frac{1}{R} \\ &+ \frac{1}{1-\nu}\delta_{pq}R_{,ij}\Bigg]\theta_{pq}(\mathbf{x}')\,d\mathbf{x}'.\end{aligned} \tag{2.9}$$

The last equation enables us to estimate the stress field produced by a continuous distribution of defects.

2.3. Discrete Defects

A defect loop is given along the boundary of a cut, denoted by S, so that one side of it, denoted by S^+, is translated by B and at the same time rotated by Ω, relative to the other side, denoted by S^-. The B and Ω are the Burgers and Frank vectors. Let $\mathbf{x}^0$ be the position vector of a point on the axis of rotation which is assumed to make the disclination. The B, Ω, and $\mathbf{x}^0$ specify the defect-loop. Let $\mathbf{l}$ be a unit vector drawn in the direction from S^- into S^+. This will be used in Section 4.

Where a defect-loop is concerned, the volume integrals appearing in (2.9); viz.,

$$\int [\cdots]\alpha_{ij}(\mathbf{x}')\,d\mathbf{x}' \quad \text{and} \quad \int [\cdots]\theta_{ij}(\mathbf{x}')\,d\mathbf{x}',$$

are substituted by the line integrals such as

$$\oint [\cdots]\{B_j + \varepsilon_{pqj}\Omega_p(x'_q - x^0_q)\}\,ds'_i \quad \text{and} \quad \oint [\cdots]\Omega_j\,ds'_i,$$

respectively. In the calculation of the line integrals we trace the loop in the direction of rotation of a right-handed screw, which is advanced in the direction from S^- to S^+. Therefore, (2.9) is transformed into

$$\sigma_{ij}(\mathbf{x}) = B_q E_{ijq}(\mathbf{x}) + \Omega_q F_{ijq}(\mathbf{x}), \tag{2.10}$$

where

$$E_{ijq}(\mathbf{x}) = \frac{G}{4\pi}\oint\left[\left(\varepsilon_{iks}\delta_{jp} + \varepsilon_{jks}\delta_{ip} - \frac{2}{1-\nu}\delta_{ij}\varepsilon_{pks}\right)\left(\frac{1}{R}\right)_{,k} + \frac{1}{1-\nu}\varepsilon_{pks}R_{,kij}\right]ds', \tag{2.11}$$

$$\begin{aligned} F_{ijq}(\mathbf{x}) = {} & \frac{G}{4\pi}\oint\left[\left(\varepsilon_{iks}\delta_{jp} + \varepsilon_{jks}\delta_{ip} - \frac{2}{1-\nu}\delta_{ij}\varepsilon_{pks}\right)\left(\frac{1}{R}\right)_{,k} + \frac{1}{1-\nu}\varepsilon_{pks}R_{,kij}\right]\varepsilon_{sqr}(x'_r - x^0_r)\,ds'_p \\ & + \frac{G}{4\pi}\oint\left[\left(\delta_{ip}\delta_{jq} + \delta_{jp}\delta_{iq} - \frac{2}{1-\nu}\delta_{ij}\delta_{pq}\right)\frac{1}{R} + \frac{1}{1-\nu}\delta_{pq}R_{,ij}\right]ds'_p. \end{aligned} \tag{2.12}$$

Equations (2.10)–(2.12) enable us to estimate the stress field produced by a discrete defect-loop.

3. Energy of Interactions Between Defects

3.1. Continuous Distributions of Defects

We shall consider two aggregates of continuously distributed defects which occupy regions $\boldsymbol{A}$ and $\boldsymbol{B}$ in the body, respectively. For the sake of simplicity, these two aggregates of defects are denoted by *defect* $\boldsymbol{A}$ and *defect* $\boldsymbol{B}$, and let $\alpha_{ij}^A(\mathbf{x})$ and $\alpha_{ij}^B(\mathbf{x})$ be their dislocation density tensors and θ_{ij}^A and θ_{ij}^B their disclination density tensors, respectively.

The elastic interaction energy between these two aggregates of defects is given by

$$E_{\text{int}} = \int_B \chi_{in}^A(\mathbf{x})\eta_{ij}^B(\mathbf{x})\, d\mathbf{x}, \tag{3.1}$$

where χ_{ij}^A is the stress function tensor for the stress field produced by defect $\boldsymbol{A}$, η_{ij}^B is the field of incompatibility caused by defect $\boldsymbol{B}$, and the volume integral is calculated over region $\boldsymbol{B}$.

The η_{ij}^B is stated in terms of α_{ij}^B and θ_{ij}^B as in (2.6), while χ_{ij}^A is stated in terms of α_{ij}^A and θ_{ij}^A as in (2.8). The substitution of those terms into (3.1) leads to

$$\begin{aligned} E_{\text{int}} = \frac{G}{8\pi} \int_B \int_A \Bigg[& \left(\varepsilon_{rki}\varepsilon_{rpq}\delta_{js} + \varepsilon_{ski}\varepsilon_{jpq} + \frac{2\nu}{1-\nu}\varepsilon_{jki}\varepsilon_{spq} \right) R_{,kp}\alpha_{ji}^A(\mathbf{x}')\alpha_{sq}^B(\mathbf{x}) \\ & - \left(\varepsilon_{tpq}\delta_{rs} + \varepsilon_{spq}\delta_{tr} + \frac{2\nu}{1-\nu}\varepsilon_{rpq}\delta_{ts} \right) R_{,p}\alpha_{rq}^A(\mathbf{x}')\theta_{ts}^B(\mathbf{x}) \\ & + \left(\varepsilon_{tpq}\delta_{sr} + \varepsilon_{spq}\delta_{tr} + \frac{2\nu}{1-\nu}\varepsilon_{rpq}\delta_{ts} \right) R_{,p}\theta_{ts}^A(\mathbf{x}')\alpha_{rq}^B(\mathbf{x}) \\ & - \left(\delta_{pt}\delta_{qs} + \delta_{ps}\delta_{qt} + \frac{2\nu}{1-\nu}\delta_{pq}\delta_{ts} \right) R\theta_{pq}^A(\mathbf{x}')\theta_{ts}^B(\mathbf{x}) \Bigg] d\mathbf{x}'\, d\mathbf{x}. \end{aligned} \tag{3.2}$$

This implies the energy of interaction between two aggregates of continuous distributions of defects.

3.2. Discrete Defects

When defect $\boldsymbol{A}$ is a single discrete defect, with $\boldsymbol{B}^A$ and $\boldsymbol{\Omega}^A$ as the Burgers and Frank vectors and with $\boldsymbol{x}^A$ as the position vector of a point on the axis of rotation, then so also for defect **B** with $\boldsymbol{B}^B$, $\boldsymbol{\Omega}^B$, and $\boldsymbol{x}^B$, respectively.

We substitute the line integrals for the volume integrals in (3.2), as mentioned in the previous section, to lead the interaction energy between these two defect-loops. The following four cases apply.

Case 1. Both $\boldsymbol{A}$ and $\boldsymbol{B}$ are single dislocation loops ($\boldsymbol{\Omega}^A = 0$, $\boldsymbol{\Omega}^B = 0$).

$$E_{\text{int}}^1 = \frac{G}{8\pi} B_i^A B_j^B \oint_A \oint_B \left(\varepsilon_{tki}\varepsilon_{tpj}\delta_{rq} + \varepsilon_{qki}\varepsilon_{rpj} + \frac{2\nu}{1-\nu}\varepsilon_{rki}\varepsilon_{qpj} \right) R_{,kp}\, ds_q\, ds_r'. \quad (3.3)$$

Case 2. $\boldsymbol{A}$ is a dislocation loop and $\boldsymbol{B}$ is a Frank disclination loop ($\boldsymbol{\Omega}^A = 0$, $\boldsymbol{B}^B = 0$).

$$\begin{aligned} E_{\text{int}}^2 = \frac{G}{8\pi} B_i^A \Omega_j^B \oint_A \oint_B \Bigg[& \varepsilon_{jnt} \left(\varepsilon_{ski}\varepsilon_{spt}\delta_{rq} + \varepsilon_{qki}\varepsilon_{rpt} \right. \\ & \left. + \frac{2\nu}{1-\nu}\varepsilon_{rki}\varepsilon_{qpt} \right) R_{,kp}(x_n - x_n^B) \\ & - \left(\varepsilon_{qpi}\delta_{rj} + \varepsilon_{jpi}\delta_{rq} + \frac{2\nu}{1-\nu}\varepsilon_{rpi}\delta_{qj} \right) R_{,p} \Bigg] ds_q\, ds_r'. \end{aligned} \quad (3.4)$$

Case 3: $\boldsymbol{A}$ is a Frank disclination loop and $\boldsymbol{B}$ a dislocation loop ($\boldsymbol{B}^A = 0$, $\boldsymbol{\Omega}^B = 0$).

$$\begin{aligned} E_{\text{int}}^2 = \frac{G}{8\pi} \Omega_i^A B_j^B \oint_A \oint_B \Bigg[& \varepsilon_{iba} \left(\varepsilon_{ska}\varepsilon_{spj}\delta_{rq} + \varepsilon_{qka}\delta_{rpj} \right. \\ & \left. + \frac{2\nu}{1-\nu}\varepsilon_{rka}\varepsilon_{qpj} \right) R_{,kp}(x_b' - x_b^A) \\ & + \left(\varepsilon_{rpj}\delta_{iq} + \varepsilon_{ipj}\delta_{rq} + \frac{2\nu}{1-\nu}\varepsilon_{qpj}\delta_{ri} \right) R_{,p} \Bigg] ds_q\, ds_r'. \end{aligned} \quad (3.5)$$

Case 4: Both $\boldsymbol{A}$ and $\boldsymbol{B}$ are single Frank disclination loops ($\boldsymbol{B}^A = 0$, $\boldsymbol{B}^B = 0$).

$$\begin{aligned} E_{\text{int}}^3 = \frac{G}{8\pi} \Omega_i^A \Omega_j^B \oint_A \oint_B \Bigg[& \varepsilon_{itm}\varepsilon_{jnu} \left(\varepsilon_{skm}\varepsilon_{spu}\delta_{rq} + \varepsilon_{qkm}\varepsilon_{rpu} \right. \\ & \left. + \frac{2\nu}{1-\nu}\varepsilon_{rkm}\varepsilon_{qpu} \right) (x_t' - x_t^A) R_{,kp}(x_n - x_n^B) \\ & - \varepsilon_{itu} \left(\varepsilon_{qpu}\delta_{rj} + \varepsilon_{jpu}\delta_{qr} + \frac{2\nu}{1-\nu}\varepsilon_{rpu}\delta_{qj} \right) (x_t' - x_t^A) R_{,p} \\ & + \varepsilon_{jnu} \left(\varepsilon_{rpu}\delta_{iq} + \varepsilon_{ipu}\delta_{qr} + \frac{2\nu}{1-\nu}\varepsilon_{qpu}\delta_{ri} \right) R_{,p}(x_n - x_n^B) \\ & - \left(\delta_{rq}\delta_{ij} + \delta_{rj}\delta_{iq} + \frac{2\nu}{1-\nu}\delta_{ri}\delta_{qj} \right) R \Bigg] ds_q\, ds_r'. \end{aligned} \quad (3.6)$$

Equations (3.3)–(3.6) enable us to estimate the interaction energies between discrete defect-loops.

4. Infinitesimal Circular Defects

4.1. Stress Field

We shall consider an infinitesimal circular defect-loop of radius a. For convenience of analysis, we introduce

$$\frac{B_i}{a}, \quad \frac{x_i}{a}, \quad \frac{\sigma_{ij}}{G},$$

as dimensionless parameters for the Burgers vector, the position vector, and stress. In this section, B_i, x_i, and σ_{ij} are used to denote these dimensionless quantities. For the sake of simplicity, we assume that the defect-loop is placed on the (x_1, x_2) coordinate plane, with its center at the origin of the coordinates.

The stress field produced by this defect-loop is given by the same equation as (2.10), although in this case σ_{ij} and B_i are dimensionless quantities and E_{ijq} and F_{ijq} are given as follows:

$$E_{ijq}(\mathbf{x}) = -\varepsilon_{3ps}\frac{1}{4x^3}\left[\left(\varepsilon_{ikq}\delta_{jp} + \varepsilon_{jkq}\delta_{ip} - \frac{1}{1-\nu}\delta_{ij}\varepsilon_{pkq}\right)\left(\delta_{ks} - \frac{3x_s x_k}{x^2}\right)\right.$$
$$- \frac{1}{1-\nu}\varepsilon_{pkq}\left\{\left(\delta_{ki} - \frac{3x_k x_i}{x^2}\right)\left(\delta_{js} - \frac{3x_j x_s}{x^2}\right)\right.$$
$$\left.\left. + \left(\delta_{kj} - \frac{3x_k x_j}{x^2}\right)\left(\delta_{is} - \frac{3x_i x_s}{x^2}\right) - \frac{12 x_i x_j x_k x_s}{x^4}\right\}\right], \tag{4.1}$$

$$F_{ijq}(\mathbf{x}) = \varepsilon_{3pk}\frac{x_k}{2x^3}\left[\delta_{qi}\delta_{jp} + \delta_{qj}\delta_{ip} - \frac{2}{1-\nu}\delta_{ij}\delta_{pq}\right] - E_{ijp}(\mathbf{x})\varepsilon_{pqr}x_r^0, \tag{4.2}$$

where the small terms are neglected and $x = |\mathbf{x}|$.

4.2. Interaction Energy

The energy of interaction between two infinitesimal circular defects can be calculated from (3.3)–(3.6). Let a, b be the radii of these defects, let $\mathbf{l}^A$, $\mathbf{l}^B$ be the unit vectors perpendicular thereto (see Section 2.3), and let $\mathring{\mathbf{x}}^A$, $\mathring{\mathbf{x}}^B$ be the position vectors of their centers.

We introduce $Ga^3/[8(1-\nu)]$ as the unit of the interaction energy, and use the same symbol, E_{int}, to denote the dimensionless interaction energy below.

From (3.3)–(3.6), we get

$$E^1_{\text{int}} = \left(\frac{b}{a}\right)^2 B_i^A B_j^B l_k^A l_s^B L_{ijks}, \tag{4.3}$$

$$E^2_{\text{int}} = \left(\frac{b}{a}\right)^2 B_i^A \Omega_j^B l_k^A l_s^B \{M_{ijks} + \varepsilon_{jnt}(\mathring{x}_n^B - x_n^B)L_{itks}, \tag{4.4}$$

$$E^3_{\text{int}} = \left(\frac{b}{a}\right)^2 \Omega_i^A \Omega_j^B l_k^A l_s^B \{N_{ijks} - \varepsilon_{iba}(\mathring{x}_b^A - x_b^A)M_{ajsk}$$
$$+ \varepsilon_{jnc}(\mathring{x}_n^B - x_n^B)M_{icks} + \varepsilon_{iba}\varepsilon_{jnc}(\mathring{x}_b^A - x_b^A)(\mathring{x}_n^B - x_n^B)L_{acks}\}, \tag{4.5}$$

where

$$L_{ijks} = L_{ijkspq}\frac{d_p d_q}{d^5}, \qquad M_{ijks} = M_{ijkspqr}\frac{d_p d_q d_r}{d^5}, \qquad N_{ijks} = N_{ijkspq}\frac{d_p d_p}{d^3}, \quad (4.6)$$

L_{ijkspq}, $M_{ijkspqr}$, and N_{ijkspq} are given as follows:

$$L_{ijkspq} = 24\delta_{ks}\delta_{ip}\delta_{jp} - 2\delta_{pq}\{2(1-\nu)\delta_{ik}\delta_{js} + (1+\nu)(\delta_{is}\delta_{jk} + \delta_{ij}\delta_{ks})\}$$
$$+ (21 - 15\nu)\delta_{ij}\varepsilon_{tps}\varepsilon_{tqk} + 6\{(1-2\nu)\varepsilon_{ips}\varepsilon_{jkq} + \nu\varepsilon_{ikp}\varepsilon_{jqs}\}, \quad (4.7)$$

$$M_{ijkspqr} = (1+\nu)\{\delta_{pq}(\delta_{is}\varepsilon_{jrk} + \delta_{ir}\varepsilon_{jks}) + 3\delta_{ir}\varepsilon_{jsq}\delta_{kp}\}$$
$$+ 2\delta_{pq}\{\delta_{ik}\varepsilon_{jrs} + \nu(\delta_{jk}\varepsilon_{irs} - \varepsilon_{ijs}\delta_{kr}) + (1-2\nu)\delta_{ij}\varepsilon_{ksr}\}$$
$$+ (3-\nu)\varepsilon_{ijr}\delta_{ks}\delta_{pq} - 3(1-\nu)\{(\varepsilon_{ijk}\delta_{sr} + \delta_{jr}\varepsilon_{iks})\delta_{pq}$$
$$+ \delta_{jr}\varepsilon_{isq}\delta_{kp} - \delta_{ir}\varepsilon_{jkq}\delta_{sp}\}, \quad (4.8)$$

$$N_{ijkspq} = (1-\nu)\{\delta_{ij}(\delta_{ks}\delta_{pq} - \delta_{pk}\delta_{qs}) + \varepsilon_{isp}\varepsilon_{jkq}\} + 2\nu\varepsilon_{ikp}\varepsilon_{jsq}, \quad (4.9)$$

where $\mathbf{d}$ means $\mathring{\mathbf{x}}^A - \mathring{\mathbf{x}}^B$, and d is $|\mathbf{d}|$.

Equations (4.3), (4.4), and (4.5) show that, as d increases, the dislocation–dislocation, dislocation–Frank disclination, and Frank disclination–Frank disclination interaction energies decrease as d^{-3}, d^{-2}, and d^{-1} decrease, respectively.

4.2. Interaction Force

The elastic interaction force exerted by defect $\boldsymbol{A}$ on defect $\boldsymbol{B}$ is given by

$$F_h = -\frac{\partial E_{\text{int}}}{\partial d_h}, \quad (4.10)$$

which is a dimensionless parameter for the interaction force, having $Gb^2/8(1-\nu)$ as the unit. By the substitution of (4.3), (4.4), and (4.5) into (4.10), the following equations are given.

A. Interaction force acting on a dislocation exerted by a dislocation

$$F_h^1 = -\left(\frac{b}{a}\right)^2 B_i^A B_j^B l_k^A l_s^B F_{ijksh}. \quad (4.11)$$

B. Interaction force acting on a Frank disclination exerted by a dislocation

$$F_h^2 = -\left(\frac{b}{a}\right)^2 B_i^A \Omega_j^B l_k^A l_s^B \{G_{ijksh} + \varepsilon_{jnt}(\mathring{x}_n^B - x_n^B)F_{itksh}\}. \quad (4.12)$$

C. Interaction force acting on a Frank disclination exerted by a Frank disclination

$$F_h^3 = -\left(\frac{b}{a}\right)^2 \Omega_i^A \Omega_j^B l_k^A l_s^B \{H_{ijksh} - \varepsilon_{iba}(\mathring{x}_b^A - x_b^A)G_{tkskh} + \varepsilon_{jnc}(\mathring{x}_n^B - x_n^B)G_{icksh}$$
$$+ \varepsilon_{iba}\varepsilon_{jnc}(\mathring{x}_b^A - x_b^A)(\mathring{x}_n^B - x_n^B)F_{acksh}\}. \quad (4.13)$$

The F_{ijksh}, G_{ijksh}, and H_{ijksh} are given as follows:

$$F_{ijksh} = \left(L_{ijkshq} + L_{ijksqh} - L_{ijkspq} \frac{5d_p d_h}{d^2} \right) \frac{d_q}{d^5}, \tag{4.14}$$

$$G_{ijksh} = \left(M_{ijkshqp} + M_{ijksphq} + M_{ijkspqh} - M_{ijkspqr} \frac{5d_r d_h}{d^2} \right) \frac{d_p d_q}{d^5}, \tag{4.15}$$

$$H_{ijksh} = \left(N_{ijkshq} + N_{ijksqh} - N_{ijkspq} \frac{3d_p d_h}{d^2} \right) \frac{d_q}{d^3}. \tag{4.16}$$

Therefore, as d increases, the dislocation–dislocation, dislocation–Frank disclination, and Frank disclination–Frank disclination interaction forces decrease as d^{-4}, d^{-3}, and d^{-2} decrease, respectively.

4.3. Basic Components of the Interaction Energy

The L_{ijks}, M_{ijks}, and N_{ijks} are the basic components of the interaction energy between infinitesimal circular defects, because the interaction energy between two infinitesimal circular defects, which are placed in arbitrary senses and having arbitrary Burgers or Frank vectors, can be stated by a linear combination of these components.

Their physical meaning is as follows: L_{ijks} is the interaction energy between two infinitesimal circular dislocations, placed perpendicularly to the x_k and x_s axes, and with the Burgers vectors parallel to the x_i and x_j axes, respectively; M_{ijks} is the interaction energy between an infinitesimal circular dislocation, which is placed perpendicularly to the x_k axis with the Burgers vector parallel to the x_i axis, and an infinitesimal circular Frank disclination placed perpendicularly to the x_s axis, with the Frank vector parallel to the x_j axis, and with an axis of rotation passing through the center of the loop; and finally, N_{ijks} is the interaction energy between two infinitesimal circular Frank disclinations, which are placed perpendicularly to the x_k and x_s axes, and having the Frank vectors parallel to the x_i and x_j axes and the axes of rotation passing through their centers, respectively.

In Tables 1 to 6, we give these basic components, which are classified into groups under the physical meaning. For example, L_{iiii} implies the dimensionless interaction energy between prismatic dislocations placed parallel to each other, and L_{ijkk} implies the dimensionless interaction energy between slip and prismatic dislocations as far as $i \neq k$ and $j \neq k$. As is stated in Table 2, no interaction is brought about between a prismatic dislocation and a twist disclination, as far as merely the terms of order d^{-2} are taken into account, nor is it between twist and wedge disclinations, as far as the terms of order d^{-1} are concerned (see Table 3).

The present work has been undertaken with an anticipation of constructing a theory of continuous distributions of infinitesimal circular defects and its application to the problem of fatigue, such as has been initiated by Kroupa (1962) and DeWit (1973).

TABLE 1. Energy of interaction between two infinitesimal circular dislocations placed parallel to each other.

Dislocation \ Dislocation	Prismatic	Slip
Prismatic	L_{iiii} $L_{1111} = (13 - 15\nu)\frac{1}{d^3} + (3 + 15\nu)\frac{d_1^2}{d^5}$	$L_{kikk};\ i \neq k$ $L_{2122} = 24\frac{d_1 d_2}{d^5}$
Slip	$L_{ikkk};\ i \neq k$ $L_{1222} = 24\frac{d_1 d_2}{d^5}$	$L_{ijkk};\ i \neq k, j \neq k$ $L_{1133} = (13 - 11\nu)\frac{1}{d^3} + 6(5 - \nu)\frac{d_1^2}{d^5} - 3(5 - 3\nu)\frac{d_3^2}{d^5}$ $L_{1233} = 6(5 - \nu)\frac{d_1 d_2}{d^5}$

TABLE 2. Energy of interaction between an infinitesimal circular dislocation and an infinitesimal circular Frank disclination placed parallel to each other.

Disclination \ Dislocation	Twist	Wedge
Prismatic	M_{iiii} No interaction	$M_{kikk};\ i \neq k$ $M_{2122} = -6\frac{d_3}{d^3}\left(1 - \frac{d_2^2}{d^2}\right)$
Slip	$M_{ikkk};\ i \neq k$ $M_{1222} = 3(1 - \nu)\frac{d_3}{d^3}\left(1 - \frac{d_2^2}{d^2}\right)$	$M_{ijkk};\ i \neq k, j \neq k$ $M_{1133} = -3(1 + \nu)\frac{d_1 d_2 d_3}{d^5}$ $M_{1233} = 6\frac{d_1^2 d_3}{d^5} + 3(1 - \nu)\frac{d_2^2 d_3}{d^5}$

TABLE 3. Energy of interaction between two infinitesimal circular Frank disclinations placed parallel to each other.

Disclination \ Disclination	Twist	Wedge
Twist	N_{iiii} $N_{1111} = (1 - \nu)\left(\frac{1}{d} - \frac{d_1^2}{d^3}\right)$	$N_{kikk};\ i \neq k$ No interaction
Wedge	$N_{ikkk};\ i \neq k$ No interaction	$N_{ijkk};\ i \neq k, j \neq k$ $N_{1133} = (1 - \nu)\left(\frac{1}{d} - \frac{d_3^2}{d^3}\right) + (1 + \nu)\frac{d_2^2}{d^3}$ $N_{1233} = -(1 + \nu)\frac{d_1 d_2}{d^3}$

TABLE 4. Energy of interaction between two infinitesimal circular dislocations placed perpendicular to each other.

Dislocation \ Dislocation	Prismatic	Slip
Prismatic	$L_{ijij};\quad i \neq j$ $L_{1212} = -(1-\nu)\dfrac{1}{d^3} + 6(1-2\nu)\dfrac{d_3^2}{d^5}$	$L_{ijik};\quad i \neq k, j \neq k$ $L_{1213} = -6(1-2\nu)\dfrac{d_1 d_2}{d^5}$
Slip	$L_{ijkj};\quad i \neq k, j \neq k$ $L_{2131} = -6(1-2\nu)\dfrac{d_2 d_3}{d^5}$	$L_{ijkl};\quad i \neq j, j \neq l, k \neq l$ $L_{1123} = -3(5-3\nu)\dfrac{d_2 d_3}{d^5}$ $L_{1223} = -6\nu\dfrac{d_1 d_3}{d^5}$

TABLE 5. Energy of interaction between an infinitesimal circular dislocation and an infinitesimal circular Frank disclination placed perpendicular to each other.

Disclination \ Dislocation	Twist	Wedge
Prismatic	$M_{ijij};\quad i \neq j$ $M_{1212} = -6(1-\nu)\dfrac{d_1 d_2 d_3}{d^5}$	$M_{ijik};\quad i \neq k, j \neq k$ $M_{1213} = -6\nu\dfrac{d_1 d_2^2}{d^5} - 6\dfrac{d_3^2 d_1}{d^5}$
Slip	$M_{ijkj};\quad i \neq k, j \neq k$ $M_{2131} = 3(1-\nu)\left(\dfrac{d_3^2 d_1}{d^5} - \dfrac{d_1 d_2^2}{d^5}\right)$	$M_{ijkl};\quad i \neq k, j \neq l, k \neq l$ $M_{1123} = -6\nu\dfrac{d_1 d_2^2}{d^5} + 3(1-\nu)\dfrac{d_3^2 d_1}{d^5}$ $M_{1223} = 6\nu\dfrac{d_1^2 d_2}{d^5} - 3(1-\nu)\dfrac{d_2 d_3^2}{d^5}$

TABLE 6. Energy of interaction between two infinitesimal circular Frank disclinations placed perpendicular to each other.

Disclination \ Disclination	Twist	Wedge
Twist	$N_{ijij};\quad i \neq j$ $N_{1212} = -(1-\nu)\dfrac{d_3^2}{d^3}$	$N_{ijik};\quad i \neq k, j \neq k$ $N_{1213} = (1-\nu)\dfrac{d_2 d_3}{d^3}$
Wedge	$N_{ijkj};\quad i \neq k, j \neq k$ $N_{2131} = (1-\nu)\dfrac{d_2 d_3}{d^3}$	$N_{ijkl};\quad i \neq k, j \neq l, k \neq l$ $N_{1123} = -2\dfrac{d_2 d_3}{d^3}$ $N_{1223} = 2\nu\dfrac{d_3 d_1}{d^3}$

References

Blin, J. (1955), Energie mutuelle de deux dislocations, *Acta Metallurgica*, **3**, 199–200.

DeWit, R. (1971), Relation between continuous and discrete disclinations, *Arch. Mech. Stos.*, **24**, 499–510.

DeWit, R. (1973a), Theory of disclinations: II. Continuous and discrete disclinations in anisotropic elasticity, *J. Res. Nat. Bur. Stand., Phys. Chem.*, **77A**, 49–100.

DeWit, R. (1973b), Continuous distribution of disclination loops, *Phys. Stat. Sol.* (a), **18**, 669–681.

Hirth, J. P. and Lothe, J. (1968), *Theory of Dislocations*, McGraw-Hill, New York.

Kondo, K. (1955, 1958, 1962, 1968), RAAG Memoirs of the Unifying Study of Basic Problems in Engineering and Physical Sciences by Means of Geometry, Gakujutsu Bunken Fukyu-kai, Tokyo.

Kossecka, E. and deWit, R. (1977a), Dislocation kinematics, *Arch. Mech.*, **29**, 633–651.

Kossecka, E. and deWit, R. (1977b), Disclination dynamics, *Arch. Mech.*, **29**, 749–767.

Kröner, E. (1958), *Kontinuumstheorie der Versetzungen und Eigenspannungen*, Springer-Verlag, Berlin.

Kroupa, F. (1962), Continuous distribution of dislocation loops, *Czech. J. Phys.*, **B12**, 191–201.

Mura, T. (1987), *Micromechanics of defects in solids*, 2nd ed., Martinus Hijhoff, Dordrecht.

Appendix

A point on the circular defect-loop of radius a, having $\mathbf{l}$ as the unit vector lying perpendicularly to the plane of the loop and $\mathbf{x}^0$ as the center, provides the coordinates as follows: For $l_3 \neq 1$,

$$x_1 = x_1^0 + a\left\{\frac{l_2}{\sqrt{1-l_3^2}}\cos t + \frac{l_1 l_3}{\sqrt{1-l_3^2}}\sin t\right\},$$

$$x_2 = x_2^0 + a\left\{-\frac{l_1}{\sqrt{1-l_3^2}}\cos t + \frac{l_2 l_3}{\sqrt{1-l_3^2}}\sin t\right\},$$

$$x_3 = x_3^0 - a\sqrt{1-l_3^2}\sin t, \tag{A1.1}$$

and for $l_3 = 1$,

$$x_1 = x_1^0 + a\cos t, \qquad x_2 = x_2^0 + a\sin t, \qquad x_3 = x_3^0, \tag{A1.2}$$

where t is a parameter such that $0 \le t \le 2\pi$. We substitute (A1.2) into (2.11) and (2.12), neglecting higher-order terms with respect to a/d, to get (4.1) and (4.2). Similar computations are made to give (4.3)–(4.6) from (3.3)–(3.6) and (A1.1).

Successive Iteration Method in the Evaluation of Average Fields in Elastically Inhomogeneous Materials

T. Mori
Department of Materials Science and Engineering,
Tokyo Institute of Technology, 4259 Nagatsuta, Midori-ku,
Yokohama 227, Japan

K. Wakashima
Research Laboratory of Precision Machinery and Electronics,
Tokyo Institute of Technology, 4259 Nagatsuta, Midori-ku,
Yokohama 227, Japan

Abstract

A successive iteration method is proposed to evaluate the average values, in inhomogeneities and a surrounding matrix, of the disturbances due to the inhomogeneities in an otherwise uniform stress field. The disturbances are reproduced by the field of the equivalent problem where the field is caused by the (elastically uniform) inclusions with equivalent eigenstrains. The method consists of the following procedures: First, the stress disturbance in a single representative inhomogeneity is evaluated. Next, the contribution of the other inhomogeneities is taken into account by considering that the representative inhomogeneity feels the additional field by the other inhomogeneities. These processes are repeated.

The equivalent eigenstrains are expressed in series forms. The series converges to a closed form under a certain condition. Using the converged value of the equivalent eigenstrains, the average values of the disturbances and the overall elastic constants are calculated. The computation involved is simple when the inhomogeneities are ellipsoidal. However, other shapes can also be treated in principle.

1. Introduction

The method for calculating the overall elastic constants of a composite is a subject which has been studied for many years. When the volume fraction of inhomogeneities (dispersoids) is small, Eshelby's method is powerful, since it can evaluate explicitly the shape effect of dispersoids (Eshelby, 1957). To account for the interaction between dispersoids in a high concentration, the so-called self-consistent method was introduced (Kroner, 1958; Budiansky, 1965; Hill, 1965). However, in the extreme case of voids or rigid dispersoids, the self-consistent method gives unsatisfactory and intuitively unacceptable overall elastic constants at finite volume fractions of these constituents.

The concept of an average field (Mori and Tanaka, 1973; Brown, 1973) in inclusions and their surrounding matrix has been used in the attempt to include the interaction between the inhomogeneous inclusions. Benveniste (1987) reviewed the literature on this attempt and applied the concept to assign the so-called concentration factors of stress and strain in a composite. The calculation involved in this application uses the result of Eshelby's celebrated work (1957), the elastic field of an ellipsoidal inclusion and the equivalency between the stress disturbance due to an inhomogeneity and the stress due to an inclusion with eigenstrains (equivalent inclusion method). The overall elastic constant of composite obtained by Benveniste is attractive, since it is free from the irregularities in the extreme cases mentioned above. It is also within the bounds proved by Hashin and Shtrikman (1963). Mura *et al.* (1981) also tried to analyze the mechanical response of a composite using the equivalent inclusion method. Their analysis was limited to an approximation suitable for the dilute dispersion of inhomogeneities. However, their paper stated that approximation would be improved for a nondilute dispersion if the result of the above approximation was further taken into account as the field to be disturbed by the inhomogeneities. If the improvement is in fact achieved, it is worthwhile repeating the approximation procedure. Thus, we have tried to examine the stress and strain disturbances due to inhomogeneities in a composite with a successive iteration method. We have found that under a certain condition, the infinite repetition of approximation leads to closed forms of the disturbances. The overall elastic constants obtained from these forms have been found to coincide with those given by Benveniste. Nevertheless, we think that the procedure and the idea used in the present study are heuristic and indicate the plausibility of the obtained results.

2. Traction-Prescribed Body Containing Inhomogeneities

2.1. Average Field in a Traction-Free Body Containing Homogeneous Inclusions with Eigenstrains

First, we review the average elastic field in the body, D, that consists of a matrix and a large number of randomly distributed inclusions. The body is free from tractions on its boundary. The elastic stiffness of the inclusions is equal to that of the matrix ($\mathbf{C}$). The inclusions have eigenstrains of $\boldsymbol{\varepsilon}^*$. For simplicity, we deal with the inclusions with an identical ellipsoidal shape, their corresponding axes being aligned. Let the internal stress and strain due to the inclusions be $\boldsymbol{\sigma}$ and $\boldsymbol{\gamma}$, respectively. The averages of these quantities are expressed by $\langle\ \rangle_\Omega$ and $\langle\ \rangle_M$ for the inclusions and matrix, respectively.

It has been shown that the averages of the stress are expressed as

$$\langle\boldsymbol{\sigma}\rangle_M = -f\boldsymbol{\sigma}^\infty, \tag{2.1}$$

$$\langle\boldsymbol{\sigma}\rangle_\Omega = (1-f)\boldsymbol{\sigma}^\infty, \tag{2.2}$$

where f is the volume fraction of the inclusions (Mori and Tanaka, 1973). Here, $\boldsymbol{\sigma}^{\infty}$ is the stress calculated for a single inclusion and is given by

$$\boldsymbol{\sigma}^{\infty} = \mathbf{C}(\boldsymbol{\gamma}^{\infty} - \boldsymbol{\varepsilon}^{*}) \tag{2.3}$$

$$\boldsymbol{\gamma}^{\infty} = \mathbf{S}\boldsymbol{\varepsilon}^{*}, \tag{2.4}$$

where $\mathbf{S}$ is the (6×6) matrix presentation of the Eshelby tensor for the inclusion and $\boldsymbol{\gamma}^{\infty}$ is the strain calculated for the inclusion. Strictly speaking, (2.3) is exact for an infinite body. However, it is valid for the present case, since we are dealing with a large number of the inclusions, and the volume of an inclusion is infinitely small compared to the volume of the finite body.

Equations (2.1) and (2.2) are derived from

$$f\langle\boldsymbol{\sigma}\rangle_{\Omega} + (1 - f)\langle\boldsymbol{\sigma}\rangle_{M} = 0, \tag{2.5}$$

$$\langle\boldsymbol{\sigma}\rangle_{\Omega} = \langle\boldsymbol{\sigma}\rangle_{M} + \boldsymbol{\sigma}^{\infty}. \tag{2.6}$$

Equation (2.5) is equivalent to the equilibrium condition for a traction-free body. Equation (2.6) is based on the following consideration. The presence of a large number of the inclusions ensures that the addition of a new inclusion does not change $\langle\boldsymbol{\sigma}\rangle_{M}$ and $\langle\boldsymbol{\sigma}\rangle_{\Omega}$. Thus, we insert an inclusion into a region where the insertion is allowed. When we repeat this process of insertion as many times as possible, this allowed region eventually covers the whole domain of the matrix. The stress in the inclusion consists of two terms. The first term is the stress which has existed before the insertion. The average of this term by the repeated insertion is $\langle\boldsymbol{\sigma}\rangle_{M}$. The second term is the stress in the inserted inclusion calculated when only this inclusion is present. It is equal to $\boldsymbol{\sigma}^{\infty}$, except when the inclusion is very near the boundary of the body. However, when the average of this term is calculated over the number of the insertion, it becomes $\boldsymbol{\sigma}^{\infty}$ for the following reason. The significant deviation of the second term from $\boldsymbol{\sigma}^{\infty}$ occurs when the inclusion is a certain distance below the boundary. This distance is of the order of the size of the inclusion, $\Omega^{1/3}$. Thus, the average of the deviation of the second term from $\boldsymbol{\sigma}^{\infty}$ is of the order of $\boldsymbol{\sigma}^{\infty}$ times $(\Omega/D)^{1/3}$. $(\Omega/D)^{1/3}$ is an extremely small quantity. Thus, the average of the stress, the sum of the first and second terms, defined in an inclusion which is inserted many times, becomes $\langle\boldsymbol{\sigma}\rangle_{M} + \boldsymbol{\sigma}^{\infty}$, (2.6).

2.2. *Stress and Strain Disturbance due to Inhomogeneities*

Next, we consider the stress field in a body, D, containing a large number of randomly distributed ellipsoidal inhomogeneities. The elastic stiffnesses of the matrix and inhomogeneities are $\mathbf{C}$ and $\mathbf{C}^{*}$, respectively. The tractions on the boundary of the body are prescribed (constant). The tractions produce uniform stress $\boldsymbol{\sigma}_0$ and strain $\boldsymbol{\gamma}_0$ when the body does not contain the inhomogeneities. $\boldsymbol{\sigma}_0$ and $\boldsymbol{\gamma}_0$ are related by

$$\boldsymbol{\sigma}_0 = \mathbf{C}\boldsymbol{\gamma}_0, \qquad \boldsymbol{\gamma}_0 = \mathbf{C}^{-1}\boldsymbol{\sigma}_0. \tag{2.7}$$

Since the body contains the inhomogeneities, the stress changes to $\boldsymbol{\sigma}_0 + \boldsymbol{\sigma}$ and the strain to $\gamma_0 + \gamma$. The terms $\boldsymbol{\sigma}$ and γ are called disturbances due to the inhomogeneities. $\boldsymbol{\sigma}_0$ is the average stress defined for D (Hill, 1963).

Following Eshelby, we seek $\boldsymbol{\sigma}$ and γ by considering the equivalent problem, where $\boldsymbol{\sigma}$ and γ are the stress and strain in the equivalent body with the equivalent inclusion (Ω) with the equivalent eigenstrains, ε^*. The equivalent inclusions have the same elastic constants as the matrix and are identically distributed as the inhomogeneities in the actual body. The equivalent body is traction free. Thus, $\boldsymbol{\sigma}$ satisfies (2.5). We will determine $\langle\boldsymbol{\sigma}\rangle_\Omega$, $\langle\boldsymbol{\sigma}\rangle_M$, $\langle\gamma\rangle_\Omega$, $\langle\gamma\rangle_M$, and ε^* in the following method.

To the zeroth-order approximation, we only consider a single inhomogeneity and ignore the other inhomogeneities. As shown by Eshelby (1957), the disturbances of the stress, $\Delta\boldsymbol{\sigma}_0$, and the strain, $\Delta\gamma_0$, in the inhomogeneity are calculated by solving the equivalency equation

$$\begin{aligned}\boldsymbol{\sigma}_0 + \Delta\boldsymbol{\sigma}_0 &= \mathbf{C}(\mathbf{C}^{-1}\boldsymbol{\sigma}_0 + \Delta\gamma_0 - \varepsilon_0^*)\\ &= \mathbf{C}^*(\mathbf{C}^{-1}\boldsymbol{\sigma}_0 + \Delta\gamma_0),\end{aligned} \tag{2.8}$$

with

$$\Delta\gamma_0 = \mathbf{S}\varepsilon_0^*. \tag{2.9}$$

Here ε_0^* is the equivalent eigenstrain in this approximation. Solving (2.8) with (2.9), we obtain

$$\varepsilon_0^* = \mathbf{Z}^{-1}(\mathbf{C} - \mathbf{C}^*)\mathbf{C}^{-1}\boldsymbol{\sigma}_0, \tag{2.10}$$

$$\Delta\boldsymbol{\sigma}_0 = \mathbf{C}(\mathbf{S} - \mathbf{I})\varepsilon_0^*, \tag{2.11}$$

with

$$\mathbf{Z} = (\mathbf{C}^* - \mathbf{C})\mathbf{S} + \mathbf{C}, \tag{2.12}$$

where $\mathbf{I}$ is the unit matrix.

Now we will give ε_0^* to all the equivalent inclusions representing the corresponding inhomogeneities. However, ε_0^* and, correspondingly, $\Delta\boldsymbol{\sigma}_0$ are overestimates or underestimates, depending on the magnitude of $\mathbf{C}^*$ and $\mathbf{C}$. For example, let us consider the case of $\mathbf{C}^* > \mathbf{C}$. The equivalency condition of (2.8) means that $\boldsymbol{\sigma}_0$ is disturbed by $\Delta\boldsymbol{\sigma}_0$ in the representative single inhomogeneity. However, the presence of the other inhomogeneities tends to reduce the stress before the disturbance from $\boldsymbol{\sigma}_0$, since the other inhomogeneities carry stress larger than $\boldsymbol{\sigma}_0$ (the average). Thus, we have to make a correction.

The first-order correction is performed by solving the equivalency equation

$$\begin{aligned}\boldsymbol{\sigma}_1 + \Delta\boldsymbol{\sigma}_1 &= \mathbf{C}(\mathbf{C}^{-1}\boldsymbol{\sigma}_1 + \Delta\gamma_1 - \varepsilon_1^*)\\ &= \mathbf{C}^*(\mathbf{C}^{-1}\boldsymbol{\sigma}_1 + \Delta\gamma_1),\end{aligned} \tag{2.13}$$

with

$$\Delta\gamma_1 = \mathbf{S}\varepsilon_1^*. \tag{2.14}$$

$\boldsymbol{\sigma}_1$ is given by

$$\boldsymbol{\sigma}_1 = \langle\boldsymbol{\sigma}_1\rangle_M = -f\Delta\boldsymbol{\sigma}_0. \tag{2.15}$$

This means that the eigenstrain of the zeroth-order approximation given to all the equivalent inclusions produces the average stress in the matrix (2.15). This average stress in the matrix is in turn disturbed by an inhomogeneity. Equation (2.15) is based on (2.1). The disturbance is reproduced by the first-order correction term of the eigenstrain, $\boldsymbol{\varepsilon}_1^*$, in (2.13) and (2.14). Solving (2.13) ~ (2.15), we have

$$\boldsymbol{\varepsilon}_1^* = -f\mathbf{A}\boldsymbol{\varepsilon}_0^*, \tag{2.16}$$

$$\Delta\boldsymbol{\sigma}_1 = \mathbf{C}(\mathbf{S}-\mathbf{I})\boldsymbol{\varepsilon}_1^*, \tag{2.17}$$

with

$$\mathbf{A} = \mathbf{Z}^{-1}(\mathbf{C}-\mathbf{C}^*)(\mathbf{S}-\mathbf{I}). \tag{2.18}$$

With the argument similar to that used to introduce the first-order correction, $\Delta\boldsymbol{\sigma}_1$, $\Delta\boldsymbol{\gamma}_1$ and $\boldsymbol{\varepsilon}_1^*$ are over- or under-corrections to the zeroth-order approximation.

The second-order correction is similarly performed, using

$$\begin{aligned}\boldsymbol{\sigma}_2 + \Delta\boldsymbol{\sigma}_2 &= \mathbf{C}(\mathbf{C}^{-1}\boldsymbol{\sigma}_2 + \Delta\boldsymbol{\gamma}_2 - \boldsymbol{\varepsilon}_2^*)\\ &= \mathbf{C}^*(\mathbf{C}^{-1}\boldsymbol{\sigma}_2 + \Delta\boldsymbol{\gamma}_2),\end{aligned} \tag{2.19}$$

$$\Delta\boldsymbol{\gamma}_2 = \mathbf{S}\boldsymbol{\varepsilon}_2^*, \tag{2.20}$$

$$\boldsymbol{\sigma}_2 = \langle\boldsymbol{\sigma}_2\rangle_{\mathrm{M}} = -f\Delta\boldsymbol{\sigma}_1. \tag{2.21}$$

These are solved as

$$\boldsymbol{\varepsilon}_2^* = -f\mathbf{A}\boldsymbol{\varepsilon}_1^* \qquad (= f^2\mathbf{A}^2\boldsymbol{\varepsilon}_0^*), \tag{2.22}$$

$$\Delta\boldsymbol{\sigma}_2 = \mathbf{C}(\mathbf{S}-\mathbf{I})\boldsymbol{\varepsilon}_2^*. \tag{2.23}$$

We can repeat similar correction procedures infinite times to obtain the disturbances in the stress and strain. The eigenstrain, $\boldsymbol{\varepsilon}_i^*$, and the stress disturbance in an inhomogeneity, $\Delta\boldsymbol{\sigma}_i$, obtained by the ith correction are given as

$$\boldsymbol{\varepsilon}_i^* = (-f)^i\mathbf{A}^i\boldsymbol{\varepsilon}_0^*, \tag{2.24}$$

$$\Delta\boldsymbol{\sigma}_i = \mathbf{C}(\mathbf{S}-\mathbf{I})\boldsymbol{\varepsilon}_i^*. \tag{2.25}$$

These iteration procedures are meaningful when the total eigenstrain, $\boldsymbol{\varepsilon}^*$, defined as

$$\begin{aligned}\boldsymbol{\varepsilon}^* &= \boldsymbol{\varepsilon}_0^* + \boldsymbol{\varepsilon}_1^* + \boldsymbol{\varepsilon}_2^* + \cdots\\ &= (\mathbf{I} - f\mathbf{A} + f^2\mathbf{A}^2 + \cdots)\boldsymbol{\varepsilon}_0^*,\end{aligned} \tag{2.26}$$

converges, see (2.24). The converging condition is that the magnitude of every eigenvalue of $-f\mathbf{A}$ is less than one. Under this condition, (2.26) is written as

$$\boldsymbol{\varepsilon}^* = (\mathbf{I} + f\mathbf{A})^{-1}\boldsymbol{\varepsilon}_0^*, \tag{2.27}$$

where $(\mathbf{I} + f\mathbf{A})^{-1}$ is the inverse of $(\mathbf{I} + f\mathbf{A})$.

The total stress disturbance, $\langle \boldsymbol{\sigma} \rangle_\Omega$, in the inhomogeneities is

$$\langle \boldsymbol{\sigma} \rangle_\Omega = \Delta\boldsymbol{\sigma}_0 + \boldsymbol{\sigma}_1 + \Delta\boldsymbol{\sigma}_1 + \boldsymbol{\sigma}_2 + \Delta\boldsymbol{\sigma}_2 + \cdots . \tag{2.28}$$

With (2.15), (2.21), and similar relations used in the iteration procedures, this is rewritten as

$$\langle \boldsymbol{\sigma} \rangle_\Omega = (1 - f)(\Delta\boldsymbol{\sigma}_0 + \Delta\boldsymbol{\sigma}_1 + \Delta\boldsymbol{\sigma}_2 + \cdots), \tag{2.29}$$

which is further changed to

$$\begin{aligned}\langle \boldsymbol{\sigma} \rangle_\Omega &= (1 - f)\mathbf{C}(\mathbf{S} - \mathbf{I})(\boldsymbol{\varepsilon}_0^* + \boldsymbol{\varepsilon}_1^* + \boldsymbol{\varepsilon}_2^* + \cdots)\\ &= (1 - f)\mathbf{C}(\mathbf{S} - \mathbf{I})(\mathbf{I} + f\mathbf{A})^{-1}\boldsymbol{\varepsilon}_0^*,\end{aligned} \tag{2.30}$$

with (2.25) to (2.27). The average disturbance of the stress, $\langle \boldsymbol{\sigma} \rangle_M$, in the matrix is calculated as

$$\begin{aligned}\langle \boldsymbol{\sigma} \rangle_M &= \boldsymbol{\sigma}_1 + \boldsymbol{\sigma}_2 + \cdots\\ &= -f\mathbf{C}(\mathbf{S} - \mathbf{I})(\mathbf{I} + f\mathbf{A})^{-1}\boldsymbol{\varepsilon}_0^*,\end{aligned} \tag{2.31}$$

using (2.15), (2.21) $\cdots$ and (2.29) and (2.30). Equation (2.31) gives the average strain disturbance, $\langle \boldsymbol{\gamma} \rangle_M$, in the matrix as

$$\langle \boldsymbol{\sigma} \rangle_M = -f(\mathbf{S} - \mathbf{I})(\mathbf{I} + f\mathbf{A})^{-1}\boldsymbol{\varepsilon}_0^*. \tag{2.32}$$

The average strain disturbance, $\langle \boldsymbol{\gamma} \rangle_\Omega$, in the inhomogeneities, is obtained by

$$\langle \boldsymbol{\gamma} \rangle_\Omega = \Delta\boldsymbol{\gamma}_0 + \mathbf{C}^{-1}\boldsymbol{\sigma}_1 + \Delta\boldsymbol{\gamma}_1 + \mathbf{C}^{-1}\boldsymbol{\sigma}_2 + \Delta\boldsymbol{\gamma}_2 + \cdots . \tag{2.33}$$

This is rewritten as

$$\begin{aligned}\langle \boldsymbol{\gamma} \rangle_\Omega &= \mathbf{S}\boldsymbol{\varepsilon}_0^* - f(\mathbf{S} - \mathbf{I})\boldsymbol{\varepsilon}_0^* + \mathbf{S}\boldsymbol{\varepsilon}_1^* - f(\mathbf{S} - \mathbf{I})\boldsymbol{\varepsilon}_1^* + \mathbf{S}\boldsymbol{\varepsilon}_1^* + \cdots\\ &= \{\mathbf{S} - f(\mathbf{S} - \mathbf{I})\}(\boldsymbol{\varepsilon}_0^* + \boldsymbol{\varepsilon}_1^* + \boldsymbol{\varepsilon}_2^* + \cdots)\end{aligned} \tag{2.34}$$

using (2.9), (2.15), (2.11), $\cdots$, or

$$\langle \boldsymbol{\gamma} \rangle_\Omega = \{\mathbf{S} - f(\mathbf{S} - \mathbf{I})\}(\mathbf{I} + f\mathbf{A})^{-1}\boldsymbol{\varepsilon}_0^* \tag{2.35}$$

from (2.27).

2.3. Overall Elastic Compliance

The strain disturbance, $\langle \boldsymbol{\gamma} \rangle$, averaged over the whole body is defined as

$$\langle \boldsymbol{\gamma} \rangle = f\langle \boldsymbol{\gamma} \rangle_\Omega + (1 - f)\langle \boldsymbol{\gamma} \rangle_M, \tag{2.36}$$

which is rewritten as

$$\langle \boldsymbol{\gamma} \rangle = f(\mathbf{I} + f\mathbf{A})^{-1}\boldsymbol{\varepsilon}_0^*, \tag{2.37}$$

or

$$\langle \boldsymbol{\gamma} \rangle = f(\mathbf{I} + f\mathbf{A})^{-1}\mathbf{Z}^{-1}(\mathbf{C} - \mathbf{C}^*)\mathbf{C}^{-1}\boldsymbol{\sigma}_0, \tag{2.38}$$

with (2.10). The body undergoes the strain which is the sum of $\boldsymbol{\gamma}_0$ and $\langle \boldsymbol{\gamma} \rangle$. Let $\bar{\mathbf{C}}^{-1}$ be the overall elastic compliance of the body. Then the above sum is equal to $\bar{\mathbf{C}}^{-1}\boldsymbol{\sigma}_0$; that is,

$$\bar{\mathbf{C}}^{-1}\boldsymbol{\sigma}_0 = \boldsymbol{\gamma}_0 + \langle\boldsymbol{\gamma}\rangle. \tag{2.39}$$

Inserting (2.7) and (2.38) into (2.39), we obtain

$$\bar{\mathbf{C}}^{-1} = \{\mathbf{I} + f(\mathbf{I} + f\mathbf{A})^{-1}\mathbf{Z}^{-1}(\mathbf{C} - \mathbf{C}^*)\}\mathbf{C}^{-1}. \tag{2.40}$$

We can rewrite this equation as

$$\bar{\mathbf{C}}^{-1} = \{\mathbf{I} + f[(1-f)(\mathbf{C}^* - \mathbf{C})(\mathbf{S} - \mathbf{I}) + \mathbf{C}^*]^{-1}(\mathbf{C} - \mathbf{C}^*)\}\mathbf{C}^{-1}. \tag{2.41}$$

3. Stress and Strain Disturbances in a Displacement-Constrained Body Containing Inhomogeneities and Overall Elastic Stiffness

Here, we consider the case where the displacement at the boundary of a body D is prescribed so that the uniform strain $\boldsymbol{\gamma}_0$ is given. When the body is uniform in elastic constants ($\mathbf{C}$), the prescribed uniform strain $\boldsymbol{\gamma}_0$ produces the uniform stress $\boldsymbol{\sigma}_0$ given by (2.7). When a large number of ellipsoidal inhomogeneities with the elastic stiffness $\mathbf{C}^*$ are introduced into D, the strain changes to $\boldsymbol{\gamma}_0 + \boldsymbol{\gamma}$ and the stress to $\boldsymbol{\sigma}_0 + \boldsymbol{\sigma}$. $\boldsymbol{\gamma}$ and $\boldsymbol{\sigma}$ are the disturbances in strain and stress, respectively. Since the strain is prescribed, we have

$$f\langle\boldsymbol{\gamma}\rangle_\Omega + (1-f)\langle\boldsymbol{\gamma}\rangle_M = 0. \tag{3.1}$$

Similar to the case of traction-prescribed loading, Section 2, we seek the equivalent problem in which the strain and stress caused by the equivalent inclusions with the equivalent eigenstrain $\boldsymbol{\varepsilon}^*$ are equal to $\boldsymbol{\gamma}$ and $\boldsymbol{\sigma}$. Repeating an argument similar to that which justifies (2.6), we can have

$$\langle\boldsymbol{\gamma}\rangle_\Omega = \langle\boldsymbol{\gamma}\rangle_M + \boldsymbol{\gamma}^\infty, \tag{3.2}$$

where $\boldsymbol{\gamma}^\infty$ is defined in (2.4). In the exact sense, (2.4) is valid for the body infinitely extended, when $\mathbf{S}$ as defined and calculated by Eshelby (1957) is used. However, the volume of the inclusion is infinitely small, compared to that of the body. Thus, the use of the Eshelby tensor is allowed in (2.4) and also in the present case.

From (3.1) and (3.2), we have

$$\langle\boldsymbol{\gamma}\rangle_M = -f\mathbf{S}\boldsymbol{\varepsilon}^* \qquad (= -f\boldsymbol{\gamma}^\infty), \tag{3.3}$$

$$\langle\boldsymbol{\gamma}\rangle_\Omega = (1-f)\mathbf{S}\boldsymbol{\varepsilon}^* \qquad (= (1-f)\boldsymbol{\gamma}^\infty). \tag{3.4}$$

Similar to the previous section, we will obtain the averages of the strain and stress disturbances and the equivalent eigenstrain by the successive iteration method. The zeroth-order approximation is obtained by solving the following equivalency equation for a single inhomogeneity:

$$\begin{aligned}\mathbf{C}\boldsymbol{\gamma}_0 + \Delta\boldsymbol{\sigma}_0 &= \mathbf{C}(\boldsymbol{\gamma}_0 + \Delta\boldsymbol{\gamma}_0 - \boldsymbol{\varepsilon}_0^*)\\ &= \mathbf{C}^*(\boldsymbol{\gamma}_0 + \Delta\boldsymbol{\gamma}_0),\end{aligned} \tag{3.5}$$

with

$$\Delta\boldsymbol{\gamma}_0 = \mathbf{S}\boldsymbol{\varepsilon}_0^*. \tag{3.6}$$

Solving (3.5), we have

$$\boldsymbol{\varepsilon}_0^* = \mathbf{Z}^{-1}(\mathbf{C} - \mathbf{C}^*)\boldsymbol{\gamma}_0. \tag{3.7}$$

Equation (2.11) also holds. We can see that (3.7) is an overestimate of the true equivalent eigenstrain for $\mathbf{C}^* < \mathbf{C}$. Equation (3.5) means that the uniform strain of $\boldsymbol{\gamma}_0$, which is present in the region to be occupied by an inhomogeneity, is disturbed by $\Delta\boldsymbol{\gamma}_0$ in the inhomogeneity. $\boldsymbol{\gamma}_0$ is the strain in the matrix. However, since the strain is prescribed, the actual strain in the matrix is smaller than $\boldsymbol{\gamma}_0$ when $\mathbf{C}^* < \mathbf{C}$: The inhomogeneities carry a larger strain and the matrix a smaller strain. Similarly, (3.7) is an underestimate for $\mathbf{C}^* > \mathbf{C}$. Thus, we will make the first-order correction, using the results of the above approximation. That is, we will solve the equivalency equation

$$\begin{aligned} \mathbf{C}\boldsymbol{\gamma}_1 + \Delta\boldsymbol{\sigma}_1 &= \mathbf{C}(\boldsymbol{\gamma}_1 + \Delta\boldsymbol{\gamma}_1 - \boldsymbol{\varepsilon}_1^*) \\ &= \mathbf{C}^*(\boldsymbol{\gamma}_1 + \Delta\boldsymbol{\gamma}_1), \end{aligned} \tag{3.8}$$

with

$$\Delta\boldsymbol{\gamma}_1 = \mathbf{S}\boldsymbol{\varepsilon}_1^*, \tag{3.9}$$

and

$$\boldsymbol{\gamma}_1 = -f\mathbf{S}\boldsymbol{\varepsilon}_0^*. \tag{3.10}$$

The idea underlying (3.8)–(3.10) is that when $\boldsymbol{\varepsilon}_0^*$ of the first-order approximation is given to all the equivalent inclusions, the matrix is elastically strained as given by (3.3) and this strain is further disturbed by a representative inhomogeneity. Equation (3.8) is solved as

$$\boldsymbol{\varepsilon}_1^* = -f\mathbf{B}\boldsymbol{\varepsilon}_0^*, \tag{3.11}$$

with

$$\mathbf{B} = \mathbf{Z}^{-1}(\mathbf{C} - \mathbf{C}^*)\mathbf{S}. \tag{3.12}$$

We will further conduct the correction procedures using the preceding corrections. The ith-order correction is obtained by solving

$$\begin{aligned} \mathbf{C}\boldsymbol{\gamma}_i + \Delta\boldsymbol{\sigma}_i &= \mathbf{C}(\boldsymbol{\gamma}_i + \Delta\boldsymbol{\gamma}_i - \boldsymbol{\varepsilon}_i^*) \\ &= \mathbf{C}^*(\boldsymbol{\gamma}_i + \Delta\boldsymbol{\gamma}_i), \end{aligned} \tag{3.13}$$

$$\Delta\boldsymbol{\gamma}_i = \mathbf{S}\boldsymbol{\varepsilon}_i^*, \tag{3.14}$$

$$\boldsymbol{\gamma}_i = -f\mathbf{S}\boldsymbol{\varepsilon}_{i-1}^*, \tag{3.15}$$

and then we obtain

$$\boldsymbol{\varepsilon}_i^* = -f\mathbf{B}\boldsymbol{\varepsilon}_{i-1}^* = (-f)^i\mathbf{B}^i\boldsymbol{\varepsilon}_0^*. \tag{3.16}$$

When the correction is repeated infinite times, the total equivalent eigenstrain becomes

$$\begin{aligned} \boldsymbol{\varepsilon}^* &= \boldsymbol{\varepsilon}_0^* + \boldsymbol{\varepsilon}_1^* + \boldsymbol{\varepsilon}_2^* + \cdots \\ &= (\mathbf{I} - f\mathbf{B} + f^2\mathbf{B}^2 + \cdots)\boldsymbol{\varepsilon}_0^*, \end{aligned} \tag{3.17}$$

from (3.16). When the magnitude of every eigenvalue of $-f\mathbf{B}$ is less than one, the series (3.17) converges to the expression of

$$\boldsymbol{\varepsilon}^* = (\mathbf{I} + f\mathbf{B})^{-1}\boldsymbol{\varepsilon}_0^*, \tag{3.18}$$

where $(\mathbf{I} + f\mathbf{B})^{-1}$ is the inverse of $(\mathbf{I} + f\mathbf{B})$.

The total stress disturbance, $\langle\boldsymbol{\sigma}\rangle_\Omega$, in the inhomogeneities is given by

$$\langle\boldsymbol{\sigma}\rangle_\Omega = \Delta\boldsymbol{\sigma}_0 + \mathbf{C}\boldsymbol{\gamma}_1 + \Delta\boldsymbol{\sigma}_1 + \mathbf{C}\boldsymbol{\gamma}_2 + \cdots, \tag{3.19}$$

which is rewritten as

$$\langle\boldsymbol{\sigma}\rangle_\Omega = \Delta\boldsymbol{\sigma}_0 - f\mathbf{CS}\boldsymbol{\varepsilon}_0^* + \Delta\boldsymbol{\sigma}_1 - f\mathbf{CS}\boldsymbol{\varepsilon}_1^* + \cdots \tag{3.20}$$

$$= \mathbf{C}\{(\mathbf{S} - \mathbf{I}) - f\mathbf{S}\}(\boldsymbol{\varepsilon}_0^* + \boldsymbol{\varepsilon}_1^* + \cdots), \tag{3.21}$$

using (3.15), (2.11), and similar equations. With (3.17), $\langle\boldsymbol{\sigma}\rangle_\Omega$ is further simplified as

$$\langle\boldsymbol{\sigma}\rangle_\Omega = \mathbf{C}\{(1 - f)\mathbf{S} - \mathbf{I}\}(\mathbf{I} + f\mathbf{B})^{-1}\boldsymbol{\varepsilon}_0^*. \tag{3.22}$$

The total stress disturbance, $\langle\boldsymbol{\sigma}\rangle_M$, in the matrix is similarly calculated as

$$\langle\boldsymbol{\sigma}\rangle_M = \mathbf{C}\boldsymbol{\gamma}_1 + \mathbf{C}\boldsymbol{\gamma}_2 + \cdots$$

$$= -f\mathbf{CS}(\mathbf{I} + f\mathbf{B})^{-1}\boldsymbol{\varepsilon}_0^*, \tag{3.23}$$

from (3.15) and (3.17). The total stress disturbance, averaged over the body, is defined as

$$\langle\boldsymbol{\sigma}\rangle = f\langle\boldsymbol{\sigma}\rangle_\Omega + (1 - f)\langle\boldsymbol{\sigma}\rangle_M, \tag{3.24}$$

which is equal to

$$\langle\boldsymbol{\sigma}\rangle = -f\mathbf{C}(\mathbf{I} + f\mathbf{B})^{-1}\boldsymbol{\varepsilon}_0^*, \tag{3.25}$$

from (3.22) and (3.23).

The overall elastic stiffness, $\bar{\mathbf{C}}$, calculated for the displacement-constrained body is defined by

$$\bar{\mathbf{C}}\boldsymbol{\gamma}_0 = \mathbf{C}\boldsymbol{\gamma}_0 + \langle\boldsymbol{\sigma}\rangle. \tag{3.26}$$

Thus, from (3.25) and (3.7), the overall stiffness is given as

$$\bar{\mathbf{C}} = \mathbf{C}\{\mathbf{I} + f(\mathbf{I} + f\mathbf{B})^{-1}\mathbf{Z}^{-1}(\mathbf{C}^* - \mathbf{C})\}, \tag{3.27}$$

or

$$\bar{\mathbf{C}} = \mathbf{C}\{\mathbf{I} + f[(1 - f)(\mathbf{C}^* - \mathbf{C})\mathbf{S} + \mathbf{C}]^{-1}(\mathbf{C}^* - \mathbf{C})\}, \tag{3.28}$$

from (2.12). The overall stiffness can be shown to be the inverse of the overall compliance, (2.41), after standard manipulations of matrix algebra. That is,

$$\bar{\mathbf{C}}^{-1}\bar{\mathbf{C}} = \mathbf{I}. \tag{3.29}$$

4. Discussion

The two approaches, following the traction-constrained and displacement-constrained conditions, yield the identical elastic constants of a body containing inhomogeneities. The present analysis also gives the same result as the work of Benveniste. First, we will see this point by examining the equivalency equation under the traction-prescribed condition. We will sum up (2.8), (2.13), (2.19) and similar equivalency equations, using (2.28), (2.31) and (2.26). Thus, we have the total equivalency equation expressed as

$$\begin{aligned}\boldsymbol{\sigma}_0 + \langle\boldsymbol{\sigma}\rangle_\Omega &= \mathbf{C}(\mathbf{C}^{-1}\boldsymbol{\sigma}_0 + \mathbf{C}^{-1}\langle\boldsymbol{\sigma}\rangle_M + \mathbf{S}\boldsymbol{\varepsilon}^* - \boldsymbol{\varepsilon}^*)\\ &= \mathbf{C}^*(\mathbf{C}^{-1}\boldsymbol{\sigma}_0 + \mathbf{C}^{-1}\langle\boldsymbol{\sigma}\rangle_M + \mathbf{S}\boldsymbol{\varepsilon}^*).\end{aligned} \tag{4.1}$$

Equation (4.1) is the same as (12) in the paper of Benveniste (1987) when the disturbance in the inhomogeneities is related to $\boldsymbol{\sigma}_0$ and the disturbance in the matrix. $\mathbf{C}^{-1}\langle\boldsymbol{\sigma}\rangle_M$ in (4.1) reads as $\boldsymbol{\varepsilon}^{(1)}$ and $\mathbf{S}\boldsymbol{\varepsilon}^*$ as $\boldsymbol{\varepsilon}^{(2)}$. Since (4.1) is a basis in Benveniste's analysis, the present result must be equal to that given by Benveniste. Second, we can see that the overall elastic constants in the present analysis are identical to those in the paper of Benveniste. The Appendix is a formal proof for this identity.

We may as well trace the meaning of (4.1) to the work of Wakashima *et al.* (1974). They have noted that the average internal stress in an inclusion is given by

$$\langle\boldsymbol{\sigma}\rangle_\Omega = \mathbf{C}\{(1-f)\mathbf{S}\boldsymbol{\varepsilon}^* + f\boldsymbol{\varepsilon}^* - \boldsymbol{\varepsilon}^*\}, \tag{4.2}$$

when there is a large number of inclusions with eigenstrain $\boldsymbol{\varepsilon}^*$ and the inclusions and matrix are equal in the elastic stiffness, $\mathbf{C}$. Equation (4.2) is obtained from (2.3)–(2.6). That is, the stress in an inclusion is calculated by considering that the total strain in the inclusions, $\langle\boldsymbol{\gamma}\rangle_\Omega$, is taken as

$$\langle\boldsymbol{\gamma}\rangle_\Omega = (1-f)\mathbf{S}\boldsymbol{\varepsilon}^* + f\boldsymbol{\varepsilon}^*. \tag{4.3}$$

Using this idea of Wakashima *et al.*, the equivalency condition to determine the stress and strain disturbances in the inhomogeneities is given as

$$\begin{aligned}\boldsymbol{\sigma}_0 + \langle\boldsymbol{\sigma}\rangle_\Omega &= \mathbf{C}\{\mathbf{C}^{-1}\boldsymbol{\sigma}_0 + (1-f)\mathbf{S}\boldsymbol{\varepsilon}^* + f\boldsymbol{\varepsilon}^* - \boldsymbol{\varepsilon}^*\}\\ &= \mathbf{C}^*\{\mathbf{C}^{-1}\boldsymbol{\sigma}_0 + (1-f)\mathbf{S}\boldsymbol{\varepsilon}^* + f\boldsymbol{\varepsilon}^*\},\end{aligned} \tag{4.4}$$

when an inhomogeneity-bearing body is subjected to the average stress of $\boldsymbol{\sigma}_0$. Inserting (2.31) into (4.1) with (2.27), we recover (4.4). The solution of $\boldsymbol{\varepsilon}^*$ satisfying (4.4) is given by (2.27).

We see that (2.41) or, equivalently, (3.28) does not show irregularities at a finite volume fraction of inhomogeneities of which $\mathbf{C}^* \to 0$ or $\mathbf{C}^* \to \infty$. Other advantages of (2.41) and (3.28) have already been discussed by Benveniste (1987). Moreover, the examination of the following example suggests that (2.41) or (3.28) is an intuitively good solution. Let us take a simple case that isotropic spherical inhomogeneities with the shear modulus of μ^* are embedded in an isotropic elastic matrix with the shear modulus of μ and the Poisson ratio of ν. First, we consider the traction-prescribed condition: The average shear stress is given. The shear component of the equivalent eigenstrain is denoted by ε^*. On the basis of the results in Section 2.2, we have

$$\varepsilon^* = \varepsilon_0^* + \varepsilon_1^* + \varepsilon_2^* + \cdots, \tag{4.5}$$

$$\varepsilon_i^* = -fA\varepsilon_{i-1}^*, \tag{4.6}$$

and

$$A = \frac{(\mu^* - \mu)(7 - 5\nu)}{\mu(7 - 5\nu) + \mu^*(8 - 10\nu)}. \tag{4.7}$$

For $\mu^* > \mu$, (4.5) is an alternating geometric series. Thus, in this case, a correction is attempted to partly cancel the overestimated term introduced by the preceding approximation or correction. That is, the ultimate value of ε^* is always between the sums of the series truncated at i and $i + 1$. If it is assumed that the attempted correction is in the right direction, the series gives a good answer to the correct value.

The above argument does not apply for the case of $\mu^* < \mu$. In this case, however, we will take another approach using the displacement prescribed condition. Under this condition, the total equivalent eigenstrain is given as

$$\varepsilon^* = \varepsilon_0^* + \varepsilon_1^* + \varepsilon_2^* + \cdots, \tag{4.8}$$

$$\varepsilon_i^* = -fB\varepsilon_{i-1}^*, \tag{4.9}$$

$$B = \frac{(\mu - \mu^*)(8 - 10\nu)}{\mu(7 - 5\nu) + \mu^*(8 - 10\nu)}, \tag{4.10}$$

by following the results in Section 3. Thus, when $\mu^* < \mu$, the eigenstrain is expressed as an alternating geometric series. Then, an argument similar to that in the preceding paragraph is repeated so that the series of (4.8) converges in the right direction to the right value.

One might argue that a correction factor corresponding to but differing from $-f\mathbf{A}$ or $-f\mathbf{B}$ can be used. It can correct the eigenstrain introduced in the preceding approximation or correction and converge the series of the eigenstrain in the iteration method. However, the fact that the inverse relation of the compliance and stiffness expressed in (3.29) is obtained strongly indicates that the present method is sound in the formulation. Of course, the factors of $-f\mathbf{A}$ and $-f\mathbf{B}$ have not been introduced arbitrarily but have been assigned on the physical basis.

We think that (2.1) and (3.3) are the most fundamental equations which play a crucial role in the present subject. We also think it fair to state that Brown and Stobbs (1971) once derived an equation similar to (2.1) with a different physical understanding of $\langle\boldsymbol{\sigma}\rangle_M$. We have given the reasoning for the validity for these equations because of their central role. Once they are taken for granted, the overall elastic constants are derived using Eshelby's idea of an equivalent inclusion. Equally fundamental are (2.3) and (2.4) which are the stress and strain calculated for a single inclusion, respectively. Only the average quantities appear in the present formulation. Thus, the present method can deal with shapes other than ellipsoids, regarding $\gamma^\infty = \mathbf{S}\boldsymbol{\varepsilon}^*$ as the average total strain in a single inclusion. In this case a numerical solution for the field inside an inclusion might be needed. Sliding inclusions are similarly treated.

5. Concluding Remarks and a Related Problem

Most of the subjects using an average field and referred to by Benveniste are analyses of an elastic field produced by inhomogeneous inclusions with the elastic stiffness $\mathbf{C}^*$ and eigenstrain $\boldsymbol{\varepsilon}^T$. The present successive iteration method can be equally applied to these subjects: The equivalent eigenstrain, $\boldsymbol{\varepsilon}^*$, can be obtained in the series form of

$$\boldsymbol{\varepsilon}^* = \boldsymbol{\varepsilon}_0^* + \boldsymbol{\varepsilon}_1^* + \boldsymbol{\varepsilon}_2^* + \cdots. \tag{5.1}$$

Here, $\boldsymbol{\varepsilon}_0^*$ satisfies the equivalency equation for the stress, $\Delta\boldsymbol{\sigma}_0$, in a single inclusion given as

$$\mathbf{C}(\mathbf{S}\boldsymbol{\varepsilon}_0^* - \boldsymbol{\varepsilon}_0^*) = \mathbf{C}^*(\mathbf{S}\boldsymbol{\varepsilon}_0^* - \boldsymbol{\varepsilon}^T)(= \Delta\boldsymbol{\sigma}_0) \tag{5.2}$$

(zeroth-order approximation). The subsequent approximation procedures are exactly the same as discussed in Section 2.2 and $\boldsymbol{\varepsilon}_i^*$ is given by (2.24) with

$$\boldsymbol{\varepsilon}_0^* = \mathbf{Z}^{-1}\mathbf{C}^*\boldsymbol{\varepsilon}^T. \tag{5.3}$$

Equations (2.27), (2.30), (2.31), (2.32), and (2.34) hold. These are identical to the expressions obtained by Pedersen (1983). The elastic energy, E, per unit volume is written as

$$E = -\tfrac{1}{2}f(1-f)(\boldsymbol{\varepsilon}^T)^t\mathbf{C}(\mathbf{S}-\mathbf{I})\{(1-f)(\mathbf{C}^*-\mathbf{C})(\mathbf{S}-\mathbf{I}) + \mathbf{C}^*\}^{-1}\mathbf{C}^*\boldsymbol{\varepsilon}^T, \tag{5.4}$$

from (2.30), (2.12), (2.18), and (5.3). $(\boldsymbol{\varepsilon}^T)^t$ is the transpose of $\boldsymbol{\varepsilon}^T$. Equation (5.4) is used in the energetic discussion of precipitation.

In conclusion, we have formulated the successive iteration method to analyze an elastic field of an elastically inhomogeneous solid. The method offers a physical basis of the average field approach which has been used in many problems of composite materials.

Acknowledgment

We are grateful to Professor N. Kinoshita, Meiji University, who suggested to us a manipulation method in matrix algebra. This research was partially supported by the Nippon Steel Corporation.

References

Benveniste, Y. (1987), A new approach to the application of Mori–Tanaka's theory in composite materials, *Mech. Materials*, **6**, 147–157.

Brown, L. M. and Stobbs, W. M. (1971), The work hardening of copper–silica, a model based on internal stress, with no plastic relaxation, *Phil. Mag.*, **23**, 1185–1199.

Brown, L. M. (1973), Back stresses, image stresses and work hardening, *Acta Metallurgica*, **21**, 879–885.

Budiansky, B. (1965), On the elastic moduli of some heterogeneous materials, *J. Mech. Phys. Solids*, **13**, 223–227.

Eshelby, J. D. (1957), The determination of the elastic field of an ellipsoidal inclusion and related problems, *Proc. Roy Soc. London*, **A241**, 376–396.

Hashin, Z. and Shtrikman, S. (1963), A variational approach to the theory of the elastic behaviour of multiphase materials, *J. Mech. Phys. Solid*, **11**, 127–140.

Hill, R. (1963), Elastic properties of reinforced solids, *J. Mech. Phys. Solids*, **11**, 357–372.

Hill, R. (1965), A self-consistent mechanics of composite materials, *J. Mech. Phys. Solids*, **13**, 213–222.

Kroner, E. (1958), Berechnung der Elastischen Konstanten des Vielkristalls aus den Konstanten des Einkristalls, *Z. Phys.*, **151**, 504–518.

Mori, T. and Tanaka, K. (1973), Average stress in matrix and average elastic energy of materials with misfitting inclusions, *Acta Metallurgica*, **21**, 571–574.

Mura, T., Furuhashi, R., and Tanaka, K. (1981), Equivalent inclusion method in composite materials, *Proceedings of the Japan–U.S. Conference on Composite Materials*, June, pp. 71–77. Applied Science, New Jersey.

Pedersen, O. B. (1983), Thermoelasticity and plasticity of composites–I, *Acta Metallurgica*, **31**, 1795–1803.

Wakashima, K., Otsuka, M., and Umekawa, S. (1974), Thermal expansions of heterogeneous solids containing aligned ellipsoidal inclusions, *J. Comp. Mater.*, **8**, 391–404.

Appendix

Here we will consider the displacement-prescribed condition and show that (3.28) is identical to equation (14a) in the paper by Benveniste (1987).

$$\bar{\mathbf{C}} = \mathbf{C}\{\mathbf{I} + f[(1-f)(\mathbf{C}^* - \mathbf{C})\mathbf{S} + \mathbf{C}]^{-1}(\mathbf{C}^* - \mathbf{C})\}, \qquad (3.28)$$

$$\mathbf{L}^* = \mathbf{L}_1 + v_2(\mathbf{L}_2 - \mathbf{L}_1)\mathbf{T}[v_1\mathbf{I} + v_2\mathbf{T}]^{-1}, \qquad (14a)$$

with

$$\mathbf{T} = [\mathbf{I} + \mathbf{S}\mathbf{L}_1^{-1}(\mathbf{L}_2 - \mathbf{L}_1)]^{-1}. \qquad (A1.1)$$

Here $\mathbf{C} = \mathbf{L}_1$, $\mathbf{C}^* = \mathbf{L}_2$, $f = v_2$, and $1 - f = v_1$. With these definitions and

$$\mathbf{N} = \mathbf{L}_2 - \mathbf{L}_1, \qquad (A1.2)$$

we rewrite (3.28) as

$$\bar{\mathbf{C}} = \mathbf{L}_1 + \nu_2\mathbf{L}_1(\nu_1\mathbf{NS} + \mathbf{L}_1)^{-1}N$$
$$= \mathbf{L}_1 + \nu_2(\nu_1\mathbf{SL}_1^{-1} + \mathbf{N}^{-1})^{-1}. \quad \text{(A1.3)}$$

Similarly, we rewrite (14a) as

$$\mathbf{L}^* = \mathbf{L}_1 + \nu_2\mathbf{N}(\mathbf{I} + \nu_1\mathbf{SL}_1^{-1}\mathbf{N})^{-1}$$
$$= \mathbf{L}_1 + \nu_2(\mathbf{N}^{-1} + \nu_1\mathbf{SL}_1^{-1})^{-1}, \quad \text{(A1.4)}$$

using (A1.2). Equation (A1.3) is identical to (A1.4). That is,

$$\bar{\mathbf{C}} = \mathbf{L}^*. \quad \text{(A1.5)}$$

Effects of Nonmetallic Inclusions on the Fatigue Strength of Metals

Y. MURAKAMI
Department of Mechanics and Strength of Solids,
Faculty of Engineering, Kyushu University, Fukuoka 812, Japan

Abstract

The equation for predicting the effects of artificial small defects on the fatigue strength of metals is introduced, and it is applied to evaluate quantitatively the effects of nonmetallic inclusions on the fatigue strength of high-strength steels. It is emphasized that the concept that nonmetallic inclusions are virtually equivalent to defects, from the viewpoint of fatigue strength and, more practically, are equivalent to small cracks, is of substantial importance.

It is shown that nonmetallic inclusions cause relatively low-fatigue strength and large scatter of the fatigue strength of steels with high static strength or high hardness. The particular characteristics of distribution of inclusion size, i.e., the distribution of extreme values, are investigated, and the statistics of extreme values is used to estimate the expected maximum size of nonmetallic inclusions. The lower limit of scatter in the fatigue strength of a high-strength steel is obtained by using the prediction equation for small defects together with the expected maximum size of nonmetallic inclusions.

1. Introduction

Nowadays it is well recognized that the problems of small defects and small cracks in fatigue cannot be solved clearly by the classical notch effect theory. The values of ΔK_{th} for long cracks which prescribe the threshold conditions of crack propagation are considered a material constant if the fatigue test is conducted under a constant stress ratio and a moderate fatigue history. However, with decreasing crack size, ΔK_{th} does not remain at a constant value and shows the trend of decrease. Although this phenomenon was indicated by Kitagawa and Takahashi (1976), and many researches on small cracks have been done since that time, the effects of crack size and geometry and also those of material properties have not been made clear.

Recently Murakami and Endo (1986a, b) proposed the equation which enables us to predict quantitatively and systematically the effects of small defects and small cracks on fatigue strength. In the equation, two representative parameters are involved. One is a geometrical parameter $\sqrt{area}$, which is defined by the square root of the projection area of defects or cracks onto the plane perpendicular to the maximum tensile stress. The other is a material parameter, i.e., the Vickers hardness H_v. They demonstrated the availability of the equation by extensive experimental data with H_v ranging from 70 to 740.

On the other hand, it is quite common that a mark named fish-eye is observed on the fatigue fracture surfaces of high-strength steels. Usually, a nonmetallic inclusion is found at the center of a fish-eye. This evidences that a nonmetallic inclusion is a crucial cuase of the fatigue fracture of high-strength steel. The role of nonmetallic inclusion in fatigue strength has been discussed from various aspects, both in metallurgy and solid mechanics. Murakami *et al.* (1988) and Murakami and Usuki (1988) proposed a concept that a nonmetallic inclusion should be considered to be equivalent to a defect, and consequently, be equivalent to a crack from the viewpoint of fatigue strength. Based on this concept, they suggested that the inclusion problems in fatigue could be treated quantitatively.

The present paper, based on the same concept, contrasts the effects of artificial small defect with those of nonmetallic inclusions and shows the equivalence of both effects. It will also be shown that the scattered distribution of inclusion size and location is the main cause of the scatter of fatigue strength of high-strength steels. Afterwards, the prediction method of the lower limit of scatter of fatigue strength will be shown.

2. Effects of Artificial Small Defects on Fatigue Strength

2.1. *Shapes and Dimension of Artificial Small Defects*

For the establishement of the law of dependence of ΔK_{th} on crack size, specimens containing sufficiently small cracks or defects must be prepared.

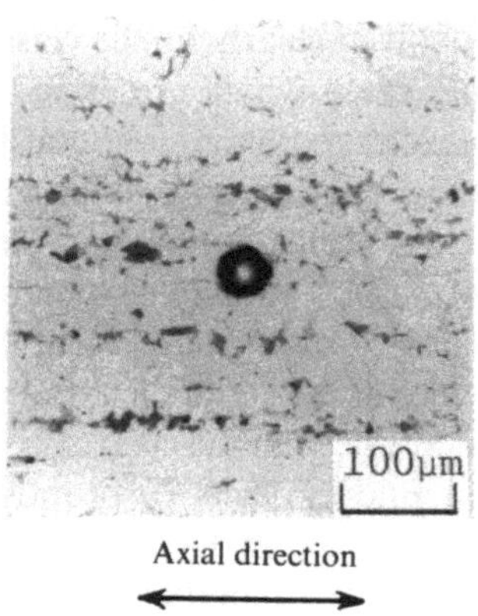

FIG. 1. Artificial small hole. Diameter = 40 μm and depth = 40 μm.

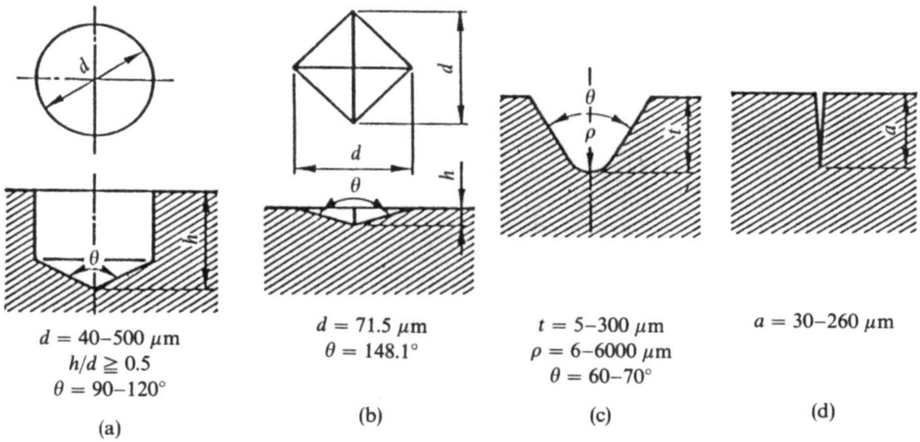

FIG. 2. Various artificial defects. (a) Hole; (b) Vickers hardness indentation; (c) notch; and (d) circumferential crack.

Figure 1 shows a hole with 40 μm diameter and 40 μm depth drilled on the surface of a low-carbon steel. The size of the hole is approximately equal to the grain size of the low-carbon steel. The propagation behaviors of the cracks emanating from this kind of artificial hole may be considered to be quite similar to those of crack nucleated at slip bands and grain boundaries.

Figure 2 shows other artificial defects considered in the present paper. Since defects occupy, in general, a three-dimensional domain, a representative geometrical parameter must be introduced in order to evaluate the effect of various defects in a unified manner.

Murakami and Endo (1983, 1986) concluded, on the basis of the three-dimensional crack analyses by Murakami and Nemat-Nasser (1983) and Murakami (1985), that the most appropriate geometrical parameter for unifying the effect of defects in fatigue is the square root of the projection area of defects, which is denoted by $\sqrt{\text{area}}$.

2.2. *Threshold Condition of Fatigue Limit*

Figure 3 shows the crack emanating from an artificial hole observed at the fatigue limit. This crack is a so-called nonpropagating crack which started from the hole edge in the early stages of stress cycle and afterward stopped propagating under the same stress level. Figure 4 illustrates the sectional view of such nonpropagating cracks. These experimental facts indicate that the fatigue limit is not the critical condition for crack initiation, but the threshold condition for nonpropagation of a crack emanating from the defect.

These facts are extremely important for the quantitative and systematic

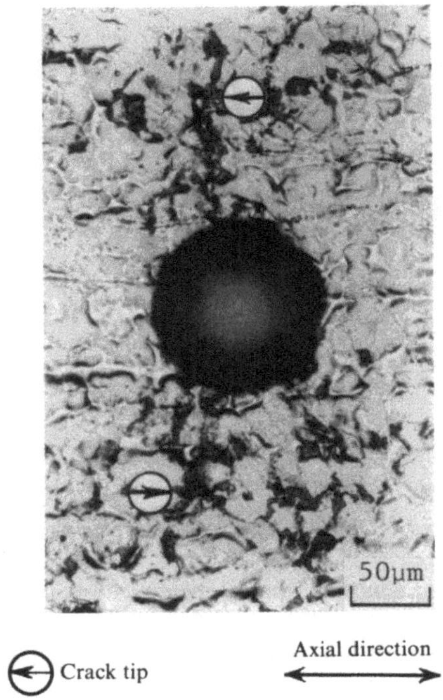

FIG. 3. Nonpropagating cracks emanating from an artificial defect and stopped propagating. Diameter of the hole = 100 μm and depth = 100 μm. The material is SAE10L45.

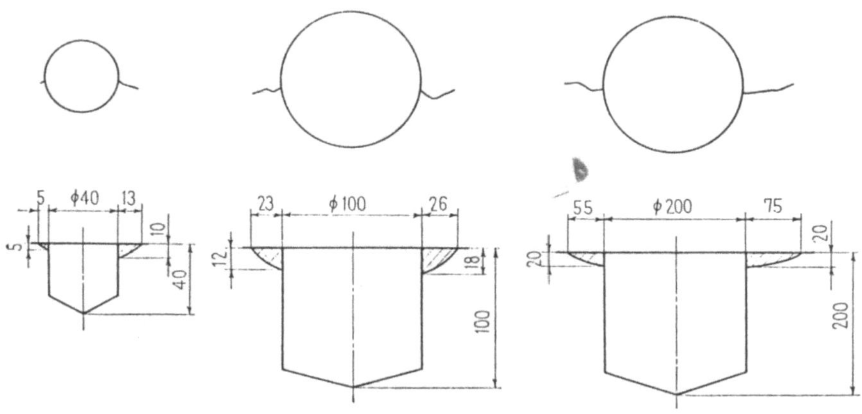

FIG. 4. Interior shapes and configurations of nonpropagating cracks. The material is 0.13% C steel. (Murakami and Endo, 1980).

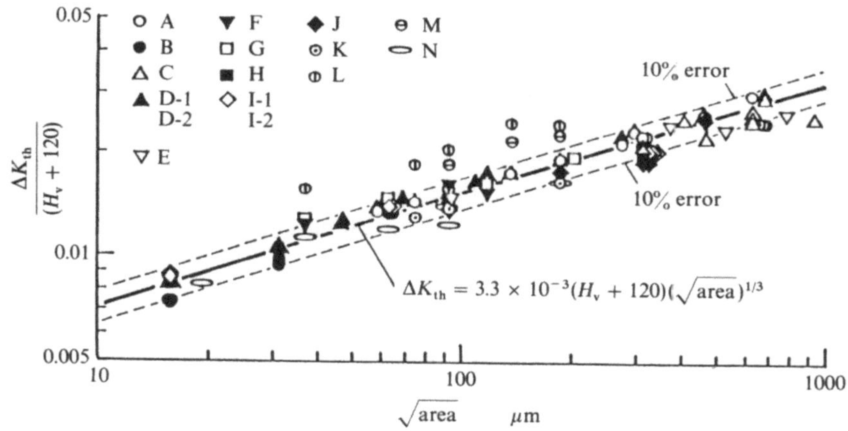

FIG. 5. Relationship between $\Delta K_{th}/(H_v + 120)$ and $\sqrt{area}$. Letters correspond to the materials listed in Table 1. (Murakami and Endo, 1986).

TABLE 1. The materials of the data plotted in Figs. 5 and 6

Material	H_V	Defect
A: S10C (annealed)	120	Notch Hole
B: S30C (annealed)	153	Notch
C: S35C (annealed)	160	Notch Hole
D-1: S45C (annealed)	180	Notch
D-2: S45C (annealed)	170	Hole
E: S50C (annealed)	177	Notch Crack
F: S45C (quenched)	650	Hole
G: S45C (quenched and tempered)	520	Hole
H: S50C (quenched and tempered)	319	Notch
I-1: S50C (quenched and tempered)	378	Notch
I-2: S50C (quenched and tempered)	375	Notch
J: 70/30 brass	70	Notch Hole
K: Aluminum alloy (2017-T4)	114	Hole
L: Stainless steel (SUS 603)	355	Hole
M: Stainless steel (YUS 170)	244	Hole
N: Maraging steel	720	Vickers hardness indentation, hole and notch

evaluation of the effect of small defects, small cracks, and nonmetallic inclusions on fatigue strength. Namely, the condition of the nonpropagation of cracks from the defect at the fatigue limit can be regarded as a crack problem rather than as a stress concentration problem of a notch in fatigue, because the final stress field in the vicinity of a nonpropagating crack from a defect is not affected by the three-dimensional volume of the defect, but is prescribed by the magnitude of the projection area of the defect. This is the reason why $\sqrt{\text{area}}$ must be introduced for the geometrical parameter.

2.3. Unifying Expression for the Fatigue Limit of Specimens Containing Small Defects

Murakami and Endo (1986) chose the Vickers hardness H_v as the most appropriate material parameter which reflects the fatigue threshold resistance of microstructures. They investigated in detail the dependence of ΔK_{th} on H_v together with the $\sqrt{\text{area}}$ of the defects.

Figure 5 demonstrates the successful unification of the extensive experimental data by H_v and $\sqrt{\text{area}}$. Figure 6 shows another expression of Fig. 5.

The equations which approximate most nearly the experimental data available at present are given as follows:

$$\Delta K_{th} = 3.3 \times 10^{-3}(H_v + 120)(\sqrt{\text{area}})^{1/3}, \tag{2.1}$$

$$\sigma_w = 1.43(H_v + 120)/(\sqrt{\text{area}})^{1/6} \tag{2.2}$$

$$\Delta K_{th}: \text{MPa}\sqrt{m}, \quad \sigma_w: \text{MPa}, \quad H_v: \text{kgf/mm}^2, \quad \sqrt{\text{area}}: \mu\text{m}$$

(*prediction equation for surface defects*).

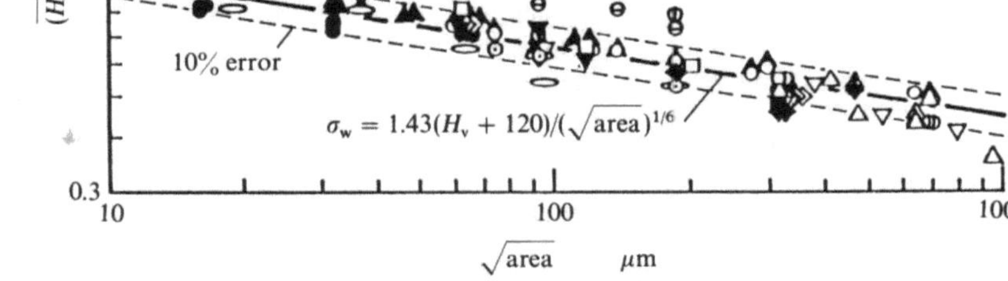

FIG. 6. Relationship between $\sigma_w/(H_v + 120)$ and $\sqrt{\text{area}}$. The same data as Fig. 5.

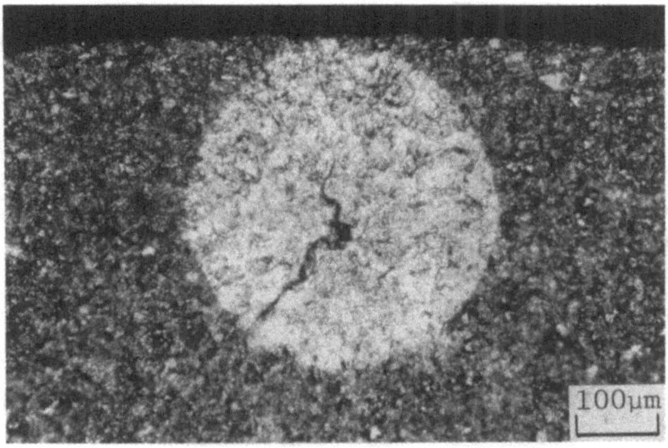

FIG. 7. An example of fish-eye observed on the fatigue fracture surface. The material is SAE10L45.

3. Effects of Nonmetallic Inclusions on the Fatigue Strength of Metals

3.1. Fish-Eye Mark on the Fatigue Fracture Surface—Inclusions Remaining on the Fracture Surface

It is quite common that a mark named fish-eye is observed on the fatigue fracture surface of high-strength steels. Figure 7 shows a typical example of a fish-eye. Usually a nonmetallic inclusion exists at the center of a fish-eye but it is sometimes not found when it is dropped out of the fracture surface. This implies that the fatigue fracture origin of high-strength steels is mainly at the nonmetallic inclusion. Various nonmetallic inclusions such as MnS, Al_2O_3, SiO_2, and TiN having various shapes and sizes have been observed (Ransom, 1954; Frith, 1955; Ramsey and Kedzie, 1957; Cummings *et al.*, 1958; Kawada *et al.*, 1963; Yokobori and Nanbu, 1966; and others).

Figure 8 shows examples of nonmetallic inclusions at the fatigue fracture origin of an ultra low-oxygen suspension spring steel tested by Saito and Ito (1985).

Somewhere on the interface between an inclusion and a matrix, at least one stress component becomes higher than the externally loaded remote stress, regardless of the difference between the elastic constants of the inclusion and the matrix. The internal stresses, which are induced at the interface by the difference in the thermal expansion coefficient and also by the transformation strain under the cooling process of steel making, have tensile components in either a radial or tangential direction at the interface and, therefore, these stresses are superimposed on the stresses caused by external stresses to make the interface or its vicinity an origin of fatigue-crack initiation. (The characteristics of these stresses have been discussed in detail by many researchers (Eshelby, 1957; Mura, 1987; and others).)

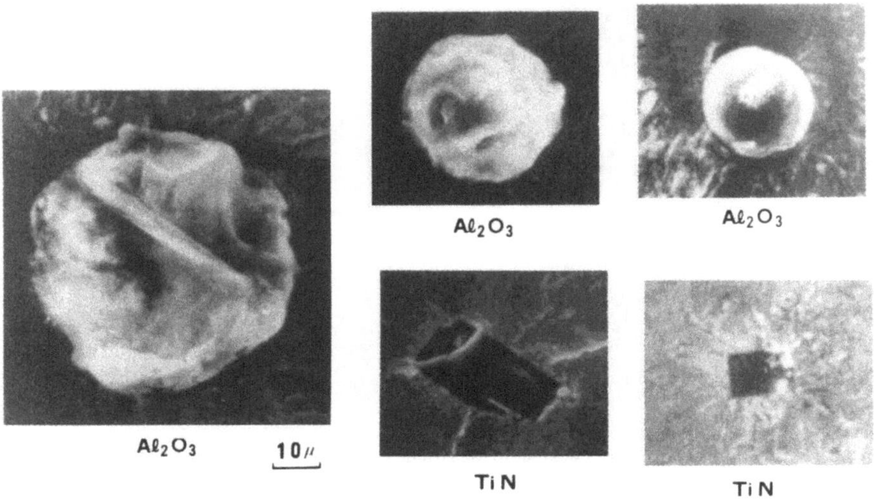

(a) lower $H_v = 534$ (b) middle $H_v = 614$ (c) upper $H_v = 588$

FIG. 8. Nonmetallic inclusions observed at the fatigue fracture origin. Pictures show typical inclusions contained in each group of specimens and the fatigue limit of each group is at (a) the lower point in the scatter band, (b) the middle point in the scatter band, and (c) the upper point in the scatter band (Saito and Ito, 1985).

Figure 9 illustrates examples of stress concentration due to remote stresses and internal stresses induced by the difference in the thermal expansion coefficient.

After a crack initiates along the interface between an inclusion and a matrix, or an inclusion itself is cracked, both the externally induced stresses and internal stresses in the inclusion are relieved, and accordingly the inclusion becomes virtualy equivalent to a hole or defect. From this consideration it is reasonable to use the equations described in Section 2 for the prediction of the fatigue strength of high-strength steels. In this case, it is natural to define $\sqrt{\text{area}}$ by the square root of the prediction area of an inclusion at the fracture origin.

Table 2 shows the sketch of the shape of inclusions (MnS) at the fracture origin and the comparison of the experimental results with the fatigue limits predicted by (3.2) which is modified for interior inclusions from (2.2).

3.2. Distribution of Inclusion Size and Scatter of the Fatigue Strength of High-Strength Steels

The fact that a nonmetallic inclusion is the fatigue fracture origin of high-strength steels means that the fatigue strength is influenced by the inclusion size ($\sqrt{\text{area}}$) and the location of the inclusion. It also follows that nonmetallic

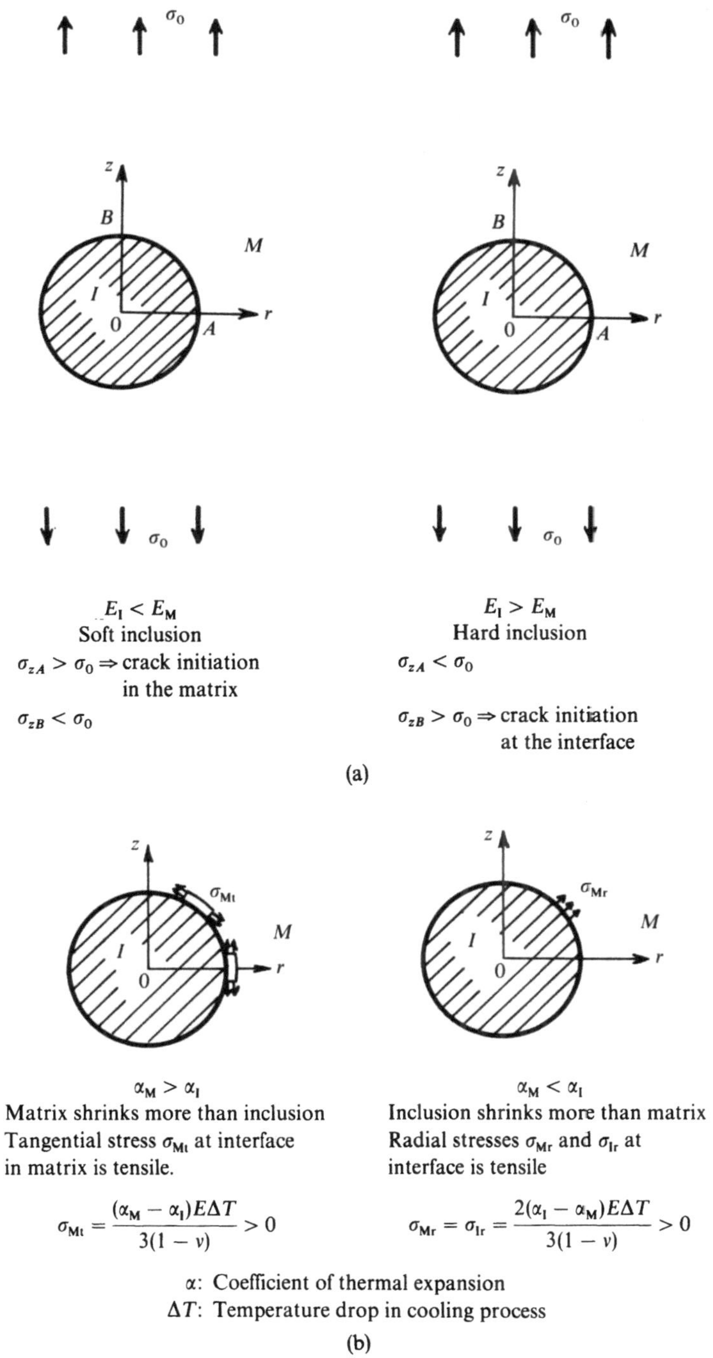

FIG. 9. Stress around a spherical inclusion. I: inclusion, M: matrix. (a) Stress around a spherical inclusion due to remote tensile stress σ_0. (b) Stress around a spherical inclusion due to the difference of thermal expansion. Young's modulus $E_I = E_M$ and Poisson's ratio $\nu_I = \nu_M$ are assumed.

TABLE 2. Size and location of the inclusions and fatigue limit predicted by (3.2).

Material and H_v	Nominal stress at surface σ (MPa)	Cycles to failure N_f ($\times 10^4$)	Inclusion size area (μm^2)	Distance from surface h (μm)	Shape of inclusions	Nominal stress at inclusion σ' (MPa)	Fatigue limit predicted by (3.2) σ'_w (MPa)	σ'/σ'_w
	981	254.36	962	316		907	752	1.21
	981	120.05	1343	370		895	731	1.22
	932	429.54	1154	390		846	740	1.14
Bearing steel	883	1280.50	962	120		858	752	1.14
	981	192.51	1343	38		971	731	1.33
$H_v \cong 734$	932	296.64	1501	420		839	724	1.16
	912	134.21	808	63		898	763	1.18
	883	277.34	416	14		879	806	1.09
	883	729.50	857	295		821	759	1.08

inclusion causes large scatter of the fatigue strength of high-strength steels. Since it is almost impossible to know inclusion sizes and locations in advance of fatigue testing, it is also impossible to predict the fatigue strength of individual specimens and machine components.

However, designers involved in practical work naturally require some information on fatigue strength. Considering these situations, Murakami *et al.* (1988) and Murakami and Usuki (1988) proposed a method to predict the lower limit of the scatter of fatigue strength on the basis of the statistical distribution of the extreme values of inclusion size ($\sqrt{\text{area}}$). In this paper, the scatter band of the fatigue strength of a high-strength steel SKH51 was predicted and was compared with the experimental results. For the prediction (2.1) and (2.2) for surface inclusion and the following equations, which were modified for *interior inclusions*, were used:

$$\Delta K_{th} = 2.77 \times 10^{-3}(H_v + 120)(\sqrt{\text{area}})^{1/3}, \tag{3.1}$$

$$\sigma_w = 1.56(H_v + 120)/(\sqrt{\text{area}})^{1/6}, \tag{3.2}$$

where $\sqrt{\text{area}}$ is the square root of the projection area of the inclusions.

Figure 10 shows the cumulative probability of the extreme values of nonmetallic inclusions for SKH51. The points in Fig. 10 indicate the value of $\sqrt{\text{area}}$ for a nonmetallic inclusion found on a fracture surface of 34 samples (specimens) under tension–compression fatigue. In this case, the unit volume for the calculation of the return period T is the volume of a test part of one specimen. Therefore, Fig. 10 can be used for the estimation of the expected maximum size $\sqrt{\text{area}_{max}}$ of inclusions contained in more specimens (say N

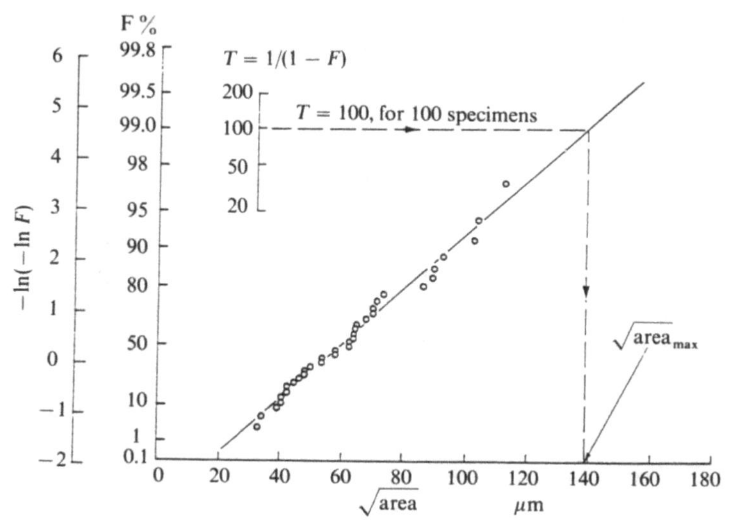

FIG. 10. Cumulative frequency of the inclusion size $\sqrt{\text{area}_{max,j}}$ of SKH-51. The unit test volume V_0, i.e., the volume of one specimen is 1272 mm^3).

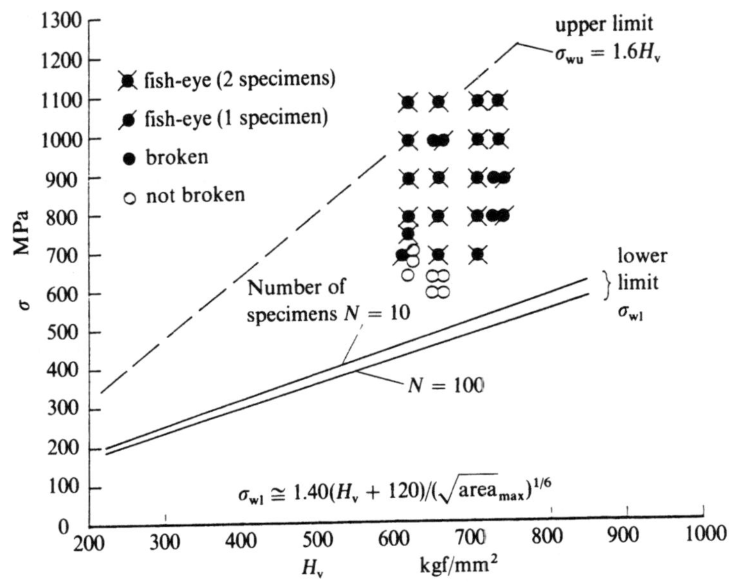

FIG. 11. Comparison between the experimental results and the lower limit of the fatigue strength which was predicted using (2.2) and the expected maximum size of the inclusion. The constant 1.43 in (2.2) is modified to 1.40, considering that the effect of the maximum inclusion becomes maximum when it exists just below the surface and, therefore, the effective value of $\sqrt{area}$ should be modified.

specimens) than those in Fig. 10. Using $\sqrt{area_{max}}$ and H_v of the matrix, the lower limit of the fatigue strength of N specimens can be predicted by (2.2).

In Fig. 11, the predicted lower limit of the fatigue strength is compared with the scattered value of the experimental data. The upper limit of the fatigue strength is obtained by an empirical equation.

4. Conclusions

(1) A unifying equation for the fatigue limit of specimens containing small defects was introduced and was applied to predict the fatigue limit of specimens containing nonmetallic inclusions. The equation includes two parameters. One is a geometrical parameter of a nonmetallic inclusion, the square root of the projection area of an inclusion $\sqrt{area}$, and the other is a material parameter, the Vickers hardness H_v. The predictions were in good agreement with the experimental results.

(2) The concept that a nonmetallic inclusion is virtually equivalent to a stress-free defect in problems of metal fatigue is of crucial importance. This concept clearly explains various problems caused by nonmetallic inclusions, such as the problems of relatively low fatigue strength and the large scatter of fatigue strength of statically high-strength steels.

(3) Combining the unifying equation with a distribution of statistics, i.e., the extreme values of inclusion size, the lower limit of fatigue strength of many specimens, machine elements, and structures can be predicted.

References

Cummings, H. N., Stulen, F. B., and Schulte, W. C. (1958), Tentative fatigue reduction factors for silicate-type inclusions in high-strength steels, *Proc. Amer. Soc. Test. Mater.*, **58**, 505–514.

Eshelby, J. D. (1957), The determination of the elastic field of an ellipsoidal inclusion, and related problems, *Proc. Roy. Soc. London*, **A241**, 376–396.

Frith, P. H. (1955), Fatigue tests on rolled alloy steels made in electric and open-hearth furnaces, *J. Iron Steel Inst.*, **180**, 26–38.

Kawada, Y., Nakazawa, H., and Kodama, S. (1963), The effects of the shapes and the distributions of inclusions on the fatigue strength of bearing steels in rotary bending, *Trans. Japan Soc. Mech. Engrs.*, **29**, 1674–1683.

Kitagawa, H. and Takahashi, S. (1976), Applicability of fracture mechanics to very small cracks or cracks in the early stage, *Proceedings of the 2nd International Conference on the Mechanical Behavior of Materials*, Boston, U.S.A., pp. 627–631.

Mura, T. (1987), *Micromechanics of Defects in Solids*, 2nd ed., Martinus Nijhoff, Dordrecht.

Murakami, Y. and Endo, T. (1980), Effects of small defects on fatigue strength of metals, *Int. J. Fatigue*, **2**, 23–30.

Murakami, Y. and Endo, M. (1983), Quantitative evaluation of fatigue strength of metals containing various small defects or cracks, *Engng. Frac. Mech.*, **17**, 1–15.

Murakami, Y. and Nemat-Nasser, S. (1983), Growth and stability of interacting surface flaws of arbitrary shape, *Engng. Frac. Mech.*, **17**, 193–210.

Murakami, Y. (1985), Analysis of stress intensity factors of modes I, II and III for inclined surface cracks of arbitrary shape, *Engng. Frac. Mech.*, **22**, 101–114.

Murakami, Y. and Endo, M. (1986a), Effects of hardness and crack geometry on ΔK_{th} of small cracks, *J. Soc. Mater. Sci., Japan*, **35**, No. 395, 911–917.

Murakami, Y. and Endo, M. (1986b), Effects of hardness and crack geometry on ΔK_{th} of small cracks emanating from small defects, in *The Behaviour of Short Fatigue Cracks*, EGF Pub. 1, edited by K. J. Miller, and E. R. de los Rios, Mechanical Engineering Publications, London, pp. 275–293.

Murakami, Y., Kodama, S., and Konuma, S. (1988), Quantitative evaluation of effects of nonmetallic inclusions on fatigue strength of high-strength steel, *Trans. Japan Soc. Mech. Engrs.*, **54**, No. 500, 688–696.

Murakami, Y. and Usuki, H. (1989), Prediction of fatigue strength of high-strength steels based on statistical evaluation of inclusion size, *Trans. Japan Soc. Mech. Engrs.*, **55**, No. 510, 213–221.

Ramsey, P. W. and Kedzie, D. P. (1957), Prot fatigue study of an aircraft steel in the ultra high-strength range, *J. Metals*, 401–407.

Ransom, J. T. (1954), The effect of inclusions on the fatigue strength of SAE4340 steels, *Trans. ASM*, **46**, 1254–1269.

Saito, M. and Itoh, Y., (1975), Some properties of ultra clean spring steel, *J. Spring, Japan Soc. Spring*, **30**, 11–19.

Yokobori, T. and Nanbu, M. (1966), Fatigue crack propagation in the high hardened steel, *Rep. Res. Inst. Strength Frac. Mater., Tohoku Univ.*, **2**, No. 2, 29–44.

Elastic Solids with Microdefects

S. NEMAT-NASSER and M. HORI
Center of Excellence for Advanced Materials,
Department of Applied Mechanics and Engineering Sciences,
University of California, San Diego, La Jolla, CA 92093, U.S.A.

Abstract

For a "representative volume element" (RVE) consisting of an elastic (linear or nonlinear) matrix and microdefects, relations between the overall stress and strain potentials and the corresponding local quantities are developed and discussed. Results are specialized to RVE's with a linearly elastic matrix. Illustrations are given for microdefects consisting of cavities, cracks, and elastic inclusions, using three averaging schemes: a scheme based on noninteracting dilute distribution of defects, the self-consistent method, and the differential scheme.

1. Introduction

In micromechanics the concept of "representative volume element" (RVE) is used to estimate the constitutive properties at a continuum material point, in terms of the microstructure and microconstituents that comprise that point and its infinitesimal neighborhood (Hashin, 1983) i.e., *to obtain the continuum constitutive properties in terms of the properties and structure of the microconstituents.* These constitutive properties, often expressed as *constitutive relations*, are then used in the balance equations to calculate the overall response of the continuum mass to applied loads and prescribed boundary data. In this approach, it is therefore required to obtain the overall averaged properties of the RVE, when subjected to the boundary data corresponding to the *uniform* fields in the continuum infinitesimal material neighborhood which the RVE is aimed to represent.

In this paper, attention is focused on the modeling of a class of materials which undergoes only small deformation, and therefore can be studied within the framework of the usual infinitesimally small deformation theory. This class includes many advanced ceramics and ceramic composites, cermets, cementitious and related materials. We consider an RVE which consists of an elastic (linear or nonlinear) matrix material containing microcavities, microcracks, microinclusions, and other possible defects. Assuming the matrix material admits stress and strain potentials, the corresponding stress and strain potentials for the overall response of the RVE are obtained, and the relation between them established. Then the results are specialized to the case when the matrix material of the RVE is linearly elastic. The overall elasticity and compliance tensors of the RVE are calculated for several practically important cases involving microcavities, microcracks, and microinclusions. Three averaging schemes are illustrated; these are: a scheme based on the noninteracting dilute distribution of defects, the self-consistent method (Hershey, 1954; Kröner, 1958; Budianksy, 1965; Hill, 1965), and the differential scheme (Roscoe, 1952, 1973; McLaughlin, 1977; Hashin, 1988).

2. Preliminaries

We refer to quantities at the local *microlevel*, i.e., within the RVE, as *microquantities*. For example, the stress and strain fields within the RVE are called *microstrain* and *microstress fields* and are denoted by $\boldsymbol{\varepsilon} = \boldsymbol{\varepsilon}(\mathbf{x})$ and $\boldsymbol{\sigma} = \boldsymbol{\sigma}(\mathbf{x})$, where $\mathbf{x}$ defines the position of the material point in the RVE. The overall or continuum quantities are called the *macroquantities*, and are denoted by uppercase letters. For example, the *macrostrain* and *macrostress fields* are $\mathbf{E} = \mathbf{E}(\mathbf{X}, t)$ and $\boldsymbol{\Sigma} = \boldsymbol{\Sigma}(\mathbf{X}, t)$, where $\mathbf{X}$ is the position of the continuum macroelement, and t is time.

For an RVE with volume V bounded by ∂V, the average strain and stess are

$$\bar{\boldsymbol{\varepsilon}} \equiv \langle \boldsymbol{\varepsilon} \rangle \equiv \frac{1}{V} \int_V \boldsymbol{\varepsilon}(\mathbf{x})\, dV, \tag{2.1a}$$

$$\bar{\boldsymbol{\sigma}} \equiv \langle \boldsymbol{\sigma} \rangle \equiv \frac{1}{V} \int_V \boldsymbol{\sigma}(\mathbf{x})\, dV. \tag{2.2a}$$

From the strain-displacement relations and the equilibrium (no body forces), it is easy to show that

$$\bar{\boldsymbol{\varepsilon}} = \frac{1}{V} \int_{\partial V} \tfrac{1}{2}(\mathbf{v} \otimes \mathbf{u} + \mathbf{u} \otimes \mathbf{v})\, dS, \tag{2.1b}$$

$$\bar{\boldsymbol{\sigma}} = \frac{1}{V} \int_{\partial V} (\mathbf{v} \cdot \boldsymbol{\sigma}) \otimes \mathbf{x}\, dS, \tag{2.2b}$$

where $\mathbf{v}$ is the exterior unit normal on ∂V, $\mathbf{u}$ is the displacement vector, and $\otimes$ denotes the dyadic product, i.e., if $\mathbf{v} = v_i \mathbf{e}_i$ and $\mathbf{w} = w_i \mathbf{e}_i$, then $\mathbf{v} \otimes \mathbf{w} = v_i w_j \mathbf{e}_i \otimes \mathbf{e}_j$, where $\mathbf{e}_i$ are the unit base vectors, $i, j = 1, 2, 3$, and the summation convention is used. It is seen from (2.1) and (2.2) that the average stress and strain are completely defined by the boundary data of the RVE (Hill, 1967).

In general, at a typical point $\mathbf{X}$ in the continuum, at a fixed instant t, the values of the macrostrain and macrostress tensors, $\mathbf{E}$ and $\boldsymbol{\Sigma}$, can be determined by the average strain and stress $\bar{\boldsymbol{\varepsilon}}$, and $\bar{\boldsymbol{\sigma}}$, over the RVE which represents the corresponding macroelement. In micromechanics it is assumed that $\mathbf{E}$ and $\boldsymbol{\Sigma}$

are equal to $\bar{\boldsymbol{\varepsilon}}$ and $\bar{\boldsymbol{\sigma}}$

$$\mathbf{E} = \bar{\boldsymbol{\varepsilon}}, \tag{2.3a}$$

$$\boldsymbol{\Sigma} = \bar{\boldsymbol{\sigma}}. \tag{2.3b}$$

Conversely, the macrostress and macrostrain tensors, $\boldsymbol{\Sigma}$ and $\mathbf{E}$, provide the uniform traction or linear displacement boundary data for the RVE. Hence, when the traction boundary data for the RVE are regarded to be prescribed, we have

$$\mathbf{v} \cdot \boldsymbol{\sigma} = \mathbf{t}^0 = \mathbf{v} \cdot \boldsymbol{\Sigma} \quad \text{on } \partial V, \tag{2.4a}$$

and when the displacements are assumed to be prescribed on ∂V, we set

$$\mathbf{u} = \mathbf{u}^0 = \mathbf{x} \cdot \mathbf{E} \quad \text{on } \partial V. \tag{2.4b}$$

Furthermore, when thermal effects are also of interest, the value of the macrotemperature, Θ, at the considered macroelement must equal the average temperature $\bar{\theta}$ over the RVE

$$\Theta = \bar{\theta} \equiv \langle \theta \rangle. \tag{2.5}$$

In general, the response of the macroelement characterized by, for example, relations among macrostrain $\mathbf{E}$, macrostress $\boldsymbol{\Sigma}$, and macrotemperature Θ, will be inelastic and history-dependent, even if the microconstituents of the corresponding RVE are elastic. This is because, in the course of deformation, flaws, microcracks, cavities, and other microdefects develop within the RVE and the microstructure of the RVE changes with changes in the overall applied loads. Therefore, the stress–strain relations for the macroelements must, in general, include additional parameters which describe the current microstructure of the corresponding RVE. We consider a class of materials whose microconstituents are elastic (linear or nonlinear) and therefore, the inelastic reponse of its macroelements stems from the generation and evolution of defects and hence, microstructural changes. For a typical macroelement we denote the current state of its microstructure, collectively, by $\mathbf{S}$, which may stand for a set of parameters, scalar or possibly tensorial, that completely defines the microstructure. For example, if the microdefects are penny-shaped cracks, $\mathbf{S}$ will stand for the sizes, orientations, and distribution of these cracks. The matrix material is elastic, and the inelasticity is produced by the growth of the cracks. If there is no change in the microstructure, e.g., no crack growth, the response of the macroelement will be elastic. Hence, we introduce the Helmholtz free energy

$$\Phi = \Phi(\mathbf{E}, \Theta; \mathbf{S}), \tag{2.6a}$$

and observe that, at *constant* $\mathbf{S}$

$$\boldsymbol{\Sigma} = \frac{\partial \Phi}{\partial \mathbf{E}}, \tag{2.6b}$$

$$H = -\frac{\partial \Phi}{\partial \Theta}, \tag{2.6c}$$

where H is the macroentropy. Similarly, the strain potential

$$\Psi = \Psi(\boldsymbol{\Sigma}, \Theta; \mathbf{S}) \tag{2.7a}$$

is introduced through the Legendre transformation

$$\Phi + \Psi = \boldsymbol{\Sigma} : \mathbf{E}, \tag{2.7b}$$

with the result that, at *constant* **S**, we have

$$\mathbf{E} = \frac{\partial \Psi}{\partial \boldsymbol{\Sigma}}, \tag{2.7c}$$

$$H = \frac{\partial \Psi}{\partial \Theta}. \tag{2.7d}$$

We now seek to express the potential functions Φ and Ψ in terms of the volume averages of the stress and strain potentials of the microconstituents.

Since the material within the RVE is assumed to be elastic, it admits a stress potential, $\phi = \phi(\mathbf{x}, \boldsymbol{\varepsilon}, \theta)$, and a strain potential, $\psi = \psi(\mathbf{x}, \boldsymbol{\sigma}, \theta)$, such that

$$\boldsymbol{\sigma} = \frac{\partial \phi}{\partial \boldsymbol{\varepsilon}}, \tag{2.8a}$$

$$\boldsymbol{\varepsilon} = \frac{\partial \psi}{\partial \boldsymbol{\sigma}}, \tag{2.8b}$$

$$\phi + \psi = \boldsymbol{\sigma} : \boldsymbol{\varepsilon}. \tag{2.8c}$$

We consider the cases of prescribed boundary tractions and prescribed boundary displacements for the RVE separately, as follows, *assuming a uniform constant temperature and a fixed microstructure for the RVE; hence, the dependence on* Θ *and* **S** *will not be displayed explicitly.*

2.1. *Stress Potential*

For the prescribed *constant* macrostrain **E**, the *variable* microstrain and microstress fields in the RVE are

$$\boldsymbol{\varepsilon} = \boldsymbol{\varepsilon}(\mathbf{x}; \mathbf{E}), \tag{2.9a}$$

$$\boldsymbol{\sigma} = \boldsymbol{\sigma}(\mathbf{x}; \mathbf{E}), \tag{2.9b}$$

where the argument **E** emphasizes that the *displacement* boundary data are prescribed through (2.4b). Hence, $\mathbf{E} = \langle \boldsymbol{\varepsilon}(\mathbf{x}; \mathbf{E}) \rangle$, and

$$\phi = \phi(\mathbf{x}, \boldsymbol{\varepsilon}(\mathbf{x}, \mathbf{E})) = \phi^E(\mathbf{x}; \mathbf{E}), \tag{2.10}$$

where the superscript E on ϕ emphasizes the fact that the stress potential is associated with the prescribed macrostrain **E**.

Now consider an infinitesimally small variation $\delta\mathbf{E}$ in the macrostrain, which produces

$$\delta\boldsymbol{\varepsilon}(\mathbf{x};\mathbf{E}) = \delta E_{ij}\frac{\partial \boldsymbol{\varepsilon}}{\partial E_{ij}}(\mathbf{x};\mathbf{E}). \tag{2.11a}$$

Then, we have

$$\langle \boldsymbol{\sigma} : \delta\boldsymbol{\varepsilon}\rangle = \left\langle \left(\frac{\partial \phi}{\partial \boldsymbol{\varepsilon}}(\mathbf{x};\boldsymbol{\varepsilon})\right) : \left(\delta E_{ij}\frac{\partial \boldsymbol{\varepsilon}}{\partial E_{ij}}(\mathbf{x};\mathbf{E})\right)\right\rangle = \left\langle \delta\mathbf{E} : \left(\frac{\partial \phi^E}{\partial \mathbf{E}}(\mathbf{x};\mathbf{E})\right)\right\rangle$$

$$= \left(\frac{\partial}{\partial \mathbf{E}}\langle \phi^E\rangle\right) : \delta\mathbf{E}. \tag{2.11b}$$

It now follows that

$$\langle \boldsymbol{\sigma}(\mathbf{x};\mathbf{E})\rangle = \frac{\partial}{\partial \mathbf{E}}\langle \phi^E\rangle. \tag{2.12}$$

We therefore *define* the macrostress potential and the corresponding macrostress by

$$\Phi^E \equiv \Phi^E(\mathbf{E}) \equiv \langle \phi^E\rangle \equiv \frac{1}{V}\int_V \phi^E(\mathbf{x};\mathbf{E})\,dV, \tag{2.13a}$$

$$\boldsymbol{\Sigma}^E \equiv \langle \boldsymbol{\sigma}(\mathbf{x};\mathbf{E})\rangle, \tag{2.13b}$$

and conclude that (Hutchinson, 1987)

$$\boldsymbol{\Sigma}^E = \frac{\partial \Phi^E}{\partial \mathbf{E}}, \tag{2.13c}$$

where the superscript E on $\boldsymbol{\Sigma}$ emphasizes that $\boldsymbol{\Sigma}^E$ is the *average stress produced by the constant macrostrain* $\mathbf{E}$. Note that the integral in (2.13a) depends on the current microstructure and hence on $\mathbf{S}$. For example, $\phi^E = 0$ in cavities and cracks. As cavities and cracks grow, local strains (microstrains) change. Hence, $\Phi^E = \langle \phi^E\rangle$ changes. We express this by writing

$$\Phi^E = \Phi^E(\mathbf{E}, \Theta; \mathbf{S}), \tag{2.13d}$$

which also includes the macrotemperature.

2.2. Strain Potential

With uniform microtemperature θ and fixed microstructure $\mathbf{S}$, let the RVE be subjected to uniform boundary tractions defined through a *constant* macrostress $\boldsymbol{\Sigma}$. The microstrain and microstress fields may be expressed as

$$\boldsymbol{\varepsilon} = \boldsymbol{\varepsilon}(\mathbf{x};\boldsymbol{\Sigma}), \tag{2.14a}$$

$$\boldsymbol{\sigma} = \boldsymbol{\sigma}(\mathbf{x};\boldsymbol{\Sigma}), \tag{2.14b}$$

where the argument $\boldsymbol{\Sigma}$ emphasizes the fact that a traction boundary value problem with constant macrostress $\boldsymbol{\Sigma}$, (2.4a), is being considered. The microstrain potential then becomes

$$\psi = \psi(\mathbf{x}, \boldsymbol{\sigma}(\mathbf{x}; \boldsymbol{\Sigma})) = \psi^{\Sigma}(\mathbf{x}; \boldsymbol{\Sigma}). \tag{2.15}$$

For an aribtrary change $\delta\boldsymbol{\Sigma}$ in the macrostress, we obtain

$$\delta\boldsymbol{\sigma}(\mathbf{x}; \boldsymbol{\Sigma}) = \delta\Sigma_{ij} \frac{\partial \boldsymbol{\sigma}}{\partial \Sigma_{ij}}(\mathbf{x}; \boldsymbol{\Sigma}), \tag{2.16a}$$

and, hence,

$$\begin{aligned} \langle \delta\boldsymbol{\sigma} : \boldsymbol{\varepsilon} \rangle &= \left\langle \left(\delta\Sigma_{ij} \frac{\partial \boldsymbol{\sigma}}{\partial \Sigma_{ij}}(\mathbf{x}; \boldsymbol{\Sigma}) \right) : \left(\frac{\partial \psi}{\partial \boldsymbol{\sigma}}(\mathbf{x}; \boldsymbol{\sigma}) \right) \right\rangle = \left\langle \delta\boldsymbol{\Sigma} : \left(\frac{\partial \psi^{\Sigma}}{\partial \boldsymbol{\Sigma}}(\mathbf{x}; \boldsymbol{\Sigma}) \right) \right\rangle \\ &= \left(\frac{\partial}{\partial \boldsymbol{\Sigma}} \langle \psi^{\Sigma} \rangle \right) : \delta\boldsymbol{\Sigma}. \end{aligned} \tag{2.16b}$$

Thus, it follows that

$$\langle \boldsymbol{\varepsilon}(\mathbf{x}; \boldsymbol{\Sigma}) \rangle = \frac{\partial}{\partial \boldsymbol{\Sigma}} \langle \psi^{\Sigma} \rangle. \tag{2.17}$$

We therefore *define* the macrostrain potential and the corresponding macrostrain by

$$\Psi^{\Sigma} \equiv \Psi^{\Sigma}(\boldsymbol{\Sigma}) \equiv \langle \psi^{\Sigma} \rangle \equiv \frac{1}{V} \int_V \psi^{\Sigma}(\mathbf{x}; \boldsymbol{\Sigma})\, dV, \tag{2.18a}$$

$$\mathbf{E}^{\Sigma} = \langle \boldsymbol{\varepsilon}(\mathbf{x}; \boldsymbol{\Sigma}) \rangle, \tag{2.18b}$$

and obtain

$$\mathbf{E}^{\Sigma} = \frac{\partial \Psi^{\Sigma}}{\partial \boldsymbol{\Sigma}}, \tag{2.18c}$$

where the superscript Σ on $\mathbf{E}$ emphasizes that $\mathbf{E}^{\Sigma}$ is the *average strain produced by the prescribed macrostress* $\boldsymbol{\Sigma}$. Like the macrostress potential $\Phi^E = \Phi^E(\mathbf{E}, \Theta; \mathbf{S})$, the macrostrain potential Ψ^{Σ} is also a function of the current macrotemperature Θ and microstructure $\mathbf{S}$. We hence write

$$\Psi^{\Sigma} = \Psi^{\Sigma}(\boldsymbol{\Sigma}, \Theta; \mathbf{S}). \tag{2.18d}$$

2.3. *Relation Between Macropotentials*

We have defined the macrostrain potential as the volume average of the microstrain potential $\psi^{\Sigma}(\mathbf{x}; \boldsymbol{\Sigma})$ when the macrostress $\boldsymbol{\Sigma}$ is prescribed, and the corresponding macrostrain

$$\mathbf{E}^{\Sigma} \equiv \langle \boldsymbol{\varepsilon}(\mathbf{x}; \boldsymbol{\Sigma}) \rangle, \tag{2.19a}$$

$$\Psi^{\Sigma} \equiv \Psi^{\Sigma}(\boldsymbol{\Sigma}) \equiv \langle \psi^{\Sigma}(\mathbf{x}; \boldsymbol{\Sigma}) \rangle, \tag{2.19b}$$

$$\mathbf{E}^{\Sigma} = \frac{\partial \Psi^{\Sigma}}{\partial \boldsymbol{\Sigma}}(\boldsymbol{\Sigma}). \tag{2.19c}$$

The notation in (2.19a) shows that $\mathbf{E}^{\Sigma}$ is the macrostrain produced by the prescribed macrostress $\boldsymbol{\Sigma}$. Now define a *new* macrostress potential function

$$\Phi^{\Sigma} \equiv \Phi^{\Sigma}(\mathbf{E}^{\Sigma}) \equiv \boldsymbol{\Sigma} : \mathbf{E}^{\Sigma} - \Psi^{\Sigma}(\boldsymbol{\Sigma}), \tag{2.20}$$

where, as usual, $\boldsymbol{\Sigma}$ is regarded a function of $\mathbf{E}^{\Sigma}$ through (2.19a).

At the local level, on the other hand, we have

$$\phi^{\Sigma} = \phi^{\Sigma}(\mathbf{x}; \boldsymbol{\Sigma}) = \phi(\mathbf{x}, \boldsymbol{\varepsilon}(\mathbf{x}; \boldsymbol{\Sigma})), \tag{2.21a}$$

$$\psi^{\Sigma} = \psi^{\Sigma}(\mathbf{x}; \boldsymbol{\Sigma}) = \psi(\mathbf{x}, \boldsymbol{\sigma}(\mathbf{x}; \boldsymbol{\Sigma})), \tag{2.21b}$$

which satisfy $\phi^{\Sigma} + \psi^{\Sigma} = \boldsymbol{\sigma}(\mathbf{x}; \boldsymbol{\Sigma}) : \boldsymbol{\varepsilon}(\mathbf{x}; \boldsymbol{\Sigma})$. Taking the average, $\langle \phi^{\Sigma} \rangle + \langle \psi^{\Sigma} \rangle = \boldsymbol{\Sigma} : \mathbf{E}^{\Sigma}$, and comparing with (2.20), we conclude that

$$\Phi^{\Sigma} = \langle \phi^{\Sigma} \rangle, \tag{2.22a}$$

$$\mathbf{E}^{\Sigma} = \frac{\partial \Psi^{\Sigma}}{\partial \boldsymbol{\Sigma}}(\boldsymbol{\Sigma}) \leftrightarrow \boldsymbol{\Sigma} = \frac{\partial \Phi^{\Sigma}}{\partial \mathbf{E}^{\Sigma}}(\mathbf{E}^{\Sigma}). \tag{2.22b}$$

In all these expressions, the superscript Σ shows that the corresponding quantity is obtained for the prescribed macrostress $\boldsymbol{\Sigma}$.

In a similar manner, when the macrostrain $\mathbf{E}$ is prescribed through a linear boundary displacement, $\mathbf{u}^0 = \mathbf{x} \cdot \mathbf{E}$ on ∂V, we have

$$\boldsymbol{\Sigma}^{E} = \langle \boldsymbol{\sigma}(\mathbf{x}; \mathbf{E}) \rangle, \tag{2.23a}$$

$$\Phi^{E} \equiv \Phi^{E}(\mathbf{E}) \equiv \langle \phi^{E}(\mathbf{x}; \mathbf{E}) \rangle, \tag{2.23b}$$

$$\boldsymbol{\Sigma}^{E} = \frac{\partial \Phi^{E}}{\partial \mathbf{E}}(\mathbf{E}). \tag{2.23c}$$

We hence *define* a new macrostrain potential Ψ^{E} by

$$\Psi^{E} \equiv \Psi^{E}(\boldsymbol{\Sigma}^{E}) \equiv \boldsymbol{\Sigma}^{E} : \mathbf{E} - \Phi^{E}(\mathbf{E}), \tag{2.24}$$

where, again, $\mathbf{E}$ is viewed as a function of $\boldsymbol{\Sigma}^{E}$. Furthermore, for the microquantities, we set

$$\phi^{E} = \phi^{E}(\mathbf{x}; \mathbf{E}) = \phi(\mathbf{x}, \boldsymbol{\varepsilon}(\mathbf{x}; \mathbf{E})), \tag{2.25a}$$

$$\psi^{E} = \psi^{E}(\mathbf{x}; \mathbf{E}) = \psi(\mathbf{x}, \boldsymbol{\sigma}(\mathbf{x}; \mathbf{E})), \tag{2.25b}$$

and $\phi^{E} + \psi^{E} = \boldsymbol{\sigma}(\mathbf{x}; \mathbf{E}) : \boldsymbol{\varepsilon}(\mathbf{x}; \mathbf{E})$. Taking the average, $\langle \phi^{E} \rangle + \langle \psi^{E} \rangle = \boldsymbol{\Sigma}^{E} : \mathbf{E}$, and comparing with (2.24), it follows that

$$\Psi^{E} = \langle \psi^{E} \rangle, \tag{2.26a}$$

$$\boldsymbol{\Sigma}^{E} = \frac{\partial \Phi^{E}}{\partial \mathbf{E}}(\mathbf{E}) \leftrightarrow \mathbf{E} = \frac{\partial \Psi^{E}}{\partial \boldsymbol{\Sigma}^{E}}(\boldsymbol{\Sigma}^{E}). \tag{2.26b}$$

2.4. *On Definition of RVE*

When the boundary tractions are given by (2.4a) with the prescribed macrostress $\boldsymbol{\Sigma}$, we have the microstrain and microstress fields, $\boldsymbol{\varepsilon} = \boldsymbol{\varepsilon}(\mathbf{x}; \boldsymbol{\Sigma})$ and $\boldsymbol{\sigma} =$

$\boldsymbol{\sigma}(\mathbf{x}; \boldsymbol{\Sigma})$, and $\mathbf{E}^{\Sigma} = \langle \boldsymbol{\varepsilon}(\mathbf{x}; \boldsymbol{\Sigma}) \rangle$ as the overall macrostrain. Suppose that, through (2.4b), the boundary displacements are defined for this macrostrain $\mathbf{E}^{\Sigma}$ by $\mathbf{u}^0 = \mathbf{x} \cdot \mathbf{E}^{\Sigma}$ on ∂V. The microstrain and microstress fields then become $\boldsymbol{\sigma} = \boldsymbol{\sigma}(\mathbf{x}; \mathbf{E}^{\Sigma})$ and $\boldsymbol{\varepsilon} = \boldsymbol{\varepsilon}(\mathbf{x}; \mathbf{E}^{\Sigma})$. In general, these fields are *not* identical with $\boldsymbol{\varepsilon}(\mathbf{x}; \boldsymbol{\Sigma})$ and $\boldsymbol{\sigma}(\mathbf{x}; \boldsymbol{\Sigma})$. Furthermore, while $\mathbf{E}^{\Sigma} = \langle \boldsymbol{\varepsilon}(\mathbf{x}; \mathbf{E}^{\Sigma}) \rangle$, there is no *a priori* reason that $\langle \boldsymbol{\sigma}(\mathbf{x}; \mathbf{E}^{\Sigma}) \rangle$ should equal $\boldsymbol{\Sigma}$ for an arbitrary heterogeneous elastic solid.

We regard the RVE to be *statistically representative* of the macroresponse of the continuum material neighborhood, if and only if any arbitrary constant macrostress $\boldsymbol{\Sigma}$ produces a macrostrain $\mathbf{E}^{\Sigma} = \langle \boldsymbol{\varepsilon}(\mathbf{x}; \boldsymbol{\Sigma}) \rangle$ such that when the displacement boundary conditions $\mathbf{u} = \mathbf{x} \cdot \mathbf{E}^{\Sigma}$ on ∂V are imposed instead, then, the macrostress

$$\langle \boldsymbol{\sigma}(\mathbf{x}; \mathbf{E}^{\Sigma}) \rangle \simeq \boldsymbol{\Sigma} \tag{2.27a}$$

is obtained, where the equality is to hold to a given degree of accuracy. Conversely, when the macrostrain $\mathbf{E}$ produces microstress and microstrain fields, $\boldsymbol{\sigma} = \boldsymbol{\sigma}(\mathbf{x}; \mathbf{E})$ and $\boldsymbol{\varepsilon} = \boldsymbol{\varepsilon}(\mathbf{x}; \mathbf{E})$, then the RVE is regarded to be statistically representative if and only if the prescribed macrostress $\boldsymbol{\Sigma}^E = \langle \boldsymbol{\sigma}(\mathbf{x}; \mathbf{E}) \rangle$ leads to a microstrain field $\boldsymbol{\varepsilon}(\mathbf{x}; \boldsymbol{\Sigma}^E)$ such that

$$\langle \boldsymbol{\varepsilon}(\mathbf{x}; \boldsymbol{\Sigma}^E) \rangle \simeq \mathbf{E}. \tag{2.27b}$$

Based on the above definitions for an RVE, the macrostrain potential, $\Psi^{\Sigma}(\boldsymbol{\Sigma})$, given by (2.19b), and the macrostress potential, $\Phi^E(\mathbf{E})$, given by (2.23b), correspond to each other in the sense that

$$\frac{\partial \Psi^{\Sigma}}{\partial \boldsymbol{\Sigma}}(\boldsymbol{\Sigma}) \simeq \mathbf{E} \leftrightarrow \frac{\partial \Phi^E}{\partial \mathbf{E}}(\mathbf{E}) \simeq \boldsymbol{\Sigma}, \tag{2.28a}$$

$$\Psi^{\Sigma}(\boldsymbol{\Sigma}) + \Phi^E(\mathbf{E}) \simeq \boldsymbol{\Sigma} : \mathbf{E}. \tag{2.28b}$$

It should be noted that $\boldsymbol{\sigma}(\mathbf{x}; \boldsymbol{\Sigma}) \neq \boldsymbol{\sigma}(\mathbf{x}; \mathbf{E})$ and $\boldsymbol{\varepsilon}(\mathbf{x}; \boldsymbol{\Sigma}) \neq \boldsymbol{\varepsilon}(\mathbf{x}; \mathbf{E})$ even for $\boldsymbol{\Sigma}$ and $\mathbf{E}$ which satisfy (2.28a, b). Moreover, in general,

$$\psi^{\Sigma}(\mathbf{x}; \boldsymbol{\Sigma}) + \phi^E(\mathbf{x}; \mathbf{E}) \neq \boldsymbol{\sigma}(\mathbf{x}; \boldsymbol{\Sigma}) : \boldsymbol{\varepsilon}(\mathbf{x}; \mathbf{E}). \tag{2.29}$$

Similarly, the complementary macropotentials, $\Phi^{\Sigma}(\mathbf{E}^{\Sigma})$ and $\Psi^{\Sigma}(\boldsymbol{\Sigma})$, are related through

$$\frac{\partial \Phi^{\Sigma}}{\partial \mathbf{E}^{\Sigma}}(\mathbf{E}^{\Sigma}) \simeq \boldsymbol{\Sigma}^E \leftrightarrow \frac{\partial \Psi^E}{\partial \boldsymbol{\Sigma}^E}(\boldsymbol{\Sigma}^E) \simeq \mathbf{E}^{\Sigma}, \tag{2.30a}$$

$$\Phi^{\Sigma}(\mathbf{E}^{\Sigma}) + \Psi^E(\boldsymbol{\Sigma}^E) \simeq \mathbf{E}^{\Sigma} : \boldsymbol{\Sigma}^E, \tag{2.30b}$$

whereas the corresponding micropotentials do not satisfy a similar relation, i.e., in general,

$$\phi^{\Sigma}(\mathbf{x}; \mathbf{E}^{\Sigma}) + \psi^E(\mathbf{x}; \boldsymbol{\Sigma}^E) \neq \boldsymbol{\sigma}(\mathbf{x}; \mathbf{E}) : \boldsymbol{\varepsilon}(\mathbf{x}; \boldsymbol{\Sigma}). \tag{2.31}$$

2.5. *Linear Versus Nonlinear Response*

When the microstructure is fixed and the material of the RVE is linearly elastic, then the corresponding overall response will also be linearly elastic. In this case, for a prescribed macrostress $\boldsymbol{\Sigma}$, the macrostrain $\mathbf{E}^{\Sigma}$ will be proportional to $\boldsymbol{\Sigma}$,

$$\mathbf{E}^{\Sigma} \equiv \bar{\boldsymbol{\varepsilon}} \equiv \langle \boldsymbol{\varepsilon}(\mathbf{x}; \boldsymbol{\Sigma}) \rangle = \bar{\mathbf{D}} : \boldsymbol{\Sigma}, \tag{2.32a}$$

where $\bar{\mathbf{D}}$ is the overall compliance tensor. Similarly, for a prescribed macrostrain $\mathbf{E}$, we obtain

$$\boldsymbol{\Sigma}^{E} \equiv \bar{\boldsymbol{\sigma}} \equiv \langle \boldsymbol{\sigma}(\mathbf{x}; \mathbf{E}) \rangle = \bar{\mathbf{C}} : \mathbf{E}, \tag{2.32b}$$

where $\bar{\mathbf{C}}$ is the overall elasticity tensor.

Now, if the RVE is statistically representative, then we must have

$$\bar{\mathbf{C}} = \bar{\mathbf{D}}^{-1} \tag{2.33a}$$

or

$$\bar{\mathbf{D}} = \bar{\mathbf{C}}^{-1}. \tag{2.33b}$$

We refer to the averaging technique, which satisfies this inverse relation, as *self-consistent* (Hershey, 1954; Kröner, 1958; Budiansky, 1965; Hill, 1965); see Mura (1982) for other references.

When the material of an RVE is nonlinearly elastic, then, for a fixed microstructure, the overall response will be nonlinearly elastic. In this case, the first gradient of the overall stress potential with respect to the overall macrostrain $\mathbf{E}$, and that of the strain potential with respect to the overall macrostress $\boldsymbol{\Sigma}$, satisfy the relation (2.28a) when the RVE is indeed statistically representative, i.e., when a consistent averaging technique is employed. However, there is no *a priori* reason to believe that a similar correspondence should exist between higher gradients of these potentials. In the remainder of this paper, attention is focused on the overall response of an RVE with linearly elastic constituent materials.

3. Overall Moduli

We now seek to obtain the overall elastic moduli of an RVE which consists of a linearly elastic matrix with uniform elasticity $\mathbf{C}$ and compliance $\mathbf{D} = \mathbf{C}^{-1}$, and a set of linearly elastic inclusions Ω_α, $\alpha = 1, 2, \ldots, n$, with elasticity $\mathbf{C}^\alpha$ and compliance $\mathbf{D}^\alpha$. The microinclusions are perfectly bonded to the matrix. Hence, the overall response of the RVE is linearly elastic, although anisotropic even if all its constituents may be isotropic. The case of an RVE containing microcavities is obtained as a limit when the elasticity tensor of each inclusion is set equal to zero, and the microcracks are considered as cavities with special geometry.

We treat the cases of prescribed constant macrostress, $\boldsymbol{\Sigma}$, and prescribed constant macrostrain, $\mathbf{E}$, separately. These cases are mutually exclusive. In the first case, the average strain $\bar{\boldsymbol{\varepsilon}}$ is given by (2.32a) for an arbitrary constant $\boldsymbol{\Sigma}$, and in the second case, the average stress $\bar{\boldsymbol{\sigma}}$ is given by (2.32b) for an arbitrary constant $\mathbf{E}$.

3.1. Macrostress Prescribed

In this case, it is easy to show that

$$(\bar{\mathbf{D}} - \mathbf{D}) : \boldsymbol{\Sigma} = \sum_{\alpha=1}^{n} f_\alpha (\mathbf{D}^\alpha - \mathbf{D}) : \bar{\boldsymbol{\sigma}}^\alpha, \tag{3.1}$$

where $f_\alpha \equiv \Omega_\alpha / V$ is the volume fraction of the αth inclusion, and $\bar{\boldsymbol{\sigma}}^\alpha$ is the average stress in the αth inclusion. Since the response of the RVE is linearly elastic,

$$\bar{\boldsymbol{\sigma}}^\alpha \equiv \langle \boldsymbol{\sigma}(\mathbf{x}; \boldsymbol{\Sigma}) \rangle_\alpha = \mathbf{h}^\alpha : \boldsymbol{\Sigma}. \tag{3.2a}$$

From $\bar{\boldsymbol{\varepsilon}}^\alpha = \mathbf{D}^\alpha : \bar{\boldsymbol{\sigma}}^\alpha$ (α not summed), it follows that

$$\bar{\boldsymbol{\varepsilon}}^\alpha \equiv \langle \boldsymbol{\varepsilon}(\mathbf{x}; \boldsymbol{\Sigma}) \rangle_\alpha = \mathbf{D}^\alpha : \mathbf{h}^\alpha : \boldsymbol{\Sigma} \qquad (\alpha \text{ not summed}). \tag{3.2b}$$

From (3.1), (3.2a), and the fact that $\boldsymbol{\Sigma}$ is arbitrary, it now follows that

$$\bar{\mathbf{D}} = \mathbf{D} + \sum_{\alpha=1}^{n} f_\alpha (\mathbf{D}^\alpha - \mathbf{D}) : \mathbf{h}^\alpha, \tag{3.3}$$

which is valid for any elastic RVE with any set of elastic inclusions of *any shape and distribution.* Note that (3.2a, b) may be regarded as defining the fourth-order tensor $\mathbf{h}^\alpha$. In many cases, this tensor may be calculated directly from (3.2a, b). For rigid inclusions, $\bar{\boldsymbol{\varepsilon}}^\alpha = \mathbf{0}$,

$$\bar{\boldsymbol{\sigma}}^\alpha \equiv \frac{1}{\Omega} \int_{\Omega_\alpha} \boldsymbol{\sigma}(\mathbf{x}; \boldsymbol{\Sigma})\, dV \equiv \frac{1}{\Omega_\alpha} \int_{\partial\Omega_\alpha} (\mathbf{n} \cdot \boldsymbol{\sigma}) \otimes \mathbf{x}\, dS = \mathbf{h}^\alpha : \boldsymbol{\Sigma}, \tag{3.4a}$$

where $\mathbf{n}$ is the outer unit normal of Ω_α. For cavities or cracks, $\bar{\boldsymbol{\sigma}}^\alpha = \mathbf{0}$,

$$\bar{\boldsymbol{\varepsilon}}^\alpha \equiv \frac{1}{\Omega_\alpha} \int_{\Omega_\alpha} \boldsymbol{\varepsilon}(\mathbf{x}; \boldsymbol{\Sigma})\, dV \equiv \frac{1}{\Omega_\alpha} \int_{\partial\Omega_\alpha} \tfrac{1}{2}(\mathbf{n} \otimes \mathbf{u} + \mathbf{u} \otimes \mathbf{n})\, dS = \mathbf{D}^\alpha : \mathbf{h}^\alpha : \boldsymbol{\Sigma}, \tag{3.4b}$$

where the tensor $\mathbf{D}^\alpha : \mathbf{h}^\alpha$ is bounded, although $\mathbf{C}^\alpha$ and $\mathbf{h}^\alpha$ go to $\mathbf{0}$ (Horii and Nemat-Nasser, 1983). Note that (3.3) yields the Reuss (1929) estimate, for $\bar{\boldsymbol{\sigma}}^\alpha = \boldsymbol{\Sigma}$,

$$\bar{\boldsymbol{\varepsilon}}^\alpha = \mathbf{D}^\alpha : \bar{\boldsymbol{\sigma}}^\alpha = \mathbf{D}^\alpha : \boldsymbol{\Sigma}, \tag{3.5a}$$

$$\bar{\mathbf{D}} = (1 - f)\mathbf{D} + \sum_{\alpha=1}^{n} f_\alpha \mathbf{D}^\alpha, \tag{3.5b}$$

where $f = \sum_{\alpha=1}^{n} f_\alpha$ is the total volume fraction of all inclusions.

3.2. Macrostrain Prescribed

In this case, instead of (3.1), we have

$$(\bar{\mathbf{C}} - \mathbf{C}) : \mathbf{E} = \sum_{\alpha=1}^{n} f_\alpha (\mathbf{C}^\alpha - \mathbf{C}) : \bar{\boldsymbol{\varepsilon}}^\alpha. \tag{3.6}$$

Again, from linearity,

$$\bar{\boldsymbol{\varepsilon}}^\alpha = \langle \boldsymbol{\varepsilon}(\mathbf{x}; \mathbf{E}) \rangle = \mathbf{j}^\alpha : \mathbf{E}, \tag{3.7a}$$

$$\bar{\boldsymbol{\sigma}}^\alpha \equiv \langle \boldsymbol{\sigma}(\mathbf{x}; \mathbf{E}) \rangle = \mathbf{C}^\alpha : \mathbf{j}^\alpha : \mathbf{E} \qquad (\alpha \text{ not summed}), \tag{3.7b}$$

$$\bar{\mathbf{C}} = \mathbf{C} + \sum_{\alpha=1}^{n} f_\alpha (\mathbf{C}^\alpha - \mathbf{C}) : \mathbf{j}^\alpha. \tag{3.8}$$

Here, also, the so-called Voigt (1889) model results for $\bar{\boldsymbol{\varepsilon}}^\alpha = \mathbf{E}$,

$$\bar{\boldsymbol{\sigma}}^\alpha = \mathbf{C}^\alpha : \bar{\boldsymbol{\varepsilon}}^\alpha = \mathbf{C}^\alpha : \mathbf{E}, \tag{3.9a}$$

$$\bar{\mathbf{C}} = (1 - f)\mathbf{C} + \sum_{\alpha=1}^{n} f_\alpha \mathbf{C}^\alpha. \tag{3.9b}$$

3.3. *Dilute Distribution of Inhomogeneities*

The tensors $\mathbf{h}^\alpha$ and $\mathbf{j}^\alpha$ depend on the material properties and geometries of the matrix and all inhomogeneities. They are symbolically written as

$$\mathbf{h}^\alpha = \mathbf{h}^\alpha(\mathbf{D}^\alpha, \Omega_\alpha; \mathbf{D}, M; \ldots; \mathbf{D}^\beta, \Omega_\beta; \ldots), \tag{3.10a}$$

$$\mathbf{j}^\alpha = \mathbf{j}^\alpha(\mathbf{C}^\alpha, \Omega_\alpha; \mathbf{C}, M; \ldots; \mathbf{C}^\beta, \Omega_\beta; \ldots), \tag{3.10b}$$

where the arguments display the functional dependency of these tensors. When inhomogeneities are far apart, we may neglect the interactions between microconstituents, and approximate (3.10a, b) by considering a single inhomogeneity embedded in the uniform *matrix* under far-field uniform stresses or strains, i.e.,

$$\mathbf{h}^\alpha \simeq \mathbf{h}^{\mathrm{DD}\alpha} \equiv \mathbf{h}^\infty(\mathbf{D}^\alpha, \Omega_\alpha; \mathbf{D}), \tag{3.11a}$$

$$\mathbf{j}^\alpha \simeq \mathbf{j}^{\mathrm{DD}\alpha} \equiv \mathbf{j}^\infty(\mathbf{C}^\alpha, \Omega_\alpha; \mathbf{C}). \tag{3.11b}$$

This we call the assumption of a *dilute distribution* (DD). The tensors $\mathbf{h}^\infty$ and $\mathbf{j}^\infty$ then have the following two properties: (1) they are independent of the size (but not the shape) of the inhomogeneity; and (2) for any inhomogeneity Ω_α,

$$\mathbf{h}^\infty = \mathbf{C}^\alpha : \mathbf{j}^\infty : \mathbf{D}, \tag{3.12a}$$

$$\mathbf{j}^\infty = \mathbf{D}^\alpha : \mathbf{h}^\infty : \mathbf{C}. \tag{3.12b}$$

In view of (3.3) and (3.8), it now follows that $\bar{\mathbf{C}}^{\mathrm{DD}}$ and $\bar{\mathbf{D}}^{\mathrm{DD}}$ obtained in this approximation are each other's inverse only to the first order in the volume fraction of inhomogeneities, i.e.,

$$\bar{\mathbf{C}}^{\mathrm{DD}} : \bar{\mathbf{D}}^{\mathrm{DD}} = \mathbf{I} + O(f^2), \tag{3.13}$$

where $\mathbf{I}$ is the fourth-order symmetric identity tensor with components $I_{ijkl} = (\delta_{ik}\delta_{jl} + \delta_{il}\delta_{jk})/2$.

3.4. Self-Consistent Estimate

In the self-consistent estimate a typical inhomogeneity Ω_α is embedded in a uniform matrix whose elasticity and compliance tensors are the, as yet unknown, overall quantities (Hershey, 1954; Kröner, 1958; Budiansky, 1965; Hill, 1965). (See also Mura (1982) for a detailed discussion and extensive references.) Hence, we approximate $\mathbf{h}^\alpha$ and $\mathbf{j}^\alpha$ for Ω_α by

$$\mathbf{h}^\alpha \simeq \mathbf{h}^{SC\alpha} \equiv \mathbf{h}^\infty(\mathbf{D}^\alpha, \Omega_\alpha; \bar{\mathbf{D}}^{SC}), \tag{3.14a}$$

$$\mathbf{j}^\alpha \simeq \mathbf{j}^{SC\alpha} \equiv \mathbf{j}^\infty(\mathbf{C}^\alpha, \Omega_\alpha; \bar{\mathbf{C}}^{SC}), \tag{3.14b}$$

where $\bar{\mathbf{D}}^{SC}$ and $\bar{\mathbf{C}}^{SC}$ are the overall moduli obtained by the *self-consistent method* (SC).

Taking advantage of $\mathbf{D}^\alpha - \mathbf{D} = -\mathbf{D} : (\mathbf{C}^\alpha - \mathbf{C}) : \mathbf{D}^\alpha$ and $\mathbf{C}^\alpha - \mathbf{C} = -\mathbf{C} : (\mathbf{D}^\alpha - \mathbf{D}) : \mathbf{C}^\alpha$, we have

$$(\mathbf{D}^\alpha - \mathbf{D}) : \mathbf{h}^{SC\alpha} = -\mathbf{D} : (\mathbf{C}^\alpha - \mathbf{C}) : \mathbf{j}^{SC\alpha} : \bar{\mathbf{D}}^{SC}, \tag{3.15a}$$

$$(\mathbf{C}^\alpha - \mathbf{C}) : \mathbf{j}^{SC\alpha} = -\mathbf{C} : (\mathbf{D}^\alpha - \mathbf{D}) : \mathbf{h}^{SC\alpha} : \bar{\mathbf{C}}^{SC}, \tag{3.15b}$$

where the identities (3.12a, b) with $\mathbf{D}$ and $\mathbf{C}$ replaced by $\bar{\mathbf{D}}^{SC}$ and $\bar{\mathbf{C}}^{SC}$, respectively, are used. Equations (3.3) and (3.8) then become

$$\mathbf{D} : (\bar{\mathbf{D}}^{SC-1} - \mathbf{C}) : \bar{\mathbf{D}}^{SC} = \mathbf{D} : \left\{ \sum_{\alpha=1}^{n} f_\alpha (\mathbf{C}^\alpha - \mathbf{C}) : \mathbf{j}^{SC\alpha} \right\} : \bar{\mathbf{D}}^{SC}, \tag{3.16a}$$

$$\mathbf{C} : (\bar{\mathbf{C}}^{SC-1} - \mathbf{D}) : \bar{\mathbf{C}}^{SC} = \mathbf{C} : \left\{ \sum_{\alpha=1}^{n} f_\alpha (\mathbf{D}^\alpha - \mathbf{D}) : \mathbf{h}^{SC\alpha} \right\} : \bar{\mathbf{C}}^{SC}. \tag{3.16b}$$

Hence, $\bar{\mathbf{D}}^{SC}$ and $\bar{\mathbf{C}}^{SC}$ obtained in this manner are each other's inverse. Furthermore, as the volume fraction of inhomogeneities decreases,

$$\bar{\mathbf{D}}^{SC} \sim \bar{\mathbf{D}}^{DD}, \tag{3.17a}$$

$$\bar{\mathbf{C}}^{SC} \sim \bar{\mathbf{C}}^{DD}. \tag{3.17b}$$

3.5. Differential Scheme

Starting with a homogeneous matrix, and considering an infinitesimally small incremental increase of the volume fraction of a typical inhomogeneity, Ω_α, until the final volume fraction is achieved, we can estimate the overall moduli by the so-called differential scheme; see Roscoe (1952, 1973), McLaughlin (1977), and Hashin (1988). The change of the overall moduli, caused when the volume fraction of Ω_α is slightly increased, can be estimated by the dilute distribution assumption. Hence, denoting the overall moduli in the differential scheme (DS) by $\bar{\mathbf{D}}^{DS}$ and $\bar{\mathbf{C}}^{DS}$, we have

$$\frac{\partial}{\partial f_\alpha} \bar{\mathbf{D}}^{DS} = \frac{1}{1 - f_\alpha} (\mathbf{D}^\alpha - \bar{\mathbf{D}}^{DS}) : \mathbf{h}^{DS\alpha}, \tag{3.18a}$$

$$\frac{\partial}{\partial f_\alpha} \bar{\mathbf{C}}^{DS} = \frac{1}{1 - f_\alpha} (\mathbf{C}^\alpha - \bar{\mathbf{C}}^{DS}) : \mathbf{j}^{DS\alpha}, \tag{3.18b}$$

where

$$\mathbf{h}^{DS\alpha} \equiv \mathbf{h}^{\infty}(\mathbf{D}^{\alpha}, \Omega_{\alpha}; \bar{\mathbf{D}}^{DS}), \tag{3.18c}$$

$$\mathbf{j}^{DS\alpha} \equiv \mathbf{j}^{\infty}(\mathbf{C}^{\alpha}, \Omega_{\alpha}; \bar{\mathbf{C}}^{DS}). \tag{3.18d}$$

Note that the shape of Ω_α is assumed to remain self-similar.

Again, from identities (3.12a, b) with $\mathbf{D}$ and $\mathbf{C}$ replaced by $\bar{\mathbf{D}}^{DS}$ and $\bar{\mathbf{C}}^{DS}$, respectively,

$$\frac{\partial}{\partial f_\alpha}\bar{\mathbf{D}}^{DS} = -\bar{\mathbf{D}}^{DS} : \left\{\frac{1}{1-f_\alpha}(\mathbf{C}^{\alpha} - \bar{\mathbf{C}}^{DS}) : \mathbf{j}^{DS\alpha}\right\} : \bar{\mathbf{D}}^{DS}, \tag{3.19a}$$

$$\frac{\partial}{\partial f_\alpha}\bar{\mathbf{C}}^{DS} = -\bar{\mathbf{C}}^{DS} : \left\{\frac{1}{1-f_\alpha}(\mathbf{D}^{\alpha} - \bar{\mathbf{D}}^{DS}) : \mathbf{h}^{DS\alpha}\right\} : \bar{\mathbf{C}}^{DS}. \tag{3.19b}$$

Initially all f_α are zero, and $\bar{\mathbf{D}}^{DS} = \mathbf{D}$, $\bar{\mathbf{C}}^{DS} = \mathbf{C}$. Hence, from (3.19), at each increment, $\bar{\mathbf{D}}^{DS} : \bar{\mathbf{C}}^{DS} = \bar{\mathbf{C}}^{DS} : \bar{\mathbf{D}}^{DS} = \mathbf{I}$, and we have

$$\frac{\partial}{\partial f_\alpha}(\bar{\mathbf{D}}^{DS} : \bar{\mathbf{C}}^{DS}) = \frac{\partial}{\partial f_\alpha}(\bar{\mathbf{C}}^{DS} : \bar{\mathbf{D}}^{DS}) = \mathbf{0}. \tag{3.20}$$

Thus, $\bar{\mathbf{D}}^{DS}$ and $\bar{\mathbf{C}}^{DS}$ are the inverse of each other, and the scheme is self-consistent. Furthermore, for a small volume fraction of inhomogeneities,

$$\bar{\mathbf{D}}^{DS} \sim \bar{\mathbf{C}}^{DD}, \tag{3.21a}$$

$$\bar{\mathbf{C}}^{DS} \sim \bar{\mathbf{C}}^{DD}. \tag{3.21b}$$

4. Microcavities

The simplest class of problems is an isotropic homogeneous linearly elastic matrix containing a random dilute distribution of circular (in two dimensions) or spherical (in three dimensions) cavities. Since the average stress in the cavities is zero,

$$\langle \boldsymbol{\varepsilon}(\mathbf{x}; \boldsymbol{\Sigma})\rangle_\alpha - \mathbf{D} : \langle \boldsymbol{\sigma}(\mathbf{x}; \boldsymbol{\Sigma})\rangle_\alpha = \bar{\boldsymbol{\varepsilon}}^{\alpha} = \mathbf{D}^{\alpha} : \mathbf{h}^{\alpha} : \boldsymbol{\Sigma} \equiv \mathbf{H}^{\alpha} : \boldsymbol{\Sigma}, \tag{4.1}$$

$$\langle \boldsymbol{\sigma}(\mathbf{x}; \mathbf{E})\rangle_\alpha - \mathbf{C} : \langle \boldsymbol{\varepsilon}(\mathbf{x}; \mathbf{E})\rangle_\alpha = -\mathbf{C} : \bar{\boldsymbol{\varepsilon}}^{\alpha} = -\mathbf{C} : \mathbf{j}^{\alpha} : \mathbf{E} \equiv -\mathbf{J}^{\alpha} : \mathbf{E} \quad (\alpha \text{ not summed}), \tag{4.2}$$

for prescribed macrostress and macrostrain cases, respectively. Hence, using (3.4b) and (3.7b), the corresponding calculations become quite simple in this case. Below we summarize some useful and interesting results obtained in this manner in two dimensions.

A fourth-order isotropic tensor can be expressed in terms of two bases, $\mathbf{E}^1$ and $\mathbf{E}^2$. In two dimensions, these bases are

$$E^1_{ijkl} = \tfrac{1}{2}\delta_{ij}\delta_{kl}, \tag{4.3a}$$

$$E^2_{ijkl} = -\tfrac{1}{2}\delta_{ij}\delta_{kl} + \tfrac{1}{2}(\delta_{ik}\delta_{jl} + \delta_{il}\delta_{jk}), \tag{4.3b}$$

which satisfy $\mathbf{E}^\alpha : \mathbf{E}^\beta = \mathbf{E}^\alpha$ for $\alpha = \beta$, and $= \mathbf{0}$ for $\alpha \neq \beta$. Then, the compliance and elasticity tensors become,

$$\mathbf{D} = \frac{\kappa - 1}{4\mu}\mathbf{E}^1 + \frac{1}{2\mu}\mathbf{E}^2, \tag{4.4a}$$

$$\mathbf{C} = \frac{4\mu}{\kappa - 1}\mathbf{E}^1 + 2\mu\mathbf{E}^2, \tag{4.4b}$$

where μ is the shear modulus, $\kappa = 3 - 4\nu$ for plane strain, and $\kappa = (3 - \nu)/(1 + \nu)$ for plane stress, with ν being the Poisson ratio.

For a circular hole, Ω_α, in an unbounded isotropic plate, $\mathbf{D}^\alpha : \mathbf{h}^\infty$ becomes

$$\mathbf{D}^\alpha : \mathbf{h}^\infty(\Omega_\alpha, \mathbf{D}^\alpha; \mathbf{D}) = \frac{\kappa + 1}{4\mu}\mathbf{E}^1 + \frac{\kappa + 1}{2\mu}\mathbf{E}^2 \qquad (\mathbf{D}^\alpha \to \infty); \tag{4.5a}$$

note that $\mathbf{D}^\alpha : \mathbf{h}^\infty$ is bounded, although $\mathbf{C}^\alpha$ and $\mathbf{h}^\infty$ are zero for cavities. The tensor $\mathbf{j}^\infty$ becomes

$$\mathbf{j}^\infty(\Omega_\alpha, \mathbf{C}^\alpha; \mathbf{C}) = \frac{\kappa + 1}{\kappa - 1}\mathbf{E}^1 + \kappa + 1\mathbf{E}^2 \qquad (\mathbf{C}^\alpha \to \mathbf{0}). \tag{4.5b}$$

Hence, the overall tensors $\bar{\mathbf{D}}^{\mathrm{DD}}$ and $\bar{\mathbf{C}}^{\mathrm{DD}}$ become,

$$\begin{aligned}\bar{\mathbf{D}}^{\mathrm{DD}} &= \frac{\bar{\kappa}^{\mathrm{DD}} - 1}{4\bar{\mu}^{\mathrm{DD}}}\mathbf{E}^1 + \frac{1}{2\bar{\mu}^{\mathrm{DD}}}\mathbf{E}^2 \\ &= \left(\frac{\kappa - 1}{4\mu} + f\frac{\kappa + 1}{4\mu}\right)\mathbf{E}^1 + \left(\frac{1}{2\mu} + f\frac{\kappa + 1}{2\mu}\right)\mathbf{E}^2,\end{aligned} \tag{4.6a}$$

$$\begin{aligned}\bar{\mathbf{C}}^{\mathrm{DD}} &= \frac{4\bar{\mu}^{\mathrm{DD}}}{\bar{\kappa}^{\mathrm{DD}} - 1}\mathbf{E}^1 + 2\bar{\mu}^{\mathrm{DD}}\mathbf{E}^2 \\ &= \left(\frac{4\mu}{\kappa - 1} - f\frac{4\mu(\kappa + 1)}{(k - 1)^2}\right)\mathbf{E}^1 + (2\mu - f2\mu(\kappa + 1))\mathbf{E}^2.\end{aligned} \tag{4.6b}$$

The self-consistent estimate $\bar{\mathbf{D}}^{\mathrm{SC}}$ $(= \bar{\mathbf{C}}^{\mathrm{SC}-1})$ becomes

$$\begin{aligned}\bar{\mathbf{D}}^{\mathrm{SC}} &= \frac{\bar{\kappa}^{\mathrm{SC}} - 1}{4\bar{\mu}^{\mathrm{SC}}}\mathbf{E}^1 + \frac{1}{2\bar{\mu}^{\mathrm{SC}}}\mathbf{E}^2 \\ &= \left(\frac{\kappa - 1}{4\mu} + f\frac{\bar{\kappa}^{\mathrm{SC}} + 1}{4\bar{\mu}^{\mathrm{SC}}}\right)\mathbf{E}^1 + \left(\frac{1}{2\mu} + f\frac{\bar{\kappa}^{\mathrm{SC}} + 1}{2\bar{\mu}^{\mathrm{SC}}}\right)\mathbf{E}^2,\end{aligned} \tag{4.7a}$$

and by the differential scheme, $\bar{\mathbf{D}}^{\mathrm{DS}}$ $(= \bar{\mathbf{C}}^{\mathrm{DS}-1})$ is the solution of the following ordinary differential equation:

$$\begin{aligned}\frac{d}{df}\bar{\mathbf{D}}^{\mathrm{DS}} &= \frac{d}{df}\left(\frac{\bar{\kappa}^{\mathrm{DS}} - 1}{4\bar{\mu}^{\mathrm{DS}}}\mathbf{E}^1 + \frac{1}{2\bar{\mu}^{\mathrm{DS}}}\mathbf{E}^2\right) \\ &= \frac{1}{1 - f}\frac{\bar{\kappa}^{\mathrm{DS}} + 1}{4\bar{\mu}^{\mathrm{DS}}}\mathbf{E}^1 + \frac{1}{1 - f}\frac{\bar{\kappa}^{\mathrm{DS}} + 1}{2\bar{\mu}^{\mathrm{DS}}}\mathbf{E}^2.\end{aligned} \tag{4.8a}$$

From (4.6a, b), (4.7a), and (4.8a), the overall elastic moduli obtained by the three approximating methods are:

$$\frac{\bar{\mu}^{DD}}{\mu} = \begin{cases} \{1 + f(\kappa + 1)\}^{-1}, & (\boldsymbol{\Sigma} \text{ prescribed}), \\ 1 - f(\kappa + 1), & (\mathbf{E} \text{ prescribed}), \end{cases} \tag{4.6c}$$

$$\frac{\bar{\kappa}^{DD}}{\kappa} = \begin{cases} \left\{1 + f\dfrac{2(\kappa + 1)}{\kappa}\right\}\{1 + f(\kappa + 1)\}^{-1}, & (\boldsymbol{\Sigma} \text{ prescribed}), \\ \left\{1 - f\dfrac{(\kappa + 1)(\kappa^2 - 2\kappa + 2)}{\kappa(\kappa - 1)}\right\}\left(1 - f\dfrac{\kappa + 1}{\kappa - 1}\right)^{-1}, & (\mathbf{E} \text{ prescribed}), \end{cases} \tag{4.6d}$$

$$\frac{\bar{\mu}^{SC}}{\mu} = (1 - 3f)\{1 + f(\kappa - 2)\}^{-1}, \tag{4.7b}$$

$$\frac{\bar{\kappa}^{SC}}{\kappa} = \left(1 - f\frac{\kappa - 2}{\kappa}\right)\{1 + f(\kappa - 2)\}^{-1}, \tag{4.7c}$$

$$\frac{\bar{\mu}^{DS}}{\mu} = \left\{1 - \frac{\kappa + 1}{3}(1 - (1 - f)^{-3})\right\}^{-1}, \tag{4.8b}$$

$$\frac{\bar{\kappa}^{DS}}{\kappa} = \frac{\kappa + 1}{\kappa}\left\{1 - \frac{\kappa + 1}{3}(1 - (1 - f)^{-3}\right\}^{-1}(1 - f)^{-3} - \frac{1}{\kappa}. \tag{4.8c}$$

Figure 1 shows graphs of these overall shear moduli given by (4.6c), (4.7b),

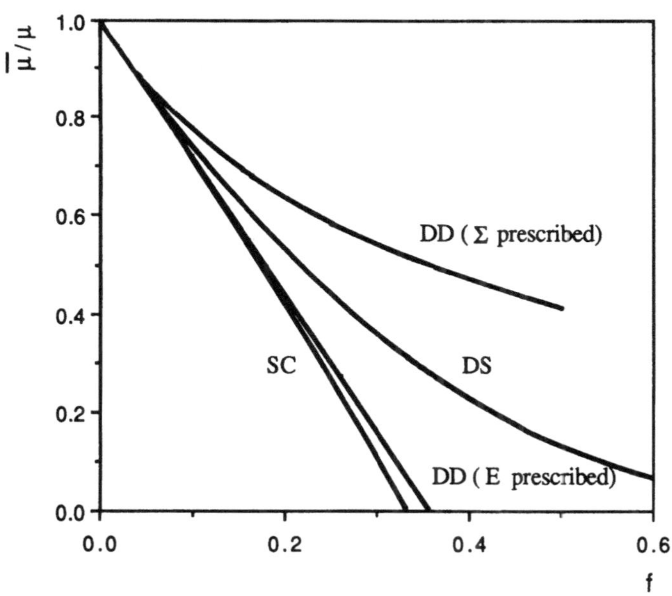

FIG. 1. Effective shear modulus of an elastic plane (plane strain) containing randomly distributed circular cavities: DD = dilute distribution, SC = self-consistent, and DS = differential scheme; the matrix Poisson ratio $\nu = 0.3$.

and (4.8b) plotted with respect to the volume fraction f; the Poisson ratio of the matrix, ν, is equal to 0.3, and $\kappa = 3 - 4\nu$ for the plane strain state. Note that these results are valid only for small void volume fractions, especially those obtained by the dilute distribution and the self-consistent assumptions.

5. Microcracks

Next, we consider three-dimensional problems of linearly elastic solids containing dilutely distributed penny-shaped microcracks. The location and size distributions are assumed to be random. For the orientation, we consider two cases: (1) all cracks are normal to, say, the x_3 axis; and (2) all cracks are parallel to the x_3 axis, having a random orientation normal to this axis. In either case, the overall elasticity and compliance tensors are transversely isotropic, with the x_1, x_2 plane being the plane of symmetry.

5.1. *Penny-Shaped Cracks*

As usual, we use a 6×6 matrix $[\mathrm{C}_{ab}]$ to represent the fourth-order symmetric elasticity tensor $\mathbf{C}$ where $\mathrm{C}_{ijkl} = \mathrm{C}_{jikl} = \mathrm{C}_{ijlk}$. The subscript a now stands for (ij), and ranges over 1 to 6 in accordance with $\{1, 2, 3, 4, 5, 6\} \equiv \{(11), (22), (33), (23), (31), (12)\}$. It is convenient to introduce the diagonal matrix $[W_{ab}]$ with $W_{aa} = 1$ for $a = 1, 2, 3$, and $W_{aa} = 2$ for $a = 4, 5, 6$. Then $\boldsymbol{\sigma} = \mathbf{C} : \boldsymbol{\varepsilon}$ corresponds to $[\sigma_a] = [\mathrm{C}_{ab}][W_{bc}][\varepsilon_c]$, where $[W_{ab}][\varepsilon_b] = [\varepsilon_{11}, \varepsilon_{22}, \varepsilon_{33}, 2\varepsilon_{23}, 2\varepsilon_{31}, 2\varepsilon_{12}]^{\mathrm{T}}$, with superscript T denoting transpose. If, in a similar manner, we define the 6×6 matrix $[\mathrm{D}_{ab}] = [\mathrm{D}_{(ij)(kl)}]$, then the compliance matrix becomes $[W_{ab}][\mathrm{D}_{bc}][W_{cd}] \equiv [D_{ab}]$. Now, since $\mathbf{I}$ corresponds to $[I_{ab}] = [W_{ab}]^{-1}$, $\mathbf{C} : \mathbf{D}$ corresponds to $[\mathrm{C}_{ab}][W_{bc}][\mathrm{D}_{cd}] = [W_{ad}]^{-1}$, and hence $[\mathrm{C}_{ab}][D_{cd}] = [\delta_{ac}]$.

For the isotropic case,

$$[\mathrm{D}_{ab}] = \frac{1}{E}\begin{bmatrix} 1 & -\nu & -\nu & 0 & 0 & 0 \\ & 1 & -\nu & 0 & 0 & 0 \\ & & 1 & 0 & 0 & 0 \\ & & & (1+\nu)/2 & 0 & 0 \\ & & & & (1+\nu)/2 & 0 \\ & & & & & (1+\nu)/2 \end{bmatrix}, \tag{5.1a}$$

$$[\mathrm{C}_{ab}] = \frac{E}{(1+\nu)(1-2\nu)}\begin{bmatrix} 1-\nu & \nu & \nu & 0 & 0 & 0 \\ & 1-\nu & \nu & 0 & 0 & 0 \\ & & 1-\nu & 0 & 0 & 0 \\ & & & (1-2\nu)/2 & 0 & 0 \\ & & & & (1-2\nu)/2 & 0 \\ & & & & & (1-2\nu)/2 \end{bmatrix}, \tag{5.1b}$$

where E and ν are the Young modulus and the Poisson ratio.

Take the $\mathbf{e}'_1, \mathbf{e}'_2, \mathbf{e}'_3$ triad such that $\mathbf{e}'_3$ coincides with the unit normal of a penny-shaped crack Ω_α, and $\mathbf{e}'_1$ is parallel to the x_1, x_2 plane. Introducing θ and ϕ, these triads can be related to the $\mathbf{e}_1, \mathbf{e}_2, \mathbf{e}_3$ triad by

$$\mathbf{e}'_1 = \cos\theta\,\mathbf{e}_1 + \sin\theta\,\mathbf{e}_2, \tag{5.2a}$$

$$\mathbf{e}'_2 = -\cos\phi\sin\theta\,\mathbf{e}_1 + \cos\phi\cos\theta\,\mathbf{e}_2 + \sin\phi\,\mathbf{e}_3, \tag{5.2b}$$

$$\mathbf{e}'_3 = \sin\phi\sin\theta\,\mathbf{e}_1 - \sin\phi\cos\theta\,\mathbf{e}_2 + \cos\phi\,\mathbf{e}_3. \tag{5.2c}$$

The strain $\bar{\boldsymbol{\varepsilon}}^\alpha$ due to the presence of a penny-shaped crack Ω_α of radius a can now be written as (Horii and Nemat-Nasser, 1983),

$$\bar{\boldsymbol{\varepsilon}}^\alpha \equiv \frac{1}{a_\alpha^3}\int_{\partial\Omega_\alpha^+} \tfrac{1}{2}\{\mathbf{n}\otimes[\mathbf{u}] + [\mathbf{u}]\otimes\mathbf{n}\}\,dS, \tag{5.3}$$

where $\partial\Omega_\alpha^+$ is the (upper) crack surface with unit normal $\mathbf{n}$, and $[\mathbf{u}]$ is the crack-opening displacement. In a manner similar to (4.1), (4.2), tensors $\mathbf{H}^{\mathrm{DD}\alpha}$ and $\mathbf{J}^{\mathrm{DD}\alpha}$ are defined by

$$\langle\boldsymbol{\varepsilon}(\mathbf{x};\boldsymbol{\Sigma})\rangle_\alpha \simeq \mathbf{D}^\alpha : \mathbf{h}^{\mathrm{DD}\alpha} : \boldsymbol{\Sigma} \equiv \mathbf{H}^{\mathrm{DD}\alpha} : \boldsymbol{\Sigma}, \tag{5.4a}$$

$$\mathbf{C} : \langle\boldsymbol{\varepsilon}(\mathbf{x};\mathbf{E})\rangle_\alpha \simeq \mathbf{C} : \mathbf{j}^{\mathrm{DD}\alpha} : \mathbf{E} \equiv \mathbf{J}^{\mathrm{DD}\alpha} : \mathbf{E} \qquad (\alpha \text{ not summed}). \tag{5.4b}$$

The tensor $\mathbf{H}^{\mathrm{DD}\alpha}$ has the following components in the $\mathbf{e}'_1, \mathbf{e}'_2, \mathbf{e}'_3$ coordinates:

$$\mathrm{H}^{\mathrm{DD}\alpha'}_{3i3i} = \frac{8(1-\nu^2)}{3E(2-\nu)}, \tag{5.5a}$$

$$\mathrm{H}^{\mathrm{DD}\alpha'}_{3333} = \frac{16(1-\nu^2)}{3E}, \qquad (i = 1, 2;\ i \text{ not summed}). \tag{5.5b}$$

From (3.12a), $\mathbf{H}^{\mathrm{DD}\alpha} = \mathbf{D}^\alpha : \mathbf{h}^{\mathrm{DD}\alpha}$ and $\mathbf{J}^{\mathrm{DD}\alpha} = \mathbf{C} : \mathbf{j}^{\mathrm{DD}\alpha}$ are related by $\mathbf{J}^{\mathrm{DD}\alpha} = \mathbf{C} : \mathbf{H}^{\mathrm{DD}\alpha} : \mathbf{C}$ (α not summed), and we have

$$\mathrm{J}^{\mathrm{DD}\alpha'}_{iijj} = \frac{16(1-\nu)\nu^2 E}{3(1+\nu)(1-2\nu)^2}, \tag{5.5c}$$

$$\mathrm{J}^{\mathrm{DD}\alpha'}_{ii33} = \frac{16(1-\nu)^2 \nu E}{3(1+\nu)(1-2\nu)^2}, \tag{5.5d}$$

$$\mathrm{J}^{\mathrm{DD}\alpha'}_{3333} = \frac{16(1-\nu)^3 E}{3(1+\nu)(1-2\nu)^2}, \tag{5.5e}$$

$$\mathrm{J}^{\mathrm{DD}\alpha'}_{3i3i} = \frac{8(1-\nu)E}{3(1+\nu)(2-\nu)}, \qquad (i, j = 1, 2;\ i, j \text{ not summed}). \tag{5.5f}$$

$\mathbf{H}^{\mathrm{DD}\alpha}$ and $\mathbf{J}^{\mathrm{DD}\alpha}$ are symmetric, although $\mathbf{j}^{\mathrm{DD}\alpha}$ is not.

Let the orientation distribution be defined by a density function $w = w(\theta, \phi)$, where

$$\frac{1}{4\pi}\int_0^\pi \int_0^{2\pi} w(\theta, \psi)\sin\phi\,d\theta\,d\phi = 1. \tag{5.6}$$

Denote the crack density by $f = N\langle a^3 \rangle$ (Budiansky and O'Connell, 1976). Then we have

$$\bar{\mathbf{D}}^{\mathrm{DD}} = \mathbf{D} + \frac{f}{4\pi}\int_0^{\pi}\int_0^{2\pi} \mathbf{H}^{\mathrm{DD}\alpha}(\theta, \phi) w(\theta, \psi) \sin\phi \, d\theta \, d\phi, \tag{5.7a}$$

$$\bar{\mathbf{C}}^{\mathrm{DD}} = \mathbf{C} - \frac{f}{4\pi}\int_0^{\pi}\int_0^{2\pi} \mathbf{J}^{\mathrm{DD}\alpha}(\theta, \phi) w(\theta, \psi) \sin\phi \, d\theta \, d\phi. \tag{5.7b}$$

5.2. *Aligned Penny-Shaped Cracks*

When all microcracks are normal to the x_3 axis, the orientation distribution function w vanishes except for $\phi = 0$, i.e., $w = 4\pi\delta(\phi)$. Hence, in the x_1, x_2, x_3 coordinates,

$$[\bar{D}_{ab}^{\mathrm{DD}}] = [D_{ab}] + f\frac{16(1-\nu^2)}{3E}\begin{vmatrix} 0 & 0 & 0 & 0 & 0 & 0 \\ & 0 & 0 & 0 & 0 & 0 \\ & & 1 & 0 & 0 & 0 \\ & & & 1/2(2-\nu) & 0 & 0 \\ & & & & 1/2(2-\nu) & 0 \\ & & & & & 0 \end{vmatrix}, \tag{5.8a}$$

$$[\bar{C}_{ab}^{\mathrm{DD}}] = [C_{ab}] - f\frac{16(1-\nu)E}{3(1+\nu)(1-2\nu)^2}$$

$$\times\begin{vmatrix} \nu^2 & \nu^2 & \nu(1-\nu) & 0 & 0 & 0 \\ & \nu^2 & \nu(1-\nu) & 0 & 0 & 0 \\ & & (1-\nu)^2 & 0 & 0 & 0 \\ & & & (1-2\nu)^2/2(2-\nu) & 0 & 0 \\ & & & & (1-2\nu)^2/2(2-\nu) & 0 \\ & & & & & 0 \end{vmatrix}, \tag{5.8b}$$

As is seen, the overall compliance and elasticity tensors are transversely isotropic with the x_1, x_2 plane being the plane of symmetry.

5.3. *Penny-Shaped Cracks Parallel to an Axis*

When all microcracks are parallel to the x_3 axis but their surface normals are randomly distributed in the x_1, x_2 plane, the orientation distribution function w becomes $2\delta(\phi - \pi/2)$. Using the rotation tensor $\mathbf{Q}$ defined by $\mathbf{e}_i' \otimes \mathbf{e}_i$, we can compute the overall moduli by the dilute distribution assumption, directly by integrating (5.7a, b). However, noting that $\bar{\mathbf{D}}^{\mathrm{DD}}$ and $\bar{\mathbf{C}}^{\mathrm{DD}}$ are transversely isotropic, their components can be obtained without such integration. To this end, let the 6 × 6 matrix for the tensor of the right-hand side of (5.7a) be

$$\left[\int(\cdots)_{ab}\right]=\begin{vmatrix} H_1 & H_2 & H_3 & 0 & 0 & 0\\ & H_1 & H_3 & 0 & 0 & 0\\ & & H_4 & 0 & 0 & 0\\ & & & H_5 & 0 & 0\\ & & & & H_5 & 0\\ & & & & & (H_1-H_2)/2 \end{vmatrix}. \tag{5.9}$$

Then, we must have

$$2H_1+2H_2+4H_3+H_4=f\frac{16(1-\nu^2)}{3E}, \tag{5.10a}$$

$$3H_1-H_2+H_4+H_5=f\frac{16(4-\nu)(1-\nu^2)}{3E(2-\nu)}, \tag{5.10b}$$

$$H_4+2H_5=f\frac{8(1-\nu^2)}{3E(2-\nu)}, \tag{5.10c}$$

$$2H_3+H_4=0, \tag{5.10d}$$

$$H_4=0. \tag{5.10e}$$

From (5.9) and (5.10a–e)

$$[\bar{D}^{DD}_{ab}]=[D_{ab}]+f\frac{2(1-\nu^2)}{3E(2-\nu)}\begin{vmatrix} 8-3\nu & -\nu & 0 & 0 & 0 & 0\\ & 8-3\nu & 0 & 0 & 0 & 0\\ & & 0 & 0 & 0 & 0\\ & & & 2 & 0 & 0\\ & & & & 2 & 0\\ & & & & & 4-\nu \end{vmatrix}. \tag{5.11a}$$

In a similar manner, $\bar{\mathbf{C}}^{DD}$ becomes

$$[\bar{C}^{DD}_{ab}]=[C_{ab}]-f\frac{2(1-\nu)E}{3(1+\nu)(2-\nu)(1-2\nu)^2}$$
$$\times\begin{vmatrix} -4\nu^3+20\nu^2-19\nu+8 & 4\nu^3-20\nu^2+15\nu & -4\nu^2+8\nu & 0 & 0 & 0\\ & -4\nu^3+20\nu^2-19\nu+8 & -4\nu^2+8\nu & 0 & 0 & 0\\ & & -8\nu^3+16\nu^2 & 0 & 0 & 0\\ & & & 2(1-2\nu)^2 & 0 & 0\\ & & & & 2(1-2\nu)^2 & 0\\ & & & & & (4-\nu)(1-2\nu)^2 \end{vmatrix}. \tag{5.11b}$$

From (5.8a, b) and (5.11a, b), the overall shear moduli in the x_1, x_2 plane, $\bar{\mu}^{DD}$, and the x_1, x_3 or x_2, x_3 directions, $\bar{\mu}^{DD}_3$, are: (1) for aligned cracks,

$$\frac{\bar{\mu}^{DD}}{\mu} = 1 \qquad (\mathbf{\Sigma} \text{ or } \mathbf{E} \text{ prescribed}), \tag{5.12a}$$

$$\frac{\bar{\mu}_3^{DD}}{\mu} = \begin{cases} \left\{1 + f\dfrac{16(1-\nu)}{3(2-\nu)}\right\}^{-1} & (\mathbf{\Sigma} \text{ prescribed}), \\ 1 - f\dfrac{16(1-\nu)}{3(2-\nu)} & (\mathbf{E} \text{ prescribed}); \end{cases} \tag{5.12b}$$

and (2) for cracks parallel to the x_3 axis,

$$\frac{\bar{\mu}^{DD}}{\mu} = \begin{cases} \left\{1 + f\dfrac{4(1-\nu)(4-\nu)}{3(2-\nu)}\right\}^{-1} & (\mathbf{\Sigma} \text{ prescribed}), \\ 1 - f\dfrac{4(1-\nu)(4-\nu)}{3(2-\nu)} & (\mathbf{E} \text{ prescribed}), \end{cases} \tag{5.13a}$$

$$\frac{\bar{\mu}_3^{DD}}{\mu} = \begin{cases} \left\{1 + f\dfrac{8(1-\nu)}{3(2-\nu)}\right\}^{-1} & (\mathbf{\Sigma} \text{ prescribed}), \\ 1 - f\dfrac{8(1-\nu)}{3(2-\nu)} & (\mathbf{E} \text{ prescribed}), \end{cases} \tag{5.13b}$$

where $\mu = E/2(1 + \nu)$. Figure 2 shows the graphs of these overall shear moduli

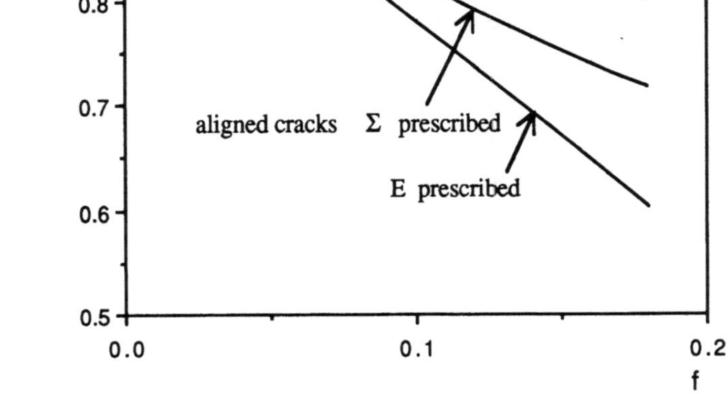

FIG. 2. Effective shear modulus of an elastic solid with random dilute distribution of penny-shaped cracks: (1) aligned normal to the x_3 direction, and (2) parallel to the x_3 direction; the matrix Poisson ratio $\nu = 0.3$.

given by (5.12b) and (5.13b) for the x_1, x_2 and x_2, x_3 directions with respect to the volume fraction f; the Poisson ratio of the matrix, ν, is equal to 0.3.

6. Micro-Inclusions

The problem of an RVE with a linearly elastic matrix and microinclusions can best be formulated and solved with the aid of Eshelby's tensor for a dilute distribution, the self-consistent method, and the differential scheme. To compare results, we consider a solid containing spherical inclusions.

For a spherical inclusion in an isotropic uniform matrix of Poisson's ratio ν, Eshelby's tensor is (Mura, 1982)

$$\mathbf{S} = s_1 \mathbf{E}^1 + s_2 \mathbf{E}^2, \tag{6.1a}$$

where

$$s_1 = \frac{1+\nu}{3(1-\nu)}, \tag{6.1b}$$

$$s_2 = \frac{2(4-5\nu)}{15(1-\nu)}, \tag{6.1c}$$

and $\mathbf{E}^1$ and $\mathbf{E}^2$ are the three-dimensional orthogonal base tensors; their components are

$$E^1_{ijkl} = \tfrac{1}{3}\delta_{ij}\delta_{kl}, \tag{6.2a}$$

$$E^2_{ijkl} = -\tfrac{1}{3}\delta_{ij}\delta_{kl} + \tfrac{1}{2}(\delta_{ik}\delta_{jl} + \delta_{il}\delta_{jk}). \tag{6.2b}$$

Note that in terms of $\mathbf{E}^1$ and $\mathbf{E}^2$ the three-dimensional isotropic compliance and elasticity tensors are expressed as

$$\mathbf{D} = \frac{1}{3K}\mathbf{E}^1 + \frac{1}{2\mu}\mathbf{E}^2, \tag{6.3a}$$

$$\mathbf{C} = 3K\mathbf{E}^1 + 2\mu\mathbf{E}^2, \tag{6.3b}$$

where K and μ are the bulk and shear moduli; $\nu = (3K - 2\mu)/2(3K + \mu)$.

For a dilute distribution of inclusions, and with *macrostress* $\boldsymbol{\Sigma}$ *assumed to be prescribed*, the overall compliance tensors are given by

$$\begin{aligned}\bar{\mathbf{D}}^{\mathrm{DD}} &= \left\{\mathbf{I} + \sum_{\alpha=1}^{n} f_\alpha (\mathbf{A}^{\mathrm{DD}\alpha} - \mathbf{S}^{\mathrm{DD}})^{-1}\right\} : \mathbf{D} \\ &= \frac{1}{3K}\left\{1 + \sum_{\alpha=1}^{n} f_\alpha \left(\frac{K}{K-K^\alpha} - s_1^{\mathrm{DD}}\right)^{-1}\right\}\mathbf{E}^1 \\ &\quad + \frac{1}{2\mu}\left\{1 + \sum_{\alpha=1}^{n} f_\alpha \left(\frac{\mu}{\mu-\mu^\alpha} - s_2^{\mathrm{DD}}\right)^{-1}\right\}\mathbf{E}^2,\end{aligned} \tag{6.4a}$$

where $\mathbf{S}^{\mathrm{DD}} = \mathbf{S}$ and

$$\mathbf{A}^{\mathrm{DD}\alpha} \equiv (\mathbf{C} - \mathbf{C}^\alpha)^{-1} : \mathbf{C} = \frac{K}{K-K^\alpha}\mathbf{E}^1 + \frac{\mu}{\mu-\mu^\alpha}\mathbf{E}^2, \tag{6.4b}$$

and K^α and μ^α are the bulk and shear moduli of isotropic microinclusion Ω_α. With *macrostrain* **E** *assumed to be prescribed*, on the other hand, we have

$$\begin{aligned}\bar{\mathbf{C}}^{\mathrm{DD}} &= \mathbf{C} : \left\{ \mathbf{I} - \sum_{\alpha=1}^{n} f_\alpha (\mathbf{A}^{\mathrm{DD}\alpha} - \mathbf{S}^{\mathrm{DD}})^{-1} \right\} \\ &= 3K \left\{ 1 - \sum_{\alpha=1}^{n} f_\alpha \left(\frac{K}{K - K^\alpha} - s_1^{\mathrm{DD}} \right)^{-1} \right\} \mathbf{E}^1 \\ &\quad + 2\mu \left\{ 1 - \sum_{\alpha=1}^{n} f_\alpha \left(\frac{\mu}{\mu - \mu^\alpha} - s_2^{\mathrm{DD}} \right)^{-1} \right\} \mathbf{E}^2. \end{aligned} \tag{6.5}$$

We thus have, for $\boldsymbol{\Sigma}$ prescribed,

$$\frac{\bar{K}^{\mathrm{DD}}}{K} = \left\{ 1 + \sum_{\alpha=1}^{n} f_\alpha \left(\frac{K}{K - K^\alpha} - s_1^{\mathrm{DD}} \right)^{-1} \right\}^{-1}, \tag{6.6a}$$

$$\frac{\bar{\mu}^{\mathrm{DD}}}{\mu} = \left\{ 1 + \sum_{\alpha=1}^{n} f_\alpha \left(\frac{\mu}{\mu - \mu^\alpha} - s_2^{\mathrm{DD}} \right)^{-1} \right\}^{-1}, \tag{6.6b}$$

and, for **E** prescribed,

$$\frac{\bar{K}^{\mathrm{DD}}}{K} = 1 - \sum_{\alpha=1}^{n} f_\alpha \left(\frac{K}{K - K^\alpha} - s_1^{\mathrm{DD}} \right)^{-1}, \tag{6.7a}$$

$$\frac{\bar{\mu}^{\mathrm{DD}}}{\mu} = 1 - \sum_{\alpha=1}^{n} f_\alpha \left(\frac{\mu}{\mu - \mu^\alpha} - s_2^{\mathrm{DD}} \right)^{-1}. \tag{6.7b}$$

In the self-consistent approach, the overall elasticity tensor $\bar{\mathbf{C}}^{\mathrm{SC}} (= \bar{\mathbf{D}}^{\mathrm{SC}-1})$ is

$$\begin{aligned}\bar{\mathbf{C}}^{\mathrm{SC}} &= \mathbf{C} + \sum_{\alpha=1}^{n} f_\alpha (\mathbf{C}^\alpha - \mathbf{C}) : \mathbf{A}^{\mathrm{SC}\alpha} : (\mathbf{A}^{\mathrm{SC}\alpha} - \mathbf{S}^{\mathrm{SC}})^{-1} \\ &= 3 \left\{ K + \sum_{\alpha=1}^{n} f_\alpha \bar{K}^{\mathrm{SC}} \frac{K^\alpha - K}{\bar{K}^{\mathrm{SC}} - K^\alpha} \left(\frac{\bar{K}^{\mathrm{SC}}}{\bar{K}^{\mathrm{SC}} - K^\alpha} - s_1^{\mathrm{SC}} \right)^{-1} \right\} \mathbf{E}^1 \\ &\quad + 2 \left\{ \mu + \sum_{\alpha=1}^{n} f_\alpha \bar{\mu}^{\mathrm{SC}} \frac{\mu^\alpha - \mu}{\bar{\mu}^{\mathrm{SC}} - \mu^\alpha} \left(\frac{\bar{\mu}^{\mathrm{SC}}}{\bar{\mu}^{\mathrm{SC}} - \mu^\alpha} - s_2^{\mathrm{SC}} \right)^{-1} \right\} \mathbf{E}^2, \end{aligned} \tag{6.8}$$

where $\mathbf{S}^{\mathrm{SC}} = s_1^{\mathrm{SC}} \mathbf{E}^1 + s_2^{\mathrm{SC}} \mathbf{E}^2$, and s_1^{SC} and s_2^{SC} are obtained from (6.1b, c) with ν replaced by $\bar{\nu}^{\mathrm{SC}}$; $\mathbf{A}^{\mathrm{SC}\alpha}$ is obtained from (6.4b) with K and μ replaced by $\bar{K}^{\mathrm{SC}}$ and $\bar{\mu}^{\mathrm{SC}}$, respectively; and $\bar{\nu}^{\mathrm{SC}} = (3\bar{K}^{\mathrm{SC}} - 2\bar{\mu}^{\mathrm{SC}})/2(3\bar{K}^{\mathrm{SC}} + \bar{\mu}^{\mathrm{SC}})$. We thus have

$$\frac{\bar{K}^{\mathrm{SC}}}{K} = 1 + \sum_{\alpha=1}^{n} f_\alpha \left(\frac{K^\alpha}{K} - 1 \right) \left\{ 1 + \left(\frac{K^\alpha}{\bar{K}^{\mathrm{SC}}} - 1 \right) s_1^{\mathrm{SC}} \right\}^{-1}, \tag{6.9a}$$

$$\frac{\bar{\mu}^{\mathrm{SC}}}{\mu} = 1 + \sum_{\alpha=1}^{n} f_\alpha \left(\frac{\mu^\alpha}{\mu} - 1 \right) \left\{ 1 + \left(\frac{\mu^\alpha}{\bar{\mu}^{\mathrm{SC}}} - 1 \right) s_2^{\mathrm{SC}} \right\}^{-1}. \tag{6.9b}$$

Finally, using the differential scheme and in view of (6.3b), we obtain

$$\frac{\partial}{\partial f_\alpha} \bar{\mathbf{C}}^{\mathrm{DS}} = -\frac{1}{1 - f_\alpha} \bar{\mathbf{C}}^{\mathrm{DS}} : (\mathbf{A}^{\mathrm{DS}\alpha} - \mathbf{S}^{\mathrm{DS}})^{-1} \qquad (\alpha \text{ not summed}), \tag{6.10}$$

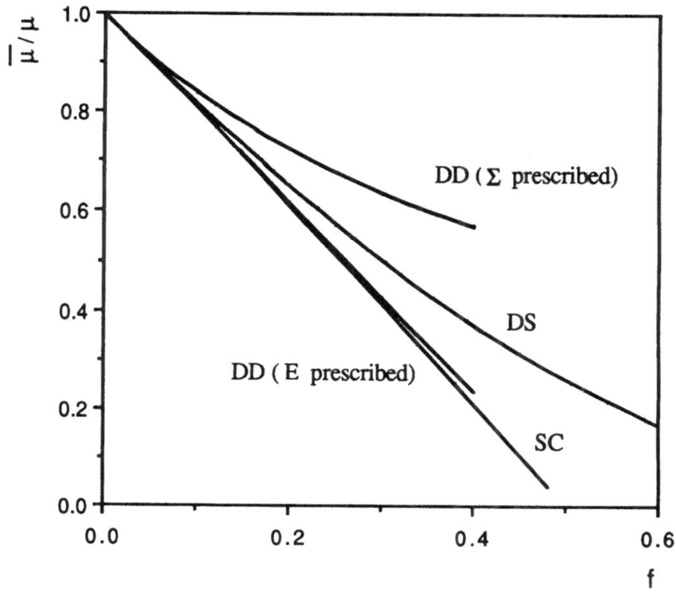

FIG. 3. Effective shear modulus of an elastic solid with randomly distributed spherical cavities: DD = dilute distribution, SC = self-consistent, and DS = differential scheme; the matrix Poisson ratio $\nu = 0.3$.

where $\mathbf{S}^{DS} = s_1^{DS}\mathbf{E}^1 + s_2^{DS}\mathbf{E}^2$, and s_1^{DS} and s_2^{DS} are obtained from (6.1b, c) with ν replaced by $\bar{\nu}^{DS}$; $\mathbf{A}^{DS\alpha}$ is obtained from (6.4b) with K and μ replaced by $\bar{K}^{DS}$ and $\bar{\mu}^{DS}$, respectively; and $\bar{\nu}^{DS} = (3\bar{K}^{DS} - 2\bar{\mu}^{DS})/2(3\bar{K}^{DS} + \bar{\mu}^{DS})$. The overall moduli are then given by the solution of the following differential equations:

$$\frac{\partial}{\partial f_\alpha}\bar{K}^{DS} = -\frac{1}{1-f_\alpha}\bar{K}^{DS}\left(\frac{\bar{K}^{DS}}{\bar{K}^{DS} - K^\alpha} - s_1^{DS}\right)^{-1}, \tag{6.11a}$$

$$\frac{\partial}{\partial f_\alpha}\bar{\mu}^{DS} = -\frac{1}{1-f_\alpha}\bar{\mu}^{DS}\left(\frac{\bar{\mu}^{DS}}{\bar{\mu}^{DS} - \mu^\alpha} - s_2^{DS}\right)^{-1} \quad (\alpha \text{ not summed}). \tag{6.11b}$$

The effective moduli of an elastic solid with spherical voids are obtained if we set $K^\alpha = \mu^\alpha = 0$. Figure 3 shows the graph of the overall shear modulus for this case, estimated by the three approximating schemes, where the Poisson ratio of the matrix, ν, is 0.3. Note that these estimates are valid for very small void volume fractions only.

Acknowledgment

This work has been supported in part by the National Science Foundation under grant No. MSM-86-15361, and in part by the U.S. Army Research Office under Contract No. DAAL-03-86-K-0169, to the University of California, San Diego.

References

Budiansky, B. (1965), On the elastic moduli of some heterogeneous materials, *J. Mech. Phys. Solids*, **13**, 223–227.

Budiansky, B. and O'Connell, R. J. (1976), Elastic moduli of a cracked solid, *Int. J. Solids Structures*, **12**, 81–97.

Hashin, Z. (1983), Analysis of composite materials, *J. Appl. Mech.*, **50**, 481–505.

Hashin, Z. (1988), The differential scheme and its application to cracked materials, *J. Mech. Phys. Solids*, **36**, 719–733.

Hershey, A. V. (1954), The elasticity of an isotropic aggregate of anisotropic cubic crystals, *J. Appl. Mech.*, **21**, 236–241.

Hill, R. (1965), A self-consistent mechanics of composite materials, *J. Mech. Phys. Solids*, **13**, 213–222.

Hill, R. (1967), The essential structure of constitutive laws for metal composites and polycrystals, *J. Mech. Phys. Solids*, **15**, 79–95.

Hutchinson, J. W. (1987), *Micro-mechanics of Damage in Deformation and Fracture*, Technical University of Denmark, Denmark.

Horii, H. and Nemat-Nasser, S. (1983), Overall moduli of solids with microcracks: load-induced anisotropy, *J. Mech. Phys. Solids*, **31**, 155–171.

Kröner, E. (1958), *Kontinuumstheorie der Versetzungen und Eigenspannungen*, Springer-Verlag, Berlin.

McLaughlin, R. (1977), A study of the differential scheme for composite materials, *Int. J. Engng. Sci.*, **15**, 237–244.

Mura, T. (1982), *Micromechanics of Defects in Solids*, Martinus Nijhoff, The Hague.

Roscoe, R. A. (1952), The viscosity of suspensions of rigid spheres, *Brit. J. Appl. Phys.*, **3**, 267–269.

Roscoe, R. A. (1973), Isotropic composites with elastic or viscoelastic phases: General bounds for the moduli and solutions for special geometries, *Rheol. Acta*, **12**, 404–411.

Reuss, A. (1929), Berechnung der Fliessgrenze von Mischkristallen auf Grund der Plastizitätsbedingung für Einkristalle, *Z. Angew. Math. Mech.*, **9**, 49–58.

Voigt, W. (1889), Über die Beziehung zwischen den beiden Elastizitätskonstanten isotroper Körper, *Wied. Ann.*, **38**, 573–587.

The Effective Moduli of Layered Media— A New Look at an Old Problem

A. N. NORRIS
Department of Mechanics and Materials Science, College of Engineering,
Rutgers University
Piscataway, NJ 08855-0909, U.S.A.

Abstract

A layered medium is one of the few composite geometries for which one can ascertain exact closed-form expressions for the effective elastic moduli relating the average stress and strain. The method of solution follows from the observation that under homogeneous loading three of the six stress components and three of the six strain components are constant from layer to layer. The other six stress and strain components can then be calculated in each layer, and by averaging over the composite, the effective moduli can be determined for an arbitrary number of anisotropic, parallel layers (e.g., Pagano, 1974). The general procedure is developed here in a coordinate-invariant manner using Hill's (1972) concept of the interior and exterior parts of second-order tensors. Two limiting types of layers can be described quite simply in this formalism: sets of aligned weak joints are treated as thin, compliant layers, and conversely, planar stiffeners are modeled as thin but very stiff layers. The net effect of adding weak joints or stiffeners is to increase the overall compliance or stiffness, respectively, by an amount that depends only upon the mechanical properties of the added thin layers. It is shown that a dilute concentration of aligned penny-shaped cracks in an isotropic medium is equivalent in effect to a set of weak joints with the same alignment. The Poisson ratio of the equivalent isotropic joints turns out to be strictly less than -1, independent of crack concentration, and therefore the equivalent layer corresponds to an unrealistic material. The general results on the effective properties are used to generate useful formulas for determining the thermoelastic moduli of fiber reinforced multi-ply laminates.

1. Introduction

Researchers in several fields of mechanics have long been interested in the effective properties of layered media. The full set of effective moduli are necessary, for example, for modeling the propagation of ultrasonic waves through laminates, and seismic waves through layered geological strata. In the theory of laminated composites, equations for the effective properties of materials composed of laminated monoclinic layers were obtained by Chou *et al.* (1972), and soon after that Pagano (1974) gave an explicit procedure for determining both the effective elastic constants and the thermal expansion coefficients for laminates with layers of arbitrary material symmetry. These results for the 21 effective elastic constants include the six moduli of the classical theory of thin laminates (Christensen, 1979), for which the out-of-plane stresses are zero.

Effective medium theory is also relevant to the propagation of long-wavelength elastic waves through stratified layers of rock. In this regard, Backus (1962) showed that stackings of transversely isotropic (TI) media with common symmetry axis in the layering direction are equivalent to an effectively homogeneous TI medium, and gave explicit formulas for calculating the effective moduli in terms of the constituent properties. The basic assumption is that each layer is thin relative to the wavelength, and that the statistical distribution of layers remains constant over many wavelengths. An effective TI medium does not, however, produce the shear-wave splitting observed for vertically propagating waves in sedimentary basins (Schoenberg and Muir, 1989, and references therein), i.e., distinct shear-wave speeds are seen. It has been suggested that this phenomenon is caused by the presence of aligned microcrack systems induced by anisotropic stress fields. Motivated by the possibility of detecting and characterizing these crack systems and other types of faults, there has been renewed interest in calculating the properties of layered geological systems (Schoenberg and Douma, 1988; Schoenberg and Muir, 1989). Some of the layers may be obliquely oriented with respect to the others, so it is necessary to devise an efficient procedure for layering when the normal to the layers is not a coordinate axis. The procedure developed here does not require the use of complicated transformation matrices associated with changes in coordinate axes, but rather uses concise, explicit tensor notation as far as possible. A consequence of the tensor formulation is that the change in properties due to thin layers of weak joints, called fractures by Schoenberg and Douma (1988), can be described in a relatively simple and transparent manner.

The mathematical preliminaries are developed in Section 2 for formulating the effective properties in coordinate invariant tensor notation. These are applied to the layered medium in Section 3, where some mention is also made to the case of two phase media and the relation of the present theory to the theory of finite rank laminates. Special limiting cases are considered in Section 4, including the joint or fracture model of Schoenberg and Douma (1988). In Section 5 the connection is made between the joint model and the change in compliance due to a dilute concentration of penny-shaped cracks. The formalism of Section 3 is recast in Section 6 using 6×6 matrix notation for the stiffness tensors, with applications to the theory of laminated fiber-reinforced composites discussed in Section 7.

2. Mathematical Preliminaries

2.1. Interior and Exterior Tensors

Regarding notation, lowercase Greek letters denote vectors, with some exceptions, such as $\boldsymbol{\sigma}$ and $\boldsymbol{\varepsilon}$. Lowercase Latin letters signify symmetric second-order tensors, and capital letters are symmetric fourth-order tensors. The components are v_i, a_{ij}, A_{ijkl}, where $i, j, k, l = 1, 2, 3$ refer to arbitrary rectangular axes. Products are understood as $(av)_i = a_{ij}v_j$, $(ab)_{ij} = a_{ik}b_{kj}$, $(Aa)_{ij} = A_{ijkl}a_{kl}$, and the summation convention on repeated subscripts is implicit. The symmetries are represented by $a_{ij} = a_{ji}$, $A_{ijkl} = A_{jikl} = A_{klij}$, and the identity tensors are $\mathbf{I}$, $I_{ijkl} = \frac{1}{2}(\delta_{ik}\delta_{jl} + \delta_{il}\delta_{jk})$, and $\mathbf{i}$, $i_{kl} = \delta_{kl}$.

Basic to our analysis are the tensor operators $\mathbf{E}$ and $\mathbf{F}$ introduced by Hill (1972) in considering the stress and strain in the neighborhood of a bonded interface of normal $\mathbf{v}$ between dissimilar materials

$$E_{ijkl} = \tfrac{1}{2}(\delta_{ik}v_jv_l + \delta_{jk}v_iv_l + \delta_{il}v_jv_k + \delta_{jl}v_iv_k) - v_iv_jv_kv_l, \tag{2.1}$$

$$F_{ijkl} = \tfrac{1}{2}(\delta_{ik} - v_iv_k)(\delta_{jl} - v_jv_l) + \tfrac{1}{2}(\delta_{jk} - v_jv_k)(\delta_{il} - v_iv_l), \tag{2.2}$$

so that

$$\mathbf{E} + \mathbf{F} = \mathbf{I}. \tag{2.3}$$

Extensive discussions of these operators and further references may be found in the review article by Walpole (1981), and also in Hill (1983). Let Ω be the space of symmetric second-order tensors, and define the subspaces

$$\Omega_E = \{\mathbf{a} \in \Omega\colon \mathbf{Ea} = \mathbf{a}\}, \tag{2.4}$$

$$\Omega_F = \{\mathbf{a} \in \Omega\colon \mathbf{Fa} = \mathbf{a}\}, \tag{2.5}$$

For any $\mathbf{a} \in \Omega$, $\mathbf{Ea}$ and $\mathbf{Fa}$ are called the *exterior* and *interior* parts of $\mathbf{a}$, respectively, and Ω_E and Ω_F are subspaces of the interior and exterior second-order tensors. The operators $\mathbf{E}$ and $\mathbf{F}$ are projection operators onto Ω_E and Ω_F, respectively; that is, for any $\mathbf{a} \in \Omega$, $\mathbf{Ea} \in \Omega_E$, $\mathbf{Fa} \in \Omega_F$,

$$\mathbf{E}^2 = \mathbf{E}, \qquad \mathbf{F}^2 = \mathbf{F}, \tag{2.6}$$

and since Ω_E and Ω_F are disjoint, i.e., $\Omega = \Omega_E \oplus \Omega_F$, it follows that

$$\mathbf{EF} = \mathbf{FE} = 0. \tag{2.7}$$

2.2. Inverse Operators

Let $\mathbf{L}$ be a symmetric fourth-order tensor, which will subsequently be used exclusively to denote the elastic stiffness. Define the inverse $\mathbf{A}\colon \Omega_E \to \Omega_E$

$$\mathbf{A} = (\mathbf{ELE})^{-1}. \tag{2.8}$$

$\mathbf{A}$ can be determined by first noting that each $\mathbf{a} \in \Omega_E$ can be expressed as

$$\mathbf{a} = \tfrac{1}{2}(\mathbf{v} \otimes \boldsymbol{\eta} + \boldsymbol{\eta} \otimes \mathbf{v}) \tag{2.9}$$

for some $\boldsymbol{\eta}$. The problem then is to find $\mathbf{a} \in \Omega_E$, given $\mathbf{b}$ where

$$\mathbf{b} = (\mathbf{ELE})\mathbf{a}. \tag{2.10}$$

But $\mathbf{Ea} = \mathbf{a}$, and using (2.9)

$$\mathbf{b}\mathbf{v} = \mathbf{f}\boldsymbol{\eta}, \tag{2.11}$$

where

$$f_{ik} = L_{ijkl} v_j v_l. \tag{2.12}$$

Equation (2.11) can be solved for $\boldsymbol{\eta}$, and thus $\mathbf{a}$ from (2.9): in summary

$$A_{ijkl} = \tfrac{1}{4}(f_{ik}^{-1} v_j v_l + f_{il}^{-1} v_j v_k + f_{jk}^{-1} v_i v_l + f_{jl}^{-1} v_i v_k). \tag{2.13}$$

The inverse to $\mathbf{L}$ is $\mathbf{M}$, which henceforth denotes the elastic compliance, and $\mathbf{LM} = \mathbf{ML} = \mathbf{I}$. The operator dual to $\mathbf{A}$ is $\mathbf{B}$ defined as $\mathbf{B}\colon \Omega_F \to \Omega_F$

$$\mathbf{B} = (\mathbf{FMF})^{-1}. \tag{2.14}$$

It can be shown without much difficulty that $\mathbf{A}$ and $\mathbf{B}$ are connected through the relations (Hill, 1983)

$$\mathbf{LA} + \mathbf{BM} = \mathbf{AL} + \mathbf{MB} = \mathbf{I}. \tag{2.15}$$

Simple, explicit expressions exist for $\mathbf{A}$ and $\mathbf{B}$ when the material is isotropic with shear modulus μ and Poisson's ratio η (Hill, 1983)

$$\mathbf{A} = \frac{1}{2\mu}\left[\mathbf{E} - \frac{\eta}{1-\eta}\mathbf{v} \otimes \mathbf{v} \otimes \mathbf{v} \otimes \mathbf{v}\right], \tag{2.16}$$

$$\mathbf{B} = 2\mu\left[\mathbf{F} + \frac{\eta}{1-\eta}(\mathbf{i} - \mathbf{v} \otimes \mathbf{v}) \otimes (\mathbf{i} - \mathbf{v} \otimes \mathbf{v})\right]. \tag{2.17}$$

2.3. *Jump Conditions at an Interface*

Let the stress, strain, and temperature deviation on either side of a bonded interface between dissimilar solids be $(\boldsymbol{\sigma}, \boldsymbol{\varepsilon}, \theta)$ and $(\boldsymbol{\sigma}^*, \boldsymbol{\varepsilon}^*, \theta^*)$. On the side with $(\boldsymbol{\sigma}, \boldsymbol{\varepsilon}, \theta)$, the constitutive relation is

$$\boldsymbol{\varepsilon} = \mathbf{M}\boldsymbol{\sigma} + \theta\boldsymbol{\alpha}, \tag{2.18}$$

where $\boldsymbol{\alpha}$ is the thermal expansion tensor. The continuity conditions across the interface, with normal $\mathbf{v}$, are

$$\mathbf{E}\boldsymbol{\sigma} = \mathbf{E}\boldsymbol{\sigma}^*, \qquad \mathbf{F}\boldsymbol{\varepsilon} = \mathbf{F}\boldsymbol{\varepsilon}^*, \qquad \theta = \theta^*. \tag{2.19}$$

These may be solved to give $(\boldsymbol{\sigma}, \boldsymbol{\varepsilon}, \theta)$ explicitly in terms of $(\sigma^*, \varepsilon^*, \theta^*)$ and the one-sided material constants $(\mathbf{L}, \boldsymbol{\alpha})$. The solution is well known (e.g., Hill, 1983) but is brief enough that we summarize it here. It follows from (2.19.2) that $(\boldsymbol{\varepsilon} - \boldsymbol{\varepsilon}^*) \in \Omega_E$, and therefore (2.19.1) may be written using (2.18) as

$$\mathbf{ELE}(\boldsymbol{\varepsilon} - \boldsymbol{\varepsilon}^*) = \mathbf{E}(\boldsymbol{\sigma}^* - \mathbf{L}\boldsymbol{\varepsilon}^* + \theta^*\mathbf{L}\boldsymbol{\alpha}). \tag{2.20}$$

The desired relation follows from (2.8)

$$\boldsymbol{\varepsilon} = \boldsymbol{\varepsilon}^* + \mathbf{A}(\boldsymbol{\sigma}^* - \mathbf{L}\boldsymbol{\varepsilon}^* + \theta^*\mathbf{L}\boldsymbol{\alpha}), \tag{2.21}$$

with obvious dual

$$\boldsymbol{\sigma} = \boldsymbol{\sigma}^* + \mathbf{B}(\boldsymbol{\varepsilon}^* - \mathbf{M}\boldsymbol{\sigma}^* - \theta^*\boldsymbol{\alpha}). \tag{2.22}$$

3. Effective Moduli of a Layered Medium

3.1. General Formulation

We ignore end effects from the sides of the composite, and assume the top and bottom surfaces are subject to uniform loadings. Then the stress and strain are piecewise constant from layer to layer, and in particular, the exterior part of the stress, $\mathbf{E}\boldsymbol{\sigma}$, and the interior part of the strain, $\mathbf{F}\boldsymbol{\varepsilon}$, are uniformly constant in all layers. The temperature is also assumed as constant throughout the composite. With $\mathbf{v}$ as the common normal to the interfaces we begin by rewriting (2.21) and (2.22), using (2.15) and (2.19), as

$$\boldsymbol{\varepsilon} = \mathbf{A}(\mathbf{E}\boldsymbol{\sigma}) + \mathbf{M}\mathbf{B}(\mathbf{F}\boldsymbol{\varepsilon}) + \theta\mathbf{A}\mathbf{L}\boldsymbol{\alpha}, \tag{3.1}$$

$$\boldsymbol{\sigma} = \mathbf{L}\mathbf{A}(\mathbf{E}\boldsymbol{\sigma}) + \mathbf{B}(\mathbf{F}\boldsymbol{\varepsilon}) - \theta\mathbf{B}\boldsymbol{\alpha}. \tag{3.2}$$

Applying these identities to the composite, layer by layer, we note that the right-hand sides involve only the constant parts of the stress and strain, viz. $\mathbf{E}\boldsymbol{\sigma}$ and $\mathbf{F}\boldsymbol{\varepsilon}$. Equations (3.1) and (3.2) thus determine the local stress and strain in each layer in terms of the global constants $\mathbf{E}\boldsymbol{\sigma}$, $\mathbf{F}\boldsymbol{\varepsilon}$, and θ. This local correction to the average stress and strain is analogous to the first-order corrections in the theory of homogenization (Auriault and Bonnet, 1987).

Let the bracket notation $\langle\cdot\rangle$ denote the volume average of a quantity over the entire composite. The macroscopic stress and strain are defined as the averages

$$\bar{\boldsymbol{\sigma}} = \langle\boldsymbol{\sigma}\rangle, \qquad \bar{\boldsymbol{\varepsilon}} = \langle\boldsymbol{\varepsilon}\rangle. \tag{3.3}$$

The global constants $\mathbf{E}\boldsymbol{\sigma}$ and $\mathbf{F}\boldsymbol{\varepsilon}$ are identical to $\mathbf{E}\bar{\boldsymbol{\sigma}}$ and $\mathbf{F}\bar{\boldsymbol{\varepsilon}}$ since $\mathbf{E}$ and $\mathbf{F}$ are constants. Therefore, averaging (3.1) and (3.2) gives

$$\bar{\boldsymbol{\varepsilon}} = \langle\mathbf{M}\mathbf{B}\rangle\bar{\boldsymbol{\varepsilon}} + \langle\mathbf{A}\rangle\mathbf{E}\bar{\boldsymbol{\sigma}} + \langle\mathbf{A}\mathbf{L}\boldsymbol{\alpha}\rangle\theta, \tag{3.4}$$

$$\bar{\boldsymbol{\sigma}} = \langle\mathbf{B}\rangle\bar{\boldsymbol{\varepsilon}} + \langle\mathbf{L}\mathbf{A}\rangle\mathbf{E}\bar{\boldsymbol{\sigma}} - \langle\mathbf{B}\boldsymbol{\alpha}\rangle\theta, \tag{3.5}$$

where we have used $\mathbf{B} = \mathbf{B}\mathbf{F}$. We note in passing that in the classical theory of thin laminates (e.g., Christensen, 1979) the exterior part of the stress, $\mathbf{E}\boldsymbol{\sigma}$, is identically zero, and so the plane-stress moduli of thin laminates follow from (3.5) as $\langle\mathbf{B}\rangle$. Eliminating $\mathbf{E}\bar{\boldsymbol{\sigma}}$ from (3.4) and (3.5) yields the macroscopic constitutive relation for arbitrary states of stress

$$\bar{\boldsymbol{\sigma}} = \bar{\mathbf{L}}\bar{\boldsymbol{\varepsilon}} - \theta\bar{\mathbf{L}}\bar{\boldsymbol{\alpha}}, \tag{3.6}$$

where the effective elastic stiffness is

$$\bar{\mathbf{L}} = \langle \mathbf{B} \rangle + \langle \mathbf{LA} \rangle \bar{\mathbf{G}} \langle \mathbf{AL} \rangle, \tag{3.7}$$

the effective thermal expansion tensor is

$$\bar{\boldsymbol{\alpha}} = \bar{\mathbf{M}}(\langle \mathbf{L}A \rangle \bar{\mathbf{G}} \langle \mathbf{AL}\boldsymbol{\alpha} \rangle + \langle \mathbf{B}\boldsymbol{\alpha} \rangle), \tag{3.8}$$

$\bar{\mathbf{M}}$ is inverse to $\bar{\mathbf{L}}$, and $\bar{\mathbf{G}}: \Omega_E \to \Omega_E$,

$$\bar{\mathbf{G}} = \langle \mathbf{A} \rangle^{-1}. \tag{3.9}$$

Equations (3.7) and (3.8) are the fundamental results of this paper, and all the quantities in these equations are well defined except for $\bar{\mathbf{G}}$ which is an operator on Ω_E. Now, all $\mathbf{a}$ and $\mathbf{b} \in \Omega_E$ can be written in the form of (2.9), i.e., $\mathbf{a} = \frac{1}{2}(\mathbf{v} \otimes \boldsymbol{\alpha} + \boldsymbol{\alpha} \otimes \mathbf{v})$ and $\mathbf{b} = \frac{1}{2}(\mathbf{v} \otimes \boldsymbol{\beta} + \boldsymbol{\beta} \otimes \mathbf{v})$, for some $\boldsymbol{\alpha}$ and $\boldsymbol{\beta}$. To find $\mathbf{G}$, $\mathbf{a}$ must be determined from $\mathbf{b} = \langle \mathbf{A} \rangle \mathbf{a}$, or equivalently, using (2.13), we must solve for $\boldsymbol{\alpha}$, given

$$\boldsymbol{\beta} = \langle \mathbf{f}^{-1} \rangle \tfrac{1}{2}(\mathbf{i} + \mathbf{v} \otimes \mathbf{v})\boldsymbol{\alpha}. \tag{3.10}$$

This is easily solved

$$\boldsymbol{\alpha} = (2\mathbf{i} - \mathbf{v} \otimes \mathbf{v})\langle \mathbf{f}^{-1} \rangle^{-1} \boldsymbol{\beta}, \tag{3.11}$$

and therefore $\mathbf{a} = \bar{\mathbf{G}}\mathbf{b}$ implies

$$\bar{\mathbf{G}}_{ijkl} = \tfrac{1}{4}(\bar{g}_{ik} v_j v_l + \bar{g}_{il} v_j v_k + \bar{g}_{jk} v_i v_l + \bar{g}_{jl} v_i v_k), \tag{3.12}$$

where

$$\bar{\mathbf{g}} = (2\mathbf{i} - \mathbf{v} \otimes \mathbf{v})\langle \mathbf{f}^{-1} \rangle^{-1}(2\mathbf{i} - \mathbf{v} \otimes \mathbf{v}). \tag{3.13}$$

It is also possible to derive an equation directly for $\bar{\mathbf{M}}$ by eliminating $\mathbf{F}\bar{\boldsymbol{\varepsilon}}$ from (3.4) and (3.5). Then instead of (3.7) we obtain the dual equation

$$\bar{\mathbf{M}} = \langle \mathbf{A} \rangle + \langle \mathbf{MB} \rangle \bar{\mathbf{H}} \langle \mathbf{BM} \rangle, \tag{3.14}$$

where $\bar{\mathbf{H}}: \Omega_F \to \Omega_F$ is the inverse of $\langle \mathbf{B} \rangle$,

$$\bar{\mathbf{H}} = \langle \mathbf{B} \rangle^{-1}. \tag{3.15}$$

The associated alternative expression for the effective thermal expansion is

$$\bar{\boldsymbol{\alpha}} = \langle \mathbf{AL}\boldsymbol{\alpha} \rangle + \langle \mathbf{MB} \rangle \bar{\mathbf{H}} \langle \mathbf{B}\boldsymbol{\alpha} \rangle. \tag{3.16}$$

This agrees with the known representation (e.g., (3.16) in Christensen, 1979) for arbitrary composite media, that

$$\bar{\boldsymbol{\alpha}} = \langle \boldsymbol{\alpha}\mathbf{S} \rangle, \tag{3.17}$$

where the stress concentration tensor $\mathbf{S}$ is defined by

$$\boldsymbol{\sigma} = \mathbf{S}\bar{\boldsymbol{\sigma}} + \mathbf{s}\theta. \tag{3.18}$$

In the present context, using (3.1) and (3.2), $\mathbf{s} = \mathbf{B}(\bar{\boldsymbol{\alpha}} - \boldsymbol{\alpha})$, and

$$\mathbf{S} = \mathbf{LA} + \mathbf{B}\bar{\mathbf{M}}. \tag{3.19}$$

Then (3.16) follows from (3.17) after simplifying $\mathbf{S}$ as $\mathbf{S} = \mathbf{LA} + \bar{\mathbf{H}}\langle \mathbf{BM} \rangle$ using

(3.14) and (3.19). We note the dual to (3.18) and (3.19) is

$$\boldsymbol{\varepsilon} = (\mathbf{A}\bar{\mathbf{L}} + \mathbf{M}\mathbf{B})\bar{\boldsymbol{\varepsilon}} + \mathbf{A}(\mathbf{L}\boldsymbol{\alpha} - \overline{\mathbf{L}\boldsymbol{\alpha}})\theta, \tag{3.20}$$

and that both the strain concentration tensor $(\mathbf{A}\bar{\mathbf{L}} + \mathbf{M}\mathbf{B})$ and the stress concentration tensor $(\mathbf{L}\mathbf{A} + \mathbf{B}\bar{\mathbf{M}})$ reduce to the identity for a homogeneous medium, by (2.15).

As an example, if there is only one phase, $\langle \mathbf{A} \rangle = \mathbf{A}$ and $\bar{\mathbf{G}} = \mathbf{A}^{-1}$, which from (2.8), gives $\bar{\mathbf{G}} = \mathbf{ELE}$. In the case of an isotropic medium of Lamé moduli λ, μ,

$$\bar{\mathbf{G}} = 2\mu\mathbf{E} + \lambda\mathbf{v} \otimes \mathbf{v} \otimes \mathbf{v} \otimes \mathbf{v}. \tag{3.21}$$

Similarly, for a single phase, $\bar{\mathbf{H}} = \mathbf{B}^{-1} = \mathbf{FMF}$, which becomes for isotropy

$$\bar{\mathbf{H}} = \frac{1}{2\mu}\left[\mathbf{F} - \frac{\lambda}{3\lambda + 2\mu}(\mathbf{i} - \mathbf{v} \otimes \mathbf{v}) \otimes (\mathbf{i} - \mathbf{v} \otimes \mathbf{v})\right]. \tag{3.22}$$

3.2. Special Case of a Two-Phase Laminate

A simpler procedure may be preferable when the layered medium consists of only two materials with moduli $\mathbf{L}_n$, $\boldsymbol{\alpha}_n$, $n = 1, 2$, and volume fractions c_1 and $c_2 = 1 - c_1$. Let $\boldsymbol{\varepsilon}_n$, $n = 1, 2$, be the local strains, then $\boldsymbol{\varepsilon}_1$ can be expressed in terms of $\boldsymbol{\varepsilon}_2$ using (2.21), and the average strain $\bar{\boldsymbol{\varepsilon}} = c_1\boldsymbol{\varepsilon}_1 + c_2\boldsymbol{\varepsilon}_2$ is

$$\bar{\boldsymbol{\varepsilon}} = [\mathbf{I} + c_1\mathbf{A}_1(\mathbf{L}_2 - \mathbf{L}_1)]\boldsymbol{\varepsilon}_2 + c_1\mathbf{A}_1(\mathbf{L}_1\boldsymbol{\alpha}_1 - \mathbf{L}_2\boldsymbol{\alpha}_2)\theta, \tag{3.23}$$

where $\mathbf{A}_1$ depends only upon $\mathbf{L}_1$. Similarly, the average stress is

$$\bar{\boldsymbol{\sigma}} = [\langle \mathbf{L} \rangle + c_1\mathbf{L}_1\mathbf{A}_1(\mathbf{L}_2 - \mathbf{L}_1)]\boldsymbol{\varepsilon}_2 - [\langle \mathbf{L}\boldsymbol{\alpha} \rangle + c_1\mathbf{L}_1\mathbf{A}_1(\mathbf{L}_2\boldsymbol{\alpha}_2 - \mathbf{L}_1\boldsymbol{\alpha}_1)]\theta. \tag{3.24}$$

Eliminating $\boldsymbol{\varepsilon}_2$ from (3.23) and (3.24), and using the definitions in (3.6) gives a simple equation for $\bar{\mathbf{L}}$ that involves $\mathbf{A}_1$ only. This relation may be rewritten in the succinct form obtained by Francfort and Murat (1986)

$$c_2(\bar{\mathbf{L}} - \mathbf{L}_1)^{-1} = (\mathbf{L}_2 - \mathbf{L}_1)^{-1} + c_1\mathbf{A}_1. \tag{3.25}$$

The effective thermal expansion $\bar{\boldsymbol{\alpha}}$ then follows from

$$\overline{\mathbf{L}\boldsymbol{\alpha}} = \langle \mathbf{L}\boldsymbol{\alpha} \rangle + c_1(\mathbf{L}_1 - \bar{\mathbf{L}})\mathbf{A}_1(\mathbf{L}_2\boldsymbol{\alpha}_2 - \mathbf{L}_1\boldsymbol{\alpha}_1), \tag{3.26}$$

which again depends only upon $\mathbf{A}_1$. Dual equations can also be derived which involve $\mathbf{B}_1$ only, thus

$$c_2(\bar{\mathbf{M}} - \mathbf{M}_1)^{-1} = (\mathbf{M}_2 - \mathbf{M}_1)^{-1} + c_1\mathbf{B}_1, \tag{3.27}$$

$$\bar{\boldsymbol{\alpha}} = \langle \boldsymbol{\alpha} \rangle + c_1(\mathbf{M}_1 - \bar{\mathbf{M}})\mathbf{B}_1(\boldsymbol{\alpha}_2 - \boldsymbol{\alpha}_1). \tag{3.28}$$

Equation (3.25) or (3.27) can be iterated to provide the effective moduli of finite rank laminates (Francfort and Murat, 1986), defined by the following sequence. Beginning with phase 2, laminate with phase 1 periodically spaced over a period L with normal in the direction $\mathbf{v}_1$, occupying volume fraction

θ_1. Then laminate the composite again with phase 1 over a period NL, with normal $\mathbf{v}_2$, and volume fraction θ_2. Repeat the process n times, such that at the jth step the periodicity is $N^{j-1}L$, the normal is $\mathbf{v}_j$, and the volume fraction is θ_j. The basic equation (3.25) or (3.27) may be used recursively in the limit of $N \to \infty$, or equivalently, when each successive lamination is into a homogeneous effective medium. After iteration, the final moduli of the rank-n laminate are given by (3.25) and (3.27) with

$$c_2 = 1 - c_1 = \prod_{j=1}^{n} (1 - \theta_j), \tag{3.29}$$

$$\mathbf{A}_1 = \sum_{j=1}^{n} \rho_j \mathbf{A}_1(\mathbf{v}_j), \tag{3.30}$$

$$\mathbf{B}_1 = \sum_{j=1}^{n} \rho_j \mathbf{B}_1(\mathbf{v}_j), \tag{3.31}$$

and

$$\rho_j = \theta_j \sum_{i=1}^{j-1} \frac{1 - \theta_i}{c_1}. \tag{3.32}$$

Note that $\sum_{j=1}^{n} \rho_j = 1$.

4. Stiffeners and Fractures

4.1. Elastic Moduli of Layered Media Containing Stiffeners

The general formalism of Section 3 simplifies somewhat if either stiffeners or fractures are present. Stiffeners are defined as materials of elastic moduli much greater than those of the other materials in the composite. Further, the volume occupied by these stiff phases is assumed to be small so that the net effect of their presence is expected to be $O(1)$. Let c denote the volume fraction of stiffeners

$$c \equiv \frac{\sum h_{\text{stiff}}}{\sum h}, \tag{4.1}$$

where h_{stiff} are the thicknesses of the stiff layers, and $\sum h$ is the thickness of all layers present. It is assumed (i) that $c \ll 1$, and (ii) that the elastic stiffness tensor $\mathbf{L}_s$ of the stiff layers is of the form

$$\mathbf{L}_s = \frac{1}{c} \tilde{\mathbf{L}}, \tag{4.2}$$

where $\tilde{\mathbf{L}}$ is a tensor of the same order of magnitude as those of the other materials present. The scaling (4.2) allows the use of asymptotic approximations in the small parameter c.

Denote the composite without the stiffeners as the background material, with subscript b. In the presence of the stiffeners we have

$$\{\langle \mathbf{A} \rangle, \langle \mathbf{LA} \rangle, \langle \mathbf{AL} \rangle, \bar{\mathbf{G}}\} = \{\langle \mathbf{A} \rangle_b, \langle \mathbf{LA} \rangle_b, \langle \mathbf{AL} \rangle_b, \bar{\mathbf{G}}_b\}[1 + O(c)], \tag{4.3}$$

but

$$\langle \mathbf{B} \rangle = \langle \mathbf{B} \rangle_b + \tilde{\mathbf{B}} + O(c), \tag{4.4}$$

where $\tilde{\mathbf{B}}$ is the tensor associated with $\tilde{\mathbf{L}}$, and defined by the orientation of the normal to the stiffened layers. For the moment let us assume all the stiffeners are parallel. The effective elastic stiffness in the presence of the stiffeners follows from (3.7), (4.3), (4.4), and $\tilde{\mathbf{B}} = c\mathbf{B}_s$, where $\mathbf{B}_s$ is the tensor for $\mathbf{L}_s$, as

$$\bar{\mathbf{L}} = \bar{\mathbf{L}}_b + c\mathbf{B}_s + O(c). \tag{4.5}$$

The effect of the stiffeners is to add to the modulus tensor the additional $c\mathbf{B}_s$, which is $O(1)$ by assumption, and the result is correct to $O(c)$. Some insight is gained by partitioning $\bar{\mathbf{L}}$ as

$$\bar{\mathbf{L}} = \mathbf{E}\bar{\mathbf{L}}\mathbf{E} + (\mathbf{F}\bar{\mathbf{L}}\mathbf{E} + \mathbf{E}\bar{\mathbf{L}}\mathbf{F}) + \mathbf{F}\bar{\mathbf{L}}\mathbf{F}, \tag{4.6}$$

from which it is clear that the stiffeners change only $\mathbf{F}\bar{\mathbf{L}}\mathbf{F}$, which may be called the interior part of $\bar{\mathbf{L}}$, and the change depends only upon the interior part of the compliance of the stiff layer, i.e., $\mathbf{F}\mathbf{M}_s\mathbf{F}$, where $\mathbf{M}_s = \mathbf{L}_s^{-1}$. It can be shown from the positive definite nature of $\mathbf{L}_s$, i.e., Tr $\mathbf{a}(\mathbf{L}_s\mathbf{a}) > 0$ for all $\mathbf{a} \neq 0$, that $\mathbf{B}_s$ is positive definite for all $\mathbf{a} \in \Omega_F$, but $\mathbf{B}_s\mathbf{a} = 0$ for all $\mathbf{a} \in \Omega_E$. Thus, the change in $\bar{\mathbf{L}}$ due to the presence of the stiffeners is positive definite on interior tensors. This is in accord with the intuitive expectation of a greater stiffness in directions parallel to the layers, but has no effect in the direction normal to the layers.

An important consequence of the additive nature of (4.5) is that oblique sets of stiffeners can be handled by adding to the background stiffness the tensors $c\mathbf{B}_s$ appropriate to each orientation.

4.2. Elastic Moduli of Fractured Media

The term fracture is used here in the same sense as defined by Schoenberg and Douma (1988), i.e., a fracture is a flat joint or thin layer of relatively small elastic stiffness, and is useful in modeling slip zones, for example. The fractures occupy total volume fraction $c \ll 1$, with elastic moduli of the form

$$\mathbf{L}_f = c\tilde{\mathbf{L}}, \tag{4.7}$$

where again $\tilde{\mathbf{L}}$ is a stiffness tensor of the same order of magnitude as the other material stiffnesses. This limit is essentially the dual of the stiffener limit (4.2). In the case of the fractures, the effect of their presence on the background medium is

$$\{\langle \mathbf{B} \rangle, \langle \mathbf{MB} \rangle, \langle \mathbf{BM} \rangle, \bar{\mathbf{H}}\} = \{\langle \mathbf{B} \rangle_b, \langle \mathbf{MB} \rangle_b, \langle \mathbf{BM} \rangle_b, \bar{\mathbf{H}}_b\}[1 + O(c)], \tag{4.8}$$

but

$$\langle \mathbf{A} \rangle = \langle \mathbf{A} \rangle_b + \tilde{\mathbf{A}} + O(c), \tag{4.9}$$

where $\tilde{\mathbf{A}}$ is defined by $\tilde{\mathbf{L}}$ and the direction normal to the fractures.

The effective compliance tensor in the presence of the fractures then follows from (3.14), (4.8), (4.9), and $\tilde{\mathbf{A}} = c\mathbf{A}_f$, as

$$\overline{\mathbf{M}} = \overline{\mathbf{M}}_b + c\mathbf{A}_f + O(c), \tag{4.10}$$

where $\mathbf{A}_f$ is associated with the actual moduli $\mathbf{L}_f$ of the fracture layer. The effect of the fractures is thus to change only the exterior part of $\overline{\mathbf{M}}$, $\mathbf{E}\overline{\mathbf{M}}\mathbf{E}$, and the change depends only upon the exterior part of the fracture stiffness tensor, $\mathbf{E}\mathbf{L}_f\mathbf{E}$. The net change in the compliance in the presence of the fractures is positive definite on exterior tensors and has no effect on interior tensors.

Just as for the stiffeners, the simple dependence of $\overline{\mathbf{M}}$ on $\mathbf{A}_f$ in (4.10) shows that oblique sets of fractures can be superposed by adding the additional compliance $c\mathbf{A}_f$ appropriate to each orientation.

5. Microcracks as Equivalent Layers

We now relate the previous results for fractures to a micromechanical model of the change in effective compliance due to the presence of a distribution of cracks. The model assumes the cracks are penny-shaped and no interaction effects are considered, i.e., we do not propose to account for finite crack density effects by some self-consistent procedure as in, for example, Budiansky and O'Connell (1976) or Horii and Nemat-Nasser (1983). Rather, we shall extend a simple theory due to Oda (1983) which is asymptotically rigorous for dilute crack concentrations. We note that Schoenberg and Douma (1988) compared their theory of fractures, which is essentially the same as ours, with the cracked medium model of Hudson (1981). The present emphasis is somewhat different from Schoenberg and Douma's (1988), and no attempt is made to compare our results with those of Hudson (1981).

We restrict our attention to isotropic solids. The starting point is the following result for the average crack opening displacement $\boldsymbol{\delta}$ of a single penny-shaped crack, radius r, subject to an applied stress $\boldsymbol{\sigma}$ (Oda, 1983; Horii and Nemat-Nasser, 1983; Mura, 1987)

$$\boldsymbol{\delta} = \frac{r}{D}\mathbf{q}\boldsymbol{\sigma}\mathbf{v}, \tag{5.1}$$

where $\mathbf{v}$ is the normal to the crack face

$$D = \frac{3}{8}\pi\left(\frac{2-\eta}{1-\eta}\right)\mu, \tag{5.2}$$

$$\mathbf{q} = 2\mathbf{i} - \eta\mathbf{v}\otimes\mathbf{v}, \tag{5.3}$$

and μ, η are the shear modulus and the Poisson ratio. Following Oda (1983), a density function $\gamma(\mathbf{v}, r)$ is introduced to account for the distribution of orientations and radii, satisfying $\int_0^\infty \int_\Omega \gamma(\mathbf{v}, r)\, d\Omega\, dr = 1$, where Ω is the unit sphere. Also, let ρ be the volume density of cracks, i.e., the number per unit volume. Then a fabric tensor density can be defined as

$$\mathbf{h}(\mathbf{v}, r) = \pi\rho r^3\gamma(\mathbf{v}, r)\mathbf{v}\otimes\mathbf{v}. \tag{5.4}$$

Oda (1983) discussed the significance of the fabric tensor and mentioned alternative definitions in the literature. Oda's fabric tensor is actually the integral $\int_0^\infty \int_\Omega 2\mathbf{h}(\mathbf{v}, r)\, d\Omega\, dr = \hat{\mathbf{h}}$, although it is more convenient for our purposes to work with $\hat{\mathbf{h}}(\mathbf{v}, r)$. We note that $\operatorname{Tr} \hat{\mathbf{h}}$ is the crack concentation parameter of Budiansky and O'Connell (1976).

It is possible, by the same arguments used by Oda (1983), to calculate the strain $\boldsymbol{\varepsilon}^{(c)}$ associated with cracks $(\mathbf{v}, r)$ in the applied stress field $\boldsymbol{\sigma}$, as

$$\boldsymbol{\varepsilon}^{(c)} = \mathbf{P}(\mathbf{v}, r)\boldsymbol{\sigma}, \tag{5.5}$$

where

$$P_{ijkl} = \frac{1}{4D}(q_{ik}h_{jl} + q_{il}h_{jk} + q_{jk}h_{il} + q_{jl}h_{ik}). \tag{5.6}$$

The result (5.5)–(5.6) is more general than that of Oda (1983) who, for some reason, chose to replace (5.1) by an empirical relation of the form $\boldsymbol{\delta} = (r/D)\boldsymbol{\sigma}\mathbf{v}$, i.e., with $\mathbf{q} = \mathbf{i}$, and his D was not well defined. In the limit of dilute concentrations $\mathbf{P}$ is the change in the bulk compliance due to the cracks $(\mathbf{v}, r)$, and can be simplfied using (2.1), (5.3), and (5.4)

$$\mathbf{P} = \pi \frac{\rho}{D} \gamma(\mathbf{v}, r) r^3 [\mathbf{E} + (1 - \eta)\mathbf{v} \otimes \mathbf{v} \otimes \mathbf{v} \otimes \mathbf{v}]. \tag{5.7}$$

In comparing these results with the change in compliance due to fractures, we immediately identify $\mathbf{P}$ with $c\mathbf{A}_f$ in (4.10), i.e., $\bar{\mathbf{M}} = \mathbf{M} + \mathbf{P}$. The analogy is strengthened by referring to (2.16), and rewriting $\mathbf{P}$ as

$$\mathbf{P} = \frac{c}{2\mu_f}\left[\mathbf{E} - \frac{\eta_f}{1 - \eta_f}\mathbf{v} \otimes \mathbf{v} \otimes \mathbf{v} \otimes \mathbf{v}\right] = c\mathbf{A}_f, \tag{5.8}$$

where

$$\eta_f = 1 - \frac{1}{\eta} \leq -1, \tag{5.9}$$

$$\mu_f = \frac{3(2 - \eta)}{16(1 - \eta)}\mu, \tag{5.10}$$

$$c = \rho r^3 \gamma(\mathbf{v}, r). \tag{5.11}$$

The cracks therefore have the same effect as thin isotropic fractures of Poisson's ratio η_f, shear modulus μ_f, volume fraction c, and normal $\mathbf{v}$. The specific choice of μ_f and c above is somewhat arbitrary; others may be chosen as long as the product c/μ_f is fixed by (5.7) and (5.8). The definition of η_f is however unambiguous, and its value of less than -1 means no physically realistic medium can be construed from the equivalent-layer moduli, which is perhaps not surprising.

The above results can be extended to consider randomly oriented cracks by averaging (5.7) over all orientations of $\mathbf{v}$; thus $\mathbf{E} \to (1/15)(8\mathbf{I} - \mathbf{i} \otimes \mathbf{i})$, and $\mathbf{v} \otimes \mathbf{v} \otimes \mathbf{v} \otimes \mathbf{v} \to (1/15)(2\mathbf{I} + \mathbf{i} \otimes \mathbf{i})$. The effective compliance tensor is then $\bar{\mathbf{M}} = \mathbf{M} + \mathbf{P}$. If we replace the moduli μ and η in $\mathbf{P}$ by the effective moduli $\bar{\mu}$

and $\bar{\eta}$, the resulting implicit equation, $2\overline{\mathbf{M}} = 2\mathbf{M} + 2\bar{\mathbf{P}}$ with $\gamma(r) = 1$ for $\bar{\mu}$ and $\bar{\eta}$, is exactly the self-consistent scheme of Budiansky and O'Connell (1976)

$$\frac{1}{\bar{\mu}}\left[\mathbf{I} - \frac{\bar{\eta}}{1+\bar{\eta}}\mathbf{i}\otimes\mathbf{i}\right] = \frac{1}{\mu}\left[\mathbf{I} - \frac{\eta}{1+\eta}\mathbf{i}\otimes\mathbf{i}\right] + \rho r^3 \frac{16}{45}\frac{(1-\bar{\eta})}{(2-\bar{\eta})}\frac{1}{\bar{\mu}} \times [(10 - 2\bar{\eta})\mathbf{I} - \bar{\eta}\mathbf{i}\otimes\mathbf{i}]. \tag{5.12}$$

6. Matrix Notation

6.1. Arbitrary Rectangular Coordinates

It is more convenient in practice to use 6×6 matrix notation for fourth-order tensors, defined according to the rule $X_{ijkl} = [X]_{IJ}$, where $ij = 11, 22, 33, 23, 13, 12$ correspond to $I = 1, 2, 3, 4, 5, 6$, respectively. The elements of the stiffness matrix $[L]$ are the standard C_{IJ} (Christensen, 1979). Second-order tensors are denoted by 6-vectors in the standard manner; thus, $\boldsymbol{\sigma} \to \{\sigma\} = (\sigma_{11}, \sigma_{22}, \sigma_{33}, \sigma_{23}, \sigma_{13}, \sigma_{12})^{\mathrm{T}}$, $\boldsymbol{\varepsilon} \to \{\varepsilon\} = (\varepsilon_{11}, \varepsilon_{22}, \varepsilon_{33}, 2\varepsilon_{23}, 2\varepsilon_{13}, 2\varepsilon_{12})^{\mathrm{T}}$, so that the anisotropic Hooke's law becomes $\{\sigma\} = [L]\{\varepsilon\}$, or $\{\varepsilon\} = [L]^{-1}\{\sigma\}$, where $[L][L]^{-1} = [L]^{-1}[L] = \mathrm{diag}(1, 1, 1, 1, 1, 1)$. Let $[M]$ be the 6×6 matrix of the compliance tensor $\mathbf{M}$, then Hooke's law is $\{\varepsilon\} = [R][M][R]\{\sigma\}$, where $[R]$ is Reuter's 6×6 matrix (Ting, 1987)

$$[R] = \mathrm{diag}(1, 1, 1, 2, 2, 2), \tag{6.1}$$

and so $[M] = [R]^{-1}[L]^{-1}[R]^{-1}$.

Define the 6-vector $\{v\}$ corresponding to the direction normal to the layering, $\mathbf{v}$, as

$$\{v\} = (v_1^2, v_2^2, v_3^2, v_2 v_3, v_1 v_3, v_1 v_2)^{\mathrm{T}}. \tag{6.2}$$

The matrix $[E]$ associated with the exterior operator $\mathbf{E}$ follows from the procedure outlined in the Appendix, and $\mathbf{F} \to [F] = [R]^{-1} - [E]$. In order to express the tensors $\mathbf{A}$ and $\mathbf{B}$ as 6×6 matrices, we first need $\{f\} = (f_{11}, f_{22}, f_{33}, f_{23}, f_{13}, f_{12})^{\mathrm{T}}$ corresponding to $\mathbf{f}$ of (2.12), which is

$$\{f\} = [\hat{L}][R]\{v\}, \tag{6.3}$$

with

$$[\hat{L}] = \begin{bmatrix} C_{11} & C_{66} & C_{55} & C_{56} & C_{15} & C_{16} \\ & C_{22} & C_{44} & C_{24} & C_{46} & C_{26} \\ & & C_{33} & C_{34} & C_{35} & C_{45} \\ & & & \frac{1}{2}(C_{44} + C_{23}) & \frac{1}{2}(C_{45} + C_{36}) & \frac{1}{2}(C_{46} + C_{25}) \\ & & & & \frac{1}{2}(C_{55} + C_{13}) & \frac{1}{2}(C_{56} + C_{14}) \\ & & & & & \frac{1}{2}(C_{66} + C_{12}) \end{bmatrix}. \tag{6.4}$$

The matrix $[A]$ follows from (2.13) and the prescription in the Appendix, and $[B]$ follows from (2.15) as

$$[B] = [L] - [L][R][A][R][L]. \tag{6.5}$$

Note the ubiquity of Reuter's matrix $[R]$, which is necessary to correctly perform tensor multiplication. In this way, a consistent, coordinate independent computational scheme can be developed to implement the results of Sections 3 and 4.

6.2. *Example: Stiffeners and Fractures*

The effect of a small volume fraction c of relatively stiff layers is to alter the macroscopic stiffness according to (4.5). For an isotropic stiff material of shear modulus μ_s and Poisson's ratio η_s, (2.17) and (4.5) imply

$$[\bar{L}] = [\bar{L}_b] + 2\mu_s c\left([F] + \frac{\eta_s}{1-\eta_s}\{\hat{v}\}\{\hat{v}\}^T\right), \tag{6.6}$$

where

$$\{\hat{v}\} = (v_1 - 1, v_2 - 1, v_3 - 1, v_4, v_5, v_6)^T. \tag{6.7}$$

Similarly, the effect of a small volume fraction of relatively compliant fracture layers follows from (4.10) which becomes, for isotropic fractures of shear modulus μ_f and Poisson's ratio η_f, using (2.16),

$$[\bar{M}] = [\bar{M}_b] + \frac{c}{2\mu_f}\left([E] - \frac{\eta_f}{1-\eta_f}\{v\}\{v\}^T\right). \tag{6.8}$$

Equations (6.6) and (6.8) provide a convenient, coordinate invariant method to estimate the change in strength due to a small concentration of relatively stiff or compliant layers.

6.3. *The Normal Direction Along a Coordinate Axis*

The standard procedure (e.g., Chou *et al.*, 1972; Pagano, 1974; Helbig and Schoenberg, 1987) is to take $\mathbf{v}$ in a coordinate direction, e.g., x_3. Thus, rearrange the elements of $[L]$ into 3×3 matrices

$$\begin{bmatrix} \mathbf{a} & \mathbf{b} \\ \mathbf{b}^T & \mathbf{d} \end{bmatrix} = \left[\begin{array}{ccc|ccc} C_{11} & C_{12} & C_{16} & C_{14} & C_{15} & C_{13} \\ C_{12} & C_{22} & C_{26} & C_{24} & C_{25} & C_{23} \\ C_{16} & C_{26} & C_{66} & C_{46} & C_{56} & C_{36} \\ \hline C_{14} & C_{24} & C_{46} & C_{44} & C_{45} & C_{34} \\ C_{15} & C_{25} & C_{56} & C_{45} & C_{55} & C_{35} \\ C_{13} & C_{23} & C_{36} & C_{34} & C_{35} & C_{33} \end{array}\right], \tag{6.9}$$

where we note that $\mathbf{b}$ is not symmetric. Then,

$$\mathbf{A} \leftrightarrow \begin{bmatrix} 0 & 0 \\ 0 & \mathbf{d}^{-1} \end{bmatrix}, \qquad \mathbf{B} \leftrightarrow \begin{bmatrix} \mathbf{a} - \mathbf{b}\mathbf{d}^{-1}\mathbf{b}^{\mathrm{T}} & 0 \\ 0 & 0 \end{bmatrix}, \tag{6.10}$$

and referring to (3.9) and (3.15),

$$\bar{\mathbf{G}} \leftrightarrow \begin{bmatrix} 0 & 0 \\ 0 & \langle \mathbf{d}^{-1} \rangle^{-1} \end{bmatrix}, \qquad \bar{\mathbf{H}} \leftrightarrow \begin{bmatrix} \langle \mathbf{a} - \mathbf{b}\mathbf{d}^{-1}\mathbf{b}^{\mathrm{T}} \rangle^{-1} & 0 \\ 0 & 0 \end{bmatrix}. \tag{6.11}$$

The effective moduli of the layered medium follow from (3.7), (6.9)–(6.11), as

$$\bar{\mathbf{a}} = \langle \mathbf{a} - \mathbf{b}\mathbf{d}^{-1}\mathbf{b}^{\mathrm{T}} \rangle + \langle \mathbf{b}\mathbf{d}^{-1} \rangle \langle \mathbf{d}^{-1} \rangle^{-1} \langle \mathbf{d}^{-1}\mathbf{b}^{\mathrm{T}} \rangle, \tag{6.12}$$

$$\bar{\mathbf{b}} = \langle \mathbf{b}\mathbf{d}^{-1} \rangle \langle \mathbf{d}^{-1} \rangle^{-1}, \tag{6.13}$$

$$\bar{\mathbf{d}} = \langle \mathbf{d}^{-1} \rangle^{-1}. \tag{6.14}$$

In the theory of thin laminates (Christensen, 1979), a state of plane stress is assumed, and it is easy to see that the effective moduli relating the in-plane stress and strain are $\bar{\mathbf{a}} - \bar{\mathbf{b}}\bar{\mathbf{d}}^{-1}\bar{\mathbf{b}}^{\mathrm{T}} = \langle \mathbf{a} - \mathbf{b}\mathbf{d}^{-1}\mathbf{b}^{\mathrm{T}} \rangle$, corresponding to $\bar{\mathbf{B}} = \langle \mathbf{B} \rangle$.

7. Examples

7.1. *All Layers Monoclinic*

Assume for simplicity that the x_3 direction is normal to the layers, and that the material in each layer possesses monoclinic symmetry about the x_1x_2 plane, implying $C_{14} = C_{15} = C_{24} = C_{25} = C_{34} = C_{35} = C_{46} = C_{56} = 0$. The composite medium then possesses monoclinic symmetry, and its 13 nonzero moduli follow from (6.9) and (6.12)–(6.14) as (Chou *et al.*, 1972)

$$\bar{C}_{IJ} = \begin{cases} \langle Q_{IJ} \rangle + \left\langle \dfrac{C_{I3}}{C_{33}} \right\rangle \left\langle \dfrac{C_{J3}}{C_{33}} \right\rangle \left\langle \dfrac{1}{C_{33}} \right\rangle^{-1}, & I, J = 1, 2, 3, 6, \\[2ex] \left\langle \dfrac{C_{IJ}}{C_{44}C_{55} - C_{45}^2} \right\rangle \left[\left\langle \dfrac{C_{44}}{C_{44}C_{55} - C_{45}^2} \right\rangle \left\langle \dfrac{C_{55}}{C_{44}C_{55} - C_{45}^2} \right\rangle \right. & \\[2ex] \qquad \left. - \left\langle \dfrac{C_{45}}{C_{44}C_{55} - C_{45}^2} \right\rangle^2 \right]^{-1}, & I, J = 4, 5, \end{cases} \tag{7.1}$$

where Q_{IJ} are

$$Q_{IJ} = C_{IJ} - \frac{C_{I3}C_{J3}}{C_{33}}, \qquad I, J = 1, 2, 3, 6, \tag{7.2}$$

which include the plane stress moduli of classical laminate theory (Christensen, 1979), $I, J = 1, 2, 6$. Let α_I be the six elements of the thermal expansion tensor, $(\alpha_{11}, \alpha_{22}, \alpha_{33}, 2\alpha_{23}, 2\alpha_{13}, 2\alpha_{12})$, then the effective tensor follows from

(3.16) and Section 6.3, as

$$\begin{aligned}
\bar{\alpha}_I &= \sum_{J,K=1,2,6} \bar{Q}_{IJ}^{-1} \langle Q_{JK}\alpha_K \rangle, \qquad I = 1, 2, 6, \\
\bar{\alpha}_3 &= \langle \alpha_3 \rangle + \sum_{I=1,2,6} \left[\left\langle \frac{C_{I3}}{C_{33}} \alpha_I \right\rangle - \left\langle \frac{C_{I3}}{C_{33}} \right\rangle \bar{\alpha}_I \right], \\
\bar{\alpha}_I &= \langle \alpha_I \rangle, \qquad I = 4, 5,
\end{aligned} \tag{7.3}$$

where $\bar{\mathbf{Q}}^{-1}$ is the inverse of the matrix of elements $\bar{Q}_{IJ}$, $I, J = 1, 2, 6$. These relatively simple expressions do not appear to have been presented previously.

7.2. *Stacks of Orthotropic Layers*

Let each layer be the same orthotropic material with coincident x_3 axis, and the other two axes in each layer are rotated from $x_1 x_2$ by an angle θ. Given the 9 material constants of the orthotropic material, the 13 nonzero elastic constants of the rotated layer can be found from known transformation rules of fourth-order tensors; explicit equations for the 13 transformed moduli are given by Christensen (1979). The composite stiffnesses then follow from (6.1) and (6.2). For simplicity, we assume equal amounts of material in the $\pm\theta$ directions. Then the composite is macroscopically orthotropic ($\bar{C}_{16} = \bar{C}_{26} = \bar{C}_{36} = \bar{C}_{45} = 0$), with effective moduli

$$\begin{bmatrix} \bar{C}_{11} \\ \bar{C}_{22} \\ \bar{C}_{12} \\ \bar{C}_{66} \end{bmatrix} = \begin{bmatrix} U_1 & U_2 & U_3 & U_6 \\ U_1 & -U_2 & U_3 & U_6 \\ U_4 & 0 & -U_3 & -U_6 \\ U_5 & 0 & -U_3 & 0 \end{bmatrix} \begin{bmatrix} 1 \\ \langle \cos 2\theta \rangle \\ \langle \cos 4\theta \rangle \\ \langle \cos 2\theta \rangle^2 \end{bmatrix}, \tag{7.4}$$

where

$$\begin{aligned}
U_1 &= \frac{1}{8}\left[3C_{11} + 3C_{22} + 2C_{12} + 4C_{66} - \frac{(C_{13} - C_{23})^2}{C_{33}} \right], \\
U_2 &= \tfrac{1}{2}(C_{11} - C_{22}), \\
U_3 &= \frac{1}{8}\left[C_{11} + C_{22} - 2C_{12} - 4C_{66} - \frac{(C_{13} - C_{23})^2}{C_{33}} \right], \\
U_4 &= \frac{1}{8}\left[C_{11} + C_{22} + 6C_{12} - 4C_{66} + \frac{(C_{13} - C_{23})^2}{C_{33}} \right], \\
U_5 &= \tfrac{1}{2}[U_1 - U_4], \\
U_6 &= \frac{1}{4}\frac{(C_{13} - C_{23})^2}{C_{33}}.
\end{aligned} \tag{7.5}$$

The remaining moduli are

$$\begin{aligned}
\bar{C}_{33} &= C_{33}, \\
\bar{C}_{13} &= \tfrac{1}{2}(C_{13} + C_{23}) + \tfrac{1}{2}(C_{13} - C_{23})\langle \cos 2\theta \rangle, \\
\bar{C}_{23} &= \tfrac{1}{2}(C_{13} + C_{23}) - \tfrac{1}{2}(C_{13} - C_{23})\langle \cos 2\theta \rangle, \\
\bar{C}_{44} &= 2C_{44}C_{55}[C_{44} + C_{55} - (C_{44} - C_{55})\langle \cos 2\theta \rangle]^{-1}, \\
\bar{C}_{55} &= 2C_{44}C_{55}[C_{44} + C_{55} + (C_{44} - C_{55})\langle \cos 2\theta \rangle]^{-1}.
\end{aligned} \tag{7.6}$$

If the layers are equally spaced at angles π/N, N an integer, it follows (Christensen, 1979) that $\langle \cos 2\theta \rangle = 0$, and the composite has at least tetragonal symmetry (six constants), since $\bar{C}_{11} = \bar{C}_{22}$, $\bar{C}_{13} = \bar{C}_{23}$, and $\bar{C}_{44} = \bar{C}_{55}$. For $N \geq 3$, $\langle \cos 4\theta \rangle = 0$, and the composite is transversely isotropic, with $\bar{C}_{66} = \tfrac{1}{2}(\bar{C}_{11} - \bar{C}_{12})$. Thus, cross-ply laminates, $N = 2$, have effective tetragonal isotropy with six elastic constants, and all higher values of N produce a TI composite with five constants. Note that the above moduli include the simpler plane-stress moduli $\bar{Q}_{IJ} = \bar{C}_{IJ} - \bar{C}_{I3}\bar{C}_{J3}/\bar{C}_{33}$, $I, J = 1, 2, 6$, normally used for thin laminates (Christensen, 1979); these follow from (7.4) and (7.6), and are seen to be independent of $\langle \cos 2\theta \rangle^2$. Also, the effective isotropy discussed here does not necessarily apply to the bending of thin laminated plates (Christensen, 1979).

The effective thermal expansion tensor can also be found directly. Let α_1, α_2, α_3 be the original principle coefficients of thermal expansion. The tensor appropriate to each rotated layer follows from the transformation properties of second-order tensors. Symmetry considerations and the assumption $\langle \theta \rangle = 0$ imply that the effective tensor is diagonal in the (x_1, x_2, x_3) system, with principal coefficients $\bar{\alpha}_1, \bar{\alpha}_2$, and $\bar{\alpha}_3$. The first two follow from (7.3), after some simplification, as

$$\begin{aligned}
\bar{\alpha}_1 = \frac{1}{2(\bar{Q}_{11}\bar{Q}_{22} - \bar{Q}_{12}^2)} \{ &(\bar{Q}_{22} - \bar{Q}_{12})[(Q_{11} + Q_{12})\alpha_1 + (Q_{22} + Q_{12})\alpha_2] \\
&+ (\bar{Q}_{22} + \bar{Q}_{12})[(Q_{11} - Q_{12})\alpha_1 \\
&+ (Q_{12} - Q_{22})\alpha_2]\langle \cos 2\theta \rangle \},
\end{aligned} \tag{7.7}$$

$$\begin{aligned}
\bar{\alpha}_2 = \frac{1}{2(\bar{Q}_{11}\bar{Q}_{22} - \bar{Q}_{12}^2)} \{ &(\bar{Q}_{11} - \bar{Q}_{12})[(Q_{11} + Q_{12})\alpha_1 + (Q_{22} + Q_{12})\alpha_2] \\
&+ (\bar{Q}_{11} + \bar{Q}_{12})[(Q_{12} - Q_{11})\alpha_1 \\
&+ (Q_{22} - Q_{12})\alpha_2]\langle \cos 2\theta \rangle \},
\end{aligned} \tag{7.8}$$

where $\bar{Q}_{IJ}$ are the effective plane-stress moduli discussed above. The expansion coefficient for the x_3 direction becomes simply

$$\bar{\alpha}_3 = \alpha_3 + \frac{C_{13}\alpha_1 + C_{23}\alpha_2 - \bar{C}_{13}\bar{\alpha}_1 - \bar{C}_{23}\bar{\alpha}_2}{C_{33}}. \tag{7.9}$$

If the layers are equally spaced at incremental angles of π/N, $N \geq 2$, then on account of the symmetry,

$$\bar{\alpha}_1 = \bar{\alpha}_2 = \frac{(Q_{11} + Q_{12})\alpha_1 + (Q_{22} + Q_{12})\alpha_2}{Q_{11} + 2Q_{12} + Q_{22}}. \tag{7.10}$$

This value is the same for cross-ply lamination ($N = 2$) and for $N = 3$ or greater, even though the elastic moduli are different in these cases.

7.3. *Fiber-Reinforced Laminates*

A practical example of a stacking of orthotropic layers is a laminate in which each layer is a unidirectional fiber-reinforced medium. If the fiber widths and separations are small compared to the layer thickness, then we can reasonably approximate the layer as an effectively homogeneous medium of transverse isotropy. The elastic constants of the layer can themselves be determined from the fiber and matrix parameters by using a suitable micromechanical model. Consider glass fibers of Lamé moduli $(\lambda, \mu) = (13.3, 29.9)$ GPa embedded in epoxy with $(\lambda, \mu) = (0.89, 1.28)$ GPa, the fibers being aligned in the x_1 direction. Then there are five independent elastic constants $(C_{11}, C_{33}, C_{13}, C_{23}, C_{66})$, with $C_{12} = C_{13}$, $C_{22} = C_{33}$, $C_{55} = C_{66}$ and $C_{44} = \frac{1}{2}(C_{33} - C_{23})$. These follow from the theory of Murakami and Hegemier (1986) as $(C_{11}, C_{33}, C_{13}, C_{23}, C_{66}) = (43.4, 10.7, 2.2, 2.2, 4.4)$ GPa for a fiber concentration of 60%. Other applicable theories give similar predictions. The anisotropic thermal expansion of the layer can also be calculated from the isotropic thermal expansion coefficients for glass and epoxy by micromechanical arguments from, for example, Tsai and Hahn (1980). The coefficients of glass and epoxy are 5.0 and 54.0, respectively, in units of $(\mu m/m)/K$. With the same fiber orientation and concentration, the theory of Tsai and Hahn (1980) gives $\alpha_1 = 6.4$ and $\alpha_2 = \alpha_3 = 28.4$.

The elastic moduli of the composite simplify if the layers are stacked at incremental angles of π/N, as discussed above. For all values of N, (7.6) gives $(\bar{C}_{33}, \bar{C}_{23}, \bar{C}_{44}) = (10.7, 2.2, 4.3)$ GPa; for transversely isotropic stacks, i.e., $N \geq 3$, (7.4) gives $(\bar{C}_{11}, \bar{C}_{12}) = (23.0, 6.2)$ GPa and $\bar{C}_{66} = \frac{1}{2}(\bar{C}_{11} - \bar{C}_{12}) = 8.4$ GPa; and for cross-ply laminates, $N = 2$, we obtain $(\bar{C}_{11}, \bar{C}_{12}, \bar{C}_{66}) = (27.0, 2.2, 4.4)$ GPa It is interesting to note that the in-plane shear modulus, $\bar{C}_{66}$, is almost twice as large for three or more plies as it is for the cross-ply laminate, $N = 2$. The thermal expansion coefficients for $N \geq 2$ follow from (7.9)–(7.10) as $\bar{\alpha}_1 = \bar{\alpha}_2 = 11.0$ and $\bar{\alpha}_3 = 31.0$, in units of $(\mu m/m)/K$.

Acknowledgments

Thanks to M. Schoenberg for discussions and encouragement. This work was supported by a grant from the National Science Foundation, Grant Number MSM 85-16256.

References

Auriault, J. L. and Bonnet, G. (1987), Surface effects in composite materials: Two simple examples, *Int. J. Engng. Sci.*, **25**, 307–323.

Backus, G. E. (1962), Long-wave elastic anisotropy produced by horizontal layering, *J. Geophys. Res.*, **67**, 4427–4440.

Budiansky, B. and O'Connell, R. J. (1976), Elastic moduli of a cracked solid, *Int. J. Solids Structures*, **12**, 81–97.

Chou, P. C., Carleone, J., and Hsu, C. M. (1972), Elastic constants of layered media, *J. Composite Materials*, **6**, 80–93.

Christensen, R. M. (1979), *Mechanics of Composite Materials*, Wiley, New York.

Francfort, G. A. and Murat, F. (1986), Homogenization and optimal bounds in linear elasticity, *Arch. Ration. Mech. Anal.*, **94**, 307–334.

Helbig, K. and Schoenberg, M. (1987), Anomalous polarization of elastic waves in transversely isotropic media, *J. Acoust. Soc. Amer.*, **81**, 1235–1245.

Hill, R. (1972), An invariant treatment of interfacial discontinuities in elastic composites, in *Continuum Mechanics and Related Problems of Analysis* (*Muskhelishvili 80th Anniversary Volume*), Moscow, pp. 597–604.

Hill, R. (1983), Interfacial operators in the mechanics of composite media, *J. Mech. Phys. Solids*, **31**, 347–357.

Horii, H. and Nemat-Nasser, S. (1983), Overall moduli of solids with microcracks: Load-induced anisotropy, *J. Mech. Phys. Solids*, **31**, 155–171.

Hudson, J. A. (1981), Wave speeds and attenuation of elastic waves in material containing cracks, *Geophys. J. Roy. Astron. Soc.*, **64**, 133–150.

Mura, T. (1987), *Micromechanics of Defects in Solids*, 2nd ed., Martinus Nijhoff, Dordrecht.

Murakami, H. and Hegemier, G. A. (1986), A mixture model for unidirectionally fiber-reinforced composites, *J. Appl. Mech.*, **53**, 765–773.

Oda, M. (1983), A method for evaluating the effect of crack geometry on the mechanical behavior of cracked rock masses, *Mech. Materials*, **2**, 163–171.

Pagano, N. J. (1974), Exact moduli of anisotropic laminates, in *Composite Materials*, Vol. 2, edited by G. P. Sendeckyj, Academic Press, New York, pp. 23–45.

Schoenberg, M. and Muir, F. (1989), A calculus for finely layered anisotropic media, *Geophys.*, **54**, 581–589.

Schoenberg, M. and Douma, J. (1988), Elastic wave propagation in media with parallel fractures and aligned cracks, *Geophys. Prospecting*, **36**, 571–590.

Ting, T. C. T. (1987), Invariants of anisotropic elastic constants, *Quart. J. Mech. Appl. Math.*, **40**, 431–448.

Tsai, S. W. and Hahn, H. T. (1980), *Introduction to Composite Materials*, Technomic Westport, CT.

Walpole, L. J. (1981), Elastic behavior of composite materials: Theoretical foundations, in *Advances in Applied Mechanics*, edited by C. S. Yih, Academic Press, New York, pp. 169–242.

Appendix: Matrix Notation

The 6 × 6 matrix $[X]$ associated with the tensor $\mathbf{X}$ of components

$$X_{ijkl} = \tfrac{1}{4}(x_{ik}v_jv_l + x_{il}v_jv_k + x_{jk}v_iv_l + x_{jl}v_iv_k), \tag{A1.1}$$

where $\mathbf{x}$ is a symmetric second-order tensor, is given by

$$[R][X][R] = \begin{bmatrix} x_1v_1 & x_6v_6 & x_5v_5 & x_5v_6 + x_6v_5 & x_1v_5 + x_5v_1 & x_1v_6 + x_6v_1 \\ & x_2v_2 & x_4v_4 & x_2v_4 + x_4v_2 & x_4v_6 + x_6v_4 & x_2v_6 + x_6v_2 \\ & & x_3v_3 & x_3v_4 + x_4v_3 & x_3v_5 + x_5v_3 & x_4v_5 + x_5v_4 \\ & & & 2x_4v_4 + x_2v_3 + x_3v_2 & x_6v_3 + x_5v_4 + x_4v_5 + x_3v_6 & x_6v_4 + x_5v_2 + x_4v_6 + x_2v_5 \\ & & & & 2x_5v_5 + x_1v_3 + x_3v_1 & x_1v_4 + x_6v_5 + x_5v_6 + x_4v_1 \\ & & & & & 2x_6v_6 + x_1v_2 + x_2v_1 \end{bmatrix} \tag{A1.2}$$

where $\{v\}$ is given by (6.3), $\{x\} = (x_{11}, x_{22}, x_{33}, x_{23}, x_{13}, x_{12})^{\mathrm{T}}$ and $[R]$ is Reuter's matrix, (6.1). In particular, the matrix $[A]$ corresponds to $\mathbf{x} = \mathbf{f}^{-1}$, and $[E]$ follows from (2.1) with $\mathbf{x} = 2\mathbf{i} - \mathbf{v} \times \mathbf{v}$, or $\{x\} = (2 - v_1, 2 - v_2, 2 - v_3, -v_4, -v_5, -v_6)^{\mathrm{T}}$.

Thermomechanical Hysteresis and Analogous Behavior of Composites

O. B. PEDERSEN
Metallurgy Department, Risø National Laboratory, DK-4000 Roskilde, Denmark

Abstract

A stress analysis gives a simple account of the observed thermal hysteresis of copper with tungsten fibers, in terms of the frictional matrix yield stress. Initial experiments at room temperature on the mechanical hysteresis of copper with high volume fractions of tungsten fibers are reported and discussed in terms of a phenomenological model, in which the frictional matrix flow stress and the reversible matrix mean stress are both split into "elastic," "plastic," and "thermal" terms. The model allows separate measurements to be made of plastic friction, and the experimental results confirm Brown and Clarke's modification of the Orowan mechanism. The possibility of extending the established mean field theory to coupled elasto-electromagnetic behavior is discussed briefly.

1. Introduction

Composites are inhomogeneous materials consisting of two or more homogeneous phases differing in their mechanical and physical properties. The theory of composites aims to predict the "effective" properties of composites from information on the phase properties and phase geometry, that is, the constitutive laws, shapes, sizes, orientations, and volume fractions of the individual phase regions. The simplest theory is the time-independent linear theory of composites (LTC), which assumes that the behaviour of the phases follows independently specified linear constitutive laws. In the 1960s and 1970s Hill and his colleagues established a mean field formulation of the LTC, which sets rigorous bounds on the effective thermoelastic moduli in terms of the phase volume fractions. In an appendix Pedersen (1983) gives a short review of Hill's theory, with special emphasis on Eshelby's contribution.

The mean field theory is a rigorous and versatile framework for studies of composite materials, but the physical assumptions of the LTC are usually too restrictive for realistic description. By itself, the theory then rarely offers a full account of the behavior of composites, it must be complemented with a diversity of specific physical models. This is one of the central themes in the general field of the micromechanics of defects in solids—to which Professor Mura has made extensive contributions. Here we first discuss the problem of thermomechanical hysteresis in metal matrix composites. The discussion is guided by the experimental behavior of a simple model system—copper with continuous tungsten fibers. We then explore the possibility of extending the mean field theory to enable it to deal with phase couplings other than the thermomechanical coupling.

2. Thermal Hysteresis

Thermal residual stresses inevitably arise during the preparation and use of metal matrix composites. Wakashima *et al.* (1974) studied the effects of thermal residual stresses in simple composites consisting of 100 μm diameter continuous tungsten fibers vacuum infiltrated with liquid copper. The composites were subjected to cyclic variations of temperature, as illustrated in Fig. 1.

It is observed that a reversal of the direction of temperature change, from heating to cooling or vice versa, is accompanied by a transient, referred to as stage I or region I, in which both fibers and matrix deform elastically. *Thermal hysteresis* results from the subsequent onset of stage II, that is, the onset of plastic flow in the matrix.

2.1. A Stress Analysis

An approximate stress analysis provides a simple model, summarizing the dominant thermal deformation characteristics of copper–tungsten at tem-

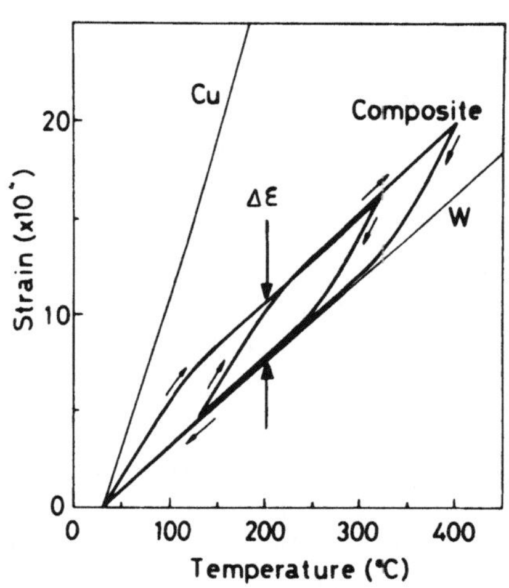

FIG. 1. Axial thermal hysteresis in the copper–tungsten system. From Wakashima *et al.*, (1974), with the definition of the strain difference $\Delta\varepsilon$ added by the present author. Courtesy of *J. Composite Materials.*

peratures below about half the absolute melting temperature of copper. In *stage I* the effective thermal expansion coefficient $\bar{m}$ in the fiber direction must satisfy the approximate stress equilibrium requirement

$$(1 - f)(\bar{m} - m_M)E_M\Delta T + f(\bar{m} - m_F)E_F\Delta T \simeq 0, \tag{2.1}$$

where f is the fiber volume fraction, E is Young's modulus, and the subscripts refer to matrix (M) and fibers (F). The approximation of (2.1) is to ignore the (small) effect of the transverse Poisson contraction, see Section 3. Thus in stage I we find

$$\bar{m} \simeq \frac{m_M E_M + f(m_F E_F - m_M E_M)}{E_M + f(E_F - E_M)}. \tag{2.2}$$

The onset of *stage II* occurs when the stress $(\bar{m} - m_M)E_M\Delta T$ in the matrix reaches the local yield stress σ_Y. If T_0 is a reference temperature at which the thermal residual stresses in the composite vanish, then the absolute value $|\Delta T_Y|$ of the temperature increment $\Delta T = T - T_0$ for the onset of stage II must satisfy

$$-(1 - f)\sigma_Y + f(\bar{m} - m_F)E_F|\Delta T_Y| \simeq 0, \tag{2.3}$$

or

$$|\Delta T_Y| \simeq \frac{(1 - f)\sigma_Y}{f(\bar{m} - m_F)E_F}. \tag{2.4}$$

Thus stage I is observed in the temperature range from $T_0 - \Delta T_Y$ to $T_0 + \Delta T_Y$, and thermal hysteresis should occur whenever this range is exceeded. In copper–tungsten, $m_F < m_M$, so the matrix is being compressed by the fibers during *heating*, and (2.3) provides the thermal strain as

$$(\bar{m}\Delta T)_{\text{heating}} \simeq m_F\Delta T + \frac{(1 - f)\sigma_Y}{fE_F}. \tag{2.5}$$

During *cooling* the matrix is being extended by the fibers, so the sign of the friction term must be reversed, that is,

$$(\bar{m}\Delta T)_{\text{cooling}} \simeq m_F\Delta T - \frac{(1 - f)\sigma_Y}{fE_F}. \tag{2.6}$$

The simplest measure of thermal hysteresis is the strain difference $\Delta\varepsilon$, added to Fig. 1 by the present author. According to (2.5) and (2.6), we find

$$\begin{aligned}\Delta\varepsilon &\simeq (\bar{m}\Delta T)_{\text{heating}} - (\bar{m}\Delta T)_{\text{cooling}} \\ &\simeq \frac{2(1 - f)\sigma_Y}{fE_F}.\end{aligned} \tag{2.7}$$

When stage I is exceeded the dimensions of the composite at a given temperature therefore depend on whether the composite is being heated or cooled, and the dimensional difference $\Delta\varepsilon$ is found to be proportional to the matrix yield stress, σ_Y. However, if the matrix is ideal plastic, with a constant σ_Y, then differentiation with respect to ΔT, whether of (2.5) or (2.6), always leads to the same thermal expansion coefficient in stage II, namely,

$$\bar{m} = m_F. \tag{2.8}$$

Thus the simple stress analysis offers a fairly comprehensive account of thermal deformation, covering the thermal expansion coefficients in stages I and II, the transition temperatures from stage I to II, and a simple measure of thermal hysteresis. The model is limited to axial deformation, it ignores Poisson contraction and it contains an adjustable parameter, the matrix yield stress σ_Y. We now discuss the model in the light of the far more sophisticated mean field theory and compare it with the available thermal cycling data on copper–tungsten.

2.2. Mean Field Theory

The established theory of thermoelasticity in composite materials is the mean field theory due to Hill (1963), Walpole (1966), Laws (1973), and others. The appendix in Pedersen (1983) gives a unified presentation of the mean field theory as formulated by Hill and his colleagues. The mean field theory refers to a general multiphase composite in which the phases are given subscripts $r = 1, 2, \ldots, N$. The mean stress in phase r is expressed in the tensor form

$$\langle\sigma\rangle_r = \bar{B}_r\langle\sigma\rangle, \tag{2.9}$$

where $\bar{B}_r$ is Hill's (1963) fourth rank *stress concentration tensor* for phase r, σ is the second rank stress tensor, and the brackets $\langle\ \rangle$ denote volume averaging. Replacing the pair of Latin indices ij with a single Greek index α, the tensors $\bar{B}_r$ and σ can be represented (Nye, 1957) by square matrices and column matrices, respectively. Stress equilibrium requires that the mean stress $\langle\sigma\rangle$ in the composite satisfies

$$\langle\sigma\rangle = \sigma^A = \sum f_r\langle\sigma\rangle_r, \tag{2.10}$$

where f_r is the volume fraction of phase r and σ^A is the tensor of applied stress. Within this general framework Hill (1963) expresses the effective elastic compliance tensor for the composite in the form $\bar{M} = \sum f_r M_r \bar{B}_r$. Even the tensor of effective thermal expansion coefficients $\bar{m}$ can be expressed in terms of $\bar{B}_r$. As shown by Laws (1973) the result is

$$\bar{m} = \sum f_r \bar{B}_r^T m_r, \tag{2.11}$$

where T denotes transposition. In the special case of the two-phase composite the mean field theory implies that

$$\bar{m} = (\bar{M} - M_2)(M_1 - M_2)^{-1}m_1 + (\bar{M} - M_1)(M_2 - M_1)^{-1}m_2. \tag{2.12}$$

Walpole (1966) showed that explicit approximations to Hill's stress concentration factors, leading to explicit bounds for $\bar{M}$, could be written in terms of Eshelby's (1957) concept of an equivalent transformation strain. In particular, he identified in his own equations Eshelby's equivalent transformation strain ε_r^T for an ellipsoidal inhomogeneity of stiffness L_r embedded in a "comparison material" of stiffness L_0 with an image stress σ^{im} applied at infinity.

If the composite consists of an isotropic elastic matrix with isotropic elastic ellipsoidal inclusions, then rigorous bounds for the effective elastic moduli are obtained by using the matrix as a comparison material, except in cases where the bulk and shear moduli (K and G) are such that $(K_1 - K_2)(G_1 - G_2) \leq 0$. It then turns out, as shown by Pedersen (1983), that the general mean field theory implies

$$\langle\sigma\rangle_{\mathrm{M}} = \sigma^{\mathrm{A}} - f_2\sigma^{\mathrm{I}}, \tag{2.13}$$

$$\langle\sigma\rangle_{\mathrm{I}} = \sigma^{\mathrm{I}} + \langle\sigma\rangle_{\mathrm{M}}. \tag{2.14}$$

Here σ^{I} is the inclusion stress provided by Eshelby's transformation theory as

$$\sigma^{\mathrm{I}} = L_{\mathrm{M}}(S - I)\varepsilon^{\mathrm{T}}, \tag{2.15}$$

where L_{M} is the stiffness tensor for the matrix, S is Eshelby's S-tensor, and I is the unit tensor.

The fact that (2.13)–(2.15) can be deduced from the general mean field theory is interesting, because it explains the initially surprising discovery (Pedersen, 1978) that some of the simple Eshelby models (e.g., Pedersen, 1978, 1979; Wakashima *et al.*, 1974, 1979; Taya and Mura, 1981) reproduce some of the rigorous bound sets by the mean field theory. Moreover, (2.13)–(2.15) are the key elements in Brown and Stobbs' (1971) model based on a simple stress analysis. In the elastically homogeneous case the Brown–Stobbs model is valid of *all* f values, as confirmed by Tanaka and Mori (1973) and Pedersen and Brown (1977). But in the *elastically heterogeneous case* the model needed further development to include elastic interactions between inclusions at high f values.

It turns out that a rigorous extension of Brown and Stobbs' stress analysis can be made simply by letting equivalent inclusions interact with the mean matrix stress and the applied stress, as in Walpole's (1966) theory (2.13)–(2.15). This type of approximation was used by Wakashima *et al.* (1974) in their energy analysis (Section 2.3) of thermal expansion and thermal hysteresis, in which equivalent inclusions are made to interact with the mean matrix stress. With an eye to studying mechanical hysteresis, the complete version of the approximation was used in an independent stress analysis (Pedersen, 1978, 1979) of the effective elastic moduli, thermal expansion coefficients, mean phase stresses, and workhardening in two-phase composites. The capability of reproducing the known and rigorously justified results of the mean field theory, including (2.12), is demonstrated and emphasized by Pedersen (1978, 1979). However, *additionally* including the interaction with the applied stress, the analysis by Pedersen (1978, 1979) also led to novel results of crucial importance for the physically based modelling of mechanical hysteresis. In particular, the mean matrix stress and the equivalent transformation strain of the fibres were both found to be *split* into "elastic" and "inelastic" components (Sections 3 and 4).

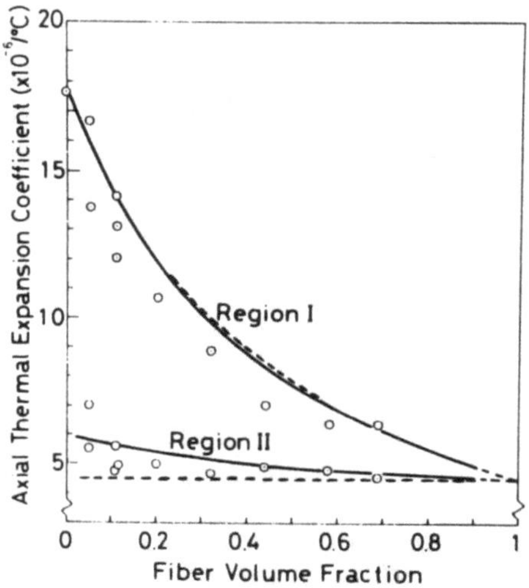

FIG. 2. Measurements of the axial thermal expansion coefficients of copper–tungsten compared with the energy calculation of Wakashima *et al.* (1974) (full curves) and the stress calculation by the present author (dashed curves). From Wakashima *et al.* (1974) with dashed curves added. Courtesy of *J. Composite Materials.*

2.3. *An Energy Analysis*

Wakashima *et al.* (1974) obtain explicit approximations for $\bar{m}$ in the cases when the inclusions are spheres, disks, or fibers, and they point out that in the case of spheres their approximation coincides with Kerner's (1956) approximation. Referring to (2.13)–(2.15) it is clear that the approximations are all part of the established mean field theory. Figure 2 shows that in the case of fibers the approximation (full curve) is in excellent agreement with the experimental behavior of copper–tungsten in stage I (region I). The dashed curve, added by the present author, represents the present stress calculation, that is, (2.2). The small difference between the curves reflects the neglect of Poisson contraction in the present model.

In dealing with stage II, Wakashima *et al.* adopt a modification of the energy minimization approach of Tanaka and Mori (1970, 1971). The modification consists of incorporating the Brown–Stobbs stress balance to account for the large f values in the elastically homogeneous case. In the elastically heterogeneous case, the Walpole approach of letting equivalent inclusions interact via the mean matrix stress is followed. The crucial feature is Ashby's (1966) continuum concept of an *effectively homogeneous* plastic strain in the matrix, $\varepsilon^P = (-\varepsilon_p/2, -\varepsilon_p/2, \varepsilon_p, 0, 0, 0)$. In reality, the plastic strain in the matrix is, of course, strongly heterogeneous, both at the level of individual dislocations and at the continuum level of slip patterns, where gradients of slip are dictated by the compatibility requirement. The concept of an effectively homogeneous ε^P

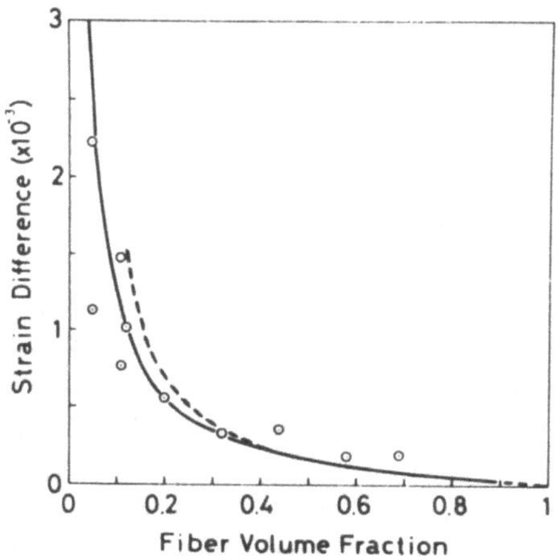

FIG. 3. Measurements of thermal hysteresis $\Delta\varepsilon$ of copper–tungsten compared with the energy calculation of Wakashima *et al.* (1974) (full curve) and the stress calculation by the present author (dashed curve). From Wakashima *et al.* (1974) with dashed curve added. Courtesy of *J. Composite Materials.*

therefore represents an enormous simplification, which apparently reduces the complex problems of stage II to nearly the level of simplicity of those of stage I. Figure 3 compares the energy calculation of thermal hysteresis, measured as $\Delta\varepsilon$, with the present stress calculation (dashed curve), i.e., (2.7).

The energy calculation exploits Eshelby's (1959) energy expressions, which deliver the elastic energy of the composite in the scalar form

$$E_{\mathrm{el}} = C_1\varepsilon_{\mathrm{p}}^2 + C_2\varepsilon_{\mathrm{p}}(m_{\mathrm{F}} - m_{\mathrm{M}})\Delta T + C_3(m_{\mathrm{F}} - m_{\mathrm{M}})^2\Delta T^2, \tag{2.16}$$

where the coefficients C_1, C_2, and C_3 depend on the elastic moduli of the phases and on the volume fraction and shape of the inclusions. The energy dissipation in the plastically deforming matrix is represented by an adjustable parameter σ_{Y}, the matrix yield stress. The equation

$$-\delta E_{\mathrm{el}} = (1 - f)\sigma_{\mathrm{Y}}|\delta\varepsilon_{\mathrm{p}}|, \tag{2.17}$$

expresses the requirement of *energy conservation*, that is, the loss, $-\delta E_{\mathrm{el}}$, of elastic energy during a virtual plastic deformation, $\delta\varepsilon_{\mathrm{p}}$, is balanced by the liberated heat $(1 - f)\sigma_{\mathrm{Y}}|\delta\varepsilon_{\mathrm{p}}|$.

In rigorous terms the stress and energy calculations are equivalent (Pedersen and Brown, 1977). Yet, we cannot expect exact agreement in the present case, because different matrix-yield criteria have been adopted and because the stress calculation is only approximate. With an ideal plastic matrix ($\sigma_{\mathrm{Y}} \equiv$ constant) we saw that the present stress calculation predicts that $\bar{m}$ is independent of f in stage II; but there is room for a possible *small* f dependence, since the Poisson effect was neglected. The energy calculation may appear to provide a somewhat better fit to the dimensional difference $\Delta\varepsilon$ (Fig. 3); but it is important to note that $\Delta\varepsilon$ is, in both calculations, proportional to σ_Y, which is merely an adjustable parameter.

3. Mechanical Hardening

The assumption of an ideal plastic matrix with an effectively homogeneous plastic strain may be examined by isothermal mechanical experiments on copper–tungsten. When the composites are pulled in the fiber direction (x_3) they display three stages: in stage I both phases stay elastic, in stage II the matrix flows plastically around the fibers, and in stage III the fibers flow or fracture. Thus in stages I and II the composite's (C) total (T) axial strain in the x_3 direction ε_{33}^{TC} is the sum $\varepsilon_{33}^{EC} + \varepsilon_{33}^{PC}$ of an elastic (E) strain and a plastic (P) strain. Experimentally, it is found that stage II is *linear*, that is, it displays a constant "modulus", defined in terms of the forward (F) flow stress and the total axial strain, as

$$E_{II} = \frac{d\sigma_{33}^{F}}{d\varepsilon_{33}^{TC}} = (E_I^{-1} + \Theta_F^{-1})^{-1}, \tag{3.1}$$

where E_I is the effective Young's modulus

$$E_I = \frac{d\sigma_{33}^{F}}{d\varepsilon_{33}^{EC}}, \tag{3.2}$$

and

$$\Theta_F = \frac{d\sigma_{33}^{F}}{d\varepsilon_{33}^{PC}}. \tag{3.3}$$

It is straightforward to calculate approximations to E_I and E_{II}, and hence Θ_F, using the stress analysis of Section 2.1. The result is

$$E_I \simeq fE_F + (1-f)E_M, \tag{3.4}$$

and, assuming that $\sigma_Y \equiv$ constant,

$$E_{II} \simeq fE_F. \tag{3.5}$$

A comparison with Hill's (1964) rigorous bounds for stage I shows that (3.4) is an excellent approximation, that is, the effect of the neglected differential Poisson contraction is small. However, it can be seen from Fig. 4 that (3.5) *cannot* account for Kelly and Lilholt's (1969) measurements of E_{II} in copper–tungsten. The dashed curve (denoted $\propto E^*f$) represents the simple model and the full curve (denoted Hill) represents Kelly and Lilholt's attempt to include differential Poisson contraction by using Hill's upper bound. Thus it turns out that the concept of an ideal plastic matrix fails completely to account for the experiments, and a surprisingly high *matrix hardening* rate has to be introduced into the model to make it fit the experiments. The top curve, which

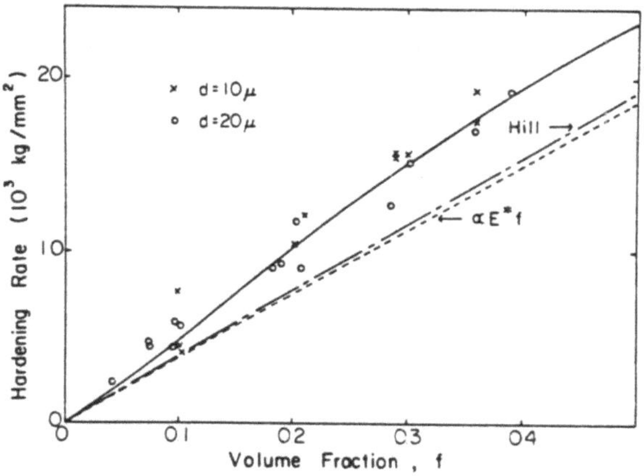

FIG. 4. Kelly and Lilholt's (1969) measurements of the axial hardening rate E_{II} of copper–tungsten compared with various calculations, see text. From Tanaka and Mori (1971). Courtesy of *Phil. Mag.*

appears to account for the measurements, is a heuristic energy analysis by Tanaka and Mori (1970). Their analysis is based on Eshelby's theory and it attempts to describe elastic fibre–fibre interaction by a variant of the self-consistent approach, while *retaining* the assumption on an ideal plastic matrix with $\sigma_Y \equiv$ constant.

However, Hill's bounds and Eshelby's theory are both rigorous continuum mechanical descriptions, which should ideally lead to identical predictions. It is indeed possible (Pedersen, 1978, 1979, 1983) to reconcile the two approaches completely. When the resulting model, which combines Walpole's theory with the simple Brown–Stobbs model, is combined with a Tresca-type flow criterion then it is found (Pedersen, 1979) to provide

$$\Theta_F = \Theta_0 = \frac{A}{(1-B)C}, \tag{3.6}$$

for an effectively homogeneous ideal plastic matrix in which stress relaxation is tentatively ignored. The scalar quantities A, B, and C are provided explicitly by the model in terms of the phase moduli and the shape and volume fraction of inclusions.

The prediction based on (3.6) agrees very accurately with that of Hill's theory, and hence it fully confirms the reality of the positive deviation in Fig. 4 and the exciting conclusion that the matrix hardening must be a genuine composite property, depending synergistically on a plastic interaction of fibres and matrix. The interesting novel feature of the mean field theory as formulated by Pedersen (1979, 1983) is that it reveals that Θ_0 is split into "*plastic*" mean stress hardening (corresponding to A/C) caused by ε_p and "*elastic*" mean stress hardening (the remainder in (3.6)) caused by the externally applied stress

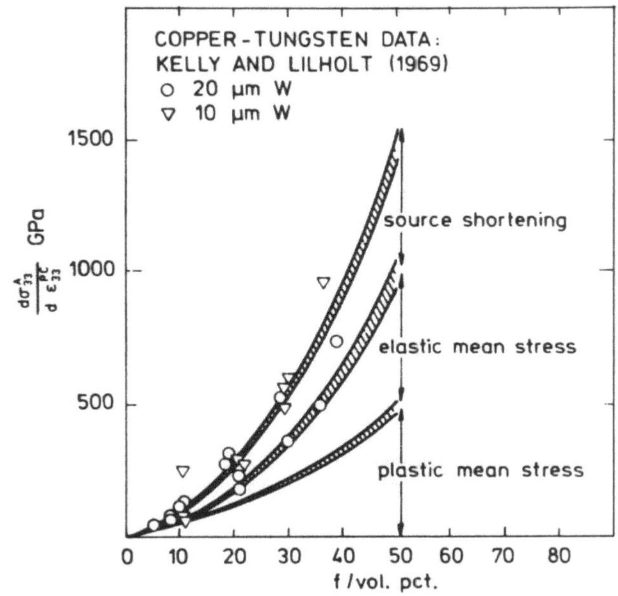

FIG. 5. When the Eshelby approach is fully reconciled with Hill's bounds it reveals that mean stress hardening is split into "elastic" and "plastic" components, see text. From Pedersen (1979).

σ^A. In Fig. 5, reproduced from Pedersen (1979), it can be seen that at high f values these two separate contributions are about equal. The positive deviation from Θ_0 may be calculated by combining the mean field theory with Brown and Clarke's (1977) modification of the Orowan model, their "*source-shortening*" model (next section). Simple arguments (Pedersen, 1988) based on dislocation mechanics exclude the various alternative models, at least at high f values.

4. Mechanical Hysteresis

Copper–tungsten composites are also used as an experimental model system in the present study of *mechanical* hysteresis in metal matrix composites. The composites are prepared by vacuum infiltrating bundles of continuous parallel tungsten fibers with 99.999% pure copper. Here we report initial results for composites with large volume fractions of 500 μm diameter fibers. A report on a similar but much more detailed study of copper with high volume fractions of 100 μm diameter tungsten fibers is in preparation.

Figure 6 shows a typical hysteresis loop measured at room temperature in tension–compression along the fiber direction. Stress readings were obtained from a tension–compression load cell and the total axial strain ε_{33}^{TC} was

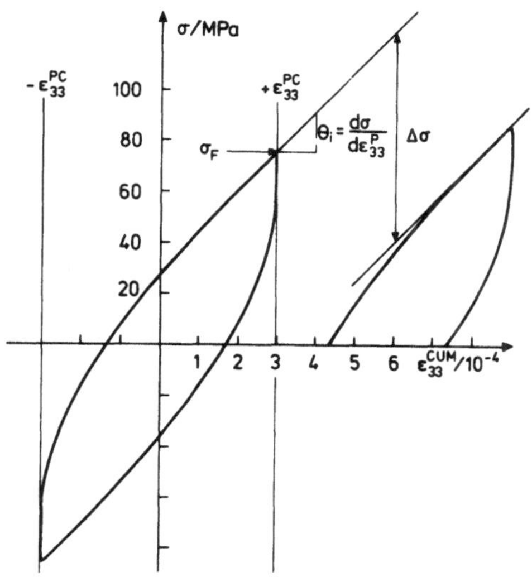

FIG. 6. Axial mechanical hysteresis in copper with 20 vol. pct. continuous 500-μm diameter tungsten fibers. The definitions of the permanent softening $\Delta\sigma$ and the "instantaneous hardening rate" Θ_i are shown.

recorded using an extensometer attached to the specimen. The stress and total strain signals were led through a plastic strain computer, as in the experiments by Pedersen and Lisiecki (1988). Hysteresis loops were recorded for each specimen at several plastic strain amplitudes, increasing from $\varepsilon_{33}^{PC} \simeq 2 \times 10^{-4}$ to 2×10^{-3}.

The mechanical hysteresis will be characterized using Orowan (1959) and Wilson's (1963) "permanent softening" $\Delta\sigma$, defined in Fig. 6. In this respect the present measurements are an extension to high f values of the experiments made by Lilholt (1977) at small f values. However, in the present study no attempt is made to measure the shape of the loops, so the "*instantaneous hardening rate*", Θ_i, which is a central feature in Lilholt's analysis of the matrix hardening, will not be used here. Instead the phenomenological model (Pedersen, 1984) extending the Orowan–Wilson interpretation of the Bauschinger effect will be applied. The model deals with measurements of $\Delta\sigma$ and σ_F, rather than with $\Delta\sigma$ and Θ_i, and is applicable at all f values.

4.1. Diffraction Studies

We saw that the linearity of stage II hardening cannot be understood as a trivial consequence of an approximately ideal plastic matrix. The positive deviation from the mean field theory (Figs. 4 and 5) reflects real physical mechanisms, and these mechanisms are nonlinear at the near-atomic level of dislocations, despite the observed linearity at the macroscopic level of matrix hardening. When modeling the mechanical hysteresis at high f values, it is essential to be able to separate the effect of elastic heterogeneity from that of plastic heterogeneity, as represented by the effectively homogeneous plastic strain ε_p in the matrix.

One important question then is whether or not ε_p does in fact behave like an Eshelby-type transformation strain, and the most direct experimental check is by X-ray (or neutron) diffraction measurement of the mean lattice strains in the deforming composite. As it turns out (Pedersen, 1984), ε_p usually does *not* behave like a transformation strain. It is necessary to allow for stress relaxation, and instead consider the unrelaxed plastic strain ε_p^* in the matrix. When the fibers are continuous then strain compatibility requires that

$$\varepsilon_p^* = \langle \varepsilon_{33}^P \rangle_M - \langle \varepsilon_{33}^P \rangle_F = \langle \varepsilon_{33}^E \rangle_F - \langle \varepsilon_{33}^E \rangle_M, \tag{4.1}$$

where a nonzero mean plastic (P) strain $\langle \varepsilon_{33}^P \rangle_F$ in the fibers represents stress relaxation, whether by flow in the fibers or by mechanisms involving cracking or interface decohesion. Cheskis and Heckel (1968) measured the mean elastic (E) strain in fibers and (polycrystalline) matrices of hot-pressed continuous fiber copper–tungsten during deformation. The corresponding values of ε_p^*, calculated from their data by Pedersen (1984), are plotted in Fig. 7 as a function of ε_p, calculated from the expression

$$\varepsilon_{33}^{PC} = \varepsilon_p - (1 - C)\varepsilon_p^*, \tag{4.2}$$

provided by the mean field theory. Clearly, (*linear*) stress relaxation accompanies deformation in stage II, but the diffraction data show that ε_p^* behaves consistently as a transformation strain ε^{T*} in the fibers.

The observation of linear stress relaxation in diffraction experiments on hot-pressed copper–tungsten agrees with a wide range of observations. Stress

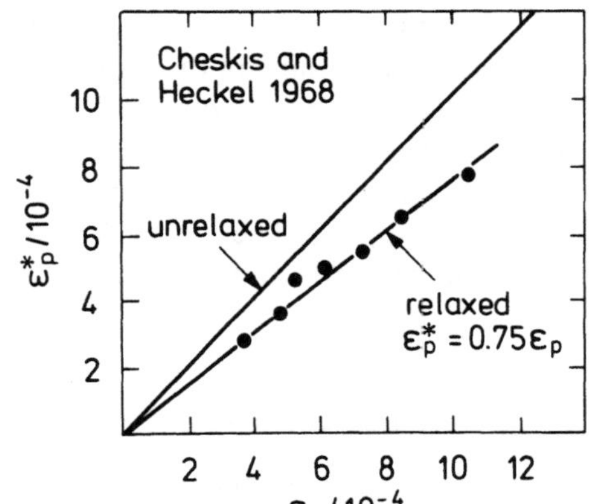

FIG. 7. Analysis of X-ray diffraction measurements of internal stresses in copper–tungsten during axial deformation provides ε_p^* as a function of ε_p within the framework of the mean field theory. The deviation from $\varepsilon_p = \varepsilon_p^*$ indicates linear stress relaxation. From Pedersen (1984).

relaxation also had to be invoked to account for cyclic experiments on copper–silica (Brown, 1979) and vacuum-infiltrated low f copper–tungsten (Lilholt, 1977).

4.2. *A Phenomenological Model*

Macroscopic linearity is perhaps the single most characteristic feature emerging from the many observational studies of stage II in continuous-fiber copper–tungsten: the composite's overall workhardening rate Θ_F and its rate $d\varepsilon_p^*/d\varepsilon_p$ of stress relaxation are both constant in stage II. The most general linear expression for the overall axial flow stress is

$$\sigma_F = (A + \alpha)\varepsilon_p^* + (B + \beta)\sigma_F + \delta + \sigma_{Th} = \frac{(A + \alpha)\varepsilon_p^* + \delta + \sigma_{Th}}{1 - B - \beta}, \quad (4.3)$$

where the mean field theory and a simple Tresca-type flow criterion for the matrix provides the coefficients A and B for mean stress hardening, see Pedersen (1984). Calculation of the phenomenological coefficients α, β, and δ for matrix friction requires a valid physical model. The term σ_{Th} is the constant contribution from thermal residual stresses present in the as-fabricated composite, it is difficult to calculate because it is caused by complex processes of differential thermal contraction and stress relaxation during fabrication. However, the presence of σ_{Th} is readily demonstrated by a cyclic experiment: The plastic mean stress term $A\varepsilon_p^*$ and σ_{Th} both have definite directions, and hence an expression for the reverse flow stress can be obtained from that for σ_F by changing the signs of $A\varepsilon_p^*$ and σ_{Th}

$$\sigma_R = \frac{(-A + \alpha)\varepsilon_p^* + \delta - \sigma_{Th}}{1 - B - \beta}. \quad (4.4)$$

It follows that the permanent softening, obtained from (4.3) and (4.4) as

$$\Delta\sigma = \sigma_F - \sigma_R = \frac{2A\varepsilon_p^* + 2\sigma_{Th}}{1 - B - \beta}, \quad (4.5)$$

has a strain-independent contribution from σ_{Th}. This contribution is easily avoided by considering the hardening *rates*, as Lilholt (1977) shows. The present model then provides

$$\Theta_F = \frac{d\sigma_F}{d\varepsilon_{33}^{PC}} = \frac{A + \alpha}{(1 - B - \beta)(d\varepsilon_p/d\varepsilon_p^* - 1 + C)}, \quad (4.6)$$

and

$$\Theta_\Delta = \frac{1}{2}\frac{d\Delta\sigma}{d\varepsilon_{33}^{PC}} = \frac{A}{(1 - B - \beta)(d\varepsilon_p/d\varepsilon_p^* - 1 + C)}, \quad (4.7)$$

which may be obtained from (4.2), (4.3), and (4.5) by differentiation. Figure 8 shows that in stage II the dependences of σ_F and $\Delta\sigma$ on strain are both linear,

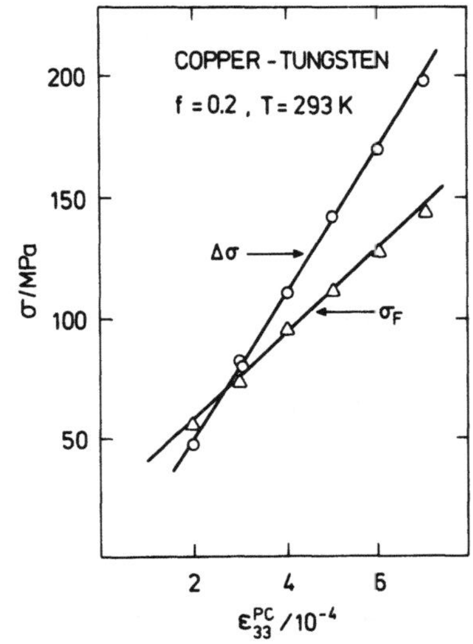

FIG. 8. The linearity of stage II is confirmed by the present cyclic experiments on copper–tungsten. Unique values of Θ_F and Θ_Δ are obtained by measuring the slopes of σ_F and $\Delta\sigma$ versus ε_{33}^{PC}.

so that unique values of Θ_F and Θ_Δ can be measured. The results of the present experiments and analysis are shown in Table 1.

When analyzed on the basis of the present model the cyclic experiments provide information on both friction and relaxation in the matrix. Comparison of (4.6) and (4.7) shows that the *plastic friction* can be measured separately as the ratio

$$\frac{\alpha}{A} = \frac{\Theta_F}{\Theta_\Delta} - 1. \tag{4.8}$$

The present cyclic experiments (Table 1) support the tensile experiments (Figs. 4 and 5) in rejecting the assumption of an ideal plastic matrix, which would mean that $\alpha/A \equiv 0$. We saw (Fig. 5) that for high f values the elastic mean stress becomes comparable with the plastic mean stress. Furthermore, we might reasonably expect that large α/A values would correspond to large β/B values. This would mean that the stress relaxation rate

$$\frac{d\varepsilon_p^*}{d\varepsilon_p} = \left[\frac{A}{(1 - B - \beta)\Theta_\Delta} + 1 - C\right]^{-1}, \tag{4.9}$$

TABLE 1. Hardening and mechanical hysteresis.

f	$d/\mu m$	T/K	Θ_F/TPa	Θ_Δ/TPa	α/A	$d\varepsilon_p^*/d\varepsilon_p$	Specimen
0.1	500	293	0.125	0.072	0.74	0.95	A11
0.2	500	293	0.180	0.150	0.20	1.18	B11

obtained by rearranging (4.7), cannot at higher f values be calculated accurately without a valid physical model of friction: According to (4.9), stress relaxation and elastic friction *cannot* be mutually separated in the Bauschinger experiment, unless $\beta \ll 1$.

4.3. *A Physical Model*

Among the many dislocation theories proposed for matrix friction hardening in copper–tungsten, Brown and Clarke's (1977) modification of the Orowan mechanism, their "source-shortening" model, appears to be the most plausible model at high f values, as argued by Pedersen (1988). Without adjustable parameters the source-shortening model predicts, for an *elastically homogeneous* composite, that

$$2\tau_M = \frac{2\mu b}{L} + \frac{5}{\pi} f\mu\varepsilon_p^*. \tag{4.10}$$

This expression immediately reproduces the dominant qualitative characteristics of matrix hardening, namely, linearity and a strong f dependence. Scale-dependence enters via the conventional Orowan stress, $2\mu b/L$, where L is the effective fiber spacing and b is the Burgers vector. An additional scale-dependence comes (Pedersen, 1985) from stress relaxation via ε_p^* in the hardening term, $(5/\pi)f\mu\varepsilon_p^*$, which is otherwise scale-independent.

In order to take *elastic heterogeneity* into account, we may replace ε_p^* in (4.10) by some suitable measure $|\varepsilon^T|$ of the equivalent transformation strain. As shown in Pedersen (1979) the deformation induced part of ε^T is split into *elastic* and *plastic* components, so that

$$\varepsilon_{11}^T = a_1\varepsilon_p^* + b_1\sigma^A, \tag{4.11}$$

$$\varepsilon_{22}^T = a_2\varepsilon_p^* + b_2\sigma^A, \tag{4.12}$$

$$\varepsilon_{33}^T = a_3\varepsilon_p^* + b_3\sigma^A, \tag{4.13}$$

where the coefficients a_i and b_i can be calculated using the mean field theory; mean field approximations to a_i and b_i for copper–tungsten at room temperature may be found in Pedersen (1988). By defining $|\varepsilon^T|$ as the average $(|\varepsilon_{22}^T| + |\varepsilon_{33}^T|)/2$ of the absolute values of ε_{22}^T and ε_{33}^T, Pedersen (1988) finds the scale-independent friction parameters

$$\alpha = \frac{5}{2\pi}\mu f(|a_2| + |a_3|), \tag{4.14}$$

$$\beta = \frac{5}{2\pi}\mu f(|b_2| + |b_3|). \tag{4.15}$$

The calculated value of α/A for copper–tungsten is identical to about 0.4 at all f values. This prediction is consistent with the present experimental values (Table 1). The values of $d\varepsilon_p^*/d\varepsilon_p$ calculated from the experimental data using

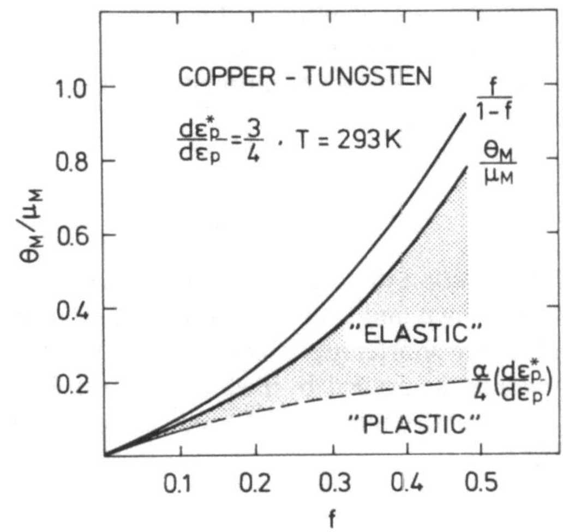

FIG. 9. The matrix hardening rate Θ predicted for copper–tungsten by combining the source-shortening model with the mean field theory. The prediction agrees accurately with the empirical curve $\mu f/(1-f)$.

(4.9) are close to unity (Table 1); but it is important to bear in mind that considerable scatter characterizes the mechanical behavior of copper–tungsten. It is clear, therefore, that more experiments are needed before the levels of friction and relaxation can be established.

A compilation (Pedersen, 1985) of the available tensile data for the copper–tungsten system shows that the derived matrix hardening rates $\Theta_m = d\tau_M/de_p$, where $e_p = 2\varepsilon_p$, are scattered around an empirical expression $f\mu/(1-f)$. Figure 9 compares this expression with the present model, which predicts

$$\Theta_M = \left(\frac{\alpha}{4} + \frac{\beta}{4}\frac{A+\alpha}{1-B-\beta}\right)\left(\frac{d\varepsilon_p^*}{d\varepsilon_p}\right), \tag{4.16}$$

as shown by Pedersen (1988). The agreement between theory and experiment is excellent, and it can be seen that the regions of high and low f values differ with respect to the cause of matrix friction: at low f values plastic friction dominates, while elastic friction dominates at high f values. Thus in the case of mechanical hysteresis it seems possible to use a physically based constitutive equation for the constrained cyclic plasticity of the matrix, rather than relying on the adjustable parameter, σ_Y, as was done in the case of thermal hysteresis.

5. Physical Analogues

Thermomechanical hysteresis displays two qualitative features with numerous general analogues in the mechanical and physical behavior of composites.

First, it is evident that in thermal hysteresis the *phase couplings* are essential. In the general context of nonstructural applications of composite materials it is necessary to consider all possible phase couplings, including those between mechanical, thermal, electrical, and magnetic behavior, as emphasized by Albers (1978). Second, hysteresis involves energy dissipation by real *physical mechanisms* of interplay between the phases, as exemplified by the Orowan mechanism and analogous interactions involving domain walls and flux line dislocations in magnetic and superconductor hardening (Haasen, 1972). It is well known that the mean field theory can be translated to cover uncoupled physical properties analogous to the dielectric constant (e.g., Hashin, 1970; Pedersen, 1978; Poulsen, 1985; Taya, 1988), and a diversity of physical mechanisms are operative in these applications. Therefore, a short formal discussion of mechanisms and couplings seems worthwhile, since it still remains to be seen whether the mean field theory can be generalized to account for couplings other than the thermomechanical coupling.

5.1. *Coupled Behavior*

We saw that nonlinear constitutive behavior of composites, like crystal plasticity, may be studied conveniently within the general framework of the mean field theory. The theory itself, however, is a formulation of the LTC which assumes that the individual phases obey simple independently specified linear constitutive laws. Following the general scheme of Nye (1957) we may express the linear moduli of homogeneous solids by a Jacobian matrix

$$J = \frac{\partial Y}{\partial X}, \tag{5.1}$$

of a set of dependent state variables Y with respect to a set of independent state variables X. For the purpose of generalizing the mean field theory to cover more general types of coupled behavior, it is essential not to mix *solenoidal* and *irrotational* variables. This becomes clear if we consider the fundamental sources of the fields in question. For example, the source of electric displacement D_i is the free charge ρ, which appears in Gauss' law of electrostatics

$$D_{i,i} = \rho. \tag{5.2}$$

In dielectrics, with no free charges, D_i is a solenoidal variable, in the sense that the divergence $D_{i,i}$ vanishes. Now, starting with the quantity $(D_j x_i)_{,j}$, it is a simple matter to show that the mean electric displacement, in a body of volume V bounded by a surface S with outward normal vector n_i, must satisfy

$$\langle D_i \rangle = \frac{1}{V} \int_S D_j x_i n_j \, dS - \frac{1}{V} \int_V x_i D_{j,j} \, dV. \tag{5.3}$$

Therefore, when the applied displacement D_i^{A} is homogeneous we get $\langle D_i \rangle = D_i^{\mathrm{A}}$. This argument may be repeated for any solenoidal field, including the magnetic field strength B_i and the stress tensor σ_{ij}, in static situations with no free currents ($B_{i,i} = 0$) and no applied body forces ($\sigma_{ij,j} = 0$), respectively. The counterpart of (5.3) for σ_{ij} may be found as eqn. 13 in Pedersen (1984). Consequently, if solenoidal fields are consistently chosen as dependent variables, then we can write a general *mean field equation*

$$\langle Y \rangle = Y^{\mathrm{A}}, \tag{5.4}$$

which applies irrespective of the constitution of the material. With our choice of variables we then have

$$J = \frac{\partial(\sigma, D, B)}{\partial(\varepsilon, E, H)} = \begin{pmatrix} \left(\dfrac{\partial \sigma_{ij}}{\partial \varepsilon_{kl}}\right)_{E,H} & \left(\dfrac{\partial \sigma_{ij}}{\partial E_l}\right)_{\varepsilon,H} & \left(\dfrac{\partial \sigma_{ij}}{\partial H_l}\right)_{\varepsilon,E} \\ \left(\dfrac{\partial D_i}{\partial \varepsilon_{kl}}\right)_{E,H} & \left(\dfrac{\partial D_i}{\partial E_l}\right)_{\varepsilon,H} & \left(\dfrac{\partial D_i}{\partial H_l}\right)_{\varepsilon,E} \\ \left(\dfrac{\partial B_i}{\partial \varepsilon_{kl}}\right)_{E,H} & \left(\dfrac{\partial B_i}{\partial E_l}\right)_{\varepsilon,H} & \left(\dfrac{\partial B_i}{\partial H_l}\right)_{\varepsilon,E} \end{pmatrix}, \tag{5.5}$$

where the independent variables ε_{kl}, E_l, and H_l are the irrotational fields of elastic strain, ε_{ij}; electric field strength, E_l; and magnetic induction, H_l. The diagonal components of J are the tensors of elastic stiffness $(\partial\sigma/\partial\varepsilon)$, dielectric permittivity $(\partial D/\partial E)$, and magnetic permeability $(\partial B/\partial H)$. Couplings are represented by the off-diagonal components which represent magnetoelectric coupling $(\partial D/\partial H)$, magnetostriction $(\partial\sigma/\partial H)$, and the piezoelectric effect $(\partial\sigma/\partial E)$.

5.2. *Ampere–Lorentz–Eshelby Theory*

There are *three* key equations in the mean field theory of the effective elastic moduli of composite materials, as discussed by Pedersen (1983). Formally, the problem of extending the theory to coupled behavior then appears to involve a search for the counterparts of these key equations. The first two equations have already been considered in the elasto-electromagnetic case: one is the *constitutive equation*, $Y = JX$, for coupled behavior, relating the irrotational variables X to the corresponding set of solenoidal variables Y. The second one is the *mean field equation*, $\langle Y \rangle = Y^{\mathrm{A}}$. We now discuss the third key equation.

As emphasized by Pedersen (1984), the transformation strain $\varepsilon_{ij}^{\mathrm{T}}$ was introduced as an analogy to the Ampere–Lorentz principle (Truesdell and Toupin, 1960) of representing distributions of current and charge by magnetizations M_i and polarizations P_i. Consequently, it should in principle be straightforward to construct an "*Ampere–Lorentz–Eshelby*" *equation*, $X^{\infty} = SX^{\mathrm{T}}$, that is, a coupled counterpart of Eshelby' solution to his transformation problem. This equation is the third key equation, and the obvious way to construct it would seem to be to repeat Eshelby's famous thought experiment. For coupled behavior the thought experiment runs as follows: Imagine that an inclusion in a linear solid undergoes a transformation which, in the absence of the surrounding infinite matrix, would be a uniform change of shape, polarization, and magnetization, as represented by the 12×1 matrix

$$X^{\mathrm{T}} = \begin{pmatrix} \varepsilon_\alpha^{\mathrm{T}} \\ E_i^{\mathrm{T}} \\ H_i^{\mathrm{T}} \end{pmatrix}. \tag{5.6}$$

Paraphrasing the experiment for the purely elastic case, we make a cut round the inclusion, take the inclusion out of the matrix, and let the transformation X^{T} of the inclusion happen in such a way that Y *remains zero* inside it. The D_i and H_i are defined in terms of the vacuum permittivity ε_0 and the vacuum permeability μ_0 as

$$D_i = \varepsilon_0 E_i + P_i, \tag{5.7}$$

and

$$H_i = \frac{1}{\mu_0} B_i - M_i. \tag{5.8}$$

Consequently, with $Y = 0$ the transformation X^{T} involves a polarization P_i^{T} and a magnetization M_i^{T}, with their corresponding layers of polarization charge and magnetization current. Now, apply equal but opposite layers of external force, $-f_i = -\sigma_{ij}^{\mathrm{T}} n_j$, free charge, $-\rho = -D_j^{\mathrm{T}} n_j$, and free "magnetization charge", $-\rho_{\mathrm{M}} = -B_j^{\mathrm{T}} n_j$, over the surface of the transformed inclusion. Put the inclusion back into its hole in the matrix and weld across the inclusion-matrix interface. Inside the inclusion the field is now $-Y^{\mathrm{T}} = -JX^{\mathrm{T}}$, and in the matrix the field is zero. Finally, cancel the applied layers of external force and free charges by applying $+f_i$, $+\rho$, and $+\rho_{\mathrm{M}}$. The cancelling layers give rise to fields of elastic displacement u_i^∞, electrostatic potential V^∞, and magnetostatic scalar potential V_{M}^∞ throughout matrix and inclusions. In the terms of Green's function

$$u_i^\infty = \int_S [G_{ij}^{11}(x - x')f_j(x') + G_i^{12}(x - x')\rho(x') + G_i^{13}(x - x')\rho_{\mathrm{M}}(x')]\, dS', \tag{5.9}$$

$$V^\infty = \int_S [G_j^{21}(x - x')f_j(x') + G^{22}(x - x')\rho(x') + G^{23}(x - x')\rho_{\mathrm{M}}(x')]\, dS', \tag{5.10}$$

$$V_{\mathrm{M}}^\infty = \int_S [G_j^{31}(x - x')f_j(x') + G^{32}(x - x')\rho(x') + G^{33}(x - x')\rho_{\mathrm{M}}(x')]\, dS', \tag{5.11}$$

where ∞ refers to the infinity of the matrix. Using the divergence theorem and the equivalence of $\partial/\partial x_i$ and $-\partial/\partial x_i'$ when operating on $|x - x'|$, we can now combine the definitions $\varepsilon_{ij} = (u_{i,j} + u_{j,i})/2$, $E_i = -V_{,i}$, and $H_i = -V_{\mathrm{M},i}$ with

$f_i = \sigma_{ij}^{\mathrm{T}} n_j$, $\rho = D_j^{\mathrm{T}} n_j$, and $\rho_{\mathrm{M}} = B_j^{\mathrm{T}} n_j$ to get

$$\varepsilon_{ij}^{\infty} = P_{ijkl}^{11}\sigma_{kl}^{\mathrm{T}} + P_{ijk}^{12}D_k^{\mathrm{T}} + P_{ijk}^{13}B_k^{\mathrm{T}}, \tag{5.12}$$

$$E_i^{\infty} = P_{ikl}^{21}\sigma_{kl}^{\mathrm{T}} + P_{ik}^{22}D_k^{\mathrm{T}} + P_{ik}^{23}B_k^{\mathrm{T}}, \tag{5.13}$$

$$H_i^{\infty} = P_{ikl}^{31}\sigma_{kl}^{\mathrm{T}} + P_{ik}^{32}D_k^{\mathrm{T}} + P_{ik}^{33}B_k^{\mathrm{T}}, \tag{5.14}$$

where the coefficients P are given by

$$P_{ijkl}^{11} = \frac{1}{2}\int_V [G_{il,kj}^{11}(x - x') + G_{jl,ki}^{11}(x - x')]\, dV', \tag{5.15}$$

and

$$P_{ij}^{22} = \int_V G_{,ij}^{22}(x - x')\, dV', \tag{5.16}$$

and the various analogous expressions. Some of these coefficients are well known: For example, using the harmonic and biharmonic potentials it is easy to verify that inside an ellipsoidal inclusion in an isotropic medium (5.16) reduces to the spatially *uniform* matrix

$$P_{ij}^{22} = \begin{pmatrix} D_a/\varepsilon & 0 & 0 \\ 0 & D_b/\varepsilon & 0 \\ 0 & 0 & D_c/\varepsilon \end{pmatrix}, \tag{5.17}$$

where ε is the dielectric constant of the medium and D_a, D_b, and D_c are the usual depolarizing or demagnetizing factors (e.g., Kittel, 1971). The coefficients P are counterparts of Hill's tensor P_{ijkl} for uncoupled elasticity and the desired Ampere–Lorentz–Eshelby equation is written formally as

$$X^{\infty} = PY^{\mathrm{T}} = PJX^{\mathrm{T}} = SX^{\mathrm{T}}, \tag{5.18}$$

where the dimensionless 12×12 S-matrix is the generalized Eshelby matrix, the Ampere–Lorentz–Eshelby matrix, for the coupled elasto-electro-magnetostic behavior of the linear medium. The actual derivation of mean field approximations in the special case of the effective compliance matrix $\overline{M}$ of a two-phase composite from the three key equations is illustrated in Pedersen (1979, 1983) and will not be repeated here. The resulting expression for $\overline{M}$ immediately translates into the expression

$$\bar{\mu} = \mu_2 + \frac{f_1}{\dfrac{1}{\mu_1 - \mu_2} + \dfrac{f_2}{3\mu_2}}, \tag{5.19}$$

for the effective magnetic permeability. We simply replace the stiffness tensor L by the (isotropic) permeability tensor, $\mu\delta_{ij}$, and Eshelby's S-tensor by the demagnetizing tensor, $P_{ij}^{22} = \delta_{ij}/3$, for a sphere. Equation (5.19) and the corresponding equation obtained by an interchange of the phase subscripts 1 and 2 coincide with the Hashin–Shtrikman bounds for the effective magnetic permeability. Thus the mean field approximation is rigorously based, and it seems clear that its evaluation for coupled behavior would be of considerable importance in the development of composites with novel physical properties.

5.3. *Irreversibility*

Through formal analogies the mean field theory then covers a vast array of physical properties of composites, which includes the steady-state transport properties described by the laws of Fick, Fourier, and Ohm. For example, by substituting the permeability μ with the conductivity λ, (5.19) can be read as an expression for the effective thermal conductivity $\bar{\lambda}$ of a two-phase composite, such as epoxy with glass powder. Figure 10 compares the bounds set by the mean field theory (shaded) with experimental values (Garrett and Rosenberg, 1974) for room-temperature conduction. Maxwell's (1892) prediction based on an independent inclusion model is also shown. The conductivity of glass–epoxy is seen to be accurately described by the mean field theory for temperatures above 10–20 K.

However, heat transport is an irreversible process, which is potentially more complicated than the reversible elasto-electromagnetostatics discussed in Section 5.2. At low temperatures (< 10 K) Garrett and Rosenberg observed a marked *scale-dependence*, which transcends the mean field theory. They could obtain values of λ, at a given temperature and f value, which were greater or smaller than the conductivity of unfilled epoxy, simply by varying the particle size of the filler. They attribute this behavior to the *physical mechanisms* of transport. Heat is carried by phonons, which interact with

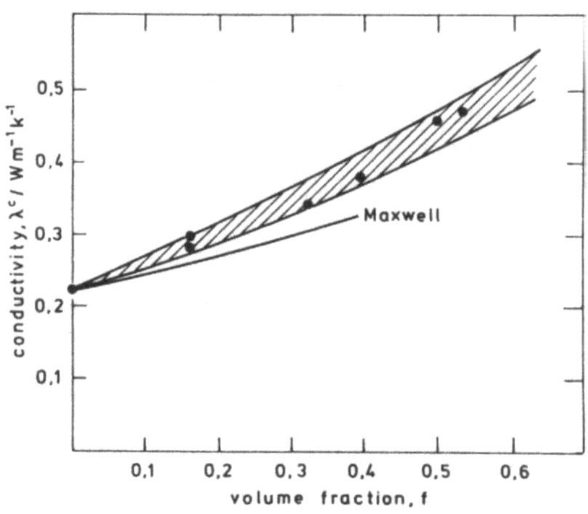

FIG. 10. Comparison of experimental data for the room temperature thermal conductivity of glass–epoxy (Garrett and Rosenberg, 1974), with theoretical predictions based on Maxwell's (1892) independent inclusion approximation and on the mean field theory (shaded area).

obstacles such as the glass–epoxy phase boundaries. At low temperatures the mean free path of phonons becomes comparable with the particle size and, consequently, the thermal conductivity of the filler is no longer an intrinsic property which may be independently specified, it becomes scale-dependent. Furthermore, the acoustic mismatch at the filler–epoxy interface causes the resistance to heat transport across the interfaces to vary roughly as T^{-3} and to become appreciable at low T values, where the effective conductivity was found to display considerable negative deviations from the bounds.

At low and intermediate f values, Garret and Rosenberg were able to account for their measurements using a model which incorporates the T-dependence of the interfacial resistance. However, at high f values ($\geq 30\%$) they observe a positive deviation from this model, which they attribute to direct contacts between the filler particles. Thus the low-temperature thermal conduction at high f values appears to occur by percolation through a connected filler phase. This mode of transport is also observed in conduction of electric charge, where the mean field theory is useful as long as the phase conductivities differ by less than a factor of 100; but for greater differences the bounds are too widely separated to be truly informative. Calculation of effective conductivity then becomes a stochastic problem in percolation theory. In general terms, the problems of thermomechanical hysteresis and effective heat conduction illustrate very clearly that the mean field theory alone cannot give a complete account of irreversible behavior. It is necessary to identify and model the actual physical mechanisms by which the continuum phase boundaries interfere with the motion of cracks, dislocations, magnetic domain walls, phonons, ions, and other discrete carriers.

6. Conclusions

(1) Thermomechanical hysteresis in copper with continuous tungsten fibres may be described on the basis of simple stress analyses, which are fully compatible with the established mean field theory of composite materials. In the case of thermal hysteresis the matrix friction stress is treated as an adjustable parameter.

(2) Initial experiments on the mechanical hysteresis in copper–tungsten confirm the existence of plastic friction hardening when analyzed on the basis of a phenomenological model of the Bauschinger effect. A modification of the Orowan mechanism accounts for the observed plastic friction hardening without the need of adjustable parameters.

(3) Theory and experiments show that elastic friction dominates matrix hardening at high fiber contents. More experiments are needed to fully establish the levels of elasto-plastic friction and relaxation during mechanical hardening of copper–tungsten.

(4) A formal discussion of the elasto-electromagnetostatic coupling in composites suggests the possibility of generalizing the mean field theory to obtain

rigorous descriptions of coupled behavior other than the thermomechanical coupling.

Acknowledgments

I want to thank Palle Nielsen and Jens Olsson for skillful experimental assistance and Lone Jørgensen for typing the manuscript.

References

Albers, W. (1978), Physical properties and design for non-structural applications of composite materials, in *Advances in Composite Materials*, edited by G. Piatti, Applied Science, London, pp. 185–207.

Ashby, M. F. (1966), Work hardening of dispersion-hardened crystals, *Phil. Mag.*, **14**, 1157–1178.

Brown, L. M. (1979), Precipitation and dispersion hardening, *Proceedings of the 5th International Conference on Strength of Metals and Alloys, Aachen, 27–31 August 1979*, edited by P. Haasen, V. Gerold, and Kostorz, G., Pergamon, Oxford 1980, pp. 1551–1571.

Brown, L. M. and Stobbs, W. M. (1971), The work-hardening of copper–silica. I. A. model based on internal stresses, with no plastic relaxation, *Phil Mag.*, **23**, 1185–1199.

Brown, L. M. and Clarke, D. R. (1977), The workhardening of fibrous composites, with particular reference to the copper–tungsten system, *Acta Metallurgica*, **25**, 563–570.

Cheskis, H. P. and Heckel, R. W. (1968), In-situ measurements of deformation behaviour of individual phases in composites by X-ray diffraction, in *Metal Matrix Composites*, ASTM-STP-438, ASTM, New York, pp. 76–91.

Eshelby, J. D. (1957), The determination of the elastic field of an ellipsoidal inclusions, and related problems, *Proc. Roy. Soc.* London, **A241**, 376–396.

Eshelby, J. D. (1959), The elastic field outside an ellipsoidal inclusion, *Proc. Roy. Soc.* London, **A252**, 561–569.

Garrett K. W. and Rosenberg H. M. (1974), The thermal conductivity of epoxy-resin/powder composite materials, *J. Phys. D: Appl. Phys.*, **7**, 1247–1258.

Haasen, P. (1972), Mechanical, magnetic and superconductor hardening by precipitates, *Mater. Sci. Engng.*, **9**, 191–196.

Hashin, Z. (1970), Theory of composite materials, in *Mechanics of Composite Materials*, edited by F. W. Wendt, H. Liebowitz, and Perrone, N., Pergamon, Oxford, pp. 201–242.

Hill, R. (1963), Elastic properties of reinforced solids: Some theoretical principles, *J. Mech. Phys. Solids*, **11**, 357–372.

Hill, R. (1964), Theory of mechanical properties of fibre-strengthened materials. I. Elastic behaviour. II. Inelastic behaviour, *J. Mech. Phys. Solids*, **12**, 199–219.

Kerner, E. H. (1956), The elastic and thermoelastic properties of composite media, *Proc. Phys. Soc.* London, **69B**, 808.

Kelly, A. and Lilholt, H. (1969), Stress–strain curve of a fibre-reinforced composite, *Phil. Mag.*, **20**, 311–328.

Kittel, C. (1971), *Introduction to Solid State Physics*, 4th ed., Wiley, New York, pp. 766.

Laws, N. (1973), On the thermostatics of composite materials, *J. Mech. Phys. Solids*, **21**, 9–17.

Lilholt, H. (1977), Hardening in two-phase materials—I. Strength contributions in fibre-reinforced copper–tungsten, *Acta Metallurgica*, **25**, 571–585.

Maxwell, J. C. (1892), *A Treatise of Electricity and Magnetism*, vol. 1, 3rd ed., Clarendon, Oxford, pp. 440.

Nye J. F. (1957), *Physical Properties of Crystals, Their Representation by Tensors and Matrices*, 4th ed., Clarendon, Oxford, pp. 322.

Orowan, E. (1959), Causes and effects of internal stresses, *Proceedings of the Symposium on Internal Stresses and Fatigue in Metals, Detroit and Warren, 4–5 September 1958*, edited by G. M. Rassweiler and W. L. Grube, Elsevier, Amsterdam, pp. 59–80.

Pedersen, O. B. (1978), Transformation theory for composites, *Z. Angew. Math. Mech.*, **58**, 227–228.

Pedersen, O. B. (1979), Thermoelasticity and plasticity of composite materials, *Proceedings of the 3rd International Conference on Mechanical Behaviour of Materials, Vol. 3, Cambridge, 20–24 August 1979*, edited by K. J. Miller and R. F. Smith, Pergamon, Oxford, pp. 263–273.

Pedersen, O. B. (1983), Thermoelasticity and plasticity of composites—I. Mean field theory, *Acta Metallurgica*, **31**, 1795–1808.

Pedersen, O. B. (1984), Mean field theory and the Bauschinger effect in composites, in *Fundamentals of Deformation and Fracture*, Proceedings of the IUTAM Eshelby Memorial Symposium, Sheffield, April 1984, edited by B. A. Bilby, K. J. Miller, and J. R. Willis, Cambridge University Press, Cambridge, 1985, pp. 263–273.

Pedersen, O. B. (1985), Residual stresses and the strength of metal matrix composites, *Proceedings 5th International Conference on Composite Materials, San Diego, 29 July–30 August 1985*, edited by W. C. Harrigan, J. Strife, and A. K. Dhingra, The Metallurgical Society of AIME, New York, pp. 1–20.

Pedersen, O. B. (1988), Dislocations and the strength of metallic composites, in *Mechanical and Physical Properties of Metallic and Ceramic Composites*, Proceedings 9th Risø International Symposium on Metallurgy and Materials Science, Risø National Laboratory, Roskilde, Denmark, 5–9 September 1988, edited by S. I. Andersen, H. Lilholt, and O. B. Pedersen, Risø National Laboratory, pp. 157–182.

Pedersen, O. B., and Brown, L. M. (1977), Equivalence of stress and energy calculations of mean stress, *Acta Metallurgica*, **25**, 1303–1305.

Pedersen, O. B. and Lisiecki, L. L. (1988), The effect of temperature on cyclic saturation in copper, *Proceedings of 8th International Conference on Strength of Metals and Alloys, Tampere, 22–26 August 1988*, edited by P. O. Kettunen, T. K. Lepistö, and M. E. Lehtonen, Pergamon, Oxford, pp. 719–724.

Poulsen, F. W. (1985), Composite electrolytes, in *Transport–Structure Relations in Fast Ion and Mixed Conductors*, Proceedings 6th Risø International Symposium on Metallurgy and Materials Science, Risø National Laboratory, Roskilde, Denmark, 9–13 September 1985, edited by F. W. Poulsen, N. H. Andersen, K. Clausen, S. Skaarup, and O. T. Sørensen, Risø National Laboratory, 1988, pp. 67–78.

Tanaka, K. and Mori, T. (1970), The hardening of crystals by non-deforming particles and fibres, *Acta Metallurgica*, **18**, 931–941.

Tanaka, K. and Mori, T. (1971), The hardening rate of fibre-strengthened materials, *Phil. Mag.*, **23**, 737–740.

Tanaka, K. and Mori, T. (1973), Average stress in matrix and average elastic energy of materials with misfitting inclusions, *Acta Metallurgica*, **21**, 571–574.

Taya, M. (1988), Modelling of physical properties of metallic and ceramic composites: Generalized Eshelby model, in *Mechanical and Physical Properties of Metallic and Ceramic Composites*, Proceedings 9th Risø International Symposium on Metallurgy and Materials Science, Risø National Laboratory, Roskilde, Denmark, 5–9 September 1988, edited by S. I. Andersen, H. Lilholt, and O. B. Pedersen, Risø National Laboratory, pp. 201–231.

Taya, M. and Mura, T. (1981), On stiffness and strength of an aligned short fiber reinforced composite containing fiber-end cracks under uniaxial applied stress, *J. Appl. Mech.*, **48**, 361–367.

Truesdell, C. and Toupin, R. (1960), The classical field theories, *Encyclopedia of Physics*, vol. III1, *Principles of Classical Mechanics and Field Theory*, edited by S. Flügge, Springer Verlag, Berlin, pp. 226–793.

Walpole, L. J. (1966), On bounds for the overall elastic moduli of inhomogeneous systems—I, *J. Mech. Phys. Solids*, **14**, 151–162.

Wakashima, K., Otsuka, M., and Umekawa, S. (1974), Thermal expansions of heterogeneous solids containing aligned ellipsoidal inclusions, *J. Composite Materials*, **8**, 391–404.

Wakashima, K., Suzuki, Y., and Umekawa, S. (1979), A micromechanical prediction of initial yield surfaces of unidirectional composites, *J. Composite Materials*, **13**, 288–302.

Wilson, D. V. (1965), Reversible work-hardening in alloys of cubic metals, *Acta Metallurgica*, **13**, 807–814.

The Effect of Voids and Inclusions on Wave Propagation in Granular Materials

M. H. SADD, A. SHUKLA, H. MEI, and C. Y. ZHU
Department of Mechanical Engineering and Applied Mechanics,
University of Rhode Island, Kingston, RI 02881, U.S.A.

Abstract

Theoretical and experimental studies have been conducted on the dynamic response of granular materials containing local discontinuities of voids and inhomogeneous inclusions. The granular medium was simulated by a specific assembly of circular disks which were subjected to explosive loadings of short duration. Voids were created by removing particular disks from the assembly, while inclusions were constructed by replacing certain disks with those of a higher impedance material. The computational simulation was accomplished through the use of the distinct element method in which the intergranular contact forces and displacements of the assembly disks are determined through a series of calculations tracing the movements of each of the individual disks. The experimental study employed the use of photoelasticity in conjunction with high-speed photography to collect photographic data of the propagation of waves in transparent assemblies of model granular media. Comparisons were made between the computational results and the experimental data for the local intergranular contact forces around each void or inclusion. Both voids and inclusions produce local wave scattering through various reflection mechanisms, and the results seem to indicate that the inclusions produced higher local wave attenuation.

1. Introduction

Granular media can be described as a collection of distinct particles which can displace independently from one another and which interact only through contact mechanisms. This type of media transmits loadings through discrete paths, and therefore the mechanical behavior of such materials under static and dynamic loading conditions is very complex and difficult to model. It is now generally accepted that the *local microstructure or fabric*, i.e., the local geometrical arrangement of particles, plays a dominant role in the transmission of mechanical loadings through these materials. The material *porosity* provides only an average estimate of microstructure, and by itself it is not sufficient to accurately predict the behavior of granular materials. Our general aim here is to understand the dynamic behavior of this type of material when it is subjected to explosive loadings of short duration which produce propagating stress waves. The discrete medium will act as a structured waveguide, providing selective paths for the waves to propagate. Amplitude attenuation will then depend strongly upon the selected path of propagation, and thus the wave propagation is strongly linked to the medium microstructure.

A considerable amount of research on the mechanical behavior of granular materials has been reported in the literature. Constitutive models have employed, for example, *elastic/plastic contact theories* (Deresiewicz, 1958; Walton, 1987; Petrakis and Dobry, 1986), *fabric tensors* (Nemat-Nasser and Mekrabadi, 1983), *distributed body models* (Goodman and Cowin, 1972), *endochronic theories* (Bazant *et al.*, 1983), *pore collapse mechanisms* (Carroll and Holt, 1972), and *probabilistic approaches* (Endley and Peyrot, 1977; Fu, 1984). The concept of modeling granular media as an array of elastic disks or spheres lead to the initial attempts at predicting wave propagation phenomena. Early wave propagation studies include Iida (1939), Takahashi and Sato (1949), Hughes and Cross (1951), Hughes and Kelly (1952), Gassman (1951), Brandt (1955), and Duffy and Mindlin (1957). This initial work investigated the propagation velocity as a function of confining pressure, particle size, and geometrical packing. This particular modeling concept led to work in determining the elastic constants of particular granular assemblies, see, for example, Hendron (1963), Petrakis and Dobry (1986), and Walton (1987). In addition, wave propagation studies for granular and porous media, by Nunziato *et al.* (1978) and Sadd and Hossain (1989), have employed the distributed body theory. Experimental studies of this problem employing the method of *dynamic photoelasticity* have been reported by Shukla *et al.* (1985, 1987, 1988). In this method, high-speed photography was used to collect photographic data of wave propagation in transparent model materials composed of assemblies of birefringent disks.

A numerical scheme developed by Cundall and Strack (1979), called the *distinct element method*, has also been used to simulate granular media by modeling the behavior of large assemblies of circular disks. In this method, the contact forces and displacements of an assembly of disks are determined through a series of calculations tracing the movements of each of the individual disks. The method is based on the use of an explicit numerical scheme in which the interaction of the granules is modeled using rigid-body dynamics, assuming each particle interaction has a particular stiffness and damping. For applications to wave propagation, the movements of each of the disks are a result of the propagation through the medium of disturbances originating at the loading points. Consequently, the wave speed and amplitude attenuation (intergranular contact force) will be a function of the physical properties of the discrete medium, i.e., the microstructure. Several successful applications of this method have been reported (Thorton and Barnes, 1986; Bathurst and Rothenburg, 1988; Sadd *et al.*, 1989), and based upon these, this method has been developed and applied to the wave propagation problems to be reported here.

The present study focuses on a specific aspect of wave propagation in granular materials, namely the effects of *voids* and *inclusions*. It is well known that actual granular media contains both voids and heterogeneous inclusions. These quantities further complicate an already complex microstructural material. Most past studies on wave propagation in these materials have been limited to looking at aggregate assemblies with uniform packing geometries. Our focus here is to investigate the local effects produced by voids and inclusions in regular arrays of circular disks, and of primary concern, is the wave scattering in the vicinity of the microstructural defect. The wave propagation phenomena is to be studied through the determination of the intergranular contact force distribution in the neighborhood of a particular void or inclusion. This problem has been studied using the computational method of distinct elements, and using the experimental technique of dynamic photoelasticity.

2. Distinct Element Method

The distinct element method is a simplified modeling concept which uses Newtonian rigid-body mechanics to model the translational and rotational motion of each disk in a model assembly. The technique establishes a discretized time stepping numerical routine, in which granule velocities and accelerations are assumed to be constant over each time interval. It is also assumed that during each time step, disturbances cannot propagate from any disk further than its immediate neighbors. Under these assumptions, the method becomes explicit, and therefore at any time increment, the resultant forces on any disk are determined solely by its interactions with the disks with which it is in contact.

In order to describe the method, consider the case of two disks in contact as shown in Fig. 1. The position, velocity, acceleration, angular velocity, angular acceleration, radius, and mass of disk 1 are labeled as: $\mathbf{r}_1$, $\mathbf{v}_1$, $\mathbf{a}_1$, ω_1, α_1, R_1, and m_1, with like notation for disk 2. The unit normal vector $\mathbf{n}$ and unit tangential vector $\mathbf{t}$ are defined as shown.

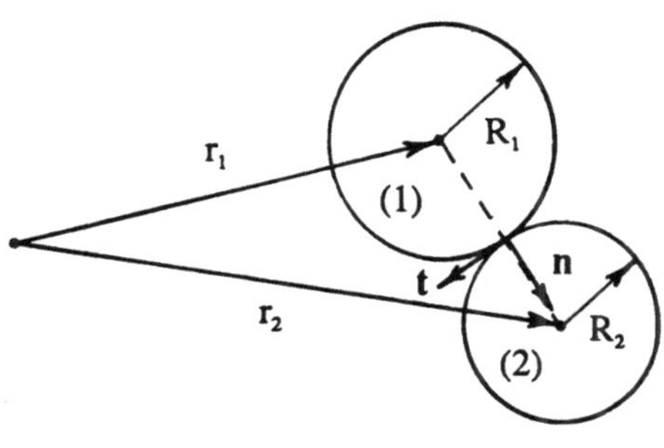

FIG. 1. Schematic of disk interaction.

The normal component of relative contact velocity between the two disks is given by

$$v_n = (\mathbf{v}_1 - \mathbf{v}_2)\cdot\mathbf{n}, \tag{2.1}$$

while the tangential relative velocity is

$$v_t = (\mathbf{v}_1 - \mathbf{v}_2)\cdot\mathbf{t} - (\omega_1 R_1 + \omega_2 R_2). \tag{2.2}$$

Using a finite difference scheme with constant properties over the time interval, the relative velocities may be integrated with respect to time to yield the incremental relative normal and tangential displacements, i.e.,

$$\begin{aligned} \Delta x_n &= v_n \Delta t = [(\mathbf{v}_1 - \mathbf{v}_2)\cdot\mathbf{n}]\Delta t, \\ \Delta x_t &= v_t \Delta t = [(\mathbf{v}_1 - \mathbf{v}_2)\cdot\mathbf{t} - (\omega_1 R_1 + \omega_2 R_2)]\Delta t. \end{aligned} \tag{2.3}$$

In a similar way, the absolute velocity may be computed from the acceleration using the relation

$$\Delta\mathbf{v} = \mathbf{a}\Delta t. \tag{2.4}$$

These relative displacement increments are to be used with a particular contact force-displacement law in order to calculate the forces on each disk in the assembly. Through allowable deformations, the disks in contact are permitted to overlap with one another such that the distance between their centers will become less than $(R_1 + R_2)$. While the general technique could include a complex nonlinear contact law, the present study incorporates a simple linear relation of the form

$$\begin{aligned} \Delta F_n &= K_n \Delta x_n, \\ \Delta F_t &= K_t \Delta x_t, \end{aligned} \tag{2.5}$$

where K_n and K_t are the normal and tangential contact stiffnesses. At each time step, the force increments ΔF_n and ΔF_t are added to the sum of the total forces F_n and F_t on each disk from previous time steps, i.e.,

$$\begin{aligned} (F_n)_N &= (F_n)_{N-1} + \Delta F_n, \\ (F_t)_N &= (F_t)_{N-1} + \Delta F_t, \end{aligned} \tag{2.6}$$

where the indices N and $N - 1$ refer to times t_N and t_{N-1}, and $\Delta t = t_N - t_{N-1}$. A Coulomb-type friction law is incorporated to deal with the tangential loading. This law is defined by

$$(F_t)_{\max} = \mu F_n + c, \tag{2.7}$$

where μ is the coefficient of friction and c is the cohesion between the two disks. If the absolute value of $(F_t)_N$ found from $(2.6)_2$ is larger than $(F_t)_{\max}$, then $(F_t)_N$ is set equal to $(F_t)_{\max}$.

Using Newton's second law of motion, the acceleration of each disk at each time interval can be determined. Now since the behavior of real granular media involves energy dissipation, the modeling introduces damping mechanisms. Two forms of such damping are therefore introduced. A *local damping* proportional to the relative disk velocities, and a *global damping* proportional to the absolute disk velocities will be included in the force balance laws. Applying Newton's law to disk 1, therefore yields

$$\mathbf{F}_1 - C_{\text{ln}} v_{\text{n}1} \mathbf{n} - C_{\text{lt}} v_{\text{t}1} \mathbf{t} - C_{\text{g}} \mathbf{v}_1 = m_1 \mathbf{a}_1,$$
$$M_1 - C_{\text{lt}} v_{\text{t}1} R_1 - C_{\text{g}} \omega_1 = I_1 \alpha_1, \tag{2.8}$$

where **F** and **M** are the resultant force and moment on the disk, C_{ln} and C_{lt} are the local damping coefficients for the normal and tangential directions, C_{g} is the global damping coefficient, and I_1 is the moment of inertia of the disk. Equations (2.8) can thus be solved for the accelerations $\mathbf{a}_1$ and α_1 over each time increment. With the accelerations known, the velocities follow from application of (2.4) and the relative displacements can then be computed from (2.3). This leads to new values of the forces through (2.5) for the next time increment, and the cycle is repeated again for each disk. In this manner, large assemblies of disks can be analyzed in a reasonable amount of computer time. Values of the stiffness and damping parameters appropriate for a given material are difficult to measure. For the static case, the stiffness properties may be computed from Hertz theory or from other elasticity analyses. However, for the dynamic case involving loadings of short duration, estimates of the model parameters are difficult to make. In order to determine estimates of these model parameters, experimental results from dynamic photoelasticity were employed. Data from simple calibration tests on a single chain of disks were used to determine values for the contact stiffness and damping for the disk assemblies under study. For the stiffness parameter, the dynamic stiffness was written as

$$K_{\text{n}} = \alpha K_{\text{n}}^{(\text{s})}, \tag{2.9}$$

where $K_{\text{n}}^{(\text{s})}$ is the static contact stiffness from Hertz theory given by

$$K_{\text{n}}^{(\text{s})} = \frac{\pi h E_1 E_2}{2(E_1 + E_2)}, \tag{2.10}$$

and α is a *dynamic stiffness coefficient* which is to be determined by the experiments, h is the disk thickness, and E_1 and E_2 are the elastic moduli of the two disks in contact. The α coefficient may be thought of as an adjustment parameter, less than unity, which accounts for the fact that the entire disk will not deform during the dynamic event.

3. Photomechanics Studies

A series of dynamic photoelastic experiments were conducted to provide experimental information on the effect of voids and inclusions on the wave propagation phenomena. For the experimental study, the granular medium was simulated with assemblies of polymeric birefringent disks of *Homalite 100*,

FIG. 2. Photoelastic fringe patterns of assembly with no voids or inclusions.

25.4 mm in diameter and 6.25 mm thick. In all the experiments reported here, the disks were assembled in an hexagonal close packing geometry as shown in Fig. 2. Voids in the assembly were created by removing disks from different locations, while inclusions were created by replacing particular disks with those of a different material (steel). Dynamic loading was achieved by detonating a small charge of PETN in a specially designed charge holder, which was mounted at the top-centerline of the model assemblies. The experimental models were placed in the optical bench of a high-speed photographic system. This high-speed photographic system operated as a series of high intensity, extremely short duration pulses of light and provided 20 photoelastic images at discrete times during the dynamic event. Framing rates of up to 10^6 frames per second are achievable, and this allows studies of wave propagation to be made, see Riley and Dally (1969).

A typical sequence of four images from each experiment are shown in Figs. 2–6. The photographic data shows the isochromatic fringe patterns at different times as the stress wave propagates through the model assemblies. The wave propagation history can thus be clearly seen from a sequence of such photographs. Figure 2 illustrates the wave patterns for the assembly with no voids or inclusions. Figs. 3 and 4 show the wave motion for the cases of single

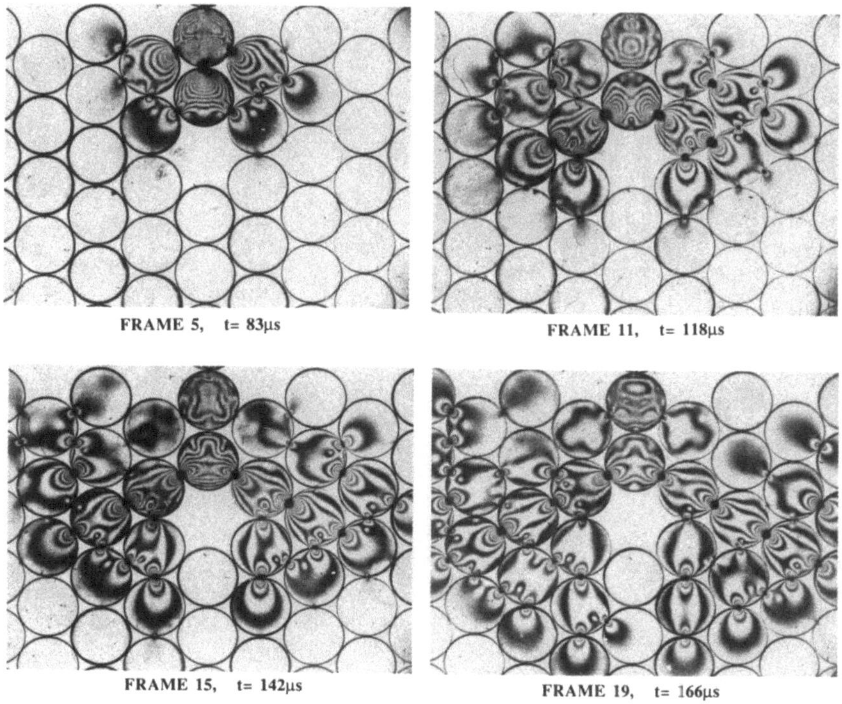

FIG. 3. Photoelastic fringe patterns of assembly with a single void.

and multiple voids, while Figs. 5 and 6 illustrate the cases with inclusions of higher density and stiffness. Inspection of the photographs reveals that the wavelength λ, of the loading pulse is much larger than the disk diameter D; in fact $\lambda \approx 4D$. Furthermore, in most cases the fringe pattern around the contact points are symmetric on either side of the contact points and are similar to the fringes obtained under static compression. Both these features indicated that around the contact zone, quasi-static loading was present during the wave propagation event. Thus Hertz theory can be used to estimate the stress field in the vicinity of the contacts.

The isochromatic fringes photographed during the experiments are lines of constant maximum shear stress, and are related to the stress field by the *stress optic law*

$$\sigma_1 - \sigma_2 = \frac{Nf_\sigma}{h}, \tag{3.1}$$

where σ_1 and σ_2 are the principal stresses, N is the fringe order, f_σ is the material fringe value, and h is the model thickness. Using relation (3.1) along with the Hertz contact stress equations, Shukla and Nigam (1985) have developed a scheme to compute the quasi-static stress field in the vicinity of the contacts between each grain. These contact stresses may then be integrated

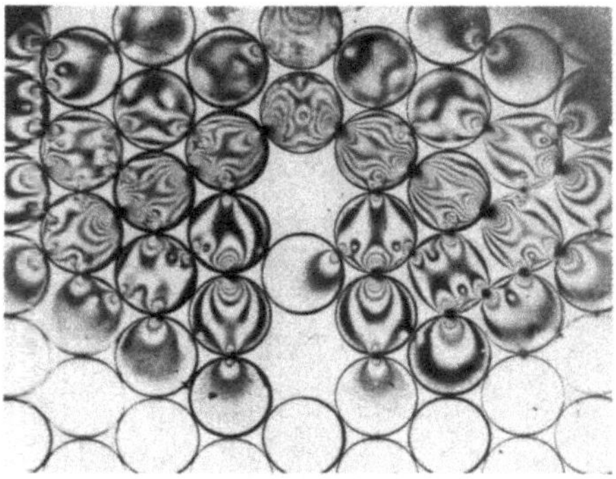

FRAME 5, t= 151μs

FRAME 10, t= 179μs

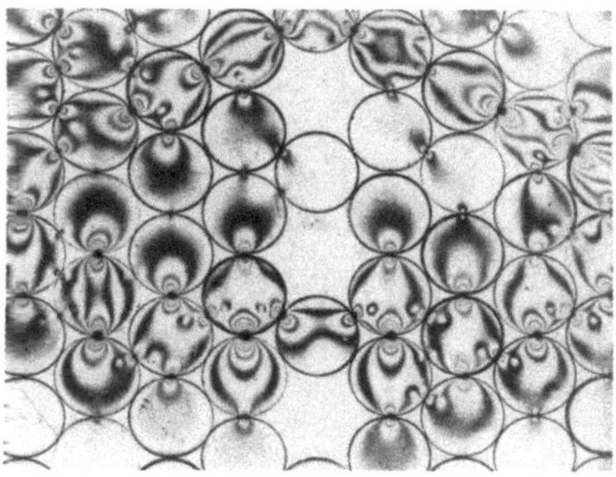

FRAME 15, t= 210μs

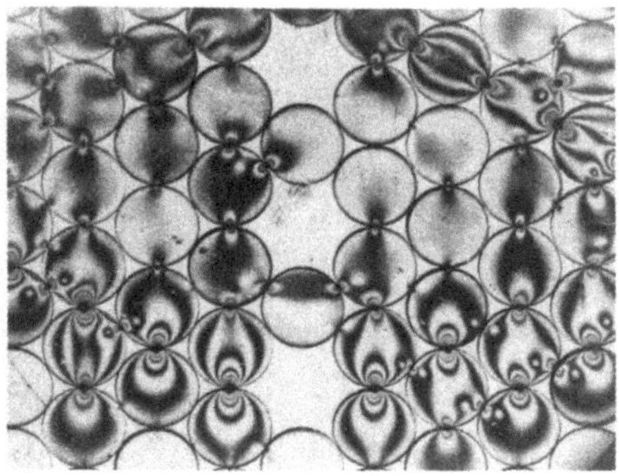

FRAME 19, t= 228μs

FIG. 4. Photoelastic fringe patterns of assembly with a series of voids.

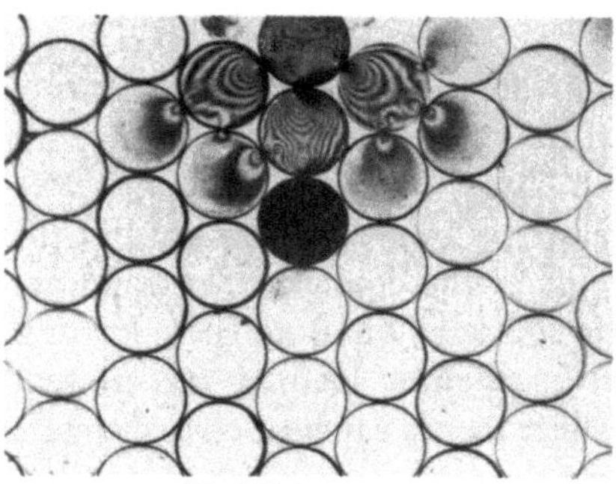

FRAME 5, t= 82μs

FRAME 9, t= 105μs

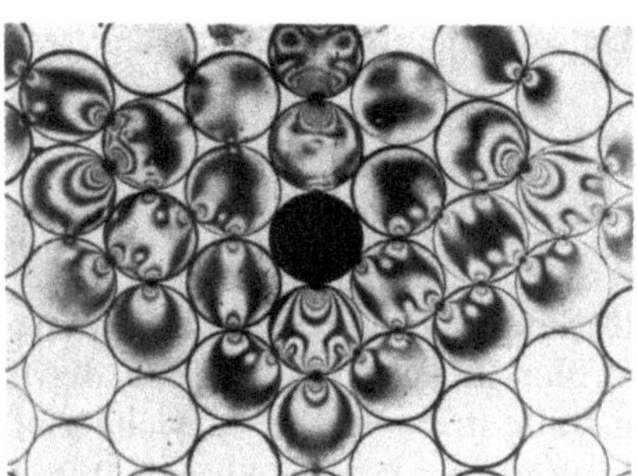

FRAME 15, t= 141μs

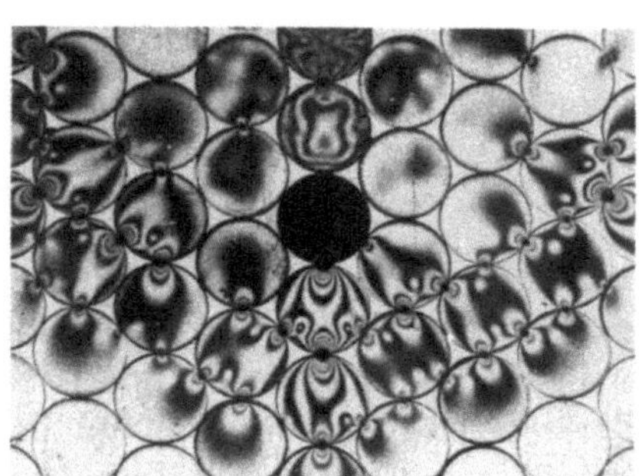

FRAME 19, t= 164μs

FIG. 5. Photoelastic fringe patterns of assembly with a single inclusion.

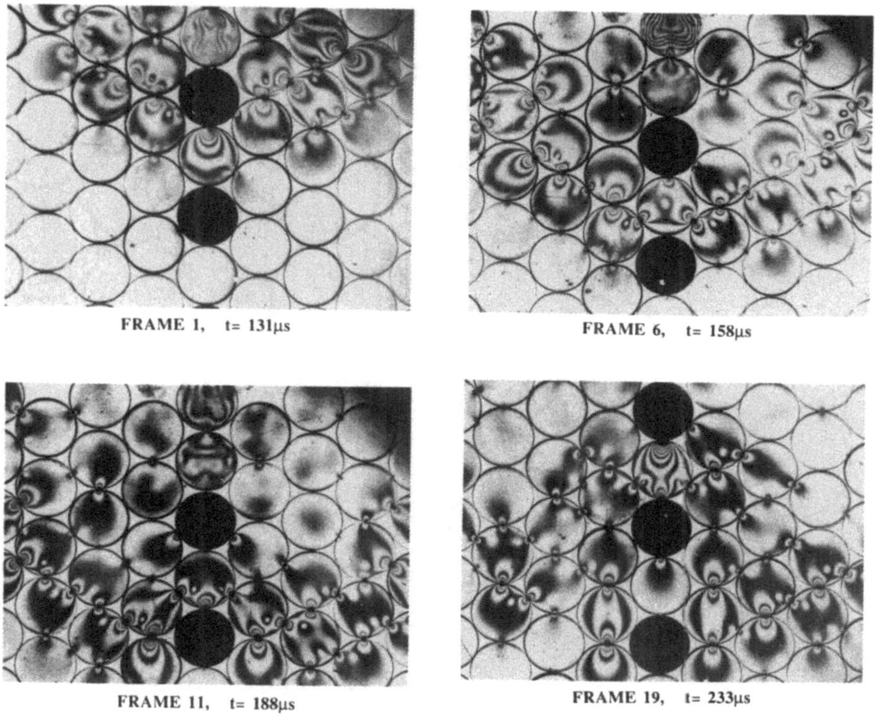

FIG. 6. Photoelastic fringe patterns of assembly with a series of inclusions.

along the contact length to determine the normal and tangential contact forces transmitted between the disks at various times during the dynamic process. Thus the integranular contact loading, which is related to the wave amplitude, can be experimentally computed. These experimental results can then be compared with the theory, and they can also be used to determine the necessary material model parameters (e.g., the contact stiffness and damping) in order to use the distinct element method.

4. Results and Comparisons

Specific results of the distinct element modeling along with comparisons to the experimental data will now be given for particular granular assemblies containing voids and inclusions. The basic assembly geometry is a hexagonal close packing arrangement shown in Fig. 7. The theoretical model used an input loading of triangular time dependence with a 60 μs duration to model the explosive loading from the experiments. The local stiffness and damping parameters K_n, K_t, C_{1n}, and C_{1t} needed in the distinct element model were

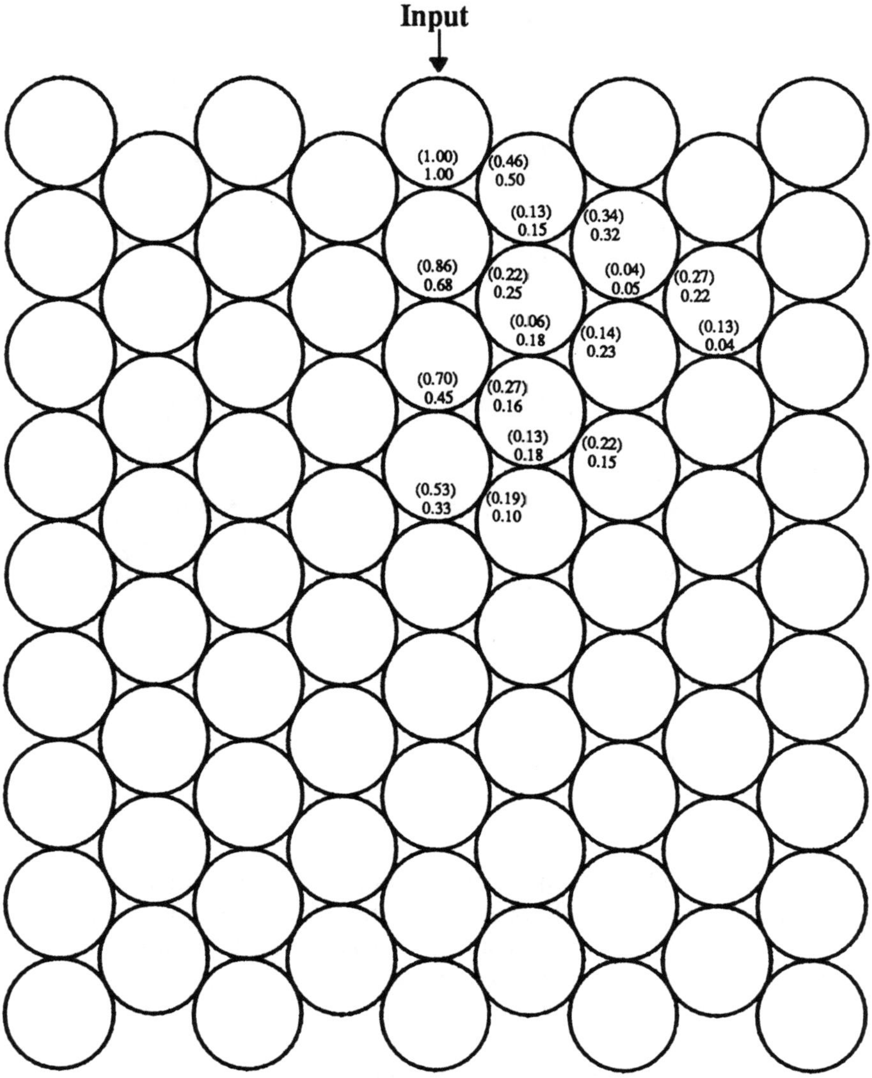

FIG. 7. Normalized maximum contact loading comparisons: No voids or inclusions. Experimental data in parentheses.

determined from experimental tests conducted on straight single chains of disks. These calibration experiments provided specific values for the wave speed and amplitude attenuation for a very simple geometry. The model parameters were thus chosen to produce the best match to the calibration data, and these values were then retained in the model to be used for calculations of more complicated geometry. The contact stiffnesses for Homalite-to-Homalite contacts were taken as $K_n = K_t = 6.4 \times 10^6$ N/m ($\alpha = 0.27$), and the local normal damping coefficient was $C_{ln} = 32$ Ns/m. Global damping was not included, and the tangential contact loading was set to zero. The time-stepping increment was taken to be 2 μs for all cases studied, and this was felt to be appropriate to calculate the essential features of the dynamic event.

Results from the distinct element technique are shown in Fig. 7 for the basic granular media with no voids or inclusions. Numerical predictions of the maximum intergranular contact loadings (normalized with respect to the input loading) are shown at various contacts in the assembly. The corre-

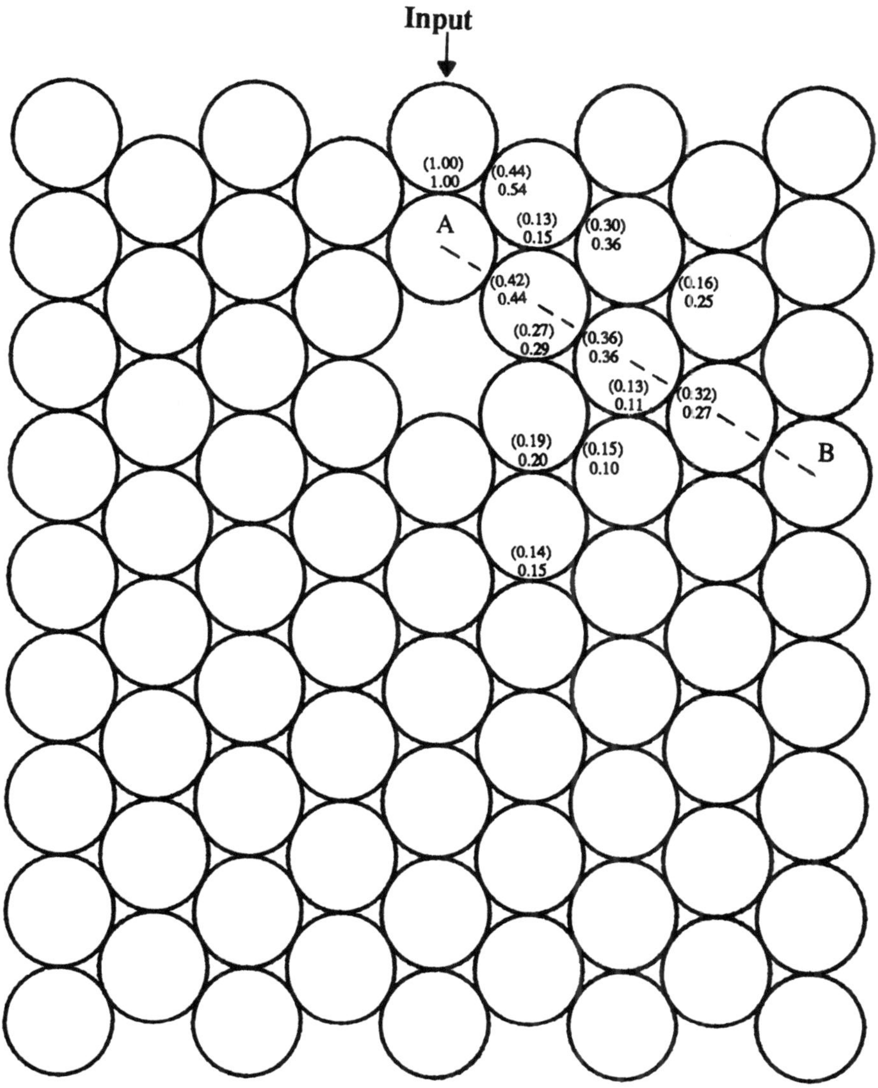

FIG. 8. Normalized maximum contact loading comparisons: Single void. Experimental data in parentheses.

sponding experimental values are also given in parentheses. The contact loading values are symmetric about the assemblies' vertical centerline, and the results indicate the rapid attenuation which occurs along various chains or paths in the assembly. Comparison of theoretical with experimental contact loadings indicates that they differ by an average amount of 25%. Experimental determination of the wave speed was accomplished from the known position of the fringe patterns in the photographs. These results indicated that the leading wave front in the Homalite assemblies propagates at approximately

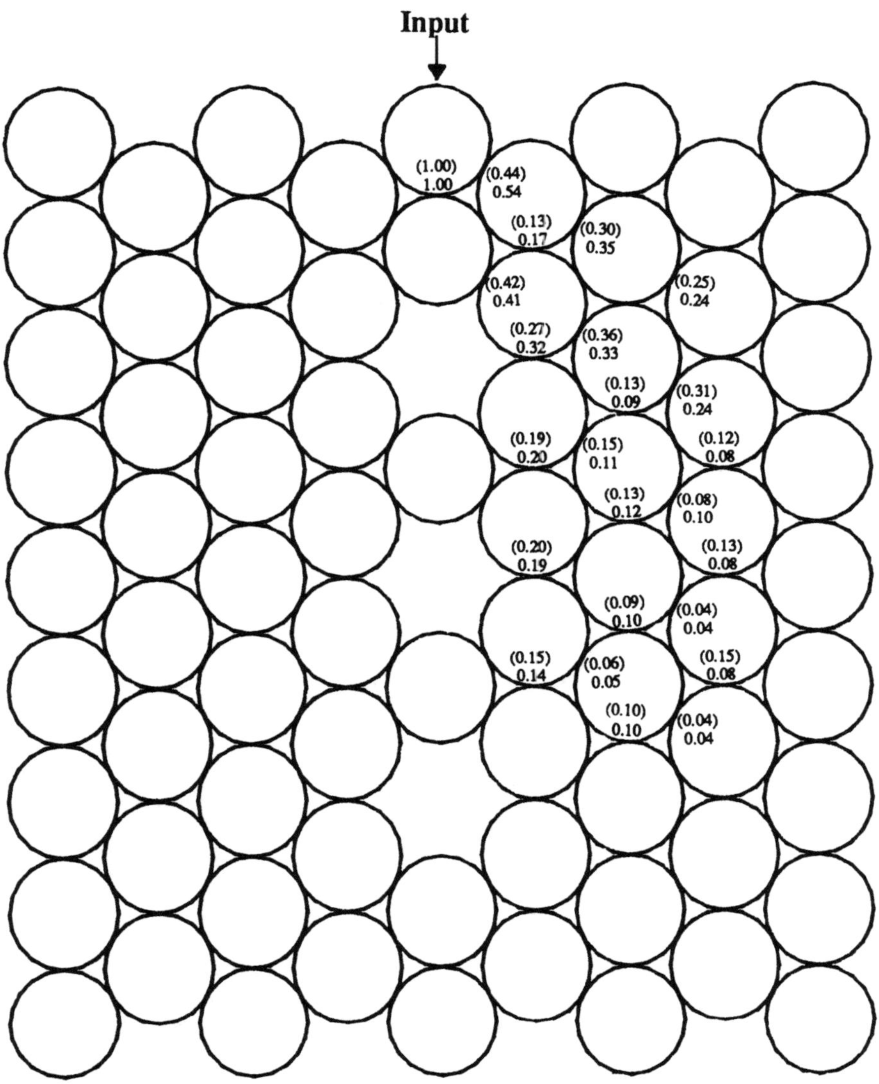

FIG. 9. Normalized maximum contact loading comparisons: Series of voids. Experimental data in parentheses.

995 m s^{-1}, which is about 50% of the P-wave speed in the virgin disk material. Wave speed predictions from the distinct element model matched well (within 10%) these measured values.

Figures 8 and 9 illustrate the case of a granular media with voids present. Figure 8 contains a single void, while Fig. 9 contains a series of voids down the vertical centerline of the assembly. The presence of a void produces significant local wave scattering and attenuation, especially along the main vertical chain down the centerline of the assembly. Local intergranular contact

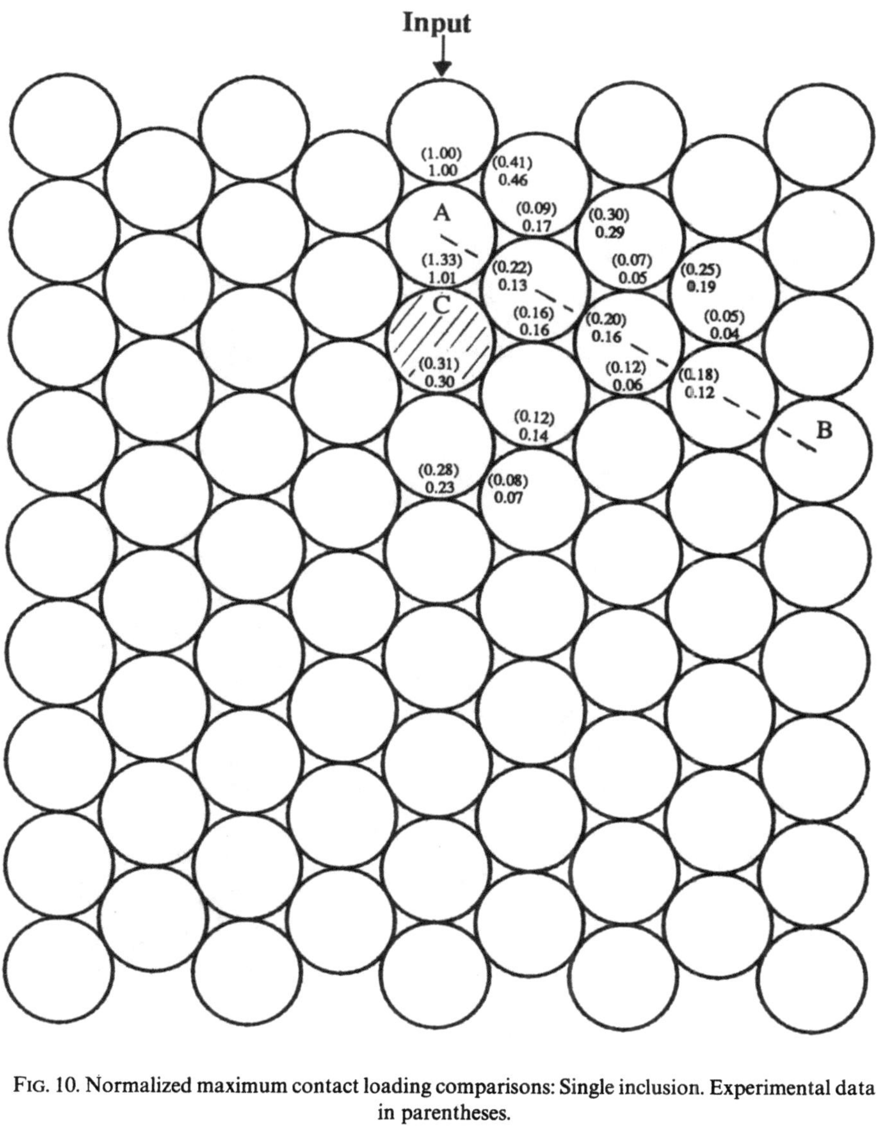

FIG. 10. Normalized maximum contact loading comparisons: Single inclusion. Experimental data in parentheses.

forces become elevated near the void; however, at contacts remote from the void, the loading values are similar in magnitude (with the exception of points on the vertical centerline) to those in Fig. 7. Computational and experimental contact loads compare to within an average difference of 15% for the assemblies with voids.

The cases of granular media with inclusions of different material are shown in Figs. 10 and 11. This situation is attempting to model the local effects of

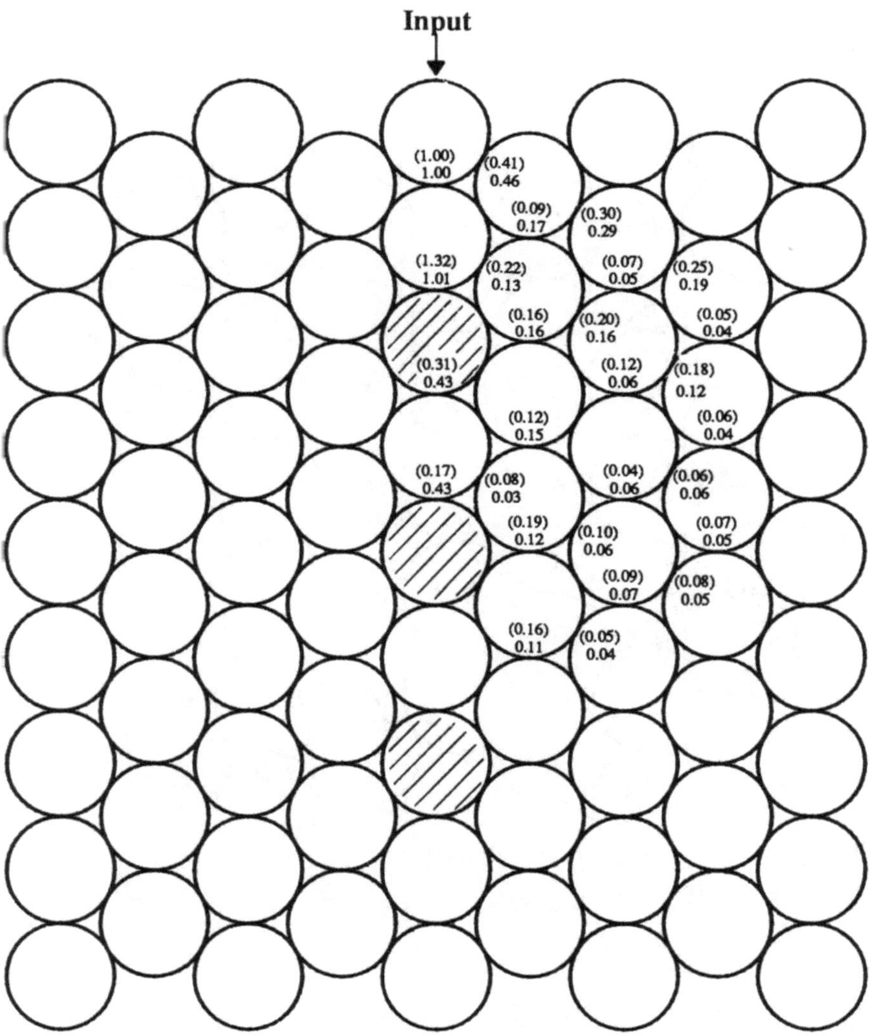

FIG. 11. Normalized maximum contact loading comparisons: Series of inclusions. Experimental data in parentheses.

heterogeneity in a granular medium. Figure 10 illustrates the case of a single inclusion, and Fig. 11 contains the case of several inclusions down the vertical centerline of the assembly. As mentioned, the inclusion is simply a disk of a different material, and in this case the inclusion material was steel, which when compared to the granular material (Homalite 100), has a much higher stiffness and density. The contact stiffness between the Homalite and steel disks was calculated using relations (2.9) and (2.10) retaining the same value of α. The local damping parameter C_{1n} was kept the same for this case, since there was only a small number of inclusions present.

There will be an impedance mismatch at the contacts of the Homalite and steel, and thus there will be a sizeable difference between the reflection and transmission phenomena at these contacts in comparison with those of the rest of the medium. For example, at point C in Fig. 10, the wave is attempting to propagate from a relatively soft material into a stiff material. Consequently, a sizeable reflection occurs at this contact producing an upward traveling wave and a large contact force. Very little wave motion is transmitted into the inclusion, and thus the inclusion acts to block the wave motion along the vertical disk chain. For the inclusion cases, the average difference between theoretical predictions and experimental data was 20–30%.

5. Conclusions

The methods of distinct elements and dynamic photomechanics have been used to study the effects of voids and heterogeneous inclusions on the wave propagation in granular materials. The granular medium was simulated by a specific assembly of circular disks arranged in an hexagonal close packing geometry. The voids were created by removing particular disks, while inclusions were constructed by replacing particular disks with those of a higher modulus material. Comparisons were made between the computational results and the experimental data for the local intergranular contact forces around each void or inclusion. Although comparisons produced some differences as high as 30%, it is felt that since the experimental data itself contain scatter of approximately 10%, the computational scheme does provide reasonable predictions. Improvements of the modeling procedures could be accomplished by incorporating a more sophisticated dynamic contact law.

Comparing the results, it is apparent that both the voids and inclusions cause local wave scattering. Compared to the case with no voids or inclusions, elevated contact forces occur near the discontinuity, and along particular paths dictated by the local microstructure, rapid wave amplitude attenuation occurs. A void produces wave scattering through free-surface reflection from the empty volume, whereas the inclusion causes sizeable reflections from the material of higher impedance. Comparing the paths AB in Figs. 8 and 10, it appears that the inclusion produces higher attenuation than the void. Current work continues on these studies, and investigations are underway on wave propagation in granular materials with additional and more complex microstructural features.

Acknowledgment

This research was supported by the Army Research Office through contract number DAAL03-86-K-0125.

References

Bathurst, R. J. and Rothenburg, L. (1988), Micromechanical aspects of isotropic granular assemblies with linear contact interactions, *J. Appl. Mech.*, **55**, 17–23.

Bazant, Z. P., Krizek, R. J., and Shieh, C. L. (1983), Hysteretic endochronic theory for sand, *J. Engng. Mech.*, **109**, 1073–1095.

Brandt, H. (1955), A study of the speed of sound of porous granular media, *J. Appl. Mech.*, **22**, 479–486.

Carroll, M. M. and Holt, A. C. (1972), Static and dynamic pore-collapse relations for ductile porous materials, *J. Appl. Phys.*, **43**, 1626–1636.

Cundall, P. and Strack, D. L. (1979), A discrete numerical model for granular assemblies, *Geotechnique*, **29**, 47–65.

Deresiewicz, H. (1958), Mechanics of granular matter, in *Advances in Applied Mechanics*, Vol. V, Academic Press, New York.

Duffy, J. and Mindlin, R. D. (1957), Stress–strain relations and vibration of granular medium", *J. Appl. Mech.*, **24**, 585–593.

Endley, S. N. and Peyrot, A. H. (1977), Load distribution in granular media, *J. Engng. Mech. Division of ASCE*, **103**, 99–111.

Fu, L. S. (1984), A new micro-mechanical theory for randomly inhomogeneous media, in *Wave Propagation in Homogeneous Media and Ultrasonic Non-Destructive Evaluation*, Vol. 62, Applied Mechanics Division, American Society of Mechanical Engineers, New York.

Gassman, F. (1951), Elastic waves through a packing of spheres, *Geophysics*, **16**, 673–685.

Goodman, M. A. and Cowin, S. C. (1972), A continuum theory for granular materials, *Arch. Rat. Mech. Anal.*, **44**, 249–266.

Hendron, A. J. (1963), The behavior of sand in one-dimensional compression, Ph.D. Thesis, University of Illinois.

Hughes, D. S. and Cross, J. H. (1951), Elastic wave velocities in rocks at high pressures and temperatures, *Geophysics*, **16**, 577–593.

Hughes, D. S. and Kelly, J. L. (1952), Variation of elastic wave velocity with saturation in sandstone, *Geophysics*, **17**, 739–752.

Iida, K. (1939), The velocity of elastic waves in sand, *Bull. Earthquake Res. Inst. Japan*, **17**, 783–808.

Nemat-Nasser, S. and Mehrabadi, M. M. (1983), Stress and fabric in granular masses, in *Mechanics of Granular Materials: New Models and Constitutive Relations*, Elsevier, Amsterdam, pp. 1–8.

Nunziato, J. W., Kennedy, J. E., and Walsh, E. (1978), The behavior of one-dimensional acceleration waves in an inhomogeneous granular solid, *Int. J. Engng. Sci.*, **16**, 637–648.

Petrakis, E. and Dobry, R. (1986), A self-consistent estimate of the elastic constants of a random array of equal spheres with application to granular soil under isotropic conditions, Report CE-86-04, Rensselaer Polytechnic Institute, Troy, NY.

Riley, W. F. and Dally, J. W. (1969), Recording dynamic fringe patterns with a Cranz–Schardin camera, *Exp. Mech.*, **9**, 27–33.

Sadd, M. H., Shukla, A., and Mei, H (1989), Computational and experimental modeling of wave propagation in granular materials, *Proceedings of the 4th International Conference on Computational Methods and Experimental Measurements, Capri, Italy, May, 1989.*

Sadd, M. H. and Hossain, M. (1989), Wave propagation in distributed bodies with applications to dynamic soild behavior, *J. Wave-Material Interaction*, to appear.

Shukla, A. and Nigam, H. (1985), A numerical-experimental analysis of the contact stress problem, *J. Strain Anal.*, **20**, 241–245.

Shukla, A. and Damania, C. (1987), Experimental investigation of wave velocity and dynamic contact stresses in an assembly of discs, *Exp. Mech.*, **27**, 268–281.

Shukla, A., Zhu, C. Y., and Sadd, M. H. (1988), Angular dependence of dynamic load transfer due to explosive loading in granular aggregate chains, *J. Strain Anal.*, **23**, 121–127.

Takahashi, T. and Sato, Y. (1949), On the theory of elastic waves in granular substance, *Bull. Earthquake Res. Inst. Japan*, **27**, 11–16.

Thornton, C. and Barnes, D. J. (1986), Computer simulated deformation of compact granular assemblies, *Acta Mech.*, **64**, 45–61.

Walton, K. (1987), The effective elastic moduli of a random packing of spheres, *J. Mech. Phys. Solids*, **35**, 213–226.

Crack, Dislocation Free Zone, and Dislocation Pile-Up Model for the Behavior of the Hall–Petch Relation in the Range of Ultrafine Grain Sizes

K. Saito, M. Iwamoto, Y. Nomura
Department of Mechanical and System Engineering, Kyoto Institute of Technology, Matugasaki, Sakyo-ku, Kyoto 606, Japan

T. Nakamura
Technical Research Laboratory, Toyo Umpanki, Co., Ltd., 3-Banchi, Ryugasaki, Ibaraki 301, Japan

Abstract

In order to elucidate the failure of the linear dependence of the yield stress of polycrystalline metals on the grain size in the range of ultrafine grain sizes, we propose the dislocation pile-up model which consists of a crack, a dislocation free zone (DFZ), and the slip band blocked at the grain boundary.

Analyzing the above model by the method of the continuously distributed theory of dislocations, we get the analytical expression which gives the relationship between the macroscopic applied stress and the grain size D. It can be shown that the behavior of the macroscopic yield stress versus $D^{-1/2}$ explains the experimental result well. Applicability of this model to the grain-size dependence of the fracture stress of engineering ceramics will also be discussed.

1. Introduction

It is confirmed experimentally that the yield stress or the fracture stress, etc., of polycrystalline metals depends on the grain sizes. This is known as the Hall–Petch relation. The theoretical background for this relation is the dislocation pile–up model that the dislocations emitted from the source at the central part of the grain pile up against the grain boundary. However, recent experiments by Armstrong (1983) show that the failure of this relation, i.e., the lowering of the yield stress or the fracture stress which might be expected from the Hall–Petch relation, occurs in the range of ultrafine grain sizes. It is concluded that the cause of failure is very probably due to the effect of the inclusions or the flaws in the grain.

In the present paper, in order to elucidate the above phenomena over both the range of normal and ultrafine grain sizes, we propose the new dislocation pile–up model which consists of a crack, a dislocation free zone (DFZ), and the slip band blocked at the grain boundary. The crack (or crack-like flaw) is brought into the model to emphasize the effect of the flaws on the yield stress or the fracture stress when the grain size becomes ultrafine.

It has recently been reported by Kobayashi and Ohr (1980) that there exists a DFZ near the crack tip. Thus, this DFZ is introduced into our model ahead of the crack tip, to render the model mathematically easy to handle, and then the dislocation source for this pile–up model is located at the end of the DFZ.

Analyzing the above model by the method of the continuously distributed theory of dislocations we get the dislocation distribution function for this model explicitly, and the stress acting on the leading dislocation can be obtained from the distribution function for the slip band. Furthermore, assuming that this stress corresponds to the stress acting on the grain boundary and then equating it with the microscopic yield stress necessary to yield the grain boundary, we get the analytical expression which gives the relationship between the macroscopic applied stress and the grain size D. It can be shown that the behavior of the macroscopic yield stress versus $D^{-1/2}$ explains the experimental results well, over both the range of normal and ultrafine grain sizes.

Finally, applicability of this model to the grain-size dependence of the fracture stress of engineering ceramics will be discussed when the friction stress of the dislocation is taken to be zero.

2. Dislocation Free Zone at the Crack Tip

As mentioned in the previous section, it was observed experimentally by Kobayashi and Ohr that there exists a DFZ near the crack tip. Hence, in this section, some considerations, on the mechanical force exerted upon the dislocation located near the crack tip, are given in order to discuss the existence of the DFZ.

Consider the semi-infinite-like body under the applied stress σ_a, which has a crack of length c and a dislocation near its tip, as shown in Fig. 1. Taking the crack tip as the origin of the coordinate, we designate the crack region as $-c < x < 0$, $y = 0$, and the position of the dislocation as $x = x_i$, $y = 0$. The direction of the dislocation is perpendicular to the paper.

Replacing the crack by the continuous distribution of infinitesimal dislocations, we obtain the following equilibrium equation from the condition that the crack surface is stress-free:

$$A \int_{-c}^{0} \frac{f(x')\,dx'}{x - x'} + \sigma_a + \frac{A}{x - x_i} = 0, \tag{2.1}$$

where $f(x)$ is the dislocation distribution function, and A is $\mu b/2\pi(1 - \nu)$ for

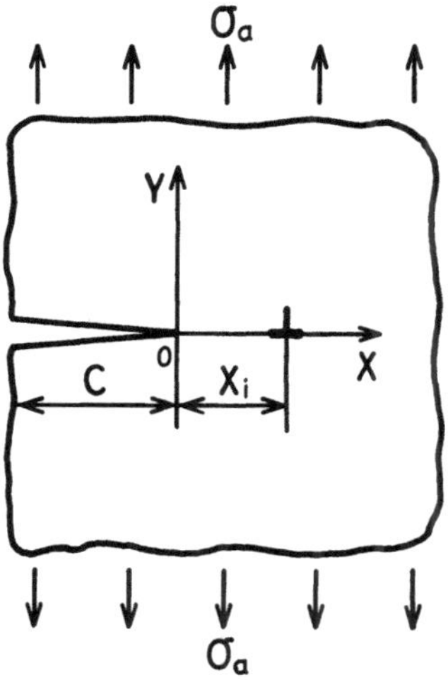

FIG. 1. Semi-infinite-like body which has a crack and a dislocation near its tip subjected to a uniform applied stress σ_a.

edge dislocations or $\mu b/2\pi$ for screw dislocations. μ and v are the shear modulus and the Poisson ratio of the material under consideration, respectively, and b is the Burgers vector of the dislocation.

Equation (2.1), which is called the singular integral equation, can be solved by the Muskhelishvili inversion theorem (Mura, 1982) as follows:

$$f(x) = \frac{\sigma_a}{\pi A}\sqrt{\frac{x+c}{-x}} - \frac{1}{\pi(x_i - x)}\sqrt{\frac{-x_i(x+c)}{x(x_i+c)}}. \tag{2.2}$$

Next, we calculate the stress acting on the real dislocation at $x = x_i$. Since the stress acting on the real dislocation is the sum of the uniform applied stress σ_a and the stress due to the continuously distributed dislocations over the crack region, we get the following expression:

$$\sigma = \sigma_a + A\int_{-c}^{0} \frac{f(x)\,dx}{x_i - x}. \tag{2.3}$$

Substituting (2.2) into (2.3), we obtain the following expression:

$$\sigma = \sigma_a\sqrt{\frac{x_i + c}{x_i}} - \frac{Ac}{2x_i(x_i + c)}. \tag{2.4}$$

Since the formula which expresses the mechanical force acting upon the dislocation at $x = x_i$ is given by $f = \sigma b$, the stress in itself (given by (2.4)) represents the force upon it.

Figure 2 shows the numerical calculations of the normalized force σ/μ. In this figure, the positive value represents the repulsive force upon the dislocation against the crack, and the negative value the attractive force toward

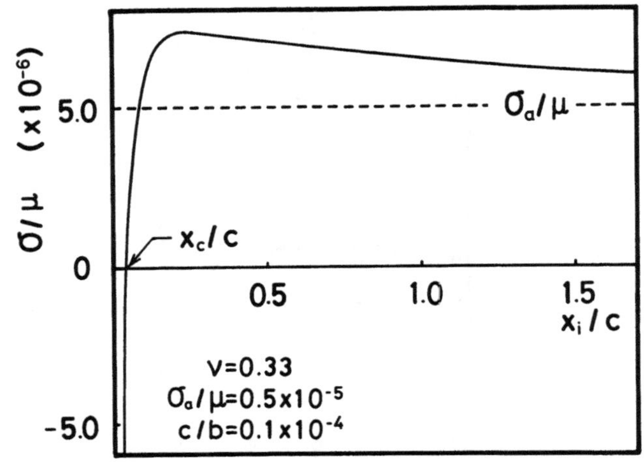

FIG. 2. Relation between the normalized force acting on the dislocation and the position of the dislocation.

it. Thus, it can be seen that there exists a critical position x_c where the resultant force is zero, which is given (from (2.4)) by

$$x_c = \frac{A^2}{4c\sigma_a^2}. \tag{2.5}$$

The above result coincides with the result given by Li (1981).

Although the above discussion is based on the behavior of one dislocation near the crack tip, it may be concluded that we can set up, near the crack tip, the DFZ which was observed by Kobayashi and Ohr.

3. Crack, Dislocation Free Zone, and Slip Band Model

3.1. *Singular Integral Equation Giving Equilibrium of Dislocations*

Consider the state shown in Fig. 3, that one of the grains of the polycrystalline aggregate has a crack in it after the application of a uniform external stress τ. The crack is introduced into the grain to emphasize the effect of the inherently existing flaws in a material on the grain-size dependence of the yield stress of polycrystalline metals when the grain size becomes ultrafine.

In addition, we also set up the DFZ near the crack tip, as discussed in the previous section. This establishment of the DFZ near the crack tip enables us to make the model easy to handle mathematically. In the grain, there is also a slip band extending from the end of the DFZ, i.e., the dislocation source, to the grain boundary.

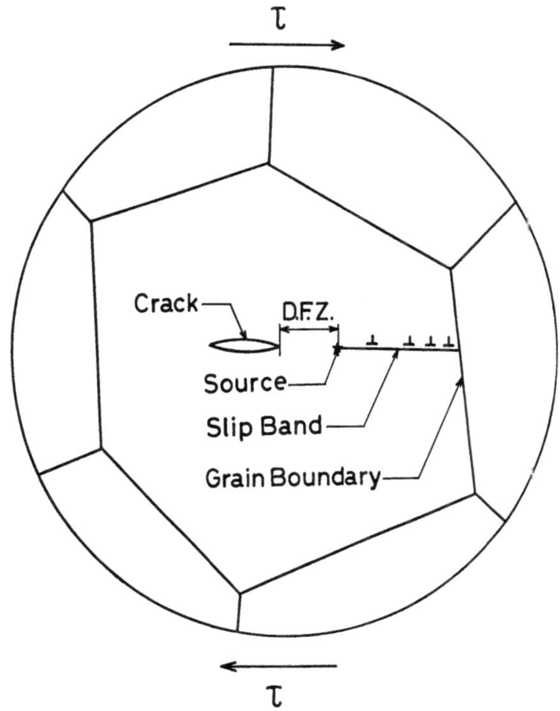

FIG. 3. Polycrystalline aggregate with a crack, a dislocation free zone, and a slip band.

The physical model shown in Fig. 3 can be transfered to the mathematical model shown in Fig. 4. If we define the origin of the coordinate at the center of the crack, then the crack is given in the region, $-c < x < c$, $y = 0$, and two DFZs are given in the regions, $-e < x < -c$, $c < x < e$, $y = 0$. The diameter of a grain is $2a$.

The model shown in Fig. 4 can be formulated by the method of the continuously distributed theory of dislocations. As before, the crack and the slip band are replaced by the continuous distribution of infinitesimal dislocations. If we define the distribution function $f(x)$, then, from the condition that the resultant stress acting on the arbitrary dislocation is zero, we obtain the following equilibrium equation:

$$A\left[\int_{-a}^{-e} + \int_{-c}^{c} + \int_{e}^{a}\right] \frac{f(x')\,dx'}{x - x'} + T = 0, \tag{3.1}$$

where

$$\left.\begin{aligned} T &= \tau, && -c < x < c, \\ T &= \tau - \tau_i, && -a < x < -e, \quad e < x < a, \end{aligned}\right\} \tag{3.2}$$

and τ_i is the friction stress of the dislocation.

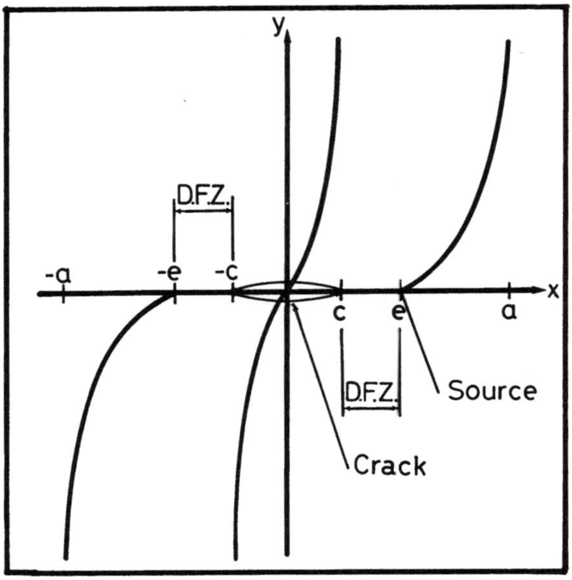

FIG. 4. Geometry of a crack, a dislocation free zone, and a slip band model and the dislocation distribution function.

3.2. *Dislocation Distribution Function*

In this section, we determine the dislocation distribution function $f(x)$ by solving the equilibrium equation (3.1). Equation (3.1) can be solved as follows by using the Muskhelishvili inversion theorem.

$$f(x) = -\pi^{-2}A^{-1}(x^2 - e^2)^{1/2}(x^2 - a^2)^{-1/2}(x^2 - c^2)^{-1/2}[(\tau - \tau_i)I_{ea} + \tau I_c] + (x^2 - e^2)^{1/2}(x^2 - a^2)^{-1/2}(x^2 - c^2)^{-1/2}Q_0. \qquad (3.3)$$

In the above equation, I_{ea} and I_c are given by

$$I_{ea} = \int_{-a}^{-e} + \int_{e}^{a} \frac{(u^2 - a^2)^{1/2}(u^2 - c^2)^{1/2}}{(u^2 - e^2)^{1/2}(u - x)}\, du, \qquad (3.4)_1$$

$$I_c = \int_{-c}^{0} + \int_{0}^{c} \frac{(u^2 - a^2)^{1/2}(u^2 - c^2)^{1/2}}{(u^2 - e^2)^{1/2}(u - x)}\, du, \qquad (3.4)_2$$

and Q_0 is the arbitrary constant.

The distribution function $f(x)$ can be determined by calculating the integrals given by $(3.4)_1$ and $(3.4)_2$ in accordance with the method presented by Chang and Ohr (1981).

3.2.1. *Evaluation of I_{ea}*

Making the proper transformation with respect to the variable in the first term of $(3.4)_1$ and combining it with the second term, we obtain the following expression:

expression:

$$I_{ea} = x[I_{ea1} + (x^2 - a^2 - c^2)I_{ea2} + (x^2 - a^2)(x^2 - c^2)I_{ea3}], \tag{3.5}$$

where

$$I_{ea1} = \int_{e^2}^{a^2} z^{-1} t\, dt, \tag{3.6$_1$}$$

$$I_{ea2} = \int_{e^2}^{a^2} z^{-1}\, dt, \tag{3.6$_2$}$$

$$I_{ea3} = \int_{e^2}^{a^2} z^{-1}(t - x^2)^{-1}\, dt, \tag{3.6$_3$}$$

$$z = [(t - a^2)(t - e^2)(t - c^2)t]^{1/2}. \tag{3.6$_4$}$$

After the tedious calculation, the integral given by $(3.6)_1$ can be reduced to

$$I_{ea1} = -2i[e^2(a^2 - c^2)]^{-1/2}\left[c^2 F\left(\frac{\pi}{2}, k\right) + (e^2 - c^2)\Pi\left(\frac{\pi}{2}, \frac{(a^2 - e^2)}{(a^2 - c^2)}, k\right)\right], \tag{3.7}$$

where k is the parameter representing the dimensions of the crack, the DFZ, and the grain diameter as follows:

$$k = \left[\frac{(a^2 - e^2)c^2}{(a^2 - c^2)e^2}\right]^{1/2}. \tag{3.8}$$

$F(\phi, k)$ and $\Pi(\phi, n, k)$ are the incomplete elliptic integrals of the first and the third kind, respectively, and $i = \sqrt{(-1)}$.

Next, the integral given by $(3.6)_2$ can be evaluated as follows:

$$I_{ea2} = -2i[e^2(a^2 - c^2)]^{-1/2} F\left(\frac{\pi}{2}, k\right). \tag{3.9}$$

Finally, the integral given by $(3.6)_3$ must be classified into the following two cases in accordance with the value of x:

(a) $0 < x^2 < c^2$. The integral is regular in this region and then can be obtained as follows:

$$\begin{aligned} I_{ea3a} = {} & -2i[e^2(a^2 - c^2)]^{-1/2}[(e^2 - x^2)(x^2 - c^2)]^{-1} \\ & \times \left[(e^2 - c^2)\Pi\left(\frac{\pi}{2}, (a^2 - e^2)(x^2 - c^2)(a^2 - c^2)^{-1}(x^2 - e^2)^{-1}, k\right)\right. \\ & \left. + (x^2 - e^2)F\left(\frac{\pi}{2}, k\right)\right]. \end{aligned} \tag{3.10}$$

(b) $e^2 < x^2 < a^2$. Since the integral is an improper integral whose integrand diverges in this region, we define this integral by its Cauchy principal value and then

$$I_{ea3b} = 2i[e^2(a^2 - c^2)]^{-1/2}[(x^2 - c^2)^{-1}F(\phi_1, k) + x^{-2}F(\phi_2, k)], \tag{3.11}$$

where

$$\phi_1 = \sin^{-1}\left[\frac{(a^2 - c^2)(x^2 - e^2)}{(a^2 - e^2)(x^2 - c^2)}\right]^{1/2}, \tag{3.12$_1$}$$

$$\phi_2 = \sin^{-1}\left[\frac{e^2(a^2 - x^2)}{(a^2 - e^2)x^2}\right]^{1/2}. \tag{3.12$_2$}$$

3.2.2. *Evaluation of I_c*

Similarly as before, making the proper transformation with respect to the variable in the first term of $(3.4)_2$ and combining it with the second term, we obtain

$$I_c = x[I_{c1} + (x^2 - a^2 - c^2)I_{c2} + (x^2 - a^2)(x^2 - c^2)I_{c3}], \tag{3.13}$$

where

$$I_{c1} = \int_0^{c^2} z^{-1}t\,dt, \tag{3.14$_1$}$$

$$I_{c2} = \int_0^{c^2} z^{-1}\,dt, \tag{3.14$_2$}$$

$$I_{c3} = \int_0^{c^2} z^{-1}(t - x^2)^{-1}\,dt, \tag{3.14$_3$}$$

and z is given by $(3.6)_4$.

The integral given by $(3.14)_1$ can be reduced to

$$I_{c1} = -2i[e^2(a^2 - c^2)]^{-1/2}\left[e^2F\left(\frac{\pi}{2}, k\right) + (e^2 - c^2)\Pi\left(\frac{\pi}{2}, \frac{c^2}{e^2}, k\right)\right]. \tag{3.15}$$

The integral given by $(3.14)_2$ can be evaluated as follows:

$$I_{c2} = -2i[e^2(a^2 - c^2)]^{-1/2}F\left(\frac{\pi}{2}, k\right). \tag{3.16}$$

The integral given by $(3.14)_3$ must be classified into the following two cases in accordance with the value of x.

(a) $e^2 < x^2 < a^2$. The integral is regular in this region and then can be obtained as follows:

$$I_{c3a} = -2i[e^2(a^2 - c^2)]^{-1/2}[x^2(x^2 - a^2)]^{-1}$$
$$\times\left[a^2\Pi\left(\frac{\pi}{2}, c^2(a^2 - x^2)(a^2 - c^2)^{-1}x^{-2}, k\right) - x^2F\left(\frac{\pi}{2}, k\right)\right]. \tag{3.17}$$

(b) $0 < x^2 < c^2$. Since the integral is an improper integral whose integrand diverges in this region, we define this integral by its Cauchy principal value and then

$$I_{c3b} = 2i[e^2(a^2 - c^2)]^{-1/2}[(x^2 - a^2)^{-1}F(\phi_3, k) + (x^2 - e^2)^{-1}F(\phi_4, k)], \tag{3.18}$$

where

$$\phi_3 = \sin^{-1}\left[\frac{(a^2 - c^2)x^2}{c^2(a^2 - x^2)}\right]^{1/2}, \tag{3.19$_1$}$$

$$\phi_4 = \sin^{-1}\left[\frac{e^2(c^2 - x^2)}{c^2(e^2 - x^2)}\right]^{1/2}. \tag{3.19$_2$}$$

3.2.3. Dislocation Distribution Function

Substituting the integrals obtained in $(3.2)_1$ and $(3.2)_2$ into (3.3), the dislocation distribution function can be determined as follows:

$0 < |x| < c$

$$\begin{aligned} f(x) = {} & 2\pi^{-2}A^{-1}[(a^2 - c^2)e^2]^{-1/2}x\left[\frac{(e^2 - x^2)}{(x^2 - a^2)(x^2 - c^2)}\right]^{1/2} \\ & \times \Bigg[(e^2 - c^2)\left[(\tau - \tau_i)\Pi\left(\frac{\pi}{2}, \frac{(a^2 - e^2)}{(a^2 - c^2)}, k\right) + \tau\Pi\left(\frac{\pi}{2}, \frac{c^2}{e^2}, k\right)\right] \\ & + \tau(x^2 + e^2 - a^2 - c^2)F\left(\frac{\pi}{2}, k\right) \\ & - (\tau - \tau_i)(e^2 - c^2)(x^2 - a^2)(x^2 - e^2)^{-1} \\ & \times \Pi\left(\frac{\pi}{2}, (a^2 - e^2)(x^2 - c^2)(a^2 - c^2)^{-1}(x^2 - e^2)^{-1}, k\right) \\ & - \tau\left\{(x^2 - c^2)F\left(\sin^{-1}\left[\frac{(a^2 - c^2)x^2}{c^2(a^2 - x^2)}\right]^{1/2}, k\right)\right. \\ & \left. + (x^2 - a^2)(x^2 - c^2)(x^2 - e^2)^{-1}F\left(\sin^{-1}\left[\frac{e^2(c^2 - x^2)}{c^2(e^2 - x^2)}\right]^{1/2}, k\right)\right\}\Bigg], \end{aligned} \tag{3.20}$$

$e < |x| < a$

$$\begin{aligned} f(x) = {} & 2\pi^{-2}A^{-1}[(a^2 - c^2)e^2]^{-1/2}x\left[\frac{(e^2 - x^2)}{(x^2 - a^2)(x^2 - c^2)}\right]^{1/2} \\ & \times \Bigg[(e^2 - c^2)\left[(\tau - \tau_i)\Pi\left(\frac{\pi}{2}, \frac{(a^2 - e^2)}{(a^2 - c^2)}, k\right) + \tau\Pi\left(\frac{\pi}{2}, \frac{c^2}{e^2}, k\right)\right] \\ & + [(\tau - \tau_i)(x^2 - a^2) + \tau(e^2 - a^2)]F\left(\frac{\pi}{2}, k\right) \\ & - (\tau - \tau_i)\left[(x^2 - a^2)F\left(\sin^{-1}\left[\frac{(a^2 - c^2)(x^2 - e^2)}{(a^2 - e^2)(x^2 - c^2)}\right]^{1/2}, k\right)\right. \\ & \left. + (x^2 - c^2)(x^2 - a^2)x^{-2}F\left(\sin^{-1}\left[\frac{e^2(a^2 - x^2)}{(a^2 - e^2)x^2}\right]^{1/2}, k\right)\right] \\ & + \tau a^2(x^2 - c^2)x^{-2}\Pi\left(\frac{\pi}{2}, \frac{c^2(a^2 - x^2)}{(a^2 - c^2)x^2}, k\right)\Bigg]. \end{aligned} \tag{3.21}$$

where the arbitrary constant Q_0 is taken to zero because $f(x)$ is symmetrical about the origin.

4. Application to the Strength Analysis of Polycrystalline Metals

4.1. Stress Acting on the Leading Dislocation of the Pile-Up

In this section, we determine the stress acting on the dislocation at the head of the pile–up from the distribution function obtained in the previous section. By approximating this stress to the stress required to yield the adjoining grain, we get the relation between the macroscopic yield stress and the grain size.

The relation between the stress acting on the leading dislocation and the distribution function $f(x)$ has been given by Bilby and Eshelby (1968) as the following expression:

$$\tau_l = -\frac{A\pi^2}{2} \lim_{x \to a} (x - a) f^2(x). \tag{4.1}$$

Substituting (3.21) into (4.1), we obtain the following expression:

$$\tau_l = \pi^{-2} A^{-1} e^{-2} a(a^2 - e^2)(e^2 - c^2)^2 (a^2 - c^2)^{-2} \times \left[(\tau - \tau_i)\Pi\left(\frac{\pi}{2}, \frac{(a^2 - e^2)}{(a^2 - c^2)}, k\right) + \tau\Pi\left(\frac{\pi}{2}, \frac{c^2}{e^2}, k\right) + \tau F\left(\frac{\pi}{2}, k\right)\right]^2. \tag{4.2}$$

4.2. Relation Between the Macroscopic Yield Stress and the Grain Sizes

By approximating the stress τ_l given by (4.2) to the stress acting on the grain boundary or the adjoining grain, and regarding this stress as the microscopic yield stress τ_c and the applied stress τ as the macroscopic yield stress τ_y, we can obtain the relationship between the macroscopic yield stress and the grain size. In (4.2), we replace τ_l and τ by τ_c and τ_y, respectively, and solve it with respect to τ_y by using the addition theorem of the elliptic integrals. Then, we obtain the following expression:

$$\tau_y = \frac{\Pi(\pi/2, (a^2 - e^2)/(a^2 - c^2), k)}{P(a, e, c)} \tau_i + \frac{Q(a, e, c)}{P(a, e, c)} (\pi^2 A \tau_c)^{1/2}, \tag{4.3}$$

where P and Q are given as follows:

$$P = 2F\left(\frac{\pi}{2}, k\right) + \left(\frac{\pi}{2}\right) e(e^2 - c^2)^{-1}(a^2 - c^2)^{1/2}, \tag{4.4}$$

$$Q = e(a^2 - c^2)(e^2 - c^2)^{-1}[(a^2 - e^2)a]^{-1/2}. \tag{4.5}$$

Figure 5 shows the normalized macroscopic yield stress τ_y/μ, calculated by (4.3) taking the inverse square root of the nondimensional half-length of the

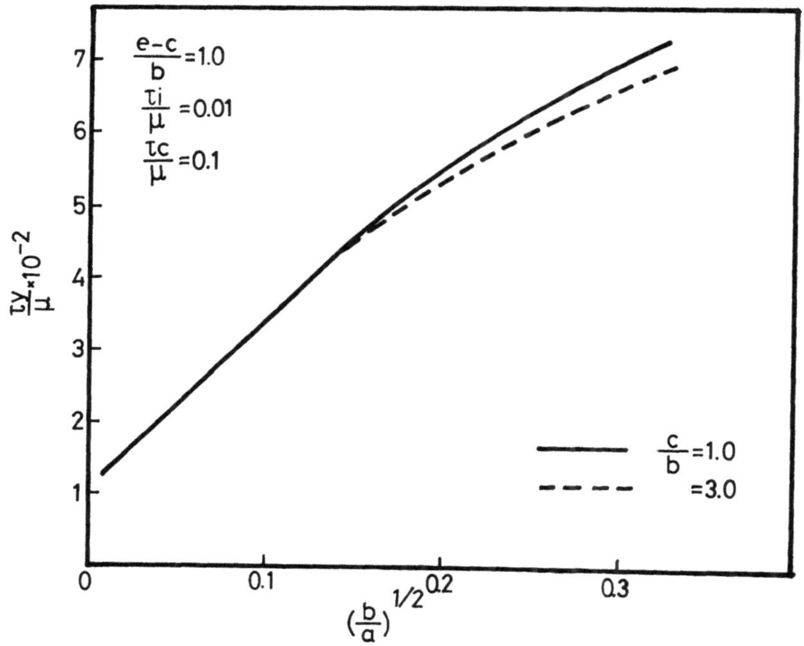

FIG. 5. Relation between the macroscopic yield stress and the grain sizes (theoretical results).

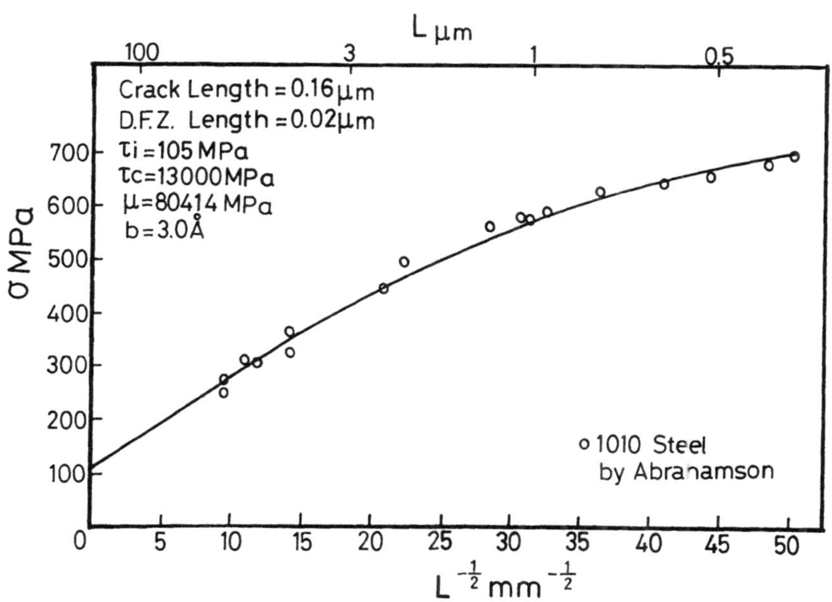

FIG. 6. Failure of the Hall–Petch relation in the range of ultrafine grain sizes (theoretical and experimental results).

grain diameter referred to the Burgers vector of the dislocation in the abscissa. In this figure, the solid line indicates the case of the shorter crack and the dotted line indicates the case of the longer crack. The Hall–Petch relation holds in the range of normal grain sizes for both cases, as shown in Fig. 5, but this relation fails in the range of ultrafine grain sizes. This explains the experimental results obtained by Armstrong (1983), as shown in Fig. 6.

In the case of normal grain sizes, even if there is a crack or a crack-like flaw in a grain, it has little effect on the yield stress and the yield stress only depends on the grain sizes. However, in case of ultrafine grain sizes, the effect of a crack or a crack-like flaw on the yield stress becomes relatively stronger, and hence the failure of the linear dependence of the yield stress on the grain size would be caused.

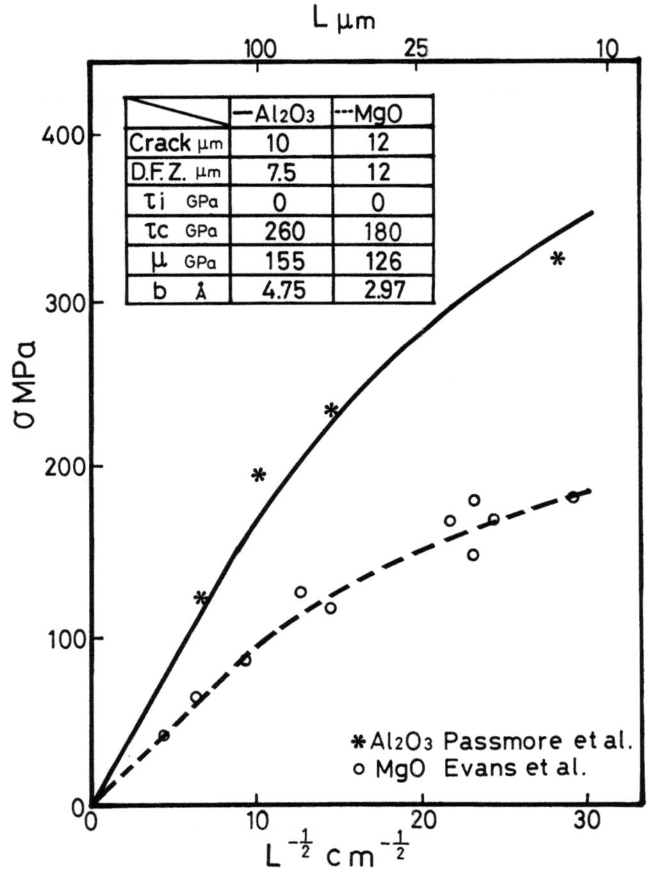

FIG. 7. Grain-size dependence of the fracture stress of engineering ceramics (theoretical and experimental results).

5. Grain-size Dependence of the Fracture Stress of Engineering Ceramics

According to Carniglia (1972), the fracture stress of different engineering ceramics depends on the grain sizes in a manner similar to the polycrstalline metals, and this relation also fails in the fine grain sizes.

Hitherto, it has been considered that this phenomenon could be explained in terms of the Griffith equation whose crack length is replaced by the grain size in the case where the grain diameter is relatively larger, and in terms of the Hall–Petch relation in the case where the grain size is relatively smaller. This relation is called the Petch–Orowan relation (Nishida, 1988). We can explain the above phenomenon in unified fashion using this model. However, since it is difficult to assume that the stress concentration, due to dislocations piled-up against the grain boundary, causes the fracture of brittle materials such as ceramics, it appears to be impossible that this model can be adopted to this phenomenon in its original form. Therefore, regarding the friction stress of the dislocation as zero, and hence the slip band as the imaginary crack, it seems possible that this model can be applied to the phenomenon. As an example, we show the experimental results on Al_2O_3 and MgO at room temperature, and the theoretical calculations by (4.3) in Fig. 7.

6. Concluding Remarks

In the present paper, we propose the new dislocation pile-up model which consists of the crack, the dislocation free zone (DFZ), and the slip band and formulate it by the method of the continuously distributed theory of dislocations, to explain the failure of the Hall–Petch relation in the range of ultrafine grain sizes confirmed experimentally by Armstrong.

We obtain the dislocation distribution function and the stress acting on the leading dislocation explicitly. Approximating this stress to the stress required to yield the adjoining grain, we finally get the relation between the macroscopic yield stress and the grain size and we can explain the experimental result.

Furthermore, regarding the friction stress of the dislocation as zero and the slip band as the imaginary crack, we can also explain similar experimental results on the fracture stress of engineering ceramics.

Acknowledgments

One of the authors (K. Saito) would like to express his sincere appreciation to Professor T. Mura of Northwestern University for his stimulating encouragement and advice in the field of micromechanics.

References

Armstrong, R. W. (1983), The yield and flow stress dependence on polycrystal grain size, in *Yield, Flow and Fracture of Polycrystals*, edited by T. N. Baker, Applied Science, London and New York, p. 1.

Bilby, B. A. and Eshelby, J. D. (1968), Dislocations and the theory of fracture, in *Fracture*, 1, edited by H. Liebowitz, Academic Press, New York, p. 99.

Carniglia, S. C. (1972), Reexamination of experimental strength versus grain-size data for ceramics, *J. Am. Ceram. Soc.*, **55**, 243–249.

Chang, S. J. and Ohr, S. M. (1981), Dislocation free zone model of fracture, *J. Appl. Phys.*, **52**, 7174–7181.

Kobayashi, S. and Ohr, S. M. (1980), In situ fracture experiments in B.C.C. metals, *Phil. Mag.*, **A42**, 763–772.

Li, J. C. M. (1981), Dislocation sources, in *Dislocation Modelling of Physical Systems*, edited by M. F. Ashby, R. Bullough, C. S. Hartley, and J. P. Hirth, Pergamon, New York, p. 498.

Mura, T. (1982), *Micromechanics of Defects in Solids*, Martinus Nijhoff, The Hague, p. 427.

Nishida, T. (1988), Ceramics no Zeisei Hakai, *Zairyou Kagaku*, **24**, 165–171 (in Japanese).

The Elastic Fields Produced by an Infinitesimal Dislocation Loop, an Interstitial Atom, and a Vacancy Moving with Uniform Velocity

H. SEKINE
Department of Engineering Science, Tohoku University, Sendai 980, Japan

Abstract

The present paper deals with, in the first place, the elastic field produced by an infinitesimal dislocation loop moving with uniform velocity in an unbounded anisotropic elastic medium. The problem is formulated on the basis of the solution of an eigenstrain problem and simple expressions are obtained for the elastic field. Next, by the use of these expressions, the elastic fields produced by an interstitial atom and a vacancy are derived. Finally, the static field produced by an infinitesimal dislocation loop is discussed.

1. Introduction

It is well known that a great number of point defects are formed in metallic materials after treatments such as quenching, irradiation, and fatigue. Such point defects move under the influence of a force field, and the motion of the point defects plays an important role in certain kinds of inelastic deformation in metallic materials. Therefore, the motion of the point defects has become thus far the subject of many investigations (Wolfer and Ashkin, 1975; Yoo and Butler, 1976; Michel, 1980).

In the present paper, the elastic field produced by an infinitesimal dislocation loop moving with uniform velocity is considered, in the first place, in the framework of unbounded anisotropic elastic continuum approximation. The problem is formulated on the basis of the solution of an eigenstrain problem. Consequently, the elastic field is expressed in line integral forms defined along a great circle on a unit sphere. Next, by use of the result, the elastic fields produced by an interstitial atom and a vacancy are derived. Finally, the static field produced by an infinitesimal dislocation loop is discussed.

2. Moving Infinitesimal Dislocation Loop

Let an eigenstrain $\varepsilon^*_{mn}(\mathbf{x}', t)$, which is a function of space and time, be distributed in an unbounded anisotropic elastic medium. Then, the induced displacement $u_i(\mathbf{x}, t)$ is given, in the coordinate form referred to a Cartesian coordinate system x_k, by (Mura, 1982)

$$u_i(\mathbf{x}, t) = -\frac{1}{(2\pi)^4}\frac{\partial}{\partial x_l}\int_{-\infty}^{\infty} d\omega \int_{-\infty}^{\infty} d\boldsymbol{\xi} \int_{-\infty}^{\infty} dt' \int_{-\infty}^{\infty} d\mathbf{x}'\, C_{klmn}\varepsilon^*_{mn}(\mathbf{x}', t')$$
$$\times N_{ik}(\boldsymbol{\xi}, \omega)D^{-1}(\boldsymbol{\xi}, \omega)\exp\{i[\boldsymbol{\xi}\cdot(\mathbf{x} - \mathbf{x}') + \omega(t - t')]\}, \tag{2.1}$$

where

$$N_{ik}(\boldsymbol{\xi}, \omega) = \frac{e_{imn}e_{kst}(C_{mjsl}\xi_j\xi_l - \rho\omega^2\delta_{ms})(C_{nptq}\xi_p\xi_q - \rho\omega^2\delta_{nt})}{2}, \tag{2.2}$$

$$D(\boldsymbol{\xi}, \omega) = e_{ijk}(C_{im1n}\xi_m\xi_n - \rho\omega^2\delta_{i1})(C_{jp2q}\xi_p\xi_q - \rho\omega^2\delta_{j2})$$
$$\times (C_{ks3t}\xi_s\xi_t - \rho\omega^2\delta_{k3}). \tag{2.3}$$

Here C_{klmn} are the elastic moduli, e_{ijk} is the permutation tensor, δ_{ij} is the Kronecker delta, ρ is the density, and t is time. Latin indices range over 1, 2, 3, and the usual summation convention is applied to every repeated index.

Consider an infinitesimal dislocation loop moving with uniform velocity V in the x_1 direction, as shown in Fig. 1. The area of the planar region bounded by the infinitesimal dislocation loop is ΔS and its unit normal is $\mathbf{v}$. When the infinitesimal dislocation loop is situated at the origin at $t = 0$, the eigenstrain for the infinitesimal dislocation loop can be written as

$$\varepsilon^*_{mn}(\mathbf{x}', t') = -b_{(n}v_{m)}\Delta S\, \delta(Vt' - x_1')\delta(x_2')\delta(x_3'), \tag{2.4}$$

where b_n is the Burgers vector and δ is the Dirac delta function.

Substituting (2.4) into (2.1) and performing the integration with respect to

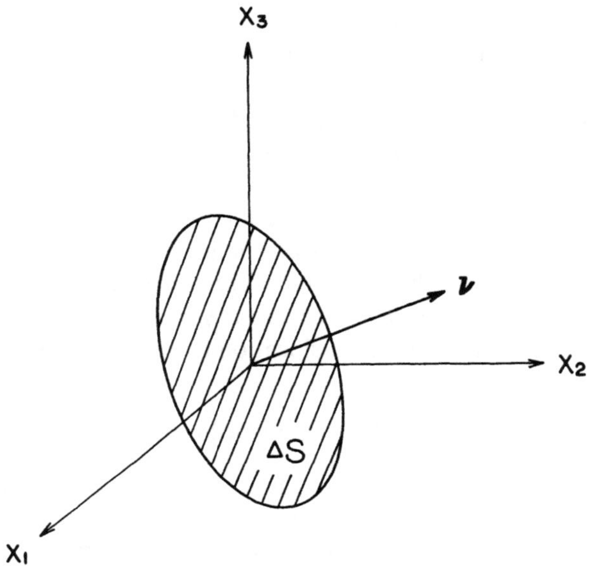

FIG. 1. An infinitesimal dislocation loop moving with uniform velocity in the x_1 direction.

$\mathbf{x}'$, we obtain

$$u_i(\mathbf{x}, t) = \frac{1}{(2\pi)^4} \frac{\partial}{\partial x_l} \int_{-\infty}^{\infty} d\omega \int_{-\infty}^{\infty} d\boldsymbol{\xi} \int_{-\infty}^{\infty} dt' \, C_{klmn} b_n v_m \Delta S \, N_{ik}(\boldsymbol{\xi}, \omega) D^{-1}(\boldsymbol{\xi}, \omega)$$
$$\times \exp\{i[\boldsymbol{\xi}\cdot\mathbf{x} + \omega t - (V\xi_1 + \omega)t']\}. \tag{2.5}$$

By the use of the formula

$$\int_{-\infty}^{\infty} \exp(i\alpha z)\, dz = \int_{-\infty}^{\infty} \exp(-i\alpha z)\, dz = 2\pi\delta(\alpha), \tag{2.6}$$

the integration of (2.5) with respect to t' and ω yields

$$u_i(\mathbf{x}, t) = \frac{1}{(2\pi)^3} \frac{\partial}{\partial x_l} \int_{-\infty}^{\infty} d\boldsymbol{\xi} \, C_{klmn} b_n v_m \Delta S \, N_{ik}(\boldsymbol{\xi}, V\xi_1) D^{-1}(\boldsymbol{\xi}, V\xi_1)$$
$$\times \exp[i\boldsymbol{\xi}\cdot(\mathbf{x} - \mathbf{x}^0)], \tag{2.7}$$

where

$$x_1^0 = Vt, \qquad x_2^0 = 0, \qquad x_3^0 = 0. \tag{2.8}$$

Introducing a new variable

$$\bar{\boldsymbol{\xi}} = \frac{\boldsymbol{\xi}}{\xi}, \tag{2.9}$$

where $\xi = (\xi_i \xi_i)^{1/2}$, we obtain

$$d\boldsymbol{\xi} = d\xi_1\, d\xi_2\, d\xi_3 = \xi^2\, d\xi\, dS(\bar{\boldsymbol{\xi}}). \tag{2.10}$$

Here $dS(\bar{\boldsymbol{\xi}})$ is the surface element on the unit sphere S^2, i.e., $\bar{\xi}_i \bar{\xi}_i = 1$.
Substituting (2.10) into (2.7), we obtain

$$u_i(\mathbf{x}, t) = \frac{1}{(2\pi)^3} \frac{\partial}{\partial x_l} \int_{0}^{\infty} d\xi \int_{S^2} dS(\bar{\boldsymbol{\xi}}) \, C_{klmn} b_n v_m \Delta S$$
$$\times N_{ik}(\bar{\boldsymbol{\xi}}, V\bar{\xi}_1) D^{-1}(\bar{\boldsymbol{\xi}}, V\bar{\xi}_1) \exp[i\xi\bar{\boldsymbol{\xi}}\cdot(\mathbf{x} - \mathbf{x}^0)]. \tag{2.11}$$

Changing $\boldsymbol{\xi}$ in (2.7) by $-\boldsymbol{\xi}$ and using (2.10), we have

$$u_i(\mathbf{x}, t) = \frac{1}{(2\pi)^3} \frac{\partial}{\partial x_l} \int_{0}^{\infty} d\xi \int_{S^2} dS(\bar{\boldsymbol{\xi}}) \, C_{klmn} b_n v_m \Delta S$$
$$\times N_{ik}(\bar{\boldsymbol{\xi}}, V\bar{\xi}_1) D^{-1}(\bar{\boldsymbol{\xi}}, V\bar{\xi}_1) \exp[-i\xi\bar{\boldsymbol{\xi}}\cdot(\mathbf{x} - \mathbf{x}^0)]. \tag{2.12}$$

From (2.6), we obtain easily the following formula:

$$\int_{0}^{\infty} [\exp(i\alpha z) + \exp(-i\alpha z)]\, dz = 2\pi\delta(\alpha). \tag{2.13}$$

Adding (2.11) and (2.12) and using formula (2.13), we get

$$u_i(\mathbf{x}, t) = \frac{1}{8\pi^2} \frac{\partial}{\partial x_l} \int_{S^2} dS(\bar{\boldsymbol{\xi}}) \, C_{klmn} b_n v_m \Delta S \, N_{ik}(\bar{\boldsymbol{\xi}}, V\bar{\xi}_1) D^{-1}(\bar{\boldsymbol{\xi}}, V\bar{\xi}_1)$$
$$\times \delta[\bar{\boldsymbol{\xi}}\cdot(\mathbf{x} - \mathbf{x}^0)]. \tag{2.14}$$

By performing the differentiation with respect to x_l, (2.14) becomes

$$u_i(\mathbf{x}, t) = \frac{1}{8\pi^2} \int_{S^2} dS(\bar{\xi})\, C_{klmn} b_n v_m \Delta S\, \bar{\xi}_l N_{ik}(\bar{\xi}, V\bar{\xi}_1) D^{-1}(\bar{\xi}, V\bar{\xi}_1) \times \delta'[\bar{\xi}\cdot(\mathbf{x}-\mathbf{x}^0)]. \tag{2.15}$$

Let us express (2.15) in a line integral form (Barnett, 1972; Mura, 1982). Define $\mathbf{y}$ and $\bar{\mathbf{y}}$ as follows

$$\mathbf{y} = \mathbf{x} - \mathbf{x}^0, \qquad \bar{\mathbf{y}} = \frac{\mathbf{y}}{y}, \tag{2.16}$$

where $y = (y_i y_i)^{1/2}$. By introducing the spherical polar coordinates θ^* and ϕ^* as shown in Fig. 2, the surface element $dS(\bar{\xi})$ is expressed as

$$dS(\bar{\xi}) = \sin\theta^*\, d\theta^*\, d\phi^* = -d(\bar{\xi}\cdot\bar{\mathbf{y}})\, d\phi^*. \tag{2.17}$$

Substituting (2.17) into (2.16) and integrating by parts with respect to $\bar{\xi}\cdot\bar{\mathbf{y}}$, we obtain

$$u_i(\mathbf{x}, t) = -\frac{1}{8\pi^2} \int_{S^1} d\phi^*\, C_{klmn} b_n v_m \Delta S\, y^{-1} \frac{\partial[\bar{\xi}_l N_{ik}(\bar{\xi}, V\bar{\xi}_1) D^{-1}(\bar{\xi}, V\bar{\xi}_1)]}{\partial(\bar{\xi}\cdot\mathbf{y})}, \tag{2.18}$$

where S^1 is the great circle on the unit sphere S^2, which is perpendicular to $\bar{\mathbf{y}}$.

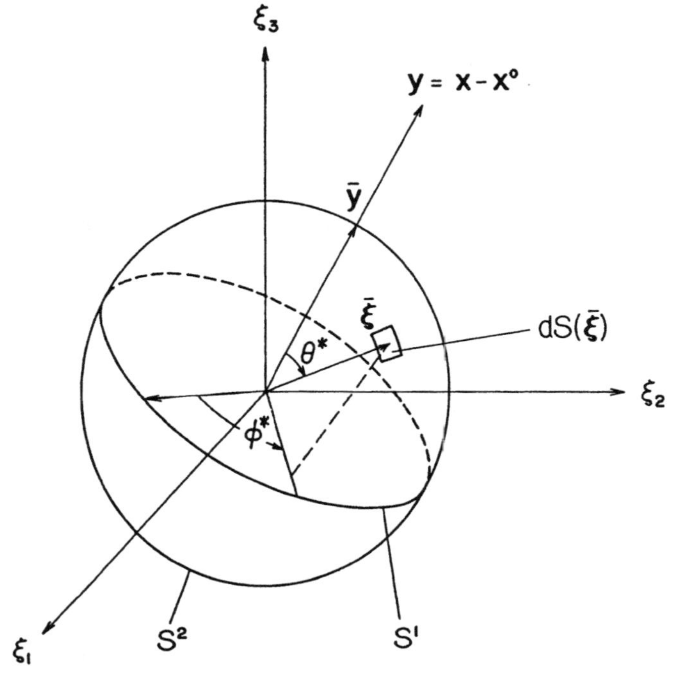

FIG. 2. The unit sphere S^2 and the definition of the spherical polar coordinates θ^* and ϕ^*.

Now, we introduce the function such that

$$K_{jk}(\bar{\xi}, V\bar{\xi}_1) = C_{jpkq}\bar{\xi}_p\bar{\xi}_q - \rho(V\xi_1)^2\delta_{jk}. \tag{2.19}$$

Since $N_{ik}(\bar{\xi}, V\bar{\xi}_1)$ and $D(\bar{\xi}, V\bar{\xi}_1)$ are, respectively, the cofactor and the determinant of the matrix $K_{jk}(\bar{\xi}, V\bar{\xi}_1)$, it follows that

$$K_{jk}N_{ik}D^{-1} = \delta_{ij}. \tag{2.20}$$

On S^1, we have

$$\frac{\partial\bar{\xi}_k}{\partial(\bar{\xi}\cdot\mathbf{y})} = y^{-1}\bar{y}_k. \tag{2.21}$$

Using (2.21), we obtain from (2.19)

$$\frac{\partial K_{jk}(\bar{\xi}, V\bar{\xi}_1)}{\partial(\bar{\xi}\cdot\mathbf{y})} = y^{-1}[C_{jpkq}(\bar{y}_p\bar{\xi}_q + \bar{\xi}_p\bar{y}_q) - 2\rho V^2\bar{\xi}_1\bar{y}_1\delta_{jk}]. \tag{2.22}$$

Therefore, in view of (2.20), we obtain the following equation on S^1:

$$\frac{\partial(N_{ik}D^{-1})}{\partial(\bar{\xi}\cdot\mathbf{y})} = -y^{-1}[C_{jpsq}(\bar{y}_p\bar{\xi}_q + \bar{\xi}_p\bar{y}_q) - 2\rho V^2\bar{\xi}_1\bar{y}_1\delta_{js}]N_{ij}N_{ks}D^{-2}. \tag{2.23}$$

By the use of (2.21) and (2.23), (2.18) becomes

$$\begin{aligned} u_i(\mathbf{x}, t) = -\frac{1}{8\pi^2}\int_{S^1} d\phi^*\, C_{klmn}b_n v_m \Delta S\, y^{-2}\{&\bar{y}_l N_{ik}(\bar{\xi}, V\bar{\xi}_1)D^{-1}(\bar{\xi}, V\bar{\xi}_1) \\ &- \bar{\xi}_l[C_{jpsq}(\bar{y}_p\bar{\xi}_q + \bar{\xi}_p\bar{y}_q) - 2\rho V^2\bar{\xi}_1\bar{y}_1\delta_{js}]N_{ij}(\bar{\xi}, V\bar{\xi}_1) \\ &\times N_{ks}(\bar{\xi}, V\bar{\xi}_1)D^{-2}(\bar{\xi}, V\bar{\xi}_1)\}. \end{aligned} \tag{2.24}$$

In a similar way, the displacement gradient $u_{i,j}(\mathbf{x}, t)$ is also obtained in a line integral form on S^1, as follows:

$$\begin{aligned} u_{i,j}(\mathbf{x}, t) = \frac{1}{8\pi^2}\int_{S^1} d\phi^*\, C_{klmn}b_n v_m \Delta S\, y^{-2}[\![&2\bar{y}_j\bar{y}_l N_{ik}(\bar{\xi}, V\bar{\xi}_1)D^{-1}(\bar{\xi}, V\bar{\xi}_1) \\ &- 2\{(\bar{y}_j\bar{\xi}_l + \bar{\xi}_j\bar{y}_l)[C_{sptq}(\bar{y}_p\bar{\xi}_q + \bar{\xi}_p\bar{y}_q) - 2\rho V^2\bar{\xi}_1\bar{y}_1\delta_{st}] \\ &+ \bar{\xi}_j\bar{\xi}_l(C_{sptq}\bar{y}_p\bar{y}_q - \rho V^2\bar{y}_1^2\delta_{st})\}N_{is}(\bar{\xi}, V\bar{\xi}_1)N_{kt}(\bar{\xi}, V\bar{\xi}_1)D^{-2}(\bar{\xi}, V\bar{\xi}_1) \\ &+ \bar{\xi}_j\bar{\xi}_l[C_{sptq}(\bar{y}_p\bar{\xi}_q + \bar{\xi}_p\bar{y}_q) - 2\rho V^2\bar{\xi}_1\bar{y}_1\delta_{st}][C_{uwvz}(\bar{y}_w\bar{\xi}_z + \bar{\xi}_w\bar{y}_z) \\ &- 2\rho V^2\bar{\xi}_1\bar{y}_1\delta_{uv}][N_{is}(\bar{\xi}, V\bar{\xi}_1)N_{ku}(\bar{\xi}, V\bar{\xi}_1)N_{tv}(\bar{\xi}, V\bar{\xi}_1) \\ &+ N_{kt}(\bar{\xi}, V\bar{\xi}_1)N_{iu}(\bar{\xi}, V\bar{\xi}_1)N_{sv}(\bar{\xi}, V\bar{\xi}_1)]D^{-3}(\bar{\xi}, V\bar{\xi}_1)]\!]. \end{aligned} \tag{2.25}$$

By differentiating (2.7) with respect to t and by comparing it with the displacement gradient $u_{i,1}(\mathbf{x}, t)$ obtained from (2.7), the velocity $\dot{u}_i(\mathbf{x}, t)$ is expressed as

$$\dot{u}_i(\mathbf{x}, t) = -Vu_{i,1}(\mathbf{x}, t). \tag{2.26}$$

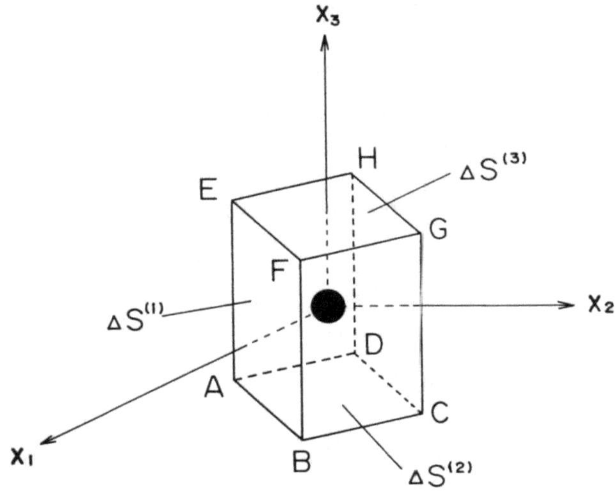

FIG. 3. An infinitesimal parallelepiped containing an interstitial atom (or a vacancy).

3. Moving Interstitial Atom and Vacancy

In this section, we are concerned with the elastic fields produced by an interstitial atom and a vacancy moving with uniform velocity.

Consider an arbitrary infinitesimal parallelepiped containing an interstitial atom (or a vacancy), as shown in Fig. 3. The parallelepiped moves with the interstitial atom (or vacancy). The length of each side of the infinitesimal parallelepiped is supposed to be of order of lattice constant. Assume that the infinitesimal parallelepiped is deformed under an affine transformation without rotations by the existence of the interstitial atom (or vacancy); that is, the side surfaces of the parallelepiped shift keeping parallel to the opposite surfaces. In this case, the interstitial atom (or vacancy) can be regarded as the superposition of three infinitesimal dislocation loops. Let the areas of the three side surfaces with outward unit normals $\mathbf{v}^{(K)}$ $(K = 1, 2, 3)$ be $\Delta S^{(K)}$ $(K = 1, 2, 3)$, as shown in Fig. 3; and let the relative movements of the opposite side surfaces to the side surfaces with the unit normals $\mathbf{v}^{(K)}$ due to the interstitial atom (or vacancy) be $\mathbf{b}^{(K)}$ $(K = 1, 2, 3)$. Then, the displacement produced by the interstitial atom (or vacancy) moving with uniform velocity V in the x_1 direction is written through (2.24) as

$$\begin{aligned} u_i(\mathbf{x}, t) = &-\frac{1}{8\pi^2}\int_{S^1} d\phi^*\, C_{klmn}\left(\sum_{K=1}^{3} b_n^{(K)} v_m^{(K)} \Delta S^{(K)}\right) y^{-2} \\ &\times \{\bar{y}_l N_{ik}(\bar{\xi}, V\bar{\xi}_1) D^{-1}(\bar{\xi}, V\bar{\xi}_1) \\ &- \bar{\xi}_l [C_{jpsq}(\bar{y}_p\bar{\xi}_q + \bar{\xi}_p\bar{y}_q) - 2\rho V^2 \bar{\xi}_1 \bar{y}_1 \delta_{js}] \\ &\times N_{ij}(\bar{\xi}, V\bar{\xi}_1) N_{ks}(\bar{\xi}, V\bar{\xi}_1) D^{-2}(\bar{\xi}, V\bar{\xi}_1)\}. \end{aligned} \tag{3.1}$$

The variation of the volume of the infinitesimal parallelepiped due to the interstitial atom (or vacancy) is given approximately by

$$\Delta W = -\sum_{K=1}^{3} b_i^{(K)} \nu_i^{(K)} \Delta S^{(K)}. \tag{3.2}$$

4. Static Field Produced by an Infinitesimal Dislocation Loop

When an infinitesimal dislocation loop is at rest at the origin, the displacement $u_i(\mathbf{x})$ is obtained by putting $V = 0$ in (2.24), as follows:

$$u_i(\mathbf{x}) = -\frac{1}{8\pi^2}\int_{S^0} d\phi^*\, C_{klmn} b_n \nu_m \Delta S\, x^{-3}[x_l N_{ik}(\bar{\xi}) D^{-1}(\bar{\xi}) - \bar{\xi}_l C_{jpsq}(x_p \bar{\xi}_q + \bar{\xi}_p x_q) N_{ij}(\bar{\xi}) N_{ks}(\bar{\xi}) D^{-2}(\bar{\xi})], \tag{4.1}$$

where

$$N_{ik}(\bar{\xi}) = \frac{e_{imn} e_{kst} C_{mjsl} C_{nptq} \bar{\xi}_j \bar{\xi}_l \bar{\xi}_p \bar{\xi}_q}{2}, \tag{4.2}$$

$$D(\bar{\xi}) = e_{ijk} C_{im1n} C_{jp2q} C_{ks3t} \bar{\xi}_m \bar{\xi}_n \bar{\xi}_p \bar{\xi}_q \bar{\xi}_s \bar{\xi}_t, \tag{4.3}$$

$$x = (x_i x_i)^{1/2}, \tag{4.4}$$

and S^0 is the great circle on the unit sphere S^2, which is perpendicular to $\mathbf{x}$, as shown in Fig. 4.

For isotropic elastic media, C_{klmn}, $N_{ik}(\bar{\xi})$, and $D(\bar{\xi})$ are simply expressed as

$$C_{klmn} = \lambda \delta_{kl} \delta_{mn} + \mu \delta_{km} \delta_{ln} + \mu \delta_{kn} \delta_{lm}, \tag{4.5}$$

$$N_{ik}(\bar{\xi}) = \mu[(\lambda + 2\mu)\delta_{ik} - (\lambda + \mu)\bar{\xi}_i \bar{\xi}_k], \tag{4.6}$$

$$D(\bar{\xi}) = \mu^2(\lambda + 2\mu), \tag{4.7}$$

where λ and μ are Lamé's elastic constants.

Let $\mathbf{m}$ and $\mathbf{n}$ be the orthogonal unit vectors lying on the plane of S^0 (see Fig. 4). Then, the unit vector $\bar{\xi}$ is expressed as

$$\bar{\xi}_i = m_i \cos\phi^* + n_i \sin\phi^*. \tag{4.8}$$

We substitute (4.5)–(4.8) into (4.1). Then, the integration is readily performed and we obtain

$$u_i(\mathbf{x}) = -\frac{\lambda + \mu}{4\pi(\lambda + 2\mu)} \frac{1}{x^3} \left\{ \frac{\mu}{\lambda + \mu} [b_i \nu_k + b_k \nu_i) x_k - b_k \nu_k x_i] + \frac{3 b_k \nu_l x_i x_k x_l}{x^2} \right\} \Delta S. \tag{4.9}$$

Equation (4.9) is equivalent to the expression which has been already obtained by Kroupa (1962).

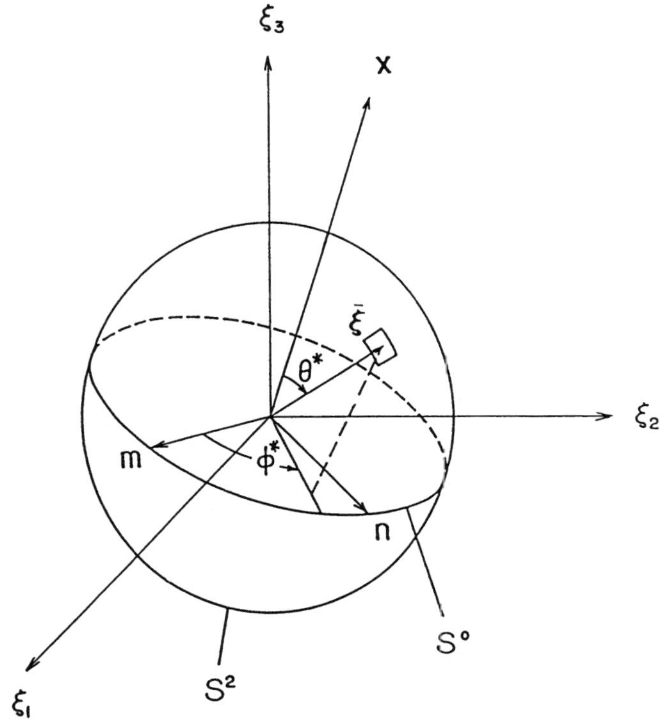

FIG. 4. The great circle S^0 on the unit sphere S^2.

References

Barnett, D. M. (1972), The precise evaluation of derivatives of the anisotropic elastic Green's functions, *Phys. Stat. Sol. (b)*, **49**, 741–748.

Kroupa, F. (1962), Continuous distribution of dislocation loops, *Czech. J. Phys.*, **B12**, 191–201.

Michel, B. (1980), Point defects and inclusions near solid surfaces and interfaces, *Arch. Mech.*, **32**, 73–79.

Mura, T. (1982), *Micromechanics of Defects in Solids*, Martinus Nijhoff, The Hague.

Wolfer, W. G. and Ashkin, M. (1975), Stress-induced diffusion of point defects to spherical sinks, *J. Appl. Phys.*, **46**, 547–557.

Yoo, M. H. and Butler, W. H. (1976), Steady-state diffusion of point defects in the interaction force field, *Phys. Stat. Sol (b)*, **77**, 181–193.

Energy Consideration on a Branched and Curved Crack Extension

Y. SUMI
Department of Naval Architecture and Ocean Engineering,
Yokohama National University, Tokiwadai Hodogaya-ku,
Yokohama 240, Japan

Abstract

An energy method is proposed to determine noncollinear crack paths in an elastic brittle solid which exhibits quasi-static crack extension under a displacement-controlled loading condition. Defining the total potential energy as the sum of the elastic potential energy stored in the body and the surface energy, its minimization with respect to crack paths leads to an energetically favorable crack growth path. In order to calculate the energies we use a second-order perturbation solution, which has been obtained for a straight crack with slightly branched and curved extensions under general far-field boundary conditions within the framework of linear fracture mechanics. As far as homogeneous materials are concerned, the minimization of the total potential energy gives rise to the crack direction that is equivalent to the one maximizing the elastic energy release rate. Possible crack paths in materials having inhomogeneous fracture toughness are also investigated by the energy method. In this case, even in a pure Mode-I loading condition, we often observe branched and curved crack extensions, which cannot be predicted by conventional stress or strain criteria. As a practical application of the present method, a brittle crack path extending along a welded joint is predicted, where the effects of applied stresses, residual stresses, and material deterioration are taken into account.

1. Introduction

The criterion of brittle fracture proposed by Griffith (1920, 1924) can be considered as a generalized equilibrium condition of the crack extension force and the crack resistance force which are, respectively, derived by differentiating the elastic potential energy and the surface energy of a brittle solid with respect to crack length. Defining the sum of the elastic potential energy and the surface energy as the total potential energy, the stationary condition of its first variation with respect to crack length thus leads to the aforementioned Griffith criterion. The characteristics of quasi-static crack extension in a brittle solid and the total potential energy of the corresponding system are investigated by Sumi *et al.* (1980), who perform the successive minimization of the total potential energy in order to determine the quasi-static growth regime of a system of interacting cracks under a displacement-controlled loading condition. Using the second-order variation of the total potential energy, stability and bifurcation analyses are also discussed, with regard to brittle fracture problems, by Potier-Ferry (1985) and by NGuyen (1987), who consider a general bifurcation and post-bifurcation behavior of rate-independent systems governed by normality laws including friction, plasticity, and brittle fracture. In the aforementioned variational studies relating to brittle fracture problems, energy consideration is made only for collinear crack extension, while a brief variational account of noncollinear crack extension problems is presented by Nemat-Nasser (1980), in which a basic concept is outlined for the minimization of the total potential energy with respect to branched and curved crack paths.

In the present paper a variational method is proposed to determine noncollinear crack paths in an elastic brittle solid which exhibits quasi-static crack extension under a displacement-controlled loading condition. The energetically most favorable crack paths are obtained by the minimization of the total potential energy, which is subjected to the subsidiary condition representing the equilibrium of the crack extension and resistance forces. In order to calculate the elastic potential energy as a function of a noncollinear crack path, the crack profile is approximated by a certain shape function which involves several shape parameters representing the noncollinearity. Since numerical difficulties exist in order to calculate accurately the first and second variations of the total potential energy with respect to a noncollinear crack extension, a second-order perturbation solution of a straight crack with slightly branched and curved extensions under general far-field boundary condition is introduced, where the variation of the energy can be expressed analytically in terms of the shape parameters. We shall examine the system governed by a single load parameter, which is proportional to stress and strain for a fixed crack geometry. Since the elastic potential energy is proportional to the square of the load parameter in this case, it has a relatively simple form which can be expanded in terms of the square root of the crack extensional length. Using the simple expression of the total potential energy, an explicit crack path criterion can be derived from the minimization condition. As far as homogeneous materials are concerned, the minimization of the total potential energy gives rise to the crack extensional direction which is equivalent to the one corresponding to the maximum energy release rate.

The present energy method is effective in order to obtain possible crack paths extending in materials with inhomogeneous fracture toughness. In the case where a pure Mode-I crack intersects a specially oriented degradation zone, the minimization of the total potential energy may lead to a branched crack extension which cannot be predicted by a conventional stress or strain criteria. As a practical application of the present method, we shall investigate a brittle crack propagation along a welded joint, which often exhibits branched and curved crack extension due to the combined effects of applied stresses, residual stresses, and material deterioration along the heat affected or thermally affected zones.

2. Variational Method for the Determination of a Noncollinear Crack Path

2.1. Statement of the Problem

Consider a two-dimensional plane-strain problem of an elastic brittle solid of unit thickness containing a noncollinear crack of length H, which extends in a quasi-statical fashion under a displacement-controlled loading condition represented by a single load parameter p. A Cartesian coordinate system $O{-}x_1, x_2$ is introduced with the origin at the tip, where the x_1 axis coincides with the tangent of the crack at the tip. A branched and curved extension of the crack is also shown in Fig. 1, in which the newly extended crack length is denoted by h, and the deviation from the x_1 axis is represented by $\lambda(x_1)$. In order to develop a variational formulation of crack path prediction, designate by U the elastic potential energy stored in the body, and by S the surface energy, and observe that

$$U = U[h, p; \lambda(x_1)] \geqq 0, \tag{2.1}$$

$$S = S[h; \lambda(x_1)] \geqq 0, \tag{2.2}$$

$$\left.\frac{\partial S}{\partial h}\right|_{+} = \lim_{\delta\to 0^{+}} \frac{S[h+\delta; \lambda] - S[h; \lambda]}{\delta} = 2\gamma[h; \lambda(x_1)] > 0, \tag{2.3}$$

in which p is assumed as an independent parameter, and $\gamma[h; \lambda(x_1)]$ is the surface energy of the unit area of the brittle material.

According to the Griffith fracture theory, an extending crack should satisfy the following equilibrium condition:

$$U_{,h}[h, p; \lambda(x_1)] + S_{,h}[h; \lambda(x_1)] = 0, \tag{2.4}$$

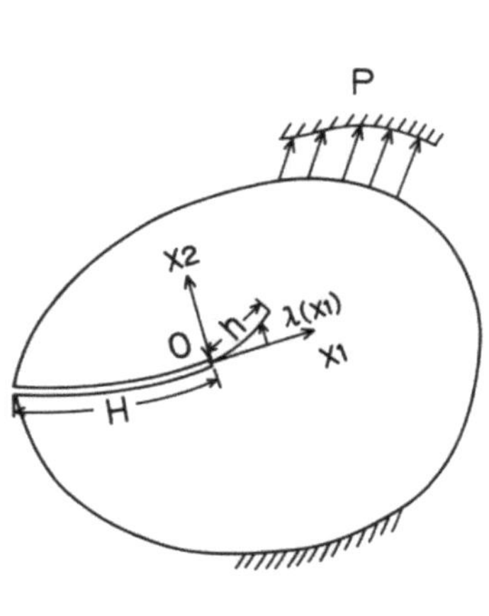

FIG. 1. A noncollinear crack with branched and curved extensions.

where a comma followed by subscript(s) indicates partial differentiation with respect to the corresponding variable(s). Equation (2.4) leads to the load parameter which preserves the equilibrium condition of the crack of extensional length h and of its shape represented by $\lambda(x_1)$. The load parameter thus obtained can be expressed as

$$p = p^{\mathrm{E}}[h; \lambda(x_1)]. \tag{2.5}$$

2.2. *Variational Principle for Crack Path Prediction*

Summing up the elastic potential energy and the surface energy, the total potential energy is defined as

$$\Pi[h, p; \lambda(x_1)] = U[h, p; \lambda(x_1)] + S[h; \lambda(x_1)]. \tag{2.6}$$

We introduce the minimum principle of the total potential energy for the crack growth path, which can be expressed as

> "Among all admissible variations in the shape, $\lambda(x_1)$, of the branched and curved extensions which correspond to a given increase of the crack length h, the ones which minimize the total potential energy Π produce the most energetically stable state, and hence are the actual ones." (2.7)

In the above statement it should be noted that the change of $\lambda(x_1)$ is admissible as far as $\lambda(x_1)$ is continuous, and that the corresponding variations of the total potential energy are calculated under the subsidiary condition (2.4). This means that the variations of the total potential energy are compared at the equilibrium states defined by (2.4).

If $\lambda(x_1)$ is selected as the actual crack path and $p^{\mathrm{E}}[h; \lambda(x_1)]$ is the corresponding equilibrium load parameter, the statement (2.7) can be written mathematically as

$$\Pi[h, p^{\mathrm{E}}(h; \lambda + \delta\lambda); \lambda + \delta\lambda] \geqq \Pi[h, p^{\mathrm{E}}(h; \lambda); \lambda]. \tag{2.8}$$

In (2.8), $\delta\lambda$ is an arbitrary admissible change of $\lambda(x_1)$, and $p^{\mathrm{E}}(h; \lambda + \delta\lambda)$ is the corresponding equilibrium load parameter. Denoting

$$\Pi^{\mathrm{E}} \equiv \Pi[h, p^{\mathrm{E}}(h; \lambda); \lambda], \tag{2.9}$$

the minimization conditions of the total potential energy (2.8) are expressed in terms of the first and second variations of Π^{E} with respect to λ, which are given as

$$\delta_\lambda \Pi^{\mathrm{E}} = 0, \tag{2.10}$$

and

$$\delta_\lambda^2 \Pi^{\mathrm{E}} > 0. \tag{2.11}$$

As long as the positive definiteness of the second variation holds, (2.10) gives rise to the energetically most favorable conditions of the noncollinear crack path $\lambda(x_1)$.

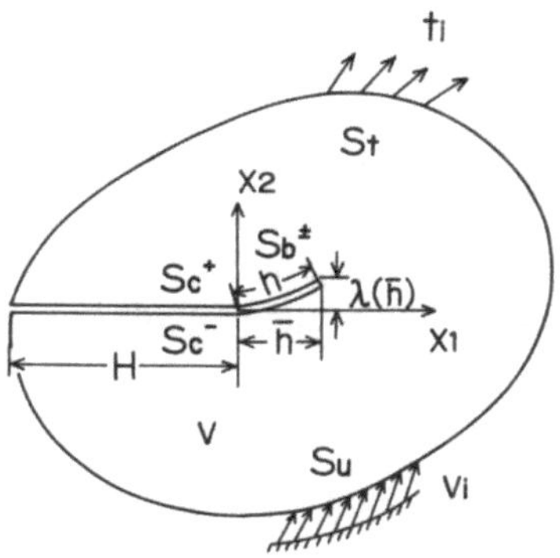

FIG. 2. A straight crack with slightly branched and curved extensions in a finite body.

3. Energy Calculation Based on a Second-Order Perturbation Solution

3.1. *A Second-Order Perturbation Solution of a Straight Crack with Slightly Branched and Curved Extensions*

In order to calculate elastic potential energy associated with branched and curved crack extensions, we introduce a second-order perturbation solution whose derivation is discussed in detail by Sumi (1989a). Let us consider a linearly elastic brittle solid containing a straight crack of length H as illustrated in Fig. 2. The boundary value problem is defined as

$$\begin{aligned} \sigma_{ij,j} &= 0 \quad \text{in } V, \\ \sigma_{ij}n_j &= t_i \quad \text{on } S_t \text{ and } S_c^{\pm}, \\ u_i &= v_i \quad \text{on } S_u, \end{aligned} \tag{3.1}$$

where σ_{ij} and u_i represent the stress tensor and the displacement vector defined in the domain V occupied by the body. The external unit normal vector on the surface of the body is denoted by n_i. Surface tractions, t_i, are prescribed on the part of the boundary, S_t, and on the crack surfaces, $S_c^{\pm}$, while surface displacements, v_i, are prescribed on the remaining outer boundary, S_u. The Cartesian coordinate system (x_1, x_2) with the origin at the crack tip has its x_1 axis on the original straight crack line. Stress distribution along the prolongation of the initial crack tip is given by

$$\begin{aligned} \sigma_{11}(x_1, 0) &= \frac{k_{\mathrm{I}}}{\sqrt{2\pi x_1}} + T + b_{\mathrm{I}}\sqrt{\frac{x_1}{2\pi}} + O(x_1), \\ \sigma_{22}(x_1, 0) &= \frac{k_{\mathrm{I}}}{\sqrt{2\pi x_1}} + b_{\mathrm{I}}\sqrt{\frac{x_1}{2\pi}} + O(x_1), \\ \sigma_{12}(x_1, 0) &= \frac{k_{\mathrm{II}}}{\sqrt{2\pi x_1}} + b_{\mathrm{II}}\sqrt{\frac{x_1}{2\pi}} + O(x_1), \end{aligned} \tag{3.2}$$

where k_{I} and k_{II} are the stress intensity factors, and the coefficients T, b_{I}, and b_{II} are also determined from the solution of the boundary value problem prior to the crack extension.

A slightly branched and curved extension of the crack is also shown in Fig. 2, in which the newly created fracture surfaces are denoted by $S_b^{\pm}$ whose projected length on the x_1 axis is $\bar{h}$, and whose deviation from the x_1 axis is represented by $\lambda(x_1)$. The shape of the noncollinear crack growth is assumed in the following form:

$$\lambda(x_1) = \alpha x_1 + \beta x_1^{3/2} + \gamma x_1^2, \tag{3.3}$$

where α, β, γ are the shape parameters, which are taken as small perturbation parameters in the analysis. Assuming the order of smallness in the following form:

$$\frac{\lambda(x_1)}{\bar{h}} = O(\varepsilon), \tag{3.4}$$

the smallness of the shape parameters are, respectively, given by

$$\alpha = O(\varepsilon), \qquad \beta\bar{h}^{1/2} = O(\varepsilon), \qquad \gamma\bar{h} = O(\varepsilon). \tag{3.5}$$

Considering predominantly Mode-I loading conditions, we also assume that the order of smallness of the near tip stress field parameters which characterize the in-plane shear mode are given by

$$\frac{k_{\mathrm{II}}}{k_{\mathrm{I}}} = O(\varepsilon), \qquad \frac{b_{\mathrm{II}}\bar{h}}{k_{\mathrm{I}}} = O(\varepsilon). \tag{3.6}$$

The second-order perturbation analysis leads to the solution of the stress intensity factors at the extended crack tip as

$$K_{\mathrm{I}} = K_{\mathrm{I}}^{(\infty)} + K_{\mathrm{I}}^{(f)} + O(\bar{h}^{3/2}), \tag{3.7}$$

$$K_{\mathrm{II}} = K_{\mathrm{II}}^{(\infty)} + K_{\mathrm{II}}^{(f)} + O(\bar{h}^{3/2}), \tag{3.8}$$

where the solution is practically applicable for the range $\lambda'(h) < 0.5$. In (3.7) and (3.8), $K_{\mathrm{I}}^{(\infty)}$ and $K_{\mathrm{II}}^{(\infty)}$ are the stress intensity factors corresponding to a semi-infinite straight crack with the branched and curved extension given by (3.3), and they are calculated as

$$\begin{aligned} K_{\mathrm{I}}^{(\infty)} = {} & (1 - \tfrac{3}{8}\alpha^2)k_{\mathrm{I}} - \tfrac{3}{2}\alpha k_{\mathrm{II}} - \left\{\tfrac{9}{4}\beta k_{\mathrm{II}} + \tfrac{9}{8}\alpha\beta k_{\mathrm{I}} - 2\sqrt{\frac{2}{\pi}}\alpha^2 T\right\}\bar{h}^{1/2} \\ & + \left\{\tfrac{1}{2}(1 + \tfrac{9}{8}\alpha^2)b_{\mathrm{I}} - \tfrac{5}{4}\alpha b_{\mathrm{II}} - 3\gamma k_{\mathrm{II}} + \frac{11}{2}\sqrt{\frac{2}{\pi}}\alpha\beta T \right. \\ & \left. - (\tfrac{27}{32}\beta^2 + \tfrac{3}{2}\alpha\gamma)k_{\mathrm{I}}\right\}\bar{h} + O(\bar{h}^{3/2}), \end{aligned} \tag{3.9}$$

$$\begin{aligned} K_{\mathrm{II}}^{(\infty)} = {} & k_{\mathrm{II}} + \tfrac{1}{2}\alpha k_{\mathrm{I}} + \left(\tfrac{3}{4}\beta k_{\mathrm{I}} - 2\sqrt{\frac{2}{\pi}}\alpha T\right)\bar{h}^{1/2} \\ & + \left\{\tfrac{1}{2}b_{\mathrm{II}} - \tfrac{1}{4}\alpha b_{\mathrm{I}} - \frac{3\sqrt{2\pi}}{4}\beta T + \gamma k_{\mathrm{I}}\right\}\bar{h} + O(\bar{h}^{3/2}), \end{aligned} \tag{3.10}$$

in which only the terms up to $O(\varepsilon^2)$ are retained. The second terms in the right-hand side of (3.7) and (3.8) are the finite body corrections given by

$$K_{\mathrm{I}}^{(f)} = [\{(1 - \tfrac{3}{8}\alpha^2)k_{\mathrm{I}} - \alpha k_{\mathrm{II}}\}\bar{k}_{11} + (k_{\mathrm{II}} - \alpha k_{\mathrm{I}})\bar{k}_{12} - \tfrac{3}{2}\alpha k_{\mathrm{I}}\bar{k}_{21} - \tfrac{3}{2}\alpha(k_{\mathrm{II}} - \alpha k_{\mathrm{I}})\bar{k}_{22}]\bar{h} + O(\bar{h}^{3/2}), \tag{3.11}$$

$$K_{\mathrm{II}}^{(f)} = [\{(1 - \tfrac{7}{8}\alpha^2)k_{\mathrm{I}} - \alpha k_{\mathrm{II}}\}\bar{k}_{21} + (k_{\mathrm{II}} - \alpha k_{\mathrm{I}})\bar{k}_{22} + \tfrac{1}{2}\alpha k_{\mathrm{I}}\bar{k}_{11} + \tfrac{1}{2}\alpha(k_{\mathrm{II}} - \alpha k_{\mathrm{I}})\bar{k}_{12}]\bar{h} + O(\bar{h}^{3/2}). \tag{3.12}$$

In (3.11) and (3.12) $\bar{k}_{11}$ and $\bar{k}_{21}$ are the stress intensity factors of Mode-I and Mode-II for $\mu = \mathrm{I}$, and $\bar{k}_{12}$ and $\bar{k}_{22}$ are those corresponding to $\mu = \mathrm{II}$ in the problems defined as

$$\begin{aligned} \sigma_{ij,j} &= 0 && \text{in } V, \\ \sigma_{ij}n_j &= -\sigma^f_{\mu ij}n_j && \text{on } S_t + S_c^{\pm}, \\ u_i &= -u^f_{\mu i} && \text{on } S_u, \quad \mu = \mathrm{I}, \mathrm{II}, \end{aligned} \tag{3.13}$$

in which $\sigma^f_{\mathrm{I}ij}$ and $\sigma^f_{\mathrm{II}ij}$ and the corresponding displacements $u^f_{\mathrm{I}i}$ and $u^f_{\mathrm{II}i}$ are, respectively, the fundamental Mode-I and Mode-II fields defined by Bueckner (1972). The fundamental stress fields are given by

$$\sigma^f_{\mathrm{I}ij} = \frac{1}{4r\sqrt{2\pi r}} \begin{cases} 2\cos(3\theta/2) - 3\sin\theta\sin(5\theta/2), & i = j = 1, \\ 2\cos(3\theta/2) + 3\sin\theta\sin(5\theta/2), & i = j = 2, \\ 3\sin\theta\cos(5\theta/2), & i \neq j, \end{cases} \tag{3.14}$$

$$\sigma^f_{\mathrm{II}ij} = \frac{1}{4r\sqrt{2\pi r}} \begin{cases} -4\sin(3\theta/2) - 3\sin\theta\cos(5\theta/2), & i = j = 1, \\ 3\sin\theta\cos(5\theta/2), & i = j = 2, \\ 2\cos(3\theta/2) - 3\sin\theta\sin(5\theta/2), & i \neq j, \end{cases} \tag{3.15}$$

where the polar coordinate system (r, θ) is taken as

$$x_1 = r\cos\theta, \qquad x_2 = r\sin\theta. \tag{3.16}$$

Following Bilby and Cardew (1975), the elastic energy release rate, G, due to a branched crack extension, can be calculated as

$$G = \frac{1 - \nu}{2\mu}(K_{\mathrm{I}}^2 + K_{\mathrm{II}}^2), \tag{3.17}$$

in which μ and ν are shear modulus and Poisson's ratio, respectively, and K_{I} and K_{II} are the stress intensity factors at the extended crack tip. Substitution of (3.7) and (3.8) into (3.17) leads to the expression of G in an ascending order of square root of h, and is given by

$$G = G_0(k_{\mathrm{I}}, k_{\mathrm{II}}; \alpha) + G_{1/2}(k_{\mathrm{I}}, k_{\mathrm{II}}, T; \alpha, \beta)h^{1/2} + G_1(k_{\mathrm{I}}, k_{\mathrm{II}}, T, b_{\mathrm{I}}, b_{\mathrm{II}}, \bar{k}_{ij}; \alpha, \beta, \gamma)h + O(h^{3/2}), \tag{3.18}$$

in which the lowest-order term is expressed as

$$G_0(k_{\mathrm{I}}, k_{\mathrm{II}}; \alpha) = \frac{1-\nu}{2\mu}\left[\left(1 - \frac{\alpha^2}{2}\right)k_{\mathrm{I}}^2 - 2\alpha k_{\mathrm{I}} k_{\mathrm{II}} + k_{\mathrm{II}}^2\right], \tag{3.19}$$

and $G_{1/2}$ and G_1 are the coefficients corresponding to the higher-order terms.

3.2. Energy Calculation and Crack Paths of a Single Load Parameter System Based on the Perturbation Solution

We shall consider the case where the elastic potential energy is represented by a single load parameter which is proportional to the stress and strain for a fixed crack geometry. In this case, the elastic potential energy and the corresponding energy release rate due to crack extension can be expressed as

$$U[h, p; \lambda(x_1)] = p^2 U^*[h; \lambda(x_1)], \tag{3.20}$$

$$G[h, p; \lambda(x_1)] = -U_{,h} = p^2[G_0^* + G_{1/2}^* h^{1/2} + G_1^* h + O(h^{3/2})], \tag{3.21}$$

in which G_0^*, $G_{1/2}^*$, and G_1^* are defined as

$$G_0^* = G_0(k_{\mathrm{I}}^*, k_{\mathrm{II}}^*; \alpha), \tag{3.22}$$

$$G_{1/2}^* = G_{1/2}(k_{\mathrm{I}}^*, k_{\mathrm{II}}^*, T^*; \alpha, \beta), \tag{3.23}$$

$$G_1^* = G_1(k_{\mathrm{I}}^*, k_{\mathrm{II}}^*, T^*, b_{\mathrm{I}}^*, b_{\mathrm{II}}^*, \bar{k}_{ij}; \alpha, \beta, \gamma). \tag{3.24}$$

In the previous equations superscript "$*$" represents the values of the corresponding quantities at the unit load parameter, $p = 1$.

Substitution of (3.20) into (2.4) leads to the load parameter of the equilibrium crack, which is given by

$$p = \left\{\frac{-S_{,h}[h; \lambda(x_1)]}{U_{,h}^*[h; \lambda(x_1)]}\right\}^{1/2}. \tag{3.25}$$

From (3.21) we have

$$U_{,h}^* = -G_0^* - G_{1/2}^* h^{1/2} - G_1^* h + O(h^{3/2}). \tag{3.26}$$

If we assume that the surface energy can be expanded in Taylor's series in terms of the crack extensional length, h, the derivative of the surface energy can be approximated by

$$S_{,h} = 2\gamma[0; \lambda(x_1)] + 2\gamma_{,h}[0; \lambda(x_1)]h + O(h^2). \tag{3.27}$$

Substitution of (3.26) and (3.27) into (3.25) leads to the following explicit representation of the equilibrium load parameter:

$$p^{\mathrm{E}}[h; \lambda(x_1)] = p_0 + p_{1/2} h^{1/2} + p_1 h + O(h^{3/2}), \tag{3.28}$$

in which

$$p_0 = \left\{\frac{2\gamma[0;\lambda]}{G_0^*}\right\}^{1/2}, \tag{3.29}$$

$$p_{1/2} = -\tfrac{1}{2}\left\{\frac{2\gamma[0;\lambda]}{G_0^*}\right\}^{1/2}\frac{G_{1/2}^*}{G_0^*}, \tag{3.30}$$

$$p_1 = \left\{\frac{2\gamma[0;\lambda]}{G_0^*}\right\}^{1/2}\left\{\frac{3}{8}\left(\frac{G_{1/2}^*}{G_0^*}\right)^2 - \frac{1}{2}\left(\frac{G_1^*}{G_0^*}\right) + \frac{\frac{1}{2}\gamma_{,h}[0;\lambda]}{\gamma[0;\lambda]}\right\}. \tag{3.31}$$

The total potential energy in an equilibrium state, Π^E, is also expanded in terms of $h^{1/2}$, and its leading terms are calculated as

$$\begin{aligned}\Pi^E[h, p^E(h;\lambda);\lambda] &= p_0^2 U_0^* + S_0 + O(h^{1/2})\\ &= \left\{\frac{2\gamma[0;\lambda]}{G_0^*}\right\}U_0^* + S_0 + O(h^{1/2}).\end{aligned} \tag{3.32}$$

In (3.32), U_0^* and S_0 are the elastic potential energy and the surface energy prior to the crack extension, and hence they are independent on the crack path, $\lambda(x_1)$. Using (3.19), (3.22), and (3.29), p_0 can be explicitly expressed in terms of the shape parameter, α, of the crack path.

Assuming that

$$\gamma[0;\lambda] = \gamma_0 = \text{const.}, \tag{3.33}$$

for a material with homogeneous fracture toughness, we observe that the minimization of (3.32) is equivalent to the condition which maximizes the elastic energy release rate, G_0, at the instant of fracture. Using (3.22) with (3.19), and taking the first and second variations of (3.32), we have

$$\delta_\lambda \Pi^E = 2\gamma_0 U_0^* \delta_\alpha\left(\frac{1}{G_0^*}\right) = 0, \tag{3.34}$$

$$\delta_\lambda^2 \Pi^E = 2\gamma_0 U_0^* \delta_\alpha^2\left(\frac{1}{G_0^*}\right) > 0. \tag{3.35}$$

Equation (3.34) leads to the branch angle α given by

$$\alpha = -\frac{2k_{II}}{k_I}. \tag{3.36}$$

The above condition is equivalent to the one predicted by the local symmetry criterion, which is discussed in detail by Cotterell and Rice (1980), Hayashi and Nemat-Nasser (1981), and Sumi *et al.* (1983).

4. Crack Paths in Materials with Inhomogeneous Fracture Toughness

*4.1. Mode-*I *Crack Intersecting a Degradation Line*

Let us consider a crack under pure Mode-I loading condition ($k_{II} = 0$), whose tip intersects a specially oriented line degradation zone at angle ω as illustrated

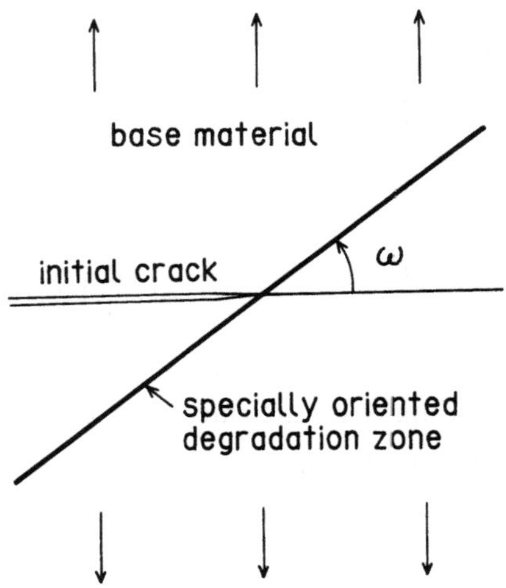

FIG. 3. A Mode-I crack intersecting a specially oriented degradation zone.

in Fig. 3, and where the critical energy release rates for base and degraded materials are G_c and $G_{c\omega}$, respectively. In this case we have

$$\gamma[0; \lambda] = \begin{cases} G_{c\omega}/2, & \alpha = \omega, \\ G_c/2, & \alpha \neq \omega. \end{cases} \tag{4.1}$$

As long as the crack extends into the base material the minimization of the total potential energy is attained at $\alpha = 0$ by using (3.36).

Substitution of (4.1) into (3.32) leads to the corresponding total potential energy, which is given by

$$\Pi^E[0; p^E(0; \alpha = 0); \alpha = 0] = \left\{ \frac{G_c}{G_0^*(k_I^*, k_{II}^* = 0; \alpha = 0)} \right\} U_0^* + S_0. \tag{4.2}$$

On the contrary, if the crack extends into the degradation zone with the branch angle ω, the total potential energy is calculated as

$$\Pi^E[0; p^E(0; \alpha = \omega); \alpha = \omega] = \left\{ \frac{G_{c\omega}}{G_0^*(k_I^*, k_{II}^* = 0; \alpha = \omega)} \right\} U_0^* + S_0. \tag{4.3}$$

Comparing (4.2) and (4.3) the branched crack extension, for which the total potential energy is minimized, occurs under the following condition:

$$\frac{G_{c\omega}}{G_0^*(k_I^*, k_{II}^* = 0; \alpha = \omega)} < \frac{G_c}{G_0^*(k_I^*, k_{II}^* = 0; \alpha = 0)}. \tag{4.4}$$

Substituting (3.19) into inequality (4.4), we have

$$G_{c\omega} < \left(1 - \frac{\omega^2}{2}\right) G_c \tag{4.5}$$

for the branched crack extension along the degradation zone, where the expression is applicable within the range $|\omega| < 0.5$.

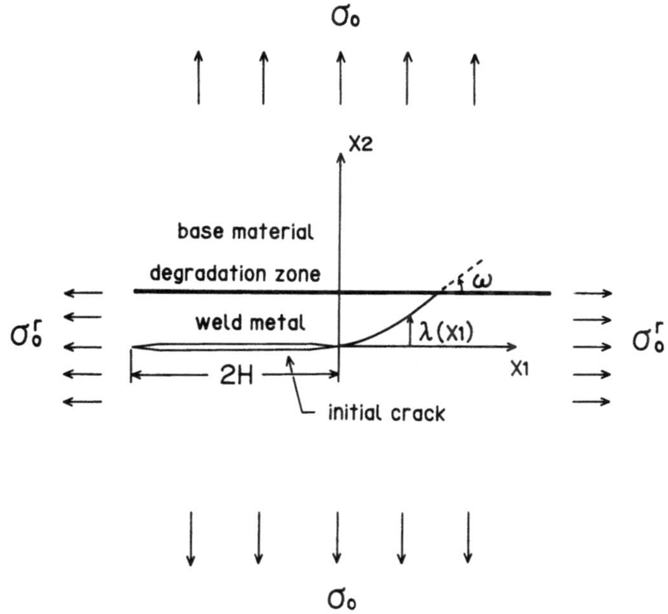

FIG. 4. A brittle crack propagating along the welded joint under the influence of applied stress σ_0, welding residual stress σ_0^r, and material deterioration along the heat affected zone.

4.2. Crack Path Prediction for Brittle Cracks Propagating Along a Welded Joint

Since welded joints are the critical parts of structures in relation to the initiation and propagation of brittle fracture, the prediction of the fracture behavior plays an essential role in the evaluation of the strength of welded joints. In this subsection we consider a transverse welded joint in an infinite plate, in which a uniform stress σ_0 is applied normal to the joint. As is shown in Fig. 4, the initial crack is assumed to be parallel to the welded joint. Material deterioration is observed along the so-called heat affected zone, which is also parallel to the welded joint with the distance L_s from the initial crack line. The longitudinal component of the welding residual stress σ_0^r which is acting parallel to the welded joint is assumed to be constant in the region under consideration.

The stress field parameters are obtained as

$$k_{\mathrm{I}} = \sigma_0\sqrt{\pi H}, \qquad T = -\sigma_0 + \sigma_0^r, \qquad b_{\mathrm{I}} = \tfrac{3}{4}\sigma_0\sqrt{\frac{\pi}{H}},$$

$$k_{\mathrm{II}} = b_{\mathrm{II}} = 0,$$

$$\bar{k}_{11} = \bar{k}_{22} = -\frac{1}{8H}, \qquad \bar{k}_{12} = \bar{k}_{21} = 0. \tag{4.6}$$

Since unstable curved crack paths are sometimes observed in the present problem, we approximate the branched and curved crack path by (3.3) with a small initial kink angle α as an imperfection parameter of the system. As long as a crack extends into a field of homogeneous fracture toughness, its path can be predicted by the local symmetry criterion, which is equivalent to the condition $K_{II} = 0$ along the curved trajectory. After substituting (4.6) into (3.10) and (3.12), we obtain the expression of the shear mode of stress intensity factor K_{II} by using (3.8). Putting (3.8) to be identically zero, the shape parameters of the crack are determined as

$$\beta = \frac{8}{3\pi}\sqrt{\frac{2}{H}}\left(\frac{\sigma_0^r}{\sigma_0} - 1\right)\alpha,$$

$$\gamma = \left\{\pi + 32\left(\frac{\sigma_0^r}{\sigma_0} - 1\right)^2\right\}\frac{\alpha}{8\pi H}. \tag{4.7}$$

In (4.7) it is observed that if the residual stress, σ_0^r, is large in comparison with the applied tensile stress, σ_0, the crack may be sharply curved due to the introduction of small imperfections.

When the crack further extends and begins to intersect the degradation zone, the question arises as to whether it intersects the zone and penetrates into the base metal with standard fracture toughness, or whether it branches to the heat affected zone, where the fracture toughness is lower than the base metal. In order to answer this question, we calculate the angle, ω, formed by the crack and the degradation zone by using (3.3) and (4.7). The condition of branched crack extension along the degradation zone is obtained in the

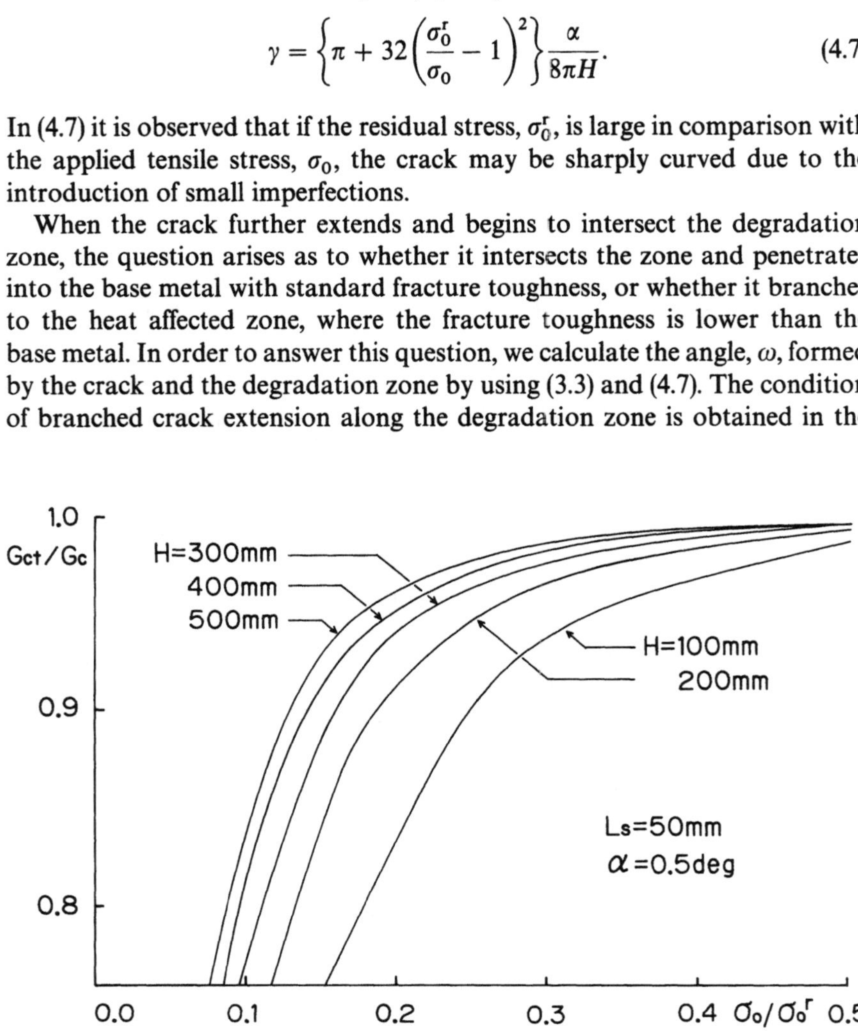

FIG. 5. Predicted crack growth behavior, taking account of the effects of material degradation and the ratio of applied stress and residual stress, where the lower right-hand sides of the respective curves represent the condition of the crack extension along the degradation zone.

following form, which is similar to inequality (4.5):

$$G_{ct} < \left(1 - \frac{\omega^2}{2}\right) G_c, \tag{4.8}$$

where G_c and G_{ct}, respectively, correspond to the critical energy release rates for the base metal and the heat affected zone. Using (4.7) and (4.8), we can examine the effects of the applied stress, the residual stress, and the fracture toughness of the materials on the crack growth behavior. Numerical examples are shown in Fig. 5, where the distance, *Ls*, between the initial crack line and the degradation zone is 50 mm and the half crack length H is selected as 100, 200, 300, 400, 500 mm with $\alpha = 0.5°$. This figure illustrates that if stress states and material properties fall on the lower right-hand side of the respective curves, a brittle crack may propagate along the degradation zone. From these results we can understand that if brittle fracture is initiated at a very low applied stress level, the crack may be sharply curved and then penetrate into the base metal, which could lead to crack arrest in the base metal (Sumi, 1989b).

References

Bilby, B. A. and Cardew, G. E. (1975), The crack with a kinked tip, *Int. J. Fracture*, **11**, 708–712.

Bueckner, H. F. (1972), Field singularities and related integral representation, in *Mechanics of Fracture*, Vol. 1, Noordhoff, Leyden, pp. 239–314.

Cotterell, B. and Rice, J. R. (1980), Slightly curved or kinked cracks, *Int. J. Fracture*, **16**, 155–169.

Griffith, A. A. (1920), Phenomena of flow and rupture in solids, *Phil. Trans. Roy. Soc. London*, **221A**, 163–198.

Griffith, A. A. (1924), Theory of Rupture, *Proceedings of the First International Congress of Applied Mechanics, Delft*, pp. 55–63.

Hayashi, K. and Nemat-Nasser, S. (1981), Energy release rate and crack kinking, *Int. J. Solids Structures*, **17**, 107–114.

Nemat-Nasser, S. (1980), Variational methods for analysis of stability of interacting cracks, in *Variational Methods in the Mechanics of Solids*, Pergamon, Oxford, pp. 249–253.

NGuyen, Q. S. (1987), Bifurcation and post-bifurcation analysis in plasticity and brittle fracture, *J. Mech. Phys. Solids*, **35**, 303–324.

Potier-Ferry, M. (1985), Towards a catastrophe theory for the mechanics of plasticity and fracture, *Int. J. Engng. Sci.*, **23**, 821–837.

Sumi, Y., Nemat-Nasser, S., and Keer, L. M. (1980), On stability and postcritical behavior of interactive tension cracks in brittle solids, *ZAMP*, **31**, 673–690.

Sumi, Y., Nemat-Nasser, S., and Keer, L. M. (1983), On crack branching and curving in a finite body, *Int. J. Fracture*, **21**, 67–79; Erratum, *Int. J. Fracture*, **24**, (1984), 159.

Sumi, Y (1989a), A second-order perturbation solution of a slightly branched and curved crack and its application to crack path prediction, *J. Soc. Naval Arch. Japan*, **165**, 245–251 (in Japanese).

Sumi, Y. (1989b), Computational crack path prediction for brittle fracture in welding residual stress fields, *Int. J. Fracture*, (in press).

Elastic/Plastic Indentation Hardness of Ceramics: The Dislocation Punching Model

K. Tanaka and H. Koguchi
Department of Mechanical Engineering, Nagaoka University of Technology, Tomiokacho 1603-1, Nagaoka 940-21, Japan

Abstract

A new model for analyzing indentation plasticity, proposed by Tanaka, Koguchi, and Mura, is discussed in relation to the plasticity in ceramics. The model is based on the punching of prismatic dislocation loops to accommodate the volume of plastic penetration at the indentation. The theory predicts that the hardness/Young's modulus ratio is almost constant irrespective of the yield strength. The predicted profiles at the unloaded state in Si_3N_4 and ZrO_2 ceramics were closely coincident with the indentation profiles showing the amount of spring back-up to 30–40% as revealed by the topographic analysis on scanning electron microscope (SEM) photographs of these materials.

1. Introduction

Micromechanics encompasses mechanics related to microstructures of materials. Mura (1982) developed a powerful and unified method to treat the problems in this field, which is called the "eigenstrain method." In particular, problems relating to inclusions and dislocations are most effectively analyzed by this method. The elastic/plastic indentation is seemingly one of the best problems for the application of the eigenstrain method. The indented plastic zone in very brittle materials is completely constrained by the surrounding elastic matrix and originates an eigenstress field.

It has been commonly accepted that the hardness test is a simple and effective means for determining the resistance of plastic flow in a material. A convenient measure of the material hardness, H_v, is taken as the indentation pressure obtained by dividing the applied peak load, P, by the area of the residual surface impression. That is, for pyramidal indentations,

$$H_v = \frac{P \sin \phi}{2c_0^2}, \tag{1.1}$$

where c_0 is half of the indentation diagonal and ϕ is half of the indenter angle (68°). For ductile metals, the Vickers hardness thus defined is about three times the yield strength, Y, in the wide range of material with yield strength/Young's ratios, Y/E, from 10^{-3} to 10^{-2} (Tanaka *et al.*, 1989). It is clear that hardness for ductile metals is an indication of the irreversible plastic deformation processes. For very brittle materials, however, the indentation pressures become much lower than would be expected from the relation for ductile metals (Marsh, 1964). This may result from the reversible deformation process or when the elastic deformation is of comparable magnitude with the plastic deformation process. The elastic/plastic behavior is mostly characterized by the impression geometries. Whereas characteristic in-surface dimensions generally remain a reasonably faithful measure of those at maximum loading, the depth does not because of the elastic recovery during unloading of the indenter (Lawn and Howes, 1981). The understanding of the mechanisms on the elastic/plastic hardness indentation is of great interest, since it not only provides insight into the nature of the plastic deformation in typically brittle polycrystallines, e.g., ceramics, but also provides the evaluation of residual stress causing attendant microcrack generation (Lawn *et al.*, 1980).

Marsh (1964) stated that the indentation process is analogous to the expansion of a spherical cavity in an elastic–plastic material (expansion cavity model). Johnson (1970) extended Marsh's suggestion by allowing the indentation pressure, p, to be transmitted via an incompressible hydrostatic core beneath the indenter replacing the cavity in Hill's theory (1950). He derived the equation

$$\frac{p}{Y} = \frac{H_v}{Y \sin \phi} = \frac{2}{3}\left[1 + \ln\left(\frac{E \cot \phi}{3Y}\right)\right]. \tag{1.2}$$

However, the expansion cavity model does not satisfy the boundary conditions at the surface. It assumes the indented geometry as a spherical cavity, and the Hill solution is applicable for an infinite body.

Recently, the problem has been solved exactly by using the continuum dislocation theory (Tanaka *et al.*, 1989; Mura *et al.*, 1989), on the assumption that the indented volume is accommodated by the punching of infinitesimal prismatic loops (dislocation punching model). This satisfies the geometrical requirement at the indented surface and includes the effect of half-space. The objective of the present study is to confirm the theoretical predictions based on the dislocation punching model by comparing them with the experimental observations. The impression geometries of two typical ceramics were observed by using a scanning electron microscope designed for the quantative measurements of surface microtopography.

2. Experimental

The ceramics observed were hot-pressed Si_3N_4 (8 mol % Y_2O_3–1 mol % ZrO_2–1 mol % Al_2O_3) and pressureless sintered ZrO_2 (3 wt % Y_2O_3). Their Young's moduli and Poisson ratios ν are shown in Table 1. The hardness test was conducted by using a conventional pyramidal indenter and a specially-

TABLE 1. Mechanical properties of Si_3N_4 and ZrO_2 and comparison of the predictions for H_v/E and Y/E with experimental results.

	Young's modulus E (GPa)	Poisson's ratio ν	Vickers hardness H_v (GPa)	H_v/E		Y/E (Estimation)	
				Experiment	Estimation	(3.13)	Fig. 5
Si_3N_4	320	0.25	15.6	0.049	0.050	0.042	0.035
ZrO_2	206	0.25	10.8	0.052	0.045 (0.051)	0.018	0.023

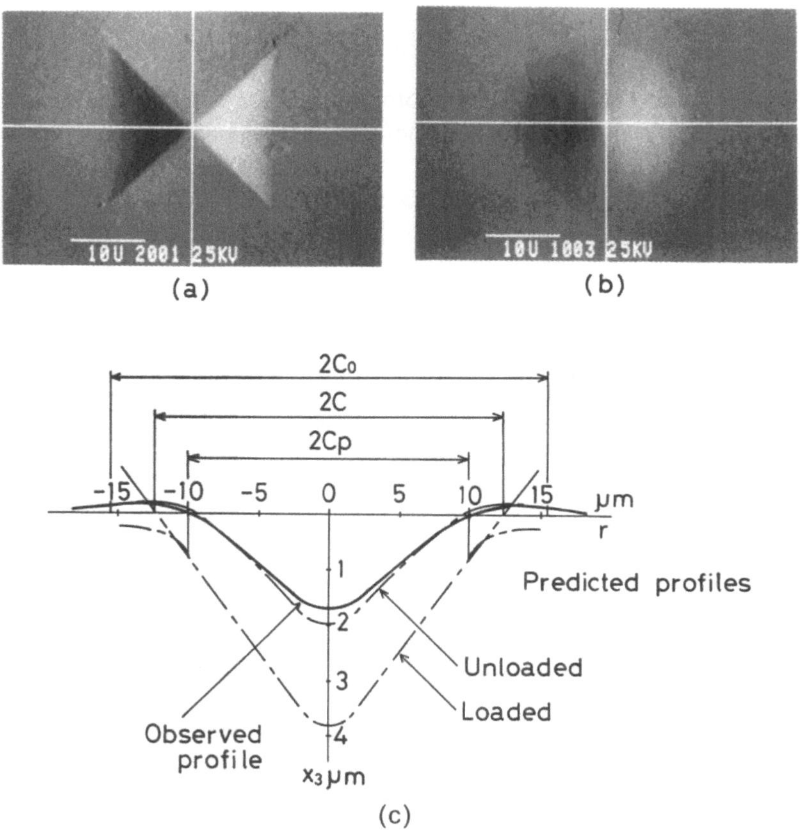

FIG. 1. SEM micrographs of (a) pyramidal and (b) conical indentations in Si_3N_4, and (c) an image diagram of the conical indentation showing cross section topography and the predicted profiles at the loaded and unloaded states.

made conical one. The hardness values of pyramidal indentations defined in (1.1) for the two ceramics are shown in Table 1. The conical indenter has the same shape as the analytical model expressed in the following secttion. Namely, it has the same volume as the Vickers pyramids; both the depths at the center and the areas at the indented surface are equalized. Hence the depth h_0, the radius c of the cone, and the hardness are described as

$$h_0 = \left(\frac{c_0}{2^{1/2}}\right) \cot \phi,$$

$$c = \left(\frac{2}{\pi}\right)^{1/2} c_0 = 0.80c_0, \tag{2.1}$$

$$H_v = \frac{P \sin \phi}{\pi c^2}.$$

Figures 1 and 2 show the SEM photographs of the indentations impressed by the load application of 9.8N with the (a) pyramidal and (b) conical indenters

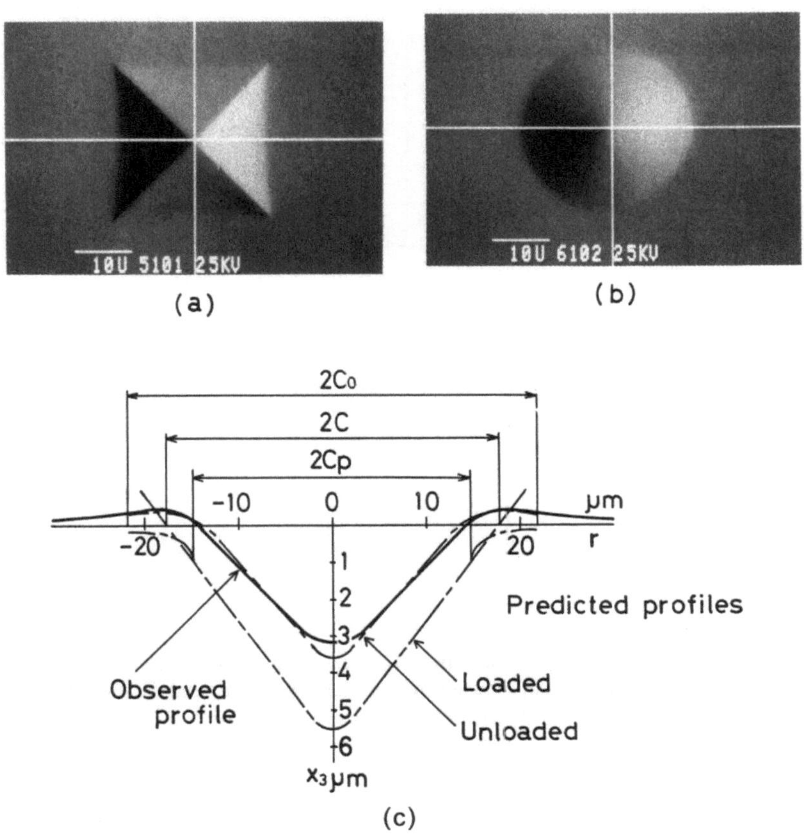

FIG. 2. SEM micrographs in ZrO_2. The illustration is the same as in Fig. 1.

for Si_3N_4 and ZrO_2, respectively. These were observed by a SEM (ELIONIX ESA-3000). The impression topography can be profiled by monitoring the difference in the signals from two different detectors at opposite positions oblique to the incident electron beam. The diagrams (c) represent the projection images of impression topography thus obtained as the cross sections along the horizontal line in the conical indentations. The abscissa of the diagrams is the horizontal direction of the photograph, while the ordinate is the depth direction of the photograph, but the latter coordinate is magnified by a factor of four against the former in order to emphasize the change in the profile of the depth direction.

3. Indentation Analysis

Figure 3 illustrates the model. We consider the case of the conical indentation, and describe the indentation displacement as a trace of the dislocation motion punched from the surface into the matrix. The prismatic dislocation loops of circular shape produce irreversible displacement, u_3^p, to make the conical indentation. Using a cylidrical coordinate, we express the irreversible indentation profile, $u_3^p(r, 0)$, as

$$\begin{aligned} u_3^p(r, 0) &= h_p\left(1 - \frac{r}{c_p}\right) \qquad \text{for} \quad r \leqq c_p, \\ &= 0 \qquad\qquad\qquad\;\; \text{for} \quad r > c_p, \end{aligned} \tag{3.1}$$

with

$$\frac{h_p}{c_p} = \frac{h_0}{c} = \left(\frac{\pi^{1/2}}{2}\right) \cot \phi. \tag{3.2}$$

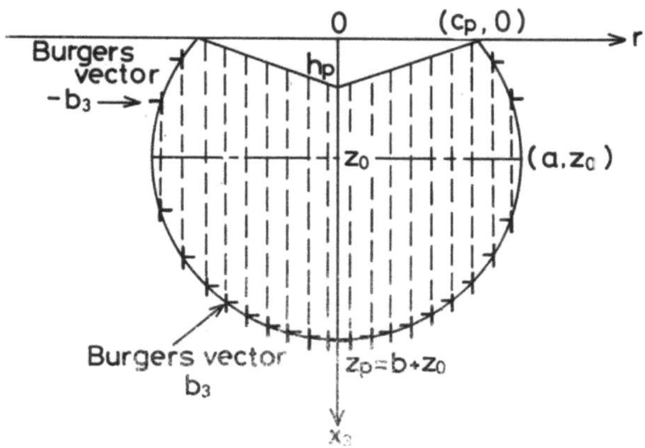

FIG. 3. Schematic illustration showing distribution of prismatic dislocation loops in an indentation plastic zone.

For the satisfication of the geometrical requirement and the simplicity of the calculation, we assume that the distribution profile of the punched dislocations or the plastic zone is semiellipsoidal with a form

$$\frac{r^2}{a^2} + \frac{(x_3 - z_0)^2}{b^2} = 1, \qquad x_3 > 0, \tag{3.3}$$

where the constants a, b, and z_0 are geometrical variables determined by the loading conditions on the contact planes and the material properties. The Burgers vectors of the prismatic dislocation loops are perpendicular to the surface and denoted by $+b_3$ on $x_3 = z_0 + b(1 - r^2/a^2)^{1/2}$ in the region $a \geqq r$, and $-b_3$ on $x_3 = z_0 - b(1 - r^2/a^2)^{1/2}$ in the region $a \geqq r \geqq c_p$. The plastic displacement at 0 is Nb_3, where N is the total number of dislocation loops in the region $r \geqq c_p$. Since $u_3^b = 0$ outside the indented portion ($a \geqq r \geqq c_p$), the dislocations with Burgers vector $-b_3$ are distributed equally to those with Burgers vector $+b_3$. The dislocations are distributed evenly in the region $r \leqq c_p$ to satisfy the condition that the plastic displacement field on the indentation plane should be piecewise linear. We assume that the density of dislocations outside the indented portion is as same as that inside, and that the number of dislocations is $N(a - c_p)/c_p$.

Now we consider the change in the indentation profile during the loading–unloading cycle. The profile at the loaded state $u_3^l(r, 0)$ is expressed by (Fig. 4a)

$$u_3^l(r, 0) = u_3^p(r, 0) + u_3^e(r, 0), \tag{3.4}$$

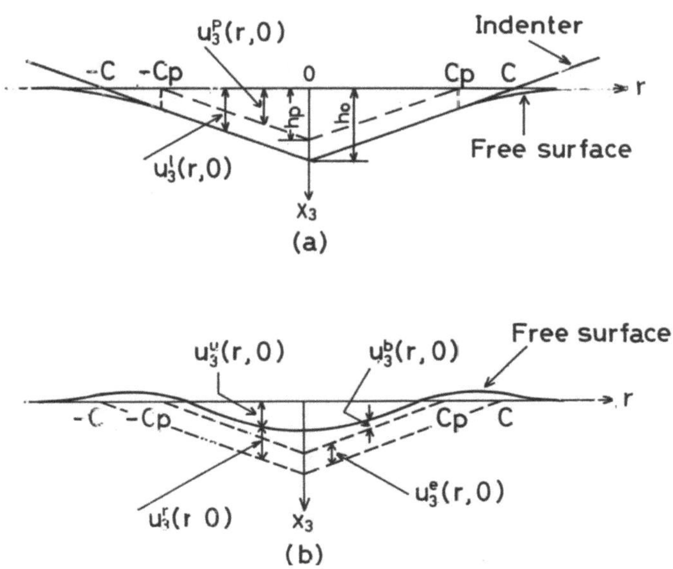

FIG. 4. Schematic illustration showing a cross section of the indentation profiles: (a) loaded state; (b) unloaded state.

where $u_3^{\mathrm{e}}(r, 0)$ is the elastic displacement. We assume that this is constant within the contact perimeter and denoted by Δu_3^{e} as

$$u_3^{\mathrm{e}}(r, 0) = \Delta u_3^{\mathrm{e}} = h_0 - h_p \qquad \text{for} \quad r \leqq c_p. \tag{3.5}$$

The profile at the unloaded state $u_3^{\mathrm{u}}(r, 0)$ is expressed by (Fig. 4b)

$$u_3^{\mathrm{u}}(r, 0) = u_3^{\mathrm{l}}(r, 0) - u_3^{\mathrm{r}}(r, 0), \tag{3.6}$$

where $u_3^{\mathrm{r}}(r, 0)$ is the elastic recovery. This is described by

$$u_3^{\mathrm{r}}(r, 0) = u_3^{\mathrm{b}}(r, 0) + u_3^{\mathrm{e}}(r, 0), \tag{3.7}$$

where $u_3^{\mathrm{b}}(r, 0)$ is the elastic recovery due to the back stress of dislocations.

The back stress is the sum of dislocation stresses of individual circular dislocation loops. Considering infinitesimal loops with the relations $b_3 = h_p/N$ and putting $c_p/N = \mathrm{d}r'$, we can change the summation into an integral form. Then we use the solution of the displacement fields for a circular prismatic loop in half-space by Salamon and Dundurs (1971), and obtain the following integrals of the Bessell functions for the elastic recovery due to the back stresses:

$$\begin{aligned} u_3^{\mathrm{b}}(\bar{r}, 0) = \frac{h_p}{\bar{c}_p} \int_0^1 d\bar{r}' \Bigg\{ & \int_0^\infty J_1(t) J_0\left(\frac{\bar{r}t}{\bar{r}'}\right) \exp - [\bar{z}_0 + \beta(1 - \bar{r}'^2)^{1/2}] \frac{t}{\bar{r}'}\, dt \\ & - \left(\frac{1}{\bar{r}'}[\bar{z}_0 + \beta(1 - \bar{r}'^2)^{1/2}]\right) \int_0^\infty J_1(t) J_0\left(\frac{\bar{r}t}{\bar{r}'}\right) \\ & \times \exp - [\bar{z}_0 + \beta(1 - \bar{r}'^2)^{1/2}] \frac{t}{\bar{r}'}\, dt \Bigg\} \\ - \left(\frac{h_p}{\bar{c}_p}\right) \int_{\bar{c}_p}^1 d\bar{r}' \Bigg\{ & \int_0^\infty J_1(t) J_0\left(\frac{\bar{r}t}{\bar{r}'}\right) \exp - [\bar{z}_0 - \beta(1 - \bar{r}'^2)^{1/2}] \frac{t}{\bar{r}'}\, dt \\ & - \left(\frac{1}{\bar{r}'}\right)[\bar{z}_0 - \beta(1 - \bar{r}'^2)^{1/2}] \int_0^\infty J_1(t) J_0\left(\frac{\bar{r}t}{\bar{r}'}\right) \\ & \times \exp - [\bar{z}_0 - \beta(1 - \bar{r}'^2)^{1/2}] \frac{t}{\bar{r}'}\, dt \Bigg\}, \end{aligned} \tag{3.8}$$

where $\bar{r} = r/a$ and $\bar{r}' = r'/a$. We can solve this equation by using the complete elliptical integrals (Tanaka *et al.*, 1989).

The elastic displacement Δu_3^{e} is related to the applied load required for the punching of the dislocation loops against the frictional stress of matrix k_0. The corresponding plastic work is irreversible and dissipated to heat. Using the virtual work conception and taking Y equal to $2k_0$, we obtain (Tanaka *et al.*, 1989)

$$\Delta u_3^{\mathrm{e}} = \left[\frac{\pi(1 - \nu)Y}{4\mu}\right]\left[\beta + \bar{z}_0\left(1 + \frac{\bar{c}_p^2}{2}\right)\right]\left(\frac{1}{\bar{c}_p^3}\right) c_p, \tag{3.9}$$

where μ is the shear modulus, $\beta = b/a$, $\bar{z}_0 = z_0/a$, and $\bar{c}_p = c_p/a$.

The total load P is the sum of the two loads relevant to the back stress and the irreversible frictional stress of dislocations, and is given by

$$P = 2\pi a^2 \int_0^{\bar{c}_p} p(\bar{r}')\bar{r}'\, d\bar{r}', \tag{3.10}$$

where $p(r)$ is a distribution of contact pressure. This is determined to satisfy the conditions that the contact plane of the indenter must keep flat during loading, that is, there is no gap between the indenter and the material surface, and that the traction on the contact plane must be compressive. Namely, we must solve an integral equation,

$$\int_0^{\bar{c}_p} \int_0^{\infty} p(\bar{r}')J_0(t\bar{r})J_0(t\bar{r}')\, dt\bar{r}'\, d\bar{r}' = \frac{[\mu/(1-\nu)][u_3^b(\bar{r},0) + \Delta u_3^e]}{c_p} \tag{3.11}$$

under the conditions that

$$\begin{aligned} p(\bar{r}') &> 0 \quad \text{for} \quad \bar{r}' \leqq \bar{c}_p, \\ p(\bar{r}') &= 0 \quad \text{for} \quad \bar{r}' > \bar{c}_p. \end{aligned} \tag{3.12}$$

4. Analytical Results and Discussions

The analytical calculation for hardness was done on the assumption that $\nu = 0.25$. The scheme of the analysis is as follows. For given values of β and $\bar{z}_0$ we can determine $\bar{c}_p$ from (3.3) by equating x_3 to 0 and r to c_p. Using the geometrical parameters, we can calculate $u_3^r(r, 0)$ from (3.11). Substituting the values into (3.11), we can obtain $p(\bar{r}')/\mu$ by solving (3.11) numerically as a function of $\Delta u_3^e/h_p$ to satisfy the condition of (3.12). In correspondence with the value of $\Delta u_3^e/h_p$, $\bar{c}$ is decided by using (3.2) and (3.5) and Y/E is determined from (3.9) since $E = 2\mu(1 + \nu)$. Finally, integrating (3.10) with $p(\bar{r}')/\mu$ thus obtained, we find $P/\pi a^2 E$ $(= P\bar{c}^2/\pi c^2 E)$ as a function of $\Delta u_3^e/h_p$ or Y/E. This yields H_v/Y from (2.1).

Figure 5 shows the analytical values of H_v/Y against E/Y in a semi-logarithmic plot as a function of β for the case that $\phi = 68°$. It is probable that the hardness value for the conical indentation is almost equivalent to that for the pyramidal one, since both the indentation volumes are identical through the relation in (2.1). Hence this may compare with the experiments obtained by the conventional indentation method. Although there are a small number of available data on the values of E/Y for ceramics, these indicate that the values range from 20 through 50 (Tanaka *et al.*, 1989). The theoretical values of H_v/Y in the appropriate range do not vary extensively with the aspect ratio of the plastic zone β.

The dotted curves in Fig. 5 show equal H_v/E value contours. It is predicted that the ratio of the hardness to the modulus should be between 0.04 and 0.06 in the range of E/Y values for ceramics. This also agrees reasonably well with

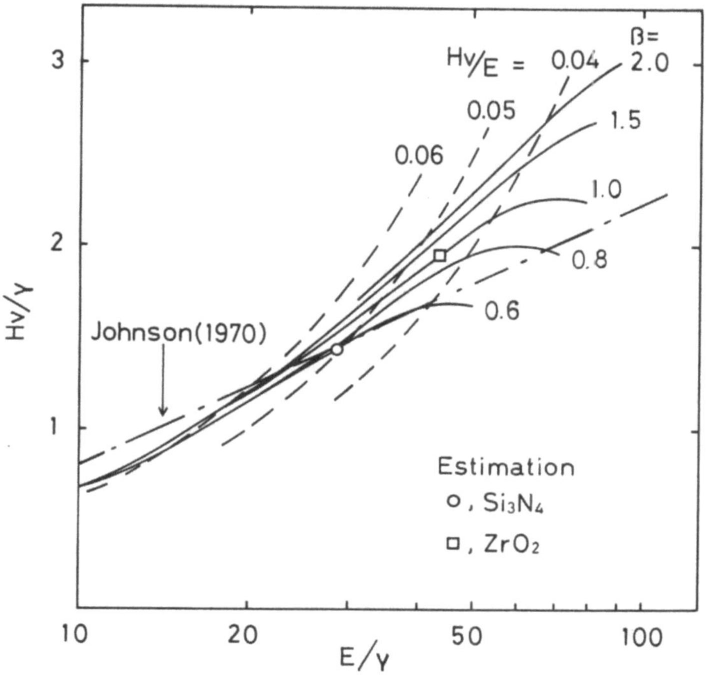

FIG. 5. Predicted hardness/yield strength ratio against the Young's modulus/yield strength ratio for ceramics as a function of β.

the experimental data on ceramics excluding glasses (Lawn and Howes, 1981; Tanaka *et al.*, 1989).

The most widely accepted analysis of indentation plasticity is that devised by Johnson (1970). Although the concept has some restrictions that reside in both the choice of the spherical cavity model and in the neglect of the free surface, Johnson's analysis in (1.2) (as shown in Fig. 5) is fairly close to the curves of the present analysis. However, the slope of his relation in the semilogarithmic plot is considerably lower, and the discrepancy between the two models becomes larger as E/Y becomes higher.

Finally, the analytical results are compared with the experimental observations. In Figs. 1(c) and 2(c) the predicted indentation profiles at both loaded and unloaded states for the two ceramics are represented by the dashed–dotted lines, respectively. These were constructed by the trial and error method to give a best fitting with the observed unloaded profiles. The estimated geometrical parameters are $\beta = 0.6$ and $\bar{z}_0 = 0.325$ for Si_3N_4, and $\beta = 1.0$ and $\bar{z}_0 = 0.42$ for ZrO_2. In the drawing of the predicted profiles the radial shape of 6 μm in radius, at the tip of the conical indenter used in the present study, is taken into account. The values of c_0 measured from the photographs of the pyramidal indenters for the two ceramics are indicated on the respective diagrams of the cone indenters. The predicted radius, c, for Si_3N_4 at the loaded state is coincident with $0.8c_0$ as expressed in (2.1), whereas for ZrO_2 it is slightly larger than $0.8c_0$ and nearly equal to $0.85c_0$.

The relations between H_v/Y and E/Y are estimated in correspondence with the best fitting profiles, and the results are exhibited in Fig. 5. The resulting H_v/E and Y/E values are also described in Table 1. The estimated values of H_v/E for the two ceramics agree with the experimental values. Particularly, the agreement for Si_3N_4 is almost perfect. For ZrO_2, by appreciating c equal to the observed value $0.85c_0$, we have a much improved H_v/E value as shown in parentheses.

The plasticity of ceramics at relatively low temperatures is limited by the Peierls force or lattice resistance. According to Frost and Ashby (1982), the rate equation for plasticity, limited by a lattice resistance, is given by

$$\dot{\gamma} = \dot{\gamma}_p \left(\frac{\tau_p}{\mu}\right)^2 \exp\left\{-\left(\frac{\Delta F}{kT}\right)\left[1 - \left(\frac{\tau_p}{\tau_0}\right)^{3/4}\right]^{4/3}\right\}, \tag{3.13}$$

where k is the Boltzmann constant, $\dot{\gamma}$ is the shear strain rate, $\dot{\gamma}_p$ is the pre-exponential parameter ($= 10^{11}\ s^{-1}$), τ_p is the shear stress, τ_0 is the flow strength at 0 K, and ΔF_k is the activation energy for lattice resistance. The latter two material parameters are roughly determined by the type of atomic bonding (Frost and Aahby, 1982); $\tau_0/\mu = 0.07$, $\Delta F_p/kT_m = 35$ for covalent bonding ceramics and $\tau_0/\mu = 0.04$, $\Delta F_p/kT_m = 10$ for oxides, where T_m is the melting temperature.

The normalized flow stresses or yield strengths Y/E for the two ceramics at room temperature were estimated from (3.13) by equating Y to $2\tau_p$ and $\dot{\gamma}$ to $10^{-3}\ s^{-1}$, and taking into account the atomic bonding types and the material parameters. The results are described in Table 1. These agree reasonably well with the values of Y/E predicted from the indentation analysis. This suggests that the hardness measurement is also a simple means for estimating the flow stress of very brittle materials.

5. Conclusions

Indentation plasticity has been analyzed on the model that the volume of plastic penetration at the indentation is accommodated by the punching of prismatic dislocation loops into the matrix and that the resulting plastic zone is semiellipsoidal. The solution is obtained by Salamon and Dumdurs (1971) on elastic fields of a dislocation loop in a two-phase material, and thus, it includes exactly the effect of the free surface. The analysis predicts that the hardness/Young's modulus ratio for ceramics is almost constant and nearly equal to 0.05. The H_v/E values of Si_3N_4 and ZrO_2 ceramics, estimated from the observed indentation profiles at the unloaded state using the model, agree closely with the experimental values.

Acknowledgments

This work was supported by a Grant-in-Aid for Science Research (No. 63460201) from the Ministry of Education, Science and Culture, Japan. The SEM observations were done by Mr. Y. Hatori at the Nagaoka University of Technology.

References

Frost, H. J. and Ashby, M. F. (1982), *Deformation Mechanism Maps*, Pergamon, Oxford.

Hill, R. (1950), *The Mathematical Theory of Plasticity*, Clarendon, Oxford.

Johnson, K. L. (1970), The correlation of indentation experiments, *J. Mech. Phys. Solids*, **18**, 429–444.

Lawn, B. R., Evans, G. A., and Marshall, D. B. (1980), Elastic/plastic indentation damage in ceramics: The median/radial crack systems, *J. Amer. Ceram. Soc.*, **63**, 574–581.

Lawn, B. R. and Howes, V. R. (1981), Elastic recovery at hardness indentations, *J. Mater. Sci.*, **16**, 2745–2752.

Marsh, D. M. (1964), Plastic flow in glass, *Proc. Roy. Soc. London*, **A270**, 420–435.

Mura, T. (1982), *Micromechanics of Defects in Solids*, Martinus Nijhoff, The Hague.

Mura, T., Yamashita, N., Mishima, T., and Hirose, Y. (1989), A dislocation model for hardness indentation problems—I, *Int. J. Engng. Sci.*, **27**, 1–10.

Salamon, N. J. and Dundurs, J. (1971), Elastic fields of a dislocation loop in two-phase material, *J. Elasticity*, **1**, 153–164.

Tanaka, K., Koguchi, H., and Mura, T. (1989), A dislocation model for hardness indentation problems—II, *Int. J. Engng. Sci.*, **27**, 11–27.

Some Thoughts on Inhomogeneous Distribution of Fillers in Composites

MINORU TAYA*
Department of Mechanical Engineering, University of Washington,
Seattle, WA 98195, U.S.A.

Abstract

The inhomogeneously distributed fillers in a composite are discussed in view of modeling of the overall properties of the composite. First, the inhomogeneous distribution of fillers is examined in its relation to fractal and percolating clusters. Then, an attempt is made to propose two parameters which can better characterize the morphology of the inhomogeneous distribution of fillers; i.e., the inhomogeneous distribution parameter ξ and the radius of gyration of the sites belonging to the clusters R_s. Finally, the methods of modeling are suggested for several cases of the inhomogeneous distribution of fillers.

1. Introduction

The macroscopic properties of a composite can be determined if the physical properties of the constituents of the composite and the microstructural morphology of fillers are known. Among the latter, several parameters are used as input data for the characterization of the macroscopic properties of the composite: the volume fraction of filler (V_f), the fiber aspect ratio (l/d), and the fiber orientation (θ). Modeling of the macroscopic properties of a composite assumes *a priori* that the sites (locations) of infinite numbers of fillers are random in space, although the orientation of fillers can be prescribed in the case of the filler being of the fiber type. The random distribution of the sites of fillers can be called "(spatially) homogeneous distribution." "The homogeneous distribution" includes the periodic distribution of fillers, which is one of the popular assumptions often used to facilitate the modeling. The vast majority of analytical models (effective medium theory) including the Eshelby method are based on the assumption of the homogeneous distribution of fillers.

There exist, however, cases where the homogeneous distribution of fillers

* Current address: Department of Materials Processing, Faculty of Engineering, Tohoku University, Sendai 980, Japan.

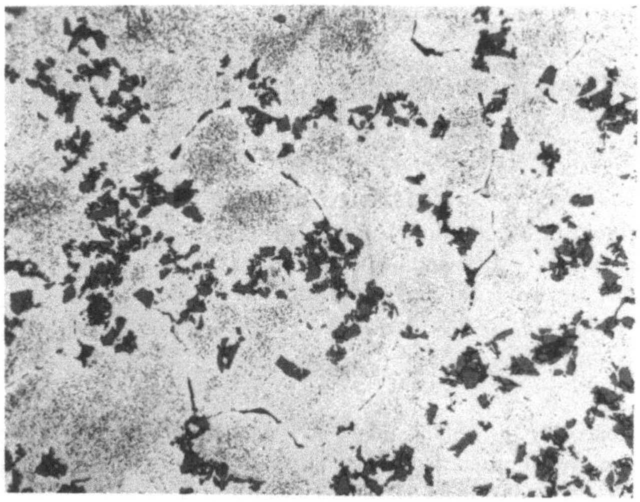

FIG. 1. A SEM photograph showing the micromorphology of SiC particles in 10% V_f Si–C_p/6061 Al composite (Lagacé and Lloyd, 1989).

cannot be realized in a composite. Figure 1 is a scanning electron microscopy (SEM) photo of as-fabricated 10% V_f SiC particulate/6061 aluminum composite where the dark islands denote SiC particulates (Lagacé and Lloyd, 1989). In spite of various sizes of SiC particulates (average size ranges from 10 to 20 μm), the distribution of particulates appears to be inhomogeneous. Moreover, some particulates are clustered to form a "clustered structure," or simply a "cluster." This "inhomogeneous distribution" of fillers may be induced by the processing of the composite where the mixing of fillers and the matrix is not complete. The inhomogeneous distribution of fillers can be looked at in terms of the morphology of "clustered structures" of fillers where a "cluster" is an aggregate of fillers (with higher density than the average value, V_f). Another example of clusters is shown in the SEM photo of a slightly etched surface of a dental composite—Fig. 2, where each cluster is made of 10–20 ultrafine sized silica particles with an average diameter 0.04 μm. These clusters are characteristic of particles of extremely small size and are caused by a Van der Waals-type force which acts between particles and is enhanced for even smaller-sized particles.

The concept of clusters is similar to that of "fractal structures" (Mandelbrot, 1982; Feder, 1988), and its characteristics can be determined by several key parameters such as cluster size (ξ) and fractal dimension (D). When the cluster size of the fillers is finite, then the macroscopic properties of such a composite would be influenced by the clusters (fillers) and also by the size of the unit cell taken, which includes clusters. This is in marked contrast to the case of the homogeneous distribution of fillers, where the macroscopic properties of the composite would not be influenced by the size of a unit cell; for the unit

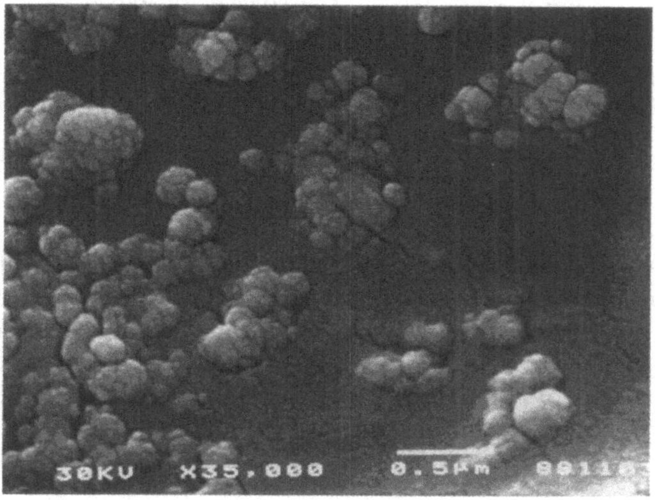

FIG. 2. A SEM photograph showing clusters made of SiO_2 particles with an average diameter of 0.04 μm in a dental composite.

cell of any size can represent a "composite." Even though the analysis of the micromorphology of fillers is made by using the "fractal" concept, it is not sufficient to predict the macroscopic properties of the composite, which requires some sort of modeling. Namely, the prediction of the macroscopic properties of a composite with such fractal structures can be made by a model based on effective medium theory (for example, the Eshelby method (Eshelby, 1957; Mura, 1987)) or a percolation model. In the case of the electrical conductivity of a composite with inhomogeneous distribution of conductivity fillers, a percolation model combined with clustered structures has been used successfully (Pike and Seager, 1974; Balberg and Binenbaum, 1983; Ueda and Taya, 1986).

Effective medium theory (EMT) is based on the assumption that the unit cell consists of a filler embedded in an infinite medium, although actual morphology of fillers requires some modifications of this unit cell model in order to account for the interactions between fillers; for example, the Mori–Tanaka theory in the case of the Eshelby model (Mori and Tanaka, 1973). This implies that the average spacing between fillers is not too small compared with the size of the filler, i.e., noncontacting fillers. In the case of the contacting fillers or the fillers with small interfiller spacing, the accuracy of EMT is expected to be poor (Taya and Ueda, 1987), except for the case of a small difference in the constituent properties.

An attempt is made in this paper to study criteria by which better modeling can be suggested for the analysis of the macroscopic properties of a composite with inhomogeneous distribution of fillers. The models to be examined are the Eshelby model representing the EMT group and a percolation model combined with the fractal concept. The concept of fractal structures and percolation will be reviewed briefly in Section 2. New parameters which better characterize the degree of the inhomogeneous distribution of fillers will be introduced in Section 3. Finally, the conclusion will be given in Section 4.

2. Fractal Structures and Percolation

Let us follow, for the time being, the method of Feder (1988) in defining the dimension of the fractal structure D, instead of following the Mandelbrot-type definition, since the former is easier to understand and more convenient for extending to the case of filler morphology. Consider a straight chain of length L which consists of N monomars (Fig. 3(a)). Denoting the radius of each monomar by R_0 and the half-length of the chain by R (i.e., $L = 2R$), we have

$$N = \left(\frac{R}{R_0}\right)^1, \tag{2.1}$$

where the superscript "1" is a power. In the case of a circular disk of radius R consisting of monomars, Fig. 3(b), N can be expressed as

$$N = \rho\left(\frac{R}{R_0}\right)^2, \tag{2.2}$$

where $\rho = \pi/(2\sqrt{3})$ for closely packed spheres. Likewise, N is obtained for a three-dimensional close packing of spherical monomars to form a spherical

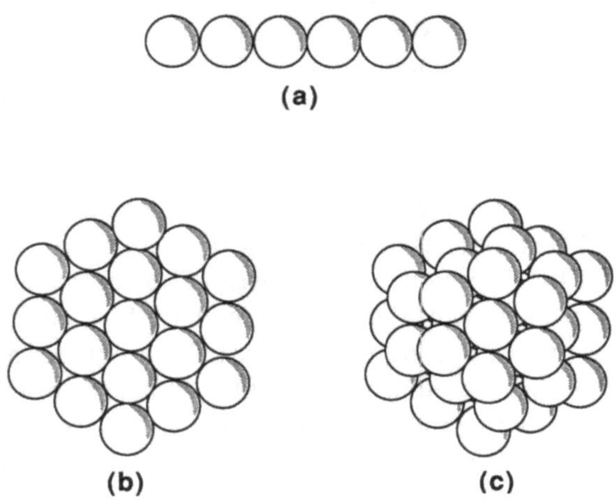

FIG. 3. An example of the Euclidean dimension E: (a) $E = 1$ for a one-dimensional chain model; (b) $E = 2$ for a two-dimensional disk model; and (c) $E = 3$ for a three-dimensional spherical model (Feder, 1988).

region of radius R, Fig. 3(c), and is given by

$$N = \rho\left(\frac{R}{R_0}\right)^3, \tag{2.3}$$

where $\rho = \pi/(3\sqrt{2})$. The values of ρ in the above equations are valid only as a limit of $R/R_0 \to \infty$. Equations (2.1)–(2.3) can be unified into a single equation

$$N = \rho\left(\frac{R}{R_0}\right)^D \quad \text{for} \quad N \to \infty. \tag{2.4}$$

Then D takes a value of integer E, which is called the "Euclidean dimension," the dimension that we usually use. All the periodic structures (simple cubic, etc.) possess integer D ($= E$), thus they are "Euclidean structures."

A question then arises: Is D always equal to E? Or, D may *not* be an integer. If not, then D is a fraction, and the structure (geometry) with such D is called a "fractal structure or geometry." If the geometry is referred to as "clustered geometry," then it is called a "fractal cluster." The determination of D is normally made by measuring the slope of the $\log N$ versus $\log(R/R_0)$ line. If the line is not straight, the corresponding structure is *not* fractal (Underwood and Banerji, 1986).

Let us calculate the value of D for an example of the chain model of 11 monomars (Fig. 4). According to Mandelbrot (1982), D can be defined as

$$D = \frac{\log N}{\log(1/r)}, \tag{2.5}$$

where N is the total number of monomars, $N = 11$ for Fig. 4, and $1/r$ is that of the monomars that would fill along the straight line, $1/r = 7$. Thus the fractal dimension of the chain of Fig. 4 is given by

$$D = \frac{\log 11}{\log 7} = 1.232. \tag{2.6}$$

In the above formulas, $N \to \infty$ was assumed, which is equivalent to the infinite size of the clustered structure. This infinite size of the clustered structure corresponds to the case where the clustered structure in a finite size unit cell extends to the opposite boundaries of the cell. This is called "percolation cluster." The percolation of a two-dimensional lattice has been well studied for regular lattice problems (Sykes and Essam, 1964; Shante and Kirkpatrick,

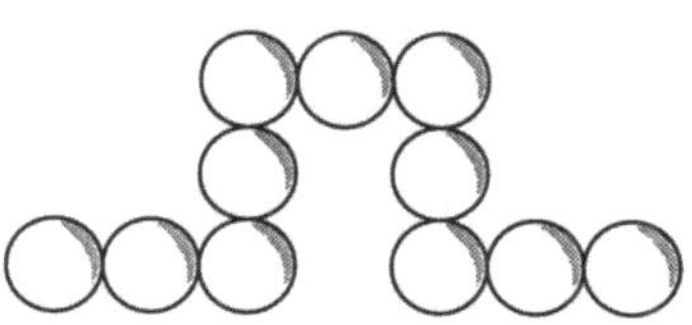

FIG. 4. An example of fractal structure with its dimension $D = 1.232$.

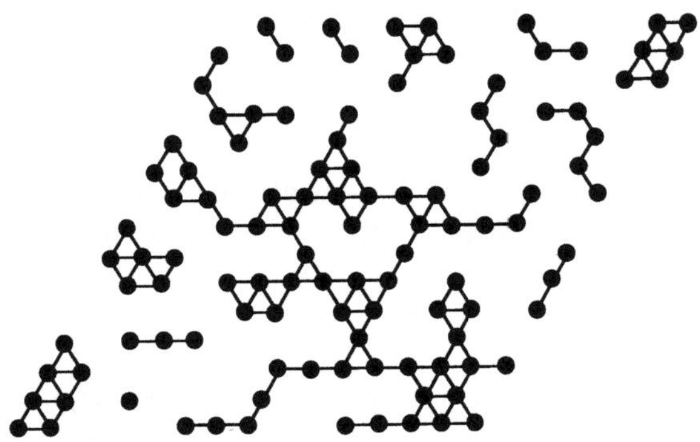

FIG. 5. An example of the percolation structure in a two-dimensional triangular lattice where only sites with $p = 1$ are shown as filled circles (Feder, 1988).

1971). An example of the percolating structure in a two-dimensional triangular lattice at $p = p_c$ is shown in Fig. 5 (Feder, 1988), where only sites with probability $p = 1$ are shown. The probability p of occupying a site is determined by the Monte Carlo method. The percolation occurs at $p = p_c = 0.5$. Then the number, $M(L)$, of the sites in the percolating cluster in a unit cell of size L is obtained as

$$M(L) \propto \begin{cases} \log L & \text{for} \quad p < p_c, \\ L^D & \text{for} \quad p = p_c, \\ L^E & \text{for} \quad p > p_c. \end{cases} \tag{2.7}$$

The powers D and E are the fractal and Euclidean dimensions, respectively; $D = 1.895$ (Stauffer, 1975) and $E = 2$. It is noted here that p can be interpreted as a fraction of filled sites in the unit cell.

In the actual distribution of fillers, sites (the centers of particulates or fibers) are not located at regular lattice points, but at random positions. Pike and Seager (1974) studied the percolation problem associated with a two-dimensional random lattice site. The locations of random sites in the Pike and Seager model are generated by using a pseudorandom number generator program. The sites so generated are planted in a unit square of 1 by 1. Figure 6(a) is an example of such sites we have generated by one of the pseudorandom number generators on a VAX computer at the University of Washington, where the total number of sites is taken to be 100 and the length of the square fillers are set equal to 0.013. The criterion for bonding two adjacent sites i and j is that the bonding function B_{ij} satisfies

$$B_{ij} = H(R_c - d_{ij}) = 1, \tag{2.8}$$

where H is the Heaviside step function, R_c is the critical radius of the circle, and d_{ij} is the interfiller spacing (distance) between the centers of the ith and

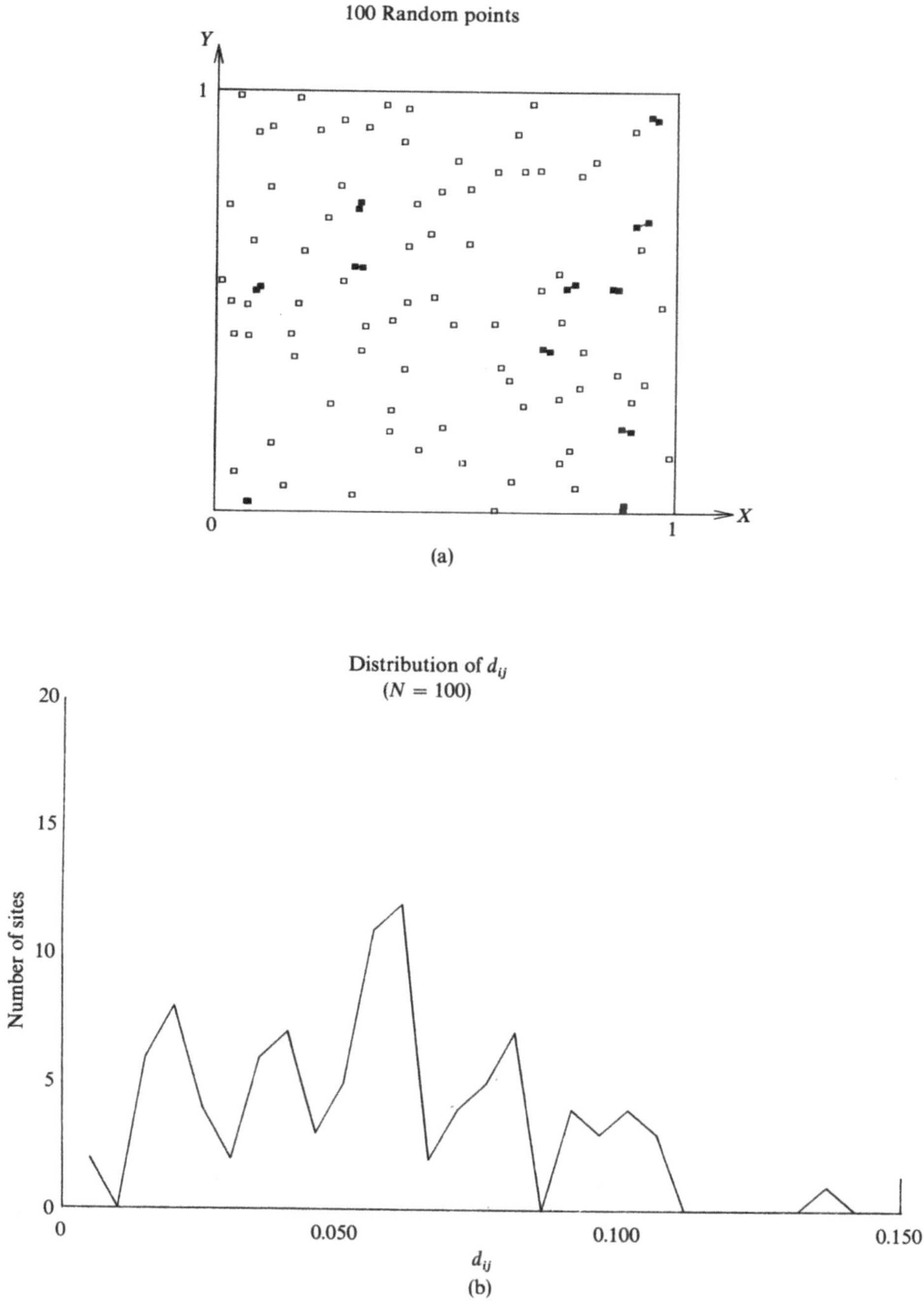

FIG. 6. (a) 100 random sites (marked by square symbols) planted in a unit square cell by a pseudorandom number generator, and (b) its distribution of interfiller (intersite) spacing d_{ij}.

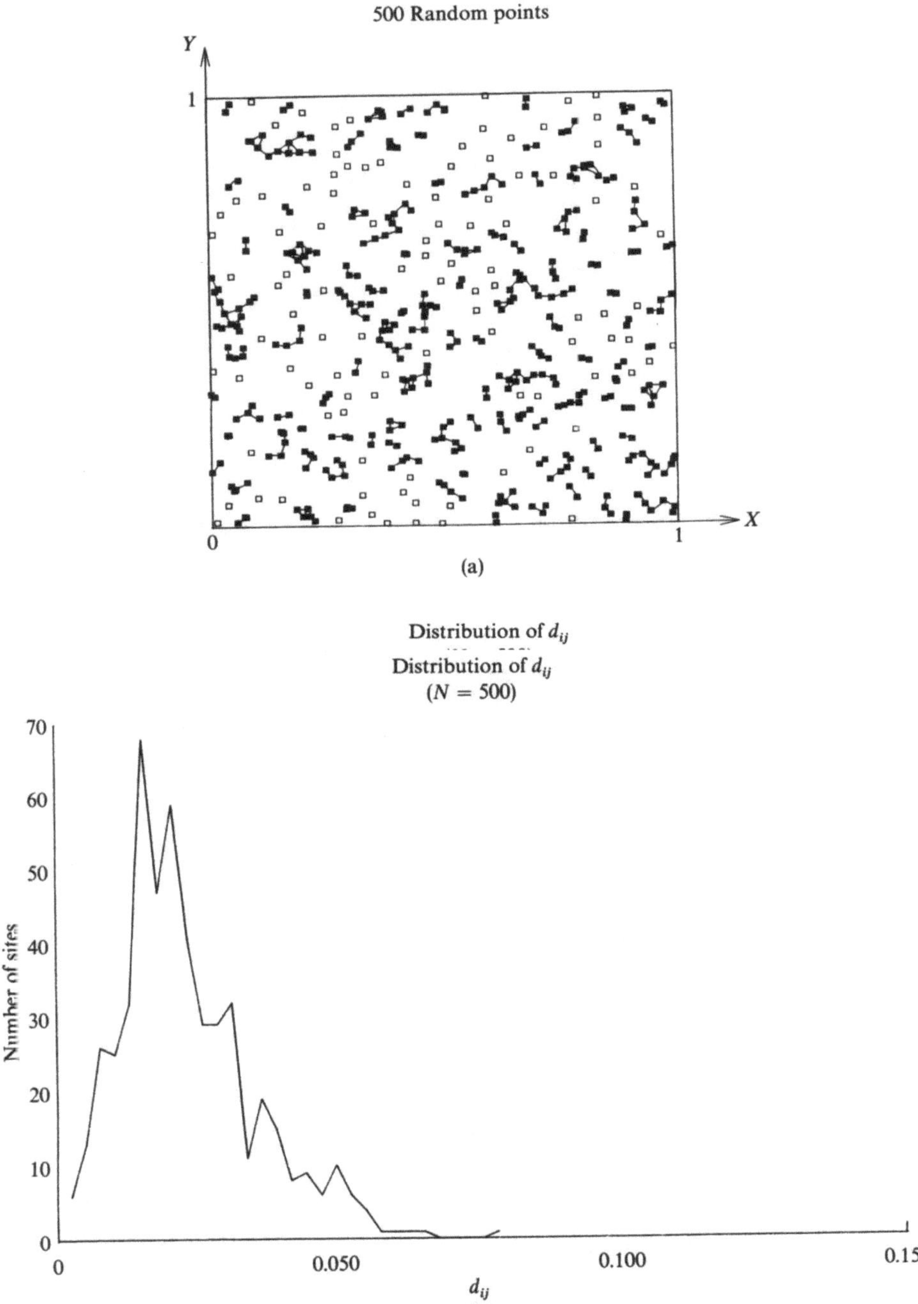

FIG. 7. (a) 500 random sites in a unit square cell, (b) its distribution of interfiller (intersite) spacing d_{ij}.

jth square fillers,

$$d_{ij} = \{(x_i - x_j)^2 + (y_i - y_j)^2\}^{1/2}, \tag{2.9}$$

and x_i and y_i are the coordinates of the center of the ith square filler. For later convenience, the distribution of interfiller spacing d_{ij} defined by (2.9) is shown in Fig. 6(b). It is clear from Fig. 6(b) that the values of d_{ij} are scattered over a wide range of spacing distance. It is also clear from the bonding criterion discussed above that whether the random sites yield a percolation stage or not depends on the value of R_c.

Suppose $R_c = 0.025$, then some sites are close to each other within R_c to form a cluster. The sites belonging to such clusters are shown as filled square marks connected by a line in Fig. 6(a). The clusters in Fig. 6(a) are still small and isolated. If the total number of random sites (N) increases to 500, the number of the clusters (N_c) and the average size of the clusters increases as seen from Fig. 7(a), although no percolation cluster is established. The distribution of d_{ij} corresponding to Fig. 7(a) is shown in Fig. 7(b), which indicates heavily localized distribution with its peak around $d_{ij} \cong 0.02$, compared to Fig. 6(b). This localized distribution of d_{ij} is a direct result of increasing N, or equivalently, the area fraction (A_f) of square fillers. Pike and Seager (1974) found that for $N = 4000$, the critical radius R_c needed to establish percolating clusters in a two-dimensional random lattice is

$$R_c = 0.000167. \tag{2.10}$$

It should be noted here that the preceding discussion on localized clusters is *not* related to fractal dimensions, unless the size of the clusters becomes infinite (1 for the unit square cell model, see (2.7)) or the clusters are of the periodic type.

3. Inhomogeneity Distribution Parameter

The examples of a random lattice discussed in the preceding section are based on the assumption that the distribution function $f(x)$ of the probability of occupying a site x on a [0, 1] interval is uniform (uniform random number)

$$f(x) = \text{const.} \qquad 0 \le x \le 1, \tag{3.1}$$

where

$$\int_0^1 f(x)\, dx = N, \tag{3.2}$$

and where N is the total number of sites. If $f(x)$ is not uniform, or $f(x)$ is uniform but bounded by $x_0 \le x \le x_1$, where $x_0 > 0$ and $x_1 < 1$ (note that a similar rule applied to the distribution along the y axis), then the distribution of sites would be "inhomogeneous," which leads to easier formation of clusters with the larger size.

The causes of the inhomogeneous distribution of fillers are many-fold: for example, the boundary effect of a viscous flow of a mixture of fillers, the matrix of a composite during the processing, and the preferred sites of precipitates during the solidification of a dispersion hardening alloy. The inhomogeneous distribution of SiC particulates in Fig. 1 is controlled by the relatively large cell size of grains, where SiC particulates tend to be located along the intercellular region (Lagacé and Lloyd, 1989). The inhomogeneous distribution of fillers can be tailored to obtain unique properties. An example of this is a thin interfacial layer between ceramic and metal, where the density of the fillers is tailored so as to diffuse the otherwise high thermal stress field at the interface (Niino *et al.*, 1987). Whatever the cause of the inhomogeneous distribution of fillers, we are concerned only with the final morphology of inhomogeneously distributed fillers, for it would determine the macroscopic properties of a composite.

Suppose the morphology of the inhomogeneous distribution of fillers ($N = 500$) is given as shown in Fig. 8(a), where several clusters of different size are formed while the remaining sites are dispersedly distributed. If those clusters are enveloped by dashed lines which form elliptical shapes, then the inhomogeneous distribution of sites within the unit square is considered to consist of two regions: an elliptic region Ω (clusters) and the remaining region $D - \Omega$ (nonclusters), where D denotes the entire region of the unit cell (Fig. 9). The distribution of interparticle spacing d_{ij} cooresponding to Fig. 8(a) is shown in Fig. 8(b), where the main peak in this figure is located over the interval of $d_{ij} = 0.015$–0.025. A comparison between Figs. 7(b) and 8(b) indicates that the more inhomogeneously distributed sites (Fig. 8(b)) yield the d_{ij} intensity peak with the smaller value. By denoting the value of the d_{ij} intensity peak by d_p, we can define a parameter ξ as

$$\xi = \frac{d_p}{l}, \tag{3.3}$$

where l is the average distance between sites and is related to the area fraction A_f

$$l = \sqrt{\frac{A_f}{N}}, \tag{3.4}$$

and N is the total number of the sites contained in the unit square cell. The value of ξ defined by (3.3) will approach 1 for the homogeneously distributed sites and 0 for the extremely localized clusters. Hence, we may call ξ an "inhomogeneity distribution parameter."

Is this inhomogeneity distribution parameter ξ complete enough for the characterization of the micromorphology of fillers? The answer to this question is "no," since we can present other examples of the inhomogeneous distribution with small ξ, which do not yield the same morphology as Fig. 8(a). Figure 10(a) and (b) illustrates such examples where the values of the d_{ij} intensity peak, d_p is assumed to be the same as that of Fig. 8(a), yet the

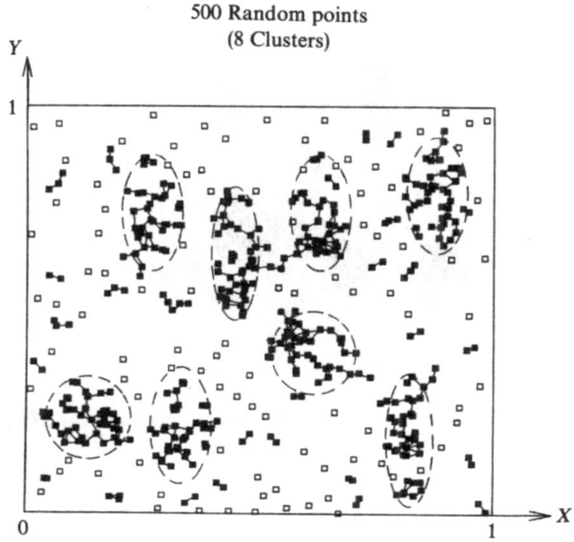

(a)

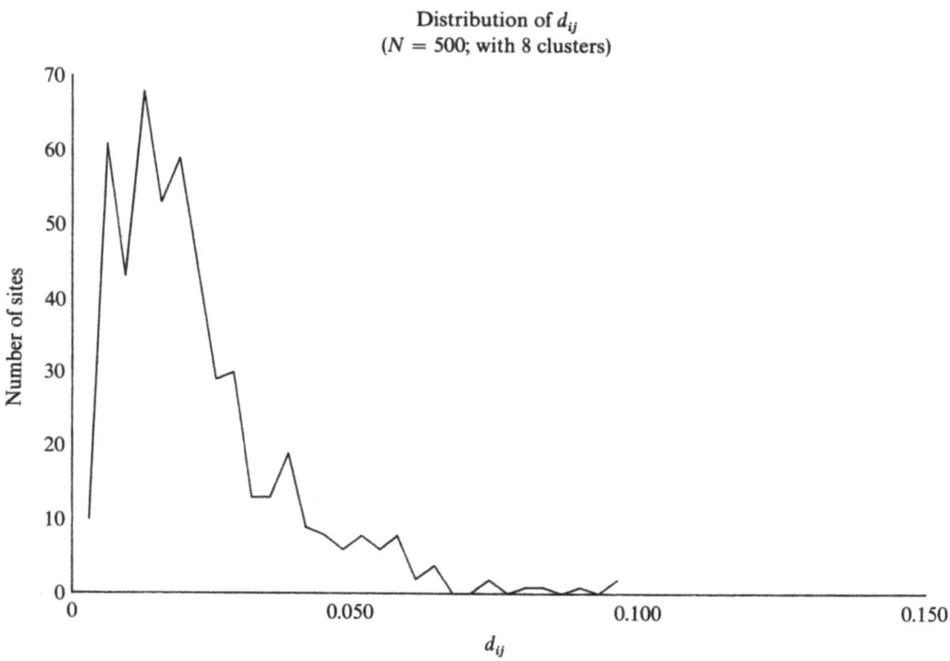

(b)

FIG. 8. (a) Inhomogeneous distribution of fillers ($N = 500$), and (b) its distribution of interfiller spacing d_{ij}.

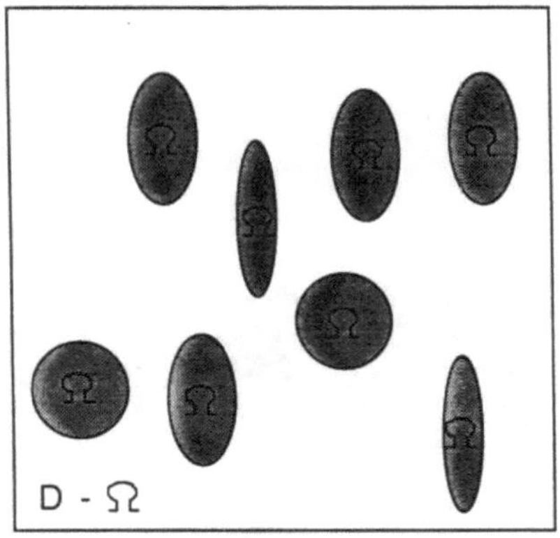

FIG. 9. Simplified morphology of Fig. 8(a) which consists of elliptic regions (Ω) and the matrix ($D - \Omega$).

morphology of the clusters is quite different from Fig. 8(a). In Fig. 10 those sites, which do not belong to the clusters, are omitted for the sake of clarity. The characterization of the extremely inhomogeneous distributions of the type shown in Fig. 10 can be, however, distinguished from Fig. 8(a) if we look at the radius of gyration, $R_g(s)$ of a cluster with s sites, which is defined by Feder (1988)

$$R_g^2(s) = \frac{1}{2s^2} \sum_{i,j=1}^{s} (\mathbf{r}_i - \mathbf{r}_j)^2, \tag{3.5}$$

where $|\mathbf{r}_i - \mathbf{r}_j|$ is the distance between the ith and the jth sites. The values of R_s of Fig. 10(a) and (b) are certainly larger than those of Fig. 8(a).

As regards the modeling of the macroscopic properties of a composite with

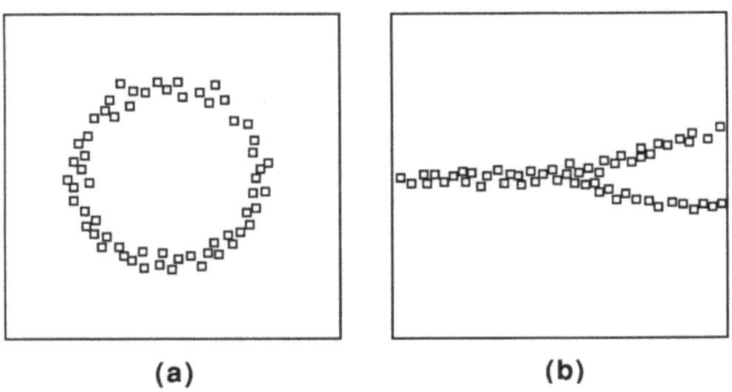

FIG. 10. Two examples of the extreme morphology of clusters: (a) a doughnut-shaped cluster, and (b) a branching cluster.

an inhomogeneous distribution of fillers, we can still use the Eshelby model for either larger or smaller values of ξ with smaller values of R_g. For in the former case, the distribution of fillers becomes close to the homogeneous distribution type including a periodic distribution, and each filler represents an inhomogeneity embedded in ths matrix. Whereas in the latter case, the Eshelby model can still be applicable if it is used twice: referring to Fig. 8(a), the first step is to homogenize the cluster domain (Ω) and the remaining domain ($D - \Omega$) to obtain the averaged properties of each domain. Thus, Fig. 8(a) can be reduced to the morphology of elliptical inhomogeneities embeded in the matrix (Fig. 9). If the shapes of the Ω's are not the same, then we must reduce these Ω's to the minimum number (M) of different elliptical inhomogeneities. The second step is to apply the Eshelby model developed for a hybrid composite where more than one kind of ellipsoidal inhomogeneity is embedded in the matrix (Taya and Chou, 1981).

The modeling of the filler morphology of the type shown in Fig. 10, however, is not as straightforward as for the type shown in Fig 8(a). As for Fig. 10(a), which may be called a "doughnut-shaped inhomogeneity," we can still treat this within the framework of the Eshelby method if the doughnut-shaped structure is considered to consist of two concentric elliptics which are embedded in the matrix (Taya and Mori, 1987). As for the morphology of the type shown in Fig. 10(b), which may be called a "branching cluster," there exists no general approach and the analysis must be tailored for a given problem; for example, a distribution of eigenstrains or dislocations may be assumed to occupy the domain of the branching cluster.

The definition of the unit cell used in the above analysis requires further consideration. If the morphology of fillers of the unit cell repeats itself in the larger sized cell, then it is the true unit cell and is called "self-similar." The clusters which are isolated within a self-similar unit cell become "finite clusters" at percolation, where the matrix is considered to be the percolating cluster (whose dimension is, of course, infinite). If the morphology of the microstructure in the unit cell is considered "fractal," then the modeling on the unit cell yields the macroscopic properties of a composite. On the other hand, if the morphology of the fillers in the unit cell represents the whole morphology of the composite without repeating itself, then it is nonfractal and the modeling requires an accurate account of the morphology. Thus, the results depend on the number of sites used. This is particularly important issue for the case of the anisotropic morphology of conducting fibers embedded in an insulating matrix (Balberg and Binenbaum, 1983; Ueda and Taya, 1986). In other words, when conducting fibers are misoriented in a two-dimensional plane (only the case of a two-dimensional misorientation is considered here, for simplicity), the overall properties of this misoriented short fiber composite can be anisotropic even though the properties of the matrix and fibers are isotropic. Balberg and Binenbaum (1983) claimed that the anisotropic electric resistivity (or conductivity) is determined by the exponent t in the following relation between the conductivity of the lattice and the probability of occupying a site p

$$\sigma \propto (p - p_c)^t. \tag{3.6}$$

The percolation threshold length, of a short fiber of a misoriented short fiber along the x axis, is different from that along the y axis unless the distribution of the fiber orientation is uniform for the entire span of the angle (i.e., completely random). This anisotropy in the percolation threshold is due to the finite number (N) of sites and it vanishes as $N \to \infty$, although the overall conductivity (or resistivity) of such a composite is anisotropic due to the exponent t (Balberg and Binenbaum, 1983). If N remains finite, which may be the case with some misoriented short fiber composites, both the anisotropy in t and the percolation threshold should exist (Ueda and Taya, 1986). It should be noted here that the anisotropic morphology of fibers yields the anisotropic properties of the composite as far as properties other than electrical conductivity are concerned.

4. Conclusion

The inhomogeneous distribution of fillers in a composite was discussed in relation to fractal and percolating clusters. A new parameter ξ which can account for the degree of the inhomogeneous distribution of fillers was introduced. It was found that an additional parameter is needed to characterize the morphology of the inhomogeneous distribution, and the radius of gyration of a cluster would suffice to cover possible extreme cases of inhomogeneous distribution.

5. Acknowledgments

This work was supported by a grant from the Alcan International Company to the University of Washington. The present author is thankful to Mr. K. Lulay for his help in the drawing of random lattice sites and the distribution of the interfiller spacing d_{ij}, and to Professor T. Mori of the Tokyo Institute of Technology for his valuable comments on some of the modelings.

References

Balberg, I. and Binenbaum, N. (1983), Computer study of the percolation threshold in a two-dimensional anisotropic system of conducting sticks, *Phys. Rev.* **B28**, 3799–3812.

Feder, J. (1988), *Fractals,* Plenum, New York.

Lagacé, H. and Lloyd, D. J. (1989), Microstructural analysis of Al–SiC composites, *Canad. Metall. Quart.*, **28**, 145–152.

Mandelbrot, B. B. (1982), *The Fractal Geometry of Nature*, W. H. Freeman, New York.

Mori, T. and Tanaka, K. (1973), Average stress in matrix and average elastic energy of materials with misfitting inclusions, *Acta Metallurgica*, **21**, 571–574.

Mura, T. (1987), *Micromechanics of Defects in Solids*, Martinus Nijhoff, The Hague.
Niino, M., Hirai, T., and Watanabe, R. (1987), Functional composites with gradient properties, *J. Japan Soc. Comp. Mater.*, **13**, no. 6, 257.
Pike, G. E. and Seager, C. H. (1974), Percolation and conductivity: A computer study, I, *Phys. Rev.*, **B10**, 1421–1434.
Shante, V. K. and Kirkpatrick, S. (1971), An introduction to percolation theory, *Adv. Phys.*, **20**, 325–357.
Sykes, M. F. and Essam, I. W. (1964), Critical percolation probabilities by the series method, *Phys. Rev.*, **113**, A310–315.
Taya, M. and Chou, T. W. (1981), On two kinds of ellipsoidal inhomogeneities in an infinite elastic body: Application to a hybrid composite, *Int. J. Solids Structures*, **17**, 553–563.
Taya, M. and Mori, T. (1987), Dislocations punched-out around a short fiber in a short fiber metal matrix composite, *Acta Metallurgica*, **35**, 155–162.
Taya, M. and Ueda, N. (1987), Prediction of in-plane electrical conductivity of a misoriented short fiber composite: Fiber percolation model vs. effective medium theory, *J. Engng. Mater. Tech.*, **109**, 252–256.
Ueda, N. and Taya, M. (1986), Prediction of electrical conductivity of misoriented short fiber composites by a percolation model, *J. Appl. Phys.* **60**, no. 1, 459–461.
Underwood, E. E. and Banerji, K. (1986), Fractals in fractography, *Mater. Sci. Engng.*, **80**, 1–14.

The Eigenvectors of the S Matrix and Their Relations with Line Dislocations and Forces in Anisotropic Elastic Solids

T. C. T. TING
Department of Civil Engineering, Mechanics and Metallurgy,
University of Illinois at Chicago, Chicago, IL 60680, U.S.A.

Abstract

The three real matrices **S**, **H**, **L** of Barnett and Lothe appear very often in the solutions to two-dimensional anisotropic elasticity problems. Of the three right eigenvectors of S, the first two are not unique. They form an ellipse on a plane called the Γ_b plane. The third right eigenvector is $\mathbf{e}_3$. The reciprocals, or the left eigenvectors, also have an ellipse formed by the first two left eigenvectors on a generally different plane called the Γ^f plane. The third left eigenvector is $\mathbf{e}^3$. For a straight line dislocation with the Burgers vector **b** and a straight line force **f** applied on the x_3 axis of the infinite anisotropic elastic medium, it is shown that if **b** is in the $\mathbf{e}_3$ direction and **f** in the $\mathbf{e}^3$ direction, the direction **h** of the infinite displacement at the origin is in the $\mathbf{e}_3$ direction, while the direction **g** of the surface traction on a radial plane (which is a constant vector independent of the choice of the radial plane) is in the $\mathbf{e}^3$ direction. On the other hand, if **b** is on the Γ_b plane while **f** is on the Γ^f plane, then **h** is on the Γ_b plane and **g** is on the Γ^f plane. There are essentially three types of geometry for the (Γ_b ellipse, $\mathbf{e}_3$) and the (Γ^f ellipse, $\mathbf{e}^3$) system. The monoclinic materials with the plane of symmetry at $x_3 = 0$ belong to Type 1 for which the Γ_b plane and the Γ^f plane coincide with the $x_3 = 0$ plane. Type 1 can also be identified by the fact that $\mathbf{S}^2$ is a symmetric matrix. The monoclinic materials with the plane of symmetry at $x_1 = 0$ or $x_2 = 0$ belong to Type 2 which has different Γ_b and Γ^f planes.

The vector **k**, which represents the line of intersection of the two planes, is the null eigenvector of the antisymmetric matrix $(\mathbf{S}^2 - \mathbf{S}^{2^T})$. For Type 2, **k** satisfies the condition $\mathbf{k}^T\mathbf{S}\mathbf{k} = 0$ and is a principal direction of the Γ_b ellipse and the Γ^f ellipse. It has the property that if **f** (or **b**) is along **k**, then so is **g** (or **h**). Type 3 applies to general anisotropic materials. It has geometry similar to Type 2 except that, since $\mathbf{k}^T\mathbf{S}\mathbf{k} \neq 0$, **k** is not a principal direction of the Γ_b or Γ^f ellipses. Therefore **f** (or **b**) along **k** does not produce **g** (or **h**) along **k**. We show that the number of independent parameters in **S**, **H**, **L** for Type 1, Type 2, and Type 3 is, four, six, and nine, respectively.

1. Introduction

In a fixed rectangular coordinate system x_1, x_2, x_3, let u_i and σ_{ij} be the displacement and stress, respectively. The stress–strain laws and the equations of equilibrium are

$$\sigma_{ij} = C_{ijks} u_{k,s}, \tag{1.1}$$

$$C_{ijks} u_{k,sj} = 0, \tag{1.2}$$

in which repeated indices imply summation, a comma stands for differentiation, and C_{ijks} are the elastic constants which possess the full symmetry

$$C_{ijks} = C_{jiks} = C_{ksij}.$$

For two-dimensional deformations in which u_i, $i = 1, 2, 3$, depend only on x_1 and x_2, the following three real matrices **S**, **H**, **L** often appear in the solutions to (1.1), (1.2). (See, e.g., Asaro *et al.*, 1973; Barnett and Lothe, 1973, 1974, 1975; Hwu and Ting, 1989a, b; Li and Ting, 1989; Ting, 1986; 1988b, c, d, 1989.) They are

$$\left.\begin{aligned} \mathbf{S} &= \frac{1}{\pi}\int_0^\pi \mathbf{N}_1(\theta)\,d\theta, \\ \mathbf{H} &= \frac{1}{\pi}\int_0^\pi \mathbf{N}_2(\theta)\,d\theta, \\ \mathbf{L} &= -\frac{1}{\pi}\int_0^\pi \mathbf{N}_3(\theta)\,d\theta, \end{aligned}\right\} \tag{1.3}$$

$$\mathbf{N}_1(\theta) = -\mathbf{T}^{-1}(\theta)\mathbf{R}^{\mathrm{T}}(\theta), \qquad \mathbf{N}_2(\theta) = \mathbf{T}^{-1}(\theta),$$

$$\mathbf{N}_3(\theta) = \mathbf{R}(\theta)\mathbf{T}^{-1}(\theta)R^{\mathrm{T}}(\theta) - \mathbf{Q}(\theta).$$

In the above, the superscript T denotes the transpose and $\mathbf{Q}(\theta)$, $\mathbf{R}(\theta)$, $\mathbf{T}(\theta)$ are 3×3 real matrices whose components are

$$\begin{aligned} Q_{ik}(\theta) &= C_{ijks} n_j(\theta) n_s(\theta), \\ R_{ik}(\theta) &= C_{ijks} n_j(\theta) m_s(\theta), \\ T_{ik}(\theta) &= C_{ijks} m_j(\theta) m_s(\theta), \\ n_i(\theta) &= (\cos\theta, \sin\theta, 0), \\ m_i(\theta) &= (-\sin\theta, \cos\theta, 0), \end{aligned}$$

in which θ is a real variable. We see that **S**, **H**, **L** depend exclusively on C_{ijks} and, as such, can be regarded as representing material property. We also see that $\mathbf{Q}(\theta)$, $\mathbf{T}(\theta)$, and $\mathbf{N}_2(\theta)$ are symmetric and positive definite if the strain energy is positive. It can be shown that $-\mathbf{N}_3(\theta)$ is symmetric and positive semidefinite (Ting, 1988a), and that **H** and **L** are symmetric and positive definite (Chadwick and Smith, 1977; Gundersen *et al.*, 1987; Ting, 1988a).

Several studies have been made on the properties of **S**, **H**, **L**. Chadwick and Smith (1977) and Bacon *et al.* (1978) proved that the matrices **H** and **L** are the prelogarithmic quadratic forms for the self-energy of dislocation lines and forces. Kirchner and Lothe (1986) stated that **S** is the prelogarithmic quadratic form controlling the interaction between dislocations and line forces. Chadwick and Ting (1987) investigated the structure and invariance properties of the three matrices. They showed that, when referred to a properly chosen basis, **S**, **H**, **L** for general anisotropic materials have the identical structure as that for isotropic materials. Kirchner and Lothe (1986) also studied the redundancy of the components of **S**, **H**, **L**. They found that, if the six components of the symmetric matrix **H** are known, **S** and **L** can be determined algebraically in terms of **H** and $\mathbf{N}_i$. Likewise, if the six components of the symmetric matrix **L** are known, **S** and **H** can be determined algebraically in terms of **L** and $\mathbf{N}_i$.

In this paper we will look at the eigenvectors of **S**. They have been presented by Chadwick and Ting (1987), but we investigate here the geometry of the ellipses traced out by the one-parameter family of two of the three eigenvectors and their reciprocal vectors. Therefore, instead of the three eigenvectors and their reciprocal eigenvectors, we study two sets of one ellipse and one eigenvector. There are essentially three types of different geometry for the ellipse–eigenvector system. We show that the ellipses and the eigenvectors play an interesting role in the solution to a straight line dislocation and a line force in the infinite anisotropic elastic material (Mura, 1987). We present the expressions of **S**, **H**, **L** for certain anisotropic materials with limited material symmetry. It is shown that the number of independent parameters for **S**, **H**, **L** is four for orthotropic materials if the planes of material symmetry coincide with the coordinate planes. For monoclinic materials, the number of independent parameters is five if $x_3 = 0$ is the plane of symmetry, and six if $x_1 = 0$ or $x_2 = 0$ is the plane of symmetry. For general anisotropic materials, the number of independent parameters is nine.

2. Representation of S, H, L

In this section we derive briefly the representation of **S**, **H**, **L**, originally due to Chadwick and Ting (1987). We will adopt a different approach in deriving the representation. We use vector notations as well as matrix notations. Thus **x** may represent a vector or a 3×1 column matrix. A scalar product of two vectors **x** and **y** may be written as

$$\mathbf{x} \cdot \mathbf{y} \qquad \text{or} \qquad \mathbf{x}^T \mathbf{y},$$

the former in vector notations while the latter is in matrix notations. The magnitude $|\mathbf{x}|$ of a vector **x** is

$$|\mathbf{x}|^2 = \mathbf{x} \cdot \mathbf{x} = \mathbf{x}^T \mathbf{x}.$$

The three matrices **S**, **H**, **L** are not independent of each other. They are related by

$$\mathbf{SH} + \mathbf{HS}^T = \mathbf{0}, \qquad \mathbf{LS} + \mathbf{S}^T\mathbf{L} = \mathbf{0}, \qquad \mathbf{HL} - \mathbf{SS} = \mathbf{I}, \tag{2.1}$$

in which $\mathbf{I}$ is the 3×3 identity matrix. Equations $(2.1)_{1,2}$ tell us that $\mathbf{SH}$ and $\mathbf{LS}$ are antisymmetric matrices.

It is shown in the Appendix of Ting (1986) that if $\mathbf{D}$ is a 3×3 symmetric positive definite real matrix and $\mathbf{W}$ is a 3×3 real antisymmetric matrix, the roots λ of the determinant

$$\|\mathbf{W} + i\lambda\mathbf{D}\| = 0$$

are $\pm s$ and 0 where s is real and positive and is given by

$$s = \{-\tfrac{1}{2}\operatorname{tr}(\mathbf{WD}^{-1})^2\}^{1/2}.$$

We now consider the eigenvalues λ' of $\mathbf{S}$,

$$\|\mathbf{S} - \lambda'\mathbf{I}\| = 0.$$

This is equivalent to

$$\|\mathbf{SH} - \lambda'\mathbf{H}\| = 0,$$

where $\mathbf{SH}$ is antisymmetric and $\mathbf{H}$ is symmetric and positive definite. If we set $\mathbf{W} = \mathbf{SH}$ and $\mathbf{D} = \mathbf{H}$, the eigenvalues of $\mathbf{S}$ are $\pm is$ and 0 where

$$s = \{-\tfrac{1}{2}\operatorname{tr}(\mathbf{S})^2\}^{1/2}.$$

Let the eigenvectors associated with the eigenvalues $\pm is$ and 0 be $(\mathbf{e}_1 \mp i\mathbf{e}_2)$, $\mathbf{e}_3$ where $\mathbf{e}_1, \mathbf{e}_2, \mathbf{e}_3$ are real vectors. That is,

$$\mathbf{S}(\mathbf{e}_1 \mp i\mathbf{e}_2) = \pm is(\mathbf{e}_1 \mp i\mathbf{e}_2), \qquad \mathbf{S}\mathbf{e}_3 = \mathbf{0}, \tag{2.2}$$

or, equating the real and imaginary parts,

$$\mathbf{S}\mathbf{e}_1 = s\mathbf{e}_2, \qquad \mathbf{S}\mathbf{e}_2 = -s\mathbf{e}_1, \qquad \mathbf{S}\mathbf{e}_3 = \mathbf{0}. \tag{2.3}$$

The $\pm$ signs employed here are different from those in Chadwick and Ting (1987). With the present choice of signs, $\mathbf{e}_1$, $\mathbf{e}_2$, $\mathbf{e}_3$ reduce to, for isotropic materials, unit vectors pointing in the positive directions of the x_1, x_2, x_3 axes, respectively.

The real vectors $\mathbf{e}^1, \mathbf{e}^2, \mathbf{e}^3$, reciprocal to $\mathbf{e}_1, \mathbf{e}_2, \mathbf{e}_3$, are defined by

$$\mathbf{e}_i \cdot \mathbf{e}^j = \mathbf{e}_i^{\mathrm{T}}\mathbf{e}^j = \delta_{ij}, \tag{2.4}$$

where δ_{ij} is the Kronecker delta. We assume that the matrix $\mathbf{S}$ is represented by

$$\mathbf{S} = S^i{}_j\mathbf{e}_i(\mathbf{e}^j)^{\mathrm{T}},$$

where repeated indices still imply summation and the matrix product $\mathbf{e}_i(\mathbf{e}^j)^{\mathrm{T}}$ produces a 3×3 matrix. Substituting into (2.3) and using (2.4) leads to the result that all $S^i{}_j$ vanish except $S^2{}_1 = -S^1{}_2 = s$. Hence $\mathbf{S}$ has the representation

$$\mathbf{S} = s\left\{\mathbf{e}_2(\mathbf{e}^1)^{\mathrm{T}} - \mathbf{e}_1(\mathbf{e}^2)^{\mathrm{T}}\right\}. \tag{2.5}$$

On the other hand, assuming that $\mathbf{H}$, $\mathbf{L}$ have the representation

$$\mathbf{H} = H^{ij}\mathbf{e}_i\mathbf{e}_j^{\mathrm{T}}, \qquad \mathbf{L} = L_{ij}\mathbf{e}^i(\mathbf{e}^j)^{\mathrm{T}},$$

and substituting, along with (2.5), into $(2.1)_{1,2}$ leads to the result that the only nonzero H^{ij}, L_{ij} are $H^{11} = H^{22}$, H^{33} and $L_{11} = L_{22}$, L_{33}. Finally, $(2.1)_3$ provides the following representations for $\mathbf{H}$ and $\mathbf{L}$:

$$\mathbf{H} = \frac{1 - s^2}{\mu\kappa}\{\mathbf{e}_1\mathbf{e}_1^{\mathrm{T}} + \mathbf{e}_2\mathbf{e}_2^{\mathrm{T}}\} + \frac{1}{\mu}\mathbf{e}_3\mathbf{e}_3^{\mathrm{T}}, \tag{2.6}$$

$$\mathbf{L} = \mu\kappa\{\mathbf{e}^1(\mathbf{e}^1)^{\mathrm{T}} + \mathbf{e}^2(\mathbf{e}^2)^{\mathrm{T}}\} + \mu\mathbf{e}^3(\mathbf{e}^3)^{\mathrm{T}}, \tag{2.7}$$

where μ and κ are constants. Due to the positive definiteness of $\mathbf{H}$ and $\mathbf{L}$ and the fact that $s > 0$, we have

$$\mu > 0, \qquad \kappa > 0, \qquad 0 < s < 1.$$

Before we close this section, we obtain from (2.5),

$$\mathbf{S}^{\mathrm{T}} = s\{\mathbf{e}^1\mathbf{e}_2^{\mathrm{T}} - \mathbf{e}^2\mathbf{e}_1^{\mathrm{T}}\}.$$

It follows from (2.4) that

$$\mathbf{S}^{\mathrm{T}}\mathbf{e}^1 = -s\mathbf{e}^2, \qquad \mathbf{S}^{\mathrm{T}}\mathbf{e}^2 = s\mathbf{e}^1, \qquad \mathbf{S}^{\mathrm{T}}\mathbf{e}^3 = \mathbf{0}. \tag{2.8}$$

Equations $(2.8)_{1,2}$ are equivalent to

$$\mathbf{S}^{\mathrm{T}}(\mathbf{e}^1 \pm i\mathbf{e}^2) = \pm is(\mathbf{e}^1 \pm i\mathbf{e}^2),$$

and hence $\mathbf{e}^1 \pm i\mathbf{e}^2$ and $\mathbf{e}^3$ are the left eigenvectors of $\mathbf{S}$ associated with the eigenvalues $\pm is$ and 0.

3. Nonuniqueness of $\mathbf{e}_1$, $\mathbf{e}_2$ and $\mathbf{e}^1$, $\mathbf{e}^2$

Most of the results we will present in this section are also contained in Chadwick and Ting (1987) but differ in detail, especially in the normalization of the vectors $\mathbf{e}_1$, $\mathbf{e}_2$, $\mathbf{e}_3$. The biorthogonal relations (2.4) do not provide us with a unique solution for $\mathbf{e}_1$, $\mathbf{e}_2$, $\mathbf{e}_3$ and their reciprocals $\mathbf{e}^1$, $\mathbf{e}^2$, $\mathbf{e}^3$. For $\mathbf{e}_3$, $\mathbf{e}^3$ of $(2.3)_3$, $(2.8)_3$, we let

$$|\mathbf{e}_3| = |\mathbf{e}^3|. \tag{3.1}$$

This and the biorthogonal relations (2.4) written for $i = j = 3$,

$$\mathbf{e}_3 \cdot \mathbf{e}^3 = 1, \tag{3.2}$$

provides us with a unique solution for $\mathbf{e}_3$, $\mathbf{e}^3$ except that if $\mathbf{e}_3$, $\mathbf{e}^3$ are the solutions, then so are $-\mathbf{e}_3$, $-\mathbf{e}^3$. Regardless of the signs, the matrices $\mathbf{e}_3\mathbf{e}_3^{\mathrm{T}}$ and $\mathbf{e}^3(\mathbf{e}^3)^{\mathrm{T}}$ are unique when conditions (3.1) and (3.2) are imposed.

As to $\mathbf{e}_1$, $\mathbf{e}_2$ and $\mathbf{e}^1$, $\mathbf{e}^2$, we let

$$(\mathbf{e}_1 \times \mathbf{e}_2) \cdot \mathbf{e}_3 = 1. \tag{3.3}$$

It follows from (2.4), (3.1), and (3.3) that

$$\left.\begin{array}{lll} \mathbf{e}^1 = \mathbf{e}_2 \times \mathbf{e}_3, & \mathbf{e}^2 = \mathbf{e}_3 \times \mathbf{e}_1, & \mathbf{e}^3 = \mathbf{e}_1 \times \mathbf{e}_2, \\ \mathbf{e}_1 = \mathbf{e}^2 \times \mathbf{e}^3, & \mathbf{e}_2 = \mathbf{e}^3 \times \mathbf{e}^1, & \mathbf{e}_3 = \mathbf{e}^1 \times \mathbf{e}^2, \end{array}\right\} \tag{3.4}$$

$$(\mathbf{e}^1 \times \mathbf{e}^2) \cdot \mathbf{e}^3 = 1,$$

$$|\mathbf{e}_1 \times \mathbf{e}_2| = |\mathbf{e}^1 \times \mathbf{e}^2| = |\mathbf{e}_3| = |\mathbf{e}^3|. \tag{3.5}$$

Conditions (2.4) and (3.3) do not provide a unique solution for $\mathbf{e}_1$, $\mathbf{e}_2$ and $\mathbf{e}^1$, $\mathbf{e}^2$. The nonuniqueness is more than the change in signs for $\mathbf{e}_1$, $\mathbf{e}_2$ and $\mathbf{e}^1$, $\mathbf{e}^2$. We see from $(2.2)_1$ that if $\mathbf{e}_1^* \mp i\mathbf{e}_2^*$ are the eigenvectors associated with the eigenvalues $\pm is$, so are

$$\mathbf{e}_1 \mp i\mathbf{e}_2 = (a \pm ib)(\mathbf{e}_1^* \mp i\mathbf{e}_2^*),$$

where a, b are arbitrary real constants. That is,

$$\mathbf{e}_1 = a\mathbf{e}_1^* + b\mathbf{e}_2^*, \qquad \mathbf{e}_2 = -b\mathbf{e}_1^* + a\mathbf{e}_2^*,$$

and, from this,

$$\mathbf{e}_1 \times \mathbf{e}_2 = (a^2 + b^2)(\mathbf{e}_1^* \times \mathbf{e}_2^*).$$

Since (3.3) applies to $\mathbf{e}_1^*$, $\mathbf{e}_2^*$ also, we let

$$a = \cos\psi, \qquad b = \sin\psi,$$

where ψ is a real parameter. We then have

$$\left.\begin{aligned} \mathbf{e}_1 &= \mathbf{e}_1^* \cos\psi + \mathbf{e}_2^* \sin\psi, \\ \mathbf{e}_2 &= -\mathbf{e}_1^* \sin\psi + \mathbf{e}_2^* \cos\psi, \end{aligned}\right\} \tag{3.6}$$

$$\mathbf{e}_1 \times \mathbf{e}_2 = \mathbf{e}_1^* \times \mathbf{e}_2^*. \tag{3.7}$$

The reciprocal vectors $\mathbf{e}^1$, $\mathbf{e}^2$ and $\mathbf{e}^{1*}$, $\mathbf{e}^{2*}$ are related by the same equations

$$\left.\begin{aligned} \mathbf{e}^1 &= \mathbf{e}^{1*} \cos\psi + \mathbf{e}^{2*} \sin\psi, \\ \mathbf{e}^2 &= -\mathbf{e}^{1*} \sin\psi + \mathbf{e}^{2*} \cos\psi, \end{aligned}\right\} \tag{3.8}$$

$$\mathbf{e}^1 \times \mathbf{e}^2 = \mathbf{e}^{1*} \times \mathbf{e}^{2*}. \tag{3.9}$$

Thus we have a one-parameter family of solutions for $\mathbf{e}_1$, $\mathbf{e}_2$, $\mathbf{e}^1$, $\mathbf{e}^2$.

Despite the nonuniqueness, it is readily shown that

$$\mathbf{e}_2(\mathbf{e}^1)^{\mathrm{T}} - \mathbf{e}_1(\mathbf{e}^2)^{\mathrm{T}} = \mathbf{e}_2^*(\mathbf{e}^{1*})^{\mathrm{T}} - \mathbf{e}_1^*(\mathbf{e}^{2*})^{\mathrm{T}},$$

$$\mathbf{e}_1\mathbf{e}_1^{\mathrm{T}} + \mathbf{e}_2\mathbf{e}_2^{\mathrm{T}} = \mathbf{e}_1^*\mathbf{e}_1^{*\mathrm{T}} + \mathbf{e}_2^*\mathbf{e}_2^{*\mathrm{T}},$$

$$\mathbf{e}^1(\mathbf{e}^1)^{\mathrm{T}} + \mathbf{e}^2(\mathbf{e}^2)^{\mathrm{T}} = \mathbf{e}^{1*}(\mathbf{e}^{1*})^{\mathrm{T}} + \mathbf{e}^{2*}(\mathbf{e}^{2*})^{\mathrm{T}}.$$

The representations of $\mathbf{S}$, $\mathbf{H}$, $\mathbf{L}$ given by (2.5)–(2.7) are therefore independent of ψ and are unique. In particular, the constants μ and κ are unique with the normalization (3.1) and (3.3). It should be pointed out that μ has the physical dimensions of stress while s and κ are dimensionless.

For isotropic materials, μ is the shear modulus and

$$\kappa = \frac{1}{1-\nu}, \qquad s = \frac{1-2\nu}{2(1-\nu)},$$

where ν is the Poisson ratio. The vectors $\mathbf{e}_1$, $\mathbf{e}_2$, $\mathbf{e}_3$ reduce to a unit vector pointing in the positive directions of the x_1, x_2, x_3 axes, respectively. We have

$$\mathbf{e}_1 = \begin{bmatrix} 1 \\ 0 \\ 0 \end{bmatrix} = \mathbf{e}^1, \qquad \mathbf{e}_2 = \begin{bmatrix} 0 \\ 1 \\ 0 \end{bmatrix} = \mathbf{e}^2, \qquad \mathbf{e}_3 = \begin{bmatrix} 0 \\ 0 \\ 1 \end{bmatrix} = \mathbf{e}^3,$$

and (2.5)–(2.7) lead to

$$\mathbf{S} = s\begin{bmatrix} 0 & -1 & 0 \\ 1 & 0 & 0 \\ 0 & 0 & 0 \end{bmatrix},$$

$$\mathbf{H} = \operatorname{diag}\left\{\frac{1-s^2}{\mu\kappa}, \frac{1-s^2}{\mu\kappa}, \frac{1}{\mu}\right\},$$

$$\mathbf{L} = \operatorname{diag}\{\mu\kappa, \mu\kappa, \mu\},$$

where diag stands for the diagonal matrix.

4. The Γ_b and Γ^f Ellipse

The one-parameter family of vectors $\mathbf{e}_1, \mathbf{e}_2$ given by (3.6) traces out an ellipse on a plane. The ellipse will be called the Γ_b ellipse and the plane will be called the Γ_b plane. The vector $\mathbf{e}^3$ is normal to the Γ_b plane. From $(2.3)_{1,2}$ we see that any vector $\mathbf{e}_b$ on the Γ_b plane satisfies the equation

$$\mathbf{S}^2\mathbf{e}_b = -s^2\mathbf{e}_b. \tag{4.1}$$

It should be noted that $\mathbf{e}_1, \mathbf{e}_2$, obtained from (4.1), may not satisfy $(2.3)_{1,2}$.

Likewise, the ellipse traced out by $\mathbf{e}^1, \mathbf{e}^2$ of (3.8) will be called the Γ^f ellipse and the plane on which the ellipse lies is the Γ^f plane. The vector $\mathbf{e}_3$ is normal to the Γ^f plane. From $(2.8)_{1,2}$, any vector $\mathbf{e}^f$ on the Γ^f plane satisfies the equation

$$(\mathbf{S}^2)^{\mathrm{T}}\mathbf{e}^f = -s^2\mathbf{e}^f. \tag{4.2}$$

Again, $\mathbf{e}^1, \mathbf{e}^2$, obtained from (4.2), may not satisfy $(2.8)_{1,2}$.

The principal axes of the Γ_b ellipse can be determined as follows. If $\mathbf{e}_1^* \cdot \mathbf{e}_2^* = 0$, $\mathbf{e}_1^*$ and $\mathbf{e}_2^*$ are the principal axes and will be denoted by $\hat{\mathbf{e}}_1, \hat{\mathbf{e}}_2$. If $\mathbf{e}_1^* \cdot \mathbf{e}_2^* \neq 0$, we obtain from (3.6),

$$\mathbf{e}_1 \cdot \mathbf{e}_2 = (\mathbf{e}_1^* \cdot \mathbf{e}_2^*) \cos 2\psi - \tfrac{1}{2}(\mathbf{e}_1^* \cdot \mathbf{e}_1^* - \mathbf{e}_2^* \cdot \mathbf{e}_2^*) \sin 2\psi.$$

The principal directions are therefore located at

$$\cot 2\hat{\psi} = \tfrac{1}{2}\left(\frac{\mathbf{e}_1^* \cdot \mathbf{e}_1^* - \mathbf{e}_2^* \cdot \mathbf{e}_2^*}{\mathbf{e}_1^* \cdot \mathbf{e}_2^*}\right),$$

and the principal axes are

$$\hat{\mathbf{e}}_1 = \mathbf{e}_1^* \cos \hat{\psi} + \mathbf{e}_2^* \sin \hat{\psi},$$

$$\hat{\mathbf{e}}_2 = -\mathbf{e}_1^* \sin \hat{\psi} + \mathbf{e}_2^* \cos \hat{\psi}.$$

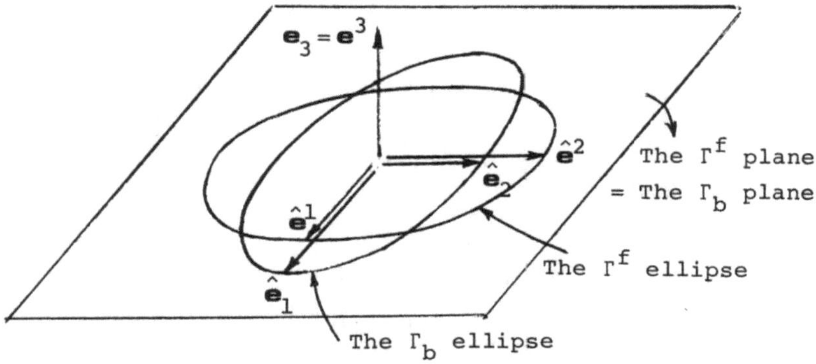

FIG. 1. Type 1: $\mathbf{e}_3 = \mathbf{e}^3$.

Using $(2.3)_1$ or $(2.3)_2$, the orthogonality of the principal axes can be written as

$$\hat{\mathbf{e}}^T \mathbf{S} \hat{\mathbf{e}} = 0, \tag{4.3}$$

which is satisfied by $\hat{\mathbf{e}}_1, \hat{\mathbf{e}}_2$. In the same manner, we can find the principal axes of the Γ^f ellipse. Again $\hat{\mathbf{e}}^1, \hat{\mathbf{e}}^2$ satisfy the equation

$$\hat{\mathbf{e}}^T \mathbf{S}^T \hat{\mathbf{e}} = 0,$$

which is equivalent to (4.3). It should be stressed that when $\mathbf{e}_1, \mathbf{e}_2$ are the principal axes of the Γ_b ellipse, the reciprocals $\mathbf{e}^1, \mathbf{e}^2$ obtained from (2.4) may not be the principal axes of the Γ^f ellipse. Regardless of whether $\mathbf{e}_1, \mathbf{e}_2$ and/or $\mathbf{e}^1, \mathbf{e}^2$ are the principal axes, it follows from $(3.5)_1$, (3.7), and (3.9) that the Γ_b ellipse and the Γ^f ellipse have the same area.

Instead of looking at the two sets of vectors $(\mathbf{e}_1, \mathbf{e}_2, \mathbf{e}_3)$ and $(\mathbf{e}^1, \mathbf{e}^2, \mathbf{e}^3)$, we will be looking at two sets of ellipse vectors (Γ_b ellipse, $\mathbf{e}_3$) and (Γ^f ellipse, $\mathbf{e}^3$) as shown in Figs. 1, 2, and 3. There is no unique choice of $\mathbf{e}_1, \mathbf{e}_2$ on the Γ_b ellipse and of $\mathbf{e}^1, \mathbf{e}^2$ on the Γ^f ellipse. However, once one of them, say $\mathbf{e}_1$, is determined, $\mathbf{e}_2$ from $(2.3)_1$ and $\mathbf{e}^1, \mathbf{e}^2$ from (2.4) are uniquely determined.

To see if $\mathbf{e}^1, \mathbf{e}^2$ are the principal axes of the Γ^f ellipse when $\mathbf{e}_1, \mathbf{e}_2$ are the principal axes of the Γ_b ellipse, we obtain from $(3.4)_{1,2}$

$$\begin{aligned} \mathbf{e}^1 \cdot \mathbf{e}^2 &= (\mathbf{e}_2 \times \mathbf{e}_3) \cdot (\mathbf{e}_3 \times \mathbf{e}_1) \\ &= (\mathbf{e}_1 \cdot \mathbf{e}_3)(\mathbf{e}_2 \cdot \mathbf{e}_3) - (\mathbf{e}_1 \cdot \mathbf{e}_2)(\mathbf{e}_3 \cdot \mathbf{e}_3). \end{aligned}$$

Therefore, when $\mathbf{e}_1 \cdot \mathbf{e}_2 = 0$, $\mathbf{e}^1 \cdot \mathbf{e}^2 = 0$ also if

$$\mathbf{e}_1 \cdot \mathbf{e}_3 = 0 \qquad \text{or} \qquad \mathbf{e}_2 \cdot \mathbf{e}_3 = 0. \tag{4.4}$$

There are two possibilities for which $(4.4)_1$ and/or $(4.4)_2$ are satisfied. They are the Type 1 and Type 2 possibilities discussed below. Before we discuss each type separately, we state the following theorem which can be proved easily.

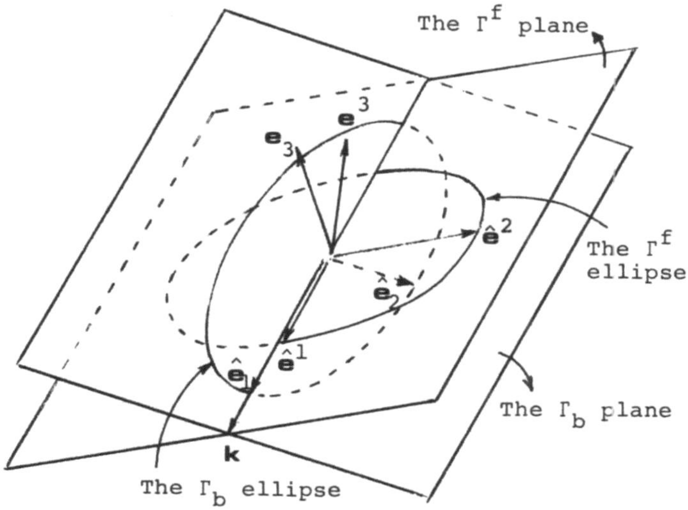

FIG. 2. Type 2: $\mathbf{e}_3 \neq \mathbf{e}^3$, $\mathbf{k}^T\mathbf{S}\mathbf{k} = 0$.

THEOREM. *Let*

$$\mathbf{W}\boldsymbol{\xi} = \mathbf{0}, \tag{4.5}$$

where $\mathbf{W}$ *is a nonzero* 3×3 *antisymmetric matrix*

$$\mathbf{W} = \begin{bmatrix} 0 & c & -b \\ -c & 0 & a \\ b & -a & 0 \end{bmatrix},$$

in which a, b, c are real. Then $\boldsymbol{\xi}$ *of* (4.5) *is unique up to a multiplicative constant given by*

$$\boldsymbol{\xi} = \begin{bmatrix} a \\ b \\ c \end{bmatrix}. \tag{4.6}$$

We now discuss each type separately.

Type 1. $\mathbf{e}_3 = \mathbf{e}^3$. When $\mathbf{e}_3 = \mathbf{e}^3$, $(4.4)_{1,2}$ are automatically satisfied in view of (2.4). Hence $\mathbf{e}_1$, $\mathbf{e}_2$ and $\mathbf{e}^1$, $\mathbf{e}^2$ can be the principal axes of their respective ellipses simultaneously. The Γ_b plane and the Γ^f plane are the same plane (Fig. 1). The three vectors $\hat{\mathbf{e}}_1$, $\hat{\mathbf{e}}_2$, $\mathbf{e}_3$ are orthogonal, so are $\hat{\mathbf{e}}^1$, $\hat{\mathbf{e}}^2$, $\mathbf{e}^3$. They have the properties

$$\left.\begin{aligned} &|\hat{\mathbf{e}}_1|\cdot|\hat{\mathbf{e}}^1| = |\hat{\mathbf{e}}_2|\cdot|\hat{\mathbf{e}}^2| = 1, \\ &|\hat{\mathbf{e}}_1|\cdot|\hat{\mathbf{e}}_2| = |\hat{\mathbf{e}}^1|\cdot|\hat{\mathbf{e}}^2| = |\mathbf{e}^3| = |\mathbf{e}_3| = 1. \end{aligned}\right\} \tag{4.7}$$

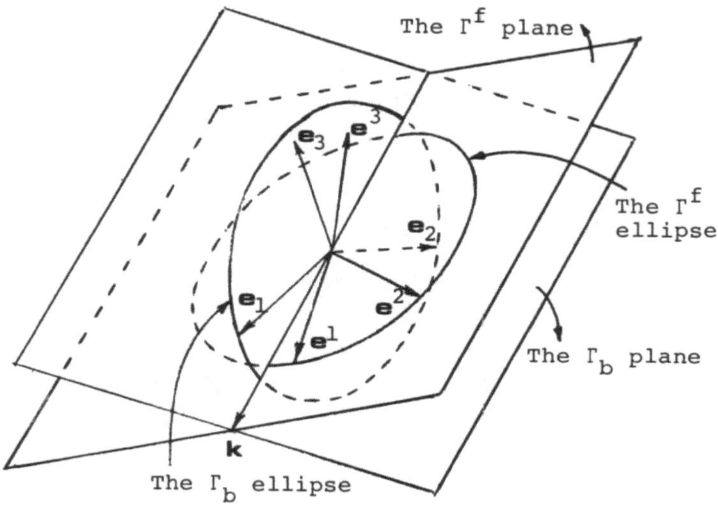

FIG. 3. Type 3: $\mathbf{e}_3 \neq \mathbf{e}^3$, $\mathbf{k}^T\mathbf{S}\mathbf{k} \neq 0$.

Since $\mathbf{e}_3 = \mathbf{e}^3$, we obtain from $(2.3)_3$ and $(2.8)_3$,

$$(\mathbf{S} - \mathbf{S}^T)\mathbf{e}_3 = \mathbf{0}. \tag{4.8}$$

This is in the form of (4.5) because $(\mathbf{S} - \mathbf{S}^T)$ is antisymmetric. Thus $\mathbf{e}_3$ can be obtained by using (4.6) subject to the normalization $(4.7)_6$.

There is an alternative way by which we can detect if $\mathbf{e}_3 = \mathbf{e}^3$ without computing $\mathbf{e}_3$, $\mathbf{e}^3$. When $\mathbf{e}_3 = \mathbf{e}^3$, the Γ_b plane and the Γ^f plane are the same plane. Any vector $\mathbf{x}$ on the plane must therefore satisfy (4.1) and (4.2) and we have

$$(\mathbf{S}^2 - \mathbf{S}^{2^T})\mathbf{x} = \mathbf{0}.$$

This is in the form of (4.5) and $\mathbf{x}$ is unique up to a multiplicative constant unless $(\mathbf{S}^2 - \mathbf{S}^{2^T})$ is a zero matrix. Since $\mathbf{x}$ is arbitrary on the Γ_b or Γ^f plane, we must have

$$\mathbf{S}^2 = \mathbf{S}^{2^T} \qquad \text{if} \quad \mathbf{e}_3 = \mathbf{e}^3.$$

Therefore, when $\mathbf{S}^2$ is symmetric (and in particular when $\mathbf{S}$ is antisymmetric), we have $\mathbf{e}_3 = \mathbf{e}^3$.

When $\mathbf{e}_3 \neq \mathbf{e}^3$, the Γ_b plane and the Γ^f plane intersect along a line which can be represented by the vector $\mathbf{k}$,

$$\mathbf{k} = a(\mathbf{e}_3 \times \mathbf{e}^3), \tag{4.9}$$

where a is an arbitrary constant. Since $\mathbf{k}$ is on the Γ_b plane as well as on the Γ^f plane, $\mathbf{k}$ satisfies (4.1) and (4.2), i.e.,

$$(\mathbf{S}^2 - \mathbf{S}^{2^T})\mathbf{k} = \mathbf{0}. \tag{4.10}$$

This is in the form of (4.5). The fact that $\mathbf{k}$ is unique assures us that $(\mathbf{S}^2 - \mathbf{S}^{2^T})$

in (4.10) cannot be a zero matrix. Hence

$$\mathbf{S}^2 \neq \mathbf{S}^{2^{\mathrm{T}}} \qquad \text{if} \quad \mathbf{e}_3 \neq \mathbf{e}^3.$$

From the theorem, $\mathbf{k}$ in (4.10) can be obtained by using (4.6). This provides us with an alternative to (4.9) for determining $\mathbf{k}$.

If $\mathbf{k}$ is a principal direction of the Γ_b ellipse, it must also be a principal direction of the Γ^f ellipse, according to (4.3), and we have

$$\mathbf{k}^{\mathrm{T}}\mathbf{S}\mathbf{k} = 0.$$

This leads to Type 2 below.

Type 2. $\mathbf{e}_3 \neq \mathbf{e}^3$ *and* $\mathbf{k}^{\mathrm{T}}\mathbf{S}\mathbf{k} = 0$. This is depicted in Fig. 2 in which $\mathbf{k}$ is the direction assumed by $\hat{\mathbf{e}}_1$ and $\hat{\mathbf{e}}^1$. The vectors $\hat{\mathbf{e}}_2, \hat{\mathbf{e}}^2$, and $\mathbf{e}_3, \mathbf{e}^3$ are on the same plane with $\mathbf{k}$ as its normal. For this case we have

$$|\hat{\mathbf{e}}_1|\cdot|\hat{\mathbf{e}}^1| = 1, \qquad |\hat{\mathbf{e}}_2|\cdot|\hat{\mathbf{e}}^2| = |\mathbf{e}_3|^2 = |\mathbf{e}^3|^2,$$
$$|\hat{\mathbf{e}}_1|\cdot|\hat{\mathbf{e}}_2| = |\hat{\mathbf{e}}^1|\cdot|\hat{\mathbf{e}}^2| = |\mathbf{e}_3| = |\mathbf{e}^3|.$$

Type 3. $\mathbf{e}_3 \neq \mathbf{e}^3$ *and* $\mathbf{k}^{\mathrm{T}}\mathbf{S}\mathbf{k} \neq 0$. This is the most general case in which $\mathbf{e}_1, \mathbf{e}_2$ and $\mathbf{e}^1, \mathbf{e}^2$ cannot be the principal axes of the respective ellipses simultaneously (Fig. 3). The vector $\mathbf{k}$ cannot be a principal direction of the Γ_b ellipse or of the Γ^f ellipse. Even though the principal axes $\hat{\mathbf{e}}_1, \hat{\mathbf{e}}_2$ of the Γ_b ellipse and the principal axes $\hat{\mathbf{e}}^1, \hat{\mathbf{e}}^2$ of the Γ^f ellipse do not satisfy (2.4), the fact that the two ellipses have the same area means that

$$|\hat{\mathbf{e}}_1|\cdot|\hat{\mathbf{e}}_2| = |\hat{\mathbf{e}}^1|\cdot|\hat{\mathbf{e}}^2| = |\mathbf{e}_3| = |\mathbf{e}^3|.$$

In summary, we can identify Type 1, Type 2, and Type 3 materials as follows. When $\mathbf{S}^2$ is symmetric, we have Type 1 in which $\mathbf{e}_3 = \mathbf{e}^3$ and $\mathbf{e}_3$ is determined from (4.8). If $\mathbf{S}^2$ is not symmetric, we determine $\mathbf{k}$ from (4.10). We have Type 2 or Type 3 depending on whether $\mathbf{k}^{\mathrm{T}}\mathbf{S}\mathbf{k}$ vanishes or not.

5. Straight Line Dislocations and Line Forces in the Infinite Medium

Let a straight line dislocation with a given Burgers vector $\mathbf{b}$ and a straight line force $\mathbf{f}$ be applied on the x_3 axis of the infinite anisotropic elastic medium. To satisfy the applied force $\mathbf{f}$ and the dislocation $\mathbf{b}$, the stress must be proportional to r^{-1} where (r, θ) are the polar coordinates

$$x_1 = r\cos\theta, \qquad x_2 = r\sin\theta.$$

The general solution to (1.1) and (1.2), for which the stress is proportional to r^{-1}, is (Kirchner, 1987; Ting, 1988b, 1989),

$$\left.\begin{aligned} 2\mathbf{u} &= -\frac{1}{\pi}(\ln r)\mathbf{h} - \hat{\mathbf{S}}(\theta)\mathbf{h} + \hat{\mathbf{H}}(\theta)\mathbf{g}, \\ 2\boldsymbol{\phi} &= \frac{1}{\pi}(\ln r)\mathbf{g} + \hat{\mathbf{S}}^{\mathrm{T}}(\theta)\mathbf{g} + \hat{\mathbf{L}}(\theta)\mathbf{h}, \end{aligned}\right\} \tag{5.1}$$

where $\mathbf{g}$ and $\mathbf{h}$ are real constant vectors, $\boldsymbol{\phi}$ is the stress function (Eshelby *et al.*, 1953; Stroh, 1958, 1962), and $\hat{\mathbf{S}}(\theta)$, $\hat{\mathbf{H}}(\theta)$, $\hat{\mathbf{L}}(\theta)$ are

$$\left.\begin{aligned}\hat{\mathbf{S}}(\theta) &= \frac{1}{\pi}\int_0^\theta \mathbf{N}_1(\theta')\,d\theta',\\ \hat{\mathbf{H}}(\theta) &= \frac{1}{\pi}\int_0^\theta \mathbf{N}_2(\theta')\,d\theta',\\ \hat{\mathbf{L}}(\theta) &= -\frac{1}{\pi}\int_0^\theta \mathbf{N}_3(\theta')\,d\theta'.\end{aligned}\right\} \tag{5.2}$$

These are the incomplete integrals which reduce to $\mathbf{S}$, $\mathbf{H}$, $\mathbf{L}$ where $\theta = \pi$. For the purpose of the present problem, a factor of 2 is added in (5.1) which is not present in Ting (1989).

Let $\mathbf{t}_r$, $\mathbf{t}_\theta$ be, respectively, the surface traction on the r = constant surface and θ = constant surface. We have

$$\mathbf{t}_\theta = \frac{\partial}{\partial r}\boldsymbol{\phi} = \frac{1}{2\pi r}\mathbf{g}, \tag{5.3}$$

$$\mathbf{t}_r = -\frac{\partial}{r\partial\theta}\boldsymbol{\phi} = \frac{1}{2\pi r}\{\mathbf{N}_3(\theta)\mathbf{h} - \mathbf{N}_1^{\mathrm{T}}(\theta)\mathbf{g}\}.$$

To balance the applied force $\mathbf{f}$ at the origin,

$$\int_0^{2\pi} \mathbf{t}_r r\,d\theta + \mathbf{f} = \mathbf{0},$$

or, noticing that $\mathbf{N}_i(\theta)$ is periodic in θ with periodicity π,

$$\mathbf{S}^{\mathrm{T}}\mathbf{g} + \mathbf{L}\mathbf{h} = \mathbf{f}. \tag{5.4}$$

On the other hand, the displacement $\mathbf{u}$ at $\theta = 2\pi$ and $\theta = 0$ must differ by the Burgers vector $\mathbf{b}$ and we have

$$\mathbf{H}\mathbf{g} - \mathbf{S}\mathbf{h} = \mathbf{b}. \tag{5.5}$$

Since $\mathbf{H}$ and $\mathbf{L}$ are positive definite, if we multiply (5.4) by $\mathbf{S}\mathbf{L}^{-1}$, we can eliminate $\mathbf{h}$ between (5.4) and (5.5) to obtain

$$\mathbf{g} = \mathbf{L}\mathbf{b} - \mathbf{S}^{\mathrm{T}}\mathbf{f}, \tag{5.6}$$

where use has been made of (2.1). Likewise, if we multiply (5.5) by $\mathbf{S}^{\mathrm{T}}\mathbf{H}^{-1}$, we have

$$\mathbf{h} = \mathbf{S}\mathbf{b} + \mathbf{H}\mathbf{f}. \tag{5.7}$$

Both $\mathbf{g}$ and $\mathbf{h}$ have a physical interpretation. From (5.3) we see that the surface traction $\mathbf{t}_\theta$ on any radial plane θ = constant is independent of θ. Thus the direction of $\mathbf{t}_\theta$ on any radial plane is given by $\mathbf{g}$. On the other hand, we obtain from $(5.1)_1$,

$$\mathbf{u} \sim \frac{1}{2\pi}\left(\ln\frac{1}{r}\right)\mathbf{h} \qquad \text{as} \quad r \to 0.$$

Therefore $\mathbf{h}$ represents the direction of the infinite displacement at the origin. It is not difficult to show from (5.6), (5.7), and (2.1) that

$$\mathbf{g}\cdot\mathbf{h} = \mathbf{b}\cdot\mathbf{f}.$$

In particular, $\mathbf{g}$ and $\mathbf{h}$ are orthogonal to each other when $\mathbf{b}$ and $\mathbf{f}$ are, or when $\mathbf{b} = \mathbf{0}$ or $\mathbf{f} = \mathbf{0}$. With the representation of $\mathbf{S}$, $\mathbf{H}$, $\mathbf{L}$ given in (2.5)–(2.7), we consider the following special cases.

(1) Let $\mathbf{b} = \mathbf{0}$. Then $\mathbf{g} = -\mathbf{S}^T\mathbf{f}$ and $\mathbf{h} = \mathbf{Hf}$. Hence,

$$\mathbf{g} = s\mathbf{e}^2, \qquad \mathbf{h} = \frac{1-s^2}{\mu\kappa}\mathbf{e}_1 \qquad \text{if} \quad \mathbf{f} = \mathbf{e}^1,$$

$$\mathbf{g} = -s\mathbf{e}^1, \qquad \mathbf{h} = \frac{1-s^2}{\mu\kappa}\mathbf{e}_2 \qquad \text{if} \quad \mathbf{f} = \mathbf{e}^2,$$

$$\mathbf{g} = \mathbf{0}, \qquad \mathbf{h} = \frac{1}{\mu}e_3 \qquad \text{if} \quad \mathbf{f} = \mathbf{e}^3.$$

(2) Let $\mathbf{f} = 0$. Then $\mathbf{g} = \mathbf{Lb}$ and $\mathbf{h} = \mathbf{Sb}$. Hence,

$$\mathbf{g} = \mu\kappa\mathbf{e}^1, \qquad \mathbf{h} = s\mathbf{e}_2 \qquad \text{if} \quad \mathbf{b} = \mathbf{e}_1,$$

$$\mathbf{g} = \mu\kappa\mathbf{e}^2, \qquad \mathbf{h} = -s\mathbf{e}_1 \qquad \text{if} \quad \mathbf{b} = \mathbf{e}_2,$$

$$\mathbf{g} = \mu\mathbf{e}^3, \qquad \mathbf{h} = \mathbf{0} \qquad \text{if} \quad \mathbf{b} = \mathbf{e}_3.$$

(3) From (5.6) we can make $\mathbf{g} = \mathbf{0}$ if we let

$$\mathbf{b} = -\mathbf{SL}^{-1}\mathbf{f}. \tag{5.8}$$

Equation (5.7) yields, when use is made of $(2.1)_3$,

$$\mathbf{h} = \mathbf{L}^{-1}\mathbf{f}. \tag{5.9}$$

Therefore if $\mathbf{b}$ and $\mathbf{f}$ are related by (5.8), the surface traction $\mathbf{t}_\theta$ on any radial plane vanishes. The representation of $\mathbf{L}^{-1}$ can be deduced from (2.7) as

$$\mathbf{L}^{-1} = \frac{1}{\mu\kappa}\{\mathbf{e}_1\mathbf{e}_1^T + \mathbf{e}_2\mathbf{e}_2^T\} + \frac{1}{\mu}\mathbf{e}_3\mathbf{e}_3^T,$$

from which we obtain, with the use of (2.5),

$$\mathbf{SL}^{-1} = \frac{s}{\mu\kappa}\{\mathbf{e}_2\mathbf{e}_1^T - \mathbf{e}_1\mathbf{e}_2^T\}. \tag{5.10}$$

From (5.8) and (5.9) we have

$$\mathbf{h} = \frac{1}{\mu\kappa}\mathbf{e}_1 \qquad \text{if} \quad \mathbf{f} = \mathbf{e}^1 \qquad \text{and} \qquad \mathbf{b} = -\frac{s}{\mu\kappa}\mathbf{e}_2,$$

$$\mathbf{h} = \frac{1}{\mu\kappa}\mathbf{e}_2 \qquad \text{if} \quad \mathbf{f} = \mathbf{e}^2 \qquad \text{and} \qquad \mathbf{b} = \frac{s}{\mu\kappa}\mathbf{e}_1,$$

$$\mathbf{h} = \frac{1}{\mu}\mathbf{e}_3 \qquad \text{if} \quad \mathbf{f} = \mathbf{e}^3 \qquad \text{and} \qquad \mathbf{b} = 0.$$

(4) Likewise, we can make $\mathbf{h} = \mathbf{0}$ if we let

$$\mathbf{f} = -\mathbf{H}^{-1}\mathbf{S}\mathbf{b}. \tag{5.11}$$

Equation (5.6) then gives

$$\mathbf{g} = \mathbf{H}^{-1}\mathbf{b}. \tag{5.12}$$

Thus when $\mathbf{b}$ and $\mathbf{f}$ are related by (5.11), the displacement at the origin is bounded. From (2.6) and (2.5) we obtain

$$\mathbf{H}^{-1} = \frac{\mu\kappa}{1-s^2}\{\mathbf{e}^1(\mathbf{e}^1)^{\mathrm{T}} + \mathbf{e}^2(\mathbf{e}^2)^{\mathrm{T}}\} + \mu\mathbf{e}^3(\mathbf{e}^3)^{\mathrm{T}},$$

$$\mathbf{H}^{-1}\mathbf{S} = \frac{\mu\kappa s}{1-s^2}\{\mathbf{e}^2(\mathbf{e}^1)^{\mathrm{T}} - \mathbf{e}^1(\mathbf{e}^2)^{\mathrm{T}}\}. \tag{5.13}$$

We thus have from (5.11) and (5.12) the following results.

$$\mathbf{g} = \frac{\mu\kappa}{1-s^2}\mathbf{e}^1 \quad \text{if} \quad \mathbf{b} = \mathbf{e}_1 \quad \text{and} \quad \mathbf{f} = -\frac{\mu\kappa s}{1-s^2}\mathbf{e}^2,$$

$$\mathbf{g} = \frac{\mu\kappa}{1-s^2}\mathbf{e}^2 \quad \text{if} \quad \mathbf{b} = \mathbf{e}_2 \quad \text{and} \quad \mathbf{f} = \frac{\mu\kappa s}{1-s^2}\mathbf{e}^1,$$

$$\mathbf{g} = \mu\mathbf{e}^3 \quad \text{if} \quad \mathbf{b} = \mathbf{e}_3 \quad \text{and} \quad \mathbf{f} = \mathbf{0}.$$

In all four special cases presented above, we see that if $\mathbf{f}$ stays on the Γ^f plane and $\mathbf{b}$ stays on the Γ_b plane, then $\mathbf{g}$ will be on the Γ^f plane while $\mathbf{h}$ will be on the Γ_b plane. If $\mathbf{f}$ is along $\mathbf{e}^3$, $\mathbf{h}$ is along $\mathbf{e}_3$, and $\mathbf{g} = \mathbf{0}$. If $\mathbf{b}$ is along $\mathbf{e}_3$, $\mathbf{g}$ is along $\mathbf{e}^3$, and $\mathbf{h} = \mathbf{0}$. While it is possible to choose $\mathbf{b}$ and $\mathbf{f}$ such that either $\mathbf{g}$ or $\mathbf{h}$ vanishes, it is not possible to have both $\mathbf{g}$ and $\mathbf{h}$ vanish.

In closing this section we note that the matrix products $\mathbf{S}\mathbf{L}^{-1}$ and $\mathbf{H}^{-1}\mathbf{S}$ of (5.10) and (5.13) also appeared in the singularity analysis of interface cracks in anisotropic bimaterials (Ting, 1986).

6. Expressions of the Matrices S, H, L for Certain Anisotropic Materials

In applications, it may occur that the coordinate planes and the planes of material symmetry do not coincide. Let x_i^* and x_i be related by

$$\mathbf{x}^* = \mathbf{\Omega}\mathbf{x},$$

$$\Omega = \begin{bmatrix} \cos\theta_0 & \sin\theta_0 & 0 \\ -\sin\theta_0 & \cos\theta_0 & 0 \\ 0 & 0 & 1 \end{bmatrix}.$$

Thus the x_i^* coordinate system is obtained by rotating the x_i coordinate system by an angle θ_0 about the x_3 axis. It can be shown (Ting, 1982) that if $\mathbf{S}^*$, $\mathbf{H}^*$,

$\mathbf{L}^*$ are the corresponding matrices in the x_i^* coordinate system, we have

$$\mathbf{S}^* = \boldsymbol{\Omega}\mathbf{S}\boldsymbol{\Omega}^{\mathrm{T}}, \qquad \mathbf{H}^* = \boldsymbol{\Omega}\mathbf{H}\boldsymbol{\Omega}^{\mathrm{T}}, \qquad \mathbf{L}^* = \boldsymbol{\Omega}\mathbf{L}\boldsymbol{\Omega}^{\mathrm{T}}. \tag{6.1}$$

We now present the expressions of $\mathbf{S}$, $\mathbf{H}$, $\mathbf{L}$ for certain anisotropic materials.

6.1. *Monoclinic Materials with the Plane of Symmetry at* $x_3 = 0$

From the discussion presented in the previous section, if we apply the force $\mathbf{f} = \mathbf{e}^3$, then the direction $\mathbf{h}$ of the displacement $\mathbf{u}$ at the origin will be along $\mathbf{e}_3$. If the material property is symmetric with the $x_3 = 0$ plane and if $\mathbf{f}$ is in the x_3 direction, then $\mathbf{h}$ must also be in the x_3 direction. Hence $\mathbf{e}_3 = \mathbf{e}^3$ and we have Type 1 geometry for the Γ_b and Γ^f ellipses. In fact, $\mathbf{e}_3$ is in the direction of the x_3 axis. According to the discussion presented in Section 4, there exist principal axes $\hat{\mathbf{e}}_1$, $\hat{\mathbf{e}}_2$, $\hat{\mathbf{e}}^1$, $\hat{\mathbf{e}}^2$ of the Γ_b and Γ^f ellipses on the $x_3 = 0$ plane. We will first assume that $\hat{\mathbf{e}}_1$ is in the direction of the x_1 axis. If not, all we have to do is apply (6.1). Hence we let

$$\mathbf{e}_1 = \begin{bmatrix} \alpha \\ 0 \\ 0 \end{bmatrix}, \qquad \mathbf{e}_2 = \begin{bmatrix} 0 \\ \alpha^{-1} \\ 0 \end{bmatrix}, \qquad \mathbf{e}_3 = \begin{bmatrix} 0 \\ 0 \\ 1 \end{bmatrix},$$

$$\mathbf{e}^1 = \begin{bmatrix} \alpha^{-1} \\ 0 \\ 0 \end{bmatrix}, \qquad \mathbf{e}^2 = \begin{bmatrix} 0 \\ \alpha \\ 0 \end{bmatrix}, \qquad \mathbf{e}^3 = \begin{bmatrix} 0 \\ 0 \\ 1 \end{bmatrix},$$

where α is a real constant. This system satisfies (2.4), (3.1), and (3.3). The representations given by (2.5)–(2.7) then have the expressions

$$\left.\begin{aligned} \mathbf{S} &= s\begin{bmatrix} 0 & -\alpha^2 & 0 \\ \alpha^{-2} & 0 & 0 \\ 0 & 0 & 0 \end{bmatrix}, \\ \mathbf{H} &= \operatorname{diag}\left\{\frac{1-s^2}{\mu\kappa}\alpha^2, \frac{1-s^2}{\mu\kappa}\alpha^{-2}, \frac{1}{\mu}\right\}, \\ \mathbf{L} &= \operatorname{diag}\{\mu\kappa\alpha^{-2}, \mu\kappa\alpha^2, \mu\}, \end{aligned}\right\} \tag{6.2}$$

where $\mathbf{H}$ and $\mathbf{L}$ are diagonal matrices. We see that $\hat{\mathbf{e}}_1$, $\hat{\mathbf{e}}_2$, $\mathbf{e}_3$ for Type 1 are eigenvectors of $\mathbf{H}$ and $\mathbf{L}$. It should be pointed out that $\mathbf{SL}^{-1}$, obtained from (6.2),

$$\mathbf{SL}^{-1} = \frac{s}{\mu\kappa}\begin{bmatrix} 0 & -1 & 0 \\ 1 & 0 & 0 \\ 0 & 0 & 0 \end{bmatrix} = \mathbf{S}^*(\mathbf{L}^*)^{-1},$$

where the second equality follows from (6.1), is independent of the choice of the x_1, x_2 axes. This property has an important application in determining whether there is an oscillation in displacement at an interface crack for monoclinic materials (Ting, 1986, 1990; Qu and Bassani, 1989).

Equations (6.2), which apply to orthotropic materials when the coordinate planes are the planes of material symmetry, have four independent parameters s, μ, κ, and α for **S**, **H**, **L** (Dongye and Ting, 1990). For monoclinic materials, for which the principal axis $\hat{\mathbf{e}}_1$ is located at an angle θ_0 from the x_1 axis, we have the additional independent parameter θ_0. For transversely isotropic materials with the axis of symmetry along the x_3 axis, (6.2) apply with $\alpha = 1$.

6.2. Monoclinic Materials with the Plane of Symmetry at $x_1 = 0$

In this case we have Type 2 geometry for the Γ_b ellipse and the Γ^f ellipse. The vector **k** is in the direction of the x_1 axis and is a principal direction of the Γ_b ellipse as well as of the Γ^f ellipse. Hence we let

$$\mathbf{e}_1 = \begin{bmatrix} \alpha \\ 0 \\ 0 \end{bmatrix}, \quad \mathbf{e}_2 = \begin{bmatrix} 0 \\ \beta\alpha^{-1}\cos\Delta \\ \beta\alpha^{-1}\sin\Delta \end{bmatrix}, \quad \mathbf{e}_3 = \begin{bmatrix} 0 \\ \beta\sin\delta \\ \beta\cos\delta \end{bmatrix},$$

$$\mathbf{e}^1 = \begin{bmatrix} \alpha^{-1} \\ 0 \\ 0 \end{bmatrix}, \quad \mathbf{e}^2 = \begin{bmatrix} 0 \\ \alpha\beta\cos\delta \\ -\alpha\beta\sin\delta \end{bmatrix}, \quad \mathbf{e}^3 = \begin{bmatrix} 0 \\ -\beta\sin\Delta \\ \beta\cos\Delta \end{bmatrix},$$

where α, β, Δ, and δ are constants with

$$\beta = [\cos(\delta + \Delta)]^{-1/2}.$$

S, **H**, **L** of (2.5)–(2.7) then have the expressions

$$\mathbf{S} = \begin{bmatrix} 0 & * & * \\ * & 0 & 0 \\ * & 0 & 0 \end{bmatrix}, \quad \mathbf{H} = \begin{bmatrix} * & 0 & 0 \\ 0 & * & * \\ 0 & * & * \end{bmatrix}, \quad \mathbf{L} = \begin{bmatrix} * & 0 & 0 \\ 0 & * & * \\ 0 & * & * \end{bmatrix},$$

in which the * represents a nonzero element. The number of independent parameters for this case is six.

We note that **k** in this case is an eigenvector of **H** and **L**.

6.3. Monoclinic Materials with the Plane of Symmetry at $x_2 = 0$

For this case, we simply apply the result of Section 6.2 to (6.1) with $\theta_0 = \pi/2$.

6.4. General Anisotropic Elastic Materials

For general anisotropic elastic materials, or for materials whose plane (or planes) of materials symmetry does not coincide with the coordinate planes, we have Type 3 geometry. There are a total of nine components in $\mathbf{e}_1$, $\mathbf{e}_2$, $\mathbf{e}_3$. We could select $\mathbf{e}_1$, $\mathbf{e}_2$ to be the principal axes of the Γ_b ellipse so that

$$\mathbf{e}_1 \cdot \mathbf{e}_2 = 0.$$

This and the two normalization conditions (3.1) and (3.2) reduce the number of independent parameters for $\mathbf{e}_1$, $\mathbf{e}_2$, $\mathbf{e}_3$ to six. The vectors $\mathbf{e}^1$, $\mathbf{e}^2$, $\mathbf{e}^3$ are then uniquely determined by (2.4). With the additional parameters s, μ, κ, the number of independent parameters in **S**, **H**, **L** for general anisotropic materials is nine.

7. Concluding Remarks

The results obtained here for **S**, **H**, **L** apply directly to the related matrices $\mathbf{S}(v)$, $\mathbf{H}(v)$, $\mathbf{L}(v)$, which occur in the steady motion of a straight line dislocation or surface waves in a semi-infinite medium in which v is the speed of the steady wave motion. For surface waves, the Γ_b ellipse is the locus of the displacement vector of the particles at the free surface.

The three matrices **S**, **H**, **L** in (1.3) require a numerical integration for most anisotropic materials. We can also express **S**, **H**, **L** in terms of the complex eigenvalues and eigenvectors of the 6×6 real matrix **N** which is obtainable from the elastic constants (Chadwick and Smith, 1977). Nevertheless, explicit expressions directly in terms of the elastic constants are available only for special materials. Some progress have been made recently in this direction. Explicit expressions for **S**, **H**, **L**, $\mathbf{S}(v)$, $\mathbf{H}(v)$, $\mathbf{L}(v)$, as well as for $\hat{\mathbf{S}}(\theta)$, $\hat{\mathbf{H}}(\theta)$, $\hat{\mathbf{L}}(\theta)$ of (5.2) have been obtained by Dongye and Ting (1990) for orthotropic materials in which the planes of symmetry coincide with the coordinate planes. For $\mathbf{S}(v)$, explicit expressions are available for orthotropic materials (Chadwick and Wilson, 1989), for cubic materials (Chadwick and Smith, 1982; Chadwick and Wilson, 1989), and for transversely isotropic materials in which the axis of symmetry is in the (x_1, x_2) or the (x_1, x_3) plane (Chadwick, 1989a). Recently, Chadwick (1989b) has determined $\mathbf{S}(v)$ for a monoclinic material for which $x_3 = 0$ is a plane of symmetry.

Acknowledgments

The work presented here has been supported by the U.S. Army Research Office, through grant DAAL 03-88-K-0079.

References

Asaro, R. J., Hirth, J. P., Barnett, D. M., and Lothe, J. (1973), A further synthesis of sextic and integral theories for dislocations and line forces in anisotropic media, *Phys. Status Solidi*, **B 60**, 261–271.

Bacon, D. J., Barnett, D. M., and Scattergood, R. O. (1978), The anisotropic continuum theory of lattice defects, *Prog. Materials Sci.*, **23**, 51–262.

Barnett, D. M. and Lothe, J. (1973), Synthesis of the sextic and the integral formalism for dislocation, Green's function and surface waves in anisotropic elastic solids, *Phys. Norv.*, **7**, 13–19.

Barnett, D. M. and Lothe, J. (1974), An image force theorem for dislocations in anisotropic bicrystals, *J. Phys. F.*, **4**, 1618–1635.

Barnett, D. M. and Lothe, J. (1975), Line force loadings on anisotropic half-spaces and wedges, *Phys. Norv.*, **8**, 13–22.

Chadwick, P. (1989a), Wave propagation in transversely isotropic elastic media. I, Homogeneous plane waves. II, Surface waves. III, The special case $a_s = 0$ and the inextensible limit., *Proc. Roy. Soc. London*, A **422**, 23–121.

Chadwick, P. (1989b), private communication.

Chadwick, P. and Smith, G. D. (1977), Foundations of the theory of surface waves in anisotropic elastic materials, *Adv. Appl. Mech.*, **17**, 303–376.

Chadwick, P. and Smith, G. D. (1982), Surface waves in cubic elastic materials, in *Mechanics of Solids*, The Rodney Hill 60th Anniversary Volume, edited by H. G. Hopkins and M. J. Sewell, Pergamon, Oxford, pp. 47–100.

Chadwick, P. and Ting, T. C. T. (1987), On the structure and invariance of the Barnett–Lothe tensors, *Quart. Appl. Math.*, **45**, 419–427.

Chadwick, P. and Wilson, N. J. (1989), Surface waves in orthorhombic and cubic elastic materials, to appear.

Dongye, Changsong and Ting, T. C. T. (1990), Explicit expressions of Barnett–Lothe tensors and their associated tensors for orthotropic materials, *Quart. Appl. Math.*, in press.

Eshelby, J. D., Read, W. T., and Shockley, W. (1953), Anisotropic elasticity with applications to dislocation theory, *Acta Metallurgica*, **1**, 251–259.

Gundersen, D. M., Barnett, D. M., and Lothe, J. (1987), Rayleigh wave existence theory. A supplementary remark, *Wave Motion*, **9**, 319–321.

Hwu, Chyanbin and Ting, T. C. T. (1989a), Solutions for the anisotropic elastic wedges at critical wedge angles, *J. Elasticity*, in press.

Hwu, Chyanbin and Ting, T. C. T. (1989b), Two-dimensional problems of the anisotropic elastic solids with an elliptic inclusion, *Quart. J. Mech. Appl. Math.*, in press.

Kirchner, H. O. K. (1987), Line defects along the axis of rotationally inhomogeneous media," *Phil. Mag.*, **A55**, 537–542.

Kirchner, H. O. K. and Lothe, J. (1986), On the redundancy of the *N* matrix of anisotropic elasticity, *Phil Mag.*, **A53**, L7–L10.

Li, Qianqian and Ting, T. C. T. (1989), Line inclusions in anisotropic elastic solids, *J. Appl. Mech.*, in press.

Mura, T. (1987), *Micromechanics of Defects in Solids*, 2nd ed, Martinus Nijhoff, Boston, MA.

Qu, Jianmin and Bassani, J. L. (1989), Cracks on bimaterial and bicrystal interfaces, *J. Mech. Phys. Solids*, in press.

Stroh, A. N. (1958), Dislocations and cracks in anisotropic elasticity, *Phil. Mag.*, **3**, 625–646.

Stroh, A. N. (1962), Steady state problems in anisotropic elasticity, *J. Math. Phys.*, **41**, 77–103.

Ting, T. C. T. (1982), Effects of change of reference coordinates on the stress analyses of anisotropic elastic materials, *Int. J. Solids Structures*, **18**, 139–152.

Ting, T. C. T. (1986), Explicit solution and invariance of the singularities at an interface crack in anisotropic composites, *Int. J. Solids Structures*, **9**, 965–983.

Ting, T. C. T. (1988a), Some identities and the structure of N_i in the Stroh formalism of anisotropic elasticity, *Quart. Appl. Math*, **46**, 109–120.

Ting, T. C. T. (1988b), Line forces and dislocations in anisotropic elastic composite wedges and spaces, *Phys. Status Solidi*, **B**, **146**, 81–90.

Ting, T. C. T. (1988c), The critical angle of the anisotropic elastic wedge subject to uniform tractions, *J. Elasticity*, **20**, 113–130.

Ting, T. C. T. (1988d), The anisotropic elastic wedge under a concentrated couple, *Quart. J. Mech. Appl. Math.*, **41**, 563–578.

Ting, T. C. T. (1989a), Line forces and dislocations in angularly inhomogeneous anisotropic elastic wedges and spaces, *Quart. Appl. Math.*, **47**, 123–128.

Ting, T. C. T. (1990), Interface cracks in anisotropic bimaterials, *J. Mech. Phys. Solids*, in press.

On the Strain Energy of Transformation Inhomogeneities in Solids

THOMAS TSAKALAKOS
Department of Mechanics and Materials Science, College of Engineering, Rutgers University, Piscataway, NJ 08855-0909, U.S.A.

Abstract

We present a theory of elastic interactions in inhomogeneous solids produced by various transformation modes. A reciprocal space formalism for the inhomogeneous modulus case is adopted, which leads to the development of an approximate closed-form solution of the elastic energy of systems, with arbitrary geometry and distribution of inhomogeneities. The Fourier representation of the elastic effect is applied to a variety of transformation modes such as spinodal decomposition, ordering, precipitation reaction in intermetallic systems, and the effect of applied stress in morphological changes. An analytical microscopic elasticity model is also discussed, to explain the origin of anomalous large contractions and expansions observed in materials with solid–solid interfaces. This model can predict relaxations at interfaces for a variety of inhomogeneous materials including metals, semiconductors, and their superlattices.

1. Introduction

Recent advances in the theory of phase transformations, developed by several authors (Khachaturyan, 1983; Krivoglaz, 1969) who incorporated the elastic strain energy into alloy thermodynamics, as well as experimental data, show that there is a real possibility of making a considerable contribution to the design of new structural materials, and of getting a new theoretical insight into inhomogeneous solids. Decomposition processes such as nucleation and growth, and coarsening, that result in the transformation of homogeneous solid solutions to a spatial distribution of multiphases in a matrix, martensitic transformations, transformation toughened ceramics, and advanced composite materials, are some examples of applications of the strain energy concepts on these morphology characteristics which principally determine materials properties such as strength and deformation mode, and may also affect toughness and fatigue resistance. Moreover, the present status of research in this area suggests that the application of external stress during decomposition produces a significant change in the precipitation process if the elastic moduli of precipitates and matrices are different (Ahn and

Tsakalakos, 1985; Jankowski *et al.*, 1985). Apart from a few theoretical studies of tractable elastic energy solutions of ellipsoidal shape inclusions, very little is known about the elastic energy of an arbitrary microstructure in the inhomogeneous modulus case.

In this paper we attempt to develop a theory of elastic interaction of elementary particles in an inhomogeneous modulus case, and the dependence of its geometry of morphology transformation on the relation between the intrinsic stress-free and the applied stress-induced crystal lattice mismatch. A reciprocal space formalism is adopted which gives a closed-form explicit equation of the elastic energy of an arbitrary geometry and of the distribution of inclusions. The theory allows for the calculation of the elastic energy in Fourier space, stress and strain distributions, and the effect of applied stress on the morphology of the precipitates.

An extension of the continuous elasticity case to the microscopic elasticity model is also developed to account for unusual phenomena in nanoscale dimensions.

2. Theoretical Foundation

We consider an anisotropic solid that contains heterogeneities such as precipitates, inclusions, composition fluctuations, or any other arbitrary distribution of new phases inside the parent matrix. In the language of micromechanics, this is equivalent to having a distribution of inhomogeneous inclusions in a continuum medium. We further assume that a force $\mathbf{f}$ is applied on the surface of the material that produces a uniform stress $\boldsymbol{\sigma}^{A}$ which in turn causes a uniform strain $\boldsymbol{\varepsilon}^{A}$ throughout the medium. Both $\boldsymbol{\sigma}^{A}$ and $\boldsymbol{\varepsilon}^{A}$ are global stresses and strains and are related to each other by the Hookian relation $\boldsymbol{\sigma}^{A} = \bar{\mathbf{C}}\boldsymbol{\varepsilon}^{A}$ where $\bar{\mathbf{C}}$ is the effective elastic constant tensor of the material. The global strain measures the overall deformation of the material neglecting the local disturbances which occur in the vicinity of the heterogeneities. If $\boldsymbol{\varepsilon}^{L}$ is the transformation strain or lattice eigenstrain and $\boldsymbol{\varepsilon}^{E}$ is the elastic strain caused by $\boldsymbol{\varepsilon}^{L}$, then the total strain ε is given by

$$\boldsymbol{\varepsilon} = \boldsymbol{\varepsilon}^{L} + \boldsymbol{\varepsilon}^{E}, \tag{2.1}$$

where $\varepsilon_{ij} = \frac{1}{2}(u_{i,j} + u_{j,i})$.

In the framework of linear elasticity we can write

$$\boldsymbol{\sigma} + \boldsymbol{\sigma}^{A} = \mathbf{C}(r)(\boldsymbol{\varepsilon} + \boldsymbol{\varepsilon}^{A} - \boldsymbol{\varepsilon}^{L}), \tag{2.2}$$

where $\boldsymbol{\sigma}$ is the internal stress and $\mathbf{C}(\mathbf{r})$ is the elastic constant tensor which depends on the position $\mathbf{r}$. In the absence of any body forces, (2.2) together with the equilibrium equation,

$$\sigma_{ij,j} = 0, \tag{2.3}$$

as well as the specific nature and configuration of the heterogeneities (boundary conditions), could be utilized to solve for the elasticity problem, i.e., to determine the stress and displacement field caused by the lattice eigenstrains and the applied stress, as well as to calculate the elastic strain energy. Such an approach is, however, unattractive because the elastic constants are functions of $\mathbf{r}$, thus involving their spatial derivatives in the equilibrium equation (2.3).

A more promising approach to this elasticity problem can be found by adopting an equivalent principle in a self-consistent scheme (Mura, 1987). Let us replace the material by one which has identical lattice heterogeneities ε^L, applied stress and strain $\boldsymbol{\sigma}^A$, $\boldsymbol{\varepsilon}^A$, internal stress $\boldsymbol{\sigma}$, and total strain $\boldsymbol{\varepsilon}$, but which maintains a constant effective elastic tensor $\bar{\mathbf{C}}$ through the material. To allow for the local disturbances caused by the variation of the elastic constants in the real material, a fictitious strain $\boldsymbol{\varepsilon}^I$ is introduced. Thus, we can now write

$$\boldsymbol{\sigma} + \boldsymbol{\sigma}^A = \bar{\mathbf{C}}[\boldsymbol{\varepsilon} + \boldsymbol{\varepsilon}^A - \boldsymbol{\varepsilon}^L - \boldsymbol{\varepsilon}^I]. \tag{2.4}$$

The inhomogeneity eigenstrain can be determined by comparing (2.2) and (2.4)

$$-\bar{\mathbf{C}}\boldsymbol{\varepsilon}^I = \Delta\mathbf{C}(\mathbf{r})[\boldsymbol{\varepsilon} + \boldsymbol{\varepsilon}^A - \boldsymbol{\varepsilon}^L]. \tag{2.5}$$

On the other hand, the total strain $\boldsymbol{\varepsilon}$ can be expressed in terms of $\boldsymbol{\varepsilon}^L$ and $\boldsymbol{\varepsilon}^I$ by first introducing (2.4) into (2.3) and taking the Fourier transform of the resulting equation. This procedure yields

$$\varepsilon(\mathbf{r}) = \int \mathbf{G}(\mathbf{r} - \mathbf{r}')[\boldsymbol{\varepsilon}^L(\mathbf{r}') + \boldsymbol{\varepsilon}^I(\mathbf{r}')]\, d^3r', \tag{2.6}$$

where $\mathbf{G}(\mathbf{r} - \mathbf{r}')$ is Green's function

$$\mathbf{G}(\mathbf{r}) = \int \mathbf{S}(\hat{n})e^{i\mathbf{k}\cdot\mathbf{r}}\, dK, \tag{2.7}$$

and $\mathbf{S}(\hat{n})$ is a function of the elastic constants $\bar{\mathbf{C}}$ and the orientation of the wave vector ($\hat{\mathbf{n}} = \mathbf{k}/|\mathbf{k}|$)

$$S_{ijkl}(\hat{n}) = C_{pqkl} n_q n_j N_{ip}(\mathbf{k}) D^{-1}(\mathbf{k}), \tag{2.8}$$

where $N_{ij} = \frac{1}{2}\varepsilon_{ikl}\varepsilon_{jmn}K_{km}K_{ln}$, ε_{ikl} = permutation tensor, $K_{ik} = C_{ijkl}n_j n_l$, and $D = \varepsilon_{mnl}K_{m1}K_{n2}K_{l3}$. By introducing (2.6) into (2.5) we obtain

$$-\bar{\mathbf{C}}\varepsilon^I(\mathbf{r}) = \Delta\mathbf{C}(\mathbf{r}) \int \mathbf{G}(\mathbf{r} - \mathbf{r}')\varepsilon^I(\mathbf{r}')\, d^3r' + \mathbf{B}(\mathbf{r}), \tag{2.9}$$

where

$$\mathbf{B}(\mathbf{r}) = \Delta\mathbf{C}(\mathbf{r})\left[\int \mathbf{G}(\mathbf{r} - \mathbf{r}')\boldsymbol{\varepsilon}^L(\mathbf{r}')\, d^3r' + \boldsymbol{\varepsilon}^A - \boldsymbol{\varepsilon}^L(\mathbf{r})\right]. \tag{2.10}$$

The inhomogeneity strain $\boldsymbol{\varepsilon}^I$ can thus be determined from the integral equation (2.9). It is apparent that an exact analytical solution of (2.9) for $\boldsymbol{\varepsilon}^I$ is formidable in the general case. Nevertheless, a few specific cases and an approximate general solution will be discussed in the following sections. It should be pointed out, however, that our presentation differs from that of

Eshelby (1957) and Mura (1987) in the case that the heterogeneities in our material are of an arbitrary shape and not of an ellipsoidal one. It can readily be shown that an ellipsoidal inhomogeneous inclusion secures a constant inhomogeneity strain $\boldsymbol{\varepsilon}^{\mathrm{I}}$ within the inclusion, and zero everywhere else. In this case, the integral equation becomes essentially an algebraic equation for $\boldsymbol{\varepsilon}^{\mathrm{I}}$. If we assume an arbitrary shape for the inclusions we can no longer have constant $\boldsymbol{\varepsilon}^{\mathrm{I}}$ within the inclusion, and $\boldsymbol{\varepsilon}^{\mathrm{I}}$ depends on $\mathbf{r}$ and the functional relationship is determined by the integral equation (2.9). The existence and uniqueness of the solution of this integral equation is beyond the scope of this paper. In the following we will consider the presentation of the elastic energy in a close Fourier space scheme and an approximate solution for the inhomogeneous case.

2.1. The Elastic Strain Energy

The total elastic potential energy of the system is given by

$$E = \frac{1}{2}\int_v (\boldsymbol{\sigma} + \boldsymbol{\sigma}^{\mathrm{A}})(\boldsymbol{\varepsilon} + \boldsymbol{\varepsilon}^{\mathrm{A}} - \boldsymbol{\varepsilon}^{\mathrm{L}})\, d^3r - \int_s \mathbf{f}\cdot(\mathbf{u} + \mathbf{u}^{\mathrm{A}})\, dS. \tag{2.11}$$

Mura (1987) has shown that the elastic potential energy of the system referred to a state of no applied stress is given by

$$\Delta E = -\int \boldsymbol{\sigma}^{\mathrm{A}}\boldsymbol{\varepsilon}^{\mathrm{L}}\, d^3r - \frac{1}{2}\int \boldsymbol{\sigma}^{\mathrm{A}}\boldsymbol{\varepsilon}^{\mathrm{I}}\, d^3r - \frac{1}{2}\int \boldsymbol{\sigma}\boldsymbol{\varepsilon}^{\mathrm{L}}\, d^3r. \tag{2.12}$$

The energy expression in (2.12) can readily be derived by applying the divergence theorem and employing the equilibrium equation. We can thus calculate the elastic energy in principle, from (2.12). This requires the solution of the integral equation (2.7) for $\boldsymbol{\varepsilon}^{\mathrm{I}}$ and the internal stress $\boldsymbol{\sigma}$ which can be computed in turn from (2.3) and (2.4). Such a representation, however, is unattractive in practical problems of phase transformations, for it is extremely difficult to obtain solutions in general cases. Moreover, the fact that it is not a closed-form representation for the energy, and the requirement of solving the integral equation in real space by using Green's function provides little or no fruitful information about the morphological changes which occur during a phase transformation.

2.2. Fourier Transform Representation

The problem can be greatly simplified by utilizing Fourier transform techniques. This has a particular significance in the method of static concentration waves in phase transformation, because the amplitude of the static lattice distortion waves are proportional to the Fourier amplitudes of the concentration waves or the long-range order parameters, where the proportionality constants can easily be obtained from the above equations. Similarly, the shape and morphology of precipitates can more easily be predicted when representing the energy in a Fourier space than in a real one. The Fourier method is especially convenient in considering diffuse X-ray or neutron scattering from distorted crystals.

The analysis is greatly simplified by introducing a characteristic function $x(\mathbf{r})$ which depends on the nature of the inhomogeneity. In the case of precipitates, it represents the shape function, $x(\mathbf{r}) = \theta(\mathbf{r})$, which is 1 within the precipitates and 0 in the matrix. In the case of concentration waves, it can be identified as the concentration variation $x(\mathbf{r}) = \Delta c(\mathbf{r}) = c(\mathbf{r}) - \bar{c}$. The Fourier transform of $x(r)$ is given by

$$x(\mathbf{k}) = \int x(\mathbf{r}) e^{i\mathbf{k}\cdot\mathbf{r}}\, d^3 r. \tag{2.13}$$

We can thus express

$$\boldsymbol{\varepsilon}^{\mathrm{L}}(\mathbf{r}) = \boldsymbol{\varepsilon}_0^{\mathrm{L}} x(\mathbf{r}), \qquad \Delta\mathbf{C}(r) = \Delta\mathbf{C}_0 x(\mathbf{r}), \tag{2.14}$$

where the zero subscript denotes the numerical values of the quantities.

Substituting for σ from (2.4) into (2.12) and Fourier transforming the resulting equation, we obtain

$$\begin{aligned}\Delta E = &-\boldsymbol{\sigma}^{\mathrm{A}}\boldsymbol{\varepsilon}^{\mathrm{L}}(\mathbf{k}=0) - \tfrac{1}{2}\boldsymbol{\sigma}^{\mathrm{A}}\boldsymbol{\varepsilon}^{\mathrm{I}}(\mathbf{k}=0)\\ &- \frac{1}{2}\int \bar{\mathbf{C}}[\mathbf{S}(\mathbf{k}) - \mathbf{I}][\boldsymbol{\varepsilon}^{\mathrm{L}}(\mathbf{k}) + \boldsymbol{\varepsilon}^{\mathrm{I}}(\mathbf{k})]\boldsymbol{\varepsilon}^{\mathrm{L}}(\mathbf{k})\, d^3 k.\end{aligned} \tag{2.15}$$

A closed-form expression for the energy is thus required for the Fourier transform of the inhomogeneity strain $\boldsymbol{\varepsilon}^{\mathrm{I}}$. The Fourier transform of (2.9) can readily be derived

$$\begin{aligned}-\bar{\mathbf{C}}\boldsymbol{\varepsilon}^{\mathrm{I}}(\mathbf{k}) = &\int \mathbf{S}(\hat{\mathbf{n}}')\Delta\mathbf{C}(\mathbf{k}' - \mathbf{k})\boldsymbol{\varepsilon}^{\mathrm{I}}(\mathbf{k}')\, d^3 k'\\ &+ \int [\mathbf{S}(\hat{\mathbf{n}}') - \mathbf{I}]\Delta\mathbf{C}(\mathbf{k}' - \mathbf{k})\boldsymbol{\varepsilon}^{\mathrm{L}}(\mathbf{k}')\, d^3 k' + \Delta\mathbf{C}(\mathbf{k})\boldsymbol{\varepsilon}^{\mathrm{A}}.\end{aligned} \tag{2.16}$$

If we assume that $\|\Delta\mathbf{C}\| \ll \|\mathbf{C}\|$, we can approximate $\boldsymbol{\varepsilon}^{\mathrm{I}}(\mathbf{k})$ by the ground state solution of the integral equation given by

$$\boldsymbol{\varepsilon}^{\mathrm{I}}(\mathbf{k}) = \int (\bar{\mathbf{C}}^{-1})\Delta\mathbf{C}(\mathbf{k}' - \mathbf{k})[\mathbf{I} - \mathbf{S}(\hat{n}')]\boldsymbol{\varepsilon}^{\mathrm{L}}(\mathbf{k}')\, d^3 k' + \Delta\mathbf{C}(\mathbf{k})\boldsymbol{\varepsilon}^{\mathrm{A}}. \tag{2.17}$$

Introducing the last equation into (2.15), we derive the elastic energy of the inhomogeneous system under the applied stress field

$$E = E_0 + E^{\mathrm{int}} + E^{\mathrm{c}} + E^{\mathrm{s-c}}, \tag{2.18}$$

where the first two terms in (2.18) express the interaction energy of the applied stress with the inhomogeneous material. E_0 is dependent on the average values of the lattice eigenstrain, $\boldsymbol{\varepsilon}^{\mathrm{L}}(\mathbf{r})$, and the elastic tensor variation, $\Delta\mathbf{C}(\mathbf{r})$, which does not have any effect on the morphology of the structure. It simply changes the kinetics of the phase transition and is given by

$$E_0 = -(\boldsymbol{\sigma}^{\mathrm{A}}\boldsymbol{\varepsilon}_0^{\mathrm{L}} + \tfrac{1}{2}\boldsymbol{\varepsilon}^{\mathrm{A}}\Delta\mathbf{C}_0\boldsymbol{\varepsilon}^{\mathrm{A}})x(\mathbf{k} = 0). \tag{2.19}$$

The second term, E^{int}, is the major part of the interaction energy and controls the morphology of the microstructure

$$E_{\mathrm{int}} = \frac{1}{2}\int Y^{\sigma}(\hat{\mathbf{n}})|x(\mathbf{k})|^2\, d^3k, \tag{2.20}$$

with

$$Y^{\sigma}(\hat{n}) = \boldsymbol{\varepsilon}^{\mathrm{A}}\Delta\mathbf{C}_0[\mathbf{I} - \mathbf{S}(\hat{\mathbf{n}})]\boldsymbol{\varepsilon}_0^{\mathrm{L}}. \tag{2.21}$$

The Fourier energy function, $Y^{\sigma}(\hat{n})$, represents the interaction of the applied stress field with the $\mathbf{k}$ wave vector through both the elastic inhomogeneity $\Delta\mathbf{C}_0$ and the lattice inhomogeneity $\boldsymbol{\varepsilon}_0^{\mathrm{L}}$.

The third term in (2.18) is the elastic coherency strain energy, E_{c}, produced by the lattice eigenstrains, $\boldsymbol{\varepsilon}^{\mathrm{L}}(\mathbf{r})$, and is given by

$$E_c = \frac{1}{2}\int Y(\hat{\mathbf{n}})|x(\mathbf{k})|^2\, d^3k, \tag{2.22}$$

where

$$Y(\hat{\mathbf{n}}) = \boldsymbol{\varepsilon}_0^{\mathrm{L}}\bar{C}[\mathbf{I} - \mathbf{S}(\hat{\mathbf{n}})]\varepsilon_0^{\mathrm{L}}. \tag{2.23}$$

This energy expression is identical to the one derived by Khachaturyan (1983) for homogeneous materials. In our treatment it contains the effective elastic constant tensor $\bar{\mathbf{C}}$.

Finally, the last term in (2.18) is a self-consistent energy term that represents the additional coherency strain energy produced by the additional internal stress field related to the variation of the elastic constant tensor and caused by the lattice eigenstrain. This energy is

$$E_{\mathrm{s\text{-}c}} = \frac{1}{2}\int_k\int_{k'} Y^{\mathrm{s\text{-}c}}(\hat{\mathbf{n}}, \hat{\mathbf{n}}')x(\mathbf{k})x(\mathbf{k}' - \mathbf{k})x^*(\mathbf{k}')\, d^3k'\, d^3k, \tag{2.24}$$

where

$$Y^{\mathrm{s\text{-}c}}(\hat{\mathbf{n}}, \hat{\mathbf{n}}') = \boldsymbol{\varepsilon}_0^{\mathrm{L}}[\mathbf{I} - \mathbf{S}(\hat{\mathbf{n}})]\Delta\mathbf{C}_0[\mathbf{I} - \mathbf{S}(\hat{\mathbf{n}}')]\boldsymbol{\varepsilon}_0^{\mathrm{L}}. \tag{2.25}$$

The self-consistent energy expression of (2.24) is a new term usually neglected when calculating the elastic energy of precipitates. $Y(\hat{n}, \hat{n}')$ has the same symmetry as the $Y(\hat{n})$ and it represents the interaction of the wave vector $\mathbf{k}$ with the $\mathbf{k}'$ vector through the elastic inhomogeneity $\Delta\mathbf{C}_0$ in the absence of the applied stress field. In most practical applications, the first and last term E_0 and $E^{\mathrm{s\text{-}c}}$ are small enough and they can be neglected (for the case of spinodal decomposition we will show that they vanish). Thus, we can write that the total elastic energy of an inhomogeneous material under applied stress is given by

$$E = \frac{1}{2} \int [Y(\hat{\mathbf{n}}) + Y^{\sigma}(\hat{\mathbf{n}})] |x(\mathbf{k})|^2 \, d^3k. \tag{2.26}$$

It is important to emphasize that this formulation in the Fourier space, of both Khachaturyan's (1983) coherency energy and the new term $Y^{\sigma}(\hat{\mathbf{n}})$ that we have derived and which represents the interaction of the applied stress and the products of phase transformations in a solid, has a superior advantage in predicting the morphological changes during a phase transition. This advantage will be manifested in a number of applications which we will consider below.

2.3. *Multiphase Generalization*

The elastic energy Fourier expression of (2.26) can easily be generalized for multiphase crystals, such as a number of different coherent precipitates in a solid. Let us designate $\boldsymbol{\varepsilon}_0^{\mathrm{L}}(\nu)$ and $\Delta\mathbf{C}_0(\nu)$ as the lattice eigenstrains and difference of the elastic constant tensor, respectively, for the νth type of phase in the crystal, and let us assume that there are N types of such phases. The total energy in this case can be written as

$$E = \frac{1}{2} \sum_{\mu,\nu} \int [Y(\hat{\mathbf{n}}, \mu, \nu) + Y^{\sigma}(\hat{\mathbf{n}}, \mu, \nu)] x_{\mu}(\mathbf{k}) x_{\nu}^{*}(\mathbf{k}) \, d^3k, \tag{2.27}$$

where

$$Y(\hat{\mathbf{n}}, \mu, \nu) + Y^{\sigma}(\hat{\mathbf{n}}, \mu, \nu) = [\boldsymbol{\varepsilon}_0^{\mathrm{L}}(\mu)\bar{\mathbf{C}} + \boldsymbol{\varepsilon}^{\mathrm{A}}\Delta\mathbf{C}_0(\mu)][\mathbf{I} - \mathbf{S}(\hat{\mathbf{n}})]\boldsymbol{\varepsilon}_0^{\mathrm{L}}(\nu), \tag{2.28}$$

and $x_{\mu}(\mathbf{k})$ is the Fourier transform of the characteristic function $x_{\mu}(\mathbf{r})$ of the μth phase.

The part of the energy for which $\mu = \nu$ are the self-energies of the individual phases under the influence of applied stress, and the remaining part represents the interaction energies of the phases. It should be noted that this interaction depends on the crystallography and shape of the phases and it is only an approximation in the pair-wide interaction sense. From the elasticity point of view, the stress and strain fields of two phases are simply superimposed.

3. Applications

3.1. *The Effect of Applied Stress on Phase Transitions*

The effect of applied stress on phase transitions can be studied by incorporating the stress interaction energy E_{int} into the general free energy expression of the crystal. The present formulation of this energy is particularly suitable for the method of static concentration waves (SCW) in order–disorder transformations and alloy decomposition. According to the SCW technique, the determination of the probabilities of occupation of sublattices of the ordered phases becomes substantially easier by formulating the problem in the reciprocal space and solving for the amplitudes of the static concentration waves. This method becomes especially powerful for long wavelengths, for it takes into account the long-range interactions which exist in the crystal. In the current treatment the linear theory of elasticity was employed which allows the study of long-range effects. Specifically, the stress interactions with long-range order, spinodal structures, precipitation reactions, and artificial composition modulations are among the phenomena to which the treatment is applicable. On the other hand, for short-range order the microscopic theory of elasticity should be used since in this system the long wavelength approximation is no longer valid. We thus confine our investigations to the above-mentioned cases.

In the absence of applied stress, the minimum work for creation of a concentration fluctuation in a crystal was derived by a number of investigators (Cahn and Hilliard, 1959; Krivloglaz, 1969) and was given by

$$R_k = V\left[\frac{\partial^2 f}{\partial c^2} + Y(\hat{\mathbf{n}}) + 2K_{ij}k_ik_j\right]|\overline{c(\mathbf{k})}|^2, \tag{3.1}$$

where f is the volume free energy, K_{ij} are the gradient energy coefficients, and $c(\mathbf{k})$ is the amplitude of the $\mathbf{k}$th fluctuation.

The applied stress introduces an additional term to the energy which can readily be expressed in terms of $|c(\bar{k})|^2$ by expanding the elastic constants about the average concentration of the material and Fourier transforming the resulting equation. Indeed, if

$$\Delta\mathbf{C}(\mathbf{r}) = \left.\frac{\partial \mathbf{C}}{\partial c}\right|_{c=\dot{c}_0} \Delta c(r),$$

from which

$$\Delta\mathbf{C}(\mathbf{k}) = \mathbf{C}'_0 c(\mathbf{k}) \qquad \text{and} \qquad \varepsilon^{\mathrm{L}}(\mathbf{r}) = \left.\frac{1}{a}\frac{da}{dc}\right|_{c=\dot{c}_0} \Delta c(\mathbf{r})$$

from which $\boldsymbol{\varepsilon}^{\mathrm{L}}(\mathbf{k}) = \boldsymbol{\varepsilon}^{\mathrm{L}}_0 c(\mathbf{k})$, where

$$\mathbf{C}'_0 = \left.\frac{\partial \mathbf{C}}{\partial c}\right|_{c=0} \qquad \text{and} \qquad \boldsymbol{\varepsilon}^{\mathrm{L}}_0 = \left.\frac{1}{d}\frac{dd}{dc}\right|_{c=0}$$

(d is the lattice spacing), we can then derive the characteristic function, $|x(\mathbf{k})|^2 = |c(\mathbf{k})|^2$, and express the total free energy of creating a fluctuation with $\mathbf{k}$ wave vector under the influence of applied stress as

$$R_k = V\left[\frac{\partial^2 f}{\partial C^2} + Y(\hat{\mathbf{n}}) + Y^{\sigma}(\hat{\mathbf{n}}) + 2K_{ij}k_ik_j\right]|\overline{c(k)}|^2, \tag{3.2}$$

where

$$Y^{\sigma}(\hat{\mathbf{n}}) = \boldsymbol{\varepsilon}^{\mathrm{A}}\mathbf{C}'_0[\mathbf{I} - \mathbf{S}(\hat{\mathbf{n}})]\boldsymbol{\varepsilon}^{\mathrm{L}}_0 \tag{3.3}$$

is the stress interaction energy function. This term is typically an order of magnitude less than $Y(Y^\sigma/Y \sim (2\varepsilon^A \Delta C)/(\varepsilon_0^L C) \sim \frac{1}{10})$, and therefore it cannot alter drastically the structure and symmetry of the fluctuation distribution in the crystal. There are two cases, however, in which $Y^\sigma(\hat{\mathbf{n}})$ plays a very important role: one near the critical temperature of an order–disorder transition and the other during spinodal decomposition. In the first case, near the critical temperature, of $\partial^2 F/\partial c^2 = 0$, the distribution of fluctuations is dominated by the symmetry of the elastic energy functions $Y(\hat{n})$ and $Y^\sigma(\hat{n})$. Since according to the SCW method the probability distribution of C_k is Gaussian, the mean square of the kth wave is given by

$$|C(k)|^2 = \frac{kT}{V[\partial^2 f/\partial c^2 + Y(\hat{n}) + Y^\sigma(\hat{n}) + 2K_{ij}k_i k_j]}. \tag{3.4}$$

Fluctuations of large amplitudes would be developed in those orientations, $\hat{\mathbf{n}}$, which minimize the above sum: $\min[Y(\hat{\mathbf{n}}) + Y^\sigma(\hat{\mathbf{n}})]$. Although the $Y^\sigma(\hat{\mathbf{n}})$ is much smaller than $Y(\hat{n})$ and cannot change the minimum value of $Y(\hat{n})$, it can still alter the development of the microstructure if there exists more than one equivalent crystallographic orientation, $\hat{\mathbf{n}}_{\min}$, which minimizes $Y(\hat{\mathbf{n}})$. Since the $Y^\sigma(\hat{\mathbf{n}})$ is proportional to the applied stress tensor $\boldsymbol{\sigma}^A$, it assumes different values at the various $\hat{\mathbf{n}}_{\min}$'s, thus altering the amplitudes of the fluctuations in those directions.

A case in which the stress interaction is quite pronounced is the spinodal decomposition. Here $d^2 f/dc^2 < 0$ in the spinodal regime and thus for some temperature and $\hat{\mathbf{n}}$, $d^2 f/dc^2 + Y(\hat{n}) \sim 0$.

It can be therefore concluded that the $Y^\sigma(\hat{\mathbf{n}})$ is the effective term that determines the growth rate of the concentration fluctuations. For temperatures well below the coherent spinodal, the orientation dependence of the fluctuations is controlled by the internal Fourier energy function $Y(\hat{\mathbf{n}})$. Again, the only way the applied stress can influence this dependence is the case of degeneracy, i.e., alter the dependence among crystallographically equivalent orientations which minimize $Y(\hat{\mathbf{n}})$. In view of (3.2), the linear theory of diffusion developed by Cahn predicts the following time evolution for the $\mathbf{k}$th fluctuation:

$$c(\mathbf{k}, t) = c(\mathbf{k}, 0)e^{R(\mathbf{k})t}, \tag{3.5}$$

where

$$R(\mathbf{k}) = -\left(\frac{M}{N_v}\right)k^2[f'' + Y(\hat{\mathbf{n}}) + Y^\sigma(\hat{\mathbf{n}}) + 2Kk^2], \tag{3.6}$$

where M is the atomic mobility, N_v is the number of atoms per unit volume, and $f'' = \partial^2 f/\partial c^2$. $R(\mathbf{k})$ is an amplification factor which differs from that derived by Cahn, by the term $Y^\sigma(\hat{\mathbf{n}})$ introduced by the applied stress. The maximum $R(\mathbf{k})$ in most cubic crystals occurs for $\mathbf{k}$ parallel to the three $\langle 100\rangle$ directions and a magnitude which is given by

$$k_m = \sqrt{-\frac{f'' + Y[100]}{4K}}. \tag{3.7}$$

Let us now define a parameter

$$\Delta T^{\sigma} = \frac{Y^{\sigma}(\hat{n})}{s''}, \tag{3.8}$$

which physically represents the difference between the coherent spinodal T_s^* and that of the stress induced coherent spinodal defined by

$$f'' + Y(001) + Y^{\sigma}(001) = 0. \tag{3.9}$$

Noting also that $f'' \cong -(T - T_s)s''$ where s is the entropy per unit volume, and combining with the results of (3.6) and (3.8) we obtain the maximum amplification factor under applied stress normalized to that without applied stress

$$\frac{R^{\sigma}_{\max}}{R^{0}_{\max}} = \left(1 - \frac{\Delta T^{\sigma}}{\Delta T}\right)^2, \tag{3.10}$$

where $\Delta T = T - T_s^*$ is the coherent spinodal temperature, and T is the annealing temperature. Since $\Delta T < 0$ and $s'' < 0$ for a spinodal system within the spinodal regime, the criterion of stress-assisted growth of the composition wave is

$$Y^{\sigma}(100) < 0. \tag{3.11}$$

From (2.21) it can be seen that the sign and magnitude $Y^{\sigma}(\hat{n})$ depends on the applied stress tensor with respect to the three $\langle 100 \rangle$ crystallographic directions.

3.2. *The Influence of Applied Stress on the Morphology of the Spinodal Structure*

To demonstrate the stress influence on the development of the microstructure, we will consider the specific example of the Au–Ni 50 a/o Au spinodal alloy. Let us also assume that a uniaxial stress is applied along the [001] of the cubic Au–Ni single crystal. In this case, the $Y(\hat{n})$ and $Y^{\sigma}(\hat{n})$ reduce to

$$Y(100) = 2\eta^2\left(C_{11} + C_{12} - \frac{2C_{12}^2}{C_{11}}\right), \tag{3.12}$$

and

$$Y^{\sigma}(001) = \sigma^{A}\eta\left[\frac{\{(C_{11} + C_{12})C'_{11} - 2C_{12}C'_{12}\}}{C_{11}(C_{11} - C_{12})} - \frac{C'_{11} + 2C'_{12}}{C_{11} + 2C_{12}}\right], \tag{3.13}$$

$$Y^{\sigma}(100) = Y^{\sigma}(010) = \sigma^{A}\eta\left[\frac{(C_{11}C'_{12} - C_{12}C'_{11}}{C_{11}(C_{11} - C_{12})} - \frac{C'_{11} + 2C'_{12}}{C_{11} + 2C_{12}}\right], \tag{3.14}$$

where σ^{A} is the uniaxial stress along the [001] orientation, $\eta = (1/a)(da/dc)$, and $C'_{ij} = (\partial C_{ij}/\partial c)|_{c=c_0}$.

The expression of the $Y(100)$ represents the Fourier elastic strain energy function produced by the internal stresses of the average lattice and is identical to the one derived by Cahn (1962). The $Y^{\sigma}(001)$ is the Fourier interaction energy function and represents the interaction of a uniaxial stress $\sigma^{A}\|[001]$ with the wave vector $\mathbf{k}\|[001]$.

The $Y^{\sigma}(100)$ and $Y^{\sigma}(010)$ are identical because of symmetry and represent the interaction of $\sigma^{A}\|[001]$ with the wave vector $\mathbf{k}\|[100]$ or $\mathbf{k}[010]$, respectively. In order to calculate these energy functions, we require the elastic constants and lattice parameter as a function of composition. These were taken from the data of Golding *et al.* (1967) and Ellwood and Bayley (1952), respectively. From these data, the following values for the energy functions were obtained

$$\begin{aligned} Y(100) &= 122.0\ \text{GPa}, \\ Y^{\sigma}(001) &= -0.237\sigma^{A}, \\ Y^{\sigma}(100) &= Y^{\sigma}(010) = 0.119\sigma^{A}. \end{aligned} \tag{3.15}$$

Thus, if a tension is applied ($\sigma^{A} > 0$) along the [001] direction, the following condition is satisfied:

$$Y^{\sigma}(001) < Y^{\sigma}(100) = Y^{\sigma}(010),$$

and as a result the [001] concentration wave will grow faster than the [100] and [010] waves. This effect becomes particularly pronounced when annealing near the coherent spinodal temperature. Indeed, from (3.8) the temperature parameter ΔT^{σ} can be estimated by assuming that the entropy of mixing is ideal

$$s'' = -\frac{N_v k_B}{c(1-c)}.$$

Introducing the above numerical value for $Y^{\sigma}(001)$ and assuming an applied tension of $\sigma = 10{,}000$ psi, we obtain

$$T = 4.2\ ^{\circ}\text{C}.$$

If we assume $T = 4.2$ from (3.9)

$$R^{\sigma}_{\max} = 4R^{0}_{\max}.$$

It is therefore concluded that by annealing the Au–Ni alloy 4.2 °C below the coherent spinodal under an applied uniaxial tension of 10,000 psi, we can increase the growth rate by a factor of four. As a result, the [001] will grow much faster and dominate the morphology of the microstructure. In fact, for $R^{0}_{\max}t \sim 1$ the [001] wave will have approximately 55 times a larger amplitude than those of the [100] and [010] waves. When $R^{0}_{\max}t \sim 2$, the amplitude of the [001] wave will be 3000 times greater than those of the two other equivalent crystallographic orientations.

The effect of applied uniaxial tension parallel to [001] for Au–Ni will thus be the development of a one-dimensional periodic structure in the [001] direction. On the other hand, if compression is applied along the [001]

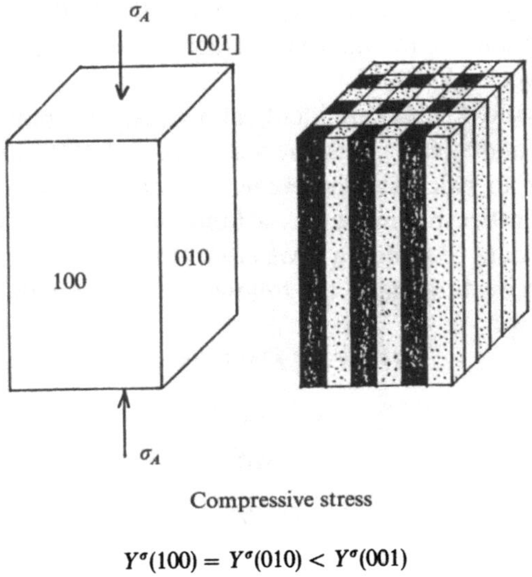

FIG. 1. The elastic interaction of a tensile stress in the [001] resulting in a plate-like formation.

direction, the calculation shows that

$$Y^\sigma(100) = Y^\sigma(010) < Y^\sigma(001),$$

which favors the growth of the [100] and [010] waves.

The resulting microstructure will consist of two perpendicular plane waves in the [100] and [010] directions, respectively, which produce rods parallel to the [001] directions. These conclusions can easily be demonstrated in Figs. 1 and 2.

In a spectacular experiment, Ahn and Tsakalakos (1985) demonstrated the effect of applied stress on the decomposition of Cu–15Ni–8Sn spinodal alloy. It was observed that the applied stress reduced the wavelength of the concentration wave shown in Fig. 3. This dependence can easily be derived from (3.7) and (3.13) and is given by

$$\frac{\lambda^\sigma}{\lambda^0} = \frac{1}{\sqrt{1 + \beta\sigma^{\mathrm{A}}}},$$

$$\beta = -\frac{1}{[S'' + Y(\hat{n})]}\left\{\varepsilon_0^{\mathrm{L}}\left[\frac{\Delta C_{11} + 2C_{12}}{C_{11} + 2C_{12}} - \frac{(C_{11} + C_{12})\Delta C_{11} - 2C_{12}\Delta C_{12}}{C_{11}(C_{11} - C_{12})}\right]\right\}.$$

3.3. Applied Stress-Induced Transformations on Coherent Precipitates

The system chosen for study is a nickel-based superalloy, in which the strengthening γ' precipitates are usually observed to form a stable coarsening micro-

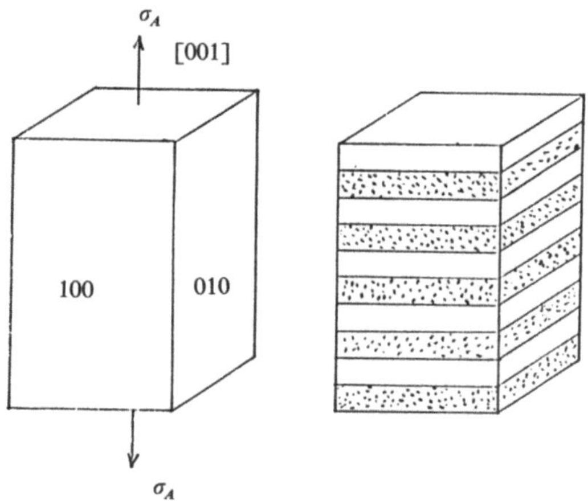

$Y^{\sigma}(001) < Y^{\sigma}(010) = Y^{\sigma}(100)$

FIG. 2. The elastic interaction of a compressive stress in the [001] resulting in rod-like formation.

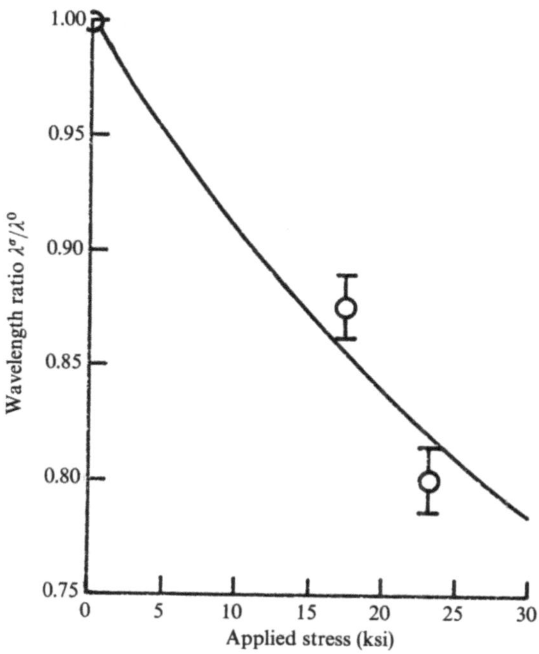

FIG. 3. The wavelength ratio, $\lambda^{\sigma}/\lambda^{0}$, of the spinodal structure, as a function of the applied stress for Cu–Ni–Sn (aging temperature, 450 °C). (Ahu and Tsakalakos (1985).)

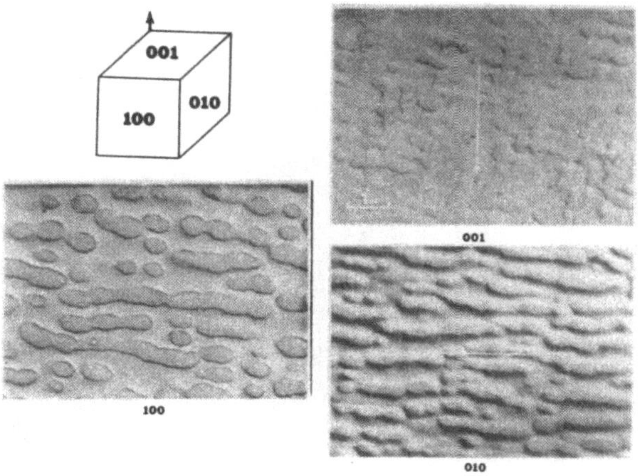

FIG. 4. Coarsening of the γ' precipitates in Udimet-700 single crystals for tensile loading. (Tien and Copley (1971).)

structure. Tien has shown that when the lattice misfit is appreciable the γ' precipitates in the form of cuboids of fairly uniform size and are aligned along the three $\langle 100 \rangle$ directions of the matrix. Upon stress annealing the morphology of the γ' precipitates was found to be dependent on stress sense. Tensile uniaxial stress resulted in precipitate plates perpendicular to the stress axis. A compressive stress provided γ' precipitates of long parallelepipeds with the long axis parallel to the stress axis. These morphologies are illustrated in Figs. 4 and 5 as observed by Tien and Copley (1971). For a uniaxial stress σ along the [001] direction the Fourier elastic energy function can be calculated

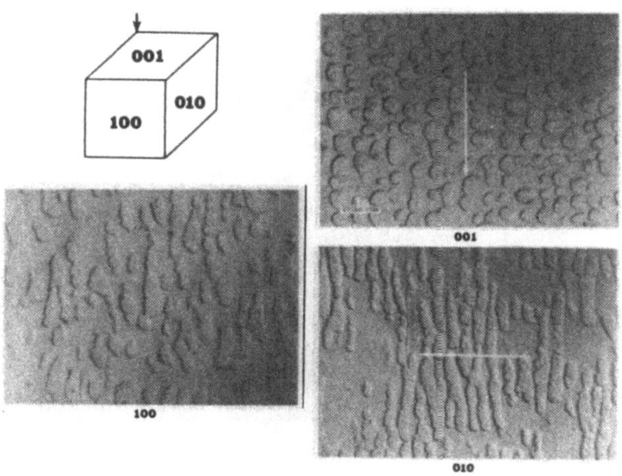

FIG. 5. Coarsening of the γ' precipitates in Udimet-700 single crystals for compressive loading. (Tien and Copley (1971).)

TABLE 1. Elastic constants.

	C_{11} $\times 10^{12}$ dyns/cm^2	C_{12}	C_{12}
γ	2.508	1.5	1.235
γ'	2.148	1.277	0.997
Δ (difference)	−0.36	−0.223	−0.238

from (2.21) in two cases

(i) $\mathbf{k} \| [001], \quad \sigma^{A} \| [001] \|,$

$$Y^{\sigma}[001] = \sigma^{A}\varepsilon_0^{L}\left\{\frac{[(C_{11} + C_{12})\Delta C_{11} - 2C_{12}\Delta C_{12}]}{C_{11}(C_{11} - C_{12})} - \frac{\Delta C_{11} + 2\Delta C_{12}}{C_{11} + 2C_{12}}\right\}, \tag{3.16}$$

and

(ii) $\mathbf{k} \| [100] \text{ or } [010] \quad \text{and} \quad \sigma^{A} \| [001],$

$$Y^{\sigma}[100] = \sigma^{A}\varepsilon_0^{L}\left\{\frac{(C_{11}\Delta C_{12} - C_{12}\Delta C_{11})}{C_{11}(C_{11} - C_{12})} - \frac{\Delta C_{11} + 2\Delta C_{12}}{C_{11} + 2C_{12}}\right\}. \tag{3.17}$$

The stress-free strain is given by $\varepsilon_0^{L} = 0.0002$, which is a positive misfit, and the elastic constants are shown in Table 1.

Using the values in Table 1, the Fourier energies for the two orientations were calculated

$$Y^{\sigma}[001] = -3.12 \times 10^{-5}\sigma^{A},$$
$$Y^{\sigma}[100] = 2.85 \times 10^{-5}\sigma^{A}.$$

Substituting the stress, $\sigma^{A} = 22{,}500$ psi, we obtain

$$\left.\begin{aligned} Y^{\sigma}[001] &= -4.836 \times 10^{4} \text{ dyns/cm}^2, \\ Y^{\sigma}[100] &= 4.42 \times 10^{4} \text{ dyns/cm}^2, \end{aligned}\right. \quad \text{for tensile,}$$

and

$$\left.\begin{aligned} Y^{\sigma}[001] &= 4.836 \times 10^{4} \text{ dyns/cm}^2, \\ Y^{\sigma}[100] &= -4.42 \times 10^{4} \text{ dyns/cm}^2, \end{aligned}\right. \quad \text{for compressive.}$$

Meanwhile, the Fourier energy function $Y[001]$ can be calculated from (2.23)

$$Y[001] = \left[C_{11} = C_{12} - \frac{2C_{12}^2}{C_{11}}\right]\eta^2,$$

which yields

$$Y[001] = 8.85 \times 10^{4} \text{ dyns/cm}^2.$$

Thus, it is seen that the interaction energy Y^{σ} has a large contribution to the total energy since it is about 50% of the coherency strain energy Y. Moreover,

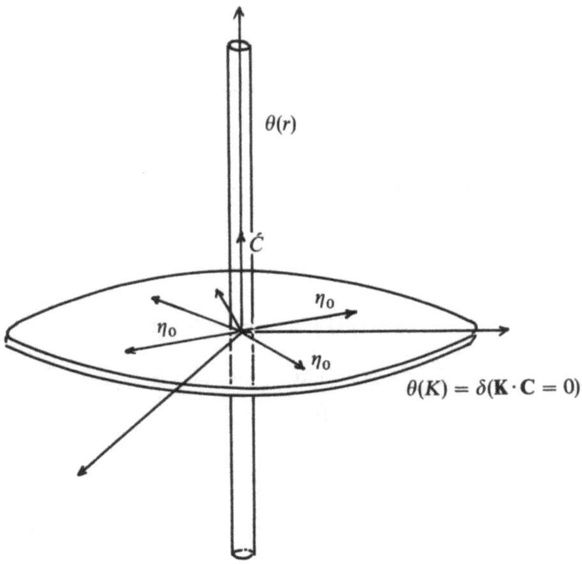

FIG. 6. Elastic interaction of an applied stress field with the [001] wave vector.

the applied stress will have a pronounced effect on the resulting morphology of the precipitates. For a tensile stress, since $Y^\sigma[001] < Y^\sigma[100] = Y^\sigma[010]$, the total energy of the system will be minimum for the [001]. This will secure faster growth in this direction and the expected morphology of the precipitates, therefore, will be plates perpendicular to the tensile axis. On the other hand, a compressive stress reverses the signs of the Y^σ's and thus yields the condition

$$Y^\sigma[100] = Y^\sigma[010] < Y^\sigma[001].$$

This condition clearly demonstrates that along the two orientations, [100] and [010], the total Fourier elastic energy is minimal, which results in rods parallel to the compressive stress axis as demonstrated in Figs. 4 and 5. The interaction of an applied stress field with various wavevectors are also shown in Figs. 6 and 7. The plane morphology of the precipitates can be easily explained by the following argument.

Since the total energy of the system is given by

$$E = \int [Y(\hat{n}) + Y^\sigma(\hat{n})]|\theta(\mathbf{k})|^2\, d\mathbf{k},$$

and $Y[001] + Y^\sigma[001] = \text{minimum}$ (for a tensile stress). E reaches an absolute minimum where $\theta(\mathbf{k})^2$ is a delta function along the [001] direction. But if $\theta(k)$ is a delta function in reciprocal space, then the shape of the precipitate defined by $\theta(\mathbf{r})$ is a delta function on the plane perpendicular to the [001] direction. The delta function representation of the Fourier transform

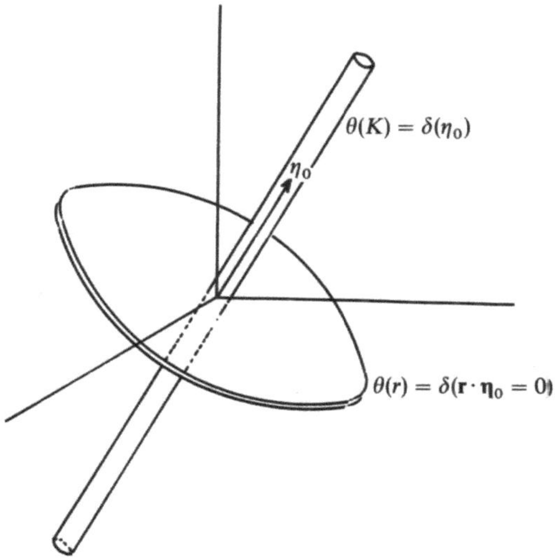

FIG. 7. Elastic interaction of an applied stress field with the [100] and [010] wave vectors.

shape function is shown schematically in Figs. 8 and 9. Computer simulations, taking into account the elastic stress-induced anisotropy, are shown in Figs. 10, 11, and 12. It can be seen that tensile stress produces plate-like morphology and compressive stress produces rod-like morphology. This is in agreement with the experimental results of Tien and Copley (1971), shown in Figs. 4 and 5.

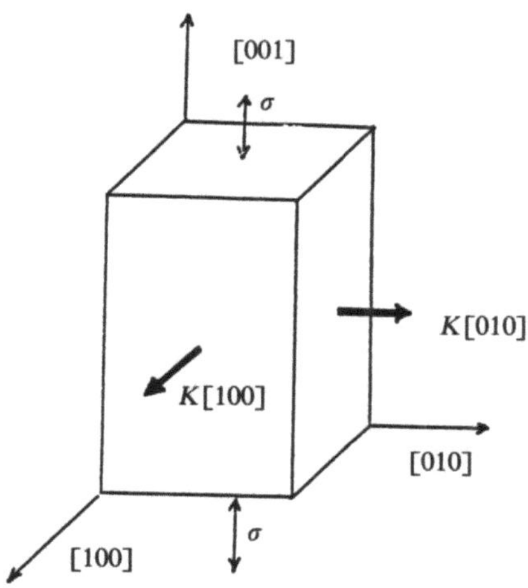

FIG. 8. A rod-like delta function distribution in the reciprocal space leading to a plate-like formation in real space.

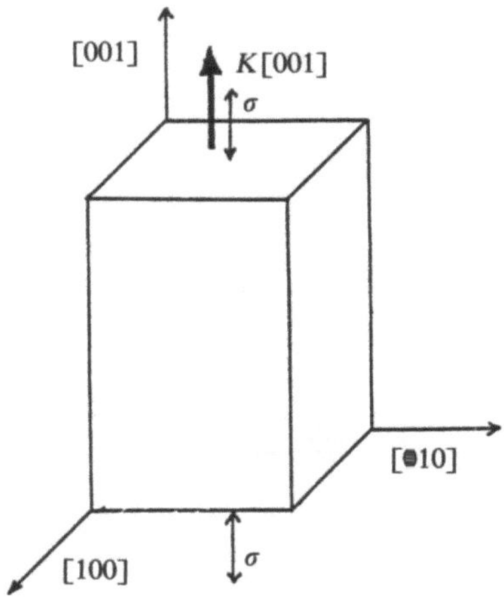

FIG. 9. A plate-like delta function distribution in the reciprocal space (a case of infinite degeneracy in hexagonal crystals on the basal plane) leading to a rod-like formation in real space.

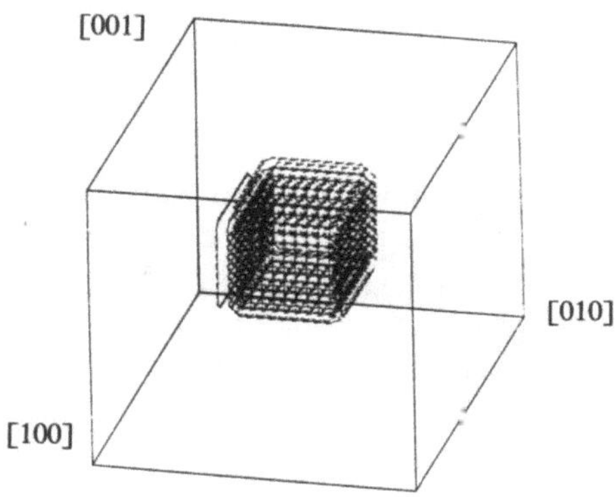

FIG. 10. A three-dimensional computer representation of the initial cuboidal structure, $t = 0$. Composition contour lines $c(x, y, z, t) = 90\%$ are shown. (Jankowski, Wingo, and Tsakalakos (1985).)

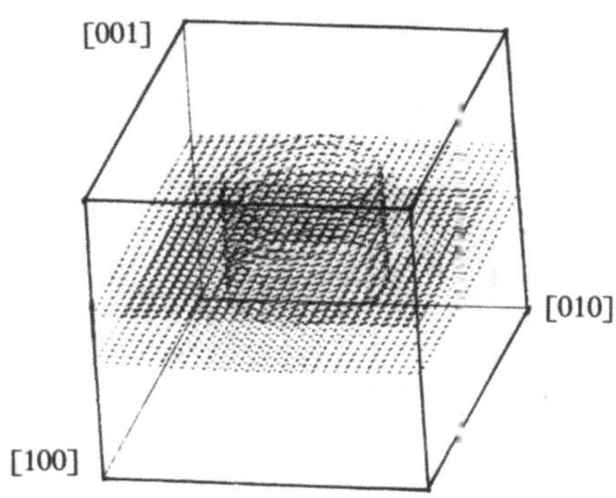

FIG. 11. Coarsening of the structure under the [001] tensile loading at time $t = 100$. Composition contour lines $c(x, y, z, t) = 90\%$ are shown. (Jankowski, Wingo, and Tsakalakos (1985).)

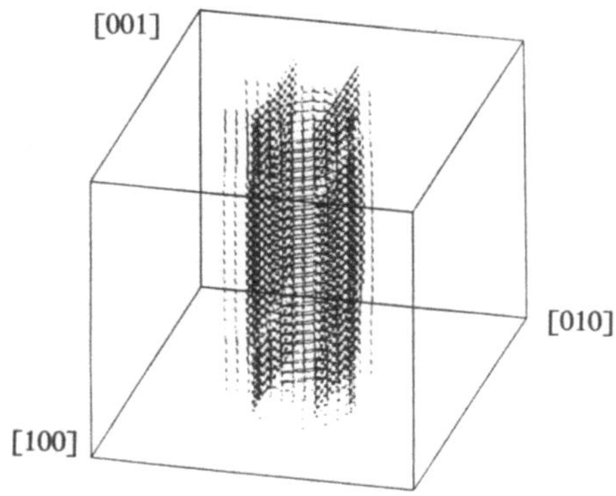

FIG. 12. Coarsening of the structure under the [001] compressive loading at time $t = 100$. Composition contour lines $c(x, y, z, t) = 90\%$ are shown. (Jankowski, Wingo, and Tsakalakos (1985).)

Similar conclusions can be drawn for uniaxial applied stresses in other crystallographic directions. Computer simulations by Jankowski *et al.* (1985) are shown in Figs. 10, 11, 12. The figures clearly demonstrate the effect of stress, i.e., plate-like formation for tension and rod-like formation for compression.

3.4. Computer Simulations in Transformation Toughened Ceramics

The transformation toughening of alumina is based on the unusual phase transformation in zirconia

$$ZrO_2 \text{ (tetragonal)} \xrightarrow{1100\,^{\circ}\mathrm{C}} ZrO_2 \text{ (monoclinic)}.$$

Accompanying this phase transformation is a 3–5% increase in volume and an 8% shear. It is known that the tetragonal to monoclinic transformation is martensitic in nature. The development of such strains on cooling through the phase transformation temperature can cause sintered zirconia in bulk to disintegrate. This problem has been largely overcome by the addition of oxides such as CaO, MgO, and Y_2O_3 to form solid solutions with zirconia, thus stabilizing the high-temperature phase over the entire temperature range.

To compute the stress and stress distributions in this ceramic material we use a computer simulation algorithm based on a distribution in the reciprocal space. The stress can readily be computed by the Fourier space formalism using (2.4) and (2.14)

$$\boldsymbol{\sigma} = \mathbf{C} \iiint_{-\infty}^{\infty} \hat{\Omega}(\hat{n})\boldsymbol{\sigma}^0\hat{\mathbf{n}} * \hat{\mathbf{n}}\theta(\mathbf{k})e^{i\mathbf{k}\cdot\mathbf{r}} \frac{d^3k}{(2\pi)^3} - \boldsymbol{\sigma}^0\theta(\mathbf{r}), \tag{3.18}$$

where $\boldsymbol{\sigma}^0 = \mathbf{C}\boldsymbol{\varepsilon}_0^{\mathrm{L}}$, $\Omega_{ik}^{-1}(\hat{\mathbf{n}}) = C_{ijkl}n_jn_l$, and $\theta(\mathbf{k}) = \int \theta(\mathbf{r})e^{i\mathbf{k}\cdot\mathbf{r}}\,d^3r$ is the Fourier

TABLE 2. Shape functions (Fourier transforms).

Shape	$\theta(k)$
(1) Sphere radius R	$3V/(kR)^3[\sin(kR) - (kR\cos(kR)]$
(2) Ellipsoid	$3V/\phi(k)^3[\sin(\phi(k)) - (\phi(k)k\cos(\phi(k)]$
	$\phi(k) = L_{ij}x_ix_j$
	$L_{ij}^{-1}x_ix_j = 1$ (ellipsoid surface)
(3) Cuboid $2L_x$, $2L_y$, $2L_z$	$V\sin(k_xL_x)/(k_x1_x)$*
	$\sin(k_yL_y)/(k_yL_y)$*
	$\sin(k_zL_z)/(k_zL_z)$
(4) Tetrahedron (L)	$i\sum_{p=1}^{p_3}\frac{e^{ik_pL}}{k_p(k_p-k_j)(k_p-k_L)} - \frac{i}{k_1k_2k_3}$
(5) Octahedron (L)	$8\sum_{p=1}^{3}\frac{k_p\sin k_pL}{(k_p^2-k_j^2)(k_p^2-k_l^2)}$
	$p \neq j \neq l$

transform of the shape function $\theta(\mathbf{r}) = \{\begin{smallmatrix}1 \text{ in}\\ 0 \text{ out}\end{smallmatrix}\}$. Table 2 shows the Fourier transforms of most common shapes of precipitates and their corresponding geometry (Table 3).

The lattice eigenstrain tensor for tetragonal to monoclinic transformation can be obtained from crystallographic data and is given by

$$\varepsilon_{ij}^{\mathrm{L}} = \begin{pmatrix} -0.000149 & 0 & 0.08188 \\ 0 & 0.02442 & 0 \\ 0.08188 & 0 & 0.02386 \end{pmatrix}.$$

The elastic constants of Al_2O_3 and ZrO_2 are also given in the literature.

Figure 13 shows the result of stress computations along the [100] directions for the aspect ratio of a cuboid ZrO_2 particle 1 : 1. Figures 14 and 15 are plots of stress versus distance along the [100] and [001] orientations, respectively. It is interesting to note that the stress is maximum near the interface along the long [001] orientations, whereas for all other directions the stresses decrease monotonically from the center of the particle. Our calculations have also shown that the stress is constant for ellipsoidal inclusions. These observations clearly demonstrate that an ellipsoidal inclusion calculation is not realistic in representing stress fields produced by precipitation or other phase transformation reactions.

These computational results are in qualitative agreement with Chiu's (1977) calculations, the one difference being that our present model has been applied to both dilatational and shear eigenstrain components introduced by the martensitic transformation. Preliminary results for more realistic shapes such as an octahedron, show that the stress decreases monotonically very rapidly within the particle. These results will be published in a future publication by Rao and Tsakalakos (1990).

TABLE 3. Geometry of shape functions.

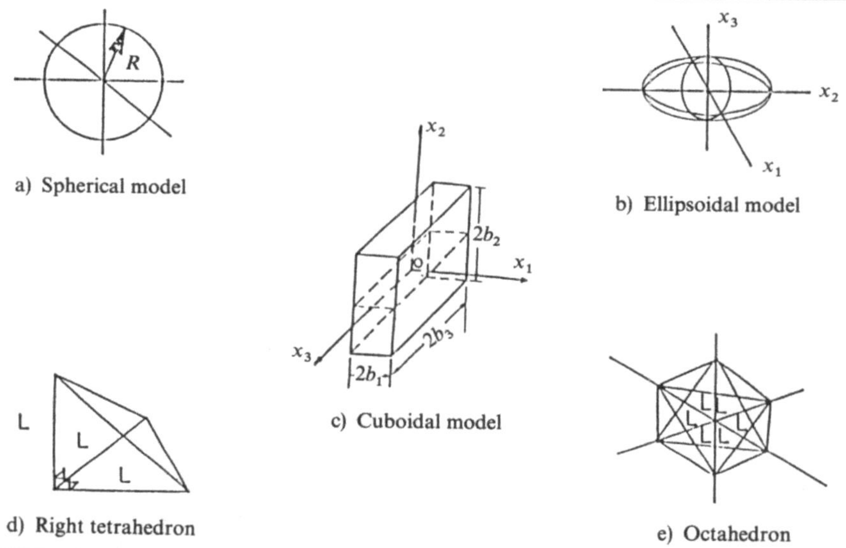

3.5. *Microscopic Elasticity Model—Interfacial Phenomena*

Coherency is an extremely important concept in materials science and certainly plays a central role in interfacial phenomena. The "supermodulus effect" in nanoscale multilayers, the strengthening mechanism in precipitation hardening alloys, and the optoelectronic properties of semiconducting materials represent some examples of the wide range of applicability of the coherency

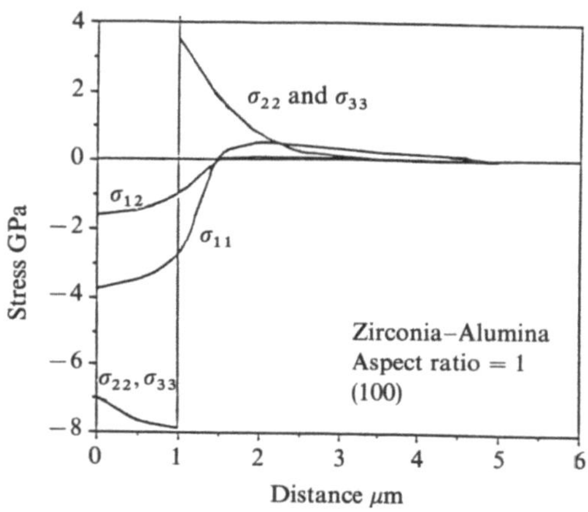

FIG. 13. Stress versus distance for a particle size of 2 μm.

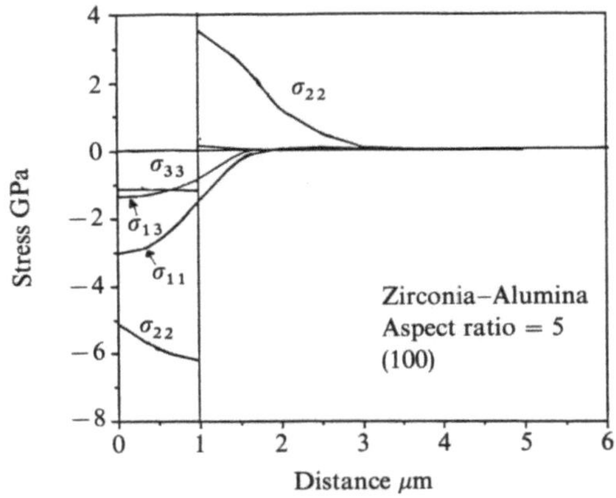

FIG. 14. Stress versus distance for a particle size of 2 μm.

concept. However, when dimensions reach nanoscale, the continuous elasticity solution for the coherency stress given above fails to address the fundamentally different situation. The structure of these interfaces are shown in Fig. 16.

In such dimensions the interface–interface interactions, the distortions of the binding energies of atoms at the interface, and the local disorder due to interdiffusion are factors which should be taken into account promptly. In fact, a number of interfacial phenomena cannot be understood on the basis of coherency. The loss of the supermodulus effect near the monolayer superlattice scale, the discrepancy of the critical bilayer thickness between experiment and theory, and a number of experimental data of anomalous inter-

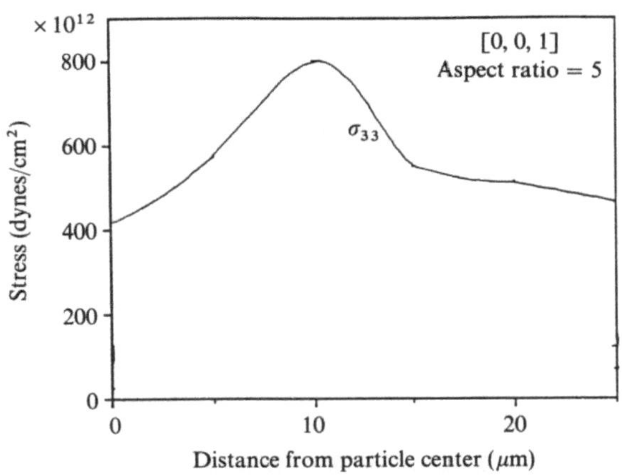

FIG. 15. Stress versus distance for a particle size of 10 μm.

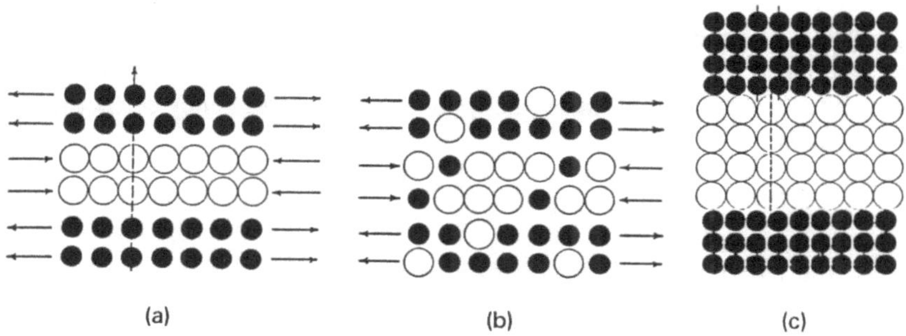

FIG. 16. Multilayered structures: (a) coherent superlattice; (b) composition modulated alloys; and (c) incoherent modulated structure.

planar expansions and contractions near interfaces, are examples directly related to the nanoscale dimensionality.

Obviously, any adequate physical and mathematical treatment of this problem should be made in terms of the changes of the binding energies near the interface. Similar models have been used on the theoretical modeling of free surfaces using the Monte Carlo method, the molecular dynamics method, the embedded atom method, or sometimes, the first principle coherent potential approximation or tight binding energy calculations.

However, in the present study we propose to use a modification of Khachaturyan's (1983) microscopic theory of elasticity which takes into account the discrete nature of the crystal lattice. This approach provides a simple and elegant analytical solution to the problem without the use of tedious and elaborate computer simulation analysis.

The case of an elastic displacement field generated by solute atoms near the interfaces becomes substantially more important when the layer thicknesses are of the same order of magnitude as the interatomic distances. The model should be viewed as a harmonic approximation of the homogeneous force constant case. Details will be published by Tsakalakos and Khachaturyan (1990) elsewhere, and only the major points are given below:

The elastic displacement field $\mathbf{u}(\mathbf{r})$ which is generated by a concentration wave $\Delta c(\mathbf{r})$ is given by

$$\mathbf{u}(\mathbf{r}) = \frac{1}{N} \sum_{k} G_{ij}(\mathbf{k}) F_j(\mathbf{k}) \Delta c(\mathbf{k}) e^{i\mathbf{kr}} \hat{e}_i. \tag{3.19}$$

where $\Delta c(\mathbf{k}) = \sum_r \Delta c(\mathbf{r}) e^{i\mathbf{kr}}$ is the Fourier transform of the concentration wave, and the summation is over N points in the first Brillouin zone of the average lattice. $\mathbf{F}(\mathbf{k})$ is the Fourier transform of the so-called Kazanki forces that describe the ability of a solute atom to distort the environmental host lattice. $G_{ij}(\mathbf{k})$ is Green's tensor of the inverse of the Fourier transform of the dynamic Born–von Karman matrix which is the fundamental characteristic of the dynamic properties of a crystal because it determines the vibration frequency spectrum. In general, the Hermitian tensor $G_{ij}(\mathbf{k})$ can be written in terms of the eigenvectors, $\hat{e}_s(\mathbf{k})$, and eigenvalues, $m\omega_s^2(\mathbf{k})$, of the dynamic matrix, where m is the mass of the host atom, ω_s is the vibration frequency of the branch s $(s = 1, 2, 3)$

$$G_{ij}(\mathbf{k}) = \sum_{s=1}^{3} \frac{e_s^i(\mathbf{k}) e_s^{*i}(\mathbf{k})}{m\omega_s^2(\mathbf{k})}. \tag{3.20}$$

Thus, the calculation of the displacement field requires knowledge of the phonon dispersion curves. On the other hand, it is possible to evaluate approximately Green's tensor from the Born–von Karman constants. Expressions of the dynamic matrix up to the eighth coordination shell are given by Khachaturyan (1983). The determination of the Kazanki forces is the most difficult problem. Expressions are also given by Khachaturyan (1983).

Application of this model in a square superlattice (square concentration wave) is considered in the following: $A[001]\|[001]$ A/B square superlattice of fcc crystal structure can be represented by a static concentration wave

$$\Delta c(r) = \begin{cases} 1 - \dfrac{L}{2}\bar{d} \leqq z \leqq \dfrac{L}{2}\bar{d}, \\ 0 - \dfrac{M}{2}\bar{d} \leqq z < \left(-\dfrac{L}{2} - 1\right)\bar{d} \ \text{ and } \ \left(\dfrac{L}{2} + 1\right)d < z < \dfrac{M}{2}\bar{d}, \end{cases}$$

where $\bar{d}$ is the average interplanar spacing and the $L + 1$ planes in layers A and M is the total number of planes in the superlattice period. It can be shown that

$$\Delta c(\mathbf{k}) = \sin\frac{\pi(L+1)}{M} \Big/ \sin\frac{\pi}{M},$$

$$G_{33}(k) = \left\{8\alpha_1\left[1 - \cos\frac{\pi}{M}\right] + 4\alpha_2 \sin\frac{2\pi}{M} + 4\alpha_3\left[1 - \cos\frac{2\pi}{M}\right]\right.$$
$$\left. = 4\alpha_3\left[1 - \cos\frac{2\pi}{M}\right]\right\}^{-1}, \tag{3.21}$$

$$F_3(k) = 8\hat{\alpha}_1\left(1 - \cos\frac{\pi}{M} + 2\hat{\alpha}_2\left(1 - \cos\frac{2\pi}{M}\right)\right.$$
$$\left. + 8\hat{\alpha}_3\left(1 - \cos\frac{2\pi}{M}\right) + 16\hat{\beta}_3\left(1 - \cos\frac{\pi}{M}\right)\right), \tag{3.22}$$

where $4\alpha_1 + 4\alpha_2 + 16\alpha_3 + 8\beta_3 = \alpha C_{11}$ (long wavelength approximation), α_1, α_2, α_3, β_3 are the Born–von Karman constants up to the third coordination shell, and $2\hat{\alpha}_1 + 2\hat{\beta}_1 + 4\hat{\beta}_2 + 4\hat{\alpha}_3 + 20\hat{\beta}_3 = \alpha C_{44}$ (long wavelength approximation) $\hat{\alpha}_1$, $\hat{\beta}_1$, $\hat{\beta}_2$, $\hat{\alpha}_3$, $\hat{\beta}_3$ are the Kazanki forces up to third neighbor approximation.

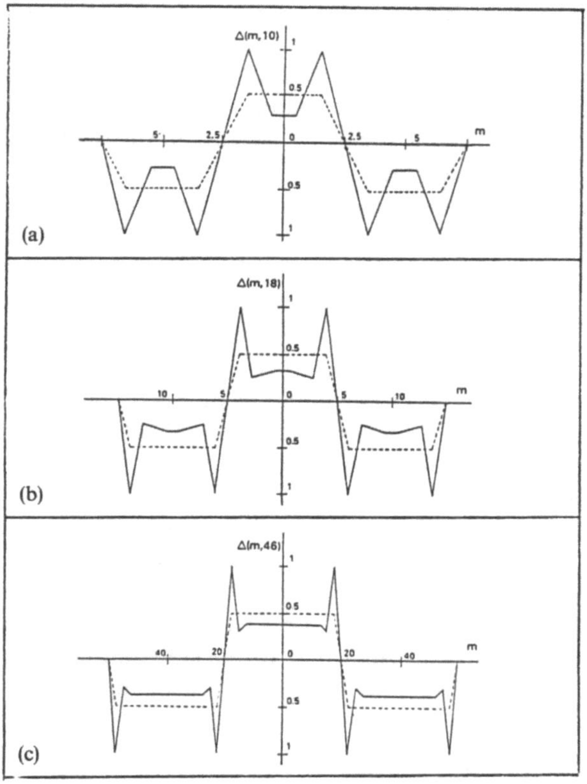

FIG. 17. (a) The percentage d-spacing change in the mth layer for a $5d_A$–$5d_B$ multilayer (solid line), $\Delta(m, 10)$. The broken line corresponds to the continuum elasticity model and the first-neighbor interaction microscopic (they are identical); (b) $\Delta(m, 18)$; and (c) $\Delta(m, 46)$.

Inserting these expressions in (3.19), the displacement field is calculated. By defining the percentage change in the mth layer of a M-layer period supper-lattice as

$$\Delta(m, M) = \frac{100\%[u(m) - u(m+1)]}{\bar{d}}, \qquad (3.23)$$

we can finally demonstrate the effect of interfaces on the interlayer relaxations in Figs. 17(a), (b), (c). Figure 17(a) is a plot of $\Delta(m, 10)$ (for five planes of the A and five planes of the B multilayers) where the square wave (curve A) shows the macroscopic coherency strain as calculated by continuous elasticity, and curve B is the three-neighbor approximation of the microscopic approach. Similarly, Fig. 17(b), (c) represents the interplanar relaxations for $9d$–$9d$ and $23d_A$–$23d_B$ superlattices of (001) layers.

The calculations are made for an Au–Ni superlattice for which approximate values of the constants are known. The most striking feature of these results is the large Δ's near the interface. The 200% increase in the strain near the interface is indicative of the strong long-range interactions in this system (Fig. 17(b), (c)). On the other hand, the 300% increase in the structural parameter Δ (Fig. 17(a)) can be understood as an interference of the relaxations from each of the neighboring interfaces. Similar behavior has been observed experimentally and predicted theoretically by Chen *et al.* (1989) in very thin slabs of Al and NiAl (110) and (210) surfaces. In the case of the Au–Ni multilayer, Imafuku *et al.* (1986) have used a molecular dynamics model to explain the origin of the supermodulus effect. It is interesting to point out that their model predicts the structural relaxations, $\Delta(m, M)$ as described above, and which are used to show the elastic modulus enhancement. Such relaxations cannot be understood on the basis of coherency itself. It should also be pointed out that the first neighbor approximation of the microscopic approach coincides with the macroscopic coherency strain shown in Fig. 17(a), (b), (c).

The nonuniform contractions and expansions in each layer near the interfaces is of the utmost importance to a variety of the mechanical, electronic, and magnetic behavior of layered structures. The recent experiments by Bizanti *et al.* (1987) on Au/Cr superlattices have shown a dramatic change in the lattices of Au and Cr perpendicular to the superlattice plane, for thicknesses less than 5 nm. Structural studies proved the epitaxial growth (100) plane of Cr layers and the (100) plane of the Au layer rotated by 45°. This habit plane orientation predicts a perfect lattice matching, i.e., $a_0(\mathrm{Au})/\sqrt{2} = 2.8838$ Å and $a_0(\mathrm{Cr}) = 2.8839$ Å. However, an elaborate X-ray diffraction *d*-spacing determination showed that the Cr interplanar spacing was stretched by 8% along the [001] direction, while Au contracted by 2% along the growth direction. Given the excellent agreement of lattice matching between Au and Cr, these changes cannot be attributed to bulk lattice misfit strain at the interfaces. At the same time, a 30% increase in the elastic constants C_{44} for the layer thicknesses that these *d*-spacing anomalies occur.

Preliminary calculations have shown that these unusual contractions and expansions can be explained by means of the Kazanki forces, and the enhancement of the C_{44} is consistent with the elastic constant calculation given above. Details of these calculations will be given in a paper to be published.

Finally, electronic and transport properties which depend on the local environment are expected to be sensitive to these interfacial structural relaxations. Preliminary work on NiAl epitaxial growth on Ga–As indicates a strong dependence, of the growth and critical thickness characteristics, on the above interfacial relaxations. Perhaps the most important example of this dependence is the magnetic behavior of Fe layers on Cu. In a spectacular experiment using Mossbauer spectroscopy, it was clearly demonstrated that α-Fe transforms into γ-Fe near the interface due to the large *d*-spacing expansions and contractions. From the paramagnetic component (central peak) and the loss of the ferromagnetic α-Fe behavior (sextet) in the Mossbauer spectrum, it is estimated that γ-Fe is present only for the first three-interplanar spacings of the Fe layer and the remaining interior layer remains α-Fe ferromagnetic. This is in excellent agreement with the interference of relaxations shown in Fig. 17(a).

4. Conclusion

We have derived a Fourier representation of the elastic strain energy of coherent precipitates in the general case, where the precipitates have different crystal structure and elastic constants. The approximate closed-form Fourier expression contains the strain energy function $Y(\hat{\mathbf{n}})$ in which average elastic constants are involved in a term $Y^{\sigma}(\hat{\mathbf{n}})$, which is the elastic interaction of an applied stress field with the inhomogeneities and is also a self-consistent Fourier energy term $Y(\hat{\mathbf{n}}, \hat{\mathbf{n}}')$ which accounts for the effect of restraining due to the elastic constant variations. We have discussed symmetry and behavior in the reciprocal space, and its influence on the development of the microstructure in alloys by cancelling or altering the minimum of the coherency energy function $Y(\hat{\mathbf{n}})$. We have successfully applied the theory to spinodal decomposition and to the γ' precipitates of nickel-based superalloys. The theory predicts the development of plate-like or rod-like precipitates for tensile or compressive stress, respectively. We have calculated the stress distribution in ZrO_2–Al_2O_3 transformation toughened ceramic in which ZrO_2 undergoes a martensitic transformation. The computations show steep stress variation as a function of distance in various orientations. Finally, we have developed a microscopic elasticity model in systems with interfaces of only a few lattice spacings apart, which demonstrate large contractions and expansions near the interfaces.

References

Ahn, S. and Tsakalakos, T. (1985), The effect of applied stress on the decomposition of Cu–15Ni–8Sn spinoidal alloys, in *Phase Transformations in Solids*, North-Holland, Amsterdam, p. 539.

Bizanti, P., Brodsky, M. B., Felcher, G. P., Grimsditch, M., and Sill, L. R. (1987), Surface waves in Au/Cr superlattices, *Phys. Rev.*, **B35**, 7813–7819.

Cahn, J. W. and Hilliard, J. E. (1958), Free energy of a nonuniform system, I. Interfacial free energy, *Acta Metallurgica*, **28**, 258.

Cahn, J. W. (1962), On spinoidal decomposition in cubic crystals, *Acta Metallurgica*, **10**, 907–913.

Chen, S. P., Voter, A. F., and Albers, R. C. (1989), Interference of surface relaxations in unsupported thin films, *Phys. Rev.*, **B39**, No. 2, 871–875.

Chiu, Y. P. (1977), On the stress-field due to initial strains in cuboid surrounded by an infinite elastic space, *J. Appl. Mech.*, **44**, 587–590.

Ellwood, E. C. and Bagley, K. Q. (1952), *J. Inst. Met.*, 617.

Eshelby, J. D. (1956), Continuum theory of defects, *Solid State Phys.*, **3**, 79.

Eshelby, J. D. (1959), The elastic field outside an ellipsoidal inclusion, *Proc. Roy. Soc.* London, **A252**, 561

Eshelby, J. D. (1959), Elastic inclusions and inhomogeneities, *Prog. Sol. Mech.*, **2**, 89.

Golding, B., Moss, S. C., and Auerbach, B. L. (1967), *Phys. Rev.*, **158**, 637.

Imafuku, M., Susajima, Y., Yamamoto, R., and Doyama, M. (1986), Computer simulations of the structures of the metallic superlattices Au/Ni and Cu/Ni and their elastic moduli, *J. Phys.* F: *Met. Phys.*, 823–829.

Jankowski, A. F., Wingo, E. M., and Tsakalakos, T. (1985), Stress induced transformations to coherent precipitates in nickel-based alloys, in *Computer Simulation of Microstructural Evolution*, AIME Conference Proceedings, edited by D. J. Srolovitz, TSM of AIME, Warrendale, PA.

Khachaturyan, A. G. (1983), *Theory of Structural Transformations in Solids*, Wiley, New York.

Khachaturyan, A. G. (1967), Some questions concerning the theory of phase transformations in solids, *J. Phys. Solid State*, **8**, 2163.

Krivoglaz, M. (1969), *Theory of X-ray and Thermal Neutron Scattering by Real Crystals*, Plenum, New York.

Mura, T. (1987), *Micromechanics of Defects in Solids*, Martinus Nijhoff, Dordrecht, p. 177.

Rao, S. and Tsakalakos, T., to be published.

Tien, J. K. and Copley, S. M. (1971), The effect of orientation and sense of applied uniaxial stress on the morphology of coherent gamma prime precipitates in stress-annealed nickel-based superalloy metals, *Met. Trans.*, **2**, 543.

Tsakalakos, T. and Jankowski, A. (1986), Mechanical properties of composition-modulated metallic foils, *Ann. Rev. Mater. Sci.*, **16**, 293–313.

Tsakalakos, T. (1981), A self-consistent model for the elastic strain energy of coherent inhomogeneous precipitates in anisotropic media, *Res. Mech. Lett.*, **1**, 409–415.

Tsakalakos, T. and Khachaturyan, A. G., to be published.

The Hemispherical Inhomogeneity Subjected to a Concentrated Force

EIICHIRO TSUCHIDA
Department of Mechanical Engineering, Saitama University,
255 Shimo-Okubo, Urawa 338, Japan

DEMITRIS KOURIS
Department of Mechanical and Aerospace Engineering,
Arizona State University, Tempe, AZ 85287-6106, U.S.A.

IWONA JASIUK
Department of Metallurgy, Mechanics and Materials Science,
Michigan State University, East Lansing, MI 48824-1226, U.S.A.

Abstract

The presence of two or more different constituents in an elastic material has a substantial effect on its mechanical behavior under thermal or mechanical loading. The overall, as well as the local, properties of such a material may bear little relation to those of the components, even though the components retain their integrity within the composite. Stress fields in composite materials under applied stresses can be simulated by the inclusion problem, when fibers in the composite are replaced by inhomogeneities. A considerable number of problems can be found in the literature involving the determination of the local stress and displacement fields in the vicinity of a single inhomogeneity or impurity, which is embedded in an infinitely extended surrounding material, i.e., the matrix. In the present paper, an analytical solution for the three-dimensional mixed boundary value problem of a hemispherical inhomogeneity is presented. The inhomogeneity is embedded at the surface of an elastic half-space and is subjected to a concentrated force. First, the problem is solved by assuming perfect bonding along the interface between the matrix and the inhomogeneity. Next, the shear-traction-free (sliding) boundary is considered and the two results are compared. For both cases, the elastic field is deduced in a series form, using Boussinesq's displacement potentials. Several calculations are performed for various combinations of elastic constants. Finally, stresses and displacements along the free surface and the matrix-inhomogeneity interface are evaluated for typical situations, in order to assess the significance of inhomogeneities in composite material applications.

1. Introduction

During the past few decades, considerable attention has been given to the study of the inclusion problem. Eshelby's important results (Eshelby, 1957) were the starting point of a considerable research effort towards better understanding of the problems involving combinations of different materials, under mechanical or thermal loading.

After extensive studies of the perfectly bonded inhomogeneity, more realistic interfacial models became necessary for the study of the micromechanical behavior of composite materials. Noble and Hussain (1969), Mura *et al.* (1985), and Mura and Furuhashi (1984), among others, considered the problem of incoherent inclusions. Most of these studies, however, involve inhomogeneities in an infinite medium.

The inclusion problem in a half-space is naturally more complicated but becomes very important in view of its applicability in composite materials studies. In addition, the availability of experimental techniques for measuring surface strains makes the analytical modeling very attractive. Tsutsui and Saito (1973), as well as Tsuchida and Mura (1983), studied the effect of the free surface for a spherical or spheroidal inclusion in a half-space. Recently, Kouris and Mura (1989) and Kouris *et al.* (1989) considered the problems of the hemispherical and hemispheroidal inhomogeneity at the free surface of an elastic half-space. In both studies, the applied load was either all-around tension at infinity, or nonshear eigenstrains sustained by the inhomogeneity.

The present paper deals with a hemispherical inhomogeneity subjected to a concentrated force; the motivation came from indentation tests used to measure the strength of the fiber–matrix interface in composite materials (Marshall, 1984). The fiberlike semiellipsoidal inhomogeneity is currently under study and the results will be presented in a future communication.

The problem is first solved under the assumption that the inclusion and the matrix are perfectly bonded. Then, the shear tractions along the interface are relaxed and the two results are compared, since they represent the lower and upper bounds of the stress concentration. The effect of the elastic properties on the stress and displacement fields is illustrated by varying the shear moduli ratio of the inhomogeneity–matrix material system.

2. Method of Solution

Consider a hemispherical inhomogeneity Ω under a concentrated force P, at the free surface of a semi-infinite medium D (Fig. 1). We denote the cylindrical and spherical coordinates by (r, θ, z) and (R, θ, ϕ), respectively. The two coordinate systems are related by the transformations, $z = R\cos\phi$, $r = R\sin\phi$.

The general solution of the displacement equations of equilibrium is expressed (in the absence of body forces and for rotational symmetry) as the sum

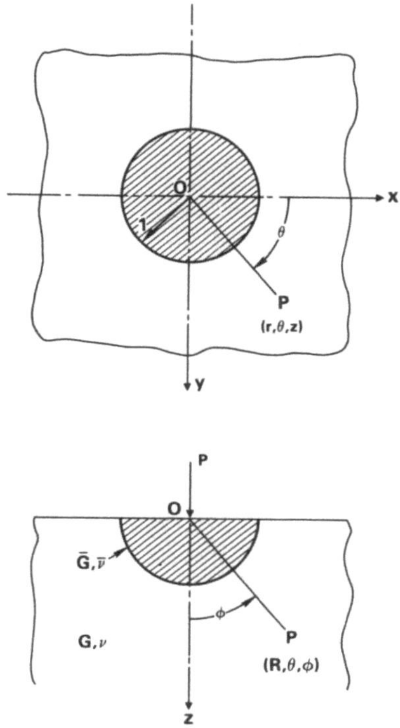

FIG. 1. Hemisperical inclusion.

of the following two fields:

$$2Gu_r = \frac{\partial \Phi_0}{\partial r}, \qquad u_\theta = 0, \qquad 2Gu_z = \frac{\partial \Phi_0}{\partial z},$$

$$\sigma_r = \frac{\partial^2 \Phi_0}{\partial r^2}, \qquad \sigma_\theta = \frac{1}{r}\frac{\partial \Phi_0}{\partial r}, \tag{2.1}$$

$$\sigma_z = \frac{\partial^2 \Phi_0}{\partial z^2}, \qquad \tau_{rz} = \frac{\partial^2 \Phi_0}{\partial r\, \partial z},$$

$$\tau_{z\theta} = \tau_{r\theta} = 0,$$

$$2Gu_r = z\frac{\partial \Phi_3}{\partial r}, \qquad u_\theta = 0,$$

$$2Gu_z = z\frac{\partial \Phi_3}{\partial z} - (3 - 4\nu)\Phi_3,$$

$$\sigma_r = z\frac{\partial^2 \Phi_3}{\partial r^2} - 2\nu\frac{\partial \Phi_3}{\partial z},$$

$$\sigma_\theta = \frac{z}{r}\frac{\partial \Phi_3}{\partial r} - 2\nu\frac{\partial \Phi_3}{\partial z}, \tag{2.2}$$

$$\sigma_z = z\frac{\partial^2\Phi_3}{\partial z^2} - 2(1-\nu)\frac{\partial\Phi_3}{\partial z},$$

$$\tau_{rz} = -(1-2\nu)\frac{\partial\Phi_3}{\partial z} + z\frac{\partial^2\Phi_3}{\partial r\,\partial z},$$

$$\tau_{z\theta} = \tau_{r\theta} = 0,$$

where

$$\nabla^2\Phi_0 = \nabla^2\Phi_3 = 0 \qquad \text{and} \qquad \nabla^2 = \frac{1}{r}\frac{\partial}{\partial r} + \frac{\partial^2}{\partial r^2} + \frac{\partial^2}{\partial z^2}.$$

Transforming (2.1), (2.2) into spherical coordinates, we arrive at the following expressions for the stress and displacement components:

$$\begin{aligned}
&2Gu_R = \frac{\partial\Phi_0}{\partial R}, \qquad u_\theta = 0, \qquad 2Gu_\phi = -\frac{\sin\phi}{R}\frac{\partial\Phi_0}{\partial\mu},\\
&\sigma_R = \frac{\partial^2\Phi_0}{\partial R^2}, \qquad \sigma_\theta = \frac{1}{R}\frac{\partial\Phi_0}{\partial R} - \frac{\mu}{R^2}\frac{\partial\Phi_0}{\partial\mu},\\
&\sigma_\phi = -\frac{\partial^2\Phi_0}{\partial R^2} - \frac{1}{R}\frac{\partial\Phi_0}{\partial R} + \frac{\mu}{R^2}\frac{\partial\Phi_0}{\partial\mu},\\
&\tau_{R\phi} = \sin\phi\left(\frac{1}{R^2}\frac{\partial\Phi_0}{\partial\mu} - \frac{1}{R}\frac{\partial^2\Phi_0}{\partial R\,\partial\mu}\right),\\
&\tau_{\phi\theta} = \tau_{R\theta} = 0,
\end{aligned} \tag{2.3}$$

and for Φ_3

$$\begin{aligned}
&2Gu_R = \mu\left[R\frac{\partial\Phi_3}{\partial R} - (3-4\nu)\Phi_3\right], \qquad u_\theta = 0,\\
&2Gu_\phi = \sin\phi\left[-\mu\frac{\partial\Phi_3}{\partial\mu} + (3-4\nu)\Phi_3\right],\\
&\sigma_R = \mu R\frac{\partial^2\Phi_3}{\partial R^2} - 2(1-\nu)\mu\frac{\partial\Phi_3}{\partial R} - 2\nu\frac{1-\mu^2}{R}\frac{\partial\Phi_3}{\partial\mu},\\
&\sigma_\theta = (1-2\nu)\mu\frac{\partial\Phi_3}{\partial R} - [2\nu + (1-2\nu)\mu^2]\frac{1}{R}\frac{\partial\Phi_3}{\partial\mu},\\
&\sigma_\phi = -\mu R\frac{\partial^2\Phi_3}{\partial R^2} - (1+2\nu)\mu\frac{\partial\Phi_3}{\partial R} + [1-(3-2\nu)(1-\mu^2)]\frac{1}{R}\frac{\partial\Phi_3}{\partial\mu},\\
&\tau_{R\phi} = \sin\phi\left[(1-2\nu)\frac{\partial\Phi_3}{\partial R} - \mu\frac{\partial^2\Phi_3}{\partial R\,\partial\mu} + 2(1-\nu)\frac{\mu}{R}\frac{\partial\Phi_3}{\partial\mu}\right],\\
&\tau_{\phi\theta} = \tau_{R\theta} = 0,
\end{aligned} \tag{2.4}$$

where $\nabla^2\Phi_0 = \nabla^2\Phi_3 = 0$,

$$\nabla^2 = \frac{1}{R^2}\frac{\partial}{\partial R}\left(R^2\frac{\partial}{\partial R}\right) + \frac{1}{R^2}\frac{\partial}{\partial\mu}\left[(1-\mu^2)\frac{\partial}{\partial\mu}\right] \qquad \text{and} \qquad \mu = \cos\phi.$$

The applied load is expressed in terms of Φ_0 and Φ_3 as follows:

$$\begin{cases} \Phi_0 = -\dfrac{a^2}{2}(1-2\bar{v})p_0 \log(R+z), \\ \Phi_3 = -\dfrac{a^2}{2} p_0 \dfrac{1}{R}, \end{cases}$$

where $P_n(\mu)$ are the Legendre functions of the first kind, $\bar{v}$ is the Poisson ratio, a is the radius of the inhomogeneity, and the quantities denoted by a bar refer to the inclusion.

The above potentials yield the following stresses and displacements:

$$\begin{aligned} \frac{2RG\bar{u}_R}{p_0 a^2} &= -2[(1-2\bar{v}) - 4(1-\bar{v})\cos\phi], \\ \frac{2RG\bar{u}_\phi}{p_0 a^2} &= -2\sin\phi\left[-\frac{1-2\bar{v}}{1+\cos\phi} + 3 - 4\bar{v}\right], \qquad \frac{2RG\bar{u}_\theta}{p_0 a^2} = 0, \\ \frac{R^2\bar{\sigma}_R}{p_0 a^2} &= -2[-(1-2\bar{v}) + 2(2-\bar{v})\cos\phi], \\ \frac{R^2\bar{\sigma}_\theta}{p_0 a^2} &= -2(1-2\bar{v})\left[\frac{1}{1+\cos\phi} - \cos\phi\right], \\ \frac{R^2\bar{\sigma}_\phi}{p_0 a^2} &= -2(1-2\bar{v})\cos\phi\left[\frac{1}{1+\cos\phi} - 1\right], \\ \frac{R^2\bar{\tau}_{R\phi}}{p_0 a^2} &= -2(1-2\bar{v})\sin\phi\left[\frac{1}{1+\cos\phi} - 1\right]. \end{aligned} \tag{2.5}$$

This solution corresponds to the field generated by a normal concentrated force, P, acting on a half-space. It gives no tractions along the free surface but yields stresses and displacements at the spherical surface, $R = a$, of the inhomogeneity.

The effects of the inhomogeneity and the requirements of equilibrium are addressed by five auxiliary sets of displacement potentials chosen to satisfy automatically the boundary conditions at the free surface and at infinity. The boundary conditions at the free surface are expressed in cylindrical coordinates by

$$(\sigma_z)_{z=0} = (\tau_{rz})_{z=0} = 0, \tag{2.6a}$$

$$(\bar{\sigma}_z)_{z=0} = (\bar{\tau}_{rz})_{z=0} = 0. \tag{2.6b}$$

At infinity

$$(\sigma_x)_{r\to\infty} = (\sigma_y)_{r\to\infty} = (\sigma_z)_{r\to\infty} = 0. \tag{2.7}$$

The required auxiliary potentials are

$$[\mathrm{I}] \quad \begin{cases} \Phi_0 = -\dfrac{a^2}{2}(1-2v)p_0 F_0 \log(R+z), \\ \Phi_3 = -\dfrac{a^2}{2} p_0 \dfrac{F_0}{R}, \end{cases}$$

$$[\mathrm{II}]\quad \begin{cases} \Phi_0 = -2(1-\nu)p_0 \sum_{n=0}^{\infty} A_n \dfrac{P_{2n}(\mu)}{R^{2n+1}}, \\ \Phi_3 = p_0 \sum_{n=0}^{\infty} (2n+1) A_n \dfrac{P_{2n+1}(\mu)}{R^{2n+2}}, \end{cases}$$

$$[\mathrm{III}]\quad \begin{cases} \Phi_0 = -(1-2\nu)p_0 \sum_{n=0}^{\infty} B_n \dfrac{P_{2n+1}(\mu)}{R^{2n+2}}, \\ \Phi_3 = p_0 \sum_{n=0}^{\infty} (2n+2) B_n \dfrac{P_{2n+2}(\mu)}{R^{2n+3}}, \end{cases}$$

for the matrix ($R > a$), and

$$[\mathrm{IV}]\quad \begin{cases} \Phi_0 = 2(1-\bar{\nu})p_0 \sum_{n=0}^{\infty} \bar{A}_n R^{2n+2} P_{2n+2}(\mu), \\ \Phi_3 = p_0 \sum_{n=0}^{\infty} (2n+2) \bar{A}_n R^{2n+1} P_{2n+1}(\mu), \end{cases}$$

$$[\mathrm{V}]\quad \begin{cases} \Phi_0 = (1-2\bar{\nu})p_0 \sum_{n=0}^{\infty} \bar{B}_n R^{2n+1} P_{2n+1}(\mu), \\ \Phi_3 = p_0 \sum_{n=0}^{\infty} (2n+1) \bar{B}_n R^{2n} P_{2n}(\mu), \end{cases}$$

for the inclusion ($R < a$).

When the matrix and inhomogeneity are perfectly bonded, the boundary conditions at the interface are

$$(\sigma_R)_{R=a} = (\bar{\sigma}_R)_{R=a}, \tag{2.8a}$$

$$(\tau_{R\phi})_{R=a} = (\bar{\tau}_{R\phi})_{R=a}, \tag{2.8b}$$

$$(u_R)_{R=a} = (\bar{u}_R)_{R=a}, \tag{2.8c}$$

$$(u_\phi)_{R=a} = (\bar{u}_\phi)_{R=a}, \tag{2.8d}$$

The unknown coefficients A_n, B_n, $\bar{A}_n$, $\bar{B}_n$, and consequently the solution, are determined by enforcing the boundary conditions (2.8a)–(2.8d).

In the case of sliding, it is assumed that the interface does not sustain any shear tractions. Consequently, the boundary conditions become

$$(\sigma_R)_{R=a} = (\bar{\sigma}_R)_{R=a}, \tag{2.9a}$$

$$(\tau_{R\phi})_{R=a} = 0, \tag{2.9b}$$

$$(\bar{\tau}_{R\phi})_{R=a} = 0, \tag{2.9c}$$

$$(u_R)_{R=a} = (\bar{u}_R)_{R=a}, \tag{2.9d}$$

In order to apply the boundary conditions along the interface, we need to express stresses and displacements due to the potentials Φ_0 and Φ_3, given in the sets [I]–[V].

Using (2.3) and (2.4), the boundary conditions (2.8a)–(2.8d) become

$$(u_R)_{R=a} = (\bar{u}_R)_{R=a} \to 2\frac{F_0}{a}\left[P_0(\mu) + 4(1-\nu)\sum_{n=1}^{\infty} w_0^{(n)}P_{2n}(\mu)\right]a^2$$

$$-\sum_{n=0}^{\infty}\frac{A_n}{a^{2n+2}}\frac{2n+1}{4n+3}[\alpha_{2n}P_{2n}(\mu) + (2n+2)(2n+5-4\nu)P_{2n+2}(\mu)]$$

$$-\sum_{n=0}^{\infty}\frac{B_n}{a^{2n+3}}\frac{2n+2}{4n+5}[\alpha_{2n+2}P_{2n+1}(\mu)$$

$$+ (2n+3)(2n+6-4\nu)P_{2n+3}(\mu)]$$

$$-\frac{1}{\Gamma}\sum_{n=0}^{\infty}\bar{A}_n a^{2n+1}\frac{2n+2}{4n+3}[(2n+1)(2n-2+4\bar{\nu})P_{2n}(\mu)$$

$$+ \bar{\alpha}_{2n+1}P_{2n+2}(\mu)]$$

$$-\frac{1}{\Gamma}\sum_{n=0}^{\infty}\bar{B}_n a^{2n}\frac{2n+1}{4n+1}[2n(2n-3+4\bar{\nu})P_{2n-1}(\mu) + \bar{\alpha}_{2n-1}P_{2n+1}(\mu)]$$

$$= \frac{2}{a}\frac{1}{\Gamma}\left[P_0(\mu) + 4(1-\bar{\nu})\sum_{n=1}^{\infty} w_0^{(n)}P_{2n}(\mu)\right]a^2, \tag{2.10}$$

where

$$\alpha_n = (n+1)^2 - 2 + 2\nu, \qquad \bar{\alpha}_n = (n+1)^2 - 2 + 2\bar{\nu},$$

$$\beta_n = (n+2)(n+5) - 2\nu, \qquad \bar{\beta}_n = (n+2)(n+5) - 2\bar{\nu}.$$

Equation (2.8d) yields

$$(u_\phi)_{R=a} = (\bar{u}_\phi)_{R=a}$$

$$\to -2\frac{F_0}{a}a^2\sum_{n=1}^{\infty}\left[(1-2\nu)\frac{4n+1}{2n(2n+1)} + (3-4\nu)w_0^{(n)}\right.$$

$$\left. -(1-2\nu)\sum_{k=0}^{\infty}\frac{4k+3}{(2k+1)(2k+2)}w_k^{(n)}\right]P'_{2n}(\mu)$$

$$-\sum_{n=0}^{\infty}\frac{A_n}{a^{2n+2}}\frac{1}{4n+3}[\alpha_{2n}P'_{2n}(\mu) + (2n+1)(2n-2+4\nu)P'_{2n+2}(\mu)]$$

$$-\sum_{n=0}^{\infty}\frac{B_n}{a^{2n+3}}\frac{1}{4n+5}[\alpha_{2n+2}P'_{2n+1}(\mu)$$

$$+ (2n+2)(2n-1+4\nu)P'_{2n+3}(\mu)]$$

$$+\sum_{n=0}^{\infty}\bar{A}_n a^{2n+1}\frac{1}{4n+3}\frac{1}{\Gamma}[(2n+2)(2n+5-4\bar{\nu})P'_{2n}(\mu)$$

$$+ \bar{\alpha}_{2n+1}P'_{2n+2}(\mu)]$$

$$+ \sum_{n=0}^{\infty} \bar{B}_n a^{2n} \frac{1}{4n+1} \frac{1}{\Gamma} [(2n+1)(2n+4-4\bar{v})P'_{2n-1}(\mu) + \bar{\alpha}_{2n-1} P'_{2n+1}(\mu)]$$

$$= -\frac{2}{a}\frac{1}{\Gamma} a^2 \sum_{n=1}^{\infty} \left[(1-2\bar{v}) \frac{4n+1}{2n(2n+1)} + (3-4\bar{v}) w_0^{(n)} - (1-2\bar{v}) \sum_{k=0}^{\infty} \frac{4k+3}{(2k+1)(2k+2)} w_k^{(n)} \right] P'_{2n}(\mu), \quad (2.11)$$

where $\Gamma = \bar{G}/G$, $\bar{G}$ is the shear modulus of the inhomogeneity, and G is the shear modulus of the matrix. For the stresses, we have

$$(\sigma_R)_{R=a} = (\bar{\sigma}_R)_{R=a} \to -2F_0 \left[(1+v)P_0(\mu) + 2(2-v) \sum_{n=1}^{\infty} w_0^{(n)} P_{2n}(\mu) \right]$$

$$+ \sum_{n=0}^{\infty} \frac{A_n}{a^{2n+3}} \frac{(2n+1)(2n+2)}{4n+3} [\alpha_{2n} P_{2n}(\mu) + \beta_{2n} P_{2n+2}(\mu)]$$

$$+ \sum_{n=0}^{\infty} \frac{B_n}{a^{2n+4}} \frac{(2n+2)(2n+3)}{4n+5} [\alpha_{2n+2} P_{2n+1}(\mu) + \beta_{2n+1} P_{2n+3}(\mu)]$$

$$- \sum_{n=0}^{\infty} \bar{A}_n a^{2n} \frac{(2n+1)(2n+2)}{4n+3} [\bar{\beta}_{2n-4} P_{2n}(\mu) + \bar{\alpha}_{2n+1} P_{2n+2}(\mu)]$$

$$- \sum_{n=0}^{\infty} \bar{B}_n a^{2n-1} \frac{2n(2n+1)}{4n+1} [\bar{\beta}_{2n-5} P_{2n-1}(\mu) + \bar{\alpha}_{2n-1} P_{2n+1}(\mu)]$$

$$= -2\left[(1+\bar{v})P_0(\mu) + 2(2-\bar{v}) \sum_{n=1}^{\infty} w_0^{(n)} P_{2n}(\mu) \right], \quad (2.12)$$

and

$$(\tau_{R\phi})_{R=a} = (\overline{\tau_{R\phi}})_{R=a}$$

$$\to -2F_0(1-2v) \sum_{n=1}^{\infty} \left[-\frac{4n+1}{2n(2n+1)} - w_0^{(n)} + \sum_{k=0}^{\infty} \frac{4k+3}{(2k+1)(2k+2)} w_k^{(n)} \right] P'_{2n}(\mu)$$

$$+ \sum_{n=0}^{\infty} \frac{A_n}{a^{2n+3}} \frac{1}{4n+3} [(2n+2)\alpha_{2n} P'_{2n}(\mu) + (2n+1)\alpha_{2n+1} P'_{2n+2}(\mu)]$$

$$+ \sum_{n=0}^{\infty} \frac{B_n}{a^{2n+4}} \frac{\alpha_{2n+2}}{4n+5} [(2n+3)P'_{2n+1}(\mu) + (2n+2)P'_{2n+3}(\mu)]$$

$$+ \sum_{n=0}^{\infty} \bar{A}_n a^{2n} \frac{1}{4n+3} [(2n+2)\bar{\alpha}_{2n} P'_{2n}(\mu) + (2n+1)\bar{\alpha}_{2n+1} P'_{2n+2}(\mu)]$$

$$+ \sum_{n=0}^{\infty} \bar{B}_n a^{2n-1} \frac{\bar{\alpha}_{2n-1}}{4n+1} [(2n+1)P'_{2n-1}(\mu) + 2nP'_{2n+1}(\mu)]$$

$$
= -2(1-2\bar{\nu}) \sum_{n=1}^{\infty} \left[-\frac{4n+1}{2n(2n+1)} - w_0^{(n)} + \sum_{k=0}^{\infty} \frac{4k+3}{(2k+1)(2k+2)} w_k^{(n)} \right] P'_{2n}(\mu). \tag{2.13}
$$

In (2.10)–(2.13), using the "half-range expansion"

$$
P_{2k+1}(\mu) = \sum_{n=0}^{\infty} w_k^{(n)} P_{2n}(\mu), \qquad P'_{2k+1}(\mu) = \sum_{n=1}^{\infty} w_k^{(n)} P'_{2n}(\mu), \tag{2.14}
$$

where

$$
w_k^{(n)} = \frac{(4n+1)(2k+1)}{(2k+1-2n)(2k+2+2n)} P_{2n}(0) P_{2k}(0),
$$

the Legendre functions of odd order are expanded in terms of the Legendre functions of even order. Then, equating the coefficients of $P_{2n}(\mu)$ and $P'_{2n}(\mu)$, we have

$$
\begin{aligned}
&-2a^2 \frac{F_0}{a} [-(1-2\nu)\delta_0^{(n)} + 4(1-\nu) w_0^{(n)}] \\
&+ \frac{A_n}{a^{2n+2}} \frac{2n+1}{4n+3} \alpha_{2n} + \frac{A_{n-1}}{a^{2n}} \frac{2n(2n-1)}{4n-1} (2n+3-4\nu) \\
&+ \sum_{k=0}^{\infty} \frac{B_k}{a^{2k+3}} \frac{2k+2}{4k+5} [\alpha_{2k+2} w_k^{(n)} + (2k+3)(2k+6-4\nu) w_{k+1}^{(n)}] \\
&+ \frac{1}{\Gamma} \left[\bar{A}_n a^{2n+1} \frac{(2n+1)(2n+2)}{4n+3} (2n-2+4\bar{\nu}) + \bar{A}_{n-1} a^{2n-1} \frac{2}{4n-1} \bar{\alpha}_{2n-1} \right. \\
&\qquad \left. + \sum_{k=0}^{\infty} \bar{B}_k a^{2k} \frac{2k+1}{4k+1} [2k(2k-3+4\bar{\nu}) w_{k-1}^{(n)} + \bar{\alpha}_{2k-1} w_k^{(n)}] \right] \\
&\qquad = -2a \frac{1}{\Gamma} [-(1-2\bar{\nu})\delta_0^{(n)} + 4(1-\bar{\nu}) w_0^{(n)}] \qquad (n = 0, 1, 2, \ldots).
\end{aligned} \tag{2.15}
$$

From (2.11),

$$
\begin{aligned}
&-2a^2 \frac{F_0}{a} \left[-(1-2\nu) \frac{4n+1}{2n(2n+1)} - (3-4\nu) w_0^{(n)} + (1-2\nu) \sum_{k=0}^{\infty} \frac{4k+3}{(2k+1)(2k+2)} w_k^{(n)} \right] \\
&+ \frac{A_n}{a^{2n+2}} \frac{1}{4n+3} \alpha_{2n} + \frac{A_{n-1}}{a^{2n}} \frac{2n-1}{4n-1} (2n-4+4\nu) \\
&+ \sum_{k=0}^{\infty} \frac{B_n}{a^{2k+3}} \frac{1}{4k+5} [\alpha_{2k+2} w_k^{(n)} + (2k+2)(2k-1+4\nu) w_{k+1}^{(n)}]
\end{aligned}
$$

$$- \bar{A}_n a^{2n+1} \frac{2n+2}{4n+3} \frac{1}{\Gamma}(2n+5-4\bar{v}) - \bar{A}_{n-1} a^{2n-1} \frac{\bar{\alpha}_{2n-1}}{4n-1} \frac{1}{\Gamma}$$

$$- \sum_{k=0}^{\infty} \bar{B}_k \frac{a^{2k}}{4k+1} \frac{1}{\Gamma} [(2k+1)(2k+4-4\bar{v}) w_{k-1}^{(n)} + \bar{\alpha}_{2k-1} w_k^{(n)}]$$

$$= -\frac{2}{\Gamma} \frac{a^2}{a} \left[-(1-2\bar{v}) \frac{4n+1}{2n(2n+1)} - (3-4\bar{v}) w_0^{(n)} \right.$$

$$\left. + (1-2\bar{v}) \sum_{k=0}^{\infty} \frac{4k+3}{(2k+1)(2k+2)} w_k^{(n)} \right] \qquad (n = 1, 2, 3, \ldots). \quad (2.16)$$

From (2.12) we get

$$- 2F_0[-(1-2v)\delta_0^{(n)} + 2(2-v) w_0^{(n)}]$$

$$+ \frac{A_n}{a^{2n+3}} \frac{(2n+1)(2n+2)}{4n+3} \alpha_{2n} + \frac{A_{n-1}}{a^{2n+1}} \frac{2n(2n-1)}{4n-1} \beta_{2n-2}$$

$$+ \sum_{k=0}^{\infty} \frac{B_k}{a^{2k+4}} \frac{(2k+2)(2k+3)}{4k+5} [\alpha_{2k+2} w_k^{(n)} + \beta_{2k+1} w_{k+1}^{(n)}]$$

$$- \left[\bar{A}_n a^{2n} \frac{(2n+1)(2n+2)}{4n+3} \bar{\beta}_{2n-4} + \bar{A}_{n-1} a^{2n-2} \frac{2n(2n-1)}{4n-1} \bar{\alpha}_{2n-1} \right.$$

$$\left. + \sum_{k=1}^{\infty} \bar{B}_k a^{2k-1} \frac{2k(2k+1)}{4k+1} [\bar{\beta}_{2k-5} w_{k-1}^{(n)} + \bar{\alpha}_{2k-1} w_k^{(n)}] \right]$$

$$= -2[-(1-2\bar{v})\delta_0^{(n)} + 2(2-\bar{v}) w_0^{(n)}] \qquad (n = 0, 1, 2, \ldots), \quad (2.17)$$

and, finally, from (2.13)

$$-2F_0(1-2v) \left[-\frac{4n+1}{2n(2n+1)} - w_0^{(n)} + \sum_{k=0}^{\infty} \frac{4k+3}{(2k+1)(2k+2)} w_k^{(n)} \right]$$

$$+ \frac{A_n}{a^{2n+3}} \frac{2n+2}{4n+3} \alpha_{2n} + \frac{A_{n-1}}{a^{2n+1}} \frac{2n-1}{4n-1} \alpha_{2n-1}$$

$$+ \sum_{k=0}^{\infty} \frac{B_k}{a^{2k+4}} \frac{\alpha_{2k+2}}{4k+5} [(2k+3) w_k^{(n)} + (2k+2) w_{k+1}^{(n)}]$$

$$+ \left[\bar{A}_n a^{2n} \frac{2n+2}{4n+3} \bar{\alpha}_{2n} + \bar{A}_{n-1} a^{2n-2} \frac{2n-1}{4n-1} \bar{\alpha}_{2n-1} \right.$$

$$\left. + \sum_{k=1}^{\infty} \bar{B}_k a^{2k-1} \frac{\bar{\alpha}_{2k-1}}{4k+1} [(2k+1) w_{k-1}^{(n)} + 2k w_k^{(n)}] \right]$$

$$= -2(1-2\bar{v}) \left[-\frac{4n+1}{2n(2n+1)} - w_0^{(n)} \right.$$

$$\left. + \sum_{k=0}^{\infty} \frac{4k+3}{(2k+1)(2k+2)} w_k^{(n)} \right] \qquad (n = 1, 2, 3, \ldots), \qquad (2.18)$$

where $\delta_i^{(n)}$ denotes the Kronecker delta.

The solution is obtained by solving the system of (2.15)–(2.18) for the unknown constants A_n, B_n, $\bar{A}_n$, $\bar{B}_n$. Then, by superposition of the expressions that can be derived from the displacement potentials [I]–[V], all stresses and displacements can be evaluated.

3. Numerical Calculations

A number of numerical calculations are carried out in order to illustrate the effect of various shear moduli ratios as well as the differences between perfect bonding and sliding. The Poisson ratios of matrix and inhomogeneity are assumed to be equal ($\nu = \bar{\nu} = 0.3$), while stresses and displacements are calculated for various shear moduli ratios $\Gamma = \bar{G}/G$. $\Gamma = 0$ represents a void (cavity) and $\Gamma \to \infty$ corresponds to a rigid inhomogeneity. The radius a was taken equal to unity.

The coefficients A_n, B_n, $\bar{A}_n$, and $\bar{B}_n$ decay monotonically with increasing n and the boundary condition requirements are met, retaining the first 15 terms of the series. It can easily be shown that the expansions (2.14) are complete and the convergence of the series is evident (Mura and Tsuchida, 1983; Mura *et al.*, 1985).

In Fig. 2, the normal stress σ_R along the interface is shown for perfect bonding and sliding. In both cases the high compressive values of σ_R at the

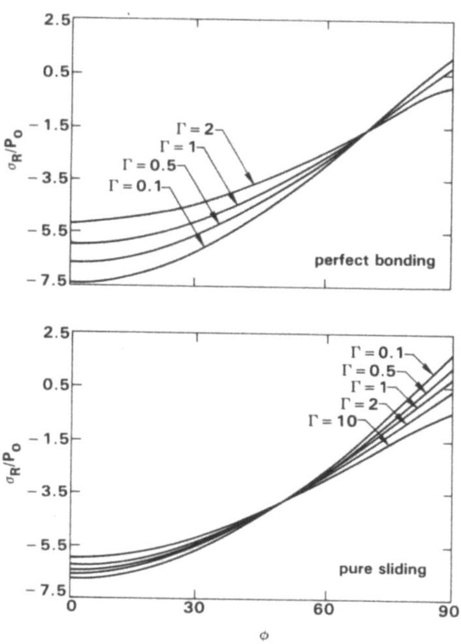

FIG. 2. Normal stress σ_R along the interface.

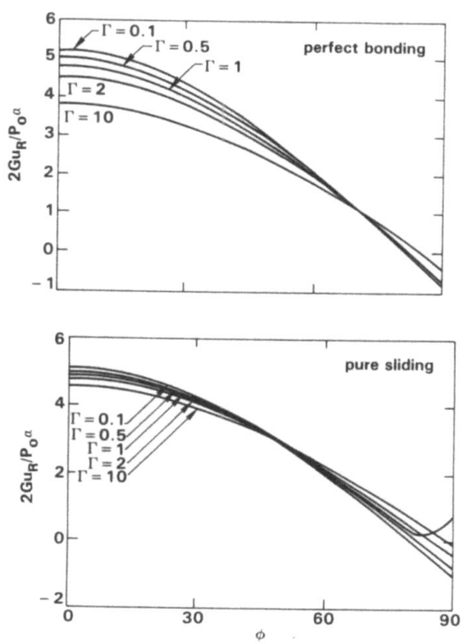

FIG. 3. Variation of the normal displacement versus ϕ.

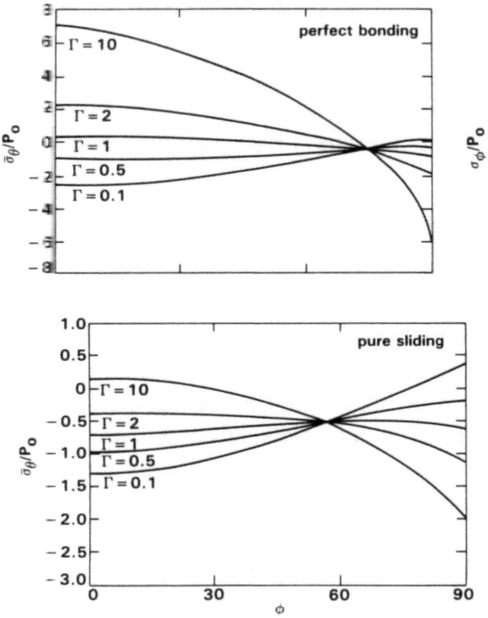

FIG. 4. Hoop stress $\bar{\sigma}_\theta$ for perfect bonding and sliding.

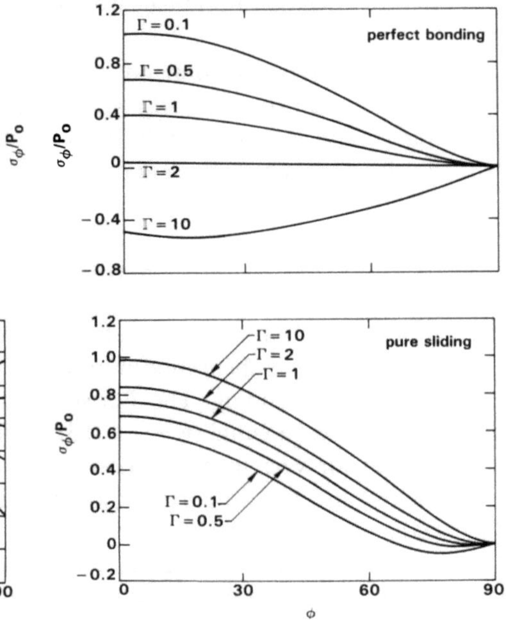

FIG. 5. Longitudinal stress σ_ϕ along the interface.

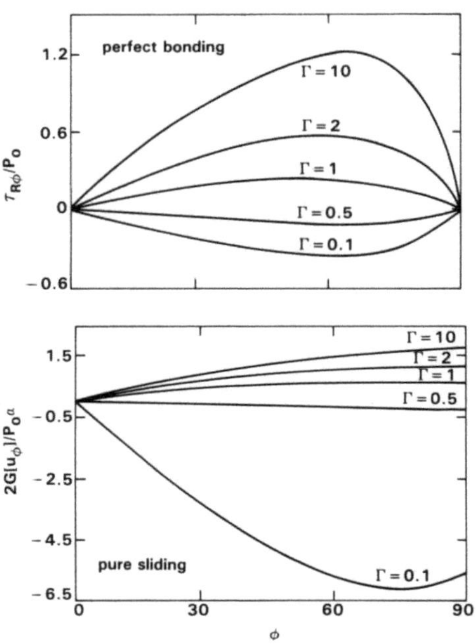

FIG. 6. Displacement discontinuity and shear stress for pure sliding and perfect bonding, respectively.

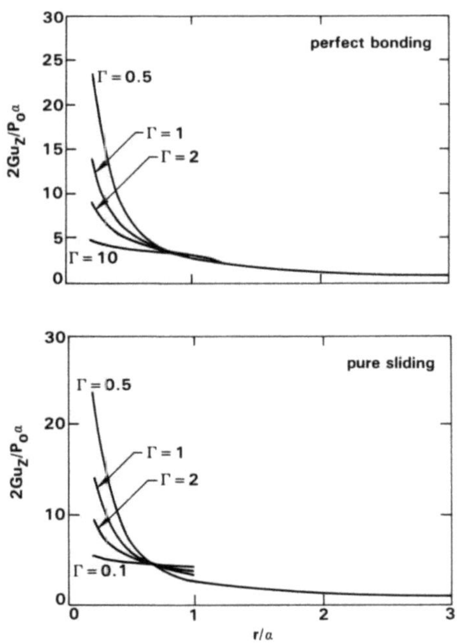

FIG. 7. Variation of the normal displacement u_z along the free surface.

bottom of the inclusion ($\phi = 0$) increase as Γ decreases ("softer" inhomogeneity). The normal displacements u_R are illustrated in Fig. 3. Both σ_R and u_R are independent of Γ when $\phi = 70$ (perfect bonding) and $\phi = 50$ (sliding). The hoop and longitudinal stresses, $\bar{\sigma}_\theta$ and σ_ϕ, respectively, are shown in Figs. 4 and 5. Stress $\bar{\sigma}_\theta$ assumes higher values as the inhomogeneity becomes "harder." The shear stress $\tau_{R\phi}$ for perfect bonding and the tangential displacement discontinuity $[u_\phi]$ for sliding are shown in Fig. 6.

Finally, the normal displacement u_z along the free surface is illustrated in Fig. 7. Away from the inhomogeneity ($r/a > 1$), the effect of Γ is negligible.

References

Boussinesq, J. (1885), *Application des Potentials*, Gauthier-Villars, Paris.

Cox, B. (1989), Surface displacements and stress field generated by a semi-ellipsoidal surface inclusion, *J. Appl. Mech.*, in press.

Eshelby, J. D. (1957), The determination of the elastic fields of an ellipsoidal inclusion and related problems, *Proc. Roy. Soc. London*, **A241**, 376–396.

Hobson, E. W. (1931), *The Theory of Spheroidal and Ellipsoidal Harmonics*, Cambridge University Press, Cambridge, England.

Kouris, D. and Mura, T. (1989), The elastic field of a hemispherical inhomogeneity at the free surface of an elastic half-space, *J. Mech. Phys. Solids*, **37**(3), 365–379.

Kouris, D., Tsuchida, E., and Mura, T. (1989), Hemispheroidal inhomogeneity at the free surface of an elastic half-space, *J. Appl. Mech.*, **56**, 70–76.

Marshall, D. (1984), An indentation method for measuring matrix–fiber frictional stresses in ceramic composites, *J. Amer. Ceram. Soc.*, **67**(12), C259–260.

Mura, T., Jasiuk, I., and Tsuchida, E. (1985), The stress field of a sliding inclusion, *Int. J. Solids Structures*, **21**, 1165–1179.

Mura, T. and Furuhashi, R. (1984), The elastic inclusion with a sliding interface, *J. Appl. Mech.*, **51**, 308–310.

Noble, B. and Hussain, M. (1969), A variational method for inclusion and indentation problems", *J. Inst. Maths. Appl.*, **5**, 194–205.

Tsuchida, E. and Mura, T. (1983), The stress field in an elastic half-space having a spheroidal inhomogeneity under all-around tension parallel to the plane boundary, *J. Appl. Mech.*, **50**, 807–816.

Tsutsui, S. and Saito, K. (1973), On the effect of a free surface and a spherical inhomogeneity on stress fields of a semi-infinite medium under axisymmetric tension, *Proceedings of the 23rd Japan National Congress on Applied Mechanics*, **23**, 547–560.

Dislocation Inhomogeneity in Cyclic Deformation

D. WALGRAEF* and E. C. AIFANTIS
Department of Mechanical Engineering and Engineering Mechanics,
Michigan Technological University, Houghton, MI 49931, U.S.A.

Abstract

The development of a dislocation inhomogeneity in the form of regular spatial patterns during cyclic deformation is examined from a dynamical instability point of view. A gradient-dependent dislocation dynamics framework is devised and the competition between mobility, interaction, and generation processes is considered. Within such a framework, it is shown that uniform dislocation distributions may become unstable versus spatial modulations leading to the formation of different types of patterns including cellular, ladderlike, and labyrinth structures. In particular, the spatially periodic wall structure of persistent slip bands (PSBs), along with the theoretical determination of the corresponding wavelength, is discussed on the basis of nonlinear dynamical equations of the reaction–diffusion type. The effect of temperature on the wavelength selection process is predicted in accordance with experimental observations. Moreover, it is shown that the possibility of secondary or multiple slip effects and the associated labyrinth structures can conveniently be described by the dynamical model. Finally, a preliminary discussion on the applicability of gradient-dependent dislocation dynamics and self-organization techniques, in describing the early stages of the deformation with emphasis on the inhomogeneous development of slip during monotonic loading, is given.

1. Introduction

The inhomogeneous character of dislocation distributions and the resulting heterogeneity of plastic deformation were known practically since the discovery of dislocation and slip (see, for example, the related chapters in the recent book by Mura (1982) and references quoted therein). No attempt has been made, however, to address the problem of the origin and evolution of this inhomogeneity, or to discuss the tendency of dislocations to form periodic and other well-organized spatial structures. At high temperatures, these regular structures consist mainly of planar networks which form the walls of

* Senior Research Associate, National Fund for Scientific Research (Belgium). Permanent address: Service de Chimie-Physique, Université Libre de Bruxelles, CP 231, B-1050 Bruxelles.

cellular planforms. At low temperatures, the dislocation cells become thicker and more diffuse. Under cyclic loading, the tendency to form ordered structures is even more pronounced. The most profound example of such self-organization is the "ladderlike" structure of persistent slip bands (PSBs), which evolves within a less regular "rodlike" vein or matrix structure. Careful transmission electron microscopy studies have repeatedly verified the occurrence of such structures during, for example, the fatigue of Cu monocrystals, and have provided precise measurements of wavelengths and dislocation densities for both the ladders and the veins. Similar observations were made for other types of single crystals and polycrystals. These results, along with the microscopic mechanisms leading to the corresponding dislocation patterning phenomena and their macroscopic manifestation on the stress–strain graph (e.g., the phase transition reminding plateau observed during the period of PSB development) are reported, for example, in the papers by Winter and co-workers (1974, 1981), Mughrabi and co-workers (1979, 1981), and Tabata *et al.* (1983), where references to previous works can also be found. Figure 1, due to Mughrabi and co-workers, shows PSBs embedded within the vein

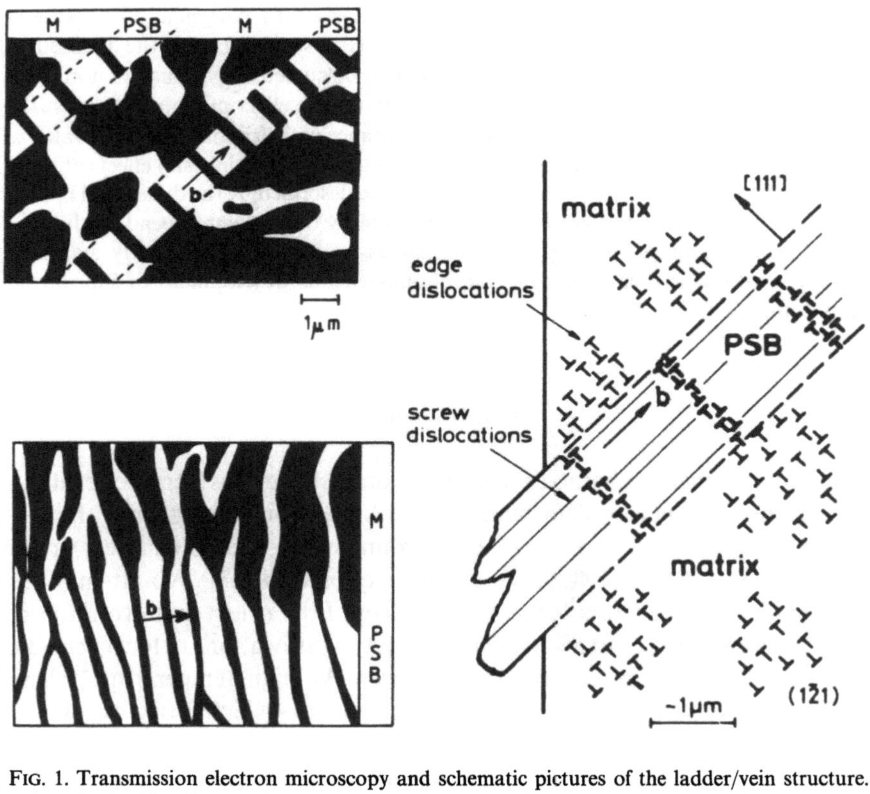

FIG. 1. Transmission electron microscopy and schematic pictures of the ladder/vein structure. (After Mughrabi and co-workers. The periodic structure of persistent slip bands (PSB) embedded within the matrix (M) region is clearly shown.)

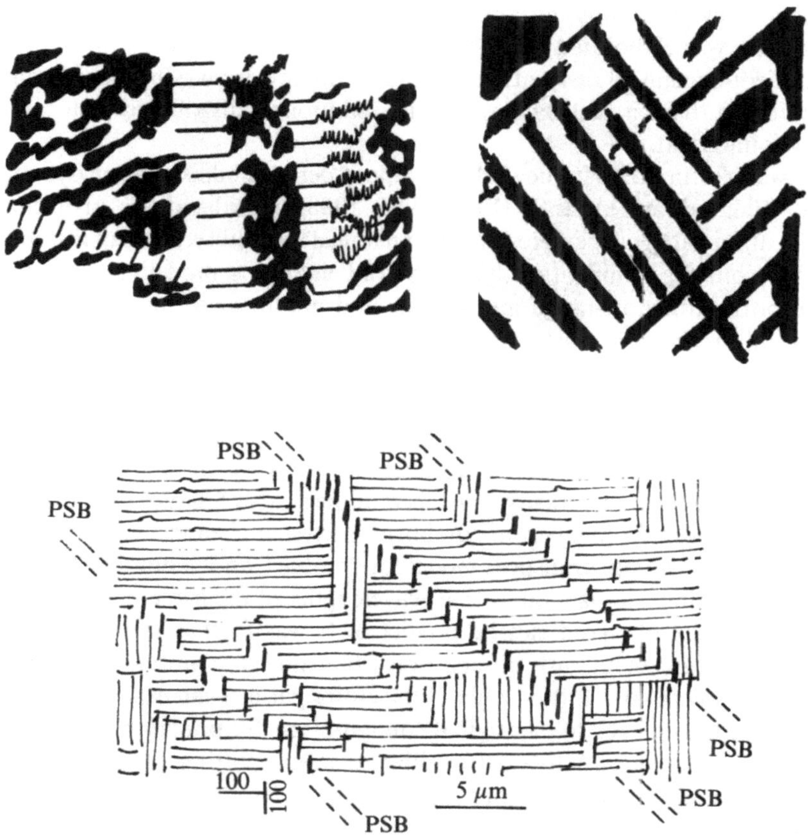

FIG. 2. Typical labyrinth structures observed in single crystals and polycrystals. (After Winter and co-workers. The possibility of the co-existence of periodic structures of different directions within separate regions of the crystal is clearly shown).

structure of Cu monocrystals oriented for single slip. In the case of multislip, PSBs with different orientations may coexist and the underlying dynamical dislocation processes may eventually lead to "labyrinth structures," as shown in Fig. 2 due to Winter and co-workers.

The first attempt to provide a description of dislocation patterns on the basis of solutions to a nonlinear partial differential equation for the dislocation densities is contained in a paper by Aifantis (1983). Essentially, this equation was an extension of the usual rate equation of dislocation dynamics obtained by the inclusion of a diffusivelike term to model the spatial interaction of dislocations. The motivation for these spatially-dependent terms came from a proposal (Aifantis, 1978) on "complete balance laws" for the internal variables containing both a "rate" and a "flux" term modeling, respectively, the growth and transport of microstructures (vacancies, dislocations, voids) within the elementary volume. For the case of dislocations, the "complete balance law" idea was initially adopted by Bammann and Aifantis (1982) and later was elaborated upon by Aifantis (1984, 1985, 1986, 1987) where a "gradient-dependent dislocation dynamics" framework was proposed and its relation to the classical framework of continuously distributed dislocations was pointed out.

It was within such a framework of gradient-dependent dislocation dynamics that Walgraef and Aifantis (1985, 1986, 1988) considered the problem of dislocation patterning as a result of a dynamical instability. Such an instability arises from the competition between gradient terms modeling dislocation transport with short-range interactions and nonlinear terms modeling dislocation production, clustering, and annihilation. This dynamical approach is quite different from the earlier static considerations based on minimization arguments for the elastic strain energy associated with the dislocation species (Holt, 1970; Kuhlman-Wilsdorf, 1982). The techniques used in the gradient-dependent dynamical theory of dislocations are those of bifurcation theory and self-organization, already developed in the study of dissipative structures and morphogenesis in far-from-equilibrium physico-chemical systems. Based on a simple reaction–transport model for "mobile" and "immobile" dislocation populations, it was found that wavelengths of the same order as those experimentally observed can be predicted. It was further shown (again in qualitative agreement with observations) that due to continuous symmetry breaking, a constant shift or glide of the entire pattern can easily be realized without relaxing back to its original configuration. Stationary phase fluctuations were also shown to exist leading to layer-splitting of the patterned structure and random development of "super defects," as experimentally observed. Moreover, the theory can provide a quantitative description of the competition between the ladderlike structure of PSBs and the rodlike structure of the surrounding veins. Both of these structures are metastable beyond threshold and the analysis indicates that PSBs can nucleate and travel as propagating fronts within the veins. Finally, it was shown that when secondary slip is taking place along with primary slip, stable labyrinth structures are generated.

The aim of this paper is to provide a generalization of the initial "elementary" model proposed by Walgraef and Aifantis (1985) in both the transport/interaction gradient terms and the reaction/clustering nonlinear terms. This is done in Section 2 where the pattern-forming instability in the weakly nonlinear regime is discussed. It is shown that the basic features and properties of the "elementary" model are preserved by the "generalized" model. Section 3 considers the effect of temperature on the wavelength f the ladder structure within the framework of the generalized model. It is shown that the model predicts the correct temperature dependence in qualitative agreement with experimental observations. Section 4 contains an extension of the generalized model to describe secondary or double (multiple) slip effects and the associated occurrence of labyrinth structures. Finally, in Section 5, we present certain preliminary results related to the applicability of these ideas in describing the early stages of deformation with emphasis on orientational stability and the inhomogeneous evolution of slip during monotonic loading.

2. A Generalized Reaction–Transport Dislocation Model

In this section we examine a generalized dynamical system of coupled nonlinear differential equations describing the collective behavior of dislocation populations during fatigue experiments, and we discuss the related patterning instabilities leading to the formation of PSB structures. As mentioned in the Introduction, this dynamical model correctly predicts the wavelength of the ladderlike structure. Furthermore, the breaking of translational and rotational symmetries at the bifurcation naturally leads to layer-splitting and imperfection development in accordance with experimental observations. Below we discuss briefly the physical basis of the model.

Consider a monocrystal oriented for single slip in a direction parallel to the x axis and submitted to cyclic loading. After an initial period where hardening occurs, the forest of immobile dislocations is already well developed and this motivates the distinction between two types of dislocations: "trapped" or nearly immobile ones of density ρ_i, and "free" or mobile ones gliding on the primary slip plane of density ρ_m. In this regime, the dislocation density is sufficiently high (10^{13}–10^{14} m^{-2}) and can be represented by a continuous concentration field on a space scale larger than a few lattice spacings. For each type of dislocation a balance equation is set up. The source and flux terms are determined by the following basic processes:

- The creation of dislocations under an applied stress is described by means of internal sources. In the case of PSB formation, a large number of dislocation clusters is already present in the crystal. Hence, the majority of the newly created dislocations will almost be immediately pinned by the forest of nearly immobile ones. Such mechanisms induce "source" terms $f(\rho_i)$ in the kinetic equation for the immobile dislocation density ρ_i.
- When the applied stress reaches certain thresholds such that thermal activation (along with internal stress (or secondary dislocation) induced activation) dominates local energy barriers, dislocations may break free and move rapidly with a stress-dependent velocity in their glide planes. The freeing processes of trapped dislocations lead to a linear source term $b\rho_m$ in the kinetic equation for the mobile dislocation density ρ_m and to a corresponding sink term in the immobile dislocation kinetics. The parameter b, specifying the freeing or mobilization rate, is vanishingly small below threshold and suddenly reaches a finite value beyond the critical stress level. This parameter will be taken as the bifurcation parameter of the system and its variation may depend on the strain rate, the temperature, and other mechanical, crystallographic, and internal structure parameters.
- The pinning of mobile dislocations by immobile dipoles or multipoles induces a sink term in their evolution equation and a corresponding source

term in the equation describing the dynamics of the latter. This term is generally nonlinear and depends on the elementary interactions between the two types of dislocation populations. It is of the form $\sum_{n>1} c_n \rho_i^n \rho_m$ where the index n models the order of the cluster participating in the pinning process. The dipole formation corresponding to quadratic couplings between mobile dislocations is mainly efficient in the early stages of the deformation process. It will not be considered in the present discussion, which considers situations where the forest of multipoles is already well developed, but we will come back to this point later in Section 5.

• The motion of free or mobile dislocations is of a drift type giving rise to plastic flow. It is described by a "complete balance law" including both a material derivative and a divergence term. For the trapped or mobile dislocations, the motion in the slip and cross-slip directions is diffusive, their mobility being mostly due to thermal effects or to the coupling with vacancies and other defects.

By taking into account the aforementioned dynamical processes, the balance equations for the two dislocation populations may be written as follows:

$$\begin{aligned} \dot{\rho}_i &= \nabla_i (D_{ij}^{(0)} - D_{ijk}^{(1)} \nabla_k^2) \nabla_j \rho_i + f(\rho_i) - b\rho_i + \sum_n c_n \rho_i^n (\rho_m^+ + \rho_m^-), \\ \dot{\rho}_m^+ &= -v \nabla_x \rho_m^+ + \frac{b}{2} \rho_i - \sum_n c_n \rho_i^n \rho_m^+, \\ \dot{\rho}_m^- &= +v \nabla_x \rho_m^- + \frac{b}{2} \rho_i - \sum_n c_n \rho_i^n \rho_m^-, \end{aligned} \tag{2.1}$$

where a further distinction between positive and negative mobile dislocation densities ρ_m^+ and ρ_m^-, of a constant average glide velocity $v = v^+ = -v^- = \bar{v} \sin \omega t$ (with $\bar{v}$ denoting the amplitude and ω the frequency of the cyclic process), has been made. A mechanical justification for the origin of the various terms, included in (2.1) on the basis of effective mass and momentum balances for dislocation populations, was given by Aifantis (1986, 1987, 1988). By expressing the dynamical equations (2.1) in terms of the sum $\rho_m = \rho_m^+ + \rho_m^-$ and the difference $\delta_m = \rho_m^+ - \rho_m^-$ we obtain

$$\begin{aligned} \dot{\rho}_i &= \nabla_i (D_{ij}^{(0)} - D_{ijk}^{(1)} \nabla_k^2) \nabla_j \rho_i + f(\rho_i) - b\rho_i + \sum_n c_n \rho_i^n \rho_m, \\ \dot{\rho}_m &= -v \nabla_x \delta_m + b\rho_i - \sum_n c_n \rho_i^n \rho_m, \\ \dot{\delta}_m &= -v \nabla_x \rho_m - \sum_n c_n \rho_i^n \delta_m. \end{aligned} \tag{2.2}$$

The terms $f(\rho_i)$ and $-b\rho_i$ in the above equations represent, respectively, creation/annihilation and mobilization of trapped dislocations. Due to the attractive character of the elastic interactions between dislocations, the diffusion coefficient tensor $\mathbf{D}^{(0)}$ may be negative definite in the high density regime while $\mathbf{D}^{(1)}$ remains positive definite. Hence, a diffusive instability may occur for high creation rates or high stress levels similar to what happens in Holt's theory of dislocation cell formation (Holt, 1970). The anisotropy of the underlying crystal structure is reflected here in the diffusion tensors. Moreover, if we take into account the explicit dependence of the glide velocity on the flow stress, and hence on the immobile dislocation density (as pointed out by Franek *et al.*, 1989), cross-diffusion terms between mobile and immobile densities appear in the dynamics. This effect should not affect qualitatively the present discussion and will be analyzed in a forthcoming publication.

On time scales larger than the period of the fatigue process and on space scales larger than the mean distance between obstables, the effective motion of mobile dislocations is diffusive. As discussed earlier by Walgraef and Aifantis (1986, 1988), this is a consequence of the combined effect of the back and forward motion of dislocations during each fatigue cycle and to the presence of a large number of pinning clusters. The effective diffusion coefficient D_M deduced from (2.2) is then proportional to the square of the mean velocity amplitude $\bar{v}$ (which may be related to the stress intensity τ via phenomenological laws of the type, for example, $\bar{v} = \bar{v}_0 \exp[-(\tau/\tau_0)^m]$) and inversely proportional to the total pinning rate. Specifically, it turns out (Walgraef and Aifantis, 1986, 1988) that $D_M \simeq \bar{v}^2/2\sum_n c_n \rho_i^{0n}$ and that its value if much larger than the values of the diffusion coefficients of the trapped dislocations. Our working version of the kinetic equation for ρ_m will then be given by

$$\dot{\rho}_m = D_M \nabla_x^2 \rho_m + b\rho_i - \sum_n c_n \rho_i^n \rho_m. \tag{2.3}$$

A less general version of (2.1) and (2.2) had been adopted and analyzed earlier by Walgraef and Aifantis (1985). In that version the diffusivity D_{ij}^0, was assumed isotropic, the diffusivity $D_{ijk}^{(1)}$ was assumed to vanish, and the index n was taken equal to 2, thus accounting only for the trapping of mobile dislocations by immobile dipoles. It turns out that the generalizations introduced in this paper do not alter the basic features and properties of the elementary model discussed earlier.

To see this, let us consider the pattern-forming instabilities associated with the dynamical system $(2.2)_1$ and (2.3). The linear stability analysis of the uniform steady state $[f(\rho_i^0) = 0,\ \rho_m^0 = b\rho_i^0/\sum_n c_n \rho_i^{0n}]$ gives, in Fourier space, the following linear evolution matrix:

$$L = \begin{bmatrix} \beta - a - q_i q_j (D_{ij}^{(0)} + q_k^2 D_{ijk}^{(1)}) & \gamma \\ -\beta & -\gamma - q_x^2 D_M \end{bmatrix}, \tag{2.4}$$

where $a = -f'(\rho_i^0)$ is positive as a result of the stability of the uniform steady state at low stress intensities, and where

$$\gamma = \sum_n c_n \rho_i^{0n} \qquad \text{and} \qquad \beta = b\left(\rho_i^0 \frac{\partial \ln \gamma}{\partial \rho_i^0} - 1\right).$$

The computation of the eigenvalues of this evolution matrix shows that, in the absence of the diffusional instability ($|D_{ij}^{(0)}| > 0$), by increasing the stress intensity (and since the motion of mobile dislocations in the primary slip direction is much faster than in any other), the uniform steady state becomes unstable versus the density modulations, first along the x direction at a bifurcation point given by

$$\beta = \beta_c = \left(\sqrt{a} + \sqrt{\frac{\gamma D_I}{D_M}}\right)^2, \qquad \mathbf{q} = q_c \mathbf{e}_x, \qquad q_c = \frac{2\pi}{\lambda_c} = \left(\frac{a\gamma}{D_I D_M}\right)^{1/4}. \quad (2.5)$$

where $D_I \equiv D_{xx}^{(0)}$ and $\mathbf{e}_x$ is the unit vector in the x direction.

It follows that the expressions given in (2.5) are exactly the same as those obtained from the initial elementary model, provided that the mobilization and pinning coefficients β and γ are properly defined.

Hence, beyond this pattern-forming instability, ladderlike structures are expected to develop with a wavelength λ_c, which is a material property depending on the dislocation mobilities as well as their multiplication and pinning rates. In fact, the wavelength selection results from a compromise between dislocation motion and interactions: the diffusion tries to increase it, while the interactions try to decrease it. By relating the parameters of the model to experimental quantities, it may be shown (Walgraef and Aifantis, 1988) that the wavelength satisfies the standard empirical or phenomenological relations such as $q_c \sim \sqrt{\rho_i^0} \sim \sqrt{\tau_s}$ where τ_s is the resolved shear stress.

When $|D_{ij}^{(0)}| < 0$, a diffusional instability occurs, leading first to the formation of cellular structures which may be associated with the vein structure of the matrix. By increasing the stress intensity, the freeing of trapped dislocations becomes more and more efficient. As a result of the highly anisotropic motion of free dislocations, this cellular structure is destabilized versus ladderlike structures with wave vectors parallel to the primary slip direction (Walgraef and Aifantis, 1985).

Having found a pattern-forming instability in the dynamical model, we must now consider the selection and stability properties of the pattern evolving beyond the instability. The nonlinear character of the dislocation dynamics does not allow, in general, analytic solutions for ρ_i and ρ_m. However, near the instability (or the weakly nonlinear regime around the bifurcation point) the dynamics may be reduced to much simpler forms, as a result of the time and space scale separations which occur between stable and unstable modes (e.g., Newell, 1974, and references quoted therein). This approximation procedure, justified by physical (slow mode dynamics (e.g., Haken, 1978)) or mathematical (center manifold theorem (e.g., Arnold, 1980)) arguments leads in our problem to the following asymptotic dynamics for the dislocation microstructure, when attention is restricted to the primary $x - y$ slip plane,

$$\tau_0 \partial_t \sigma = [\varepsilon - d_x(q_c^2 + \nabla^2)^2 + d_y \nabla_y^2]\sigma - v\sigma^2 - u\sigma^3, \qquad (2.6)$$

where $\varepsilon = (\beta - \beta_c)/\beta_c$, $d_x = D_I/\beta_c q_c^2$, $d_y = D_I/\beta_c(1 + \gamma/q_c^2 D_M)$, and $\nabla^2 = \nabla_x^2 + \nabla_y^2$. The order parameterlike variable σ is a linear combination of the two dislocation densities while u and v may be obtained explicitly in terms of the phenomenological coefficients of the model. (The "characteristic time"-like constant τ_0 is also directly related to the original parameters of the model and should not be confused with the reference yieldlike or threshold stress τ_0 used in other places of the text.)

In isotropic cases ($d_y = 0$), the only stable steady states beyond threshold ($\varepsilon > 0$) correspond to hexagonal or layered patterns, while in highly anisotropic media the cellular structures become unstable. The situation is similar to the one discussed for the simpler model by Walgraef and Aifantis (1985) and is depicted in Fig. 3. It thus turns out again the layered structures with walls perpendicular to the primary slip direction will be the only possible ones. Their amplitude equation is given by

$$\tau_0 \partial_t A = \varepsilon A + (\xi_x^2 \nabla_x^2 + \xi_y^2 \nabla_y^2) A - 3u|A|^2 A, \qquad (2.7)$$

where $\sigma(\mathbf{r}, t) = A(\mathbf{r}, t) \exp[iq_c x] + A^*(\mathbf{r}, t) \exp[-iq_c x]$ with * denoting complex conjugate, $|\ |$ magnitude, and $\xi_x^2 = 4q_c^2 d_x$, $\xi_y^2 = d_y$. This equation describes the slow amplitude modulations on the critical length scale of the pattern.

The preferred solution (with respect to the associated Lyapounov function), for steady state modulations of uniform amplitude in a medium with natural boundary conditions, is then given by

$$\sigma_0 = 2R_0 \cos(q_c x + \phi_0), \qquad R_0 = \sqrt{\frac{\varepsilon}{3u}}, \qquad \phi_0 = \text{constant}. \qquad (2.8)$$

3. Temperature Effects on the PSB Microstructures

In this section we discuss how the results of the previous dynamical analysis of dislocation patterning are affected by the temperature of the fatigue process. We show that it is possible, within the present dynamical framework, to interpret the existing experimental observations. In this connection, we remark that while the qualitative aspects of PSB structures obtained during fatigue experiments for Cu (Basinski *et al.*, 1980) and Mg (Kwadjo and Brown, 1978) crystals at various temperatures are the same, their quantitative properties are temperature-dependent. The saturation stress corresponding to the plateau of the stress–strain curve decreases for increasing temperatures, while the wavelength of the ladderlike structure increases with temperature. Basinski *et al.* (1980) also found (by decreasing the temperature after saturation at room temperature), that the smaller wavelength low-temperature ladders occupy the same sites as the high-temperature ladders. On the other hand, for fatigue experiments performed at room temperature after preliminary saturation at 77.4 K, the low-temperature ladders remain unchanged while new high-temperature ladders form in the matrix. As will be shown below, this phenomenon is consistent with the dynamical description of structure formation near a pattern-forming instability.

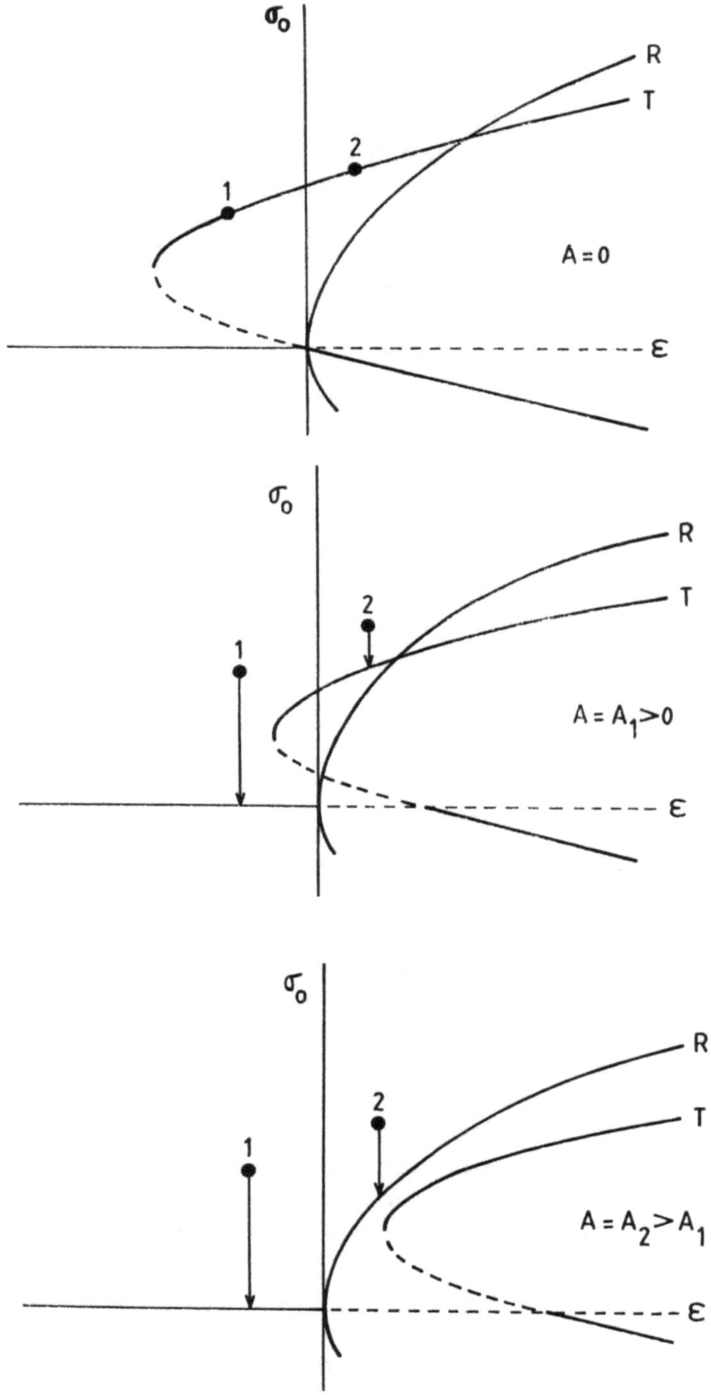

FIG. 3. Bifurcation diagram for two-dimensional anisotropic systems with increasing anisotropy. (The branch labeled with R represents roll or layered structures of the wave vector $\mathbf{q} = q_c\mathbf{e}_x$, while T represents structures of triangular symmetry. States 1 and 2, corresponding to initially stable triangular or honeycomb patterns, relax to new states when the anisotropy is increased.)

In order to study the effect of temperature on the ladderlike structure beyond the patterning instability within the present dynamical framework, it is necessary to describe the temperature dependence of the various parameters of the dynamical system $(2.2)_1$ and (2.3). The nonlinear couplings corresponding to annihilation, creation, and pinning represent athermal effects in the plastic regime and will thus be described by constant parameters with respect to temperature. On the other hand, the terms associated with dislocation mobilities and freeing rates are highly sensitive to temperature, as can be concluded from the following discussion:

- The mobility of trapped dislocations is associated with thermal effects describable in terms of the elastic free energy or the coupling with vacancies. It is thus assumed to be proportional to the temperature.
- The average velocity of free dislocations is also expected to increase with temperature. This can be justified in view of the Arrhenius-type rate equations describing thermally activated dislocation motion through a random network of obstacles. Alternatively, such an effect can be understood on the basis of the constitutive description of plastic deformation via the temperature dependence of a reference yieldlike stress. This is apparent, for example, in typical expressions for the mean dislocation velocity of the form $v = v_0 \exp[-(\tau_0/\tau)^m]$ where the threshold stress τ_0 is a decreasing function of temperature.
- The freeing of mobile dislocations from the nearly immobile forest of multipoles may occur via different mechanisms such as stress induced destabilization of dipoles or via fragmentation of dipolar loops by secondary dislocations (Ackerman *et al.*, 1984; Buchinger *et al.*, 1984). In the present model, these effects are described by the production of mobile dislocations at the expense of the trapped ones. The corresponding rate coefficient b is then assumed to vary linearly with the local rate of plastic deformation. This, in turn, may be taken proportional to the Arrhenius-like function $\exp[(A/T)(\tau - \tau_0 - B\sqrt{\rho_i^0})]$ where τ_0 is a threshold yieldlike stress associated with the freeing mechanism and $B\sqrt{\rho_i^0}$ is an internal or back stress due to the forest of immobile dislocations.

On the basis of the above discussion it follows that the wavelength of the ladderlike structure λ_c of PSBs is an increasing function of the temperature since λ_c is proportional to $(D_I/D_M)^{1/4}$. Subsequently, the critical value of the stress associated with the pattern-forming instability (or the plateau stress τ_s associated with the PSB formation) is a decreasing function of the temperature since τ_s is proportional to $1/\lambda_c^2$ (recall that $q_c \sim \sqrt{\rho_i^0} \sim \sqrt{\tau_s}$).

Another interesting observation is the fact that above threshold, i.e., beyond the patterning instability, a band of wave vectors parallel to the primary slip direction may correspond to stable structures. Let us further elaborate on this point by considering the stability of ladderlike structures of the wave vector $q\mathbf{e}_x$ with q close to q_c. On writing the slow mode as

$$\sigma = A \exp(iqx) + A^* \exp(-iqx), \tag{3.1}$$

the kinetic equation for the amplitude $A(x, y, t)$ of this pattern is given by

$$\tau_0 \partial_t A = [\varepsilon - d_x(q_c^2 - q^2 + 2iq\nabla_x + \nabla^2)^2 + d_y \nabla_y^2]A - 3u|A|^2 A. \quad (3.2)$$

Hence, a structure of amplitude $[(\varepsilon - d_x(q_c^2 - q^2)^2)/3u]^{1/2}$ may in principle exist provided that $\varepsilon > d_x(q_c^2 - q^2)^2$. The stability analysis, however, suggests that such a structure is only stable in the domain of the parameter space defined by the relations

$$\varepsilon > 3d_x(q_c^2 - q^2)^2 \quad \text{and} \quad q^2 > q_c^2 - \frac{d_y}{2d_x}. \quad (3.3)$$

It is a matter of algebra to express these relations in terms of the temperature-dependent variables, i.e., the threshold stress or strain amplitudes and critical wave number q_c. From (2.5) one effectively obtains:

$$\beta > \beta_c + 3\sqrt{\frac{a\gamma D_I}{D_M}}\left(1 - \left(\frac{q}{q_c}\right)^2\right)^2, \quad (3.4a)$$

or

$$\tau > \tau_c + 3\sqrt{\frac{a\gamma D_I}{D_M}}\left(1 - \left(\frac{q}{q_c}\right)^2\right)^2 \frac{1}{\left.\dfrac{\partial \ln \beta(\tau)}{\partial \tau}\right|_{\tau=\tau_c}}, \quad (3.4b)$$

and

$$\frac{q}{q_c} > \frac{\bar{q}}{q_c} = \frac{1}{\sqrt{2}}\sqrt{1 - \sqrt{\frac{\gamma D_I}{a D_M}}}. \quad (3.5)$$

We now propose the following scenario for the effect of sudden temperature variations: The stability domains associated with the initial and final temperature states are given in Fig. 4. Patterns with wave numbers belonging to the intersection of these two domains remain stable and the corresponding wavelength is not expected to change when a temperature change occurs. On the other hand, if the wave number of the initial pattern (corresponding to the initial temperature) is outside the stability domain associated with the final temperature, the system will evolve to a structure having a wavelength corresponding to a stable state with respect to the final temperature.

Let us discuss this process explicitly in the cases of either a sudden decrease or sudden increase in temperature after saturation.

(i) *Temperature decrease after saturation*: Let us suppose that the structure obtained at saturation for the initial temperature T_i has the corresponding preferred wavelength $q_c(T_i)$ and that the temperature decrease is such that $q_c(T_i) < \bar{q}(T_f)$ where T_f denotes the final temperature. The initial pattern is then unstable and should relax to the vein structure for stress amplitudes such that $\tau < \tau_c(T_f)$, or for strain amplitudes such that $\beta < \beta_c(T_f)$, depending on whether the experiment is stress or strain controlled. If $\tau > \tau_c(T_f)$ or $\beta > \beta_c(T_f)$, the patterns relax to a wall structure with a smaller wave-

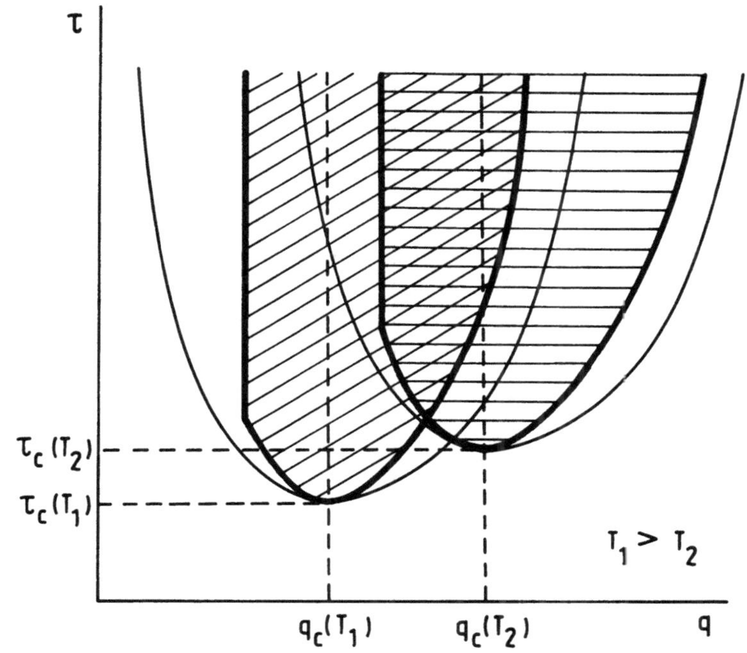

FIG. 4. Stability limits for wall structures with different temperatures T_1 and T_2. (Patterns in the doubly hatched regions are not affected by temperature jumps between T_1 and T_2. (M is the marginal stability curve, E is the Eckhaus stability limit versus longitudinal fluctuations, and Z is the zigzag limit versus transverse fluctuations.))

length (see Fig. 5). Since the distance from threshold is reduced in the final state, we should also observe a reduction in the amplitude of the pattern, i.e., a reduction in its dislocation content, and the walls should become thicker. In this connection, it is pointed out that numerical analysis (Walgraef *et al.*, 1987, 1988; Schiller and Walgraef, 1988) suggests that a reduction of the wall fraction occurs when the bifurcation parameter increases (e.g., from nearly 0.5 close to threshold to 0.09 for ε larger than 3).

(ii) *Temperature increase after saturation*: In this case, two possibilities arise as illustrated in Fig. 6. According to its saturation stress or strain, the initial pattern may correspond to either: (a) an unstable pattern with respect to the final conditions leading to an increase in its wavelength; or (b) a stable pattern with respect to the final conditions. No modification of the pattern wavelength is expected in this case. However, since the final state is farther above threshold than the initial one, the system is forced either to increase the amplitude of the existing patterns (leading to thinner walls) or to create new PSBs having, however, the preferred wavelength associated with the final state, i.e., $q_c(T_f)$. Hence, a final state corresponding to PSBs with wall structures having different wavelengths is compatible with the predictions of the dynamical model.

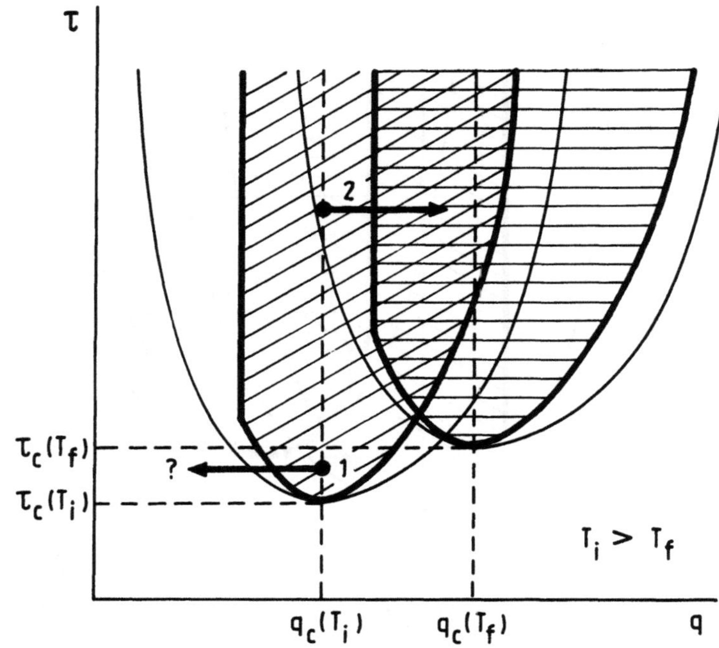

FIG. 5. The effect of temperature decrease after saturation. (The initial pattern 1 is marginally unstable with respect to the final conditions and should relax to the homogeneous steady state. On the other hand, pattern 2 lies within the marginal curve corresponding to the final temperature but is still unstable and should relax to a pattern with a smaller wavelength.)

The effect of temperature variations on the selection of PSB microstructures was qualitatively discussed in the framework of the amplitude equation for the dislocation pattern near the bifurcation point. The predictions of this type of dynamical model are consistent with experimental observations. Near threshold, the qualitative aspects of the discussion outlined here are in fact independent of the mode of testing, i.e., controlled stress or strain amplitude, since they only depend on the parabolic shape of the marginal stability curve. Farther from threshold, however, deviations between the two situations may arise. In fact, direct comparison between the predictions of the present model and experimental observations may be more convenient in the case of stress-controlled tests. In the case of strain-controlled experiments, a macroscopic constraint is exerted on the system, forcing the stress to adjust itself continuously according to the local strain and internal stress associated with the forest of nearly immobile dislocations, and this could continuously influence the pattern formation and evolution mechanisms.

Hence, by systematically performing variable-temperature fatigue tests after saturation, it would be possible to determine experimentally the stability curves of the wall structure in either stress- or strain-controlled experiments. The shape of these curves could then be used in the determination of the

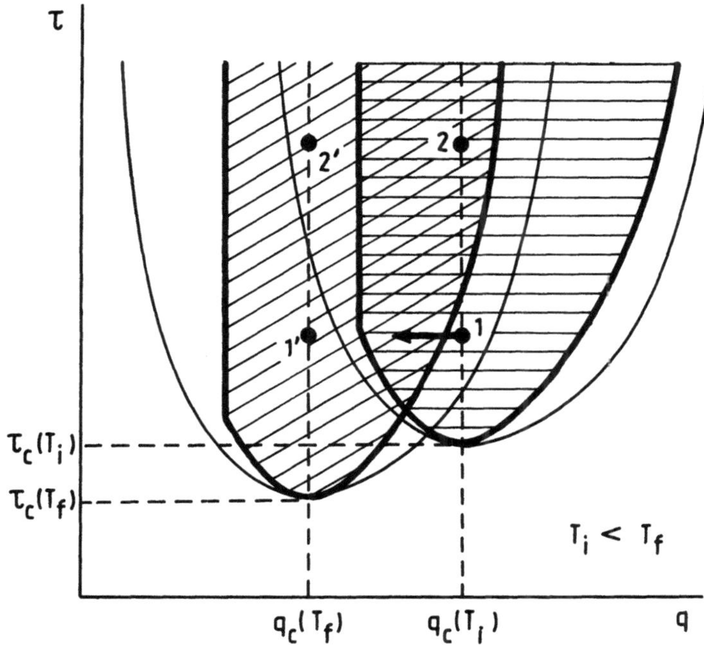

FIG. 6. The effect of temperature increase after saturation. (The initial pattern 1 lies within the marginal curve corresponding to the final temperature but is unstable and should relax to a pattern with a larger wavelength. The initial pattern 2 is in the stable region for both initial and final conditions and its wavelength should not vary. On the other hand, newly created patterns (1′, 2′) should nucleate with a wave number corresponding to the maximum growth rate associated with the final temperature $q_c(T_f)$.)

parameters of the model, especially in relation to the validity of the theoretical arguments about their temperature dependence.

In concluding this section, we provide a brief discussion on the evolution of the pattern far from the threshold. In this case, the weakly nonlinear approximation breaks down and numerical analysis is needed to study the behavior of the patterns and test the validity of the results obtained in the near-bifurcation regime. Currently, only the one-dimensional aspects of the reaction–diffusion model corresponding to patterning in the primary slip direction have been tested numerically (Walgraef *et al.*, 1987, 1988; Schiller and Walgraef, 1988). The various kinetic rates were obtained from experimentally accessible quantities such as the mean dislocation density, the mean dipole annihilation length, and the typical plastic deformation rates. Stable periodic structures were found above the predicted threshold with wavelengths between 0.5 μm and 1.5 μm, and maximum amplitudes for ρ_i of the order of $2 \cdot 10^{15}$ m^{-2} with the wall fraction being around 0.09. Different nucleation processes have been used as initial conditions, including slight dislocation density inhomogeneities in the bulk and stress inhomogeneities at

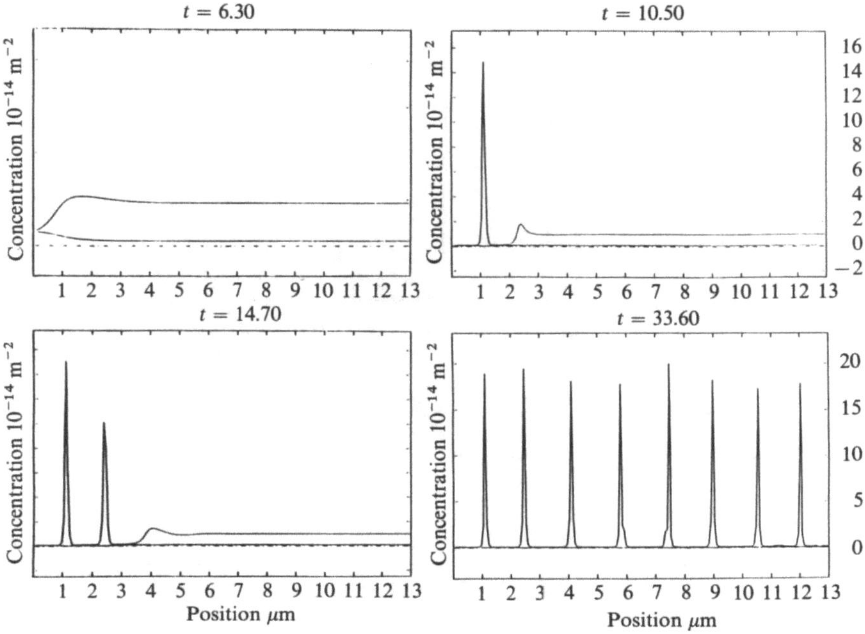

FIG. 7. The development of a wall structure from a stress inhomogeneity at the left border of the sample. (Snapshots of the time evolution of the free and trapped dislocation densities in space are shown. The time is given in periods of the fatigue process, and with the bifurcation parameter $\varepsilon = 3.5$.)

the surface. In particular, the inwards propagation of the wall structure from a surface inhomogeneity is illustrated in Fig. 7. The results of the numerical analysis suggest that the pattern properties remain qualitatively the same over a wide range of values of the bifurcation parameter. Quantitative variations are indeed observed and the comparison between numerical and experimental results will allow better estimates for the model parameters. Moreover, it will provide deeper insight into the role of various nonlinear interaction processes. A two-dimensional analysis is further required to capture the competition between cellular and wall structures, the stability properties of the patterns, and to compute the shape of the macroscopic stress–strain curve.

4. Secondary/Multiple Slip and Labyrinth Structures

Experimental observations of dislocation patterning in fatigued crystals oriented for double slip show interesting additional features (see, e.g., Jin and Winter, 1984; Ackerman *et al.*, 1984). According to the interactions between the dislocations associated with each slip system, different types of structures emerge. For example, if the interaction is strong, leading to locking effects of the Lomer–Cottrel type, PSB and ladderlike structures associated with the two slip systems develop in separate domains. On the other hand, if the interaction produces a third type of dislocation with a different Burgers vector, cellular structures may emerge. Similar observations have been made, in general, when secondary slip is activated, as well as in the case of fatigued polycrystals.

In order to describe such situations of double slip in terms of the proposed dynamical approach, we need to introduce at least two families of mobile dislocations. With them we associate two diffusion coefficients modeling, in effect, dislocation transport along two directions. For simplicity, we assume that these "effective slip" directions (x and y) are mutually orthogonal and express the relevant balance equations in terms of the new system of coordinates as follows:

$$\begin{aligned} \dot{\rho}_{\mathrm{i}} &= \nabla_i(D_{ij}^{(0)} - D_{ijk}^{(1)}\nabla_k^2)\nabla_j\rho_{\mathrm{i}} + f(\rho_{\mathrm{i}}) - (b_1 + b_2)\rho_{\mathrm{i}} + \sum_n c_n\rho_{\mathrm{i}}^n(\rho_{\mathrm{m1}} + \rho_{\mathrm{m2}}), \\ \dot{\rho}_{\mathrm{m1}} &= D_1\nabla_x^2\rho_{\mathrm{m1}} + b_1\rho_{\mathrm{i}} - \sum_n c_n\rho_{\mathrm{i}}^n\rho_{\mathrm{m1}}, \\ \dot{\rho}_{\mathrm{m2}} &= D_2\nabla_y^2\rho_{\mathrm{m2}} + b_2\rho_{\mathrm{i}} - \sum_n c_n\rho_{\mathrm{i}}^n\rho_{\mathrm{m2}}, \end{aligned} \tag{4.1}$$

where b_1 and b_2 are the freeing rates corresponding to the two families of free dislocations which are interacting with each other indirectly, via their interaction with the trapped dislocations. Direct interaction between the two mobile populations could be included in this description without causing a qualitative change in the results. The linear stability analysis around the uniform steady state yields the following evolution matrix:

$$L = \begin{pmatrix} (\beta_1 + \beta_2) - a - q_iq_j(D_{ij}^{(0)} + q_k^2D_{ijk}^{(1)}) & \gamma & \gamma \\ -\beta_1 & -\gamma - q_x^2D_1 & 0 \\ -\beta_2 & 0 & -\gamma - q_y^2D_2 \end{pmatrix}. \tag{4.2}$$

Next, we assume that the slow mobility of the trapped dislocations is almost random ($D_{ij}^{(0)} = D_I\delta_{ij}$; $D_I > 0$), that each freeing rate is proportional to the stress projection on the corresponding slip plane ($b_i \propto \beta_i \propto \beta \cos\theta_i$), and that the effective diffusivities of the mobile states are similar ($D_1 \simeq D_2 \equiv D_M$). It follows that the uniform steady state becomes unstable versus modulations in the x or y directions at a bifurcation point given by

$$\begin{aligned} \beta = \beta_c &= \left(\sqrt{a} + \sqrt{\frac{\gamma D_I}{D_M}}\right)^2, \qquad \mathbf{q} = \mathbf{q}\mathbf{e}_x \quad \text{or} \quad \mathbf{q} = q_c\mathbf{e}_y, \\ q_c &= \frac{2\pi}{\lambda_c} = \left(\frac{a\gamma}{D_ID_M}\right)^{1/4}. \end{aligned} \tag{4.3}$$

As a result, the slow mode dynamics, in the three-dimensional space, is of the form

$$\tau_0 \partial_t \sigma = [\varepsilon - d(q_c^2 + \nabla^2)^2 - \kappa \nabla_x^2 \nabla_y^2 + d_z \nabla_z^2]\sigma - v\sigma^2 - u\sigma^3, \qquad (4.4)$$

and, hence, the steady state solutions corresponding to modulations of uniform amplitude are given by the relations

$$\begin{aligned} \sigma_x &= 2R_0 \cos(q_c x + \phi_0), \qquad \sigma_y = 2R_0 \cos(q_c y + \psi_0), \\ R_0 &= \sqrt{\frac{\varepsilon}{3u}}, \qquad \phi_0, \psi_0 = \text{constants}. \end{aligned} \qquad (4.5)$$

Since the two slip directions are equivalent as far as the pattern forming instability is concerned, we could also expect the formation of square patterns of the type

$$\sigma_0 = A_x \exp(iq_c x) + A_x^* \exp(-iq_c x) + A_y \exp(iq_c y) + A_y^* \exp(-iq_c y), \qquad (4.6)$$

with their corresponding amplitude equations given by

$$\begin{aligned} \tau_0 \partial_t A_x &= \varepsilon A_x + (\xi_\parallel^2 \nabla_x^2 + \xi_\perp^2 \nabla_y^2 + \xi_z^2 \nabla_z^2) A_x - 3uA_x(|A_x|^2 + 2|A_y|^2), \\ \tau_0 \partial_t A_y &= \varepsilon A_y + (\xi_\parallel^2 \nabla_y^2 + \xi_\perp^2 \nabla_x^2 + \xi_z^2 \nabla_z^2) A_y - 3uA_y(|A_y|^2 + 2|A_x|^2). \end{aligned} \qquad (4.7)$$

The linear stability analysis of the uniform amplitude steady state, $|A_x^0| = |A_y^0| = \sqrt{\varepsilon/9u}$, shows that square planforms are unstable. Hence, the only possible structures correspond to PSBs associated with one or other slip systems, or to mixed structures corresponding to regions where the two slip systems dominate alternatively. But as a consequence of the instability of square patterns, when one wall structure nucleates in some region of the crystal it impedes the other one developing in the same region. For example, if we consider amplitude variations in the x direction only, chains of alternating kink–antikink solutions of the amplitude equation

$$\begin{aligned} \tau_0 \partial_t A_x &= \varepsilon A_x + \xi_\parallel^2 \nabla_x^2 A_x - 3uA_x(|A_x|^2 + 2|A_y|^2), \\ \tau_0 \partial_t A_y &= \varepsilon A_y + \xi_\perp^2 \nabla_x^2 A_y - 3uA_y(|A_y|^2 + 2|A_x|^2), \end{aligned} \qquad (4.8)$$

exist, corresponding to a succession of domains where A_x and A_y are alternatively zero and nearly equal to $\sqrt{\varepsilon/3u}$. As recently discussed by Coullet *et al.* (1987), these solutions are dynamically unstable with very long transients. But random pinning of these kink and antikink solutions may occur due to the interactions of these "defects" with inhomogeneities of the medium, or due to oscillatory interactions between the defects themselves. Since the wall structures are initiated within the relatively regular vein structure of the matrix, the experimentally observed alternating domains of PSBs associated with one or other slip system are consistent with their description, within the amplitude equation formalism. A similar analysis may justify the existence of labyrinth structures when secondary slip is activated (see also Walgraef and Aifantis, 1986).

If the freeing rates are the same for the two slip systems, the wave vectors of the dislocation microstructures are oriented along the principal axes of the global diffusion tensor, leading to the observed fact that deformation bands do not always develop along slip lines.

According to the orientation of the tensile axis, the interaction between mobile dislocations may also trigger the formation of new dislocations (Jin and Winter, 1984). When the angles between the Burgers vectors associated with the three families are equal, such as in cyclically deformed [100] copper, the resulting amplitude equation for the microstructures is isotropic. In this case, the presence of quadratic nonlinearities may induce the formation of cellular structures.

5. Early Stages of Deformation and Dislocation Current Effects

As shown earlier, the dynamics of the gliding dislocations may be effectively diffusional when a crystal with a highly developed forest of immobile dislocations is subjected to cyclic loading. In the early stages of deformation, however, the situation may be different, according to the mode of loading (monotonic or cyclic). In fact, anisotropic plastic flow effects due to the current- or driftlike motion of dislocations are dominant in this case and directly affect the orientational stability of the corresponding deformation patterns.

Let us consider a crystal oriented for single slip at the beginning of the plastic deformation. In this case it is not necessary to distinguish between "free" and "trapped" dislocations since pronounced immobile dislocation structures have not developed yet. One average dislocation density, $\rho = \rho^+ + \rho^-$, may thus be introduced to describe the dynamics of the process. Due to the interactions with other defects, and crystal imperfections, small mobilities in the glide and climb directions result, as before, into an effective diffusivity D to be assigned to both the ρ^+ and ρ^- populations. Moreover, a large glide current, $\mathbf{v}^+ = -\mathbf{v}^- \equiv \mathbf{v}$, in the primary slip direction (associated with the plastic flow) must now be introduced for both the ρ^+ and ρ^- populations. By writing separate balance equations for the density of dislocations, with positive and negative Burgers vectors, and by adding source and sink terms representing dislocation creation and annihilation, we obtain the following dynamical system:

$$\begin{aligned} \dot{\rho}^+ &= D\nabla^2\rho^+ - \mathbf{v}\cdot\nabla\rho^+ - NL^+(\rho^+, \rho^-), \\ \dot{\rho}^- &= D\nabla^2\rho^- + \mathbf{v}\cdot\nabla\rho^- - NL^-(\rho^+, \rho^-), \end{aligned} \tag{5.1}$$

where $\mathbf{v}$ is the mean glide velocity. The linear stability analysis around the uniform steady state (denoted by $|_0$) leads to the following evolution matrix in Fourier space:

$$L = \begin{pmatrix} -\alpha - q^2D - i\mathbf{q}\cdot\mathbf{v} & -\beta \\ -\beta & -\alpha - q^2D + i\mathbf{q}\cdot\mathbf{v} \end{pmatrix}, \tag{5.2}$$

where

$$q = |\mathbf{q}|, \qquad \alpha = \frac{\partial NL^+}{\partial \rho^+}\bigg|_0 = \frac{\partial NL^-}{\partial \rho^-}\bigg|_0 \quad \text{and} \quad \beta = \frac{\partial NL^+}{\partial \rho^-}\bigg|_0 = \frac{\partial NL^-}{\partial \rho^+}\bigg|_0.$$

In the case of monotonic loading, the dominant contributions to the dislocation dynamics during the early stages of plastic deformation correspond to dislocation multiplication and annihilation. Thus $-NL^{\pm}(\rho^+, \rho^-) = a\rho^{\pm} - b\rho^+\rho^-$ where a and b can be related to phenomenological quantities (Essmann and Mughrabi, 1979; Essmann *et al.*, 1981). It turns out that in this case the steady state $\rho^{+0} = \rho^{-0} = a/b$ is always unstable since the linear evolution matrix is

$$L = \begin{pmatrix} -q^2D - i\mathbf{q}\cdot\mathbf{v} & -a \\ -a & -q^2D + i\mathbf{q}\cdot\mathbf{v} \end{pmatrix} \tag{5.3}$$

with eigenvalues $\omega = -q^2D \pm \sqrt{a^2 - (\mathbf{q}\cdot\mathbf{v})^2}$. Hence, local fluctuations in the difference between positive and negative dislocations densities have a tendency to grow with a diffusion limited expansion perpendicularly to the glide direction, as well as to propagate in the slip direction. This may be easily seen by expressing the dynamical system (5.3) in terms of the sum (ρ) and the difference (δ) of the positive and negative dislocation densities

$$\begin{aligned} \dot{\delta} &= D\nabla^2\delta - \mathbf{v}\cdot\nabla\rho + a\delta, \\ \dot{\rho} &= D\nabla^2\rho - \mathbf{v}\cdot\nabla\delta + a\rho - \frac{b}{2}(\rho^2 - \delta^2), \end{aligned} \tag{5.4}$$

and is consistent with the experimental observation of the formation of slip lines of increasing thickness (Neuhauser, 1986).

In cyclic loading, due to the periodic reversal of the stress and the symmetry of the problem with respect to positive and negative dislocations, it is reasonable to assume that $-NL^{\pm}(\rho^+, \rho^-) = (a/2)(\rho^+ + \rho^-) - b\rho^+\rho^-$ and the steady state is given by $\rho^{+0} = \rho^{-0} = a/b$. In this case, the linear evolution matrix is

$$L = \begin{pmatrix} -\alpha/2 - q^2D - i\mathbf{q}\cdot\mathbf{v} & -\alpha/2 \\ -\alpha/2 & -\alpha/2 - q^2D + i\mathbf{q}\cdot\mathbf{v} \end{pmatrix}, \tag{5.5}$$

with eigenvalues $\omega = -\alpha/2 - q^2D \pm \sqrt{\alpha^2/2 - (\mathbf{q}\cdot\mathbf{v}^2)}$. Since the dynamics may also be written as

$$\begin{aligned} \dot{\delta} &= D\nabla^2\delta - \mathbf{v}\cdot\nabla\rho, \\ \dot{\rho} &= D\nabla^2\rho - \mathbf{v}\cdot\nabla\delta + a\rho - \frac{b}{2}(\rho^2 - \delta^2), \end{aligned} \tag{5.6}$$

uniform distributions are expected with eventually long-lived clusters resulting from the diffusional character of the evolution of δ.

The diffusion coefficient D may be negative as a consequence of the attractive character of dislocation interactions (Holt, 1970; Aifantis, 1985). In this case, we must consider higher-order space derivatives in the dynamics and write

$$\begin{aligned}\dot{\rho}^+ &= D\nabla^2\rho^+ - E\nabla^4\rho^+ - \mathbf{v}\cdot\nabla\rho^+ - NL^+(\rho^+,\rho^-),\\ \dot{\rho}^- &= D\nabla^2\rho^- - E\nabla^4\rho^- + \mathbf{v}\cdot\nabla\rho^- - NL^-(\rho^+,\rho^-),\end{aligned} \tag{5.7}$$

A pattern-forming instability may occur depending on the sign of the eigenvalues of the linear evolution matrix of the homogeneous steady state, which reads

$$L = \begin{pmatrix} -\alpha + q^2|D| - q^4E - i\mathbf{q}\cdot\mathbf{v} & -\beta \\ -\beta & -\alpha + q^2|D| - q^4E + i\mathbf{q}\cdot\mathbf{v}\end{pmatrix}. \tag{5.8}$$

For perturbations with a wave vector orthogonal to the glide velocity, the larger eigenvalue is $\omega = q_y^2|D| - q_y^4E$, leading to the development of patterns having the usual preferred wavelength, $\lambda_c = 2\pi\sqrt{2E/|D|}$. On the other hand, the eigenvalues associated with perturbations of wave vectors parallel to the glide direction are $\omega = -\alpha/2 + q_x^2|D| - q_x^4E \pm \sqrt{\alpha^2/2 - (q_x v_x)^2}$, and we see that the plastic flow has a stabilizing effect on these perturbations. The resulting patterns will then correspond to layered structures having their layers parallel to the slip planes. These structures, being nucleated preferentially at inhomogeneities, will then propagate in the crystal orthogonally to the slip planes, in a way which is reminiscent of the nucleation process of the Lüders bands (Neuhauser, 1986).

Hence, we see that the plastic flow in the form of a current- or driftlike motion of dislocations, at the early stages of deformation, may have different effects on the orientational stability of the corresponding deformation patterns. When it induces the destabilization process, it triggers structures with their walls perpendicular to the slip direction. On the other hand, when instability is induced by other processes such as the balance between dislocation multiplication and annihilation, its presence tends to orient the layers parallel to the slip planes, as in the case of the Lüders bands.

Acknowledgments

Financial support through a NATO grant for international collaboration in research is gratefully acknowledged. E. C. A. also acknowledges support from the U.S. National Science Foundation under grant CES-8800459.

References

Ackermann, F., Kubin, L. P., Lepinoux, J., and Mughrabi, H. (1984), The dependence of dislocation microstructure on plastic strain amplitude in cyclically strained copper single crystals, *Acta Metallurgica* **32**, 715–725.

Aifantis, E. C. (1978), A proposal for continuum with microstructure, *Mech. Res. Comm.*, **5**, 139–145.

Aifantis, E. C. (1983), Dislocation kinetics and the formation of deformation bands, in *Defects, Fracture and Fatigue*, edited by G. C. Sih and J. W. Provan, Martinus-Nijhoff, The Hague, pp. 75–84.

Aifantis, E. C. (1984a), On the mechanics of modulated structures, in *Modulated Structure Materials*, NATO ASI Series 83, edited by T. Tsakalakos, Martinus-Nijhoff, The Hague, pp. 357–385.

Aifantis, E. C. (1984b), On the microstructural origin of certain inelastic models, *Transactions of ASME, J. Mat. Engng. Tech.*, **106**, 326–330.

Aifantis, E. C. (1984c) Towards a continuum approach to dislocation patterning, in *Dislocations in Solids—Recent Advances*, AMD-63, edited by X. Markenscoff, ASME, New York, pp. 23–33.

Aifantis, E. C. (1985a), Continuum models for dislocated states and media with microstructures, in *Mechanics of Dislocations*, edited by E. C. Aifantis and J. P. Hirth, ASM, Metals Park, New York, pp. 127–146.

Aifantis, E. C. (1985b), On dislocation patterning, in *Dislocations in Solids*, edited by H. Suzuki, T. Ninomiya, K. Sumino, and S. Takeuchi, Tokyo University Press, Tokyo, pp. 41–47.

Aifantis, E. C. (1986a), Mechanics of microstructures—I, II, III, in *Mechanical Properties and Behaviour of Solids: Plastic Instabilities*, Proceedings ICTP Enrico Fermi School of Theoretical Physics, Trieste, Italy, edited by V. Balakrishnan and C. E. Bottani, World Scientific, Sinpagore, pp. 314–353.

Aifantis, E. C. (1986b), On the dynamical origin of dislocation patterns, *Mat. Sci. Engng.*, **81**, 563–574.

Aifantis, E. C. (1987), The physics of plastic deformation, *Int. J. Plasticity*, **3**, 211–247.

Aifantis, E. C. (1988), On the problem of dislocation patterning the persistent slip bands, in *Nonlinear Phenomena in Materials Science* (*Solid State Phenomena*, Vols. 3 and 4), edited by L. Kubin and G. Martin, Trans. Tech. Publ., pp. 397–406.

Arnold, V. (1980), *Chapitres supplémentaires de la théorie des equations differentielles ordinaires*, MIR, Mowcow.

Bammann, D.J. and Aifantis, E.C. (1982), On a proposal for a continuum with microstructure, *Acta Mechanica*, **45**, 91–121.

Basinski, Z. S., Korbel, A. S., and Basinski, S. J. (1980), The temperature dependence of the saturation stress and dislocation substructure in fatigued copper single crystals, *Acta Metallurgica*, **28**, 191–207.

Buchinger, L., Stanzl, S., and Laird, C. (1984), Dislocation structures in copper single crystals fatigued at low amplitudes, *Phil. Mag.*, **A50**, 275–298.

Coullet, P., Elphick C., and Repaux, D. (1987), *Phys. Rev. Lett.* **28**, 431.

Essmann, U. and Mughrabi, H. (1979), Annihilation of dislocations during tensile and cyclic deformation and limits of dislocation densities, *Phil. Mag.*, **A40**, 731–756.

Essmann, U., Gösele, U., and Mughrabi, H. (1981), A model of extrusions and intrusions in fatigued metals, I. Point-defect production and the growth of extrusions", *Phil. Mag.*, **A44**, 406–426.

Franek, A., Kalus, R., and Kratochvil, J. (1989), On stability of dislocation structure in cyclically deformed metal crystals and formation of persistent slip bands, preprint.

Haken, H. (1978), *Synergetics—An Introduction*, 2nd ed., Springer-Verlag, Berlin–Heidelberg–New York.

Holt, L. (1970), Dislocation cell formation in metals, *J. Appl. Phys.*, **41**, 3197–3201.

Jin, N. Y. and Winter, A. T. (1984), Dislocation structures in cyclically deformed [001] copper crystals, *Acta Metallurgica*, **32**, 1173–1176. (See also: Cyclic deformation of copper single crystals oriented for double slip, *Acta Metallurgica*, **32**, 989–995.)

Kuhlmann-Wilsdorf, D. (1982), Theory of dislocation cell sizes in deformed metals, *Mat. Sci. Engng.*, **55**, 79–83.

Kwadjo, R. and Brown, L. M. (1978), Cyclic hardening of magnesuim single crystals, *Acta Metallurgica*, **26**, 1117–1132.

Mughrabi, H. (1981), Cyclic plasticity of matrix and persistent slip bands in fatigued metals, in *Continuum Models for Discrete Systems*, Vol. 4, edited by O. Brulin and R. K. T. Hsieh, North-Holland, pp. 241–257.

Mughrabi, H., Ackermann, F., and Herz, K. (1979), Persistent slipbands in fatigued face-centered and body-centered cubic metals, in *Fatigue Mechanisms*, edited by J. T. Fong, ASTM-STP 675, ASTM, New York, pp. 69–105.

Mura, T. (1982), *Micromechanics of Defects in Solids*, Martinus-Nijhoff, The Hague.

Neuhauser, H. (1986), Physical manifestations of instabilities in plastic flow, in *Mechanical Behaviour of Solids: Plastic Instabilities*, Proceedings ICTP Enrico Fermi School of Theoretical Physics Trieste, Italy, edited by V. Balakrishnan and C. Bottani, World Scientific, Singapore, pp. 209–252.

Newell, A. C. (1974), Envelope equations, in *Nonlinear Wave Motion*, Lectures in Applied Mathematics, Vol. 15, edited by A. C. Newell, AMS, Providence, pp. 157–163.

Schiller, C. and Walgraef, D. (1988), Numerical simulation of persistent slip band formation, *Acta Metallurgica*, **A36**, 563–574.

Tabata, T., Fujita, H., Hiraoka, M., and Onishi, K. (1983), Dislocation behavior and the formation of persistent slip bands in fatigued copper single crystals observed by high-voltage electron microscopy. *Phil. Mag.*, **A47**, 841–857.

Walgraef, D. and Aifantis, E. C. (1985a), On the formation and stability of dislocation patterns—I, II, III, *Int. J. Engng. Sci.*, **23**, 1351–1358, 1359–1364, 1365–1372.

Walgraef, D. and Aifantis, E. C. (1985b), Dislocation patterning in fatigued metals as a result of dynamical instabilities, *J. Appl. Phys.*, **58**, 688–691.

Walgraef, D. and Aifantis, E. C. (1986), Dislocation patterning in fatigued metals: Labyrinth structures and rotational effects, *Int. J. Engng. Sci.*, 1789–1798.

Walgraef, D., Schiller, C., and Aifantis, E. C. (1987), Reaction–diffusion approach for dislocation patterns, in *Patterns, Defects, and Microstructures in Nonequilibrium Systems*, edited by D. Walgraef, Martinus-Nijhoff, The Hague, pp. 257–269.

Walgraef, D. and Aifantis, E. C. (1988), Plastic instabilities, dislocation patterns and nonequilibrium phenomena, *Res. Mechanica*, **23**, 161–195.

Winter, A. T. (1974), A model for the fatigue of copper at low plastic strain amplitudes, *Phil. Mag.*, **30**, 719–738.

Winter, A. T., Pedersen, O. B., and Rasmussen, K. V. (1981), Dislocation microstructures in fatigued copper polycrystals, *Acta Metallurgica*, **29**, 735–748.

Equivalence of Green's Function and the Fourier Series Representation of Composites with Periodic Microstructure

KEVIN P. WALKER
Engineering Science Software, Inc., Smithfield, RI 02917, U.S.A.

ERIC H. JORDAN
University of Connecticut, Storrs, CT 06268, U.S.A.

ALAN D. FREED
NASA–Lewis Research Center, Cleveland, OH 44135, U.S.A.

Abstract

This work is concerned with modeling the nonlinear mechanical deformation of composites comprised of a periodic microstructure under small displacement conditions at elevated temperatures. The practical motivation for such work stems from the need to design and optimize new multiphase materials and to predict their micromechanical and bulk material behavior under in-service thermomechanical loading conditions.

Two different methods, one based on a Fourier series approach and the other on a Green's function approach, are used in modeling the micromechanical behavior of the composite material. These two methods are shown to be equivalent to each other via the Poisson sum formula. Although the constitutive formulations are based on a micromechanical approach, it should be stressed that the resulting equations are volume averaged to produce overall "effective" constitutive relations which relate the bulk, volume averaged, stress increment to the bulk, volume averaged, strain increment. As such, they are macromodels which can be used directly in nonlinear finite element structural analysis programs.

1. Introduction

The ultimate objective of this work is to produce a computer program to analyze the heterogeneous stress and strain history variation at the "damage critical" locations of a composite structure operating at elevated temperatures. This paper describes some of the theoretical foundations for that program. A mesomechanics (Haritos *et al.*, 1988) approach is adopted which relates the micromechanical behavior of the heterogeneous composite to its in-service macroscopic behavior.

A comprehensive application of micromechanics to mechanical deformation problems is given by Mura (1982) in his book *Micromechanics of Defects in Solids.* The composite materials in which we are interested have fibers which are closely packed together in periodic arrays. Pictures of metal matrix composites (tungsten–fiber-reinforced superalloys) which exhibit a periodic microstructure can be found in the article by Petrasek *et al.* (1986).

Some composites are actually comprised of a periodic microstructure whilst others are possessed of an essentially randomly distributed microstructure. When the fibers in a composite material occupy a large volume fraction of the material, the induced deformation in one fiber interacts with and alters the induced deformation in the neighboring fibers. When the fibers are densely packed the interaction effect becomes dominant and must be accounted for in the constitutive formulation.

Iwakuma and Nemat-Nasser (1983) and Nemat-Nasser *et al.* (1981, 1982, 1983) have exploited the mathematical simplicity of a periodic microstructure in order to develop elastic, plastic, and creep constitutive models for composite materials. The assumption of periodicity allows the heterogeneous stress, strain, and displacement fields to be expanded in a Fourier series, which greatly simplifies the ensuing computations. This technique fully accounts for the interaction effects between neighboring fibers. Even when the composite is comprised of closely packed fibers distributed at random, the method gives accurate results (Nemat-Nasser *et al.*, 1982) for the "effective" elasticity tensor. When densely packed fibers form a large volume fraction of the composite material, these interaction effects play a dominant role and must be included in the calculations. It appears that inclusion of the interaction effects can be as, or more, important than inclusion of the random nature of the microstructure when the fibers occupy a large volume fraction of the composite material.

The nonlinear constitutive behavior of composites with a periodic microstructure can also be treated with a Green's function approach as shown in the expositions by Eshelby (1957), Gubernatis and Krumhansl (1975), Korringa (1973), Zeller and Dederichs (1973), and Barnett (1971, 1972). Here, the periodic heterogeneous material property variation—due to the fibers—is treated as a fictitious body force in the matrix material. The Green's function is used to evaluate the displacement due to a unit point force in the matrix material, and the actual displacement at any point in the composite can then be determined by summing (integrating) the effect due to a volume distribution of fictitious periodic body forces.

Dvorak (1986), Dvorak *et al.* (1982, 1988) and Bahei-El-Din *et al.* (1987) have also made great progress in modeling the micromechanical behavior of nonlinear periodic composite materials and are embarked on a combined experimental and theoretical effort.

Work on the theoretical foundations behind the homogenization of micromechanical constitutive models to produce bulk macroscopic models has been underway in France by Devriès and Léné (1987), Duvaut (1988), Renard and Marmonier (1987), Léné and Leguillon (1982), Léné (1986, 1987), and Sanchez-Palencia (1980, 1985).

Aboudi (1987) has recently developed a macroscopic formulation for periodic composites based on volume averaging Bodner's viscoplastic constitutive model (Bodner, 1987) over the unit periodic cell, but the method is general and is not restricted to any particular constitutive model. This work expands the heterogeneous displacement throughout the constituent phases of the composite material as linear and higher-order functions of the coordinates. Good agreement with experimental results was obtained by this method.

A more general approach is adopted in the present work, where the displacement is not restricted to linear or quadratic variations throughout each constituent phase, but varies according to the "exact" theory of an infinite periodic composite.

The purpose of the present paper is to outline briefly the Fourier series and Green's function formulations for the nonlinear constitutive behavior of viscoplastic composites comprised of a periodic microstructure, and to show that the formulations are equivalent by deriving the Green's function representation from the Fourier series representation using the Poisson sum formula. Further details concerning the formulations can be found in a recent report (Walker *et al.*, 1989).

2. Theoretical Modeling Approaches

A periodic composite material is supposedly acted upon by an imposed strain increment $\Delta\varepsilon_{ij}^0$ and responds in bulk with a stress increment $\Delta\sigma_{ij}^0$. These values are then equated to the respective volume averaged quantities in order to obtain the "effective" constitutive relation for the composite material, i.e.,

$$\Delta\sigma_{ij}^0 = \frac{1}{V_c}\iiint_{V_c} \Delta\sigma_{ij}(\mathbf{r})\, dV(\mathbf{r}) \qquad \text{and} \qquad \Delta\varepsilon_{ij}^0 = \frac{1}{V_c}\iiint_{V_c} \Delta\varepsilon_{ij}^{\mathrm{T}}(\mathbf{r})\, dV(\mathbf{r}). \quad (2.1)$$

In Section 3 it is shown that the volume averaged or "effective" constitutive relation for the composite material can be written as

$$\Delta\sigma_{ij}^0 = D_{ijkl}^{\mathrm{m}}\Delta\varepsilon_{kl}^0 - \frac{1}{V_c}\iiint_{V_c}\left\{D_{ijkl}^{\mathrm{m}}\Delta c_{kl}(\mathbf{r}) - \delta D_{ijkl}(\mathbf{r})\left[\Delta\varepsilon_{kl}^{\mathrm{T}}(\mathbf{r}) - \Delta c_{kl}(\mathbf{r})\right]\right\} dV(\mathbf{r}), \tag{2.2}$$

where V_c is the volume of a unit periodic cell in the composite material, $\Delta\varepsilon_{kl}^{\mathrm{T}}(\mathbf{r})$ is the total strain increment at point $\mathbf{r}$ in the periodic cell due to the imposed uniform total strain increment $\Delta\varepsilon_{kl}^0$ at the surface of the composite, and $\Delta c_{kl}(\mathbf{r})$ is the strain increment at point $\mathbf{r}$ in the periodic cell representing the deviation from isothermal elastic behavior. The fourth rank tensor $\delta D_{ijkl}(\mathbf{r})$ is defined by the relation

$$\delta D_{ijkl}(\mathbf{r}) = \vartheta(\mathbf{r})(D^{\mathrm{f}}_{ijkl} - D^{\mathrm{m}}_{ijkl}), \tag{2.3}$$

where $\vartheta(\mathbf{r}) = 1$ in the fiber and $\vartheta(\mathbf{r}) = 0$ in the matrix, with D^{f}_{ijkl} denoting the elasticity tensor of the fiber and D^{m}_{ijkl} that of the matrix.

In the expression for the average or "effective" constitutive relation in (2.2), the quantities $\Delta\varepsilon^0_{kl}$, D^{m}_{ijkl} and $\delta D_{ijkl}(\mathbf{r})$ are given. The deviation strain increment $\Delta c_{kl}(\mathbf{r})$ can be obtained throughout the periodic cell as a function of position $\mathbf{r}$ by using an explicit Euler forward difference method, since the stress and state variables in a viscoplastic formulation will be known functions of position at the beginning of the increment. Everything is therefore known explicitly except the total strain increment $\Delta\varepsilon^{\mathrm{T}}_{kl}(\mathbf{r})$.

In the Fourier series approach described in Section 3 we find that the total strain increment is determined by solving the integral equation

$$\Delta\varepsilon^{\mathrm{T}}_{kl}(\mathbf{r}) = \Delta\varepsilon^0_{kl} + \frac{1}{V_{\mathrm{c}}}\sum\sum_{n_p=0}^{\pm\infty}\sum{}' g_{klij}(\boldsymbol{\zeta})$$

$$\times \iiint_{V_{\mathrm{c}}} e^{i\boldsymbol{\xi}\cdot(\mathbf{r}-\mathbf{r}')}\{D^{\mathrm{m}}_{ijrs}\Delta c_{rs}(\mathbf{r}') - \delta D_{ijrs}(\mathbf{r}')[\Delta\varepsilon^{\mathrm{T}}_{rs}(\mathbf{r}') - \Delta c_{rs}(\mathbf{r}')]\}\, dV(\mathbf{r}'), \tag{2.4}$$

where the fourth rank tensor $g_{klij}(\zeta)$ is given by

$$g_{klij}(\zeta) = \tfrac{1}{2}(\zeta_j\zeta_l M^{-1}_{ik}(\zeta) + \zeta_j\zeta_k M^{-1}_{il}(\zeta)), \tag{2.5}$$

in which the Christoffel stiffness tensor $M_{ij}(\zeta)$, with inverse $M^{-1}_{ij}(\zeta)$, is defined by the relation

$$M_{ij}(\zeta) = D^{\mathrm{m}}_{pijq}\zeta_p\zeta_q, \tag{2.6}$$

with $\zeta_p = \xi_p/\sqrt{\xi_m\xi_m} = \xi_p/\xi$ being a unit vector in the direction of the Fourier wave vector $\boldsymbol{\xi}$, and $\xi = \sqrt{\xi_m\xi_m}$ denoting the magnitude of the vector $\boldsymbol{\xi}$. In (2.4) the sum is taken over integer values in which

$$\xi_1 = \frac{2\pi n_1}{L_1}, \qquad \xi_2 = \frac{2\pi n_2}{L_2}, \qquad \xi_3 = \frac{2\pi n_3}{L_3}, \tag{2.7}$$

and L_1, L_2, L_3 are the dimensions of the unit periodic cell in the x_1, x_2, x_3 directions, so that $V_{\mathrm{c}} = L_1L_2L_3$. The values of n_1, n_2, n_3 are given by

$$n_p = 0, \pm 1, \pm 2, \pm 3, \ldots, \text{etc.}, \quad \text{for} \quad p = 1, 2, 3; \tag{2.8}$$

and the prime on the triple summation signs indicates that the term with $n_1 = n_2 = n_3 = 0$ is excluded from the sum.

In the Green's function approach the total strain increment $\Delta\varepsilon^{\mathrm{T}}_{kl}(\mathbf{r})$ is determined by solving a different integral equation, viz.,

$$\Delta\varepsilon^{\mathrm{T}}_{kl}(\mathbf{r}) = \Delta\varepsilon^{0}_{kl} + \iiint\limits_{V} U_{klmn}(\mathbf{r}-\mathbf{r}')\{D^{\mathrm{m}}_{mnrs}\Delta c_{rs}(\mathbf{r}') - \delta D_{mnrs}(\mathbf{r}')[\Delta\varepsilon^{\mathrm{T}}_{rs}(\mathbf{r}') - \Delta c_{rs}(\mathbf{r}')]\}\, dV(\mathbf{r}'), \tag{2.9}$$

where the fourth rank tensor $U_{klmn}(\mathbf{r}-\mathbf{r}')$ gives the kl component of the total strain increment at point $\mathbf{r}$ due to the mn component of a stress increment applied at point $\mathbf{r}'$ in the infinite matrix with elasticity tensor D^{m}_{mnrs}, i.e.,

$$U_{klmn}(\mathbf{r}-\mathbf{r}') = -\frac{1}{2}\left(\frac{\partial^2 G_{km}(\mathbf{r}-\mathbf{r}')}{\partial x_l \partial x_n} + \frac{\partial^2 G_{lm}(\mathbf{r}-\mathbf{r}')}{\partial x_k \partial x_n}\right), \tag{2.10}$$

and the volume integration in (2.9) extends over all the periodic cells in the composite material, i.e., over the entire composite.

The Green's function tensor is defined in Barnett (1971, 1972) and Mura (1982) by the Fourier integral

$$G_{ij}(\mathbf{r}-\mathbf{r}') = \iiint\limits_{-\infty}^{\infty} \frac{d^3\mathbf{K}}{(2\pi)^3}\frac{M^{-1}_{ij}(\zeta)}{K^2} e^{-i\mathbf{K}\cdot(\mathbf{r}-\mathbf{r}')}, \tag{2.11}$$

in which the vector ζ is now defined by the relation $\zeta_i = K_i/K$ with $K = \sqrt{K_q K_q}$ denoting the magnitude of the vector $\mathbf{K} = (K_1, K_2, K_3)$.

In Section 5 it is shown, by applying the Poisson sum formula, that (2.4) and (2.9) are identical, although the summation extends over the integer values n_1, n_2, n_3 in (2.4) and extends over the periodic cells in (2.9).

Both (2.4) and (2.9) are implicit integral equations for the determination of the total strain increment $\Delta\varepsilon^{\mathrm{T}}_{kl}(\mathbf{r})$, since this unknown quantity appears both on the left-hand sides of the equations and on the right-hand sides under the volume integrations.

The "effective" constitutive relation given in (2.2) and the total strain increment relation, given by either (2.4) or (2.9), contain the volume integration of the deviation strain increment $\Delta c_{kl}(\mathbf{r})$. In the periodic cell the deviation strain increment at any point $\mathbf{r}$ will be determined from a unified viscoplastic constitutive relation (Lemaitre and Chaboche, 1985) appropriate to the constituent phase in which the point $\mathbf{r}$ resides. If a constituent phase is included at the fiber–matrix interface, a constitutive relation can also be proposed for this chemically degrading phase, and the resulting inelastic strain increment determined for inclusion in the volume integrals. This may be important for metal matrix composites where boron, carbon, and silicon carbide react chemically with superalloy matrices at elevated temperatures.

Equations (2.2), (2.4), and (2.9) form the basic incremental constitutive equations for determining the "effective" overall deformation behavior of a composite material with a periodic microstructure. In order to update the stress state in each of the constituent phases in preparation for integrating the "effective" constitutive relation over the next increment, the constitutive relation

$$\Delta\sigma_{ij}(\mathbf{r}) = D_{ijkl}(\mathbf{r})[\Delta\varepsilon^{\mathrm{T}}_{kl}(\mathbf{r}) - \Delta c_{kl}(\mathbf{r})] \tag{2.12}$$

is used, where $D_{ijkl}(\mathbf{r}) = D^{\mathrm{f}}_{ijkl}$ or D^{m}_{ijkl} according as the point $\mathbf{r}$ is in the fiber or matrix. The stress $\sigma_{ij}(\mathbf{r})$ and state variables $q_i(\mathbf{r})$ can now be updated at each point $\mathbf{r}$ in preparation for computing $\Delta c_{kl}(\mathbf{r})$ in the next increment.

The derivation of the preceding equations and some methods for their solution are discussed in the succeeding sections of this paper.

3. Fourier Series Approach

The application of Fourier series to the calculation of the "effective" overall constitutive behavior of periodic composites has been dealt with in detail by Iwakuma and Nemat-Nasser (1983) and Nemat-Nasser *et al.* (1981, 1982, 1983). This work is used in this section to develop constitutive relationships for viscoplastic composite materials under small displacement conditions.

The periodic composite is supposedly acted upon at its surface by a spatially linear displacement increment, $\Delta u_i^0(\mathbf{r})$, given by

$$\Delta u_i^0(\mathbf{r}) = x_j \Delta\varepsilon_{ij}^0 + x_j \Delta\omega_{ij}^0, \tag{3.1}$$

where $\Delta\varepsilon_{ij}^0$ and $\Delta\omega_{ij}^0$ are the spatially uniform strain and rotation increments at the surface of the composite.

If the matrix material was homogeneous and had no fibers embedded in it, the strain increment would be homogeneous and given by

$$\Delta\varepsilon_{ij}^0 = \frac{1}{2}\left(\frac{\partial(\Delta u_i^0)}{\partial x_j} + \frac{\partial(\Delta u_j^0)}{\partial x_i}\right). \tag{3.2}$$

Since this is constant, we may trivially volume average $\Delta\varepsilon_{ij}^0$ over the volume V of the homogeneous matrix material to obtain

$$\Delta\varepsilon_{ij}^0 = \frac{1}{V}\iiint_V \frac{1}{2}\left(\frac{\partial(\Delta u_i^0(\mathbf{r}))}{\partial x_j} + \frac{\partial(\Delta u_j^0(\mathbf{r}))}{\partial x_i}\right) dV(\mathbf{r}), \tag{3.3}$$

which, by Gauss' divergence theorem, may be written as

$$\Delta\varepsilon_{ij}^0 = \frac{1}{V}\iint_S \tfrac{1}{2}(n_j(\mathbf{r})\Delta u_i^0(\mathbf{r}) + n_i(\mathbf{r})\Delta u_j^0(\mathbf{r}))\, dS(\mathbf{r}), \tag{3.4}$$

where the integral extends over the surface of the material and $n_i(\mathbf{r})$ denotes the outwardly directed unit normal vector at point $\mathbf{r}$ on the surface. Thus, by applying the displacement increment $\Delta u_i^0(\mathbf{r})$ in (3.1) over the surface of the material to produce the surface strain increment given in (3.4), (3.2) and (3.3) show that the strain increment in the matrix material is spatially uniform.

If the displacement increment $\Delta u_i^0(\mathbf{r})$ in (3.1) is applied to the actual composite material, the total displacement increment within the material, $\Delta u_i^{\mathrm{T}}(\mathbf{r})$, will vary in a periodic manner due to the assumed geometric periodicity of the composite material, so that

$$\Delta u_i^{\mathrm{T}}(\mathbf{r}) = \Delta u_i^0(\mathbf{r}) + \Delta u_i(\mathbf{r}), \tag{3.5}$$

where $\Delta u_i^0(\mathbf{r})$ is the displacement increment which would be induced in the homogeneous matrix if the fiber phase were absent, and $\Delta u_i(\mathbf{r})$ is the perturbation or deviation from the homogeneous value due to the presence of the fibers.

Corresponding to these displacement increments, the total strain increment at any point $\mathbf{r}$ in the composite, $\Delta\varepsilon_{kl}^{\mathrm{T}}(\mathbf{r})$, is given by the relation

$$\Delta\varepsilon_{kl}^{\mathrm{T}}(\mathbf{r}) = \Delta\varepsilon_{kl}^0 + \Delta\varepsilon_{kl}(\mathbf{r}), \tag{3.6}$$

where

$$\Delta\varepsilon_{kl}^0 = \frac{1}{2}\left(\frac{\partial(\Delta u_k^0)}{\partial x_l} + \frac{\partial(\Delta u_l^0)}{\partial x_k}\right) \quad \text{and} \quad \Delta\varepsilon_{kl}(\mathbf{r}) = \frac{1}{2}\left(\frac{\partial(\Delta u_k)}{\partial x_l} + \frac{\partial(\Delta u_l)}{\partial x_k}\right), \tag{3.7}$$

with $\Delta\varepsilon_{kl}^0$ representing the spatially constant total strain increment which would be produced on the surface and in the interior of the homogeneous matrix if the fibers were absent, and with $\Delta\varepsilon_{kl}(\mathbf{r})$ representing the deviation from the uniform value due to the presence of the fibers. Both the total strain increment $\Delta\varepsilon_{kl}^{\mathrm{T}}(\mathbf{r})$ and the perturbed strain increment $\Delta\varepsilon_{kl}(\mathbf{r})$ vary throughout the composite in a periodic manner.

We *define* the volume averaged stress and strain increments as $\langle\Delta\sigma_{ij}\rangle$ and $\langle\Delta\varepsilon_{ij}\rangle$, respectively. The required "effective" constitutive equation for the composite material is then an expression relating the volume averaged stress and strain increments when these are equated to the respective values, $\Delta\sigma_{ij}^0$ and $\Delta\varepsilon_{ij}^0$, applied at the surface. For a function $f(\mathbf{r})$, which varies with position, the volume average is defined by the relation

$$\langle f\rangle = \frac{1}{V}\iiint_V f(\mathbf{r})\, dV(\mathbf{r}). \tag{3.8}$$

Since the composite is assumed to be comprised of a periodic aggregate of identical unit cells, we may write

$$\langle f\rangle = \frac{1}{V_c}\iiint_{V_c} f(\mathbf{r})\, dV(\mathbf{r}), \tag{3.9}$$

where V_c denotes the volume of the unit periodic cell.

If we volume average the total strain increment in (3.6), we obtain

$$\langle \Delta\varepsilon_{kl}^{\mathrm{T}} \rangle = \frac{1}{V_c} \iiint_{V_c} \Delta\varepsilon_{kl}^{\mathrm{T}}(\mathbf{r})\, dV(\mathbf{r}) = \Delta\varepsilon_{kl}^{0} + \frac{1}{V_c} \iiint_{V_c} \Delta\varepsilon_{kl}(\mathbf{r})\, dV(\mathbf{r}), \quad (3.10)$$

or

$$\langle \Delta\varepsilon_{kl}^{\mathrm{T}} \rangle = \Delta\varepsilon_{kl}^{0} + \langle \Delta\varepsilon_{kl} \rangle. \quad (3.11)$$

But the volume averaged total strain increment is equated with the value applied at the surface, so that $\langle \Delta\varepsilon_{kl}^{\mathrm{T}} \rangle = \Delta\varepsilon_{kl}^{0}$ and

$$\langle \Delta\varepsilon_{kl} \rangle = 0, \quad (3.12)$$

which shows that the volume average of the perturbation strain increment, $\Delta\varepsilon_{kl}(\mathbf{r})$, is equal to zero.

If the elasticity tensor is denoted by $D_{ijkl}(\mathbf{r})$ and the inelastic strain tensor by $\varepsilon_{kl}^{\mathrm{P}}(\mathbf{r})$, then the constitutive equation at any point $\mathbf{r}$ in the composite material can be written as

$$\sigma_{ij}(\mathbf{r}) = D_{ijkl}(\mathbf{r})(\varepsilon_{kl}^{\mathrm{T}}(\mathbf{r}) - \varepsilon_{kl}^{\mathrm{P}}(\mathbf{r}) - \alpha_{kl}(\mathbf{r})(T - T_0)), \quad (3.13)$$

where

$$\alpha_{kl}(\mathbf{r})(T - T_0) = \int_{T_0}^{T} \alpha_{kl}^{*}(\mathbf{r}, T)\, dT \quad (3.14)$$

is the thermal strain and $\alpha_{kl}(\mathbf{r})$, $\alpha_{kl}^{*}(\mathbf{r})$ are the average and instantaneous coefficients of thermal expansion.

The incremental form of Hooke's law is

$$\Delta\sigma_{ij}(\mathbf{r}) = D_{ijkl}(\mathbf{r})(\Delta\varepsilon_{kl}^{\mathrm{T}}(\mathbf{r}) - \Delta c_{kl}(\mathbf{r})), \quad (3.15)$$

where $\Delta c_{kl}(\mathbf{r})$ denotes the incremental strain representing the deviation from isothermal elastic conditions and is given by

$$\Delta c_{kl}(\mathbf{r}) = \Delta\varepsilon_{kl}^{\mathrm{P}}(\mathbf{r}) + \alpha_{kl}^{*}(\mathbf{r})\Delta T - D_{klij}^{-1}(\mathbf{r})\Delta D_{ijmn}(\mathbf{r})(\varepsilon_{mn}^{\mathrm{T}}(\mathbf{r}) - \varepsilon_{mn}^{\mathrm{P}}(\mathbf{r}) - \alpha_{mn}(\mathbf{r})(T - T_0)), \quad (3.16)$$

in which the tensor $\Delta D_{ijkl}(\mathbf{r})$ represents the incremental change in the elasticity tensor due to the temperature increment ΔT.

In a unified viscoplastic constitutive formulation (Lemaitre and Chaboche, 1985) which is integrated by an explicit Euler forward difference method, the inelastic strain increment $\Delta\varepsilon_{kl}^{\mathrm{P}}(\mathbf{r})$ is a function of the current stress (at the beginning of the increment), $\sigma_{ij}(\mathbf{r})$, and the current values of the state variables, $q_i(\mathbf{r})$. For example, if

$$\dot{\varepsilon}_{ij}^{\mathrm{P}} = f_{ij}(\sigma_{rs}, q_s), \quad (3.17)$$

then $\Delta\varepsilon_{ij}^{\mathrm{P}} = f_{ij}(\sigma_{rs}, q_s)\Delta t$, and the inelastic strain increment is independent of the total strain increment $\Delta\varepsilon_{kl}^{\mathrm{T}}(\mathbf{r})$. This independence of the inelastic strain increment on the total strain increment is no longer true if an implicit integration method (e.g., backward difference) or subincrementation method is used.

The elasticity tensor $D_{ijkl}(\mathbf{r})$ may be written as

$$D_{ijkl}(\mathbf{r}) = D^{\mathrm{m}}_{ijkl} + \delta D_{ijkl}(\mathbf{r}), \tag{3.18}$$

where

$$\delta D_{ijkl}(\mathbf{r}) = \vartheta(\mathbf{r})(D^{\mathrm{f}}_{ijkl} - D^{\mathrm{m}}_{ijkl}), \tag{3.19}$$

with $\vartheta(\mathbf{r}) = 1$ in the fiber and $\vartheta(\mathbf{r}) = 0$ in the matrix, the superscripts f and m referring to the elasticity tensor of the fiber and matrix, respectively. The constitutive equation at any point $\mathbf{r}$ can then be written, from (3.15), as

$$\Delta\sigma_{ij}(\mathbf{r}) = (D^{\mathrm{m}}_{ijkl} + \delta D_{ijkl}(\mathbf{r}))(\Delta\varepsilon^{0}_{kl} + \Delta\varepsilon_{kl}(\mathbf{r}) - \Delta c_{kl}(\mathbf{r})), \tag{3.20}$$

or

$$\begin{aligned}\Delta\sigma_{ij}(\mathbf{r}) = {} & D^{\mathrm{m}}_{ijkl}(\Delta\varepsilon^{0}_{kl} + \Delta\varepsilon_{kl}(\mathbf{r})) \\ & -\{D^{\mathrm{m}}_{ijkl}\Delta c_{kl}(\mathbf{r}) - \delta D_{ijkl}(\mathbf{r})[\Delta\varepsilon^{0}_{kl} + \Delta\varepsilon_{kl}(\mathbf{r}) - \Delta c_{kl}(\mathbf{r})]\}.\end{aligned} \tag{3.21}$$

If the quantity in braces is set equal to $D^{\mathrm{m}}_{ijkl}\Delta\varepsilon^{*}_{kl}(\mathbf{r})$, that is, if

$$D^{\mathrm{m}}_{ijkl}\Delta\varepsilon^{*}_{kl}(\mathbf{r}) = D^{\mathrm{m}}_{ijkl}\Delta c_{kl}(\mathbf{r}) - \delta D_{ijkl}(\mathbf{r})[\Delta\varepsilon^{0}_{kl} + \Delta\varepsilon_{kl}(\mathbf{r}) - \Delta c_{kl}(\mathbf{r})], \tag{3.22}$$

then (3.21) can be written in the form

$$\Delta\sigma_{ij}(\mathbf{r}) = D^{\mathrm{m}}_{ijkl}(\Delta\varepsilon^{\mathrm{T}}_{kl}(\mathbf{r}) - \Delta\varepsilon^{*}_{kl}(\mathbf{r})) = D^{\mathrm{m}}_{ijkl}(\Delta\varepsilon^{0}_{kl} + \Delta\varepsilon_{kl}(\mathbf{r}) - \Delta\varepsilon^{*}_{kl}(\mathbf{r})). \tag{3.23}$$

From the preceding equation it is evident that the eigenstrain increment, $\Delta\varepsilon^{*}_{kl}(\mathbf{r})$, represents the incremental deviation from isothermal elastic behavior in the composite material when the elasticity tensor is taken to be a spatially constant tensor appropriate to that of the matrix phase.

Newton's law for continuing static equilibrium throughout the strain increment requires that

$$\frac{\partial(\Delta\sigma_{ij}(\mathbf{r}))}{\partial x_j} = 0. \tag{3.24}$$

Equations (3.23) and (3.24) then require that

$$\frac{\partial\{D^{\mathrm{m}}_{ijkl}(\Delta\varepsilon^{0}_{kl} + \Delta\varepsilon_{kl}(\mathbf{r}) - \Delta\varepsilon^{*}_{kl}(\mathbf{r}))\}}{\partial x_j} = 0, \tag{3.25}$$

or, if $\Delta\varepsilon^{0}_{kl}$ is constant,

$$D^{\mathrm{m}}_{ijkl}\frac{\partial(\Delta\varepsilon_{kl}(\mathbf{r}))}{\partial x_j} = D^{\mathrm{m}}_{ijkl}\frac{\partial(\Delta\varepsilon^{*}_{kl}(\mathbf{r}))}{\partial x_j}. \tag{3.26}$$

Due to the geometric periodicity of the composite we may expand $\Delta u_k(\mathbf{r})$ and $\Delta\varepsilon^{*}_{kl}(\mathbf{r})$ in a Fourier series (Mura, 1982, Appendix 3). This gives

$$\begin{aligned}\Delta u_k(\mathbf{r}) = {} & \sum_{n_1=0}^{\pm\infty}\sum_{n_2=0}^{\pm\infty}\sum_{n_3=0}^{\pm\infty}{}' \Delta\hat{u}_k(n_1, n_2, n_3) \\ & \times \exp\left[i\left(\frac{2\pi n_1}{L_1}x_1 + \frac{2\pi n_2}{L_2}x_2 + \frac{2\pi n_3}{L_3}x_3\right)\right],\end{aligned} \tag{3.27}$$

where L_1, L_2, L_3 are the dimensions of a unit cell in the x_1, x_2, x_3 directions. The coefficients $\Delta\hat{u}_k$ in the Fourier expansion are determined by multiplying each side of (3.27) by $\exp\left[-i\left(\frac{2\pi m_1}{L_1}x_1 + \frac{2\pi m_2}{L_2}x_2 + \frac{2\pi m_3}{L_3}x_3\right)\right]$ and integrating over the volume of the unit cell to give

$$\Delta\hat{u}_k(n_1, n_2, n_3) = \frac{1}{L_1 L_2 L_3}\int_{x_1=0}^{L_1}\int_{x_2=0}^{L_2}\int_{x_3=0}^{L_3} \Delta u_k(\mathbf{r}) \times \exp\left[-i\left(\frac{2\pi n_1}{L_1}x_1 + \frac{2\pi n_2}{L_2}x_2 + \frac{2\pi n_3}{L_3}x_3\right)\right] dx_1\, dx_2\, dx_3, \tag{3.28}$$

where only the terms with $m_i = n_i$ survive in the summations.

Equations (3.27) and (3.28) can be written in shortened form as

$$\Delta u_k(\mathbf{r}) = \sum\sum_{n_p=0}^{\pm\infty}\sum{}' \Delta\hat{u}_k(\boldsymbol{\xi})e^{i\boldsymbol{\xi}\cdot\mathbf{r}}, \tag{3.29}$$

with coefficients $\Delta\hat{u}_k(\boldsymbol{\xi})$ determined by the inverse relation

$$\Delta\hat{u}_k(\boldsymbol{\xi}) = \frac{1}{V_c}\iiint_{V_c} \Delta u_k(\mathbf{r})e^{-i\boldsymbol{\xi}\cdot\mathbf{r}}\, dV(\mathbf{r}), \tag{3.30}$$

where

$$\boldsymbol{\xi} = (\xi_1, \xi_2, \xi_3), \qquad \mathbf{r} = (x_1, x_2, x_3), \qquad V_c = L_1 L_2 L_3, \tag{3.31}$$

with

$$\xi_i = \frac{2\pi n_i}{L_i} \quad \text{(no sum on } i\text{)} \qquad \text{for } i = 1, 2, 3. \tag{3.32}$$

The strain increment $\Delta\varepsilon_{kl}^*(\mathbf{r})$ can also be expanded in a Fourier series to give

$$\Delta\varepsilon_{kl}^*(\mathbf{r}) = \sum\sum_{n_p=0}^{\pm\infty}\sum{}' \Delta\hat{\varepsilon}_{kl}^*(\boldsymbol{\xi})e^{i\boldsymbol{\xi}\cdot\mathbf{r}}, \tag{3.33}$$

with coefficients $\Delta\hat{\varepsilon}_{kl}^*$ determined by the inverse relation

$$\Delta\hat{\varepsilon}_{kl}^*(\boldsymbol{\xi}) = \frac{1}{V_c}\iiint_{V_c} \Delta\varepsilon_{kl}^*(\mathbf{r})e^{-i\boldsymbol{\xi}\cdot\mathbf{r}}\, dV(\mathbf{r}). \tag{3.34}$$

In (3.29) and (3.33) the prime indicates that the term with $n_1 = n_2 = n_3 = 0$ is excluded from the summations, since $\Delta\hat{u}_k$ $(n_1 = 0, n_2 = 0, n_3 = 0)$ represents a rigid body displacement increment and $\Delta\hat{\varepsilon}_{kl}^*$ $(n_1 = 0, n_2 = 0, n_3 = 0)$ represents a spatially uniform strain increment.

By substituting (3.29) into (3.7); (3.7) into the left-hand side of (3.26); and (3.34) into the right-hand side of (3.26), the equilibrium relationship becomes

$$D_{ijkl}^{\mathrm{m}}\sum\sum_{n_p=0}^{\pm\infty}\sum{}' \tfrac{1}{2}(\Delta\hat{u}_k(\boldsymbol{\xi})\xi_l\xi_j + \Delta\hat{u}_l(\boldsymbol{\xi})\xi_k\xi_j)e^{i\boldsymbol{\xi}\cdot\mathbf{r}} = -iD_{ijkl}^{\mathrm{m}}\sum\sum_{n_p=0}^{\pm\infty}\sum{}' \Delta\hat{\varepsilon}_{kl}^*(\boldsymbol{\xi})\xi_j e^{i\boldsymbol{\xi}\cdot\mathbf{r}}, \tag{3.35}$$

or

$$D^{\mathrm{m}}_{ijkl}\xi_l\xi_j\Delta\hat{u}_k(\boldsymbol{\xi}) = -iD^{\mathrm{m}}_{ijkl}\xi_j\Delta\hat{\varepsilon}^*_{kl}(\boldsymbol{\xi}). \tag{3.36}$$

If $\xi = \sqrt{\xi_i\xi_i}$ denotes the magnitude of the vector $\boldsymbol{\xi}$, a unit vector $\boldsymbol{\zeta}$ in the direction of $\boldsymbol{\xi}$ can be written as $\zeta_i = \xi_i/\xi$. Equation (3.36) can therefore be written in the form

$$\xi^2 D^{\mathrm{m}}_{ijkl}\left(\frac{\xi_l}{\xi}\right)\left(\frac{\xi_j}{\xi}\right)\Delta\hat{u}_k(\boldsymbol{\xi}) = -iD^{\mathrm{m}}_{ijkl}\xi_j\Delta\hat{\varepsilon}^*_{kl}(\boldsymbol{\xi}), \tag{3.37}$$

or

$$\xi^2(D^{\mathrm{m}}_{ijkl}\zeta_l\zeta_j)\Delta\hat{u}_k(\boldsymbol{\xi}) = -iD^{\mathrm{m}}_{ijkl}\xi_j\Delta\hat{\varepsilon}^*_{kl}(\boldsymbol{\xi}). \tag{3.38}$$

The second rank tensor,

$$M_{ik}(\boldsymbol{\zeta}) = M_{ki}(\boldsymbol{\zeta}) = D^{\mathrm{m}}_{ijkl}\zeta_l\zeta_j, \tag{3.39}$$

is called the Christoffel stiffness tensor and (3.38) can be written as

$$\xi^2 M_{ik}(\boldsymbol{\zeta})\Delta\hat{u}_k(\boldsymbol{\xi}) = -iD^{\mathrm{m}}_{ijrs}\xi_j\Delta\hat{\varepsilon}^*_{rs}(\boldsymbol{\xi}). \tag{3.40}$$

This equation can be inverted by premultiplying each side by the inverse tensor $\xi^{-2}\mathbf{M}^{-1}$ to give the Fourier expansion coefficients

$$\Delta\hat{u}_k(\boldsymbol{\xi}) = -iM^{-1}_{ik}(\boldsymbol{\zeta})D^{\mathrm{m}}_{ijrs}\xi_j\Delta\hat{\varepsilon}^*_{rs}(\boldsymbol{\xi})\xi^{-2}. \tag{3.41}$$

The expansion coefficients can now be substituted into the Fourier expansion of $\Delta u_k(\mathbf{r})$ in (3.29) to give

$$\Delta u_k(\mathbf{r}) = -\sum\sum_{n_p=0}^{\pm\infty}\sum{}' \, i\xi^{-2}M^{-1}_{ik}(\boldsymbol{\zeta})D^{\mathrm{m}}_{ijrs}\xi_j\Delta\hat{\varepsilon}^*_{rs}(\boldsymbol{\xi})e^{i\boldsymbol{\xi}\cdot\mathbf{r}}. \tag{3.42}$$

This result may now be substituted into (3.7), so that the perturbation strain increment may be written as

$$\Delta\varepsilon_{kl}(\mathbf{r}) = \sum\sum_{n_p=0}^{\pm\infty}\sum{}' \, \tfrac{1}{2}(\xi^{-2}M^{-1}_{ik}(\boldsymbol{\zeta})\xi_j\xi_l + \xi^{-2}M^{-1}_{il}(\boldsymbol{\zeta})\xi_j\xi_k)D^{\mathrm{m}}_{ijrs}\Delta\hat{\varepsilon}^*_{rs}(\boldsymbol{\xi})e^{i\boldsymbol{\xi}\cdot\mathbf{r}}. \tag{3.43}$$

If we define the fourth rank tensor $g_{klij}(\boldsymbol{\zeta})$ by the relation

$$g_{klij}(\boldsymbol{\zeta}) = \tfrac{1}{2}(M^{-1}_{ik}(\boldsymbol{\zeta})\zeta_j\zeta_l + M^{-1}_{il}(\boldsymbol{\zeta})\zeta_j\zeta_k), \tag{3.44}$$

then the perturbation strain increment can be written in the form

$$\Delta\varepsilon_{kl}(\mathbf{r}) = \sum\sum_{n_p=0}^{\pm\infty}\sum{}' \, g_{klij}(\boldsymbol{\zeta})D^{\mathrm{m}}_{ijrs}\Delta\hat{\varepsilon}^*_{rs}(\boldsymbol{\xi})e^{i\boldsymbol{\xi}\cdot\mathbf{r}}, \tag{3.45}$$

and by inserting the relation for the Fourier expansion coefficients $\Delta\hat{\varepsilon}^*_{rs}$ from (3.34), we obtain

$$\Delta\varepsilon_{kl}(\mathbf{r}) = \frac{1}{V_c}\sum\sum_{n_p=0}^{\pm\infty}\sum{}' \, g_{klij}(\boldsymbol{\zeta})\iiint_{V_c} D^{\mathrm{m}}_{ijrs}\Delta\varepsilon^*_{rs}(\mathbf{r}')e^{i\boldsymbol{\xi}\cdot(\mathbf{r}-\mathbf{r}')}\,dV(\mathbf{r}'), \tag{3.46}$$

where the integration extends over the volume, $V_c = L_1L_2L_3$, of the unit periodic cell.

From (3.6) the total strain increment is given by

$$\Delta\varepsilon_{kl}^{\mathrm{T}}(\mathbf{r}) = \Delta\varepsilon_{kl}^{0} + \frac{1}{V_c}\sum\sum_{n_p=0}^{\pm\infty}\sum{}' g_{klij}(\zeta)\iiint_{V_c} D_{ijrs}^{\mathrm{m}}\Delta\varepsilon_{rs}^{*}(\mathbf{r}')e^{i\xi\cdot(\mathbf{r}-\mathbf{r}')}\,dV(\mathbf{r}'), \quad (3.47)$$

which, from the definition of $D_{ijrs}^{\mathrm{m}}\Delta\varepsilon_{rs}^{*}(\mathbf{r}')$ in (3.22), may be written in the final form

$$\Delta\varepsilon_{kl}^{\mathrm{T}}(\mathbf{r}) = \Delta\varepsilon_{kl}^{0} + \frac{1}{V_c}\sum\sum_{n_p=0}^{\pm\infty}\sum{}' g_{klij}(\zeta)\iiint_{V_c} e^{i\xi\cdot(\mathbf{r}-\mathbf{r}')}$$
$$\times\{D_{ijrs}^{\mathrm{m}}\Delta c_{rs}(\mathbf{r}') - \delta D_{ijrs}(\mathbf{r}')[\Delta\varepsilon_{rs}^{\mathrm{T}}(\mathbf{r}') - \Delta c_{rs}(\mathbf{r}')]\}\,dV(\mathbf{r}'). \quad (3.48)$$

This implicit integral equation—(2.4) in Section 2—must be solved to yield the total strain increment $\Delta\varepsilon_{kl}^{\mathrm{T}}(\mathbf{r})$ at each point $\mathbf{r}$ in the unit periodic cell.

Instead of solving for $\Delta\varepsilon_{kl}^{\mathrm{T}}(\mathbf{r})$ from this implicit integral equation, we could use (3.6) and (3.22) to eliminate $\Delta\varepsilon_{kl}^{\mathrm{T}}(\mathbf{r})$ from (3.48) to give an equivalent integral equation for $\Delta\varepsilon_{kl}^{*}(\mathbf{r})$, viz.,

$$D_{ijkl}^{\mathrm{m}}\Delta\varepsilon_{kl}^{*}(\mathbf{r}) = D_{ijkl}^{\mathrm{m}}\Delta c_{kl}(\mathbf{r}) - \delta D_{ijkl}(\mathbf{r})[\Delta\varepsilon_{kl}^{0} - \Delta c_{kl}(\mathbf{r})]$$
$$- \delta D_{ijkl}(\mathbf{r})\frac{1}{V_c}\sum\sum_{n_p=0}^{\pm\infty}\sum{}' g_{klmn}(\zeta)\iiint_{V_c} D_{mnrs}^{\mathrm{m}}\Delta\varepsilon_{rs}^{*}(\mathbf{r}')e^{i\xi\cdot(\mathbf{r}-\mathbf{r}')}\,dV(\mathbf{r}'). \quad (3.49)$$

The incremental constitutive relation at any point $\mathbf{r}$ is given in (3.23), and this relation can be used to update the stress state at any point $\mathbf{r}$ in the unit cell once (3.49) is solved for $\Delta\varepsilon_{kl}^{*}(\mathbf{r})$. Alternatively, (3.48) can be solved for $\Delta\varepsilon_{kl}^{\mathrm{T}}(\mathbf{r})$ and inserted into (3.22) and (3.23). The overall "effective" constitutive relation for the composite material can be obtained by averaging (3.23) over the unit periodic cell. This gives

$$\langle\Delta\sigma_{ij}\rangle = \langle D_{ijkl}^{\mathrm{m}}(\Delta\varepsilon_{kl}^{0} + \Delta\varepsilon_{kl} - \Delta\varepsilon_{kl}^{*})\rangle \quad (3.50)$$

or

$$\langle\Delta\sigma_{ij}\rangle = D_{ijkl}^{\mathrm{m}}\Delta\varepsilon_{kl}^{0} + D_{ijkl}^{\mathrm{m}}\langle\Delta\varepsilon_{kl}\rangle - D_{ijkl}^{\mathrm{m}}\langle\Delta\varepsilon_{kl}^{*}\rangle. \quad (3.51)$$

If we define $\Delta\sigma_{ij}^{0} = \langle\Delta\sigma_{ij}\rangle$ as the volume averaged stress increment, $\overline{\Delta\varepsilon_{kl}^{*}} = \langle\Delta\varepsilon_{kl}^{*}\rangle$ as the volume averaged eigenstrain increment, and note from (3.12) that the volume averaged perturbation strain increment is zero, i.e. $\langle\Delta\varepsilon_{kl}\rangle = 0$, then the overall "effective" constitutive relationship is

$$\Delta\sigma_{ij}^{0} = D_{ijkl}^{\mathrm{m}}\Delta\varepsilon_{kl}^{0} - D_{ijkl}^{\mathrm{m}}\overline{\Delta\varepsilon_{kl}^{*}}, \quad (3.52)$$

or, from (3.22),

$$\Delta\sigma_{ij}^{0} = D_{ijkl}^{\mathrm{m}}\Delta\varepsilon_{kl}^{0} - \frac{1}{V_c}\iiint_{V_c}\{D_{ijkl}^{\mathrm{m}}\Delta c_{kl}(\mathbf{r}) - \delta D_{ijkl}(\mathbf{r})[\Delta\varepsilon_{kl}^{\mathrm{T}}(\mathbf{r}) - \Delta c_{kl}(\mathbf{r})]\}\,dV(\mathbf{r}). \quad (3.53)$$

The procedure for integrating the overall "effective" constitutive relation then proceeds as follows:

(1) From a knowledge of the stress state throughout the unit periodic cell at the current time, t, calculate the inelastic strain increment $\Delta\varepsilon_{kl}^{\mathrm{P}}(\sigma_{rs}, q_s, \mathbf{r})$ from an appropriate unified viscoplastic constitutive relation. The viscoplastic constitutive relation will vary according as $\mathbf{r}$ is in the fiber or matrix phase, respectively.
(2) Compute the eigenstrain $\Delta\varepsilon_{kl}^{*}(\mathbf{r})$ throughout the unit periodic cell from either the implicit integral (3.49) or from (3.48) and (3.22).
(3) Compute the stress increment throughout the unit periodic cell from (3.23) and update the stress, strain, and viscoplastic state variables according to the relations

$$\sigma_{ij}(\mathbf{r}, t + \Delta t) = \sigma_{ij}(\mathbf{r}, t) + \Delta\sigma_{ij}(\mathbf{r}),$$

$$\varepsilon_{ij}^{\mathrm{T}}(\mathbf{r}, t + \Delta t) = \varepsilon_{ij}^{\mathrm{T}}(\mathbf{r}, t) + \Delta\varepsilon_{ij}^{\mathrm{T}}(\mathbf{r}),$$

$$q_i(\mathbf{r}, t + \Delta t) = q_i(\mathbf{r}, t) + \Delta q_i(\mathbf{r}).$$

(4) Calculate the overall "effective" stress and strain increment for the composite from (3.53) and update the overall "effective" stress and strain from the relations

$$\sigma_{ij}^{0}(t + \Delta t) = \sigma_{ij}^{0}(t) + \Delta\sigma_{ij}^{0},$$

$$\varepsilon_{ij}^{0}(t + \Delta t) = \varepsilon_{ij}^{0}(t) + \Delta\varepsilon_{ij}^{0}.$$

(5) Repeat the preceding calculations for each incremental load step.

The preceding algorithm makes use of the fact that the inelastic strain increment $\Delta\varepsilon_{kl}^{\mathrm{P}}(\mathbf{r})$ is independent of the total strain increment $\Delta\varepsilon_{kl}^{\mathrm{T}}(\mathbf{r})$ if an explicit Euler forward difference method is used to integrate the unified viscoplastic relations for the fiber and matrix phases. If an implicit method—such as backward difference or subincrementation—is used, the inelastic strain increment depends on the total strain increment. In this case the total strain increment must be obtained by iterating (3.48) in the form

$$\Delta\varepsilon_{kl}^{\mathrm{T}}(\mathbf{r}) = \Delta\varepsilon_{kl}^{0} + \frac{1}{V_{\mathrm{c}}}\sum\sum_{n_p=0}^{\pm\infty}\sum{}' g_{klij}(\zeta)\iiint_{V_{\mathrm{c}}} e^{i\boldsymbol{\xi}\cdot(\mathbf{r}-\mathbf{r}')}\{D_{ijrs}^{\mathrm{m}}\Delta c_{rs}(\mathbf{r}', \Delta\varepsilon_{pq}^{\mathrm{T}}(\mathbf{r}'))$$

$$- \delta D_{ijrs}(\mathbf{r}')[\Delta\varepsilon_{rs}^{\mathrm{T}}(\mathbf{r}') - \Delta c_{rs}(\mathbf{r}', \Delta\varepsilon_{pq}^{\mathrm{T}}(\mathbf{r}'))]\}\, dV(\mathbf{r}'). \tag{3.54}$$

The first iterative guess can be taken as $\Delta\varepsilon_{kl}^{\mathrm{T}}(\mathbf{r}) = \Delta\varepsilon_{kl}^{0}$, and the right-hand side evaluated to give an improved guess for $\Delta\varepsilon_{kl}^{\mathrm{T}}(\mathbf{r})$. This process is then continued with

$$\{\Delta\varepsilon_{kl}^{\mathrm{T}}(\mathbf{r})\}_{\lambda+1} = \Delta\varepsilon_{kl}^{0} + \frac{1}{V_{\mathrm{c}}}\sum\sum_{n_p=0}^{\pm\infty}\sum{}' g_{klij}(\zeta)\iiint_{V_{\mathrm{c}}} e^{i\boldsymbol{\xi}\cdot(\mathbf{r}-\mathbf{r}')}\{D_{ijrs}^{\mathrm{m}}\Delta c_{rs}(\mathbf{r}', \{\Delta\varepsilon_{pq}^{\mathrm{T}}(\mathbf{r}')\}_{\lambda}$$

$$- \delta D_{ijrs}(\mathbf{r}')[\{\Delta\varepsilon_{rs}^{\mathrm{T}}(\mathbf{r}')\}_{\lambda} - \Delta c_{rs}(\mathbf{r}', \{\Delta\varepsilon_{pq}^{\mathrm{T}}(\mathbf{r}')\}_{\lambda})]\}\, dV(\mathbf{r}'), \tag{3.55}$$

until the λth and $(\lambda + 1)$th iterates of $\Delta\varepsilon_{kl}^{\mathrm{T}}(\mathbf{r})$ converge.

Equation (3.49) is not so convenient for iteration as (3.48) when the inelastic strain increment depends on the total strain increment. It is always necessary to know the total strain increment $\Delta\varepsilon^{\mathrm{T}}_{kl}(\mathbf{r})$ in order to calculate the inelastic strain increment $\Delta\varepsilon^{\mathrm{P}}_{rs}(\mathbf{r}, \Delta\varepsilon^{\mathrm{T}}_{pq}(\mathbf{r}))$. But (3.22), viz.,

$$D^{\mathrm{m}}_{ijkl}\Delta\varepsilon^{*}_{kl}(\mathbf{r}) = D^{\mathrm{m}}_{ijkl}\Delta c_{kl}(\mathbf{r}, \Delta\varepsilon^{\mathrm{T}}_{pq}(\mathbf{r})) - \delta D_{ijkl}(\mathbf{r})[\Delta\varepsilon^{\mathrm{T}}_{kl}(\mathbf{r}) - \Delta c_{kl}(\mathbf{r}, \Delta\varepsilon^{\mathrm{T}}_{pq}(\mathbf{r}))] \quad (3.56)$$

is an implicit equation for $\Delta\varepsilon^{\mathrm{T}}_{kl}(\mathbf{r})$ when the iterated quantity, $\Delta\varepsilon^{*}_{kl}(\mathbf{r})$, is given. Equation (3.48) is therefore the appropriate equation to iterate when the inelastic strain increment depends on the total strain increment. For further details, see Walker *et al.*, (1989).

4. Green's Function Approach

The equation of continuing static equilibrium for the composite material throughout an applied strain increment is given by

$$\frac{\partial(\Delta\sigma_{ij}(\mathbf{r}))}{\partial x_j} + \Delta f_i(\mathbf{r}) = 0, \quad (4.1)$$

where $\Delta f_i(\mathbf{r})$ is the incremental body force per unit volume of the composite material. From (3.23) and (4.1) we obtain

$$D^{\mathrm{m}}_{ijkl}\frac{\partial(\Delta\varepsilon^{\mathrm{T}}_{kl}(\mathbf{r}))}{\partial x_j} = \frac{\partial}{\partial x_j}(D^{\mathrm{m}}_{ijkl}\Delta\varepsilon^{*}_{kl}(\mathbf{r})) - \Delta f_i(\mathbf{r}). \quad (4.2)$$

From this equation it is clear that the divergence of the stress variation produced by $\Delta\varepsilon^{*}_{kl}(\mathbf{r})$ may be formally regarded as a fictitious body force increment, analogous to $\Delta f_i(\mathbf{r})$, which is applied to the homogeneous matrix material with elasticity tensor D^{m}_{ijkl}. The theory of elasticity for homogeneous materials is generally concerned with the solution of the homogeneous differential equation (4.2)—Navier's equation—when the right-hand side is zero. When body forces are present, the standard method of solution is to obtain the displacement solution at $\mathbf{r}$ due to a unit body force applied at $\mathbf{r}'$. This solution is given by the Green's function $G_{ij}(\mathbf{r} - \mathbf{r}')$ which gives the displacement in the ith direction at $\mathbf{r}$ due to a unit point force applied in the jth direction at $\mathbf{r}'$. For a distributed incremental body force $\Delta f_i(\mathbf{r}')$ the displacement increment at $\mathbf{r}$ is obtained by summing the results for the distribution in the form

$$\Delta u_i(\mathbf{r}) = \iiint_V G_{ij}(\mathbf{r} - \mathbf{r}')\Delta f_j(\mathbf{r}')\, dV(\mathbf{r}'). \quad (4.3)$$

The integration extends over the whole volume, V, of the composite material which may be regarded as being of infinite extent.

When $\Delta f_j(\mathbf{r}') = 0$ we know that the displacement solution is $\Delta u^{\mathrm{T}}_i(\mathbf{r}) = \Delta u^{0}_i(\mathbf{r})$, corresponding to an applied uniform strain increment $\Delta\varepsilon^{0}_{ij}$ on the infinite boundary of the homogeneous matrix. For an effective distributed body force increment, given by the right-hand side of (4.2), with $\Delta f_j(\mathbf{r}') = 0$, the solution for the total displacement increment $\Delta u_i^{\mathrm{T}}(\mathbf{r})$ can be written as

$$\Delta u_i^{\mathrm{T}}(\mathbf{r}) = \Delta u_i^0(\mathbf{r}) - \iiint_V G_{ik}(\mathbf{r}-\mathbf{r}')\frac{\partial}{\partial x_l'}(D^{\mathrm{m}}_{klmn}\Delta\varepsilon^*_{mn}(\mathbf{r}'))\,dV(\mathbf{r}'). \tag{4.4}$$

This corresponds to (3.5), the volume integral representing the perturbed displacement increment $\Delta u_i(\mathbf{r})$ in (3.5) and (3.42).

For a material which is homogeneous with elasticity tensor D^{m}_{ijkl} the Green's function satisfies the differential relation (Mura, 1982, p. 10)

$$D^{\mathrm{m}}_{ijkl}\frac{\partial^2 G_{km}(\mathbf{r}-\mathbf{r}')}{\partial x_j\,\partial x_l} + \delta_{im}\delta(\mathbf{r}-\mathbf{r}') = 0, \tag{4.5}$$

where δ_{im} is the Kronecker delta tensor given by $\delta_{im} = 1$ if $i = m$ and $\delta_{im} = 0$ if $i \neq m$, and $\delta(\mathbf{r}-\mathbf{r}')$ is the three-dimensional Dirac delta function defined by the relation

$$\delta(\mathbf{r}-\mathbf{r}') = \delta(x_1 - x_1')\delta(x_2 - x_2')\delta(x_3 - x_3'). \tag{4.6}$$

By applying Fourier integral transform techniques the Green's tensor is shown (Barnett, 1971, 1972) to have the Fourier integral form

$$G_{ij}(\mathbf{r}-\mathbf{r}') = \iiint_{-\infty}^{\infty}\frac{d^3\mathbf{K}}{(2\pi)^3}\frac{M_{ij}^{-1}(\zeta)}{K^2}e^{-i\mathbf{K}\cdot(\mathbf{r}-\mathbf{r}')}, \tag{4.7}$$

in which the inverse Christoffel stiffness tensor $M_{ij}^{-1}(\zeta)$ is defined by

$$M_{ij}^{-1}(\zeta) = (D^{\mathrm{m}}_{pijq}\zeta_p\zeta_q)^{-1}, \tag{4.8}$$

with $\zeta_p = K_p/\sqrt{K_mK_m} = K_p/K$ being a unit vector in the direction of the Fourier wave vector $\mathbf{K}$, and $K = \sqrt{K_mK_m}$ denoting the magnitude of the wave vector $\mathbf{K}$.

Making use of the relation

$$G_{ik}(\mathbf{r}-\mathbf{r}')\frac{\partial}{\partial x_l'}(D^{\mathrm{m}}_{klmn}\Delta\varepsilon^*_{mn}(\mathbf{r}')) = \frac{\partial}{\partial x_l'}(G_{ik}(\mathbf{r}-\mathbf{r}')D^{\mathrm{m}}_{klmn}\Delta\varepsilon^*_{mn}(\mathbf{r}')) - \frac{\partial G_{ik}(\mathbf{r}-\mathbf{r}')}{\partial x_l'}D^{\mathrm{m}}_{klmn}\Delta\varepsilon^*_{mn}(\mathbf{r}'), \tag{4.9}$$

we may write (4.4) in the form

$$\Delta u_i^{\mathrm{T}}(\mathbf{r}) = \Delta u_i^0(\mathbf{r}) - \iiint_V \frac{\partial}{\partial x_l'}(G_{ik}(\mathbf{r}-\mathbf{r}')D^{\mathrm{m}}_{klmn}\Delta\varepsilon^*_{mn}(\mathbf{r}'))\,dV(\mathbf{r}') + \iiint_V \frac{\partial G_{ik}(\mathbf{r}-\mathbf{r}')}{\partial x_l'}D^{\mathrm{m}}_{klmn}\Delta\varepsilon^*_{mn}(\mathbf{r}')\,dV(\mathbf{r}'). \tag{4.10}$$

The first volume integral can be transformed into a surface integral via Gauss' divergence theorem, viz.,

$$\iiint\limits_V \frac{\partial}{\partial x_l'}(G_{ik}(\mathbf{r}-\mathbf{r}')D^{\mathrm{m}}_{klmn}\Delta\varepsilon^*_{mn}(\mathbf{r}'))\,dV(\mathbf{r}')$$

$$= \iint\limits_S n_l(\mathbf{r}')G_{ik}(\mathbf{r}-\mathbf{r}')D^{\mathrm{m}}_{klmn}\Delta\varepsilon^*_{mn}(\mathbf{r}')\,dS(\mathbf{r}'). \tag{4.11}$$

The surface integral extends over the entire outer surface of the "infinite" matrix material. Since this is assumed to be at an infinite distance, all the integration points $\mathbf{r}'$ in the surface integral are at an infinite distance from the field point $\mathbf{r}$ and $G_{ik}(\mathbf{r}-\mathbf{r}') = 0$. Thus, for an infinite body the first volume integral in (4.10) vanishes. This would not be the case for a finite body in which the field point $\mathbf{r}$ is close to the surface integration point $\mathbf{r}'$, and the volume (or surface) integral would need to be retained for these situations. In this case other surface integrals would arise (Korringa, 1973; Walker *et al.*, 1989) due to the application of boundary incremental displacements or surface tractions on the surface of the material.

From the properties of Green's function,

$$\frac{\partial G_{ik}(\mathbf{r}-\mathbf{r}')}{\partial x_l'} = -\frac{\partial G_{ik}(\mathbf{r}-\mathbf{r}')}{\partial x_l}, \tag{4.12}$$

which follows since G_{ik} is a function of

$$\mathbf{r}-\mathbf{r}' = (x_1 - x_1', x_2 - x_2', x_3 - x_3'). \tag{4.13}$$

Equation (4.10) may then be written alternatively as

$$\Delta u_i^{\mathrm{T}}(\mathbf{r}) = \Delta u_i^0(\mathbf{r}) - \iiint\limits_V \frac{\partial G_{ik}(\mathbf{r}-\mathbf{r}')}{\partial x_l} D^{\mathrm{m}}_{klmn}\Delta\varepsilon^*_{mn}(\mathbf{r}')\,dV(\mathbf{r}'). \tag{4.14}$$

But $\Delta\varepsilon^{\mathrm{T}}_{ij}(\mathbf{r}) = \frac{1}{2}(\partial(\Delta u_i^{\mathrm{T}}(\mathbf{r}))/\partial x_j + \partial(\Delta u_j^{\mathrm{T}}(\mathbf{r}))/\partial x_i)$, so that by differentiating (4.14) with respect to x_i and x_j and taking half the sum, we obtain

$$\Delta\varepsilon^{\mathrm{T}}_{ij}(\mathbf{r}) = \Delta\varepsilon^0_{ij} + \iiint\limits_V U_{ijkl}(\mathbf{r}-\mathbf{r}')D^{\mathrm{m}}_{klmn}\Delta\varepsilon^*_{mn}(\mathbf{r}')\,dV(\mathbf{r}'), \tag{4.15}$$

which, by means of (3.22), may be written as

$$\Delta\varepsilon^{\mathrm{T}}_{ij}(\mathbf{r}) = \Delta\varepsilon^0_{ij} + \iiint\limits_V U_{ijkl}(\mathbf{r}-\mathbf{r}')\{D^{\mathrm{m}}_{klrs}\Delta c_{rs}(\mathbf{r}')$$

$$-\delta D_{klrs}(\mathbf{r}')[\Delta\varepsilon^{\mathrm{T}}_{rs}(\mathbf{r}') - \Delta c_{rs}(\mathbf{r}')]\}\,dV(\mathbf{r}'). \tag{4.16}$$

An equivalent integral equation, involving the eigenstrain increment $\Delta\varepsilon^*_{ij}(\mathbf{r})$, can also be obtained by using (3.22) to eliminate $\Delta\varepsilon^{\mathrm{T}}_{ij}(\mathbf{r})$ from (4.15), which gives

$$D^{\mathrm{m}}_{ijkl}\Delta\varepsilon^*_{kl}(\mathbf{r}) = D^{\mathrm{m}}_{ijkl}\Delta c_{kl}(\mathbf{r}) - \delta D_{ijkl}(\mathbf{r})[\Delta\varepsilon^0_{kl} - \Delta c_{kl}(\mathbf{r})]$$
$$- \delta D_{ijkl}(\mathbf{r}) \iiint_V U_{klmn}(\mathbf{r}-\mathbf{r}')D^{\mathrm{m}}_{mnrs}\Delta\varepsilon^*_{rs}(\mathbf{r}')\,dV(\mathbf{r}'). \tag{4.17}$$

In the preceding equations the operator,

$$U_{ijkl}(\mathbf{r}-\mathbf{r}') = -\frac{1}{2}\left(\frac{\partial^2 G_{ik}(\mathbf{r}-\mathbf{r}')}{\partial x_j\,\partial x_l} + \frac{\partial^2 G_{jk}(\mathbf{r}-\mathbf{r}')}{\partial x_i\,\partial x_l}\right), \tag{4.18}$$

gives the ij component of the strain increment at point $\mathbf{r}$ due to an applied stress increment component kl at point $\mathbf{r}'$ in an infinite homogeneous medium with elasticity tensor D^{m}_{ijkl} and Green's function given by (4.7).

From (3.6), (3.46), and (4.15) we see that the perturbed strain increment, $\Delta\varepsilon_{kl}(\mathbf{r}) = \Delta\varepsilon^{\mathrm{T}}_{kl}(\mathbf{r}) - \Delta\varepsilon^0_{kl}$, is given by the equivalent relations

$$\Delta\varepsilon_{kl}(\mathbf{r}) = \frac{1}{V_c}\sum\sum_{n_p=0}^{\pm\infty}\sum{}' g_{klmn}(\zeta) \iiint_{V_c} D^{\mathrm{m}}_{mnrs}\Delta\varepsilon^*_{rs}(\mathbf{r}')e^{i\boldsymbol{\xi}\cdot(\mathbf{r}-\mathbf{r}')}\,dV(\mathbf{r}'), \tag{4.19}$$

or

$$\Delta\varepsilon_{kl}(\mathbf{r}) = \iiint_V U_{klmn}(\mathbf{r}-\mathbf{r}')D^{\mathrm{m}}_{mnrs}\Delta\varepsilon^*_{rs}(\mathbf{r}')\,dV(\mathbf{r}'). \tag{4.20}$$

The volume integral in the Fourier series representation extends over the volume, V_c, of the unit periodic cell and the summation extends over the integers $n_p = 0, \pm1, \pm2, \ldots$, etc., where $p = 1, 2, 3$. In the Green's function approach the volume integral extends over the entire infinite medium, i.e., over all the periodic cells comprising the material. It is shown in Section 5 that the Fourier summation expression in (4.19) can be converted into the Green's function expression in (4.20) by means of the Poisson sum formula.

From (3.22) it is evident that if the elastic properties of the fiber are the same as that of the matrix, then $\delta D_{ijkl}(\mathbf{r}) = \vartheta(\mathbf{r})(D^{\mathrm{f}}_{ijkl} - D^{\mathrm{m}}_{ijkl}) = 0$, in which case

$$\Delta\varepsilon^*_{kl}(\mathbf{r}) = \Delta c_{kl}(\mathbf{r}) \tag{4.21}$$

is known explicitly without having to solve the integral equation. From (3.48) and (4.16) it can also be observed that $\Delta\varepsilon^{\mathrm{T}}_{kl}(\mathbf{r})$ is known explicitly when $\delta D_{ijkl}(\mathbf{r}) = 0$. The explicit relation in (4.21) holds only when an explicit Euler forward difference method is used to integrate the viscoplastic constitutive relations. For implicit integration methods in which the inelastic strain increment $\Delta\varepsilon^{\mathrm{P}}_{kl}(\mathbf{r})$ depends on the total strain increment $\Delta\varepsilon^{\mathrm{T}}_{kl}(\mathbf{r})$, (3.48) and (4.16) show that even when $\delta D_{ijkl}(\mathbf{r}) = 0$, the equation to determine $\Delta\varepsilon^{\mathrm{T}}_{kl}(\mathbf{r})$ is still an implicit integral equation.

5. Relationship Between Fourier Series and Green's Function Approaches

In the composite material the total strain increment $\Delta\varepsilon^{\mathrm{T}}_{kl}(\mathbf{r})$ is periodic in $\mathbf{r}$ and is defined by the relationship

$$\Delta\varepsilon^{\mathrm{T}}_{kl}(\mathbf{r}) = \Delta\varepsilon^{0}_{kl} + \Delta\varepsilon_{kl}(\mathbf{r}), \tag{5.1}$$

where $\Delta\varepsilon^{0}_{kl}$ is the strain increment applied to the composite's boundary and is equal to the volume average of $\Delta\varepsilon^{\mathrm{T}}_{kl}(\mathbf{r})$ over the unit periodic cell, and $\Delta\varepsilon_{kl}(\mathbf{r})$ is the deviation or perturbation from the average value due to the presence of the fibers.

From (4.19) and (4.20) the perturbed strain increment is given in the Fourier series and Green's function approaches by the equivalent relations

$$\Delta\varepsilon_{kl}(\mathbf{r}) = \frac{1}{V_c}\sum\sum_{n_p=0}^{\pm\infty}\sum{}' g_{klij}(\boldsymbol{\zeta})\iiint_{V_c} D^{\mathrm{m}}_{ijrs}\Delta\varepsilon^{*}_{rs}(\mathbf{r}')e^{i\boldsymbol{\xi}\cdot(\mathbf{r}-\mathbf{r}')}\,dV(\mathbf{r}'), \tag{5.2}$$

or

$$\Delta\varepsilon_{kl}(\mathbf{r}) = \iiint_{V} U_{klij}(\mathbf{r}-\mathbf{r}')D^{\mathrm{m}}_{ijrs}\Delta\varepsilon^{*}_{rs}(\mathbf{r}')\,dV(\mathbf{r}'). \tag{5.3}$$

We now show that these equations are equivalent and that the Green's function relation is the Poisson sum transformation of the Fourier series relation.

From the definition of $g_{klmn}(\boldsymbol{\zeta})$ in (3.44) we may write

$$g_{klij}(\boldsymbol{\zeta}) = \tfrac{1}{2}(M^{-1}_{ik}(\boldsymbol{\zeta})\zeta_j\zeta_l + M^{-1}_{il}(\boldsymbol{\zeta})\zeta_j\zeta_k), \tag{5.4}$$

or

$$g_{klij}(\zeta_1,\zeta_2,\zeta_3) = \tfrac{1}{2}(M^{-1}_{ik}(\zeta_1,\zeta_2,\zeta_3)\zeta_j\zeta_l + M^{-1}_{il}(\zeta_1,\zeta_2,\zeta_3)\zeta_j\zeta_k), \tag{5.5}$$

where

$$\zeta_i = \frac{\xi_i}{\xi} = \frac{\dfrac{2\pi n_i}{L_i}}{\sqrt{\left(\dfrac{2\pi n_1}{L_1}\right)^2 + \left(\dfrac{2\pi n_2}{L_2}\right)^2 + \left(\dfrac{2\pi n_3}{L_3}\right)^2}} \quad \text{(no sum on } i\text{)}$$
$$\text{for} \quad i = 1, 2, 3. \tag{5.6}$$

We may therefore write

$$g_{klij}(\boldsymbol{\zeta}) = g_{klij}(\zeta_1(n_1,n_2,n_3), \zeta_2(n_1,n_2,n_3), \zeta_3(n_1,n_2,n_3)) = f_{klij}(n_1,n_2,n_3), \tag{5.7}$$

and the perturbation strain increment can then be written in the form

$$\Delta\varepsilon_{kl}(\mathbf{r}) = \frac{1}{L_1L_2L_3}\sum_{n_1=0}^{\pm\infty}\sum_{n_2=0}^{\pm\infty}\sum_{n_3=0}^{\pm\infty}{}' f_{klij}(n_1,n_2,n_3)\iiint_{V_c} D^{\mathrm{m}}_{ijrs}\Delta\varepsilon^{*}_{rs}(\mathbf{r}')$$
$$\times \exp\left\{i\left[\frac{2\pi n_1}{L_1}(x_1 - x_1') + \frac{2\pi n_2}{L_2}(x_2 - x_2') + \frac{2\pi n_3}{L_3}(x_3 - x_3')\right]\right\} dx_1'\,dx_2'\,dx_3', \tag{5.8}$$

or as

$$\Delta\varepsilon_{kl}(\mathbf{r}) = \frac{1}{L_1 L_2 L_3} \sum_{n_1=0}^{\pm\infty} \sum_{n_2=0}^{\pm\infty} \sum_{n_3=0}^{\pm\infty}{}' h_{kl}(n_1, n_2, n_3), \tag{5.9}$$

where

$$h_{kl}(n_1, n_2, n_3) = f_{klij}(n_1, n_2, n_3) \iiint_{V_c} D^{\mathrm{m}}_{ijrs} \Delta\varepsilon^*_{rs}(\mathbf{r}')$$
$$\times \exp\left\{i\left[\frac{2\pi n_1}{L_1}(x_1 - x_1') + \frac{2\pi n_2}{L_2}(x_2 - x_2') + \frac{2\pi n_3}{L_3}(x_3 - x_3')\right]\right\} dx_1'\, dx_2'\, dx_3'. \tag{5.10}$$

By the Poisson sum formula (Morse and Feshbach, 1958) we may write

$$\sum_{n_1=0}^{\pm\infty} \sum_{n_2=0}^{\pm\infty} \sum_{n_3=0}^{\pm\infty}{}' h_{kl}(n_1, n_2, n_3)$$
$$= \sum_{m_1=0}^{\pm\infty} \sum_{m_2=0}^{\pm\infty} \sum_{m_3=0}^{\pm\infty} \frac{L_1 L_2 L_3}{(2\pi)^3} \iiint_{-\infty}^{\infty} d^3\mathbf{K}\, e^{i(m_1 K_1 L_1 + m_2 K_2 L_2 + m_3 K_3 L_3)}$$
$$\times h_{kl}\left(\frac{K_1 L_1}{2\pi}, \frac{K_2 L_2}{2\pi}, \frac{K_3 L_3}{2\pi}\right), \tag{5.11}$$

where the sum over the integers n_1, n_2, n_3 is replaced by the sum over the integers m_1, m_2, m_3 in the Fourier integrals, the sum over m_i including the case where $m_1 = m_2 = m_3 = 0$.

We now have the alternative sum

$$\Delta\varepsilon_{kl}(\mathbf{r}) = \frac{1}{L_1 L_2 L_3} \sum_{n_1=0}^{\pm\infty} \sum_{n_2=0}^{\pm\infty} \sum_{n_3=0}^{\pm\infty}{}' h_{kl}(n_1, n_2, n_3)$$
$$= \sum_{m_1=0}^{\pm\infty} \sum_{m_2=0}^{\pm\infty} \sum_{m_3=0}^{\pm\infty} \iiint_{-\infty}^{\infty} \frac{d^3\mathbf{K}}{(2\pi)^3} e^{i(m_1 K_1 L_1 + m_2 K_2 L_2 + m_3 K_3 L_3)}$$
$$\times f_{klij}\left(\frac{K_1 L_1}{2\pi}, \frac{K_2 L_2}{2\pi}, \frac{K_3 L_3}{2\pi}\right)$$
$$\times \iiint_{V_c} D^{\mathrm{m}}_{ijrs} \Delta\varepsilon^*_{rs}(\mathbf{r}') e^{i[K_1(x_1 - x_1') + K_2(x_2 - x_2') + K_3(x_3 - x_3')]}\, dx_1'\, dx_2'\, dx_3', \tag{5.12}$$

or

$$\Delta\varepsilon_{kl}(\mathbf{r}) = \sum_{m_1=0}^{\pm\infty} \sum_{m_2=0}^{\pm\infty} \sum_{m_3=0}^{\pm\infty} \iiint_{-\infty}^{\infty} \frac{d^3\mathbf{K}}{(2\pi)^3} \times f_{klij}\left(\frac{K_1L_1}{2\pi}, \frac{K_2L_2}{2\pi}, \frac{K_3L_3}{2\pi}\right) e^{i(K_1x_1+K_2x_2+K_3x_3)} \times \iiint_{V_c} D^{\mathrm{m}}_{ijrs}\Delta\varepsilon^*_{rs}(\mathbf{r}') \times e^{-i[K_1(x_1'-m_1L_1)+K_2(x_2'-m_2L_2)+K_3(x_3'-m_3L_3)]}\, dx_1'\, dx_2'\, dx_3'. \tag{5.13}$$

Due to the geometric periodicity of the unit cell we may write

$$\Delta\varepsilon^*_{rs}(\mathbf{r}') = \Delta\varepsilon^*_{rs}(x_1', x_2', x_3') = \Delta\varepsilon^*_{rs}(x_1' - m_1L_1, x_2' - m_2L_2, x_3' - m_3L_3), \tag{5.14}$$

and

$$dx_1'\, dx_2'\, dx_3' = d(x_1' - m_1L_1)d(x_2' - m_2L_2)d(x_3' - m_3L_3), \tag{5.15}$$

so that by making the change of variable

$$(x_1' - m_1L_1, x_2' - m_2L_2, x_3' - m_3L_3) = (x_1'', x_2'', x_3'') = \mathbf{r}'', \tag{5.16}$$

the perturbation strain increment is

$$\Delta\varepsilon_{kl}(\mathbf{r}) = \iiint_{-\infty}^{\infty} \frac{d^3\mathbf{K}}{(2\pi)^3} f_{klij}\left(\frac{K_1L_1}{2\pi}, \frac{K_2L_2}{2\pi}, \frac{K_3L_3}{2\pi}\right) e^{i(K_1x_1+K_2x_2+K_3x_3)} \times \sum_{m_1=0}^{\pm\infty} \sum_{m_2=0}^{\pm\infty} \sum_{m_3=0}^{\pm\infty} \iiint_{V_c(m_1,m_2,m_3)} D^{\mathrm{m}}_{ijrs}\Delta\varepsilon^*_{rs}(\mathbf{r}'')e^{-i\mathbf{K}\cdot\mathbf{r}''}\, dV(\mathbf{r}''), \tag{5.17}$$

where the volume integration extends over the volume $V_c(m_1, m_2, m_3)$ of the unit cell whose center is at the point (m_1L_1, m_2L_2, m_3L_3). Since m_1, m_2, m_3 range over all integer values, the summation of the volume integrals extends to *all* the cells in the periodic lattice, i.e., it extends over the entire volume, V, of the composite medium. The expression for $\Delta\varepsilon_{kl}(\mathbf{r})$ thus takes the form

$$\Delta\varepsilon_{kl}(\mathbf{r}) = \iiint_{-\infty}^{\infty} \frac{d^3\mathbf{K}}{(2\pi)^3} f_{klij}\left(\frac{K_1L_1}{2\pi}, \frac{K_2L_2}{2\pi}, \frac{K_3L_3}{2\pi}\right) e^{i\mathbf{K}\cdot\mathbf{r}} \times \iiint_{V} D^{\mathrm{m}}_{ijrs}\Delta\varepsilon^*_{rs}(\mathbf{r}'')e^{-i\mathbf{K}\cdot\mathbf{r}''}\, dV(\mathbf{r}''). \tag{5.18}$$

By interchanging the order of the volume and wave vector integrals and noting that $\mathbf{r}''$ can be replaced by $\mathbf{r}'$ since it is a dummy integration variable, we obtain

$$\Delta\varepsilon_{kl}(\mathbf{r}) = \iiint_V dV(\mathbf{r}') \iiint_{-\infty}^{\infty} \frac{d^3\mathbf{K}}{(2\pi)^3}$$

$$\times f_{klij}\left(\frac{K_1L_1}{2\pi}, \frac{K_2L_2}{2\pi}, \frac{K_3L_3}{2\pi}\right) e^{i\mathbf{K}\cdot(\mathbf{r}-\mathbf{r}')} D^{\mathrm{m}}_{ijrs} \Delta\varepsilon^*_{rs}(\mathbf{r}'). \tag{5.19}$$

Introducing $(K_1L_1/2\pi,\ K_2L_2/2\pi,\ K_3L_3/2\pi)$ in place of (n_1, n_2, n_3) in the expression for

$$f_{klij}(n_1, n_2, n_3) = g_{klij}(\zeta_1(n_1, n_2, n_3), \zeta_2(n_1, n_2, n_3), \zeta_3(n_1, n_2, n_3)), \tag{5.20}$$

then gives

$$g_{klij}\left(\zeta_1\left(\frac{K_1L_1}{2\pi}, \frac{K_2L_2}{2\pi}, \frac{K_3L_3}{2\pi}\right), \zeta_2\left(\frac{K_1L_1}{2\pi}, \frac{K_2L_2}{2\pi}, \frac{K_3L_3}{2\pi}\right),\right.$$

$$\left.\zeta_3\left(\frac{K_1L_1}{2\pi}, \frac{K_2L_2}{2\pi}, \frac{K_3L_3}{2\pi}\right)\right) = \frac{1}{2}\left(\frac{M^{-1}_{ik}(\zeta)}{K^2} K_jK_l + \frac{M^{-1}_{il}(\zeta)}{K^2} K_jK_k\right), \tag{5.21}$$

with

$$\zeta_i = \frac{K_i}{K} = \frac{K_i}{\sqrt{K_qK_q}}, \tag{5.22}$$

and the perturbed strain increment takes the form

$$\Delta\varepsilon_{kl}(\mathbf{r}) = \iiint_V dV(\mathbf{r}') \iiint_{-\infty}^{\infty} \frac{d^3\mathbf{K}}{(2\pi)^3} \frac{1}{2}\left(\frac{M^{-1}_{ik}(\zeta)}{K^2} K_jK_l + \frac{M^{-1}_{il}(\zeta)}{K^2} K_jK_k\right)$$

$$\times e^{i\mathbf{K}\cdot(\mathbf{r}-\mathbf{r}')} D^{\mathrm{m}}_{ijrs} \Delta\varepsilon^*_{rs}(\mathbf{r}'). \tag{5.23}$$

But, from (4.7),

$$G_{ik}(\mathbf{r}-\mathbf{r}') = \iiint_{-\infty}^{\infty} \frac{d^3\mathbf{K}}{(2\pi)^3} \frac{M^{-1}_{ik}(\zeta)}{K^2} e^{-i\mathbf{K}\cdot(\mathbf{r}-\mathbf{r}')}$$

$$= \iiint_{-\infty}^{\infty} \frac{d^3\mathbf{K}}{(2\pi)^3} \frac{M^{-1}_{ik}(\zeta)}{K^2} e^{i\mathbf{K}\cdot(\mathbf{r}-\mathbf{r}')}, \tag{5.24}$$

since $G_{ik}(\mathbf{r}-\mathbf{r}') = G_{ik}(\mathbf{r}'-\mathbf{r})$, and therefore

$$\frac{\partial^2 G_{ik}(\mathbf{r}-\mathbf{r}')}{\partial x_j\, \partial x_l} = -\iiint_{-\infty}^{\infty} \frac{d^3\mathbf{K}}{(2\pi)^3} \frac{M^{-1}_{ik}(\zeta)}{K^2} K_jK_l e^{i\mathbf{K}\cdot(\mathbf{r}-\mathbf{r}')}. \tag{5.25}$$

Inserting the last relation into the expression for $\Delta\varepsilon_{kl}(\mathbf{r})$ then shows that

$$\Delta\varepsilon_{kl}(\mathbf{r}) = -\iiint_V dV(\mathbf{r}') \frac{1}{2}\left(\frac{\partial^2 G_{ik}(\mathbf{r}-\mathbf{r}')}{\partial x_j\, \partial x_l} + \frac{\partial^2 G_{il}(\mathbf{r}-\mathbf{r}')}{\partial x_j\, \partial x_k}\right) D^{\mathrm{m}}_{ijrs} \Delta\varepsilon^*_{rs}(\mathbf{r}'). \tag{5.26}$$

From the definition of the tensor $U_{klmn}(\mathbf{r} - \mathbf{r}')$ in (4.18), we see that

$$\Delta\varepsilon_{kl}(\mathbf{r}) = \iiint_V U_{klij}(\mathbf{r} - \mathbf{r}') D^{\mathrm{m}}_{ijrs} \Delta\varepsilon^*_{rs}(\mathbf{r}')\, dV(\mathbf{r}'), \tag{5.27}$$

which is the result obtained with Green's function approach.

The Fourier series expression for the perturbation strain increment is thus identical to Green's function expression and the two are linked via the Poisson sum formula.

6. Concluding Remarks

The Fourier series and Green's function representations have been shown to be equivalent approaches by means of the Poisson sum formula. This method is well known in mathematical physics and is used extensively to turn slowly convergent Fourier series into a series of rapidly converging Fourier integrals. Both representations offer promising approaches to modeling the viscoplastic behavior of metal matrix composites at elevated temperatures. Having shown their equivalence we are free to choose between them based on mathematical and/or numerical convenience. Each is expected to be suited to different situations with respect to convergence of the series with increasing fiber volume fraction. Future work will explore the relative advantages of each formulation and the overall usefulness of these approaches in modeling the nonlinear viscoplastic deformation behavior of metal matrix composites.

Acknowledgment

This work was supported by Grant NAG3-882 from the NASA–Lewis Research Center and by Contract DE-AC02-88ER13895 from the United States Department of Energy.

References

Aboudi, J. (1987), Closed form constitutive equations for metal matrix composites, *Int. J. Engng. Sci.*, **25**, No. 9, 1229–1240.

Bahei-El-Din, Y. A., Dvorak, G. J., Lin, J., Shah, R., and Wu, J.-F. (1987), Local fields and overall response of fibrous and particulate metal matrix composites, Final Technical Report to Alcoa Laboratories under Contract No. 379(52R)053(22L), Department of Civil Engineering, Rensselaer Polytechnic Institute, Troy, New York 12181, November.

Barnett, D. M. (1971), The elastic energy of a straight dislocation in an infinite anisotropic elastic medium, *Phys. Stat. Sol.*, (b), **48**, 419–428.

Barnett, D. M. (1972), The precise evaluation of derivatives of the anisotropic elastic Green's functions, *Phys. Stat. Sol.*, (b), **49**, 741–748.

Bodner, S. R. (1987), Review of a unified elastic-viscoplastic theory, in *Unified Constitutive Equations for Plastic Deformation and Creep of Engineering Alloys*, edited by A. K. Miller, Elsevier Applied Science, Amsterdam.
Devriès, F. and Léné, F. (1987), Homogenization at set macroscopic stress: Numerical implementation and application. *La Reserche Aérospatiale*, No. 1, 33–51.
Duvaut, G. (1988), Functional and mechanical analysis of continuous media–application to the study of elastic composites with periodic structure: Homogenization, NASA Technical Translation Report, NASA TT-20188.
Dvorak, G. J. (1986), Thermal expansion of elastic–plastic composite materials, *ASME J. Appl. Mech.*, **53**, 737–743.
Dvorak, G. J. and Bahei-El-Din, Y. A. (1982), Plasticity analysis of fibrous composites, *ASME J. Appl. Mech.*, **49**, 327–335.
Dvorak, G. J., Bahei-El-Din, Y. A., Macheret, Y., and Liu, C. H. (1988), An experimental study of elastic–plastic behavior of a fibrous boron–aluminum composite, Technical Report to the Office of Naval Research under Contract No. N00014-85-K-0247 and to the U.S. Army Research Office under Contract No. DAAG29-85-K-0011 from Rensselaer Polytechnic Institute, Troy, New York 12181, February.
Eshelby, J. D. (1957), The determination of the elastic field of an ellipsoidal inclusion, and related problems, *Proc. Roy. Soc. London*, **A241**, 376–396.
Gubernatis, J. E. and Krumhansl, J. A. (1975), Macroscopic engineering properties of polycrystalline materials: Elastic properties, J. Appl. Phys., **46**, No. 5, 1875–1883.
Haritos, G. K., Hager, J. W., Amos, A. K., Salkind, M. J., and Wang, A.S.D. (1988), Mesomechanics: The microstructure–mechanics connection, *Int. J. Solids Structures*, **24**, No. 11, 1081–1096.
Iwakuma, T. and Nemat-Nasser, S. (1983), Composites with periodic microstructure, *Computers and Structures*, **16**, Nos. 1–4, 13–19.
Korringa, K. (1973), Theory of elastic constants of heterogeneous media, *J. Math. Phys*, **14**, No. 4, 509–513.
Lemaitre, J. and Chaboche, J. L. (1985), *Mécanique des matériaux solides*, Dunod, Paris.
Léné, F. (1986), Damage constitutive relations for composite materials, *Engng. Fracture Mech.*, **25**, Nos. 5/6, 713–728.
Léné, F. (1987), Contribution to the study of composite materials and their damage, Translation of Thesis presented to Université Pierre et Marie Curie, Paris VI, NASA Technical Translation Report, NASA TT-20107.
Léné, F. and Leguillon, D. (1982), Homogenized constitutive law for a partially cohesive composite material, *Int. J. Solids Structures*, **18**, No. 5, 443–458.
Morse, P. M. and Feshbach, H. (1958), *Methods of Theoretical Physics*, McGraw-Hill, New York.
Mura, T. (1982), *Micromechanics of Defects in Solids*, Martinus-Nijhoff, The Hague/Boston/London.
Nemat-Nasser, S. and Iwakuma, T. (1983), Micromechanically based constitutive relations for polycrystalline solids, NASA Conference Publication NASA CP 2271, pp. 113–136.
Nemat-Nasser, S., Iwakuma, T., and Hejazi, M. (1982), On composites with periodic structure, *Mech. Mater.*, **1**, 239–267.
Nemat-Nasser, S. and Taya, M. (1981), On effective moduli of an elastic body containing periodically distributed voids. *Quart. Appl. Math.*, **39**, 43–59.

Petrasek, D. W., McDanels, L. J., Westfall, L. J., and Stephens, J. R. (1986), Fiber-reinforced superalloy composites provide an added performance edge, *Metal Progress Magazine*, August.

Renard, J. and Marmonier, M. F. (1987), Study of damage initiation in the matrix of a composite material by a homogenization method, *La Reserche Aérospatiale*, No. 6, 43–51.

Sanchez-Palencia, E. (1980), *Nonhomogeneous Media and Vibration Theory*, Lecture Notes in Physics, No. 127, Springer-Verlag, Berlin.

Sanchez-Palencia, E. (1985), *Homogenization Techniques for Composite Media*, Lecture Notes in Physics, No. 272, edited by E. Sanchez-Palencia and A. Zaoui, Springer-Verlag, Berlin.

Teodosiu, C. (1982), *Elastic Models of Crystal Defects*, Springer-Verlag, Berlin–Heidelberg–New York, p. 90.

Walker, K. P., Jordan, E. H., and Freed, A. D. (1989), *Nonlinear Mesomechanics of Composites with Periodic Microstructure*: First Report, NASA TM-102051.

Zeller, R. and Dederichs, P. H. (1973), Elastic constants of polycrystals, *Phys. Stat. Sol.*, (b), **55**, 831–842.

Polarization, Virtual Mass, and Analogous Elastic Properties

L. J. WALPOLE
School of Mathematics, University of East Anglia, Norwich NR4 7TJ, England

Abstract

Polarization and virtual mass are dual, electrostatic, and hydrodynamical, physical quantities calculated for a given solid body by means of the appropriate exterior potential fields. Analogous physical quantities can be derived from appropriate exterior elastostatic fields. We establish some new isoperimetric inequalities, which each bound a physical quantity in terms of geometrical parameters of an arbitrarily shaped body, and convert to an exact calculation for a particular shape.

1. Introduction

In the classical problems of potential theory for the region exterior to a solid body, paramount physical quantities (such as exterior electrostatic capacitance and polarization, hydrodynamic virtual mass) are to be evaluated in terms of geometrical parameters of the particular body. The surveys of Schiffer and Szegö (1949), Polya and Szegö (1951), and Payne (1967) are relevant particularly to our present standpoint. Calculations can be completed usually for ellipsoidal and sometimes for a few other special body shapes. But for a general shape, the analysis can hope only to bound the physical quantity usefully in terms of the geometrical ones. Such an inequality is described as isoperimetric if it converts to an equality for a particular, possibly degenerate, body shape. There are analogous problems for an exterior elastic medium which associate another set of physical quantities with the body, as indicated recently by Villaggio (1986).

We plan to examine these electrostatic, hydrodynamical, and elastic problems particularly to establish some new isoperimetric inequalities, together with the special solutions that achieve equality. We give the gravitational (Newtonian) potential of the body a recurring role. It can be calculated by integration for a given body shape. Moreover, it has helpful general properties that serve for all shapes.

The given solid body is supposed to occupy a region B of volume V, of arbitrary shape, in three-dimensional infinite space. The components of the unit outward normal at a typical point of its boundary ∂B are denoted by n_i. The region exterior to the body is denoted by E. W denotes the whole region made up of B and E together. The coordinate origin O is fixed at any chosen point within B, such as the center of volume. Any other point in B or E has Cartesian coordinates x_i and position vector $\mathbf{r}$ relative to O, and is at distance $|\mathbf{r}| = r$ from O.

We shall refer frequently to symmetric second-rank, three-dimensional Cartesian tensors, and to their well-known properties. Such a tensor a_{ij} ($= a_{ji}$) is described as positive definite if the scalar $a_{ij}\lambda_i\lambda_j$ is positive for every nonzero vector λ_i, or as positive semidefinite if this scalar is merely nonnegative, possibly zero. We have then the particular properties

$$a_{ii} > 0, \qquad a_{ii} \geq 0, \tag{1.1}$$

of a positive or nonnegative trace, respectively. A positive definite tensor a_{ij} has an inverse, b_{ij} say (denoted also by a_{ij}^{-1}), which is likewise symmetric and positive definite, and which produces the second-rank unit tensor, the Kronecker delta, by means of the multiplications

$$a_{ik}b_{kj} = b_{ik}a_{kj} = \delta_{ij}.$$

A positive lower bound is set consequently on the product of the traces by the inequality

$$a_{ii}b_{jj} \geq 9, \tag{1.2}$$

with equality when a_{ij} (and likewise b_{ij}) is simply a positive multiple of the Kronecker delta. This inequality can be proved by assigning the "spectral representation" to a_{ij} and by appealing to the well-known inequality that obliges the arithmetic mean of three positive numbers to be no smaller than the reciprocal of their harmonic mean. When we meet subsequently the property that a difference $c_{ij} - d_{ij}$ is positive definite (or semidefinite), it will be displayed more briefly as the tensor "inequality" $c_{ij} > d_{ij}$ (or $c_{ij} \geq d_{ij}$). This property is passed on likewise to the reversed difference of the inverses if c_{ij} and d_{ij} are both positive definite. That is,

$$c_{ij} > 0, \qquad d_{ij} > 0, \qquad c_{ij} \geq d_{ij} \rightarrow d_{ij}^{-1} \geq c_{ij}^{-1}. \tag{1.3}$$

In connection with a body which is symmetrical about the x_3 axis, we meet the particular construction

$$a_{ij} = a\delta_{i3}\delta_{j3} + b(\delta_{ij} - \delta_{i3}\delta_{j3}) \tag{1.4}$$

in which the scalar coefficients a ($= a_{33}$) and b ($= a_{11} = a_{22}$) are those that appear in the main diagonal of a matrix representation. These two scalars are both positive when their tensor is positive definite, and each is simply replaced by its reciprocal in order to construct the inverse tensor.

2. Gravitational Potential

The gravitational potential, $\varphi(\mathbf{r})$, of attracting matter of uniform density imagined to fill B, is defined in suitable units by the volume integration

$$\varphi(\mathbf{r}) = -\int_B \frac{dx_1'\,dx_2'\,dx_3'}{4\pi|\mathbf{r}-\mathbf{r}'|},$$

over B with respect to the primed coordinates. We insert the negative factor $(-1/4\pi)$ here in preference to letting it obtrude more frequently in other places. Then the equations of Poisson and Laplace

$$\nabla^2\varphi = 1, \qquad \nabla^2\varphi = 0, \tag{2.1}$$

are satisfied at all points of B and E, respectively, where ∇^2 denotes the Laplacian operator $\partial^2/\partial x_i\,\partial x_i$ (with the summation convention for repeated suffixes). At points in E remote from B, φ tends to $-V/4\pi r$. There is a continuous variation of φ and each of its first derivatives across the boundary ∂B. However, as anticipated by (2.1), there is an abrupt jump in each second derivative which can be expressed precisely as

$$\varphi_{,ij}(B) - \varphi_{,ij}(E) = n_i n_j \tag{2.2}$$

at a typical boundary point. In addition to these well-known properties, we may show first that

$$\int_W \varphi_{,ki}\varphi_{,kj}\,dV = \int_B \varphi_{,ij}\,dv = V\langle\varphi_{,ij}\rangle,$$

where $dV = dx_1\,dx_2\,dx_3$, where the angular brackets denote the volume average of their contents over the body B, and where the integrand has been expressed first as the divergence $(\varphi_{,ki}\varphi_{,j})_{,k}$ in order to convert volume integrations from E to B, via intermediate surface integrals into which the jump (2.2) is inserted. The body-averaged second derivatives of φ form therefore a symmetric, positive definite, dimensionless, tensor. We can show further that

$$\int_E \varphi_{,ki}\varphi_{,kj}\,dV = V(\langle\varphi_{,ij}\rangle - \langle\varphi_{,ki}\varphi_{,kj}\rangle), \tag{2.3}$$

and also immediately that

$$\int_B (\delta_{ki} - \varphi_{,ki})(\delta_{kj} - \varphi_{,kj})\,dV = V(\langle\varphi_{,ki}\varphi_{,kj}\rangle - 2\langle\varphi_{,ij}\rangle + \delta_{ij}), \tag{2.4}$$

$$\int_B (\varphi_{,ki} - \langle\varphi_{,ki}\rangle)(\varphi_{,kj} - \langle\varphi_{,kj}\rangle)\,dV = V(\langle\varphi_{,ki}\varphi_{,kj}\rangle - \langle\varphi_{,ki}\rangle\langle\varphi_{,kj}\rangle),$$

$$(\langle\varphi_{,ki}\rangle - \tfrac{1}{3}\delta_{ki})(\langle\varphi_{,kj}\rangle - \tfrac{1}{3}\delta_{kj}) = \langle\varphi_{,ki}\rangle\langle\varphi_{,kj}\rangle - \tfrac{2}{3}\langle\varphi_{,ij}\rangle + \tfrac{1}{9}\delta_{ij}.$$

The tensors on the right-hand sides of these last four equations receive the symmetric and positive semidefinite properties from the left-hand sides, to establish the chain of tensor inequalities

$$\langle \varphi_{,ij} \rangle \geq \langle \varphi_{,ki}\varphi_{,kj} \rangle \geq \langle \varphi_{,ki} \rangle \langle \varphi_{,kj} \rangle \geq \tfrac{2}{3}\langle \varphi_{,ij} \rangle - \tfrac{1}{9}\delta_{ij}, \tag{2.5}$$

which contract, on taking their trace, to

$$1 \geq \langle \varphi_{,ki}\varphi_{,ki} \rangle \geq \langle \varphi_{,ki} \rangle \langle \varphi_{,ki} \rangle \geq \tfrac{1}{3}. \tag{2.6}$$

By addition of (2.3) and (2.4), and by appeal to the theorem (1.3), we infer that each of the three differences

$$\delta_{ij} - \langle \varphi_{,ij} \rangle, \qquad \langle \varphi_{,ij} \rangle^{-1} - \delta_{ij}, \qquad (\delta_{ij} - \langle \varphi_{,ij} \rangle)^{-1} - \delta_{ij},$$

is a symmetric, positive definite tensor.

In particular, when the body B has an ellipsoidal shape, there is a well-known calculation of φ at interior and exterior points in terms of elliptic integrals, or of elementary integrals for spheroidal and spherical shapes. In the interior of B, φ is simply a quadratic function of the Cartesian coordinates and so every individual second derivative (not just their Laplacian combination) is constant there, and is denoted still by the angular brackets. The middle inequalities of (2.5) and (2.6) revert to equalities for an ellipsoidal shape.

When the body has a general axisymmetric shape, with respect to the x_3 axis, its body-averaged second derivatives can be expressed in terms of a single bounded parameter α as

$$\langle \varphi_{,ij} \rangle = \alpha\delta_{i3}\delta_{j3} + \tfrac{1}{2}(1 - \alpha)(\delta_{ij} - \delta_{i3}\delta_{j3}), \qquad 0 \leq \alpha \leq 1, \tag{2.7}$$

in view of the Poisson equation (2.1) and the positive definite property. In particular, for a spheroid of semiaxes εb, εb, and b, the integration can be simplified as

$$\alpha = \int_0^1 \frac{x^2\, dx}{x^2 + \varepsilon^2(1 - x^2)}, \tag{2.8}$$

and can be evaluated consequently as

$$\alpha - \frac{\varepsilon^2}{\varepsilon^2 - 1} = \frac{\varepsilon^2 \ln[(1 + \sqrt{1 - \varepsilon^2})/(1 - \sqrt{1 - \varepsilon^2})]}{2(1 - \varepsilon^2)^{3/2}}$$
$$= -\frac{\varepsilon^2(\tan^{-1}\sqrt{\varepsilon^2 - 1})}{(\varepsilon^2 - 1)^{3/2}},$$

for the prolate ($\varepsilon < 1$) and oblate ($\varepsilon > 1$) spheroids, respectively. The limiting values

$$\alpha = \varepsilon^2 \ln(2/\varepsilon) - \varepsilon^2, \qquad \alpha = 1 - \frac{\pi}{2\varepsilon}, \tag{2.9}$$

are reached for the extremely prolate (small ε) and the extremely oblate (large ε) spheroids, respectively. The upper bound of unity is approached by α in the oblate limit, and the first inequalities of (2.5) and (2.6) approach equalities as well. The lower bound of zero is approached by α in the prolate limit. For a sphere, α equates to one-third and we have the calculation

$$\langle \varphi_{,ij} \rangle = \tfrac{1}{3}\delta_{ij}, \tag{2.10}$$

namely as the particular isotropic tensor that conforms to the Poisson equation (2.1). The third inequalities of (2.5) and (2.6) revert to equalities for the spherical shape.

3. Polarization and Virtual Mass

In the exterior region E, we seek separately the two vector potentials v_i and w_i which satisfy there the Laplace equations

$$\nabla^2 v_i = 0, \qquad \nabla^2 w_i = 0,$$

on condition that each is of the order $1/r^2$ at great distances r from B, and which also satisfy the respective boundary conditions

$$v_i = x_i + c_i, \qquad n_j w_{i,j} = n_i,$$

at points on ∂B. The constants c_i are to be determined so as to ensure that v_i diminishes at the required rate at remote points. The polarization tensor P_{ij} and the virtual mass tensor M_{ij} are evaluated by the integrations

$$P_{ij} = \int_E v_{i,k} v_{j,k} \, dV, \qquad M_{ij} = \int_E w_{i,k} w_{j,k} \, dV,$$

which leave them both symmetric and positive definite. The ratios P_{ij}/V and M_{ij}/V [the latter called α_{ij} by Batchelor (1967)] are dimensionless and depend only on the shape of the body B.

For an ellipsoidal body, we may construct at once the solutions

$$v_i = \langle \varphi_{,ij} \rangle^{-1} \varphi_{,j}, \qquad w_i = -(\delta_{ij} - \langle \varphi_{,ij} \rangle)^{-1} \varphi_{,j},$$

if we recall the general properties of the gravitational potential φ [continuity of first derivatives and a jump (2.2) of second derivatives across ∂B] and the particular properties for the ellipsoidal B (first derivatives of φ are linear functions of x_i within B). Equivalent expressions are quoted by Schiffer and Szegö (1949) in terms of the elliptic integrals but the verifications of those solutions is lengthier. Consequently, we proceed to the evaluations

$$\frac{P_{ij}}{V} = \langle \varphi_{,ij} \rangle^{-1} - \delta_{ij}, \tag{3.1}$$

$$\frac{M_{ij}}{V} = (\delta_{ij} - \langle \varphi_{,ij} \rangle)^{-1} - \delta_{ij} = (\langle \varphi_{,ij} \rangle^{-1} - \delta_{ij})^{-1}, \tag{3.2}$$

as positive definite tensors, by which it is shown that P_{ij}/V and M_{ij}/V are inverse to each other for an ellipsoid. The theorem (1.2) and the Poisson equation (2.1) imply that

$$P_{ii} \geq 6V, \qquad M_{ii} \geq \frac{3V}{2}, \tag{3.3}$$

for every ellipsoidal body of volume V. For a spherical body, it follows with the help of (2.10) that

$$P_{ij} = 2V\delta_{ij}, \qquad M_{ij} = \tfrac{1}{2}V\delta_{ij}.$$

The inequalities (3.3) are therefore of the isoperimetric type which revert to equalities when the ellipsoidal shape becomes spherical, as was shown by Schiffer and Szegö (1949) and by Polya (1947), by explicit examination of the elliptic integrals.

The tensor inequality

$$P_{ij} = \int_E v_{i,k} v_{j,k}\, dV$$
$$\geq \int_E (t_{i,k} v_{j,k} + t_{j,k} v_{i,k} - t_{i,k} t_{j,k})\, dV,$$

is valid when the vector field t_i is equated to $a_{ij}\varphi_{,j}$ in particular, for constant coefficients a_{ij}, since equality would be restored by the addition of the positive semidefinite product $(v_{i,k} - t_{i,k})(v_{j,k} - t_{j,k})$ to the bracketed integrand. After each term of the integrand is written as the appropriate divergence, $t_{i,k}v_{j,k}$ as $(t_{i,k}v_j)_{,k}$ and so on, the integrations can be all converted to ones over the interior region B, via intermediate surface integrals over ∂B, into which the boundary values of v_i and the jump (2.2) are substituted. We find consequently that

$$\frac{P_{ij}}{V} \geq a_{ij} + a_{ji} + \langle \varphi_{,km}\varphi_{,kn}\rangle a_{im}a_{jn}$$
$$-\langle \varphi_{,mn}\rangle a_{im}a_{jn} - a_{im}\langle \varphi_{,mj}\rangle - a_{jm}\langle \varphi_{,im}\rangle.$$

We may proceed at once to the best choice of the constants a_{ij} if all the angular bracketed averages are calculated. But we are content to weaken this inequality first by incorporating the middle one of (2.5), after which the best choice of a_{ij} is $\langle \varphi_{,ij}\rangle^{-1}$. After a corresponding analysis for the virtual mass tensor, we are brought to the chain of tensor inequalities

$$\frac{P_{ij}}{V} \geq \langle \varphi_{,ij}\rangle^{-1} - \delta_{ij} \geq \left(\frac{M_{ij}}{V}\right)^{-1}, \tag{3.4}$$

the last of which can be recast as

$$\frac{M_{ij}}{V} \geq (\delta_{ij} - \langle \varphi_{,ij}\rangle)^{-1} - \delta_{ij}.$$

An isoperimetric status can be conferred here since the equalities (3.1) and (3.2) are restored whenever the body B is ellipsoidal. With the help of the theorems (1.1) and (1.2), and the Poisson equation (2.1), we confirm at once that the pair of scalar isoperimetric inequalities (3.3) apply to every (ellipsoidal and nonellipsoidal) body of volume V, as was shown otherwise by Schiffer (1957). Furthermore, after (3.4) is rearranged slightly by appeal to the theorem (1.3), and after the Poisson equation (2.1) is recalled, we derive the further pair

$$e_{ii} \leq 1, \qquad f_{ii} \leq 2, \tag{3.5}$$

where

$$e_{ij} = \left(\delta_{ij} + \frac{P_{ij}}{V}\right)^{-1}, \qquad f_{ij} = \left(\delta_{ij} + \frac{M_{ij}}{V}\right)^{-1}.$$

These bracketed tensors are the dipole coefficients (apart from multiplicative factors) introduced by Schiffer and Szegö (1949). The equality signs are attained in (3.5) by any ellipsoidal body B, not just by a spherical one.

For the general axisymmetric body B, the polarization and virtual mass tensors are constructed as

$$\frac{P_{ij}}{V} = p\delta_{i3}\delta_{j3} + q(\delta_{ij} - \delta_{i3}\delta_{j3}),$$

$$\frac{M_{ij}}{V} = m\delta_{i3}\delta_{j3} + n(\delta_{ij} - \delta_{i3}\delta_{j3}),$$

in terms of dimensionless scalar coefficients $p, q, m,$ and n which are all positive, in view of the positive definite status of the tensors. The inequalities (3.4) simplify to the scalar isoperimetric ones

$$p \geq \frac{1}{\alpha} - 1 \geq \frac{1}{m}, \tag{3.6}$$

$$q \geq \frac{2}{1-\alpha} - 1 \geq \frac{1}{n}, \tag{3.7}$$

in terms of the coefficient α of (2.7). As α is confined between zero and unity, it is shown by (3.7) that q is not merely positive but is always greater than unity. For any simply connected axisymmetric body B, Payne (1956) has found the relation

$$q = 1 + 2m, \tag{3.8}$$

between polarization and virtual mass coefficients. When the body has the spheroidal shape that evaluates α by (2.8), all the equality signs are attained in (3.6) and (3.7), and the relation (3.8) is obeyed. Although q and m each tend to infinite values as α approaches its upper limit of unity in (2.9) for the extremely oblate spheroid, the corresponding components of polarization and virtual mass are kept finite as the volume V diminishes, to show in the degenerate limit of a circular disk of radius a that

$$P_{11} = P_{22} = 2M_{33} = \frac{16a^3}{3}.$$

Likewise p approaches infinity but P_{33} remains finite when α reaches its lower limit of zero for the extremely prolate spheroid.

4. Analogous Elastic Properties

We seek in the exterior region E a vector field u_i, the elastic displacement, together with the accompanying symmetric tensor fields

$$e_{ij} = \tfrac{1}{2}(u_{i,j} + u_{j,i}), \qquad \sigma_{ij} = \lambda e_{kk}\delta_{ij} + 2\mu e_{ij},$$

of strain e_{ij} and stress σ_{ij}, where λ and μ are given scalar constants (the Lamé elastic constants of the solid medium that fills E), subject to the physical constraints that μ and $3\lambda + 2\mu$ are always positive. The necessary condition for elastic equilibrium is that $\sigma_{ij,j}$ should be zero at every point of E. At great distances r from B, u_i is to be of the order $1/r^2$.

In the elastic analogue of the polarization problem, it is stipulated further that u_i reduces to a linear function on the boundary of B, so that

$$u_i = f_{ij}x_j + g_i$$

on ∂B, with arbitrarily prescribed constant coefficients f_{ij} (components of a generally unsymmetric tensor). The constant vector components g_i are to be determined so as to make u_i vanish at remote points at the required rate. The body is subject to a resultant couple, in general, but not to a resultant force. For an ellipsoidal body, Daniele (1911) has constructed the solution u_i in a manner described by Eshelby (1961). When the ellipsoidal body is required to have no imposed couple, and consequently when a relation between the symmetric and skew-symmetric parts of f_{ij} is enforced, the solution was given by Eshelby (1961).

On the other hand, in the counterpart of the virtual mass problem, surface traction forces are imposed on the exterior elastic medium by the condition

$$\sigma_{ij}n_j = t_{ij}n_j$$

on ∂B, with given constant coefficients t_{ij}. The solution is constructed again by Eshelby (1961) for the ellipsoidal body when the components t_{ij} are symmetric ($= t_{ji}$), but not when they are unsymmetric and contributive then to a resultant couple.

We confine attention in the sequel to the dual cases where f_{ij} and t_{ij} are both skew-symmetric (each expressible generally in terms of its axial vector), and to the dual cases where f_{ij} is dilatational and t_{ij} is hydrostatic (each a multiple of the Kronecker delta). We define the analogues of the polarization and virtual mass tensors together with some isoperimetric inequalities and accompanying special solutions.

4.1. Prescribed Rotation of Rigid Inclusion

The conditions at the boundary ∂B are appropriate for a rotated rigid inclusion B bonded to the exterior elastic medium E. On ∂B

$$u_i = \varepsilon_{ijk}\omega_j x_k + g_i, \tag{4.1}$$

where ω_i is an arbitrarily prescribed constant vector (of angular displacement) and ε_{ijk} is the alternating symbol (taking the value 1 when i, j, k is a cyclic permutation of 1, 2, 3, the value -1 for a noncyclic permutation, and the value zero in the other cases where i, j, k are not all different). By means of the two equivalent definitions

$$R_{ij}\omega_i\omega_j = \frac{1}{2}\int_E \sigma_{ij}e_{ij}\,dV, \qquad M_i = 2R_{ij}\omega_j,$$

in terms first of the positive elastic strain energy in the exterior E and secondly of the resultant couple M_i acting on the body, we introduce the symmetric, positive definite second-rank "rotation" tensor R_{ij}. The equivalence can be verified by converting the volume integral into a surface one over ∂B, to take account of the boundary condition (4.1) and the condition of no resultant force. The ratio R_{ij}/V depends on the shape of the body and on the elastic constants.

We can construct the inequality

$$\int_E \sigma_{ij}e_{ij}\,dV \geq \int_E \Sigma_{ij}(2e_{ij} - E_{ij})\,dV, \tag{4.2}$$

where

$$E_{ij} = \tfrac{1}{2}(U_{i,j} + U_{j,i}), \qquad \Sigma_{ij} = \lambda E_{kk}\delta_{ij} + 2\mu E_{ij}, \tag{4.3}$$

in terms of a vector field U_i (of order $1/r^2$ for large r), since equality could be restored by adding the positive quantity $(\sigma_{ij} - \Sigma_{ij})(e_{ij} - E_{ij})$ to the right-hand integrand. It suffices here to make the particular choice

$$U_i = c\varepsilon_{ijk}\omega_j\varphi_{,k}$$

in terms of a constant c and the gravitational potential φ of B. At all points of E, it presents the simplifying properties

$$U_{i,i} = 0, \qquad \Sigma_{ij,j} = 0, \tag{4.4}$$

the first of which removes the elastic constant λ from the analysis, while the second allows the integrand in (4.2) to be expressed as a divergence. By means of the divergence theorem, the integrations can be all converted eventually to ones over the interior region B, via intermediate surface integrals over ∂B, into which the boundary values of u_i are substituted. We arrive at the tensor inequality

$$\frac{2R_{ij}}{\mu V} \geq [4c + 2c^2(\langle\varphi_{,mn}\varphi_{,mn}\rangle - 1)]\delta_{ij} + 3c^2(\langle\varphi_{,ij}\rangle - \langle\varphi_{,ki}\varphi_{,kj}\rangle),$$

which can be contracted and simplified further, on reference to (1.1) and to the final inequality of (2.6), as

$$R_{ii} \geq 9\mu V, \tag{4.5}$$

after finally making the best choice of the constant c ($= 3$). When the body B

has a spherical shape, with radius a and center at the origin, it can be verified that

$$u_i = U_i = \frac{a^3 \varepsilon_{ijk} \omega_j x_k}{r^3}$$

and, consequently, that

$$R_{ij} = 3\mu V \delta_{ij}.$$

The inequality (4.5) is therefore an isoperimetric one whose equality sign is attained by the spherical shape.

We specialize next to an axisymmetric rigid body B whose rotation tensor R_{ij} has the structure (1.4), in order to simplify the calculation of the component R_{33}. It suffices to take an angular rotation ω about the x_3 axis of symmetry, so that

$$\omega_1 = \omega_2 = 0, \qquad M_1 = M_2 = 0,$$

$$\omega_3 = \omega, \qquad M_3 = 2R_{33}\omega.$$

The displacement field u_i can be expressed then in terms of the polarization components v_1 and v_2 for the body, as was shown by Kanwal (1961) in the corresponding context of a slow rotation in a fluid of viscosity μ. Thus we can set

$$u_1 = -\omega v_2, \qquad u_2 = \omega v_1, \qquad u_3 = 0,$$

so as to satisfy the boundary conditions (4.1) at ∂B and those at infinity, and the equilibrium conditions. It is to be observed that u_1 and u_2 each satisfy Laplace's equation and the solenoidal condition $u_{i,i} = 0$, which eliminates the elastic constant λ. Consequently, by the analysis of Kanwal (1961), the connection

$$R_{33} = \mu V(1 + q)$$

is made with the polarization coefficient q of B, and in (3.8) with the virtual mass coefficient m.

For the spheroidal shape, it only remains to transfer our previous evaluations of the polarization fields v_1 and v_2, and of the coefficient q, to the present analysis. An equivalent outcome was reached by Villaggio (1986) by a complicated use of harmonic functions set in prolate and oblate spheroidal coordinate systems. This method was employed in an identical mathematical context (of the viscous fluid) by Jeffery (1915). For a few other nonspheroidal, axisymmetric body shapes, we may refer to Schiffer and Szegö (1949), Payne (1952), and Richardson (1977).

Lastly, with the help of (3.7), we establish the isoperimetric inequalities

$$R_{33} \geq \frac{2\mu V}{1 - \alpha} \geq 2\mu V, \tag{4.6}$$

for any axisymmetric body of volume V, in terms of its gravitational parameter α of (2.7). The first equality sign is attained by every spheroid while the second is approached in the extremely prolate limit. For the circular disk of radius a, in particular, α tends to unity and R_{33} tends to the finite limit $16\mu a^3/3$.

4.2. Prescribed Torsional Traction

The condition at the boundary ∂B is that

$$\sigma_{ij}n_j = \varepsilon_{ijk}n_j\tau_k, \tag{4.7}$$

in terms of the arbitrarily prescribed, constant components τ_k of a vector (whose direction is that of the resultant couple). By means of the two equivalent definitions

$$S_{ij}\tau_i\tau_j = \frac{1}{2}\int_E \sigma_{ij}e_{ij}\,dV,$$

$$\varepsilon_{ijk}S_{km}\tau_m = -\frac{1}{2}\int_{\partial B} (u_in_j - u_jn_i)\,dS,$$

in terms of the positive strain energy in E, and of a surface integral over ∂B, we introduce a symmetric, positive definite, second-rank tensor S_{ij}. By recourse again to a vector field U_i, and to the definitions (4.3), we construct the inequality

$$\int_E \sigma_{ij}e_{ij}\,dV \geq \int_E (2\sigma_{ij} - \Sigma_{ij})E_{ij}\,dV \tag{4.8}$$

by dropping a positive quantity from the right-hand integrand. It is appropriate to specify that

$$U_i = c\varepsilon_{ijk}\varphi_{,j}\tau_k$$

in terms of a constant c. The properties (4.4) are obtained again, and the integrations can be transferred to the interior region, after substitution of the boundary condition (4.7) in the intervening surface integrals. We derive consequently the tensor inequality

$$\frac{2S_{ij}}{V} \geq \mu c^2[3(\langle\varphi_{,ij}\rangle - \langle\varphi_{,ki}\varphi_{,kj}\rangle) - 2(1 - \langle\varphi_{,mn}\varphi_{,mn}\rangle)\delta_{ij}] - 2c(\delta_{ij} - \langle\varphi_{,ij}\rangle),$$

which contracts and simplifies further, with the help of (1.1) and the final inequality of (2.6), to

$$S_{ii} \geq \frac{V}{\mu}, \tag{4.9}$$

after, finally, making the best choice of c as $-1/\mu$. For a spherical body B, it may be verified now that U_i is identifiable as the exact solution u_i, that

$$S_{ij} = V\frac{\delta_{ij}}{3\mu},$$

and hence that S_{ij}/V is inverse to R_{ij}/V. The inequality (4.9) is identified as another isoperimetric one whose equality sign is attained by the spherical shape. In the particular axisymmetric case where the body has the spheroidal shape and where τ_1 and τ_2 both vanish, to leave the couple acting only about the x_3 axis, it may be verified that U_i stands again as the exact solution u_i, to hence allow the calculation

$$S_{33} = \frac{(1-\alpha)V}{2\mu},$$

in terms of the gravitational parameter (2.8) of the spheroid. It is thereby confirmed, after taking the first equality sign in (4.6), that S_{33}/V is the reciprocal of R_{33}/V for every spheroid.

4.3. *Prescribed Pressure*

The condition at the boundary ∂B is that

$$\sigma_{ij}n_j = -\beta n_i,$$

where β is a given constant scalar (the pressure). A scalar parameter Q is to be evaluated by either of the two equivalent definitions

$$Q\beta^2 = \tfrac{1}{2}\int_E \sigma_{ij}e_{ij}\,dV, \qquad 2Q\beta = \int_{\partial B} u_i n_i\,dS,$$

in terms first of the positive elastic strain energy in E and secondly of a surface integral over ∂B, which is equal (for $\beta > 0$) to the increase in the volume of the region B due to the elastic straining. We construct the vector field

$$U_i = c\varphi_{,i} \tag{4.10}$$

where c is a constant, and we appeal again to the definitions (4.3) and to the inequality (4.8). After the integrations are transferred once again to the interior region, it is found that

$$\frac{Q\beta^2}{V} \geq \beta c - \mu c^2(1 - \langle\varphi_{,ij}\varphi_{,ij}\rangle),$$

and, consequently, by incorporating the final inequality of (2.6) and the best choice of c as $3\beta/4\mu$, that

$$Q \geq \frac{3V}{8\mu}.$$

The equality sign is attained in this isoperimetric inequality, and U_i coincides with the exact solution u_i, when the region B is spherical. That is, for given β, V, and μ, the smallest increase in the volume of B, and the smallest total strain energy in E, is achieved by the spherical shape.

4.4. *Prescribed Expansion*

The condition at the boundary ∂B is that

$$u_i = \gamma x_i + g_i,$$

where γ is a given constant scalar. The region B thereby increases in volume (if $\gamma > 0$) by the amount $3\gamma V$. A scalar T can be evaluated by either of the equivalent definitions

$$T\gamma^2 = \frac{1}{2}\int_E \sigma_{ij}e_{ij}\,dV, \qquad T\gamma = -\frac{1}{4}\int_{\partial B} \sigma_{ij}(n_i x_j + n_j x_i)\,dS.$$

A vector field U_i is defined again by (4.10) in terms of a new constant c, and the inequality (4.2) and definitions (4.3) are recalled. We find first that

$$\frac{T\gamma^2}{V} \geq 4\mu c\gamma - \mu c^2(1 - \langle \varphi_{,ij}\varphi_{,ij}\rangle)$$

and finally that

$$T \geq 6\mu V,$$

after incorporating the final inequality of (2.6) and the best choice of c as 3γ. The equality sign is attained in this final isoperimetric inequality, and U_i coincides with the exact solution u_i, when the region B is spherical.

References

Batchelor, G. K. (1967), *An Introduction to Fluid Dynamics*, Cambridge University Press, London, p. 403.

Daniele, E. (1911), Sul problema dell equilibrio elastico nello spazio esterno ad un ellissoide per dati spostamenti in superficie, *Nuovo Cimento* [6], **1**, 211–230.

Eshelby, J. D. (1961), Elastic inclusions and inhomogeneities, in *Progress in Solid Mechanics*, edited by I. N. Sneddon and R. Hill, Vol. II, North-Holland, Amsterdam, pp. 87–140.

Jeffery, G. B. (1915), On the steady rotation of a solid of revolution in a viscous fluid, *Proc. London Math. Soc.*, **14**, 327–338.

Kanwal, R. P. (1961), Slow steady rotation of axially symmetric bodies in a viscous fluid, *J. Fluid Mech.*, **10**, 17–24.

Payne, L. E. (1952), On axially symmetric flow and the method of generalized electrostatics, *Quart. Appl. Math.* **10**, 197–204.

Payne, L. E. (1956), New isoperimetric inequalities for eigenvalues and other physical quantities, *Comm. Pure Appl. Math.*, **9**, 531–542.

Payne, L. E. (1967), Isoperimetric inequalities and their applications, *SIAM Rev.*, **9**, 453–488.

Polya, G. (1947), A minimum problem about the motion of a solid through a fluid", *Proc. Nat. Acad. Sci. U.S.A.*, **33**, 218–221.

Polya, G. and Szegö, G. (1951), *Isoperimetric Inequalities in Mathematical Physics*, Princeton University Press, Princeton, NJ.

Richardson, S. (1977), Axisymmetric slow viscous flow about a body of revolution whose section is a cardioid, *Quart. J. Mech. Appl. Math.*, **30**, 369–374.

Schiffer, M. (1957), Sur la polarization et al masse virtuelle, *C. R. Acad. Sci. Paris*, **244**, 3118–3121.

Schiffer, M. and Szegö, G. (1949), Virtual mass and polarization, *Trans. Amer. Math. Soc.*, **67**, 130–205.

Villaggio, P. (1986), The main elastic capacities of a spheroid, *Arch. Ration. Mech. Analysis*, **92**, 337–353.

Edge Crack Solution Through Use of Dislocation Shielding/Antishielding

J. Weertman
Department of Materials Science and Engineering,
Department of Geological Sciences,
Northwestern University, Evanston, IL 60208, U.S.A.

Abstract

The edge crack problem in an elastic solid has been treated by a number of investigators. The crack tip stress intensity factor, under uniform loading, of a mode I edge crack of length *a* has been shown to be 12% greater than a crack of half-length *a* within an infinite solid. In this paper a new method of solution for this old problem is presented. The crack within an infinite solid can be considered to be a distribution of infinitesimal dislocations on the (horizontal) crack plane. The edge crack is made up of both an infinitesimal edge dislocation distribution on the (horizontal) crack plane and an infinitesimal edge dislocation distribution on the (vertical) free surface when the edge crack is treated by the image dislocation method. (That is, the edge crack is considered to be within an infinite solid and image dislocations are used to cancel surface tractions on the free surface.) The image dislocations on the free surface shield or antishield the crack tip. The dislocation crack tip stress shielding/antishielding stress intensity factor L for a dislocation (of arbitrary Burgers vector) situated at an arbitrary position about a crack was derived earlier. The edge crack solution is derived with use of this L. The use of L is the novel feature of the solution. The crack intensity factor for the mode I edge crack under uniform loading is found to be 12% greater than the equivalent crack within an infinite solid.

1. Introduction

In this paper the edge crack problem is treated using distributions of infinitesimal dislocations. This way of viewing a crack is now common for a crack in an infinite solid. For the crack in an infinite solid the dislocation distribution exists only on the crack plane. For the edge crack an additional dislocation distribution must be considered—a distribution on the free surface. The dislocations in this second distribution will shield or antishield the crack tip. (The general expression for dislocation crack tip stress shielding/antishielding stress intensity factor L for a dislocation (of arbitrary Burgers vector) situated at an arbitrary position about a crack was derived earlier by Weertman (1984).

The edge crack problem, of course, has been considered by a number of investigators (see Hartranft and Sih (1973) and references cited by them). Their solutions do not make use of dislocation crack tip shielding or antishielding or explicit use of dislocation distributions.

The edge crack considered in this paper is a mode I crack. The results apply too to the mode II crack. (The mode III edge crack is essentially the same as a mode III crack in an infinite solid.)

2. Image Dislocations

The stress field of a discrete edge dislocation of Burgers vector b in a climb orientation (that is, the Burgers vector is in the y direction, which is perpendicular to the horizontal crack plane) situated at position x, $y = 0$ near a free surface (see Fig. 1) is the sum of the stress field (in an infinite solid) of this same dislocation—the stress field of a discrete image dislocation of Burgers vector $-b$ situated at $-x$, $y = 0$, and the stress field of a distribution of infinitesimal edge dislocations in a glide orientation (that is, their Burgers vectors are in the x direction, which is parallel to the crack plane). The distribution of infinitesimal dislocation is on the vertical, $x = 0$, free surface plane (see Fig. 1). The dislocation distribution function $B_{fs}(y)$ is given by (Weertman, 1964)

$$B_{fs}(y) = \left[\frac{4yx^2}{\pi(y^2 + x^2)^2}\right] b. \tag{2.1}$$

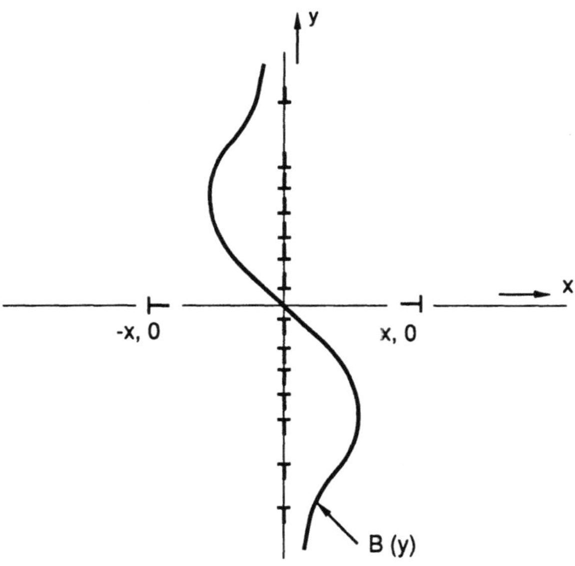

FIG. 1. Climb edge dislocation near a free surface, its image of opposite sign, and an image dislocation distribution of infinitesimal dislocations on the free surface.

The sum of the stress fields of the real dislocation, its image, and the infinitesimal dislocation gives rise to no traction stress on the vertical free surface plane.

3. Basic Equations

Suppose a crack in an infinite solid is subjected to a remotely applied constant tensile stress σ_{ra} and directly applied to the crack plane tensile stress $\sigma_{ca}(x)$, which may be a function of position x. To have a solution the stress across the crack plane produced by the dislocation distribution $B_{cp}(x)$ on the crack plane must be equal to

$$\sigma_{B_{cp}}(x) = -\sigma_{ra} + \sigma_{ca}(x) + \sigma_{if}(x), \tag{3.1}$$

in the region on the crack plane where $B_{cp}(x) \neq 0$. Here $\sigma_{if}(x)$ is the interatomic stress across the crack plane. For a crack of half-width a the stress $\sigma_{if}(x) = 0$ where $-a < x < a$. In the region $|x| > a$, for the Bilby–Cottrell–Swinden–Dugdale (BCSD) type crack, $\sigma_{if}(x) = \pm\sigma_0$,where σ_0 is a finite constant, wherever $B_{cp}(x) \neq 0$. The solution for the crack in a linear elastic solid is obtained by letting $|\sigma_0| \to \infty$ in the BCSD solution.

The distribution $B_{cp}(x)$ of infinitesimal climb edge dislocations is given by the Muskhelishvili equation

$$B_{cp}(x) = -\frac{2\alpha(c^2 - x^2)^{1/2}}{\pi G} \int_{-c}^{c} \left\{ \frac{\sigma_{B_{cp}}(x')}{(x - x')(c^2 - x'^2)^{1/2}} \right\} dx', \tag{3.2}$$

where G is the shear modulus. The c is determined by the equation

$$\int_{-c}^{c} \left\{ \frac{\sigma_{B_{cp}}(x')}{(c^2 - x'^2)^{1/2}} \right\} \left\{ 1 - q + q\left(\frac{x'}{c}\right) \right\} dx' = \frac{Gb_T}{2\alpha c}. \tag{3.3}$$

Here $q = 0$ if $\sigma_{B_{cp}}(x) = \sigma_{B_{cp}}(-x)$, and $q = 1$ if $\sigma_{B_{cp}}(x) = -\sigma_{B_{cp}}(-x)$. The term b_T is the total Burgers vectors of the dislocations in the distribution (that is, $b_T = \int_{-c}^{c} B_{cp}(x)$). (Note that $c > a$ and that $B_{cp}(x) = 0$ where $|x| > c$.)

In the limit $|\sigma_0| \to \infty$, and thus $(c - a)/a \ll 0$, the crack tip stress intensity factor K can be shown (Weertman, 1984) from (3.1)–(3.3), to be given by

$$K = -\left(\frac{a}{\pi}\right)^{1/2} \int_{-a}^{a} \left\{ \frac{\sigma_{B_{cp}}(x)}{(a^2 - x^2)^{1/2}} \right\} \left\{ 1 - q + q\left(\frac{x}{a}\right) \right\} dx. \tag{3.4}$$

Under a constant applied stress (that is, when $\sigma_{ra} \neq 0$ and $\sigma_{ca}(x) = 0$, and thus $b_T = 0$), $B_{cp}(x) = K[2\alpha/G(\pi a)^{1/2}][(x/(c^2 - x^2)^{1/2}]$ and $K = \sigma_{ra}(\pi a)^{1/2}$ when $|\sigma_0| \to \infty$ and, therefore, $(c - a)/a \to 0$.

A discrete glide edge dislocation of Burgers vector b (the Burgers vector parallel to the horizontal crack plane) which is situated at position $x = 0$, y on the vertical $x = 0$ plane that bisects the crack gives rise (when $(c - a)/a \ll 0$) to a mode I shielding or antishielding stress intensity factor L given by

(Weertman, 1984)

$$L = \left[\frac{bG}{2\alpha(\pi a)^{1/2}}\right]\left[\frac{ay|y|}{(c^2 + y^2)^{3/2}}\right]. \tag{3.5}$$

4. Edge Crack

The edge crack can be modeled as a crack of half-width a in an infinite solid for a stress $\sigma_{B_{cp}}(x) = \sigma_{B_{cp}}(-x)$ with a dislocation distribution $B_{ecp}(x) = -B_{ecp}(-x)$ on the horizontal crack plane, and the distribution $B_{fs}(y)$ of infinite-simal glide edge dislocations on the vertical, traction free surface, plane that bisects the crack. This latter distribution, found from (2.1), is

$$B_{fs}(y) = \int_0^c \left[\frac{4yx'^2}{\pi(y^2 + x'^2)^2}\right] B_{ecp}(x')\,dx'. \tag{4.1}$$

The distribution $B_{fs}(y)$ gives rise to a tensile stress $\sigma_{B_{fs}}(x)$ that acts across the crack plane. The stress $\sigma_{B_{fs}}(x)$ can be found from the stress field of a discrete dislocation. It is given by the equation

$$\begin{aligned}\sigma_{B_{fs}}(x) &= \frac{G}{2\pi\alpha}\int_{-\infty}^{\infty}\left[\frac{y(y^2 - x^2)}{(x^2 + y^2)^2}\right] B_{fs}(y)\,dy \\ &= \left(\frac{G}{2\pi\alpha}\right)\int_{-c}^{c}\left[\frac{x'(|x'| - |x|)}{(|x| + |x'|)^3}\right] B_{ecp}(x')\,dx'.\end{aligned} \tag{4.2}$$

Equation (3.2) must be modified for the presence of the stress $\sigma_{B_{fs}}(x)$ to the form

$$B_{ecp}(x) = -\frac{2\alpha(c^2 - x^2)^{1/2}}{\pi G}\int_{-c}^{c}\left\{\frac{\sigma_{B_{cp}}(x') - \sigma_{B_{fs}}(x')}{(x - x')(c^2 - x'^2)^{1/2}}\right\} dx'. \tag{4.3}$$

Let the constant (with units of stress) σ_a^* be defined by the equation

$$\sigma_a^* = \frac{1}{\pi}\int_{-c}^{c}\left[\frac{\sigma_{B_{fs}}(x)}{(c^2 - x^2)^{1/2}}\right] dx. \tag{4.4}$$

Equation (3.3) for the determination of c becomes

$$\int_{-c}^{c}\left\{\frac{\sigma_{B_{cp}}(x) - \sigma_a^*}{(c^2 - x'^2)^{1/2}}\right\} dx' = 0, \tag{4.5}$$

and (3.4), modified for the edge crack stress intensity factor K^*, is

$$\begin{aligned}K^* &= -\left(\frac{a}{\pi}\right)^{1/2}\int_{-a}^{a}\left\{\frac{\sigma_{B_{cp}}(x) - \sigma_a^*}{(a^2 - x^2)^{1/2}}\right\}\left\{1 - q + q\left(\frac{x}{a}\right)\right\} dx \\ &= K + \sigma_a^*(\pi a)^{1/2}.\end{aligned} \tag{4.6}$$

The value of c given by (4.5) is the same as for an ordinary BCSD crack in an infinite solid which is subjected to an applied stress $\sigma_{ra} - \sigma_{ca} + \sigma_a^*$ rather than an applied stress $\sigma_{ra} - \sigma_{ca}$ (or an applied stress intensity factor $K^* = K + \sigma_a^*(\pi a)^{1/2}$ rather than an applied stress intensity factor K, where K is the stress intensity factor of a crack in an infinite solid subjected to the stress $\sigma_{ra} - \sigma_{ca}$). In other words, the image dislocation distribution on the free surface has the same effect on changing the stress singularity at the crack tip as does increasing the applied stress by the amount σ_a^* on the stress singularities at the tips of a crack within an infinite solid.

The stress intensity factor K^* of the edge crack is also equal to $K^* = K + L$, where K is the stress intensity factor of a crack in an infinite solid subjected to the stress $\sigma_{ra} - \sigma_{ca}$, and L is the total crack tip dislocation shielding or antishielding stress intensity factor from the dislocation distribution $B_{fs}(y)$ on the free surface. An equation for L is found by combining (3.5) and (4.1)

$$L = \frac{2G(\pi a)^{1/2}}{\pi^2 \alpha} \int_{-\infty}^{\infty} \left\{ \int_0^c \left\{ \frac{|y|^3}{(c^2 + y^2)^{3/2}} \right\} \left[\frac{x'^2}{(y^2 + x'^2)^2} \right] B_{ecp}(x')\, dx' \right\} dy. \quad (4.7)$$

Let $B_{cep}(x)$ be expressed as $B_{ecp}(x) = B_{cp}^*(x) + \{B_{ecp}(x) - B_{cp}^*(x)\}$, where $B_{cp}^*(x)$ is the dislocation distribution solution for the crack in an infinite solid when a stress σ_a^* is added to the applied stresses already present. (Note that the crack tip stress intensity arising from the dislocation distribution $B_{cp}^*(x)$ is K^*, the same as from the dislocation distribution $B_{ecp}(x)$.

Setting $B_{cp}^*(x)$ into (4.1) and placing the result into (4.7) gives

$$\begin{aligned} L = {} & \frac{G(\pi a)^{1/2}}{2\pi\alpha} \int_{-\infty}^{\infty} \left\{ \frac{y|y|}{(c^2 + y^2)^{3/2}} \right\} B_{fs}^*(y)\, dy \\ & + \frac{2G(\pi a)^{1/2}}{\pi^2 \alpha} \int_{-\infty}^{\infty} \left\{ \int_0^c \left\{ \frac{|y|^3}{(c^2 + y^2)^{3/2}} \right\} \left[\frac{x'^2}{(y^2 + x'^2)^2} \right] \right. \\ & \left. \times \{B_{ecp}(x') - B_{cp}^*(x')\}\, dx' \right\} dy. \end{aligned} \quad (4.8)$$

In (4.8) the dislocation distribution $B_{fs}^*(y)$ is the distribution required to make the surface traction free only if the distribution $B_{cp}^*(x)$ were to exist on the crack plane.

When the total applied stress is the constant remotely applied stress σ_{ra} and $(c - a)/a \to 0$, then $B_{cp}^*(x) = B_{BCS}^*(x) \equiv K^*[2\alpha/G(\pi a)^{1/2}][x/(c^2 - x^2)^{1/2}]$ and $B_{fs}^*(y) = B_{fsBCS}^*(y)$ where

$$\begin{aligned} B_{fsBCS}^*(y) = {} & \left[\frac{4\alpha K^*}{\pi G(\pi a)^{1/2}} \right] \left[\frac{y}{(c^2 + y^2)^{3/2}} \right] \\ & \times \left\{ (y^2 + 2c^2) \ln \left\{ \frac{c + (c^2 + y^2)^{1/2}}{|y|} \right\} - c(c^2 + y^2)^{1/2} \right\}. \end{aligned} \quad (4.9)$$

Equation (4.7) can be expressed in the form

$$L = \gamma_0 K^* + \frac{2G(\pi a)^{1/2}}{\pi^2 \alpha} \int_0^c I(c, x') x'^2 \{B_{ecp}(x') - B_{cp}^*(x')\}\, dx'. \quad (4.10a)$$

The term $\gamma_0 K^*$ is

$$\gamma_0 K^* = \frac{G(\pi a)^{1/2}}{2\pi\alpha} \int_{-\infty}^{\infty} \left\{ \frac{y|y|}{(c^2 + y^2)^{3/2}} \right\} B^*_{\text{fs}}(y)\, dy$$
$$= \frac{2G(\pi a)^{1/2}}{\pi^2 \alpha} \int_0^c I(c, x') x'^2 B^*_{\text{cp}}(x')\, dx'. \tag{4.10b}$$

The constant γ_0 is equal to $\gamma_0 = 0.09471\ldots$, when $B^*_{\text{fs}}(y) = B^*_{\text{fsBCS}}(y)$ (that is, when the applied stress is a constant). The term $I(c, x')$ is

$$I(c, x') = 2\int_0^{\infty} \left\{ \frac{y^3}{(c^2 + y^2)^{3/2}(y^2 + x'^2)^2} \right\} dy$$
$$= \left\{ 4 \frac{\partial^2}{\partial z\, \partial(x'^2)} \int_0^{\infty} \left\{ \frac{y}{(zy^2 + c^2)^{1/2}(y^2 + x'^2)} \right\} dy \right\}_{z=1}$$
$$= \left\{ \frac{\partial^2}{\partial z\, \partial(x'^2)} \left(\frac{2}{(c^2 - zx'^2)^{1/2}} \right\} \log \left\{ \frac{zx'^2}{[c - (c^2 - zx'^2)^{1/2}]^2} \right) \right\}_{z=1}$$
$$= \frac{2c^2 + x'^2}{(c^2 - x'^2)^{5/2}} \log \left\{ \frac{x'}{c - (c^2 - x'^2)^{1/2}} \right\} - 3 \left\{ \frac{c}{(c^2 - x'^2)^2} \right\}. \tag{4.10c}$$

The term $\{B_{\text{ecp}}(x) - B^*_{\text{cp}}(x)\}$ in (4.10a), using (4.3)–(4.5) can be expressed as

$$B_{\text{ecp}}(x) = B^*_{\text{cp}}(x) + \frac{2\alpha(c^2 - x^2)^{1/2}}{\pi G} \int_{-c}^{c} \left\{ \frac{\sigma_{B_{\text{fs}}}(x') - \sigma^*_{\text{a}}}{(x - x')(c^2 - x'^2)^{1/2}} \right\} dx'$$
$$= B^*_{\text{cp}}(x) + \frac{2\alpha(c^2 - x^2)^{1/2}}{\pi G} \int_{-c}^{c} \left\{ \frac{\sigma_{B_{\text{fs}}}(x')}{(x - x')(c^2 - x'^2)^{1/2}} \right\} dx'$$
$$= B^*_{\text{cp}}(x) + \frac{(c^2 - x^2)^{1/2}}{\pi^2} \int_{-c}^{c} \left\{ \int_{-c}^{c} \left\{ \frac{t(|t| - |x'|)}{(|x'| + |t|)^3} \right\} \right.$$
$$\left. \times \left\{ \frac{1}{(x - x')(c^2 - x'^2)^{1/2}} \right\} B_{\text{ecp}}(t)\, dt \right\} dx'. \tag{4.11}$$

(Note that $\int_{-c}^{c} \{1/(x - x')(c^2 - x'^2)^{1/2}\}\, dx' = 0$.)

A subtle point, but a very important one, to be appreciated about (4.11): The constant c for the dislocation distribution $B^*_{\text{cp}}(x)$ for the crack in an infinite solid is identical to the constant c for the edge crack. It is because of this fact that (4.11) can take on the form it does. For example, it is not possible to rewrite (4.11) by replacing in it the distribution $B^*_{\text{cp}}(x)$ with the distribution $B_{\text{cp}}(x)$ because the value of the constant c associated with the distribution $B_{\text{cp}}(x)$ is not the same as the one associated with $B^*_{\text{cp}}(x)$. It is for this reason that $B_{\text{ecp}}(x)$ is expressed as $B_{\text{ecp}}(x) = B^*_{\text{cp}}(x) + \{B_{\text{ecp}}(x) - B^*_{\text{cp}}(x)\}$ instead of as $B_{\text{ecp}}(x) = B_{\text{cp}}(x) + \{B_{\text{ecp}}(x) - B_{\text{cp}}(x)\}$.

With use of (4.11), (4.10a) becomes

$$L = \gamma_0 K^* + \frac{2G(\pi a)^{1/2}}{\pi^4 \alpha} \int_0^c \left\{ \int_{-c}^{c} \left(\int_{-c}^{c} I(c, x') \right.\right.$$
$$\left.\left. \times \left\{ \frac{x'^2(c^2 - x'^2)^{1/2}}{(x' - x)(c^2 - x^2)^{1/2}} \right\} \left\{ \frac{t(|t| - |x|)}{(|t| + |x|)^3} \right\} B_{\text{ecp}}(t)\, dt \right) dx \right\} dx'. \tag{4.12a}$$

Setting $B_{ecp}(t) = B^*_{cp}(t) + \{B_{ecp}(t) - B^*_{cp}(t)\}$ in (4.12a) results in

$$L = (\gamma_0 + \gamma_1)K^* + \frac{2G(\pi a)^{1/2}}{\alpha\pi^4}\int_0^c \left\{\int_{-c}^c \left(\int_{-c}^c I(c, x')\{B_{ecp}(t) - B^*_{cp}(t)\}\right.\right.$$
$$\times \left.\left.\left\{\frac{x'^2(c^2 - x'^2)^{1/2}}{(x' - x)(c^2 - x^2)^{1/2}}\right\}\left\{\frac{t(|t| - |x|)}{(|t| + |x|)^3}\right\} dt\right) dx\right\} dx'. \tag{4.12b}$$

The term $\gamma_1 K^*$ is

$$\gamma_i K^* = \frac{2G(\pi a)^{1/2}}{\pi^4\alpha}\int_0^c \left\{\int_{-c}^c \left(\int_{-\infty}^\infty I(c, x')\left\{\frac{x'^2(c^2 - x'^2)^{1/2}}{(x' - x)(c^2 - x^2)^{1/2}}\right\}\right.\right.$$
$$\times \left.\left.\frac{y(y^2 - x^2)}{(x^2 + y^2)^2} B^*_{fs}(y)\, dy\right) dx\right\} dx'$$
$$= \frac{2G(\pi a)^{1/2}}{\pi^3\alpha}\int_0^c \left\{\int_{-\infty}^\infty I(c, x')\left\{\left(\frac{y}{|y|}\right)x'^3(c^2 - x'^2)^{1/2}\right\} B^*_{fs}(y)\right.$$
$$\times \left.\left[\frac{y^2(x'^2 + 2c^2 + 3y^2)}{(c^2 + y^2)^{3/2}(x'^2 + y^2)^2}\right] dy\right\} dx'. \tag{4.13}$$

The constant γ_1 is equal to $\gamma_1 = 0.0106\ldots$, when $B^*_{fs}(y) = B^*_{fsBCS}(y)$ (that is, when the applied stress is a constant).

The next higher-order form of (4.10) is found by again replacing $\{B_{ecp} - B^*_{cp}\}$ with the integral which contains $\sigma_{B_{fs}}$ and in turn replacing $\sigma_{B_{fs}}$ with an integral containing the distribution B_{ecp} or B_{fs}. The term γ_2 is given by the equation

$$\gamma_2 K^* = \frac{2G(\pi a)^{1/2}}{\pi^6\alpha}\int_0^c \left\{\int_{-c}^c \left(\int_{-c}^c \left[\int_{-c}^c \left\{\int_{-\infty}^\infty I(c, x')\left\{\frac{x'^2(c^2 - x'^2)^{1/2}}{(x' - x)(c^2 - x^2)^{1/2}}\right\}\right.\right.\right.\right.$$
$$\times \left\{\frac{t(|t| - |x|)}{(|t| + |x|)^3}\right\}\left\{\frac{(c^2 - t^2)^{1/2}}{(t - s)(c^2 - s^2)^{1/2}}\right\}\left\{\frac{u(u^2 - s^2)}{(u^2 + s^2)^2}\right\}\Bigg\}$$
$$\times \left.\left.\left.\left. B^*_{fs}(u)\, du\right\} ds\right] dt\right) dx\right\} dx'$$
$$= \frac{2G(\pi a)^{1/2}}{\pi^5\alpha}\int_0^c \left\{\int_{-c}^c \left(\int_{-c}^c \left\{\int_{-\infty}^\infty I(c, x')\left\{\frac{x'^2(c^2 - x'^2)^{1/2}}{(x' - x)(c^2 - x^2)^{1/2}}\right\}\right.\right.\right.$$
$$\times \left[\frac{u^2(t^2 + 2c^2 + 3u^2)}{(c^2 + u^2)^{3/2}(t^2 + u^2)^2}\right]$$
$$\times \left.\left.\left.\left\{\frac{t^2(c^2 - t^2)^{1/2}(|t| - |x|)}{(|t| + |x|)^3}\right\}\left(\frac{u}{|u|}\right) B^*_{fs}(u)\, du\right\} dt\right) dx\right\} dx'. \tag{4.14}$$

The constant γ_2 is equal to $\gamma_2 = 0.0025\ldots$, when $B^*_{fs}(y) = B^*_{fsBCS}(y)$.

Repeating the process still gives high-order forms for this equation. Each repetition gives a more accurate value of $L = (\gamma_0 + \gamma_1 + \gamma_2 + \cdots)K^*$. In each repetition the constant in front of the multiple integral decreases by the factor

$1/\pi^2$. As a consequence, the successive higher-order constants γ_i are smaller and smaller.

Since $K^* = K + L$,

$$K^* = \frac{K}{1 - \gamma_0 - \gamma_1 - \gamma_2 - \cdots}. \tag{4.15}$$

The value of K^* obtained using just the constants γ_0, γ_1, and γ_2 is $K^* \simeq K/(1 - \gamma_0 - \gamma_1 - \gamma_2) \simeq 1.121K$ when $B^*_{fs}(y) = B^*_{fsBCS}(y)$ (that is, when the applied stress is a constant). The value in the literature (Hartranft and Sih, 1973), which has been carried out with different methods by different investigators to a higher number of significant figures, is $K^* \simeq 1.1215K$.

The solution for the dislocation density function $B_{ecp}(x)$ is found in the same manner. The zero-order approximate solution $B_{ecp0}(x)$ is $B_{ecp0}(x) = B^*_{cp}(x)$. The first-order approximate solution $B_{ecp1}(x)$, found from (4.11), is

$$B_{ecp1}(x) = B_{ecp0}(x) + \frac{(c^2 - x^2)^{1/2}}{\pi^2} \int_{-c}^{c} \left\{ \int_{-c}^{c} \left\{ \frac{t(|t| - |x'|)}{(|x'| + |t|)^3} \right\} \times \left\{ \frac{1}{(x - x')(c^2 - x'^2)^{1/2}} \right\} B^*_{cp}(t)\, dt \right\} dx'. \tag{4.16}$$

The second-order approximate solution $B_{ecp2}(x)$, also found from (4.11), is

$$B_{ecp2}(x) = B_{ecp1}(x) + \frac{(c^2 - x^2)^{1/2}}{\pi^4} \int_{-c}^{c} \left\{ \int_{-c}^{c} \left(\int_{-c}^{c} \left[\int_{-c}^{c} \left\{ \frac{t(|t| - |x'|)}{(|x'| + |t|)^3} \right\} \times \left\{ \frac{u(|u| - |v|)}{(|v| + |u|)^3} \right\} \left\{ \frac{(c^2 - t^2)^{1/2}}{(x - x')(t - u)(c^2 - x'^2)^{1/2}(c^2 - u^2)^{1/2}} \right\} \times B^*_{cp}(u)\, du \right\} dv \right) dt \right] dx'. \tag{4.17}$$

And so on.

Acknowledgments

This research was supported by the National Science Foundation under NSF Grant DMR-86-07896.

References

Hartranft, R. J. and Sih G. C. (1973), Alternating method applied to edge and surface crack problems, in *Mechanics of Fracture* 1, *Methods of Analysis and Solutions of Crack Problems*, edited by G. C. Sih, Noordhoff, Leyden, pp. 179–238.

Weertman, J. (1964), Continuum distribution of dislocations on faults with finite friction, *Bull. Seismological Soc. Amer.* **54**, 1035–1058.

Weertman, J. (1984), Crack tip stress intensity factor of the double slip plane crack model: Short cracks and short short-cracks, *Int. J. Fracture*, **26**, 31–42.

Variational Estimates for the Overall Behavior of a Nonlinear Matrix–Inclusion Composite

J. R. WILLIS
School of Mathematical Sciences, University of Bath,
Bath BA2 7AY, England

Abstract

A variational structure developed over the last few years by the author is applied to the estimation of the overall behavior of a composite material, in the regime of small strains but physically nonlinear response. The composite is modeled as nonlinearly elastic and the results take the form of stationary estimates for the overall energy and complementary energy functions. Simple explicit formulas are derived for the cases of a nonlinear matrix reinforced by rigid inclusions or weakened by voids, distributed with any two-point statistics. Detailed implications are developed for an incompressible isotropic matrix containing an isotropic distribution of voids. The uniaxial stress–strain behavior of the matrix is arbitrary (apart from mild restrictions). Numerical results are presented for "linear plus power-law" matrix behavior.

1. Introduction

This paper addresses the problem of determining the overall behavior of a composite material which comprises a nonlinear matrix containing inclusions made of a single different nonlinear material. The discussion is restricted to small strains and the materials are nonlinearly elastic, at least formally: with a different interpretation, the same analysis applies to steady nonlinear creep. The work employs a variational structure, introduced by Willis (1983) and developed further by Talbot and Willis (1985), which reduces to that of Hashin and Shtrikman (1962) in the special case of linear constitutive behavior. It is developed in a form suitable to the study of statistically anisotropic materials and constitutes a generalization to nonlinear behavior of work by Willis (1977, 1981, 1982).

Two basic formulations are discussed based, respectively, on energy and complementary energy. They display a precise duality so that only one has to be developed in detail. After considering a generally inhomogeneous material, specialization is made, first to an n-component composite and then to a two-component composite, for which simple general equations can be given, for any two-point statistics. The formulation based on energy is shown to simplify dramatically in the case of a matrix reinforced by rigid inclusions, to allow an explicit closed-form formula to be given, from which lower bounds, and self-consistent estimates, for the overall energy function, $\tilde{W}(\bar{e})$, can be developed. A corresponding formula, relating to the overall complementary energy, $\tilde{W}^*(\bar{\sigma})$, for a matrix containing voids, follows immediately from the duality of the two formulations.

The remainder of the work concentrates on an incompressible matrix containing voids. Equations are developed for fairly general isotropic behavior of the matrix and their implications are explored numerically for "linear plus power-law" behavior. Two lower bounds for $\tilde{W}^*(\bar{\sigma})$ and three self-consistent estimates are studied. One bound is found to be better than the other. The self-consistent estimates are in reasonably good agreement when the mean normal stress is low, and thus they are expected to provide a good indication of overall behavior. High mean normal stress, however, tests two of the self-consistent estimates to destruction: they violate the bound, while the remaining self-consistent estimate remains above it by a substantial amount. In this case, the behavior of the composite remains an open problem, except for the establishment of the lower bound.

2. The General Theory

2.1. Formulation

The general problem that will be addressed relates to a medium which occupies a domain Ω and has boundary $\partial\Omega$. Small deformations are envisaged and its local constitutive relation is taken as

$$\sigma_{ij} = \frac{\partial W}{\partial e_{ij}}(e_{kl}; x_m),$$

or more concisely,

$$\sigma = W'(e; x), \tag{2.1}$$

where σ represents Cauchy stress and W is taken to be a convex function of the infinitesimal strain e. The medium is inhomogeneous (and will later be specialized to a composite) so the potential W depends explicitly on position x. An alternative interpretation is to regard the medium as one undergoing nonlinear creep, in which case e becomes the strain-rate tensor. This viewpoint was taken by Ponte Castañeda and Willis (1988), who performed analysis for the special case of power-law creep but, for ease of exposition, e will be regarded, throughout this work, as infinitesimal strain.

The overall response of the medium is defined, following Hill (1963, 1972), in terms of the mean energy per unit volume, $\tilde{W}(\bar{e})$, which is induced when the displacement boundary condition

$$u_i = \bar{e}_{ij}x_j \tag{2.2}$$

is applied over $\partial\Omega$, $\bar{e}$ being a constant tensor. The mean strain in the medium is then $\bar{e}$ exactly and it follows from the minimum energy principle that

$$\tilde{W}(\bar{e}) = \inf \frac{1}{|\Omega|} \int_\Omega W(\bar{e} + e'; x)\, dx, \tag{2.3}$$

the infimum being taken over a suitable set of strain fields e' whose associated displacement u' is zero on $\partial\Omega$. If $\bar{\sigma}$ is defined to be the mean value of σ over Ω, application of the principle of virtual work yields the result

$$\bar{\sigma} = \tilde{W}'(\bar{e}). \tag{2.4}$$

This was demonstrated by Hill (1963, 1972).

An overall complementary potential $\tilde{W}^*$ is defined from $\tilde{W}$ as

$$\tilde{W}^*(\bar{\sigma}) = \sup_{\bar{e}} \{\bar{\sigma}\cdot\bar{e} - \tilde{W}(\bar{e})\}, \tag{2.5}$$

the dot product implying the double contraction $\bar{\sigma}_{ij}\bar{e}_{ij}$. The supremum (2.5) is attained when $\bar{e}$ and $\bar{\sigma}$ are related by (2.4). It follows also that

$$\tilde{W}(\bar{e}) = \sup_{\bar{\sigma}} \{\bar{\sigma}\cdot\bar{e} - \tilde{W}^*(\bar{\sigma})\}. \tag{2.6}$$

This supremum is attained when

$$\bar{e} = \tilde{W}^{*\prime}(\bar{\sigma}). \tag{2.7}$$

Equations (2.4) and (2.7) imply each other.

The inequality to follow was first noted by Willis (1989). If W^* is defined as the local complementary energy density, (2.3) can be expanded to give

$$\tilde{W}(\bar{e}) = \inf_{e'} \frac{1}{|\Omega|} \int_\Omega \sup_{\sigma} \{\sigma\cdot(\bar{e} + e') - W^*(\sigma; x)\}\, dx. \tag{2.8}$$

The operations inf and sup can be interchanged at the expense of replacing the equality in (2.8) by an inequality. Evaluating the infimum first, with σ fixed, yields $-\infty$ unless the mean value of $\sigma\cdot e'$ over Ω is zero; that is, unless σ has zero divergence (in a generalized sense). It follows therefore that

$$\tilde{W}(\bar{e}) \geq \sup \frac{1}{|\Omega|} \int_\Omega \{\sigma\cdot\bar{e} - W^*(\sigma; x)\}\, dx, \tag{2.9}$$

the supremum being taken over fields σ with $\operatorname{div}\sigma = 0$. The result (2.9) can be written

$$\tilde{W}(\bar{e}) \geq \sup_{\bar{\sigma}} \{\bar{\sigma}\cdot\bar{e} - \tilde{W}_*(\bar{\sigma})\}, \tag{2.10}$$

where

$$\tilde{W}_*(\bar{\sigma}) = \inf \int_\Omega W^*(\sigma; x)\, dx, \tag{2.11}$$

with the infimum in (2.11) taken over fields σ with $\operatorname{div}\sigma = 0$ and which have mean value $\bar{\sigma}$. Equality is usually attained in (2.10). Then, $\tilde{W}_* =$

$\tilde{W}^*$ and (2.4) and (2.7) follow. A more complete discussion, in the context of nonlinear electrostatics, is given by Toland and Willis (1989).

It should be noted that the definition (2.11) for $\tilde{W}_*$ involves prescribing a mean value for σ but not detailed boundary conditions. It will be assumed in the sequel that $\tilde{W}_* = \tilde{W}^*$, though it may be noted that (2.10) provides a lower bound for $\tilde{W}$ even when equality is not attained. Conversely, the inequality (2.10) provides a bound for $\tilde{W}_*$, in terms of $\tilde{W}$.

2.2. *Hashin–Shtrikman Structure*

This subsection summarizes a variational structure developed by Willis (1983) and Talbot and Willis (1985), which reduces to that of Hashin and Shtrikman (1962) in the case of linear material behavior. With the definition

$$U(\tau; x) = \sup_e \{\tau \cdot e - W(e; x) + W_0(e)\}, \tag{2.12}$$

it follows that, for any e, τ,

$$W(e; x) \geq \tau \cdot e + W_0(e) - U(\tau; x), \tag{2.13}$$

and hence, using (2.3), that

$$\tilde{W}(\bar{e}) \geq \inf_{e'} \frac{1}{|\Omega|} \int_\Omega \{\tau \cdot (\bar{e} + e') + W_0(\bar{e} + e') - U(\tau; x)\}\, dx, \tag{2.14}$$

for any τ and any "comparison energy function" W_0. The inequality is trivial unless W_0 is chosen to grow no faster than W when $\|e\|$ is large.

Slightly more generally, if U is defined so that

$$U(\tau; x) = \operatorname{stat}_e \{\tau \cdot e - W(e; x) + W_0(e)\}, \tag{2.15}$$

then

$$\tilde{W}(\bar{e}) \approx \inf_{e'} \frac{1}{|\Omega|} \int_\Omega \{\tau \cdot (\bar{e} + e') + W_0(\bar{e} + e') - U(\tau; x)\}\, dx. \tag{2.16}$$

If the stationary value in (2.15) is a global maximum, then (2.14) is valid. Conversely, if it is a global minimum, an upper bound for $\tilde{W}$ results. When (2.15) is only a stationary value, then (2.16) provides an approximation for $\tilde{W}$.

A dual structure can be developed similarly: let

$$V(\eta; x) = \operatorname{stat}\{\sigma \cdot \eta - W^*(\sigma; x) + W_0^*(\sigma)\}. \tag{2.17}$$

Then, using (2.11) with $\tilde{W}_*$ identified with $\tilde{W}^*$,

$$\tilde{W}^*(\bar{\sigma}) \approx \inf_{\sigma'} \frac{1}{|\Omega|} \int_\Omega \{(\bar{\sigma} + \sigma') \cdot \eta + W_0^*(\bar{\sigma} + \sigma') - V(\eta; x)\}\, dx, \tag{2.18}$$

the infimum being taken over fields σ' with div $\sigma' = 0$ and having zero mean value over Ω.

Further progress depends upon being able to find the infima in (2.16) and (2.18). These are problems for a nonlinear "comparison body," unless W_0 is chosen to correspond to linear behavior. Assuming, therefore, that

$$W_0(e) = \tfrac{1}{2}e \cdot L_0 e, \tag{2.19}$$

where L_0 is a constant fourth-order "tensor of moduli," the infimum in (2.16) is attained when

$$\operatorname{div}(L_0 e') + \operatorname{div} \tau = 0, \qquad x \in \Omega, \tag{2.20}$$

with e' associated with a displacement u' for which

$$u' = 0, \qquad x \in \partial\Omega. \tag{2.21}$$

The linear problem (2.20), (2.21) can be solved in terms of a Green's function, as outlined by Willis (1977), for instance: formally,

$$e' = -\Gamma\tau, \tag{2.22}$$

where Γ is a linear operator whose kernel involves two derivatives of the Green's function.

For the problem (2.18), let

$$W_0^*(\sigma) = \tfrac{1}{2}\sigma \cdot M_0 \sigma, \tag{2.23}$$

where M_0 is the "tensor of compliances" inverse to L_0. For given η, let

$$\tau = -L_0 \eta \tag{2.24}$$

and let e' be given by (2.22). Then

$$\sigma' = L_0 e' + (\tau - \bar{\tau}) \tag{2.25}$$

has zero divergence and zero mean value and, in fact, attains the infimum in (2.18), because (2.16) and (2.18) define the same problem. Thus, expressing τ in terms of η,

$$\sigma' = -\Delta\eta, \tag{2.26}$$

where the linear operator Δ is defined so that

$$\Delta\eta = [L_0(\eta - \bar{\eta}) - L_0 \Gamma L_0 \eta]. \tag{2.27}$$

An elementary property of Γ (see, for example, Willis, 1981), is obtained by noting that, for any τ, (2.20) and (2.22) are satisfied simultaneously. Thus, τ may be replaced by $-L_0\Gamma\tau$ in (2.22), to give

$$e' = \Gamma L_0 e'$$

for any τ. Thus, using (2.22) on both sides, and noting that τ is arbitrary,

$$\Gamma L_0 \Gamma = \Gamma. \tag{2.28}$$

Then also, from (2.27),

$$\Delta M_0 \Delta = \Delta. \tag{2.29}$$

3. Specialization to a Composite

3.1. A General Composite

The material contained in Ω is now regarded as a mixture of n different constituents, or phases, firmly bonded across interfaces. The rth phase has energy function W_r, so that

$$W(e; x) = \sum_{r=1}^{n} W_r(e) f_r(x), \tag{3.1}$$

where f_r takes the value 1 if point x is in material r and zero otherwise. In obvious notation, $U_r(\tau)$ corresponds to $W_r(e)$ and it is natural to choose $\tau(x)$ to be piecewise constant, taking the value τ_r in material r. Then, from (2.15),

$$U_r(\tau_r) = \operatorname{stat}\{\tau \cdot e - W_r(e) + W_0(e)\}, \tag{3.2}$$

and the stationary value is attained when $e = \gamma_r$, where γ_r satisfies

$$\tau_r = W_r'(\gamma_r) - W_0'(\gamma_r), \qquad \gamma_r = U_r'(\tau_r). \tag{3.3}$$

Now substitute

$$e' = -\Gamma\tau = -\sum_{r=1}^{n} \Gamma\tau_r f_r$$

into (2.16) to give the infimum. Upon use of (2.28), the result simplifies to

$$\tilde{W}(\bar{e}) \approx \tfrac{1}{2}\bar{e} \cdot L_0 \bar{e} + \sum c_r[\tau_r \cdot \bar{e} - U_r(\tau_r)] - \tfrac{1}{2}\sum\sum \tau_r \cdot A_{rs}\tau_s, \tag{3.4}$$

where c_r is the volume fraction of material r, given by

$$c_r = \frac{1}{|\Omega|}\int_\Omega f_r \, dx, \tag{3.5}$$

and

$$A_{rs} = \frac{1}{|\Omega|}\int_\Omega f_r(\Gamma f_s) \, dx. \tag{3.6}$$

3.2. A Two-Phase Composite

In the special case $n = 2$, use of the relation $f_1 + f_2 = 1$ allows A_{rs} to be expressed in the form

$$A_{rs} = c_r(\delta_{rs} - c_s)P \quad \text{(no sum on } r\text{)}, \tag{3.7}$$

where

$$c_1 c_2 P = \frac{1}{|\Omega|}\int_\Omega f_1(\Gamma f_1) \, dx. \tag{3.8}$$

Then, (3.4) reduces to

$$\tilde{W}(\bar{e}) \approx \tfrac{1}{2}\bar{e} \cdot L_0 \bar{e} + \sum c_r\{\tau_r \cdot \bar{e} - U_r(\tau_r) - \tfrac{1}{2}\tau_r \cdot P\tau_r\} + \tfrac{1}{2}\sum\sum c_r c_s \tau_r \cdot P\tau_s. \tag{3.9}$$

This is stationary with respect to τ_r when

$$U_r'(\tau_r) + P(\tau_r - \bar{\tau}) = \bar{e}, \tag{3.10}$$

with $\bar{\tau} = \sum c_r \tau_r$.

3.3. *A Matrix Containing Rigid Inclusions*

Suppose phase 2 corresponds to rigid material. Then

$$\begin{aligned} W_2(e) &= 0, \qquad & e &= 0, \\ &= +\infty, \qquad & e &\neq 0, \end{aligned} \tag{3.11}$$

and correspondingly,

$$U_2(\tau_2) = 0, \qquad \gamma_2 = 0. \tag{3.12}$$

Equations (3.10) thus yield (for this two-phase composite)

$$\begin{aligned} \gamma_1 + c_2 P(\tau_1 - \tau_2) &= \bar{e}, \\ c_1 P(\tau_2 - \tau_1) &= \bar{e}, \end{aligned} \tag{3.13}$$

with γ_1 and τ_1 related by (3.3). Equations (3.13) can be solved explicitly

$$\gamma_1 = \frac{\bar{e}}{c_1},$$

$$\tau_1 = W_1'\left(\frac{\bar{e}}{c_1}\right) - L_0 \frac{\bar{e}}{c_1}, \qquad \tau_2 = W_1'\left(\frac{\bar{e}}{c_1}\right) + (P^{-1} - L_0)\frac{\bar{e}}{c_1}. \tag{3.14}$$

Also,

$$U_1(\tau_1) = \tau_1 \cdot \gamma_1 - W_1(\gamma_1) + W_0(\gamma_1). \tag{3.15}$$

Substituting into (3.9) therefore yields the explicit formula

$$\tilde{W}(\bar{e}) \approx \frac{1}{2}\frac{c_2}{c_1}\bar{e}\cdot(P^{-1} - L_0)\bar{e} + c_1 W_1\left(\frac{\bar{e}}{c_1}\right). \tag{3.16}$$

The right-hand side of (3.16) is always a "stationary estimate" for $\tilde{W}(\bar{e})$ and is a lower bound so long as

$$\tfrac{1}{2} e \cdot L_0 e \leq W_1(e) \qquad \text{when} \quad \|e\| \to \infty, \tag{3.17}$$

so that U_1 is defined as the supremum, (2.12). No upper bound can be obtained because U_2 can never be defined as a finite infimum.

3.4. *A Matrix Containing Voids*

Suppose phase 2 corresponds to empty space. Then $W_2 = 0$ but

$$\begin{aligned} W_2^*(\sigma) &= 0, \qquad & \sigma &= 0, \\ &= +\infty, \qquad & \sigma &\neq 0. \end{aligned} \tag{3.18}$$

The reasoning given above for rigid inclusions therefore follows through exactly, if the dual formulation is adopted. In particular, the result corresponding to (3.16) is

$$\tilde{W}^*(\bar{\sigma}) \approx \frac{1}{2}\frac{c_2}{c_1}\bar{\sigma}\cdot(Q^{-1} - M_0)\bar{\sigma} + c_1 W_1^*\left(\frac{\bar{\sigma}}{c_1}\right), \tag{3.19}$$

where Q is defined by (3.8) except that Δ replaces Γ. From (2.27), therefore,

$$Q = L_0 - L_0 P L_0, \tag{3.20}$$

and a complete duality between (3.16) and (3.19) may be noted, since (3.20) implies

$$Q^{-1} - M_0 = (P^{-1} - L_0)^{-1}. \tag{3.21}$$

4. Example

The example considered here is that of an incompressible nonlinear matrix containing a statistically uniform array of voids. These need not be spherical but their two-point statistics will be taken as isotropic, so that the expectation value

$$p_{11}(x, x') = \langle f_1(x) f_1(x') \rangle, \tag{4.1}$$

depends on $|x - x'|$ only. The composite overall will be isotropic but compressible so it is natural to take the tensor L_0 to be isotropic, with bulk modulus K_0 and shear modulus μ_0. The notation

$$L_0 = (3K_0, 2\mu_0), \tag{4.2}$$

is employed, following Hill (1965), since products of isotropic tensors may then be calculated simply by multiplying corresponding components. The tensor P in this case is well known. From Willis (1977), for example,

$$P = \left(\frac{1}{3K_0 + 4\mu_0}, \frac{3(K_0 + 2\mu_0)}{5\mu_0(3K_0 + 4\mu_0)}\right), \tag{4.3}$$

regardless of the detailed form of $p_{11}(|x - x'|)$.

The behavior of the matrix material is taken to be such that, under uniaxial tension σ_e, the axial strain e_e is given by

$$e_e = F(\sigma_e). \tag{4.4}$$

More generally, consistently with (4.4), σ_e is defined as the "equivalent stress" $(3\sigma'_{ij}\sigma'_{ij}/2)^{1/2}$, where $\sigma'_{ij} = \sigma_{ij} - \sigma_m \delta_{ij}$ and σ_m is the mean normal stress, and e_e is defined as the "equivalent strain" $(2e'_{ij}e'_{ij}/3)^{1/2}$; since the matrix is incompressible, $e'_{ij} = e_{ij}$ but the general definition is needed elsewhere. The potential W_1^* is taken to depend on σ only through σ_e; thus,

$$W_1^*(\sigma) = \int_0^{\sigma_e} F(s)\, ds. \tag{4.5}$$

The general expression (3.19) for $\tilde{W}^*$ then becomes

$$\tilde{W}^*(\bar{\sigma}) \approx \frac{1}{2}\frac{c_2}{c_1}\left\{\frac{3}{4\mu_0}\bar{\sigma}_m^2 + \frac{2(K_0 + 2\mu_0)}{\mu_0(9K_0 + 8\mu_0)}\bar{\sigma}_e^2\right\} + c_1 \int_0^{\bar{\sigma}_e/c_1} F(s)\,ds. \quad (4.6)$$

Convexity of W_1^* is assured if $F(s)$ is monotone increasing; here, the stronger assumption will be made that $F(s)$ is convex and grows at least linearly when s is large.

4.1. Lower Bounds

The expression (4.6) provides a lower bound for $\tilde{W}^*(\bar{\sigma})$, so long as the function $V_1(\eta_1)$ which was buried in its derivation in fact is defined as a supremum. Now from (2.17),

$$V_1(\eta) = \text{stat}\left\{3\sigma_m\eta_m + \sigma'_{ij}\eta'_{ij} - \int_0^{\sigma_e} F(s)\,ds + \frac{1}{2K_0}\sigma_m^2 + \frac{1}{6\mu_0}\sigma_e^2\right\}, \quad (4.7)$$

and the stationary value is attained when

$$\eta_m + \frac{1}{3K_0}\sigma_m = 0, \qquad \eta'_{ij} - \frac{3}{2}\left\{F(\sigma_e) - \frac{1}{3\mu_0}\sigma_e\right\}\frac{\sigma'_{ij}}{\sigma_e} = 0. \quad (4.8)$$

The second equation of (4.8) shows that σ'_{ij} is aligned with η'_{ij},

$$\eta_e = F(\sigma_s) - \frac{1}{3\mu_0}\sigma_e, \quad (4.9)$$

and

$$\sigma'_{ij}\eta'_{ij} = \sigma_e\eta_e, \quad (4.10)$$

η_e being normalized like e_e.

The stationary value (4.7) cannot be a supremum for any finite K_0. However, allowing $K_0 \to \infty$, and defining $V_1(\eta)$ as the supremum,

$$\begin{aligned} V_1(\eta) &= +\infty, \qquad \eta_m \neq 0, \\ &= \sigma_e\eta_e - \int_0^{\sigma_e} F(s)\,ds + \frac{1}{6\mu_0}\sigma_e^2, \qquad \eta_m = 0, \end{aligned} \quad (4.11)$$

where σ_e and η_e are related by (4.9). This definition is valid so long as the stationary value (4.9), with $\sigma_m = \eta_m = 0$, in fact is the supremum. With the assumptions made for F, the requirement is that

$$F(\sigma_e) - \frac{1}{3\mu_0}\sigma_e > 0, \quad (4.12)$$

at the stationary point. It is also known, from the analysis of Section 3.3, that the required value of η is η_1, given by

$$\eta_1 = W_1^{*\prime}\left(\frac{\bar{\sigma}}{c_1}\right) - M_0\frac{\bar{\sigma}}{c_1}, \quad (4.13)$$

while the value of σ_e corresponding to the stationary point is $\bar{\sigma}_e/c_1$: equations (3.14) give this information for the "dual" formulation for rigid inclusions. The constraint of incompressibility of the matrix can be handled either by considering first a compressible matrix and then taking a limit or directly: in either case, the conclusion is that $(\eta_1)_m = 0$ while (4.13) correctly relates $(\eta_1)_e$ and $\bar{\sigma}_e$. Thus, (4.6) yields a lower bound for $\tilde{W}^*(\bar{\sigma})$, so long as

$$\mu_0 > \frac{s}{(3F(s))}; \qquad s = \frac{\bar{\sigma}_e}{c_1}. \tag{4.14}$$

For a given $\bar{\sigma}$, the best bound of this type is attained by taking equality in (4.14). Thus,

$$\tilde{W}^*(\bar{\sigma}) \geq \tilde{W}_1^*(\bar{\sigma}) = \frac{c_2 F(\bar{\sigma}_e/c_1)}{2\bar{\sigma}_e}\{\tfrac{9}{4}\bar{\sigma}_m^2 + \tfrac{2}{3}\bar{\sigma}_e^2\} + c_1 \int_0^{\bar{\sigma}_e/c_1} F(s)\, ds. \tag{4.15}$$

In the sequel, this bound will be referred to as the bound of type 1. A similar analysis could be performed starting from the dual formulation, to obtain an upper bound for $\tilde{W}$ and hence, by duality, a lower bound for $\tilde{W}^*$. To do this directly would involve complicated algebra but this can be avoided by noting that Talbot and Willis (1985) proved generally that the bound for $\tilde{W}$ so obtained is precisely the convex dual of that obtained from the formulation for $\tilde{W}^*$, provided that both are obtained from the same fixed linear comparison medium. Thus, the upper bound is the Legendre transform of (4.6), taken with K_0, μ_0 constant (and $K_0 \to \infty$). The result (with K_0 finite) is

$$\tilde{W}(\bar{e}) \approx \frac{2c_1\mu_0}{3c_2}(3\bar{e}_m)^2 + \frac{c_1c_2(K_0 + 2\mu_0)}{\mu_0(9K_0 + 8\mu_0)}s^2 + c_1 sF(s) - c_1 \int_0^s F(s')\, ds', \tag{4.16}$$

where s satisfies

$$\bar{e}_e = \frac{2c_2(K_0 + 2\mu_0)}{\mu_0(9K_0 + 8\mu_0)}s + F(s). \tag{4.17}$$

The expression (4.16), with $K_0 \to \infty$, is a bound so long as μ_0 satisfies (4.14), with s now given by (4.17), and the best bound, for given $\bar{e}$, follows by taking equality in (4.14). Thus,

$$\tilde{W}(\bar{e}) \leq \frac{2c_1}{3c_2}\left(\frac{s}{3F(s)}\right)(3\bar{e}_m)^2 + c_1\left(1 + \frac{c_2}{3}\right)sF(s) - c_1 \int_0^s F(s')\, ds', \tag{4.18}$$

where s is related to $\bar{e}_e$ by

$$\bar{e}_e = \left(1 + \frac{2c_2}{3}\right)F(s). \tag{4.19}$$

A bound for $\tilde{W}^*$, corresponding to (4.18), can be obtained by taking Legendre transforms. The result is

$$\tilde{W}^*(\bar{\sigma}) \geq \frac{9c_2}{8c_1}\left(\frac{F(s)}{s}\right)\bar{\sigma}_m^2 + \left(1 + \frac{2c_2}{3}\right)\bar{\sigma}_e F(s) - c\left(1 + \frac{c_2}{3}\right)sF(s) + c_1 \int_0^s F(s')\, ds', \tag{4.20}$$

with

$$\bar{\sigma}_e = \frac{s}{(1 + 2c_2/3)}\left[\frac{c_1c_2}{3n} + c_1\left(1 + \frac{c_2}{3}\right) - \frac{9c_2}{8c_1}\left(\frac{\bar{\sigma}_m}{s}\right)^2\left(\frac{n-1}{n}\right)\right], \quad (4.21)$$

where

$$n(s) = \frac{sF'(s)}{F(s)}. \quad (4.22)$$

The relation (4.21) allows (4.20) to be rewritten

$$\tilde{W}^*(\bar{\sigma}) \geq \tilde{W}_2^*(\bar{\sigma}) = \frac{9c_2}{8c_1}\left(\frac{sF(s)}{n}\right)\left(\frac{\bar{\sigma}_m}{s}\right)^2 + \frac{c_1c_2}{3n}sF(s) + c_1\int_0^s F(s')\,ds'. \quad (4.23)$$

The bound (4.23) will be referred to as of type 2.

Although it is known (Ponte Castañeda and Willis, 1988) that $\tilde{W}$ and $\tilde{W}^*$ are convex, and the expressions (4.6) and (4.16) are convex, the "optimal" bounding functions (4.15) and (4.18) are not. Correspondingly, the lower bound (4.23), obtained by Legendre transformation of (4.18), is not always a single-valued function of $\bar{\sigma}_e$, if it is generated by varying the parameter s. The required bound is, of course, the branch of the function that defines the supremum, and this is convex.

4.2. "Self-Consistent" Estimates

The expression (4.6) provides an approximation to $\tilde{W}^*(\bar{\sigma})$, for any choice of K_0, μ_0. It is natural to speculate on the nature of the "best" values for K_0, μ_0, to ensure that the approximation is a good one. Two possible prescriptions are now considered. The first involves the *postulate* that, if K_0, μ_0 are chosen so that $W_0^*(\bar{\sigma}) = \tilde{W}^*(\bar{\sigma})$, then the approximation (4.6) would be exact. The assumption is thus that

$$\tilde{W}^*(\bar{\sigma}) = \frac{\bar{\sigma}_m^2}{2K_0} + \frac{\bar{\sigma}_e^2}{6\mu_0} = \frac{1}{2}\frac{c_2}{c_1}\left\{\frac{3}{4\mu_0}\bar{\sigma}_m^2 + \frac{2(K_0 + 2\mu_0)}{\mu_0(9K_0 + 8\mu_0)}\bar{\sigma}_e^2\right\} + c_1\int_0^{\bar{\sigma}_e/c_1} F(s)\,ds,$$

and so, taking $K_0 = 4\mu_0 c_1/(3c_2)$ and then eliminating μ_0,

$$\tilde{W}_3^*(\bar{\sigma}) = \frac{c_1^2(1 - c_2/3)}{(1 - 2c_2)}\left\{1 + \frac{9c_2}{4c_1}\left(\frac{\bar{\sigma}_m}{\bar{\sigma}_e}\right)^2\right\}\int_0^{\bar{\sigma}_e/c_1} F(s)\,ds. \quad (4.24)$$

The suffix 3 is placed on $\tilde{W}^*$ to define it as the approximation of type 3.

An alternative self-consistent prescription is obtained by postulating that (4.6) would be exact if K_0, μ_0 were chosen so that the mean polarization, $\bar{\eta}$, is zero. As shown by Talbot and Willis (1987), this is equivalent to the requirement that

$$\tilde{W}^{*\prime}(\bar{\sigma}) = W_0^{*\prime}(\bar{\sigma}). \quad (4.25)$$

Implementing this prescription leads to the result

$$\tilde{W}_4^*(\bar{\sigma}) = \frac{3c_2(1-c_2/3)}{2(1-2c_2)}\bar{\sigma}_e F\left(\frac{\bar{\sigma}_e}{c_1}\right)\left\{\frac{3}{4}\left(\frac{\bar{\sigma}_m}{\bar{\sigma}_e}\right)^2 + \frac{2(1+c_2/2)}{9(1-c_2/3)}\right\}$$
$$+ c_1 \int_0^{\bar{\sigma}_e/c_1} F(s)\, ds. \qquad (4.26)$$

Other prescriptions can be generated, starting from the dual formulation involving $\tilde{W}(\bar{e})$. The one corresponding to (4.25) requires that

$$\tilde{W}'(\bar{e}) = W_0'(\bar{e}). \qquad (4.27)$$

This leads, after some algebra, to an approximation for $\tilde{W}$ whose Legendre transform yields

$$\tilde{W}_5^* = \frac{9c_2}{8c_1}\left(\frac{1-c_2/3}{1-2c_2}\right)\left(\frac{sF(s)}{n}\right)\left(\frac{\bar{\sigma}_m}{s}\right)^2 + \frac{c_1c_2(1+c_2/2)}{3(1-2c_2)}\left(\frac{sF(s)}{n}\right)$$
$$+ c_1 \int_0^s F(s')\, ds', \qquad (4.28)$$

with

$$\bar{\sigma}_e = s\left\{\frac{c_2(1+c_2/2)}{3(1-c_2/3)}\left(\frac{n+1}{n}\right) + \left(\frac{1-2c_2}{1-c_2/3}\right) - \frac{9c_2}{8c_1^2}\left(\frac{n-1}{n}\right)\left(\frac{\bar{\sigma}_m}{s}\right)^2\right\}. \qquad (4.29)$$

This is not single-valued as it stands but the required branch is the one that yields the supremum.

4.3. *Results*

Some computations have been performed, for matrix behavior corresponding to

$$F(s) = e_0\left\{\frac{s}{\sigma_0} + \left[\left(\frac{s}{\sigma_0}\right)^m - \left(\frac{\sigma_y}{\sigma_0}\right)^m\right]H(s-\sigma_y)\right\}, \qquad (4.30)$$

with $m > 1$. Thus, behavior is linear up to a yield stress σ_y, after which the strain has an "inelastic" component which follows a power law. The parameters σ_0 and e_0 are used to normalize the results; the ratio σ_0/e_0 defines Young's modulus for the matrix material. In all of the results presented, $\sigma_y/\sigma_0 = 0.2$, $m = 10$, and the porosity $c_2 = 0.3$. Stresses are scaled by σ_0, strains by e_0, and potentials (W^*) are scaled by $\sigma_0 e_0$.

Figure 1(a) shows normalized plots of all of the estimates $\tilde{W}_r^*$ (for $r = 1$ to 5) against $\bar{\sigma}_e$, with mean normal stress chosen so that $\bar{\sigma}_m/\bar{\sigma}_0 = 0.25$. The lower bounds ($r = 1$ and 2) coincide at low values of $\bar{\sigma}_e$, corresponding to linear behavior, and the three self-consistent estimates ($r = 3$, 4, and 5) likewise coincide in this regime. At higher levels of $\bar{\sigma}_e$, the bound $\tilde{W}_2^*$ lies above the bound $\tilde{W}_1^*$, so $\tilde{W}_2^*$ is the better. The self-consistent estimate $\tilde{W}_3^*$

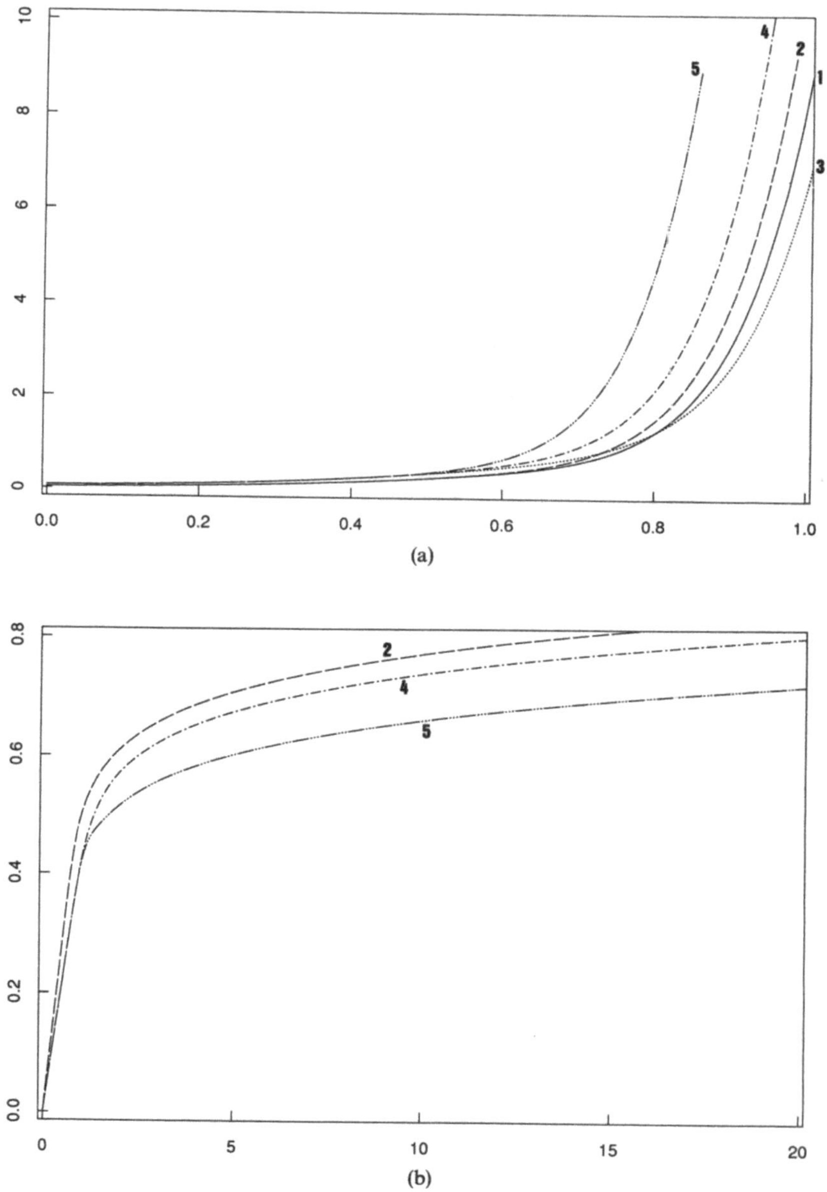

FIG. 1. Plots of (a) $\tilde{W}_r^*/(\sigma_0 e_e)$ against $\bar{\sigma}_e/\sigma_0$, and (b) $\bar{\sigma}_e/\sigma_0$ against $\bar{e}_e/e_0$, for a medium with porosity $c_2 = 0.3$, with $\bar{\sigma}_m/\sigma_0 = 0.25$. (b) shows only those curves corresponding to $r = 2$, 4, and 5.

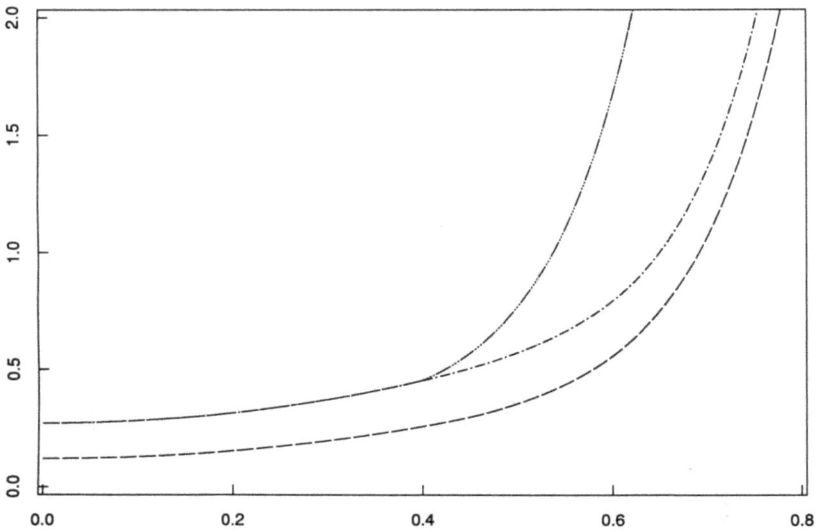

(a)

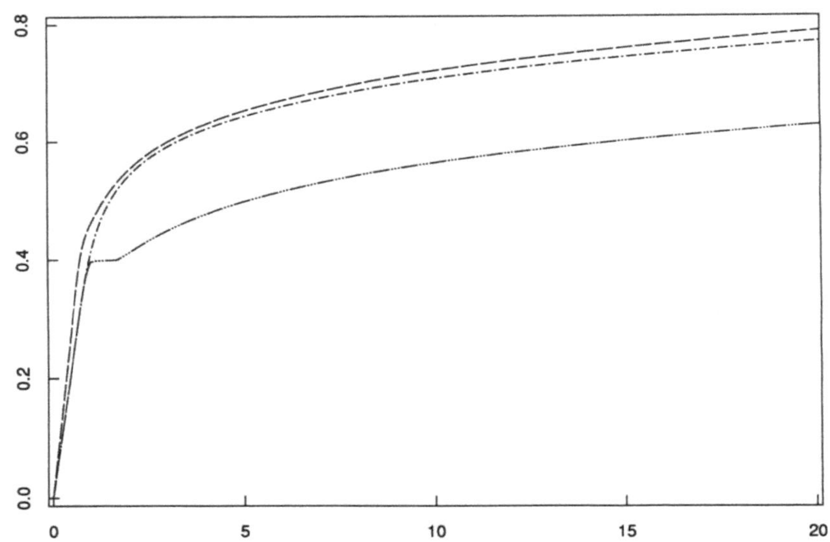

(b)

FIG. 2. Plots of (a) $\tilde{W}_r^*/(\sigma_0 e_0)$ against $\bar{\sigma}_e/\sigma_0$, and (b) the associated $\bar{\sigma}_e/\sigma_0$ against $\bar{e}_e/e_0$, with $\bar{\sigma}_m/\sigma_0 = 0.5$ for $r = 2, 4,$ and 5.

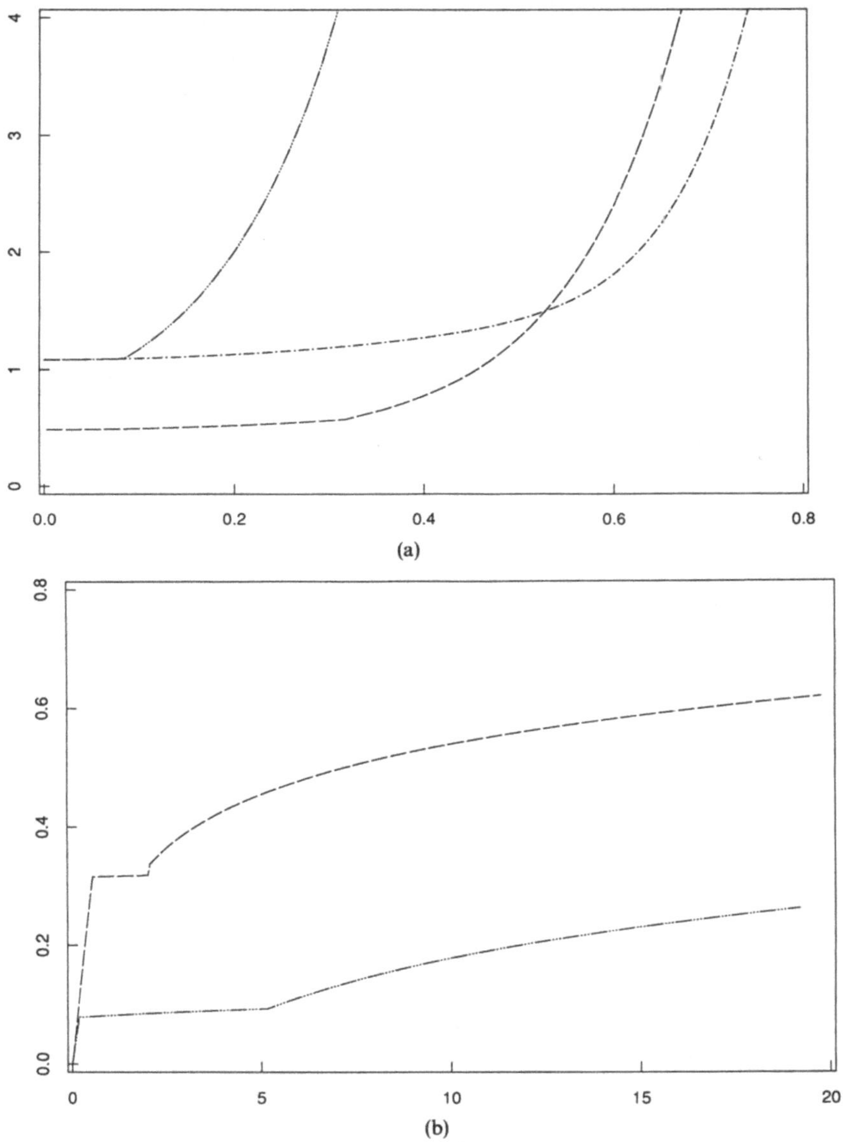

FIG. 3. Plots of (a) $\tilde{W}_r^*/(\sigma_0 e_0)$ against $\bar{e}_e/e_0$ for $r = 2$, 4, and 5, and (b) $\bar{\sigma}_e/\sigma_0$ against $\bar{e}_e$ for $r = 2$ and 5 only, with $\bar{\sigma}_m/\sigma_0 = 1.0$.

violates the bound $\tilde{W}_2^*$—and even the bound $\tilde{W}_1^*$—and so must be rejected. Similar behavior occurs at other values of $\bar{\sigma}_m$, so attention is henceforth restricted to the bound $\tilde{W}_2^*$ and the polarization-based self-consistent estimates $\tilde{W}_4^*$ and $\tilde{W}_5^*$. Figure 1(b) gives the estimates for $\bar{\sigma}_e$ versus $\bar{e}_e$, corresponding to $\tilde{W}_2^*$, $\tilde{W}_4^*$, and $\tilde{W}_5^*$. The upper curve corresponds to $\tilde{W}_2^*$ but it is not guaranteed to be a bound, except in the linear regime. Figure 2(a) shows plots of $\tilde{W}_2^*$, $\tilde{W}_4^*$, and $\tilde{W}_5^*$ against $\bar{\sigma}_e$, for $\bar{\sigma}_m/\sigma_0 = 0.5$. In contrast to behavior at $\bar{\sigma}_m/\sigma_0 = 0.25$, formula (4.28) for $\tilde{W}_5^*$ yields a function with more than one branch, corresponding to failure of convexity of the energy function $\tilde{W}_5$ from which it was derived. Only the upper branch is shown and this gives rise to the corner on the curve. This feature is reflected in Fig. 2(b), which gives the associated estimates for $\bar{\sigma}_e$ against $\bar{e}_e$. The curve derived from $\tilde{W}_5^*$ has a plateau which corresponds to the corner. Figure 3(a) gives $\tilde{W}_r^*$ for $r = 2$, 4, and 5, plotted against $\bar{\sigma}_e$, with $\bar{\sigma}_m/\sigma_0 = 1.0$. This time, the formula (4.20) for $\tilde{W}_2^*$ also has more than one branch and the curve shown in the figure has a corner. The bounds and self-consistent estimates differ substantially and no reliable inference can be made for the actual stress–strain behavior. The self-consistent estimate $\tilde{W}_4^*$, however, violates the bound and so has to be rejected. The two remaining estimates for $\bar{\sigma}_e$ versus $\bar{e}_e$, corresponding to $\tilde{W}_2^*$ and $\tilde{W}_5^*$ are shown in Fig. 3(b).

Similar observations can be made for other "hardening exponents" m and porosities c_2: larger values of m tend to test the formulas to destruction at lower values of $\bar{\sigma}_m/\sigma_0$, and the self-consistent estimates deviate further from the bounds as c_2 increases, diverging when $c_2 = 0.5$, as in the linear case.

5. Concluding Remarks

It is evident from the results presented that the problem of determining the overall behavior of nonlinear composites has still not been settled definitively. The general formulas, (3.16) for a matrix with rigid inclusions and (3.19) for a matrix with cavities, have the virtue of simplicity and have not been given previously, for general constitutive response and general statistics, though bounds and self-consistent estimates for linear behavior and isotropic statistics have been available for many years; generalizations to power-law behavior and isotropic statistics were given by Ponte Castañeda and Willis (1988). The writer is not aware that the duality between (3.16) and (3.19) has been noted before, even in special cases.

The difficulty with the nonlinear problem is that the form of the overall functions $\tilde{W}$, $\tilde{W}^*$ is not known in advance, and this severely limits the methods that are available for studying the problem. The major result of this work is the bound $\tilde{W}_2^*$: it is given by an explicit formula, valid for a wide range of material behavior, and provides the only rigorous statement that can so far be made. It is noteworthy that two of the three "self-consistent" estimates that were constructed, all of which coincide for linear behavior, violated the bound, even though they were derived from the same variational structure. It is suspected that the behavior of the medium is given fairly well by any of the self-consistent estimates at low values of $\bar{\sigma}_m/\sigma_0$, when they are in reasonable agreement with one another; the behavior at high values of $\bar{\sigma}_m/\sigma_0$ remains to be established, except that $\tilde{W}^*$ must lie above $\tilde{W}_2^*$.

References

Hashin, Z. and Shtrikman, S. (1962), On some variational principles in anisotropic and nonhomogeneous elasticity, *J. Mech. Phys. Solids*, **10**, 335–342.

Hashin, Z. and Shtrikman, S. (1963), A variational approach to the theory of the elastic behavior of multiphase materials, *J. Mech. Phys. Solids*, **11**, 127–140.

Hill, R. (1963), Elastic properties of reinforced solids: Some theoretical principles, *J. Mech. Phys. Solids*, **11**, 357–372.

Hill, R. (1965), Continuum micro-mechanics of elastoplastic polycrystals, *J. Mech. Phys. Solids*, **13**, 89–101.

Hill, R. (1972), On constitutive macro-variables for heterogeneous solids at finite strain, *Proc. Roy. Soc. London*, **A326**, 131–147.

Ponte Castañeda, P. and Willis, J. R. (1988), On the overall properties of nonlinearly viscous composites, *Proc. Roy. Soc. London*, **A416**, 217–244.

Talbot, D. R. S. and Willis, J. R. (1985), Variational principles for inhomogeneous nonlinear media, *IMA J. Appl. Math.*, **35**, 39–54.

Talbot, D. R. S. and Willis, J. R. (1987), Bounds and self-consistent estimates for the overall properties of nonlinear composites, *IMA J. Appl. Math.*, **39**, 215–240.

Toland, J. F. and Willis, J. R. (1989), Duality for families of natural variational principles in nonlinear electrostatics, *SIAM J. Math. Anal.* (to appear).

Willis, J. R. (1977), Bounds and self-consistent estimates for the overall properties of anisotropic composites, *J. Mech. Phys. Solids*, **25**, 185–202.

Willis, J. R. (1981), Variational and related methods for the overall properties of composites, in *Advances in Applied Mechanics*, Vol. 21, edited by C. S. Yih, Academic Press, New York, pp. 1–48.

Willis, J. R. (1982), Elasticity theory of composites, in *Mechanics of Solids*, the Rodney Hill 60th Anniversary Volume, edited by H. G. Hopkins and M. J. Sewell, Pergamon Press, Oxford, pp. 653–686.

Willis, J. R. (1983), The overall elastic response of composite materials, *J. Appl. Mech.*, **50**, 1202–1209.

Willis, J. R. (1989), The structure of overall constitutive relations for a class of nonlinear composites, *IMA J. Appl. Math.* (to appear).

Theory of Plasticity for a Class of Inclusion and Fiber-Reinforced Composites

Y. H. ZHAO* and G. J. WENG
Department of Mechanics and Materials Science, Rutgers University, New Brunswick, NJ 08903, U.S.A.

Abstract

Based on the theoretical framework recently established by Tandon and Weng (1988), a multiaxial theory of plasticity is developed for a class of composites containing unidirectionally aligned spheroidal inclusions. The aspect ratio of inclusions may range from that of a thin disk all the way to that of a continuous fiber, and its influence on the transversely isotropic elastoplastic behavior of the composite is investigated. Under a combined stress the secant moduli of the composite and the average stress concentration factors of the matrix are derived. It is shown that the yield condition of the composite generally depends on the hydrostatic pressure, contributing most significantly for the disk and fiber-reinforced cases. Both the initial yield stress and the work-hardening modulus of the anisotropic composite are also seen to be strongly influenced by the shape of inclusions; the disks are found to be most effective in the transverse plane but the fibers are so along the axial direction. Finally, a rather detailed account on the elastoplastic behavior of fiber-reinforced composites is given, and a comparison of the stress–strain curve with experiment is also demonstrated.

1. Introduction

This paper is concerned with the determination of the overall elastoplastic behavior for a class of unidirectionally aligned composites. The homogeneously dispersed inclusions or fibers are assumed to be spheroidal in shape, with a common aspect ratio α. These inclusions, as is usually the case, are taken to be elastic, but the surrounding matrix is ductile and may undergo plastic deformation. Both phases are perfectly bonded together, without any void nucleation or growth. This class of composites covers a wide range of inclusions; as illustrated in Fig. 1 they may range from thin disks to spherical particles, and all the way to continuous fibers. In this case, the composite as a whole is transversely isotropic and five independent stress–strain relations are required to characterize its elastic–plastic response. The theory to be

* Permanent address: Department of Civil Engineering, Shenyang Institute of Architectural Engineering, Liaoning, People's Republic of China.

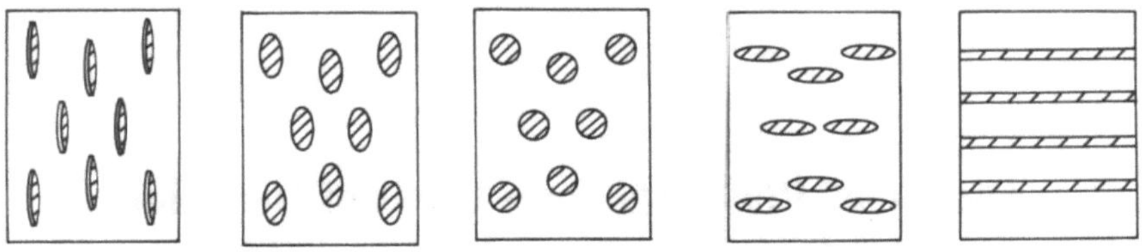

FIG. 1. Microgeometries for a class of inclusion and fiber-reinforced composites: (a) disks, (b) oblate inclusions, (c) spheres, (d) prolate inclusions and short fibers, and (e) continuous fibers.

presented is directed towards this end, so that the effect of the volume fraction and the aspect ratio of inclusions can be accounted for under a proportionally increasing combined stress.

The study is based upon the theoretical framework recently established by Tandon and Weng (1988) for particle-reinforced plasticity. The idea there was to combine the Mori–Tanaka (1973) concept of mean-stress in elasticity and Hill's (1965) discovery of a decreasing constraint power of the matrix in polycrystal plasticity. This concept of mean-stress has proven to be reliable in the determination of the effective moduli of composites; the results are consistent with the Hashin–Shtrikman (1963) bounds and Hill's (1963) exact solution for the isotropic composites (see Weng, 1984; Tandon and Weng, 1986) and also with the Hill (1964a) and Hashin (1965) bounds for the fiber-strengthened case (Zhao *et al.*, 1989). Under the special condition of monotonic, proportional loading the weakening constraint power of the matrix can be represented, approximately, by the secant moduli of the matrix (Berveiller and Zaoui, 1979). This simple approximation was used by Weng (1982) to calculate the tensile stress–strain curve of a polycrystal; the result was very close to Hutchinson's (1970) more elaborate incremental calculation using Hill's (1965) original model.

The theory of a particle-reinforced solid thus derived readily shows that the yield stress and work-hardening modulus of the composite are inversely proportional to the deviatoric part of average stress concentration factors of the matrix, and therefore these properties will increase (or decrease) with increasing hard (or soft) particle concentration. When applied to a silica particles/epoxy matrix system, the predicted stress–strain curves of the composite, up to 47% of particle volume fraction, are shown to be in good agreement with experimental data. This theory seems to give a more accurate prediction than those using an elastic constraint of the matrix (e.g., Kroner, 1961; Pedersen, 1983; Arsenault and Taya, 1987), which appears to be more suitable for the class of composite in which the matrix phase remains elastic but the particles are plastically deforming (see Weng, 1988, for a discussion on two kinds of composite plasticity). Taking the ductile matrix to be of the von Mises type, Tandon and Weng (1988) also proved that the plastic response of the particle-reinforced composite can be accompanied by a plastic volume expansion.

One novel feature of the theory is that despite the nonlinear work-hardening behavior of the matrix, the stress–strain relations of the composite can be readily generated from those of its constituents without any iteration. Thus under a monotonic proportional loading this approach appears to offer both simplicity and accuracy, and is capable of capturing the essential features of the elastic–plastic response for the two-phase composite.

2. The Mean-Field Theory

Let us first specify the properties of both constitutents. The elastic inclusion phase will be referred to as phase 1 and the ductile matrix phase as phase 0, with κ_r and μ_r representing the bulk and shear moduli of the rth phase. The tensile stress and plastic strain of the matrix is assumed to follow the modified Ludwik equation

$$\sigma = \sigma_y + h\cdot(\varepsilon^{p})^n, \tag{2.1}$$

where σ is the flow stress at the plastic strain, ε^p, and σ_y, h, and n are the initial yield stress, the strength coefficient, and the work-hardening exponent, respectively. The three material constants can readily be determined from a simple tensile test.

Under a monotonic, proportional loading this equation can be generalized to a combined stress state

$$\sigma^* = \sigma_y + h\cdot(\varepsilon^{p*})^n, \tag{2.2}$$

where, following the von Mises theory, the effective stress and strain are given by

$$\sigma^* = (\tfrac{3}{2}\sigma'_{ij}\sigma'_{ij})^{1/2}, \qquad \varepsilon^{p*} = (\tfrac{2}{3}\varepsilon^{p}_{ij}\varepsilon^{p}_{ij})^{1/2}, \tag{2.3}$$

in terms of the deviatoric stress σ'_{ij} and the plastic strain ε^p_{ij}, and the flow rule is

$$\varepsilon^{p}_{ij} = \frac{3}{2}\cdot\frac{\varepsilon^{p*}}{\sigma^*}\sigma'_{ij}. \tag{2.4}$$

Taking the total strain to be the sum of the elastic and plastic components, the "secant" Young's modulus of the matrix is

$$E^s_0 = \frac{1}{\dfrac{1}{E_0} + \dfrac{\varepsilon^{p*}}{\sigma_y + h\cdot(\varepsilon^{p*})^n}}, \tag{2.5}$$

at a given ε^{p*}, where E_0 is its usual Young's modulus. The plastic incompressibility and the assumption of overall isotropy further lead to the connections

$$\nu^s_0 = \tfrac{1}{2} - (\tfrac{1}{2} - \nu_0)\frac{E^s_0}{E_0},$$

$$\kappa^s_0 = \frac{E^s_0}{3(1-2\nu^s_0)} = \kappa_0, \qquad \mu^s_0 = \frac{E^s_0}{2(1+\nu^s_0)}, \tag{2.6}$$

for the "secant" Poisson ratio, and the secant bulk and shear moduli, in turn. The quantities ν_0 and κ_0 are the ordinary Poisson ratio and bulk modulus; the "secant" Lamé constant λ_0^s will also be used later in the theory.

In this light the plastic state of the matrix can be characterized by ε^{p*} or ν_0^s, or symbolically by the secant moduli tensor L_0^s.

The properties of the boron/aluminum system will be used throughout the paper for numerical illustrations; their elastic and plastic constants are (see Adams, 1970):

$$
\begin{aligned}
&\text{Boron fibers or inclusions:} \quad && E_1 = 379\ \text{GPa}, \quad && \nu_1 = 0.20, && \\
&\text{Aluminum matrix:} && E_0 = 55.8\ \text{GPa}, && \nu_0 = 0.32, && \\
& && \sigma_y = 87.8\ \text{MPa}, && h = 972\ \text{MPa}, && n = 0.55.
\end{aligned}
\tag{2.7}
$$

We now proceed to recapitulate briefly the mean-field theory recently established by Tandon and Weng (1988), but with a slightly different concept on the comparison material. We shall also prove that both the traction and the displacement-prescribed conditions will lead to an identical elastoplastic response for the composite. Here the familiar symbolic notations will be adopted; a second-order tensor will be denoted by a boldface Greek letter and a fourth-order tensor by an ordinary capital letter.

2.1. A Traction-Prescribed Condition

When the composite is subjected to a boundary traction which gives rise to a certain uniform stress $\bar{\boldsymbol{\sigma}}$, we may denote the (as yet unknown) secant moduli tensor of its matrix state by L_0^s. At this instant we conceptually introduce an identically shaped, linearly elastic comparison material, also with a moduli tensor L_0^s, and subject it to the same traction; its strain is then given by

$$\boldsymbol{\varepsilon}^0 = L_0^{s^{-1}} \bar{\boldsymbol{\sigma}}. \tag{2.8}$$

Since L_0^s continues to decrease as $\bar{\boldsymbol{\sigma}}$ increases, the stiffness of the comparison material also decreases accordingly. This choice of comparison material differs from the previous one suggested by Tandon and Weng (1988), but is believed to be able to reflect the current state of the matrix more accurately. Both choices, however, lead to the same overall elastoplastic response of the composite in the end.

The average stress and strain of the matrix in the composite system generally differs from those of the comparison material, say by $\tilde{\boldsymbol{\sigma}}$ and $\tilde{\boldsymbol{\varepsilon}}$, respectively. Its mean stress is

$$\boldsymbol{\sigma}^{(0)} = \bar{\boldsymbol{\sigma}} + \tilde{\boldsymbol{\sigma}} = L_0^s(\boldsymbol{\varepsilon}^0 + \tilde{\boldsymbol{\varepsilon}}). \tag{2.9}$$

Similarly further perturbations, denoted by $\boldsymbol{\sigma}^{pt}$ and $\boldsymbol{\varepsilon}^{pt}$, also exist in the inclusions with respect to the surrounding matrix; their average stress can thus be written as

$$
\begin{aligned}
\boldsymbol{\sigma}^{(1)} = \bar{\boldsymbol{\sigma}} + \tilde{\boldsymbol{\sigma}} + \boldsymbol{\sigma}^{pt} &= L_1(\boldsymbol{\varepsilon}^0 + \tilde{\boldsymbol{\varepsilon}} + \boldsymbol{\varepsilon}^{pt}) \\
&= L_0^s(\boldsymbol{\varepsilon}^0 + \tilde{\boldsymbol{\varepsilon}} + \boldsymbol{\varepsilon}^{pt} - \boldsymbol{\varepsilon}^*),
\end{aligned}
\tag{2.10}
$$

where L_1 is the elastic moduli tensor of phase 1, or inclusions, and $\boldsymbol{\varepsilon}^*$ Eshelby's (1957) equivalent transformation strain (or Mura's eigenstrain, 1987).

The perturbed strain $\boldsymbol{\varepsilon}^{pt}$ is taken to be related to $\boldsymbol{\varepsilon}^*$ through Eshelby's (1957) S tensor, but calculated in the comparison material. Denoting this tensor by S_0^s, we write

$$\boldsymbol{\varepsilon}^{pt} = S_0^s \boldsymbol{\varepsilon}^*. \tag{2.11}$$

The nonvanishing components of the S_0^s tensor for a spheroidal inclusion in general, and a circular cylinder in particular, are listed in the Appendix.

Since the weighted mean of $\boldsymbol{\sigma}^{(r)}$ must be in balance with $\bar{\boldsymbol{\sigma}}$, we have

$$\tilde{\boldsymbol{\sigma}} = -c_1 \boldsymbol{\sigma}^{pt} \quad \text{or} \quad \tilde{\boldsymbol{\varepsilon}} = -c_1(\boldsymbol{\varepsilon}^{pt} - \boldsymbol{\varepsilon}^*)$$
$$= -c_1(S_0^s - I)\boldsymbol{\varepsilon}^*, \tag{2.12}$$

where c_1 is the volume function of phase 1, and I is the symmetric fourth-order identity tensor. Substitution of (2.11) and (2.12) into the last of (2.10) leads to

$$\boldsymbol{\varepsilon}^* = A^{*(\sigma)} \boldsymbol{\varepsilon}^0, \tag{2.13}$$

where

$$A^{*(\sigma)} = -[(L_1 - L_0^s)(c_1 I + c_0 S_0^s) + L_0^s]^{-1}(L_1 - L_0^s), \tag{2.14}$$

and the symbol $[\cdot]^{-1}$ represents the inverse of $[\cdot]$, with the superscript σ indicating the traction-prescribed condition.

Similarly, from the weighted mean $\bar{\boldsymbol{\varepsilon}} = c_1 \boldsymbol{\varepsilon}^{(1)} + c_0 \boldsymbol{\varepsilon}^{(0)}$, we have

$$\bar{\boldsymbol{\varepsilon}} = \boldsymbol{\varepsilon}^0 + c_1 \boldsymbol{\varepsilon}^*. \tag{2.15}$$

Since the secant moduli tensor of the composite L_s, or the compliances tensor M_S, is defined by $\bar{\boldsymbol{\varepsilon}} = M_s \bar{\boldsymbol{\sigma}}$, it is given by

$$M_s = (I + c_1 A^{*(\sigma)}) M_0^s, \tag{2.16}$$

at a given M_0^s of the matrix.

Unlike the elastic problem, M_s alone cannot fully account for the overall elastoplastic behavior of the composite; indeed, the flow-stress and plastic–strain relation of the matrix given by (2.2) must also be satisfied. To determine such a condition we need to find $\boldsymbol{\sigma}^{(0)}$, in terms of $\bar{\boldsymbol{\sigma}}$, as

$$\boldsymbol{\sigma}^{(0)} = B_0 \bar{\boldsymbol{\sigma}}, \tag{2.17}$$

where B_0 is the fourth-order, average-stress concentration tensor of the matrix and is given by

$$B_0 = I + c_1 L_0^s (I - S_0^s) A^{*(\sigma)} L_0^{s^{-1}}, \tag{2.18}$$

from (2.9).

2.2. A Displacement-Prescribed Condition

Similarly, we now subject the composite and the comparison material to the same boundary displacement which will give rise to a uniform strain $\bar{\boldsymbol{\varepsilon}}$. The stress required for the comparison material is now given by

$$\boldsymbol{\sigma}^0 = L_0^s \bar{\boldsymbol{\varepsilon}}, \tag{2.19}$$

where L_0^s is also the current secant moduli tensor of the matrix. Using the same notations we have

$$\boldsymbol{\sigma}^{(0)} = \boldsymbol{\sigma}^0 + \tilde{\boldsymbol{\sigma}} = L_0^s(\bar{\boldsymbol{\varepsilon}} + \tilde{\boldsymbol{\varepsilon}}), \tag{2.20}$$

for the matrix, and

$$\begin{aligned}\boldsymbol{\sigma}^{(1)} = \boldsymbol{\sigma}^0 + \tilde{\boldsymbol{\sigma}} + \boldsymbol{\sigma}^{pt} &= L_1(\bar{\boldsymbol{\varepsilon}}_1 + \tilde{\boldsymbol{\varepsilon}} + \boldsymbol{\varepsilon}^{pt}) \\ &= L_0^s(\bar{\boldsymbol{\varepsilon}} + \tilde{\boldsymbol{\varepsilon}} + \boldsymbol{\varepsilon}^{pt} - \boldsymbol{\varepsilon}^*),\end{aligned} \tag{2.21}$$

for the inclusions. The balance of the weighted mean strains with the external $\bar{\boldsymbol{\varepsilon}}$ now leads to

$$\tilde{\boldsymbol{\varepsilon}} = -c_1\boldsymbol{\varepsilon}^{pt} = -c_1 S_0^s \boldsymbol{\varepsilon}^*, \tag{2.22}$$

as opposed to (2.12) for the traction-prescribed condition. Also keeping (2.11), we now have

$$\boldsymbol{\varepsilon}^* = A^{*(\varepsilon)}\bar{\boldsymbol{\varepsilon}}, \tag{2.23}$$

from (2.21), where

$$A^{*(\varepsilon)} = -[c_0(L_1 - L_0^s)S_0^s + L_0^s]^{-1}(L_1 - L_0^s), \tag{2.24}$$

as opposed to (2.14).

The stress of the composite, from $\bar{\boldsymbol{\sigma}} = c_1\boldsymbol{\sigma}^{(1)} + c_0\boldsymbol{\sigma}^{(0)}$, can be written as

$$\bar{\boldsymbol{\sigma}} = L_0^s(\bar{\boldsymbol{\varepsilon}} - c_1\boldsymbol{\varepsilon}^*). \tag{2.25}$$

It follows that the secant moduli tensor of the composite, defined by $\bar{\boldsymbol{\sigma}} = L_s\bar{\boldsymbol{\varepsilon}}$, is given by

$$L_s = L_0^s(I - c_1 A^{*(\varepsilon)}). \tag{2.26}$$

2.3. Equivalency Between the Two Boundary Conditions

The secant compliances tensor of the composite M_s is given by (2.16) under a traction-prescribed condition, and its secant moduli tensor L_s is given by (2.26) under a displacement-prescribed condition. Although not proved in Tandon and Weng (1988) we now show that both considerations lead to an identical elastoplastic property for the composite.

To see this we denote the M_s in (2.16) by $M_s^{(\sigma)}$, and L_s in (2.26) by $L_s^{(\varepsilon)}$. Then multiplying the $[\cdot]$ term in (2.14) on both ends of (2.16), we have

$$\begin{aligned}&[(L_1 - L_0^s)(c_1 I + c_0 S_0^s) + L_0^s]M_s^{(\sigma)} \\ &\qquad = [(L_1 - L_0^s)(c_1 I + c_0 S_0^s) + L_0^s]M_0^s - c_1(L_1 - L_0^s)M_0^s \\ &\qquad = [c_0(L_1 - L_0^s)S_0^s + L_0^s]M_0^s.\end{aligned} \tag{2.27}$$

Similarly, rewriting (2.26) as $M_0^s L_s^{(\varepsilon)} = I - c_1 A^{*(\varepsilon)}$ first and then multiplying it by the $[\cdot]$ term in (2.24) on both ends, we find

$$\begin{aligned}[c_0(L_1 - L_0^s)S_0^s + L_0^s]M_0^s L_s^{(\varepsilon)} &= c_0(L_1 - L_0^s)S_0^s + L_0^s + c_1(L_1 - L_0^s) \\ &= [(L_1 - L_0^s)(c_1 I + c_0 S_0^s) + L_0^s].\end{aligned} \tag{2.28}$$

Comparison between (2.27) and (2.28) reflects the desired reciprocal relation

$$M_s^{(\sigma)} L_s^{(\varepsilon)} = I. \tag{2.29}$$

Such an equivalency for the effective elastic moduli was first disclosed by Weng (1984) for the isotropic, particle-reinforced, composite and then by Benveniste (1987) for the anisotropic case. Now that this relation still stands for the elastic–plastic response, we only need to proceed under one type of boundary condition, for this we choose the commonly adopted, prescribed traction.

3. Determination of $A^{*(\sigma)}$, L_s, and B_0

The determination of M_s and B_0, in view of (2.16) and (2.18), requires the information of $A^{*(\sigma)}$. With the principal or symmetric axis of the spheroids aligned along the x_1 direction, tensor $A^{*(\sigma)}$ is transversely symmetric and the equation $\boldsymbol{\varepsilon}^* = A^{*(\sigma)}\boldsymbol{\varepsilon}^0$ of (2.13) can be expanded in terms of the components as

$$\begin{aligned}
\varepsilon_{11}^* &= a_1^*\varepsilon_{11}^0 + a_2^*\varepsilon_{22}^0 + a_2^*\varepsilon_{33}^0,\\
\varepsilon_{22}^* &= a_3^*\varepsilon_{11}^0 + a_4^*\varepsilon_{22}^0 + a_5^*\varepsilon_{33}^0,\\
\varepsilon_{33}^* &= a_3^*\varepsilon_{11}^0 + a_5^*\varepsilon_{22}^0 + a_4^*\varepsilon_{33}^0,\\
\varepsilon_{12}^* &= a_6^*\varepsilon_{12}^0, \qquad \varepsilon_{13}^* = a_6^*\varepsilon_{13}^0,\\
\varepsilon_{23}^* &= a_7^*\varepsilon_{23}^0,
\end{aligned} \tag{3.1}$$

where, due to the transverse isotropy, $a_7^* = a_4^* - a_5^*$. The two shear components are uncoupled, and can readily be found from (2.14) as

$$\begin{aligned}
a_6^* &= -\frac{1}{c_1 + 2c_0 S_{1212}^s + \mu_0^s/(\mu_1 - \mu_0^s)},\\
a_7^* &= -\frac{1}{c_1 + 2c_0 S_{2323}^s + \mu_0^s/(\mu_1 - \mu_0^s)}.
\end{aligned} \tag{3.2}$$

The normal components are coupled; following Tandon and Weng's (1984) notations of D_i, B_i, and A, but modified for the secant moduli, we have

$$\begin{aligned}
a_1^* &= \frac{D_1(B_4 + B_5) - 2B_2}{A},\\
a_2^* &= -\frac{(1 + D_1)B_2 - (B_4 + B_5)}{A},\\
a_3^* &= \frac{B_1 - D_1 B_3}{A},\\
a_4^* &= \frac{(1 + D_1)B_1 - 2B_3}{2A} + \frac{a_7^*}{2},\\
a_5^* &= \frac{(1 + D_1)B_1 - 2B_3}{2A} - \frac{a_7^*}{2},
\end{aligned} \tag{3.3}$$

where

$$D_1 = 1 + \frac{2(\mu_1 - \mu_0^s)}{(\lambda_1 - \lambda_0^s)},$$

$$D_2 = \frac{(\lambda_0^s + 2\mu_0^s)}{(\lambda_1 - \lambda_0^s)}, \tag{3.4}$$

$$D_3 = \frac{\lambda_0^s}{(\lambda_1 - \lambda_0^s)},$$

and

$$\begin{aligned} B_1 &= c_1 D_1 + D_2 + c_0(D_1 S_{1111}^s + 2S_{2211}^s), \\ B_2 &= c_1 + D_3 + c_0(D_1 S_{1122}^s + S_{2222}^s + S_{2233}^s), \\ B_3 &= c_1 + D_3 + c_0[S_{1111}^s + (1 + D_1)S_{2211}^s], \\ B_4 + B_5 &= c_1(1 + D_1) + (D_2 + D_3) + c_0[2S_{1122}^s + (1 + D_1)(S_{2222}^s + S_{2233}^s)], \\ A &= 2B_2B_3 - B_1(B_4 + B_5). \end{aligned} \tag{3.5}$$

The five secant compliances of the composite can then be found from (2.16). The two shear components are again uncoupled, and can readily be written as

$$\frac{\mu_{12}^s}{\mu_0^s} = \frac{1}{1 + c_1 a_6^*} = 1 + \frac{c_1}{\mu_0^s/(\mu_1 - \mu_0^s) + 2c_0 S_{1212}^s},$$

$$\frac{\mu_{23}^s}{\mu_0^s} = \frac{1}{1 + c_1 a_7^*} = 1 + \frac{c_1}{\mu_0^s/(\mu_1 - \mu_0^s) + 2c_0 S_{2323}^s}. \tag{3.6}$$

The normal components, following Tandon and Weng (1984) for the elastic case, are given by

$$\frac{E_{11}^s}{E_0^s} = \frac{1}{1 + c_1(a_1^* - 2\nu_0^s a_2^*)},$$

$$\frac{E_{22}^s}{E_0^s} = \frac{1}{1 + c_1[a_4^* - \nu_0^s(a_3^* + a_5^*)]}, \tag{3.7}$$

$$\nu_{12}^s = \frac{\nu_0^s - c_1[a_3^* - \nu_0^s(a_4^* + a_5^*)]}{1 + c_1(a_1^* - 2\nu_0^s a_2^*)},$$

or

$$\frac{\kappa_{23}^s}{\overline{\kappa}_0^s} = \frac{(1 + \nu_0^s)(1 - 2\nu_0^s)}{(1 - \nu_0^s - 2\nu_0^s\nu_{12}^s) + c_1[2(\nu_{12}^s - \nu_0^s)a_3^* + (1 - \nu_0^s - 2\nu_0^s\nu_{12}^s)(a_4^* + a_5^*)]},$$

where ν_{12}^s is the secant major Poisson ratio of the composite, and $\overline{\kappa}_0^s$ is the secant plane–strain bulk modulus of the matrix.

Similarly, $\boldsymbol{\sigma}^{(0)} = B_0\bar{\boldsymbol{\sigma}}$ in (2.17) can be expanded into

$$\sigma_{11}^{(0)} = b_1\bar{\sigma}_{11} + b_2\bar{\sigma}_{22} + b_2\bar{\sigma}_{33},$$

$$\sigma_{22}^{(0)} = b_3\bar{\sigma}_{11} + b_4\bar{\sigma}_{22} + b_5\bar{\sigma}_{33},$$

$$\sigma_{33}^{(0)} = b_3\bar{\sigma}_{11} + b_5\bar{\sigma}_{22} + b_4\bar{\sigma}_{33},$$
$$\sigma_{12}^{(0)} = b_6\bar{\sigma}_{12}, \qquad \sigma_{13}^{(0)} = b_6\bar{\sigma}_{13}, \tag{3.8}$$
$$\sigma_{23}^{(0)} = b_7\bar{\sigma}_{23},$$

where again due to the transverse symmetry $b_7 = b_4 - b_5$. The two shear components follow readily from (2.18) as

$$b_6 = 1 + c_1(1 - 2S_{1212}^s)a_6^* = \frac{2(\mu_1 - \mu_0^s)S_{1212}^s + \mu_0^s}{(c_1 + 2c_0S_{1212}^s)(\mu_1 - \mu_0^s) + \mu_0^s},$$
$$b_7 = 1 + c_1(1 - 2S_{2323}^s)a_7^* = \frac{2(\mu_1 - \mu_0^s)S_{2323}^s + \mu_0^s}{(c_1 + 2c_0S_{2323}^s)(\mu_1 - \mu_0^s) + \mu_0^s}. \tag{3.9}$$

The normal components can be written as

$$b_1 = 1 + c_1(p_1q_1 + 2p_2q_3),$$
$$b_2 = c_1[-p_1q_2 + p_2(q_4 + q_5)],$$
$$b_3 = c_1[p_3q_1 + (p_4 + p_5)q_3], \tag{3.10}$$
$$b_4 = 1 + c_1(-p_3q_2 + p_4q_4 + p_5q_5),$$
$$b_5 = c_1(-p_3q_2 + p_4q_5 + p_5q_4),$$

where

$$p_1 = \frac{(1 - \nu_0^s)(1 - S_{1111}^s) - 2\nu_0^s S_{2211}^s}{(1 + \nu_0^s)(1 - 2\nu_0^s)},$$
$$p_2 = \frac{\nu_0^s(1 - S_{2222}^s - S_{2233}^s) - (1 - \nu_0^s)S_{1122}^s}{(1 + \nu_0^s)(1 - 2\nu_0^s)},$$
$$p_3 = \frac{\nu_0^s(1 - S_{1111}^s) - S_{2211}^s}{(1 + \nu_0^s)(1 - 2\nu_0^s)}, \tag{3.11}$$
$$p_4 = \frac{(1 - \nu_0^s)(1 - S_{2222}^s) - \nu_0^s(S_{1122}^s + S_{2233}^s)}{(1 + \nu_0^s)(1 - 2\nu_0^s)},$$
$$p_5 = \frac{\nu_0^s(1 - S_{1122}^s - S_{2222}^s) - (1 - \nu_0^s)S_{2233}^s}{(1 + \nu_0^s)(1 - 2\nu_0^s)},$$

and

$$q_1 = a_1^* - 2\nu_0^s a_2^*,$$
$$q_2 = \nu_0^s a_1^* - (1 - \nu_0^s)a_2^*,$$
$$q_3 = a_3^* - \nu_0^s(a_4^* + a_5^*), \tag{3.12}$$
$$q_4 = a_4^* - \nu_0^s(a_3^* + a_5^*),$$
$$q_5 = a_5^* - \nu_0^s(a_3^* + a_4^*).$$

It can readily be checked that since $S_{2222}^s - S_{2233}^s = 2S_{2323}^s$, the relation $b_7 = b_4 - b_5$ is satisfied.

4. Effective Stress of the Composite

When a composite is subject to certain combinations of the macroscopic stress $\bar{\sigma}_{ij}$, the average stress of the matrix is now given by (3.8) through the stress concentration factors $b_1, b_2, \ldots, b_7$. The effective stress of the matrix σ^* can then be calculated from the first of (2.3), and it is this σ^* that determines the average elastic–plastic state of the matrix and, therefore, of the composite material. The quantity σ^* thus translates from the matrix to the global level and also represents the effective stress on the plastic deformation of the composite. Although its magnitude may be normalized to $\bar{\sigma}_{11}$ or $\bar{\sigma}_{22}$ as desired, the anisotropic nature of the composite does not seem to make either choice a particularly desirable one. We therefore decide to keep its magnitude as it is.

Then the effective stress of the composite σ^* is given by the combination

$$\begin{aligned} 2\sigma^{*2} = {} & [(b_1 - b_3)\bar{\sigma}_{11} + (b_2 - b_4)\bar{\sigma}_{22} + (b_2 - b_5)\bar{\sigma}_{33}]^2 \\ & + [(b_1 - b_3)\bar{\sigma}_{11} + (b_2 - b_5)\bar{\sigma}_{22} + (b_2 - b_4)\bar{\sigma}_{33}]^2 \\ & + (b_4 - b_5)^2(\bar{\sigma}_{22} - \bar{\sigma}_{33})^2 + 6b_6^2(\bar{\sigma}_{12}^2 + \bar{\sigma}_{13}^2) + 6b_7^2\bar{\sigma}_{23}^2. \end{aligned} \tag{4.1}$$

This quadratic function, which is transversely isotropic, now represents the yield function of the composite. The condition $\sigma^* = \sigma_y$ marks the onset of plastic deformation and, upon continuous loading, the relation $\sigma^* = \sigma_y + h(\varepsilon^{p*})^n$ and the secant moduli given by (3.6) and (3.7) allow us to determine the work-hardening behavior of the two-phase system.

The yield function thus derived is pressure-dependent, and therefore can not be reduced to Hill's (1948) anisotropic form. To see this let us impose a hydrostatic pressure on the composite such that $\bar{\sigma}_{11} = \bar{\sigma}_{22} = \bar{\sigma}_{33} = \bar{\sigma}_H$; we find that

$$\frac{\sigma^*}{\bar{\sigma}_H} = c_1[(p_1 - p_3)(q_1 - 2q_2) + (2p_2 - p_4 - p_5)(q_3 + q_4 + q_5)]. \tag{4.2}$$

This function, trivially vanishing at $c_1 = 0$, in general depends on the aspect ratio of the inclusion and the extent of plastic deformation. For the boron/aluminum system given in (2.7), its aspect-ratio dependence at $v_s = 0.4$ for three selected volume fractions is depicted in Fig. 2(a) on a log scale. The composite with disk-shaped inclusions (aspect ratio $\alpha = 0.001$, on the left) is seen to be most affected by the hydrostatic pressure, and the fiber-reinforced composite ($\alpha > 100$) also shows a considerable amount of pressure sensitivity (σ^* is 10% of $\bar{\sigma}_H$ at $c_1 = 0.3$). A particle-reinforced composite ($\alpha = 1$), as shown by Tandon and Weng (1988), remains insensitive to it. (This is a consequence of the mean-field approach. Local yielding in the matrix of course can occur, but this plastic strain, averaged out over the entire matrix, should vanish.) In all cases the influence of the hydrostatic pressure, as shown in Fig. 2(b) at $c_1 = 0.3$, decreases upon continuous plastic deformation.

The initial yield surface of the composite can be obtained by setting $\sigma^* = \sigma_y$.

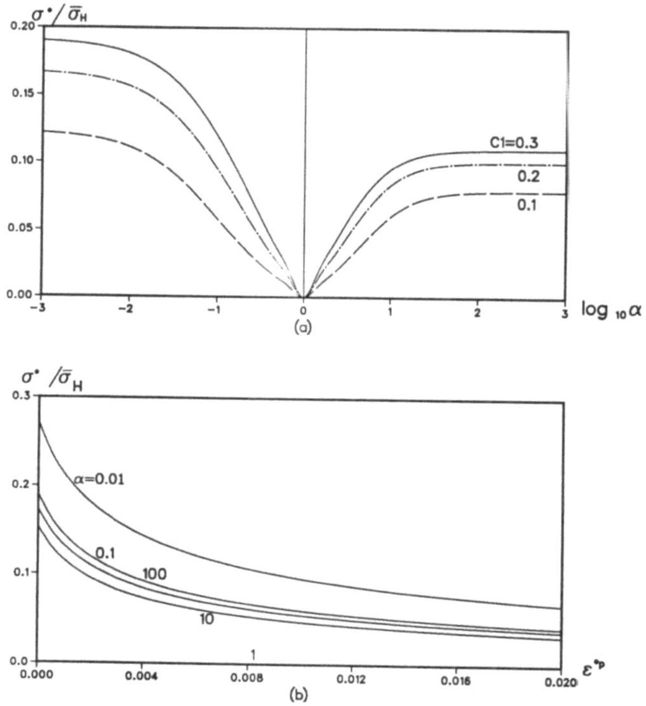

FIG. 2. The contribution of a pure hydrostatic pressure on the effective stress of the composite as a function of: (a) aspect-ratio, and (b) plastic strain.

In this case the elastic moduli of the matrix are used to evaluate the stress concentration factors $b_1, b_2, \ldots, b_7$. The yield surface of the composite is also aspect-ratio dependent. For instance, at $c_1 = 0.3$, the yield surface in the $(\bar{\sigma}_{11}, \bar{\sigma}_{12})$ space shows an extended elastic region along the $\bar{\sigma}_{11}$ axis for prolate inclusions (Fig. 3(a)), and in the $(\bar{\sigma}_{22}, \bar{\sigma}_{12})$ space it exhibits a similar effect along the $\bar{\sigma}_{22}$ axis for oblate ones (Fig. 3(b)). Yielding under the axial shear $\bar{\sigma}_{12}$ is generally less sensitive to the aspect ratio of inclusions. Since yielding is not defined locally these yield surfaces should be viewed in terms of a measurable, though infinitesimally small, overall plastic strain.

Although the hydrostatic pressure may contribute to σ^* of the composite, an axisymmetric loading characterized by $\bar{\sigma}_{22} = \bar{\sigma}_{33} = m\bar{\sigma}_{11}$ may provide a state under which the composite will not have an overall plastic strain. This is marked by the condition $\sigma^* = 0$ in (4.1); it occurs when

$$m = \frac{b_1 - b_3}{b_4 + b_5 - 2b_2}, \tag{4.3}$$

which reduces to 1 when the inclusions are spherical.

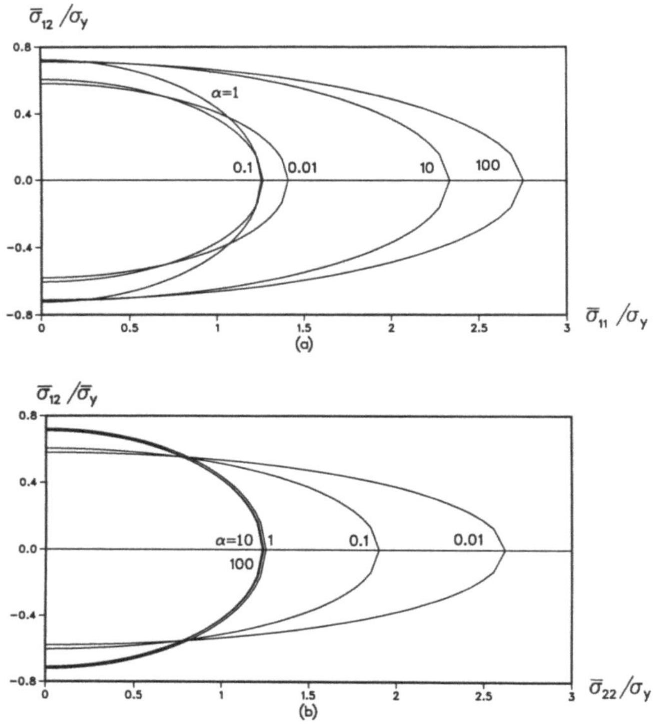

FIG. 3. The effect of aspect-ratio of inclusions on the initial yield surface in the (a) $(\bar{\sigma}_{11}, \bar{\sigma}_{12})$ space, and (b) $(\bar{\sigma}_{22}, \bar{\sigma}_{12})$ space.

5. Aspect-Ratio Dependence of the Stress–Strain Relations

Now that both the secant moduli and the effective stress of the composite have been established, its stress–strain relations can be *generated* from (2.2) of the matrix without any interation.

The elastic moduli are given by (3.6) and (3.7), but with the superscript s deleted for the elastic condition. The initial yield point is marked by setting $\sigma^* = \sigma_y$ as already discussed. Then, as ε^{p*} increases from 0 to a certain value, the secant moduli of the matrix also changes according to (2.5) and (2.6). This is accompanied by a corresponding change for D_1, D_2, and D_3 in (3.4), and, at a given aspect ratio α, a change for the S^s_{ijkl} components in the Appendix. The latter change leads to the aspect-ratio dependence for $B_1, \ldots, B_5$ in (3.5) and $a_1^*, \ldots, a_5^*$, and therefore the aspect-ratio dependence for $E^s_{11}, E^s_{22}, \ldots$, etc., in (3.7). On the other hand, the corresponding stress-concentration factors $b_1, \ldots, b_7$, with the help of (3.11) and (3.12), can be determined from (3.10), and this leads to the effective stress σ^* from (4.1) for the desired combination of $\bar{\sigma}_{ij}$. This σ^* must be set equal to $\sigma_y + h\cdot(\varepsilon^{p*})^n$ for the considered ε^{p*}, thereby providing the magnitude for the combined $\bar{\sigma}_{ij}$. This magnitude of $\bar{\sigma}_{ij}$ must lie on the linear line with the corresponding secant modulus E^s_{11}, or E^s_{22}, etc., and it gives rise to a point for the stress–strain curve of the composite. At any given point the total strain and the plastic strain of the composite are given symbolically by

$$\bar{\boldsymbol{\varepsilon}} = L_s^{-1}\bar{\boldsymbol{\sigma}} \qquad \text{and} \qquad \bar{\boldsymbol{\varepsilon}}^p = (L_s^{-1} - L^{-1})\bar{\boldsymbol{\sigma}}, \tag{5.1}$$

respectively, L being its elastic moduli tensor given by (3.6) and (3.7) without the superscript s. As ε^{p*} increases the entire stress–strain curve of the composite is generated.

For example, under a uniaxial stress $\bar{\sigma}_{11}$, the secant modulus E^s_{11} at a given ε^{p*} is given by (3.7), and the magnitude of $\bar{\sigma}_{11}$ can be determined from (4.1) and (2.2), as

$$\sigma^* = (b_1 - b_3)\bar{\sigma}_{11} = \sigma_y + h\cdot(\varepsilon^{p*})^n,$$

or (5.2)

$$\bar{\sigma}_{11} = \frac{\sigma_y + h\cdot(\varepsilon^{p*})^n}{b_1 - b_3},$$

where b_1 and b_3 are calculated from (3.10) for the considered aspect ratio α and volume fraction c_1. The calculated E^s_{11} and $\bar{\sigma}_{11}$ provide a point on the $\bar{\sigma}_{11}$ versus $\bar{\varepsilon}_{11}$ curve, and more points may similarly be found by increasing the ε^{p*} value.

Under a combined proportional loading, say with the ratio $\bar{\sigma}_{ij} = \alpha_{ij}\bar{\sigma}$, where α_{ij} are the desired proportional constants and $\bar{\sigma}$ is a monotonically increasing stress, the multiaxial stress–strain relations can be established similarly. At a given ε^{p*}, the secant moduli are again calculated from (3.6) and (3.7), and the magnitude of $\bar{\sigma}$ is calculated from (4.1) and (2.2), namely,

$$\begin{aligned}\frac{1}{\sqrt{2}}&\{[(b_1 - b_3)\alpha_{11} + (b_2 - b_4)\alpha_{22} + (b_2 - b_5)\alpha_{33}]^2\\ &+ [(b_1 - b_3)\alpha_{11} + (b_2 - b_5)\alpha_{22} + (b_2 - b_4)\alpha_{33}]^2\\ &+ (b_4 - b_5)^2(\alpha_{22} - \alpha_{33})^2 + 6[b_6(\alpha_{12}^2 + \alpha_{13}^2) + b_7\alpha_{23}^2]\}^{1/2}\bar{\sigma}\\ = {}&\sigma_y + h\cdot(\varepsilon^{p*})^n.\end{aligned} \tag{5.3}$$

This value of $\bar{\sigma}$ leads to the individual components of $\bar{\sigma}_{ij}$ and, together with the derived secant moduli, also gives rise to the points on the stress–strain curves.

To demonstrate the aspect-ratio dependence of the stress–strain relations we followed this procedure to calculate the five independent stress–strain relations; the results for the uniaxial tension, transverse tension, plane–strain biaxial loading, axial shear, and transverse shear at $c_1 = 0.30$ are shown, respectively, in Fig. 4. The initial linear portions in each figure represent the elastic response, and the subsequent nonlinear portions are the elastic–plastic response of the composites. Fibrous composites ($\alpha = 100$) are generally more

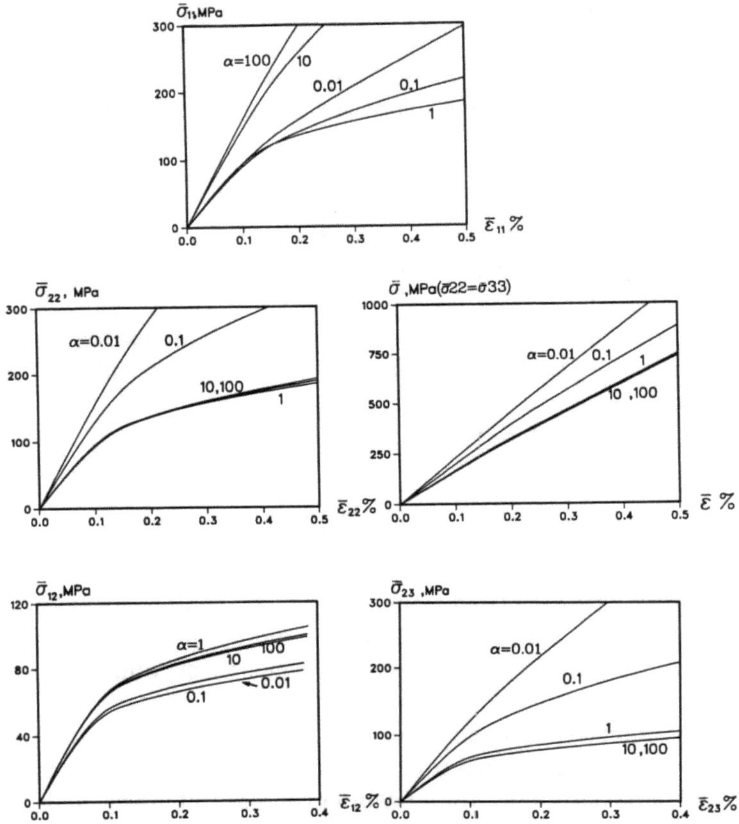

FIG. 4. The aspect-ratio dependence of the stress–strain relations under five respective loadings.

superior for the axial response ($\bar{\sigma}_{11}$), and the disk ones ($\alpha = 0.01$) are stiffer in the transverse direction ($\bar{\sigma}_{22}$ and $\bar{\sigma}_{23}$). The responses under the plane–strain biaxial loading are almost all elastic for all types of inclusions.

6. Continuous Fiber-Reinforced Composites

We now give a more detailed account for the technologically important continuous fiber-reinforced composite. Such a problem has also been examined by Hill (1964b) and Accorsi and Nemat-Nasser (1986) in the context of incremental deformation. Primarily motivated by the experimental observations (Dvorak *et al.*, 1988), Dvorak and Bahei-El-Din (1987) introduced a bimodal theory which is capable of accounting for the behavior at the flat portions of the yield surface. Using a periodic hexagonal array to describe the microgeometry of fibers, Teply and Dvorak (1988) also recently considered the local field to establish the bounds of the instantaneous moduli. These theories are also suitable for nonradial loadings and interested readers should refer to these original works. For now, the S-tensor depends only on ν_0^s, and the preceding results associated with the general spheroidal inclusions can be greatly simplified. For instance, the $B_1, B_2, \ldots, B_5$ components in (3.5) are now given by

$$
\begin{aligned}
B_1 &= c_1 D_1 + D_2 + \frac{c_0 \nu_0^s}{1-\nu_0^s}, \\
B_2 &= c_1 + D_3 + \frac{c_0}{2(1-\nu_0^s)}, \\
B_3 &= c_1 + D_3 + \frac{c_0 \nu_0^s (1 + D_1)}{2(1-\nu_0^s)}, \\
B_4 + B_5 &= (D_2 + D_3) + (1 + D_1)\left[c_1 + \frac{c_0}{2(1-\nu_0^s)}\right].
\end{aligned}
\tag{6.1}
$$

These simple expressions make it possible to write the five independent secant moduli in (3.6) and (3.7) as

$$
\begin{aligned}
E_{11}^s &= c_1 E_1 + c_0 E_0^s + \frac{4 c_1 c_0 (\nu_1 - \nu_0^s)^2}{c_1/\bar{\kappa}_0^s + c_0/\bar{\kappa}_1 + 1/\mu_0^s}, \\
\nu_{12}^s &= c_1 \nu_1 + c_0 \nu_0^s + \frac{c_1 c_0 (\nu_1 - \nu_0^s)(1/\bar{\kappa}_0^s - 1/\bar{\kappa}_1)}{c_1/\bar{\kappa}_0^s + c_0/\bar{\kappa}_1 + 1/\mu_0^s}, \\
\frac{\mu_{12}^s}{\mu_0^s} &= 1 + \frac{c_1}{\mu_0^s/(\mu_1 - \mu_0^s) + c_0/2}, \\
\frac{\mu_{23}^s}{\mu_0^s} &= 1 + \frac{c_1}{\dfrac{\mu_0^s}{(\mu_1 - \mu_0^s)} + \dfrac{c_0(\bar{\kappa}_0^s + 2\mu_0^s)}{2(\bar{\kappa}_0^s + \mu_0^s)}}, \\
\kappa_{23}^s &= \bar{\kappa}_0^s + \frac{c_1}{1/(\bar{\kappa}_1 - \bar{\kappa}_0^s) + c_0/(\bar{\kappa}_0^s + \mu_0^s)},
\end{aligned}
\tag{6.2}
$$

and

$$
E_{22}^s = \frac{4\kappa_{23}^s}{\kappa_{23}^s/\mu_{23}^s + 1 + 4\nu_{12}^{s2}\kappa_{23}^s/E_{11}^s}.
$$

The simplification processes leading to (6.2) from (3.6) and (3.7) are carried out following the same steps shown by Zhao *et al.* (1989) for the elastic moduli. In the latter case the five independent moduli—with the superscript s deleted in all expressions—are immediately recognized as Hill's (1964a) and Hashin's (1965) lower bounds when the matrix is the softer phase, and their upper bounds if the matrix is the harder.

The simpler B_i values give rise to simpler a_i^* in (3.3). Similarly, the p_i values in (3.11) become

$$p_1 = \frac{1}{1 - \nu_0^{s2}}, \qquad p_2 = p_3 = \frac{\nu_0^s}{2(1 - \nu_0^{s2})},$$
$$p_4 = \frac{3}{8(1 - \nu_0^{s2})}, \qquad p_5 = \frac{1}{8(1 - \nu_0^{s2})}, \tag{6.3}$$

and these, together with the simpler a_i^* in (3.12), also provide a simpler expression for the stress concentration factors b_i in (3.10).

This leads to a simpler effective stress σ^* in (4.1), and the initial yield surface, given by $\sigma^* = \sigma_y$, can be calculated conveniently. Such a yield surface has also been derived by Wakashima *et al.* (1979) in their study of thermomechanical loading. The dependence of σ^* on the pure hydrostatic pressure $\bar{\sigma}_H$ in (4.2) now can be simplified to

$$\frac{\sigma^*}{\bar{\sigma}_H} = \frac{c_1(1 - 2\nu_0^s)}{2(1 - \nu_0^{s2})}\left[(2 - \nu_0^s)\left(a_1^* - \frac{2a_2^*}{1 - 2\nu_0^s}\right) - (1 - 2\nu_0^s)(a_3^* + a_4^* + a_5^*)\right]. \tag{6.4}$$

This dependence is accurately reflected in Fig. 2 with $\alpha \geq 100$.

Of some theoretical importance is the dependence of the five initial yield points on the volume fraction of fibers c_1. These points, or proportional limits on the macroscopic stress–strain curves, are determined from the condition of $\sigma^* = \sigma_y$ under the five respective loadings, and the results, for the same boron/aluminum system given by (2.7), are depicted as solid lines in Fig. 5.

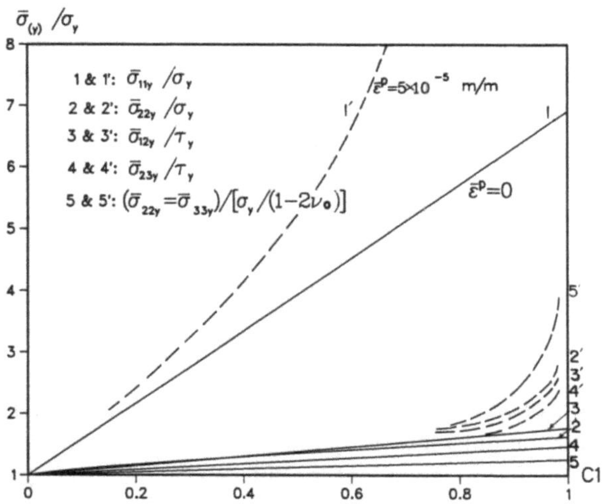

FIG. 5. The c_1 dependence of five initial yield stresses of continuous fiber-reinforced composites. The solid lines indicate those defined by the proportional limit, whereas the dashed lines correspond to a small amount of plastic strain of the composite.

These curves are normalized by the corresponding yield stresses of the matrix under respective loadings, where τ_y represents the yield stress under a pure shear ($\sigma_y = \sqrt{3}\tau_y$), and $\sigma_y/(1-2\nu_0)$ is the yield stress under the biaxial plane–strain loading ($\bar{\sigma}_{22} = \bar{\sigma}_{33}$, $\bar{\varepsilon}_{11} = 0$). This picture indicates that the axial yield stress $\bar{\sigma}_{11}$ can be very significantly improved by the fiber reinforcements, with others showing only a moderate increase.

It is to be noted from this figure that the yield stress of the composite at $c_1 = 1$ remains finite in all five directions. Such a consequence results from the fact that, as $c_1 \to 1$, there still exists a vanishingly small volume of matrix carrying a finite magnitude of stesss. Since the average stress in the matrix is finite, the condition $\sigma^* = \sigma_y$ also leads to a finite yield stress, or proportional limit, for the composite. Yet, its volume is so small that its contribution to the plastic strain of the composite is totally insignificant. Thus, if the yield point of the composite is defined with a finite—no matter how small—amount of plastic strain, its yield stress will approach infinity as $c_1 \to 1$. Such a definition

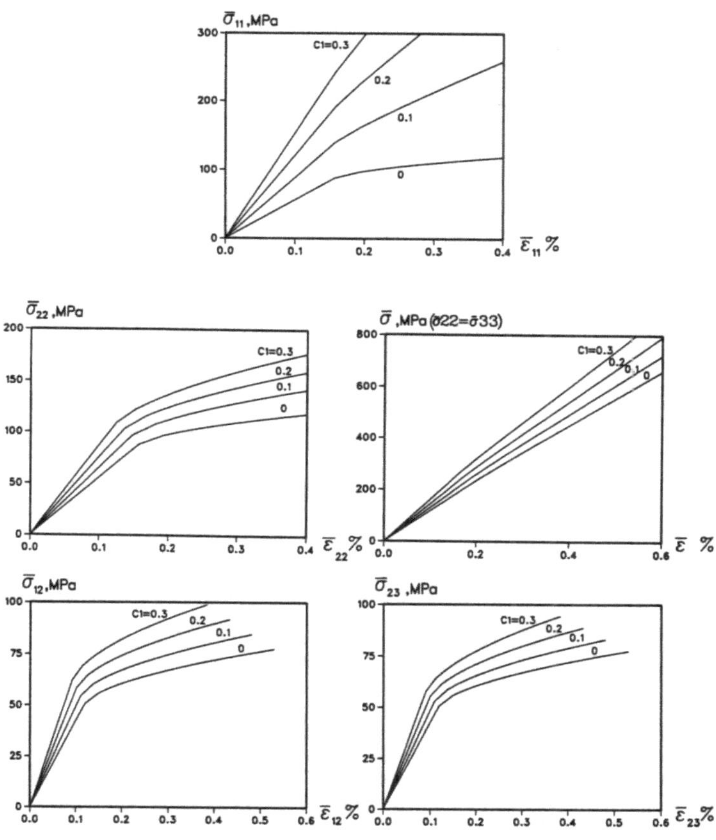

FIG. 6. Individual stress–strain relations of continuous fiber-reinforced composites under five respective loadings.

will lead to a significantly higher yield stress compared to that defined by the proportional limit ($\bar{\varepsilon}^p = 0$) at higher c_1. For instance, taking $\bar{\varepsilon}^p = 5 \times 10^{-5}$ m/m (not ε^p of the matrix) as the definition under respective loadings, the normalized yield stresses will be those depicted by the dashed lines in the same figure. Such a new definition does not affect the yield points significantly at low c_1, but apparently can result in a rahter drastic departure at high concentration of fibers.

The stress–strain relations of the continuous fiber-reinforced composites are shown in Fig. 6 at three fiber concentrations, along with that of the pure matrix, under the five respective loadings. While the behavior of the composite has shown a generally stiffer response in all directions, the improvement is apparently most noticeable under uniaxial tension.

At this point it is perhaps desirable that the theoretical prediction be compared with some experimental data. For the particle-reinforced composite such a comparison was given in Tandon and Weng (1988) for a silica/epoxy system, and the results show that up to $c_1 = 0.47$ the comparison between the two was reasonably good. The authors have not been able to find any experimental data with the disk or short-fiber reinforcement, but were able to locate one set of results for the boron/aluminum fiber-reinforced composite. As shown in Fig. 7, three independent tests were carried out under $\bar{\sigma}_{22}$ at General Dynamics (see Adams, 1970) for $c_1 = 0.34$. These three sets of data are marked

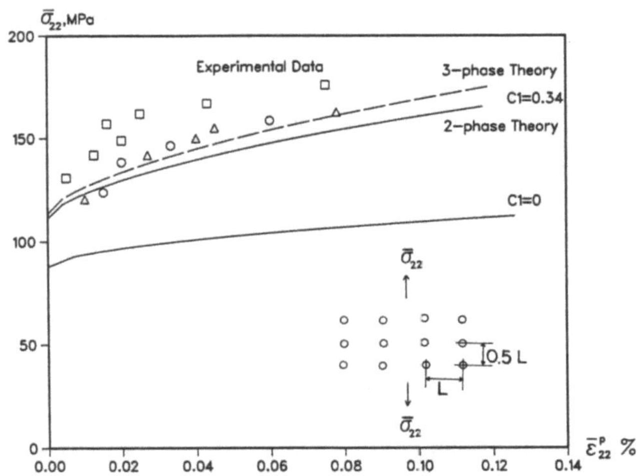

FIG. 7. Comparison between the theoretical prediction and the experimental data under a transverse normal loading. The line marked with "2-phase Theory" was calculated using Eshelby's (1957) original S tensor whereas the line marked with "3-phase Theory" was calculated using Eshelby's type of S tensor derived by Luo and Weng (1989) in a 3-phase concentric solid. The actual microgeometry of the filaments in the test specimen was more favorably aligned along the 2-axis, giving rise to a stiffer response.

by the squares, triangles, and circles, respectively, with the last two showing a better reproducibility. The material constants of boron and aluminum given in (2.7) were essentially derived from this report, and these values were used in the theory to calculate the $\bar{\sigma}_{22}$ versus $\bar{\varepsilon}_{22}^{p}$ relation. The theoretical result is plotted as solid line in Fig. 2, with a mark "2-phase Theory." It was reported that deformation beyond the indicated strain range began to show matrix cracking and, therefore, the data beyond 0.14% of plastic strain were not suitable for the present comparison.

The theoretical prediction appears to be somewhat softer than the experimental data. But, as pointed out by Adams (1970), the filament distribution across the transverse plane was not ideally isotropic; in fact, a direct measurement from a photomicrograph of the actual test specimen cross section shows a rectangular array of fiber as shown in the inset of Fig. 7. The testing direction ($\bar{\sigma}_{22}$) is about twice as closely packed as its perpendicular direction (3 axis). It is therefore anticipated that the test data along the 2 axis will show a slightly stronger response.

7. Final Remarks

The simple theory thus derived for the aspect-ratio dependence of two-phase plasticity is based upon Mori and Tanaka's (1973) mean-field approximation and Hill's (1965) discovery of a decreasing constraint power in the polycrystal matrix. A recent examination of the Mori–Tanaka theory by Luo and Weng (1987) suggests that, while their theory is fundamentally sound, the connection between $\boldsymbol{\varepsilon}^{pt}$ and $\boldsymbol{\varepsilon}^*$ in (2.11) through Eshelby's original S tensor may not be the best possible. A better connection perhaps can be provided by using the S tensor derived from a three-phase concentric solid. In this case a stress-free strain $\boldsymbol{\varepsilon}^*$ is introduced into the inclusion region which is embedded in the matrix, and the matrix, with a thickness appropriate for its volume fraction, is further surrounded by an infinity extended outer phase which has the property of the composite. The components of the S tensor for such a three-phase configuration can then be derived from the difference of mean strains in the inclusion and the intermediate matrix. Such a problem was also solved by Luo and Weng (1987) for a three-phase spherically concentric solid, and more recently (Luo and Weng, 1989) for a three-phase cylindrically concentric material.

When these new components of the S tensor were used in the Mori-Tanaka theory to predict the effective bulk and shear moduli of a particle-reinforced composite, Luo and Weng (1987) found that the bulk modulus remains unchanged, but the shear modulus becomes stiffer (or softer) if the inclusion is the harder (or softer) phase. When used to predict the five elastic moduli E_{11}, ν_{12}, μ_{12}, μ_{23}, and κ_{23} of a fiber-reinforced composite, it was found (Luo and Weng, 1989) that the only change occurs in μ_{23}, and it also becomes stiffer (or softer in the same way). As is evident from the last of (6.2), a higher μ_{23} also leads to a higher E_{22} and, when extended to the elastoplastic analysis, it will result in a stiffer $\bar{\sigma}_{22}$ versus $\bar{\varepsilon}^p_{22}$ curve.

The dashed line in Fig. 7 marked by "3-phase Theory" was calculated using such an S tensor (Luo and Weng, 1989, with the elastic moduli replaced by the corresponding secant moduli); some improvement is observed.

Hill's discovery of a weakening constraint power in a plastically deforming matrix was originally expressed in terms of the tangent moduli. Its representation by the secant moduli has proved to be appropriate under a proportional loading. When the composite is subject to a proportional external loading, however, the mean stress components of the matrix may or may not increase proportionally and, if not, it will be necessary to examine to what extent the departure, from a strict proportionality, really is.

Such a property is controlled by the variations of the stress concentration factors $b_1, b_2, \ldots, b_7$ in (3.8) as ε^{p*} increases under any desired proportional $\bar{\sigma}_{ij}$. For instance, under a monotonically increasing tensile stress $\bar{\sigma}_{11}$, the ratio of $\sigma^{(0)}_{22}/\sigma^{(0)}_{11}$ in the matrix is given by b_3/b_1, which, according to (3.10)–(3.12), depends on the aspect ratio of the inclusions α and the extent of plastic deformation in the matrix. To illustrate these effects we depict in Fig. 8 the loading paths of the average stress in the matrix for five selected aspect ratios. The initial linear portions of these curves are in the elastic state, representing a strict proportional loading irrespective of the aspect ratio. The subsequent portions after the turns correspond to the plastic state. In the case of prolate inclusions ($\alpha > 1$) the departure from strict linearity is very slight indeed, although in the oblate case it is more visible. However, such a departure, even

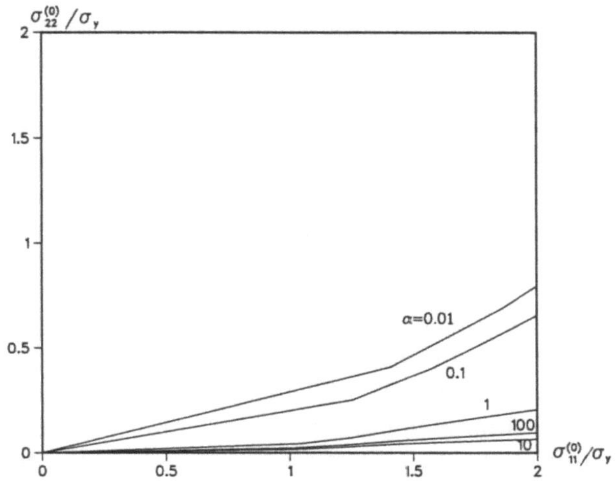

FIG. 8. Departure of mean-stress components of the matrix from a strict proportionality under a monotonically increasing external $\bar{\sigma}_{11}$ for five selected inclusion shapes.

in the latter case, still falls well within Budiansky's (1959) criterion for physical soundness in the application of deformation theory under nonproportional loading.

As a final remark, it is often noted that the in-situ, local behavior of the ductile matrix in a composite system is often harder due to the presence of strong inclusions. Such an assertion is perhaps generally true, but its extent will vary from point to point. For instance, in a metal matrix composite dislocations moving towards a particle or fiber may experience a strong resistance, but those moving away from it may even benefit from the possible repulsive force exerted on them. Such a point-to-point account of the in-situ hardening in the constitutive equation is clearly beyond the scope of this paper. The present mean-field theory, in effect, smears these local variations out and, in this context, although the average in-situ behavior of the matrix may still be a bit harder, its increase probably is not very significant. As demonstrated by comparison with experiments, the present theory is believed to have captured the essence of the prediction of the stress–strain relations for the particle- and fiber-strengthened composites.

Acknowledgment

This work was supported by the National Science Foundation, Solid and Geo-Mechanics program, under Grant MSM 86-14151.

References

Accorsi, M. L. and Nemat-Nasser, S. (1986), Bounds on the overall elastic and instantaneous elastoplastic moduli of periodic composites, *Mech. Materials*, **5**, 209–220.

Adams, D. F. (1970), Inelastic analysis of a unidirectional composite subjected to transverse normal loading, *J. Composite Materials*, **4**, 310–328.

Arsenault, R. J. and Taya, M. (1987), Thermal residual stress in metal matrix composite, *Acta Metallurgica*, **35**, 651–659.

Benveniste, Y. (1987), A new approach to the application of Mori–Tanaka theory in composite meterials, *Mech. Materials*, **6**, 147–157.

Berveiller, M. and Zaoui, A. (1979), An extension of the self-consistent scheme to plastically-flowing polycrystals, *J. Mech. Phys. Solids*, **26**, 325–344.

Budiansky, B. (1959), A reassessment of deformation theories of plasticity, *J. Appl. Mech.*, **26**, 259–264.

Dvorak, G. J. and Bahei-El-Din, Y. A. (1987), A bimodal plasticity theory of fibrous composite materials, *Acta Mech.*, **69**, 219–241.

Dvorak, G. J., Bahei-El-Din, Y. A., Macheret, Y., and Liu, C. H. (1988), An experimental study of elastic–plastic behavior of a fibrous boron–aluminum composite, *J. Mech. Phys. Solids*, **36**, 655–687.

Eshelby, J. D. (1957), The determination of the elastic field of an ellipsoidal inclusion, and related problems, *Proc. Roy. Soc. London*, **A241**, 376–396.

Hashin, Z. and Shtrikman, S. (1963), A variational approach to the theory of the elastic behavior of multiphase materials, *J. Mech. Phys. Solids*, **11**, 127–140.

Hashin, Z. (1965), On elastic behavior of fibre reinforced materials of arbitrary transverse phase geometry, *J. Mech. Phys. Solids*, **13**, 119–134.

Hill, R. (1948), A theory of the yielding and plastic flow of anisotropic metals, *Proc. Roy. Soc. London*, **A193**, 281–297.

Hill, R. (1963), Elastic properties of reinforced solids: Some theoretical principles, *J. Mech. Phys. Solids*, **11**, 357–372.

Hill, R. (1964a), Theory of mechanical properties of fibre-strengthened materials: I. Elastic behavior, *J. Mech. Phys. Solids*, **12**, 199–212.

Hill, R. (1964b), Theory of mechanical properties of fibre-strengthened materials: II. Inelastic behavior, *J. Mech. Phys. Solids*, **12**, 213–218.

Hill, R. (1965), Continuum micro-mechanics of elastoplastic polycrystals, *J. Mech. Phys. Solids*, **13**, 89–101.

Hutchinson, J. W. (1970), Elastic–plastic behavior of polycrystalline metals and composites, *Proc. Roy. Soc. London*, **A319**, 247–272.

Kröner, E. (1961), Zur plastischen verformung des vielkristalls, *Acta Metallurgica*, **9**, 155–161.

Luo, H. A. and Weng, G. J. (1987), On Eshelby's inclusion problem in a three-phase spherically concentric solid, and a modification of Mori–Tanaka's method, *Mech. Materials*, **6**, 347–361.

Luo, H. A. and Weng, G. J. (1989), On Eshelby's S-tensor in a three-phase cylindrically concentric solid, and the elastic moduli of fiber reinforced composites, *Mech. Materials*, **8** (in press).

Mori, T. and Tanaka, K. (1973), Average stress in the matrix and average elastic energy of materials with misfitting inclusions, *Acta Metallurgica*, **21**, 571–574.

Mura, T. (1987), *Micromechanics of Defects in Solids*, 2nd ed., Martinus Nijhoff, Dordrecht.

Pedersen, O. B. (1983), Thermoelasticity and plasticity of composites—I. Mean field theory, *Acta Metallurgica*, **13**, 1795–1808.

Tandon, G. P. and Weng, G. J. (1984), The effect of aspect ratio of inclusions on the elastic properties of unidirectionally aligned composites, *Polymer Composites*, **5**, 327–333.

Tandon, G. P. and Weng, G. J. (1986), Average stress in the matrix and effective moduli of randomly oriented composites, *Composite Sci. Tech.*, **27**, 111–132.

Tandon, G. P. and Weng, G. J. (1988), A theory of particle-reinforced plasticity, *J. Appl. Mech.*, **55**, 126–135.

Teply, J. L. and Dvorak, G. J. (1988), Bounds on overall instantaneous properties of elastic–plastic composites, *J. Mech. Phys. Solids*, **36**, 29–58.

Wakashima, K., Suzuki, Y., and Umekawa, S. (1979), A micromechanical prediction of initial yield surfaces of unidirectional composites, *J. Composite Materials*, **13**, 288–302.

Weng, G. J. (1982), A unified, self-consistent theory for the plastic-creep deformation of metals, *J. Appl. Mech.*, **49**, 728–734.

Weng, G. J. (1984), Some elastic properties of reinforced solid, with special reference to isotropic ones containing spherical inclusions, *Int. J. Engng. Sci.*, **22**, 845–856.

Weng, G. J. (1988), Theoretical principles for the determination of two kinds of composite plasticity, in *Mechanics of Composite Materials*, edited by Dvorak, G. J. and Laws, N., AMD-Vol. **92**, ASME, New York, pp. 193–208.

Zhao, Y. H., Tandon, G. P., and Weng, G. J. (1989), Elastic moduli for a class of porous materials, *Acta Mech.*, **76**, 105–130.

Appendix: Eshelby's *S* Tensor in Terms of the Secant Moduli

For a spheroidal inclusion with the symmetric axis identified as x_1, the components of Eshelby's tensor S^s_{ijkl} are

$$S^s_{1111} = \frac{1}{2(1-\nu^s_0)}\left\{1 - 2\nu^s_0 + \frac{3\alpha^2 - 1}{\alpha^2 - 1} - \left[1 - 2\nu^s_0 + \frac{3\alpha^2}{\alpha^2-1}\right]g\right\},$$

$$S^s_{2222} = S^s_{3333} = \frac{3}{8(1-\nu^s_0)}\frac{\alpha^2}{\alpha^2-1} + \frac{1}{4(1-\nu^s_0)}\left[1 - 2\nu^s_0 - \frac{9}{4(\alpha^2-1)}\right]g,$$

$$S^s_{2233} = S^s_{3322} = \frac{1}{4(1-\nu^s_0)}\left\{\frac{\alpha^2}{2(\alpha^2-1)} - \left[1 - 2\nu^s_0 + \frac{3}{4(\alpha^2-1)}\right]g\right\},$$

$$S^s_{2211} = S^s_{3311} = -\frac{1}{2(1-\nu^s_0)}\frac{\alpha^2}{\alpha^2-1} + \frac{1}{4(1-\nu^s_0)}\left\{\frac{3\alpha^2}{\alpha^2-1} - (1-2\nu^s_0)\right\}g,$$

$$S^s_{1122} = S^s_{1133} = -\frac{1}{2(1-\nu^s_0)}\left[1 - 2\nu^s_0 + \frac{1}{\alpha^2-1}\right] \tag{A1.1}$$

$$+ \frac{1}{2(1-\nu^s_0)}\left[1 - 2\nu^s_0 + \frac{3}{2(\alpha^2-1)}\right]g,$$

$$S^s_{2323} = \frac{1}{4(1-\nu^s_0)}\left\{\frac{\alpha^2}{2(a^2-1)} + \left[1 - 2\nu^s_0 - \frac{3}{4(\alpha^2-1)}\right]g\right\},$$

$$S^s_{1212} = S^s_{1313} = \frac{1}{4(1-\nu^s_0)}\left\{1 - 2\nu^s_0 - \frac{\alpha^2+1}{\alpha^2-1}\right.$$

$$\left. - \frac{1}{2}\left[1 - 2\nu^s_0 - \frac{3(\alpha^2+1)}{\alpha^2-1}\right]g\right\},$$

where ν^s_0 is the secant Poisson ratio of the matrix, α is the aspect ratio of the inclusion ($= l/d$), and g is given by

$$g = \frac{\alpha}{(\alpha^2-1)^{3/2}}\{\alpha(\alpha^2-1)^{1/2} - \cosh^{-1}\alpha\}, \qquad \text{prolate shape,} \tag{A1.2}$$

$$= \frac{\alpha}{(1-\alpha^2)^{3/2}}\{\cos^{-1}\alpha - \alpha(1-\alpha^2)^{1/2}\}, \qquad \text{oblate shape.} \tag{A1.3}$$

For a spherical inclusion, they simplify to

$$
\begin{aligned}
S^s_{1111} &= S^s_{2222} = S^s_{3333} = \frac{7 - \nu^s_0}{15(1 - \nu^s_0)}, \\
S^s_{1122} &= S^s_{2233} = S^s_{3311} = \frac{5\nu^s_0 - 1}{15(1 - \nu^s_0)}, \\
S^s_{1212} &= S^s_{2323} = S^s_{3131} = \frac{4 - 5\nu^s_0}{15(1 - \nu^s_0)}.
\end{aligned}
\tag{A1.4}
$$

For a circular cylinder, we have

$$
\begin{aligned}
S^s_{1111} &= 0, \\
S^s_{2222} &= S^s_{3333} = \frac{5 - 4\nu^s_0}{8(1 - \nu^s_0)}, \\
S^s_{2233} &= S^s_{3322} = \frac{4\nu^s_0 - 1}{8(1 - \nu^s_0)}, \\
S^s_{2211} &= S^s_{3311} = \frac{\nu^s_0}{2(1 - \nu^s_0)}, \\
S^s_{1122} &= S^s_{1133} = 0, \\
S^s_{2323} &= \frac{3 - 4\nu^s_0}{8(1 - \nu^s_0)}, \\
S^s_{1212} &= S^s_{1313} = \tfrac{1}{4}.
\end{aligned}
\tag{A1.5}
$$

Author Index

Subject Index

MIX
Papier aus verantwortungsvollen Quellen
Paper from responsible sources
FSC® C105338

If you have any concerns about our products,
you can contact us on
ProductSafety@springernature.com

In case Publisher is established outside the EU,
the EU authorized representative is:
Springer Nature Customer Service Center GmbH
Europaplatz 3, 69115 Heidelberg, Germany

Printed by Libri Plureos GmbH
in Hamburg, Germany